CAMBRIDGE STUDIES IN
ADVANCED MATHEMATICS : 43

ABSOLUTELY SUMMING OPERATORS

Already published

1 W.M.L. Holcombe *Algebraic automata theory*
2 K. Petersen *Ergodic theory*
3 P.T. Johnstone *Stone spaces*
4 W.H. Schikhof *Ultrametric calculus*
5 J.-P. Kahane *Some random series of functions, 2nd edition*
6 H. Cohn *Introduction to the construction of class fields*
7 J. Lambek & P.J. Scott *Introduction to higher-order categorical logic*
8 H. Matsumura *Commutative ring theory*
9 C.B. Thomas *Characteristic classes and the cohomology of finite groups*
10 M. Aschbacher *Finite group theory*
11 J.L. Alperin *Local representation theory*
12 P. Koosis *The logarithmic integral I*
13 A. Pietsch *Eigenvalues and s-numbers*
14 S.J. Patterson *An introduction to the theory of the Riemann zeta-function*
15 H.J. Baues *Algebraic homotopy*
16 V.S. Varadarajan *Introduction to harmonic analysis on semisimple Lie groups*
17 W. Dicks & M. Dunwoody *Groups acting on graphs*
18 L.J. Corwin & F.P. Greenleaf *Representations of nilpotent Lie groups and their applications*
19 R. Fritsch & R. Piccinini *Cellular structures in topology*
20 H Klingen *Introductory lectures on Siegel modular forms*
21 P. Koosis *The logarithmic integral II*
22 M.J. Collins *Representations and characters of finite groups*
24 H. Kunita *Stochastic flows and stochastic differential equations*
25 P. Wojtaszczyk *Banach spaces for analysts*
26 J.E. Gilbert & M.A.M. Murray *Clifford algebras and Dirac operators in harmonic analysis*
27 A. Frohlich & M.J. Taylor *Algebraic number theory*
28 K. Goebel & W.A. Kirk *Topics in metric fixed point theory*
29 J.F. Humphreys *Reflection groups and Coxeter groups*
30 D.J. Benson *Representations and cohomology I*
31 D.J. Benson *Representations and cohomology II*
32 C. Allday & V. Puppe *Cohomological methods in transformation groups*
33 C. Soulé et al *Lectures on Arakelov geometry*
34 A. Ambrosetti & G. Prodi *A primer of nonlinear analysis*
35 J. Palis & F. Takens *Hyperbolicity and sensitive chaotic dynamics at homoclinic bifurcations*
36 M. Auslander, I. Reiten & S. Smalo *Representation theory of Artin algebras*
37 Y. Meyer *Wavelets and operators*
38 C. Weibel *An introduction to homological algebra*
39 W. Bruns & J. Herzog *Cohen-Macaulay rings*
40 V. Snaith *Explicit Brauer induction*
42 E.B. Davies *Spectral theory and differential operators*
43 J. Diestel, H. Jarchow & A. Tonge *Absolutely summing operators*
44 P. Mattila *Geometry of sets and measures in Euclidean spaces*
45 R. Pinsky *Positive harmonic functions and diffusion*

Absolutely Summing Operators

Joe Diestel
Department of Mathematics and Computer Science
Kent State University

Hans Jarchow
Mathematisches Institut
Universität Zürich

Andrew Tonge
Department of Mathematics and Computer Science
Kent State University

CAMBRIDGE UNIVERSITY PRESS
Cambridge, New York, Melbourne, Madrid, Cape Town, Singapore, São Paulo

Cambridge University Press
The Edinburgh Building, Cambridge CB2 8RU, UK

Published in the United States of America by Cambridge University Press, New York

www.cambridge.org
Information on this title: www.cambridge.org/9780521431682

First published 1995
This digitally printed version 2008

A catalogue record for this publication is available from the British Library

ISBN 978-0-521-43168-2 hardback
ISBN 978-0-521-06493-4 paperback

CONTENTS

INTRODUCTION

The roots of the theory of p-summing operators lie in work undertaken by Alexandre Grothendieck from the 1950s. However, it was only in 1967 that Albrecht Pietsch clearly isolated this class of operators and established many of their fundamental properties. Within a year Pietsch's work gained recognition thanks to the appearance of the seminal paper "Absolutely summing operators in $\mathcal{L}_p$-spaces and their applications" of Joram Lindenstrauss and Aleksander Pełczyński.

The time was ripe for these ideas to gain ascendance. They soon made serious inroads into other areas of analysis – witness Pełczyński's influential monograph "Banach Spaces of Analytic Functions and Absolutely Summing Operators" which was published within a decade of Pietsch's introduction – and this trend has continued unabated.

Unfortunately, students and specialists in subjects other than Banach space theory have scant sources to learn about our topic. P. Wojtaszczyk's excellent and broad ranging "Banach Spaces for Analysts" is pitched at a level high enough to be uncomfortable for many students. Indeed, with the possible exception of G.J.O. Jameson's elegant text "Summing and Nuclear Norms in Banach Space Theory", which carefully avoids the measure theoretical underpinnings of the subject, no general introduction to the area exists. We hope our book goes some way towards filling this gap.

Our aim is to introduce the student of analysis to the basic facts about p-summing operators and their relatives and to show how their theory can be applied in a variety of situations. We have tried to follow the dictum espoused by C.A. Rogers:

"As the book is largely based on lectures, and as I like my students to follow my lectures, proofs are given in detail; this may bore the mature mathematician, but it will I believe be a great help to anyone trying to learn the subject *ab initio*."

We expect that those with a sound background in real analysis, complex analysis and functional analysis will be ready and able to gain from the study of this text. As indicated above, we have presented much of the material in classes and seminars, and this has greatly influenced our ordering.

The first six chapters give a thorough and, we hope, accessible introduction which should be suitable for presentation in classes, seminars and the like. Naturally, the material in later chapters is often much tougher, and sometimes we find it convenient to call on substantial results from other areas without proof. On occasion, theorems are given two different proofs; as one might expect, this is to illustrate how various developments lend new insights.

Despite the length of the text, we have striven to avoid being encyclopaedic. However, at the end of each chapter, we have notes and remarks in which we relate historical information, indicate directions to be pursued, provide results that are a bit beyond our perceived scope, and generally pander to our encyclopaedic urges without poisoning the text. We make no claim of completeness (although we always work with Banach spaces). Hopefully, we have been accurate in allocating priorities.

As is usual in such ventures, we have benefitted from the support of many people and institutions. We take the opportunity to thank some of those to whom we are indebted.

The Mathematics Departments at Kent State University and Universität Zürich have provided us with stages where we tried our material on the most valuable of audiences – graduate students and colleagues. To all these people we extend our thanks for their patience and perceptive advice.

We have also had the opportunity to test our wares at a number of other institutions, most particularly, Centrale (Caracas, Venezuela), Complutense (Madrid, Spain), Los Andes (Merida, Venezuela), New England (Armidale, Australia), PAN (Warsaw, Poland), Pretoria (Republic of South Africa), São Paulo (Brazil), Sevilla (Spain) and University College (Dublin, Ireland).

In addition, we have had access through the years to the wit and wisdom of numerous mathematicians and friends who have greatly influenced our thinking. Among them we mention R. Alencar, R.M. Aron, D. Barcenas, G. Bennett, F. Bombal, C. Cardassi, B. Carl, G. Curbera, W.J. Davis, A. Defant, S. Dineen, S.W. Drury, T. Figiel, C. Finet, K. Floret, D.J.H. Garling, K. John, W.B. Johnson, V.M. Kadets, S. Kaijser, N.J. Kalton, H. König, S. Kwapień, L.E. Labuschagne, K. Lermer, D.R. Lewis, V.D. Mascioni, B. Maurey, V.D. Milman, A. Pełczyński, A. Pietsch, G. Pisier, J.R. Retherford, B. Sims, A. Ströh, J. Swart, N. Tomczak-Jaegermann, N.T. Varopoulos and G. West. To each of these individuals and institutions we acknowledge our debt and gratitude.

Our thanks also go to S. Kyburz for preparing a substantial part of the manuscript.

In addition, we wish to thank the National Science Foundation, Schweizerischer Nationalfonds zur Förderung der wissenschaftlichen Forschung, the Polish Academy of Sciences and the Kent State Office of Research and Sponsored Programs for important financial support.

Kent and Zürich

April 1994

NOTATION

We shall stick to standard definitions and terminology. **N**, **Q**, **R**, **C** will be the systems of natural, rational, real and complex numbers; **Z** will denote the ring of integers, and occasionally we shall write $\mathbf{N}_0 := \mathbf{N} \cup \{0\}$. We shall also use **K** to stand for **R** or **C**.

Our main interest is in Banach spaces over **K** and in their operators. Throughout this text, X, Y, Z,... will be Banach spaces; typical members will be $x, y, z, ...$, respectively, perhaps with indices. When we deal with finite dimensional spaces, E, F, H,... may be used to denote the space in question. By span M and conv M we mean the linear span and the convex hull of a subset M of a Banach space.

The norm of a Banach (or just normed) space X will usually be denoted by $\|\cdot\|$, but when more precision is desirable, we may also use $\|\cdot\|_X$. B_X will denote the closed unit ball $\{x \in X : \|x\| \leq 1\}$ of our Banach space X, whereas $S_X = \{x \in X : \|x\| = 1\}$ is its unit sphere.

As usual, bounded linear maps between Banach spaces are referred to as operators; they will be denoted by u, v, w,.... Let X and Y be Banach spaces. The collection $\mathcal{L}(X,Y)$ of all operators $u : X \to Y$ is a Banach space with respect to the (uniform) norm $\|u\| = \sup_{x \in B_X} \|ux\|$. If $X = Y$, then $\mathcal{L}(X)$ is used instead of $\mathcal{L}(X,X)$. We shall extend this convention to subsystems of operators as well. For example, $\mathcal{F}(X,Y)$ will be used to denote the collection of all operators in $\mathcal{L}(X,Y)$ which have a finite dimensional range, and $\mathcal{F}(X)$ replaces $\mathcal{F}(X,X)$.

A Banach space operator $u : X \to Y$ is *isometric* if $\|ux\| = \|x\|$ for all $x \in X$; it is an *isometry* if it is in addition onto.

The continuous dual of a Banach space X is $X^* := \mathcal{L}(X, \mathbf{K})$; its typical member will be denoted by x^*, and for $x \in X$, we shall write $\langle x^*, x\rangle$ (or $\langle x, x^*\rangle$) for the action of x^* on x. The bidual of X is the space $X^{**} = (X^*)^*$, and $x \mapsto \langle \cdot, x\rangle$ establishes an isometric embedding $X \hookrightarrow X^{**}$; we denote it by k_X.

Only in Chapter 18 do we deviate from our notation concerning elements; we do this to avoid potential confusion with the notation of $*$-algebras.

Hilbert spaces will generally be denoted by H or H_k, and if x and y are elements of a Hilbert space their inner product will be written $(x|y)$.

The classical function spaces $C(K)$ of continuous **K**-valued functions on a compact Hausdorff space K as well as the spaces $L_p(\mu)$ relative to a measure μ need no further comment. The latter are Banach spaces when $1 \leq p \leq \infty$.

Occasionally, we shall also take into account the case $0 < p < 1$; then we are dealing with complete metrizable topological vector spaces. If μ is the counting measure on some given set J, we write ℓ_p^J in place of $L_p(\mu)$. Thus ℓ_∞^J just consists of all bounded functions $f : J \to \mathbf{K}$, and $\|f\|_\infty = \sup_{j\in J} |f(j)|$. If p is finite, then $f : J \to \mathbf{K}$ belongs to ℓ_p^J if and only if $\{j \in J : f(j) \neq 0\}$ is countable and $\sum_{j\in J} |f(j)|^p$ is finite; this sum is then $\|f\|_p^p$.

By c_0^J we mean the closed subspace of ℓ_∞^J consisting of all functions $f : J \to \mathbf{K}$ which 'vanish at infinity': given $\varepsilon > 0$ there is a finite subset M of J such that $f(j) \leq \varepsilon$ for all $j \in J \setminus M$.

The classical sequence spaces ℓ_p and c_0 are obtained by taking $J = \mathbf{N}$. If J is a finite set, $J = \{1,...,n\}$ say, then we prefer to write ℓ_p^n in place of ℓ_p^J.

At certain important junctures we shall also encounter the classical Lorentz spaces $L_{p,q}(\mu)$ relative to a measure space (Ω, Σ, μ); we shall only be interested in the case $1 \leq p, q \leq \infty$. $L_{p,q}(\mu)$ consists of all (μ-a.e. equivalence classes of) measurable functions $f : \Omega \to \mathbf{K}$ whose *decreasing rearrangement*

$$f^* : [0, \mu(\Omega)) \longrightarrow \mathbf{K} : t \mapsto \inf\{s > 0 : \mu(|f| > s) \leq t\}$$

satisfies

$$\|f\|_{p,q} := \Big(\int_0^{\mu(\Omega)} \big(t^{(1/p)-1} f^*(t)\big)^q dt\Big)^{1/q} < \infty$$

if $q < \infty$, and

$$\|f\|_{p,\infty} := \sup_{t\geq 0} t^{1/p} |f^*(t)| < \infty$$

if $q = \infty$. The spaces obtained in this way are Banach spaces though $\|\cdot\|_{p,q}$ is, in general, only equivalent to a Banach space norm. But we always have $L_{p,p}(\mu) = L_p(\mu)$ isometrically. Details can be found in many books, for example in E.M. Stein and G. Weiss [1971].

If μ is counting measure on $\mathbf{N}$ (or $\{1,...,n\}$) we write $\ell_{p,q}$ (or $\ell_{p,q}^n$) instead of $L_{p,q}(\mu)$. When we take the absolute values of the members of a scalar sequence (a_k) and arrange them in non-increasing order, we obtain the decreasing rearrangement (a_k^*), and then

$$\|(a_k)\|_{p,q} = \Big(\sum_n \big(n^{(1/p)-1} a_n^*\big)^q\Big)^{1/q}$$

if $q < \infty$, whereas

$$\|(a_k)\|_{p,\infty} = \sup_n n^{1/p} \cdot a_n^*.$$

Let $1 \leq p < \infty$. If $(X_j)_{j\in J}$ is a family of Banach spaces, then its ℓ_p *direct sum*,

$$\big(\bigoplus_{j\in J} X_j\big)_p,$$

(also written $\big(\bigoplus_{j\in J} X_j\big)_{\ell_p^J}$) consists of all $(x_j)_{j\in J} \in \prod_{j\in J} X_j$ such that $(\|x_j\|)_{j\in J}$ belongs to ℓ_p^J. This is a Banach space with respect to the norm

$$\|(x_j)_{j\in J}\| = \|(\|x_j\|)_{j\in J}\|_p = \big(\sum_{j\in J} \|x_j\|^p\big)^{1/p}.$$

The meaning of $\left(\bigoplus_{j\in J} X_j\right)_\infty$ and $\left(\bigoplus_{j\in J} X_j\right)_{c_0}$ is self-explanatory. As in the scalar case, we make the canonical identification of the dual of $\left(\bigoplus_{j\in J} X_j\right)_{c_0}$ with $\left(\bigoplus_{j\in J} X_j^*\right)_1$; the duality is given by $\langle (x_j^*), (x_j)\rangle = \sum_{j\in J}\langle x_j^*, x_j\rangle$. In the same way, for $1 \le p < \infty$, we shall identify the dual of $\left(\bigoplus_{j\in J} X_j\right)_p$ with $\left(\bigoplus_{j\in J} X_j^*\right)_{p^*}$. The ℓ_p direct sum of n Banach spaces $X_1,\ldots, X_n$ will also be denoted by $X_1 \oplus_p \ldots \oplus_p X_n$.

1. Unconditional and Absolute Summability in Banach Spaces

The Dvoretzky - Rogers Theorem

Recall that a sequence (x_n) in a normed space is *absolutely summable* if $\sum_n \|x_n\| < \infty$, and is *unconditionally summable* if $\sum_n x_{\sigma(n)}$ converges, regardless of the permutation σ of the indices. It is traditional to say that the series $\sum_n x_n$ is absolutely (unconditionally) convergent if the sequence (x_n) is absolutely (unconditionally) summable.

A theorem of Dirichlet from elementary analysis asserts that a scalar sequence is absolutely summable precisely when it is unconditionally summable. Simple natural adjustments to the proof show that this theorem extends to the setting of any finite dimensional normed space.

What happens in infinite dimensional spaces? Without completeness we can get nowhere.

1.1 Proposition: *A normed space is a Banach space if and only if every absolutely summable sequence is unconditionally summable.*

This elementary old standby finds frequent use in proofs of completeness, and a brief indication of its proof is worthy of our attention.

Proof. To show completeness we need to prove that every Cauchy sequence (x_n) is convergent. For this it suffices to find a convergent subsequence, a task which is not difficult since any 'sufficiently rapid' Cauchy subsequence will do the trick. For example, choose an increasing sequence of positive integers (n_k) so that if $y_k = x_{n_{k+1}} - x_{n_k}$, then $\|y_k\| \leq 2^{-k}$. As (y_k) is absolutely summable, it is (unconditionally) summable. The convergence of (x_{n_k}) now follows from the identity $x_{n_1} + y_1 + ... + y_k = x_{n_{k+1}}$.

Conversely, it is plain that in a Banach space the sequence of partial sums of an absolutely summable sequence is a Cauchy sequence, and so convergent. An application of Dirichlet's Theorem to the sum of the norms allows us to obtain *unconditional* convergence. QED

In standard infinite dimensional spaces there are usually easy examples of unconditionally summable sequences which fail to be absolutely summable. For instance, in ℓ_2 the sequence $(\lambda_n e_n)$ – where e_n is the n-th unit coordinate vector – is unconditionally summable when $(\lambda_n) \in \ell_2$, but not absolutely summable unless $(\lambda_n) \in \ell_1$.

Remarkably enough, this example can be transplanted, after suitable modification, to any infinite dimensional Banach space. The operation requires a modicum of solid effort.

1.2 Dvoretzky-Rogers Theorem: *Let X be an infinite dimensional Banach space. Then no matter how we choose $(\lambda_n) \in \ell_2$ there is always an unconditionally summable sequence (x_n) in X with $\|x_n\| = |\lambda_n|$ for all n.*

So, choosing (λ_n) in ℓ_2 but not in ℓ_1, we have sequences in infinite dimensional Banach spaces which are unconditionally summable but not absolutely summable.

The heart of the Dvoretzky-Rogers Theorem is a striking geometrical lemma.

1.3 Lemma: *Let E be a $2n$-dimensional Banach space. There exist n vectors $x_1,\dots,x_n \in B_E$, each of norm $\geq 1/2$, such that regardless of the scalars $\lambda_1,\dots,\lambda_n$ we have*

$$\Big\|\sum_{j\leq n} \lambda_j x_j\Big\| \leq \Big(\sum_{j\leq n} |\lambda_j|^2\Big)^{1/2}.$$

Proof. First we fix some (standard) notations. If w is a linear map between two k-dimensional vector spaces, each with a chosen basis, then $\det(w)$ and $\operatorname{tr}(w)$ denote the determinant and trace of the matrix representing w with respect to these chosen bases. Changing the bases will change $\det(w)$ by a constant factor but will leave $\operatorname{tr}(w)$ unchanged.

Our objective is to find a norm one isomorphism $u : \ell_2^{2n} \to E$ satisfying

$$(*) \qquad |\operatorname{tr}(u^{-1}v)| \leq 2n\cdot\|v\|$$

for all operators $v : \ell_2^{2n} \to E$.

Locating u is no problem: take any u for which

$$\det(u) = \max\big\{|\det(v)| : v \in \mathcal{L}(\ell_2^{2n}, E),\ \|v\| = 1\big\}.$$

The compactness of the unit sphere of $\mathcal{L}(\ell_2^{2n}, E)$ and the continuity of $\det(\cdot)$ guarantee the existence of such an operator u.

Establishing $(*)$ takes more care; we use a perturbation argument. Let ε be a non-zero scalar and $v \in \mathcal{L}(\ell_2^{2n}, E)$. Then, by the choice of u and the nature of determinants,

$$\frac{|\det(u+\varepsilon v)|}{\|u+\varepsilon v\|^{2n}} \leq \det(u),$$

so that

$$|\det(u+\varepsilon v)| \leq \det(u)\cdot\|u+\varepsilon v\|^{2n} \leq \det(u)\cdot(1+|\varepsilon|\cdot\|v\|)^{2n}.$$

Let id denote the identity on ℓ_2^{2n}. The invertibility of u gives

$$\begin{aligned}|\det(u+\varepsilon v)| &= \det(u)\cdot|\det(id+\varepsilon u^{-1}v)|\\ &= \det(u)\cdot|1+\varepsilon\cdot\operatorname{tr}(u^{-1}v)+c(\varepsilon)|\end{aligned}$$

where $|c(\varepsilon)| = O(|\varepsilon|^2)$ as $\varepsilon \to 0$. In tandem, these observations tell us that

$$|1+\varepsilon\cdot\mathrm{tr}\,(u^{-1}v)+c(\varepsilon)| \le (1+|\varepsilon|\cdot\|v\|)^{2n} = 1+2n\cdot|\varepsilon|\cdot\|v\|+O(|\varepsilon|^2)$$

for small ε. Choosing ε judiciously enough to satisfy $\varepsilon\cdot\mathrm{tr}\,(u^{-1}v) = |\varepsilon\cdot\mathrm{tr}\,(u^{-1}v)|$, we soon have for small $|\varepsilon|$

$$1+|\varepsilon|\cdot|\mathrm{tr}\,(u^{-1}v)| \le 1+|\varepsilon|\cdot 2n\cdot\|v\|+O(|\varepsilon|^2),$$

from which

$$|\mathrm{tr}\,(u^{-1}v)| \le 2n\cdot\|v\|+O(|\varepsilon|).$$

Passing with ε to zero we obtain $(*)$.

Now if P is any orthogonal projection on ℓ_2^{2n} with m-dimensional range, we can exploit $(*)$ to get

$$m = \mathrm{tr}\,(P) = \mathrm{tr}\,(u^{-1}uP) \le 2n\cdot\|uP\|.$$

In other words, $\|uP\| \ge m/2n$.

At last we can address the problem of selecting suitable $x_1,\dots,x_n$ in E. The key is to choose appropriate orthonormal vectors $y_1,\dots,y_n$ in ℓ_2^{2n} and then to set $x_j = u(y_j)$ for each j.

Since $\|u\| = 1$, there is a $y_1 \in \ell_2^{2n}$ with $\|y_1\| = 1$ and $\|uy_1\| = 1$. Let P_1 be the orthogonal projection of ℓ_2^{2n} onto the orthogonal complement $[y_1]^\perp$ of (the span of) y_1. Then $\|uP_1\| \ge (2n-1)/2n$, so there is a $y_2 \in [y_1]^\perp$ with $\|y_2\| = 1$ and $\|uy_2\| = \|uP_1y_2\| \ge (2n-1)/2n$. Let P_2 be the orthogonal projection of ℓ_2^{2n} onto the orthogonal complement $[y_1,y_2]^\perp$ of y_1 and y_2. Then $\|uP_2\| \ge (2n-2)/2n$, so there is a $y_3 \in [y_1,y_2]^\perp$ with $\|y_3\| = 1$ and $\|uy_3\| = \|uP_2y_3\| \ge (2n-2)/2n$. — Continue.

After n steps we have orthonormal vectors $y_1,\dots,y_n$ in ℓ_2^{2n}. Set $x_j = uy_j$ for $1 \le j \le n$. Then $\|x_j\| \ge (2n-j+1)/2n \ge 1/2$ for all j, and at the same time, if $\lambda_1,\dots,\lambda_n$ are scalars then

$$\Big\|\sum_{j\le n}\lambda_jx_j\Big\| = \Big\|u\big(\sum_{j\le n}\lambda_jy_j\big)\Big\| \le \|u\|\cdot\Big\|\sum_{j\le n}\lambda_jy_j\Big\| = \big(\sum_{j\le n}|\lambda_j|^2\big)^{1/2},$$

by the orthonormality of the y_j's. QED

This lemma will recur in several important improvements; see 14.2 and 19.15.

To prove the Dvoretzky - Rogers Theorem, it is convenient to use an alternative formulation of unconditional summability.

1.4 Lemma: *A sequence (x_n) in a Banach space is unconditionally summable if and only if it is sign summable, that is $\sum_n \varepsilon_nx_n$ converges for all choices of signs $\varepsilon_n = \pm 1$.*

The proof of this lemma requires some rather formal manipulations, which are best seen in the context of other equivalent variations of unconditional summability. We shall confront these shortly (1.5 through 1.9) but for now we take the lemma on trust.

Proof of the Dvoretzky-Rogers Theorem. Fix $(\lambda_n) \in \ell_2$ and choose positive integers $n_1 < n_2 < \ldots$ such that, for each $k \in \mathbf{N}$,

$$\sum_{n \geq n_k} |\lambda_n|^2 \leq 2^{-2k}.$$

Since X is infinite dimensional, the geometrical lemma 1.3 applies with impunity to all dimensions. We can therefore find a sequence of vectors (y_n) in B_X, each of norm $\geq 1/2$, such that for any scalar sequence (α_n) and any k we have

$$\Big\| \sum_{n=n_k}^{N} \alpha_n y_n \Big\| \leq \Big(\sum_{n=n_k}^{N} |\alpha_n|^2 \Big)^{1/2},$$

no matter how we select $n_k \leq N < n_{k+1}$. We weight the y_j's by setting $x_j = \lambda_j y_j / \|y_j\|$. Take note: regardless of the signs $\varepsilon_n = \pm 1$ and regardless of $n_k \leq N < n_{k+1}$ we have

$$\Big\| \sum_{n=n_k}^{N} \varepsilon_n x_n \Big\| \leq \Big(\sum_{n=n_k}^{N} \frac{|\lambda_n|^2}{\|y_n\|^2} \Big)^{1/2} \leq 2^{-k+1}.$$

It follows that the partial sums of $(\varepsilon_n x_n)$ are Cauchy. Hence (x_n) is sign summable, and so unconditionally summable. Of course, our weighty scaling of the y_n's ensures that $\|x_n\| = |\lambda_n|$ for all n. QED

Unconditional Convergence and the Orlicz-Pettis Theorem

We have now seen that in infinite dimensional Banach spaces the notions of unconditional and absolute summability are never equivalent. Just what *can* we say about unconditional summability?

In the next few pages we amass an extensive arsenal of useful equivalences to unconditional summability. Though some of these are elementary in nature others, like the Orlicz-Pettis Theorem 1.8, are subtle and surprising results.

To start, we shall rely only on manipulative skill, and we shall confine ourselves to careful reworkings of the scalar-valued case.

1.5 Theorem: *For a sequence (x_n) in a Banach space, the following are equivalent:*

(i) *(x_n) is unconditionally summable.*

(ii) *(x_n) is unordered summable, that is, for any $\varepsilon > 0$ there is a positive n_ε such that whenever M is a finite subset of $\mathbf{N}$ with $\min M > n_\varepsilon$ we have $\|\sum_{n \in M} x_n\| < \varepsilon$.*

(iii) *(x_n) is subseries summable, that is, for any strictly increasing sequence (k_n) of positive integers, $\sum_n x_{k_n}$ is convergent.*

(iv) *(x_n) is sign summable.*

Proof. (i)⇒(ii): Arguing contrapositively, we assume that (ii) is false and look for a permutation σ of $\mathbf{N}$ which makes the partial sums of $(x_{\sigma(n)})$ fail to be Cauchy. There is an exceptional $\delta > 0$ such that, regardless of $m \in \mathbf{N}$, there is always a finite set $M \subset \mathbf{N}$ with $\min M > m$ and $\|\sum_{k\in M} x_k\| \geq \delta$. This allows us to construct a sequence (M_n) of finite subsets of $\mathbf{N}$ such that, for all $n \in \mathbf{N}$,

$$\max M_n < \min M_{n+1} \qquad \text{and} \qquad \Big\|\sum_{k\in M_n} x_k\Big\| \geq \delta.$$

But then, if σ is a permutation of $\mathbf{N}$ which maps each (integer) interval $[\min M_n, \min M_n + |M_n|)$ to M_n, the partial sums of $(x_{\sigma(n)})$ cannot be Cauchy.
(ii)⇒(i): Let σ be a permutation of $\mathbf{N}$. Fix $\varepsilon > 0$ and choose $n_\varepsilon \in \mathbf{N}$ as in the definition of unordered summability. Certainly there is an $m_\varepsilon \in \mathbf{N}$ such that $\{1,\ldots,n_\varepsilon\} \subset \sigma(\{1,\ldots,m_\varepsilon\})$, and this tells us that if $q > p > m_\varepsilon$, then $\|\sum_{n=p}^{q} x_{\sigma(n)}\| < \varepsilon$. In other words, $(x_{\sigma(n)})$ is summable.
(ii)⇒(iii): Fix $\varepsilon > 0$ and choose $m_\varepsilon \in \mathbf{N}$ so that $\|\sum_{n\in M} x_n\| < \varepsilon$ for every finite subset M of $\mathbf{N}$ with $\min M > m_\varepsilon$. Now if (k_n) is a strictly increasing sequence of natural numbers we have $k_n \geq n$ for each n, so that if $q > p > m_\varepsilon$, then $\|\sum_{n=p}^{q} x_{k_n}\| < \varepsilon$. In other words, (x_{k_n}) is summable.
(iii)⇒(iv): Let (ε_n) be a sequence of ± 1's. If (x_n) is subseries summable and we set $S^+ = \{n \in \mathbf{N} : \varepsilon_n = 1\}$ and $S^- = \{n \in \mathbf{N} : \varepsilon_n = -1\}$, then it must be the case that both series $\sum_{n\in S^\pm} x_n$ are convergent. Fix $\varepsilon > 0$. Notice that if $p < q$ are natural numbers and $M^\pm = \{n \in S^\pm : p \leq n \leq q\}$, then

$$\sum_{n=p}^{q} \varepsilon_n x_n = \sum_{n\in M^+} x_n - \sum_{n\in M^-} x_n.$$

It follows from the convergence of $\sum_{n\in S^\pm} x_n$ that $\|\sum_{n\in M^\pm} x_n\| < \varepsilon/2$ for p sufficiently large. Then $\|\sum_{n=p}^{q} \varepsilon_n x_n\| < \varepsilon$ for such p, and this is enough to ensure sign summability.
(iv)⇒(ii): For this final step, we will again take the contrapositive route. Assuming that (x_n) is not unordered summable, we have no option but to admit the existence of an exceptional $\delta > 0$ and a sequence (M_k) of finite subsets of $\mathbf{N}$ for which $\max M_k < \min M_{k+1}$ and $\|\sum_{n\in M_k} x_n\| \geq \delta$. Assign ε_n the value $+1$ if $n \in \bigcup_k M_k$, and -1 otherwise. The partial sums of $((1+\varepsilon_n)x_n)$ then fail to be Cauchy, so at least one of the series $\sum x_n$ and $\sum \varepsilon_n x_n$ will find it impossible to converge. QED

We have just seen that unconditional summability of (x_n) yields summability of $(b_n x_n)$ for certain very special sequences in ℓ_∞ – the sequences of signs. These special (b_n)'s are, at least in the real case, just the extreme points of B_{ℓ_∞}, so, in view of the Krein-Milman Theorem, the next result should not come as too much of a surprise.

1.6 Bounded Multiplier Test: *In any Banach space, the sequence (x_n) is unconditionally summable if and only if the sequence $(b_n x_n)$ is summable for every $(b_n) \in \ell_\infty$.*

Proof. One direction is trivial since we know sign summability implies unconditional summability. Accordingly, we fix an unconditionally summable sequence (x_n) in the Banach space X and show that $(b_n x_n)$ is summable when $b = (b_n) \in \ell_\infty$. The completeness of X means that we need only show that $\|\sum_{k=m}^n b_k x_k\|$ converges to zero as $m, n \to \infty$. Using duality,

$$\Big\|\sum_{k=m}^{n} b_k x_k\Big\| = \sup_{x^* \in B_{X^*}} \Big|\Big\langle x^*, \sum_{k=m}^{n} b_k x_k\Big\rangle\Big| \leq \|b\|_\infty \cdot \sup_{x^* \in B_{X^*}} \sum_{k=m}^{n} |\langle x^*, x_k\rangle|,$$

and this leads us to try to show that

$$(*) \qquad \lim_{m\to\infty} \sup_{x^* \in B_{X^*}} \sum_{k \geq m} |\langle x^*, x_k\rangle| = 0.$$

Unordered summability is the appropriate medium. If $\varepsilon > 0$ is given, there is an $m_\varepsilon \in \mathbf{N}$ such that for any finite set $M \subset \mathbf{N}$ with $\min M > m_\varepsilon$ we have $\|\sum_{k\in M} x_k\| < \varepsilon$. Let us concentrate on $\mathrm{Re}\,\langle x^*, x_k\rangle$ for a fixed $x^* \in B_{X^*}$. Choose $n > m > m_\varepsilon$ and set

$$M^+ = \{m \leq k \leq n : \mathrm{Re}\,\langle x^*, x_k\rangle \geq 0\}$$

and

$$M^- = \{m \leq k \leq n : \mathrm{Re}\,\langle x^*, x_k\rangle < 0\}.$$

Then

$$\begin{aligned}\sum_{k=m}^{n} |\mathrm{Re}\,\langle x^*, x_k\rangle| &= \Big|\mathrm{Re}\,\Big\langle x^*, \sum_{k\in M^+} x_k\Big\rangle\Big| + \Big|\mathrm{Re}\,\Big\langle x^*, \sum_{k\in M^-} x_k\Big\rangle\Big| \\ &\leq \Big\|\sum_{k\in M^+} x_k\Big\| + \Big\|\sum_{k\in M^-} x_k\Big\| < 2\cdot\varepsilon.\end{aligned}$$

Similarly, $\sum_{k=m}^n |\mathrm{Im}\,\langle x^*, x_k\rangle| < 2\varepsilon$. All this is clearly enough to prove $(*)$.

QED

We have now an extensive array of norm convergence conditions equivalent to unconditional summability. It is remarkable that most of these conditions are equivalent to an analogous weak convergence condition, and our next aim is to set up the transfer process by proving two subtle interchange of limits theorems, the first an old gem of Schur, its sequel the Orlicz - Pettis Theorem.

1.7 Schur's ℓ_1 Theorem: *In ℓ_1, weak convergence and norm convergence of sequences are the same.*

Proof. For the non-trivial implication there is no loss of generality in working with a weakly null sequence $(x^{(n)})$ in ℓ_1. We need to show that $\sum_k |x_k^{(n)}| \to 0$ as $n \to \infty$.

The weakly null nature of $(x^{(n)})$ ensures that for any fixed integer N, $\sum_{k=1}^N |x_k^{(n)}| \to 0$ as $n \to \infty$. The hard part is to control the tails. To achieve this, recall that B_{ℓ_∞} is weak$*$compact, and is even weak$*$metrizable, since ℓ_∞ is the dual of the separable space ℓ_1. It is perhaps worth noting that a suitable metric on B_{ℓ_∞} is given by $d(f,g) = \sum_k 2^{-k}|f_k - g_k|$. It is certainly worth recalling that a basis of weak$*$neighbourhoods about some $\hat{f} \in B_{\ell_\infty}$ is given by all

$$U(\hat{f}, \delta, N) = \{f \in B_{\ell_\infty} : |f_k - \hat{f}_k| < \delta,\ 1 \le k \le N\},$$

where $\delta > 0$ and $N \in \mathbf{N}$.

The combination of the weak$*$topology – which is finitary in nature – and the compact metrizability – which will enable us to invoke Baire's Category Theorem – turns out to be just what we need to gain uniform control of the tails.

Fix $\varepsilon > 0$, and for each $m \in \mathbf{N}$, set

$$B_m = \bigcap_{n \ge m} \{f \in B_{\ell_\infty} : |\langle f, x^{(n)} \rangle| \le \frac{\varepsilon}{3}\}.$$

Each B_m is a weak$*$closed subset of B_{ℓ_∞}, since it is an intersection of such sets. The B_m's clearly grow with m, and the weak nullity of $(x^{(n)})$ gives

$$B_{\ell_\infty} = \bigcup_m B_m.$$

Baire's Category Theorem now tells us that one of the B_m's, say B_{m_0}, has non-empty interior. This ensures that we can find $\hat{f} \in B_{m_0}$, $\delta > 0$ and $N \in \mathbf{N}$ such that

$$U(\hat{f}, \delta, N) \subset B_{m_0} \subset B_m$$

for all $m \ge m_0$. By adjusting m_0 if necessary, we can also assume

$$\sum_{k=1}^{N} |x_k^{(m)}| < \frac{\varepsilon}{3}$$

for all $m \ge m_0$.

Now fix any $m \ge m_0$. Take advantage of the finitary constraints governing membership of $U(\hat{f}, \delta, N)$, and define $f \in B_{\ell_\infty}$ by

$$f_k = \hat{f}_k \text{ if } 1 \le k \le N \quad \text{and} \quad f_k = \operatorname{sign} x_k^{(m)} \text{ if } k > N.$$

Then $f \in U(\hat{f}, \delta, N) \subset B_{m_0}$, so $|\langle f, x^{(m)} \rangle| \le \varepsilon/3$. It follows that

$$\begin{aligned}\sum_k |x_k^{(m)}| &= \sum_{k \le N} |x_k^{(m)}| + \Big|\sum_k f_k x_k^{(m)} - \sum_{k \le N} f_k x_k^{(m)}\Big| \\ &\le \sum_{k \le N} (1 + |f_k|)|x_k^{(m)}| + |\langle f, x^{(m)} \rangle| \le \frac{2\varepsilon}{3} + \frac{\varepsilon}{3} = \varepsilon,\end{aligned}$$

which was what we wanted. QED

With this theorem in mind we say that a Banach space has the *Schur property* if all of its weakly convergent sequences are norm convergent.

Schur's ℓ_1 Theorem is a cornerstone for our proof of the Orlicz - Pettis Theorem. We shall also make critical use of another fundamental result about weak topologies, due to S. Mazur:

- *for convex subsets, in particular subspaces, the weak and norm closures coincide.*

1.8 Orlicz - Pettis Theorem: *For sequences in a Banach space, weak subseries summability and (norm) subseries summability are the same.*

Naturally, the definition of weak subseries summability is almost a carbon copy of subseries summability: just replace the norm topology by the weak topology.

Proof. Obviously the norm property implies the weak property, but the converse requires work.

Let (x_n) be weakly subseries summable. We may as well assume that our Banach space is separable since all the action happens inside the (weakly) closed linear span of the x_n's. The proof proceeds by creating a natural operator $v : X^* \to \ell_1$ given by $v(x^*) = (\langle x^*, x_n \rangle)$. We shall use Schur's ℓ_1 Theorem to show that v is compact. From there it will only be a short step to obtain subseries summability of (x_n).

We must first justify v's existence as a bounded linear operator. Since (x_n) is weakly subseries summable, weak-$\lim_n \sum_{j=1}^n x_{k_j}$ exists for all increasing sequences (k_j) of positive integers. It follows that the *scalar* sequence $(\langle x^*, x_n \rangle)$ is unconditionally summable, so absolutely summable, for each x^* in X^*. Thus the map $v : X^* \to \ell_1$ is born. Easily, v is linear and has closed graph; hence v is a bounded linear operator.

To show that v is compact, we start with a sequence (x_m^*) in B_{X^*}. The supposed separability of X makes B_{X^*} compact and metrizable in the weak$*$topology. This allows us to extract a weak$*$convergent subsequence $(x_{m_k}^*)$ from (x_m^*) with weak$*$limit x_0^*, say. If we can show that vx_0^* is the norm limit of $(vx_{m_k}^*)$, we shall have established the compactness of v. But, since v takes values in ℓ_1, Schur's ℓ_1 Theorem is at our disposal, and so all we need to show is that vx_0^* is the weak limit of $(vx_{m_k}^*)$. For this it is enough to prove that $\langle f, vx_0^* \rangle = \lim_k \langle f, vx_{m_k}^* \rangle$ for any f in some norm dense subset of ℓ_∞. Recall that the simple functions form such a dense subset of ℓ_∞, so linearity allows us to restrict our testing set to the collection of characteristic functions $f = 1_M$ of subsets M of $\mathbf{N}$. For such an f, the weak subseries summability of (x_n) gives

$$\begin{aligned} \langle f, vx_0^* \rangle &= \sum_{n\in M} \langle x_0^*, x_n \rangle = \langle x_0^*, \sum_{n\in M} x_n \rangle = \lim_{k\to\infty} \langle x_{m_k}^*, \sum_{n\in M} x_n \rangle \\ &= \lim_{k\to\infty} \sum_{n\in M} \langle x_{m_k}^*, x_n \rangle = \lim_{k\to\infty} \langle f, vx_{m_k}^* \rangle. \end{aligned}$$

The proof could now be completed in a variety of ways. Perhaps the most elementary begins by recalling how to identify the relatively compact subsets of ℓ_1: these are the bounded subsets K with uniformly small tails, by which we mean that regardless of $\varepsilon > 0$ there is an $n_\varepsilon \in \mathbf{N}$ such that $\sum_{n>n_\varepsilon} |a_n| \le \varepsilon$ for every $(a_n) \in K$. We shall take $v(B_{X^*})$ as our K. Then, for any finite subset M of $\mathbf{N}$ with $\min M > n_\varepsilon$,

$$\| \sum_{n\in M} x_n \| \le \sup_{x^*\in B_{X^*}} \sum_{n\in M} |\langle x^*, x_n \rangle| \le \varepsilon.$$

Our sequence is unordered summable. QED

Stop and think for a moment. We have actually proved more than what was stated in the Orlicz-Pettis Theorem: the compactness of the natural operator $v : X^* \to \ell_1$ is also equivalent to subseries summability of (x_n).

The Orlicz-Pettis Theorem has opened the door to the weak topology. Using the proofs of the elementary equivalences to (norm) subseries summability (1.5) as a model, it is routine to derive a similar list of weak properties equivalent to weak subseries summability. Let's draw together all that we know.

1.9 Omnibus Theorem on Unconditional Summability: *For a sequence (x_n) in a Banach space X, the following are equivalent:*

(i) (x_n) *is unconditionally summable.*

(ii) (x_n) *is unordered summable.*

(iii) (x_n) *is subseries summable.*

(iv) (x_n) *is sign summable.*

(v) $(b_n x_n)$ *is summable for every* (b_n) *in* ℓ_∞.

(vi) (x_n) *is weakly subseries summable.*

(vii) (x_n) *is weakly sign summable, that is* $\sum_n \varepsilon_n x_n$ *converges weakly in* X *for every choice of signs* $\varepsilon_n = \pm 1$

(viii) $(b_n x_n)$ *is weakly summable in* X *for every* $(b_n) \in \ell_\infty$.

(ix) $v : X^* \to \ell_1 : x^* \mapsto (\langle x^*, x_n \rangle)_n$ *is a compact operator.*

(x) $(b_n) \mapsto \sum_n b_n x_n$ *defines a compact operator* $\ell_\infty \to X$.

(xi) $(b_n) \mapsto \sum_n b_n x_n$ *defines a compact operator* $c_0 \to X$.

(xii) $(b_n) \mapsto \sum_n b_n x_n$ *defines a bounded operator* $\ell_\infty \to X$.

Our discussion has shown so far that (i) through (ix) are equivalent. Thanks to (v), the operator in (x) is just (that induced by) the adjoint of the one in (ix), and v in (ix) is the adjoint of the operator in (xi). Since (x)⇒(xi) and (x)⇒(xii)⇒(v) are trivial, the remaining equivalences are a consequence of Schauder's Theorem: a Banach space operator is compact precisely when its (bi)adjoint is.

The omission of a weak analogue of (i) from our list is inevitable. For an example, let (e_n) be the standard unit vector basis in c_0. The sequence given by $x_1 = e_1$ and $x_n = e_n - e_{n-1}$ for $n \geq 2$ shows that $\sum_n x_{\sigma(n)}$ may very well converge weakly in a Banach space for every permutation σ of $\mathbf{N}$ without (x_n) being unconditionally summable.

Khinchin's Inequality

Having seen that unconditional and absolute summability are *not* the same in general, it behoves us to seek out analytic conclusions that *can* be drawn about unconditionally summable sequences. This section serves both as an appetizer for future mathematical treats and as a provider of important ballast for times of need.

Everything hangs on our next topic, the remarkable Rademacher functions and their rôle in Khinchin's Inequality. Formally, the *Rademacher functions*

$$r_n : [0,1] \longrightarrow \mathbf{R},\ n \in \mathbf{N},$$

are defined by setting

$$r_n(t) := \operatorname{sign}(\sin 2^n \pi t).$$

It is useful to observe that if we extend $r_1 = 1_{(0,1/2]} - 1_{[1/2,1)}$ periodically to the whole line, then $r_{n+1}(t) = r_1(2^n t)$ for all n. Occasionally, it will be convenient to refer to the constant one function as the zero'th Rademacher function r_0.

To get an understanding of the Rademacher functions, we recommend picturing their graphs rather than struggling with the formulae:

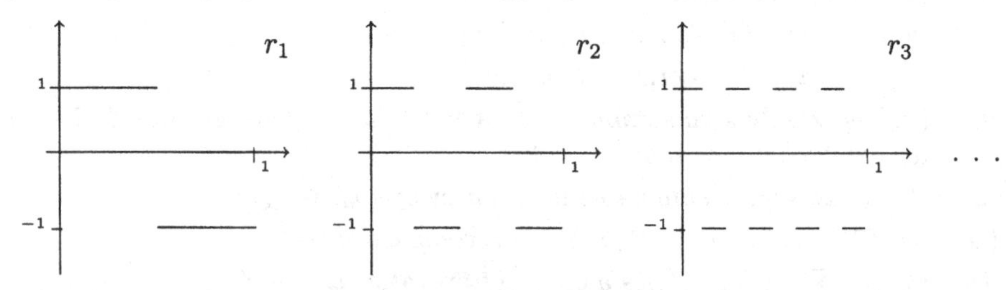

The most important feature of the Rademacher functions is that they have nice orthogonality properties. If $0 < n_1 < n_2 < ... < n_k$ and $p_1,...,p_k \geq 0$ are integers, then

$$\int_0^1 r_{n_1}^{p_1}(t)\cdot \ldots \cdot r_{n_k}^{p_k}(t)\,dt = \begin{cases} 1 & \text{if each } p_j \text{ is even} \\ 0 & \text{otherwise}\ . \end{cases}$$

This can easily be seen from the pictures; an analytical proof would be supremely boring.

An immediate consequence is that the r_n's form an orthonormal sequence in $L_2[0,1]$, and so

$$\int_0^1 \left| \sum a_n r_n(t) \right|^2 dt = \sum |a_n|^2$$

for all $(a_n) \in \ell_2$. Beware! They do *not* form an orthonormal basis: $\cos(2\pi t)$ and $r_1 \cdot r_2$, for example, are orthogonal to all the r_n's.

The main result about the Rademacher functions is a powerful inequality.

1.10 Khinchin's Inequality: *For any $0 < p < \infty$, there are positive constants A_p, B_p such that regardless of the scalar sequence (a_n) in ℓ_2 we have*

$$A_p\cdot\Big(\sum_n |a_n|^2\Big)^{1/2} \leq \Big(\int_0^1 \Big|\sum_n a_n r_n(t)\Big|^p\,dt\Big)^{1/p} \leq B_p\cdot\Big(\sum_n |a_n|^2\Big)^{1/2}.$$

Notice that the statement can be rephrased to say that on the span of the Rademacher functions all the L_p metrics are equivalent. The best possible constants A_p and B_p depend on the choice of the scalar field, but we defer any discussion of this topic to Notes and Remarks. It will be enough to prove Khinchin's Inequality for real numbers; the complex case follows readily enough when all scalars are decomposed into their real and imaginary parts.

Bearing in mind the monotonicity of the L_p metrics, we could establish Khinchin's Inequality by concentrating on powers $p = 2^n$ and using induction. The general inductive step is repulsive, but it is informative and elementary to present the case $p=4$ in isolation. It is this case that we shall use in the course of proving another notable inequality, due to Grothendieck. An independent argument, valid for all $0 < p < \infty$, will follow hard on the heels of the special case.

Proof of Khinchin's Inequality for p=4. A simple limit argument shows that it is sufficient to work with a finite sequence $(a_1,\dots,a_m)$ of real scalars. Then

$$\begin{aligned}\int_0^1 \Big|\sum_{n\le m} a_n r_n(t)\Big|^4 dt &= \int_0^1 \Big(\sum_{i\le m} a_i r_i(t)\Big)\cdot\Big(\sum_{j\le m} a_j r_j(t)\Big)\cdot\Big(\sum_{k\le m} a_k r_k(t)\Big)\cdot\Big(\sum_{\ell\le m} a_\ell r_\ell(t)\Big)\,dt \\ &= \sum_{i,j,k,\ell\le m} a_i a_j a_k a_\ell \int_0^1 r_i(t) r_j(t) r_k(t) r_\ell(t)\,dt = 3\cdot\sum_{i,j\le m} a_i^2 a_j^2 - 2\cdot\sum_{i\le m} a_i^4\end{aligned}$$

since the orthogonality properties of the r_n's kill off the integrals unless the indices are equal in pairs. The negative term is present since we must take care not to count the cases $i = j = k = \ell$ more than once. It follows that

$$\int_0^1 \Big|\sum_{n\le m} a_n r_n(t)\Big|^4 dt \le 3\cdot\sum_i a_i^2\cdot\sum_j a_j^2 = 3\cdot\Big(\sum_n a_n^2\Big)^2,$$

and so

$$\Big\|\sum_{n\le m} a_n r_n\Big\|_2 \le \Big\|\sum_{n\le m} a_n r_n\Big\|_4 \le 3^{1/4}\cdot \Big\|\sum_{n\le m} a_n r_n\Big\|_2 .$$

We have shown $B_4 \le 3^{1/4}$ and $A_4 \ge 1$; in fact, it is easily seen that $A_4 = 1$.

QED

It should be clear, in principle, how to tackle a general inductive step to treat the cases where p is a power of 2. But it should also be clear that such a procedure would inevitably lead us to countenance considerably convoluted combinatorial contortions. We back away from this challenge and follow another path.

Proof of Khinchin's Inequality for general p. The procedure is to start by establishing the result for integer values of p. An argument using Hölder's Inequality and the monotonicity of the L_p norms will then dispose of the remaining cases. As before, it suffices to deal with finitely many real numbers $a_1, \dots, a_m$, not all zero.

If $p \in \mathbf{N}$ and $y \in \mathbf{R}$, then $|y|^p < p! \cdot (1 + |y|^p/p!) \leq p! \cdot e^{|y|}$. So, if we set $f(t) = \sum_{n \leq m} a_n r_n(t)$, we have

$$\int_0^1 |f(t)|^p dt \leq p! \cdot \int_0^1 e^{|f(t)|} dt \leq p! \cdot \int_0^1 \left(e^{f(t)} + e^{-f(t)}\right) dt.$$

Normalize to arrange that $\|f\|_2 = (\sum_{n \leq m} a_n^2)^{1/2} = 1$, and observe that $\int_0^1 e^{f(t)}\, dt = \int_0^1 \Pi_{n \leq m} \exp(a_n r_n(t))\, dt$. Expanding the integrand as a product of power series, and taking note of the orthogonality properties of the r_n's, we may interchange the integral and the product. This can also be seen, if you prefer, by thinking of the r_n's as independent random variables. In any event,

$$\int_0^1 e^{f(t)} dt = \prod_{n \leq m} \int_0^1 \exp\big(a_n r_n(t)\big)\, dt = \prod_{n \leq m} \cosh(a_n),$$

and a quick comparison of power series reveals that this is

$$\leq \prod_{n \leq m} \exp\left(\frac{a_n^2}{2}\right) = \exp\Big(\sum_{n \leq m} \frac{a_n^2}{2}\Big) = e^{1/2}.$$

By symmetry, $\int_0^1 e^{-f(t)} dt \leq e^{1/2}$, and consequently

$$\int_0^1 |f(t)|^p dt \leq 2p! \cdot e^{1/2}.$$

For $2 \leq p < \infty$, we take account of monotonicity and homogeneity of the L_p norms to conclude that, for arbitrary $a_1, \dots, a_m \in \mathbf{R}$,

$$\Big(\sum_{n \leq m} a_n^2\Big)^{1/2} \leq \Big\| \sum_{n \leq m} a_n r_n \Big\|_p \leq \big(2k! \cdot e^{1/2}\big)^{1/k} \cdot \Big(\sum_{n \leq m} a_n^2\Big)^{1/2},$$

k being the integer following p.

We are left with the case $0 < p < 2$. We settle this by invoking a now standard trick using Hölder's Inequality. Given $0 < p < 2$, define $0 < \theta < 1$ by $\theta = (2 - (p/2))^{-1}$, so that $p\theta + 4(1 - \theta) = 2$. Then

$$\int_0^1 |f(t)|^2 dt = \int_0^1 |f(t)|^{p\theta} \cdot |f(t)|^{4(1-\theta)} dt \leq \Big(\int_0^1 |f(t)|^p dt\Big)^{\theta} \cdot \Big(\int_0^1 |f(t)|^4 dt\Big)^{1-\theta}.$$

We know that $\|f\|_4 \leq B_4 \cdot \|f\|_2$, whence $B_4^{2-4/p} \cdot \|f\|_2 \leq \|f\|_p$. On the other hand, another appeal to monotonicity yields $\|f\|_p \leq \|f\|_2$. QED

The device used to treat the case $0 < p < 2$ can be employed to establish a slightly more general fact: if L_p metrics are equivalent on some common subspace for $0 < p_1 < p \leq p_2$, then they must also be equivalent for $0 < p \leq p_2$.

The Dvoretzky-Rogers Theorem (1.2) indicates that regardless of the infinite dimensional Banach space X and regardless of the scalar sequence (λ_n) in ℓ_2 there will always be an unconditionally summable sequence (x_n) in X for which $\|x_n\| = |\lambda_n|$ for all n. Can we do better? For instance, can we replace

ℓ_2 by the bigger space ℓ_q when $q > 2$? The next result puts paid to all such speculations and affords us our first opportunity to use Khinchin's Inequality.

1.11 Orlicz's Theorem: *If (f_n) is an unconditionally summable sequence in $L_1[0,1]$, then $\sum_n \|f_n\|_1^2 < \infty$.*

Proof. As usual, by considering real and imaginary parts separately, it is enough to work with real-valued functions. Recall from our Omnibus Theorem 1.9 that (f_n) gives rise to the (compact) operator $v: L_\infty[0,1] \to \ell_1 : g \mapsto (\langle g, f_n \rangle)$. By sign summability, any sequence of signs $\varepsilon_n = \pm 1$ satisfies

$$\Big\| \sum_n \varepsilon_n f_n \Big\|_1 \le \sup_{g \in B_{L_\infty}} \sum_n |\langle g, f_n \rangle| = \|v\|.$$

We observe that regardless of the natural number m, we can use Minkowski's Inequality followed by Khinchin's Inequality to deduce that

$$\Big(\sum_{n \le m} \|f_n\|_1^2\Big)^{1/2} \le \int_0^1 \Big(\sum_{n \le m} f_n(s)^2\Big)^{1/2} ds \le A_1^{-1} \cdot \int_0^1 \int_0^1 \Big| \sum_{n \le m} f_n(s) r_n(t) \Big| \, dt \, ds.$$

Switching the order of integration, the last expression becomes

$$A_1^{-1} \cdot \int_0^1 \Big\| \sum_{n \le m} r_n(t) f_n \Big\|_1 \, dt,$$

and this is bounded above by $A_1^{-1} \cdot \|v\|$ since $r_n(t) = \pm 1$ for all t outside the countable set of dyadic rationals $\{k \cdot 2^{-n} : 0 \le k \le 2^n, \, n \in \mathbf{N}\}$. As a result,

$$\Big(\sum_n \|f_n\|_1^2\Big)^{1/2} \le A_1^{-1} \cdot \|v\| < \infty. \qquad \text{QED}$$

Later we shall prove much more general versions of Orlicz's Theorem; see 3.12, 10.7 and 11.17.

Another powerful consequence of Khinchin's Inequality is the existence of easily described Hilbert subspaces of $L_p[0,1]$. Let us introduce the notation

$$Rad_p$$

for the closed linear span of the Rademacher functions in $L_p[0,1]$, for any $0 < p \le \infty$.

1.12 Theorem: (a) *For $0 < p < \infty$, Rad_p is isomorphic to ℓ_2.*

(b) *Rad_∞ is isomorphic to ℓ_1, isometrically in the real case.*

(c) *For $1 < p < \infty$, Rad_p is complemented in $L_p[0,1]$.*

Proof. (a) This is an immediate consequence of Khinchin's Inequality. The isomorphism is given by $(a_n) \mapsto \sum_n a_n r_n$.

(b) First the real case. If $c_1, \ldots, c_n$ are real numbers, we can certainly find a dyadic interval of length 2^{-n} on which each r_j $(1 \le j \le n)$ agrees in sign with the corresponding c_j. It is now clear that

$$\Big\| \sum_{j=1}^n c_j r_j \Big\|_\infty = \sum_{j=1}^n |c_j|.$$

The complex case is a quick consequence. If $c_j = a_j + i \cdot b_j$ $(1 \le j \le n)$ are complex numbers,

$$\frac{1}{2} \cdot \sum_{j=1}^{n} |c_j| \le \max\Big\{\sum_{j=1}^{n} |a_j|, \sum_{j=1}^{n} |b_j|\Big\} = \max\Big\{\Big\|\sum_{j=1}^{n} a_j r_j\Big\|_\infty, \Big\|\sum_{j=1}^{n} b_j r_j\Big\|_\infty\Big\}$$
$$\le \Big\|\sum_{j=1}^{n} c_j r_j\Big\|_\infty \le \sum_{j=1}^{n} |c_j|.$$

In each case, the passage to infinite sums is elementary.

(c) For each $1 < p < \infty$, we have to find a bounded linear projection $P_p : L_p[0,1] \to L_p[0,1]$ with range Rad_p.

When $p = 2$, all is straightforward: just let P_2 be the orthogonal projection on $L_2[0,1]$ defined by $P_2 f = \sum_n \langle f, r_n\rangle r_n$, where $\langle f, r_n\rangle = \int_0^1 f(t) r_n(t)\, dt$.

When $p > 2$, we make use of the fact that $L_p[0,1]$ embeds continuously into $L_2[0,1]$; with Khinchin's Inequality, this ensures that $f \mapsto \sum_n \langle f, r_n\rangle r_n$ defines a bounded operator $P_p : L_p[0,1] \to L_p[0,1]$ with

$$\|P_p f\|_p = \Big\|\sum \langle f, r_n\rangle r_n\Big\|_p \le B_p \cdot \Big\|\sum \langle f, r_n\rangle r_n\Big\|_2 \le B_p \cdot \|f\|_2 \le B_p \cdot \|f\|_p.$$

This map is patently a projection onto Rad_p.

When $1 < p < 2$, duality comes to the rescue. The map $P_p = (P_{p^*})^*$ is definitely a bounded linear projection on $L_p[0,1]$. To see that it behaves the way we want, first note that Rad_p is in the range of P_p, since $\langle P_p r_n, g\rangle = \langle r_n, g\rangle$ for all natural numbers n and all g in $L_{p^*}[0,1]$; we even have

$$\langle P_p f, g\rangle = \langle f, P_{p^*} g\rangle = \Big\langle f, \sum_n \langle g, r_n\rangle r_n\Big\rangle = \lim_{n\to\infty} \sum_{k=1}^{n} \langle f, r_k\rangle \cdot \langle g, r_k\rangle$$
$$= \lim_{n\to\infty} \Big\langle \sum_{k=1}^{n} \langle f, r_k\rangle r_k, g\Big\rangle$$

for $f \in L_p[0,1]$ and $g \in L_{p^*}[0,1]$.

This shows that $\sum_{n=1}^\infty \langle f, r_n\rangle r_n$ converges weakly to $P_p f$, and so Rad_p is weakly dense, and thus norm dense in the range of P_p by convexity. Since we are dealing with Banach spaces, Rad_p must coincide with the range of P_p.

QED

Note that we also get $P_p f = \sum_{n=1}^\infty \langle f, r_n\rangle r_n$ for $f \in L_p[0,1]$, since P_p obviously extends P_2 and $L_2[0,1]$ embeds densely into $L_p[0,1]$.

Simply by playing around with two two-dimensional vectors we could have shown that, in the complex case, Rad_∞ is not isometrically isomorphic to ℓ_1.

We emphasize that Rad_1 is *not* complemented in $L_1[0,1]$. More machinery is needed to justify this assertion, and once this has been developed, we shall even be able to show that no subspace isomorphic to Hilbert space can ever be complemented in $L_1[0,1]$. This will become clearer by the end of this chapter.

GROTHENDIECK'S INEQUALITY

Though Orlicz's Theorem dashes any hope of substantial refinements of the Dvoretzky - Rogers Theorem, a change of viewpoint opens up new horizons.

Let $u : X \to Y$ be a continuous linear operator between Banach spaces. Reflect for a moment: u takes absolutely summable sequences into absolutely summable sequences and unconditionally summable sequences into unconditionally summable sequences. A fortiori, u takes absolutely summable sequences into unconditionally summable sequences. But for *any* infinite dimensional Banach space the identity *fails* to take *all* unconditionally summable sequences to absolutely summable ones.

Those operators $u : X \to Y$ which *do* take unconditionally summable sequences (x_n) in X to absolutely summable sequences (ux_n) in Y are called *absolutely summing operators*. They and their many remarkable ramifications form the subject matter of this book.

We present a fundamental and elegant result of Grothendieck to whet the appetite for what follows.

1.13 Grothendieck's Theorem: *Every continuous linear operator $u : \ell_1 \to \ell_2$ is absolutely summing.*

The theorem will appear as a consequence of a profound matrix inequality which will recur in many guises throughout these proceedings. Our proof of this inequality hinges on the simplest form of Khinchin's Inequality (1.10).

1.14 Grothendieck's Inequality: *There is a universal constant κ_G for which, given any Hilbert space H, any $n \in \mathbf{N}$, any $n \times n$ scalar matrix (a_{ij}) and any vectors $x_1, \ldots, x_n$, $y_1, \ldots, y_n$ in B_H, we have*

$$(G) \qquad \Big|\sum_{i,j} a_{ij}(x_i|y_j)\Big| \le \kappa_G \cdot \max\Big\{\Big|\sum_{i,j} a_{ij}s_it_j\Big| : |s_i| \le 1,\ |t_j| \le 1\Big\}.$$

Two comments are in order. First, the maximum occuring on the right hand side of (G) can be viewed as the norm of (a_{ij}) considered as an operator from ℓ_∞^n to ℓ_1^n. Since any norming subset of $B_{\ell_\infty^n}$ serves to evaluate the norm of this operator, we may, in the real case, replace (G) by

$$(G') \qquad \Big|\sum_{i,j} a_{ij}(x_i|y_j)\Big| \le \kappa_G \cdot \max\Big\{\Big|\sum_{i,j} a_{ij}\varepsilon_i\varepsilon'_j\Big| : \varepsilon_i, \varepsilon'_j = \pm 1\Big\}.$$

Second, the best possible κ_G – currently unknown – is generally called *Grothendieck's constant*; it depends on the chosen scalar field.

Proof of Grothendieck's Inequality. Here we are not going to worry about getting good estimates for κ_G. It is therefore enough to concern ourselves only with real matrices and real Hilbert spaces. The complex case follows, with a possible doubling of the constant, by the usual decomposition into real and imaginary parts.

For simplicity we write

$$\|a\| = \sup\big\{\big|\sum_{i,j} a_{ij}s_it_j\big| : |s_i| \leq 1,\ |t_j| \leq 1\big\}$$

and

$$|\!|\!|a|\!|\!| = \sup\big|\sum_{i,j} a_{ij}(x_i|y_j)\big|$$

where the latter supremum is taken over all Hilbert spaces H and all vectors $x_1,\ldots,x_n$, $y_1,\ldots,y_n$ in the unit ball B_H of the ambient Hilbert space. It is plain that separable Hilbert spaces suffice to obtain this supremum, since at each stage we only use a finite number of vectors.

A central idea in the proof is the use of the Rademacher functions to obtain an embedding of a separable Hilbert space H into $L_2[0,1]$ which respects the inner product. If (e_n) is an orthonormal basis for H, any $x \in H$ has a Fourier series expansion $x = \sum_n (x|e_n)e_n$. Using $\xi_n = (x|e_n)$ we define $X : [0,1] \to \mathbf{R}$ by

$$X(t) := \sum_n \xi_n r_n(t).$$

Thanks to the orthonormality of the r_n's, each X belongs to $L_2[0,1]$, with $\|X\|_2 = \|x\|$, and for any x, $y \in H$

$$(x\,|\,y) = \int_0^1 X(t)Y(t)\,dt.$$

Were it the case (it isn't, by the way) that, for all the X's in $L_2[0,1]$ resulting from the x's in B_H, the function $|X|$ had a fixed upper bound, say M, then Grothendieck's Inequality would be a triviality. Indeed, in such a case we would have for any $x_1,\ldots,x_n$, $y_1,\ldots,y_n$ in B_H

$$\begin{aligned}\big|\sum_{i,j} a_{ij}(x_i|y_j)\big| &= \big|\int_0^1 \sum_{i,j} a_{ij}X_i(t)Y_j(t)\,dt\big| \\ &\leq M^2 \cdot \int_0^1 \big|\sum_{i,j} a_{ij}\frac{X_i(t)}{M}\cdot\frac{Y_j(t)}{M}\big|\,dt \leq M^2 \cdot \|a\|.\end{aligned} \tag{1}$$

It would follow that $|\!|\!|a|\!|\!| \leq M^2 \cdot \|a\|$.

Something is needed to circumvent the lack of uniform boundedness displayed by the image of B_H in $L_2[0,1]$. The functional representation affords us two advantages: we can exploit the simple device of truncation, and we gain access to Khinchin's Inequality.

Given $x \in H$ and $M > 0$, define X^L and X^U in $L_2[0,1]$ by

$$X^L(t) := \begin{cases} X(t), & \text{if } |X(t)| \leq M \\ M \cdot \operatorname{sign} X(t), & \text{otherwise} \end{cases}$$

and

$$X^U(t) := X(t) - X^L(t).$$

We have split each X into two components. The first, X^L, is uniformly bounded by M, so arguments like those in (1) are applicable; the second, X^U, has a readily estimable L_2 norm. In fact, we show that for any $x \in B_H$

$$\|X^U\|_2 \leq \frac{\sqrt{3}}{4M}\,. \tag{2}$$

Observe that for a given $t \in [0,1]$, $X^U(t)$ may or may not be zero. If not, then $|X(t)| > M$, and in fact $|X^U(t)| = |X(t)| - M$. The elementary inequality $s \leq m + (s^2/4m)$ $(m, s > 0)$ reveals that, regardless of $X^U(t)$'s status, $|X^U(t)| \leq X(t)^2/4M$, so that

$$\|X^U\|_2^2 \leq \frac{1}{16M^2} \cdot \|X\|_4^4 \leq \frac{3}{16M^2}$$

thanks to the simplest form of Khinchin's Inequality.

With this in hand we can complete our proof. If $x_1, \ldots, x_n$, $y_1, \ldots, y_n$ are in B_H, then by (1) and (2)

$$\begin{aligned}
\Big|\sum_{i,j} a_{ij}(x_i|y_j)\Big| &= \Big|\int_0^1 \sum_{i,j} a_{ij} X_i(t) Y_j(t)\, dt\Big| \\
&\leq \Big|\int_0^1 \sum_{i,j} a_{ij} X_i^L(t) Y_j^L(t)\, dt\Big| + \Big|\int_0^1 \sum_{i,j} a_{ij} X_i^U(t) Y_j^L(t)\, dt\Big| \\
&\qquad + \Big|\int_0^1 \sum_{i,j} a_{ij} X_i(t) Y_j^U(t)\, dt\Big| \\
&\leq M^2 \cdot \|a\| + \frac{\sqrt{3}}{4M} \cdot \Big|\int_0^1 \sum_{i,j} a_{ij} \cdot \Big(\frac{X_i^U(t) Y_j^L(t)}{\|X_i^U\|_2} + \frac{X_i(t) Y_j^U(t)}{\|Y_j^U\|_2}\Big)\, dt\Big| \\
&\leq M^2 \cdot \|a\| + \frac{\sqrt{3}}{2M} \cdot |\!|\!|a|\!|\!|.
\end{aligned}$$

We have seen that for each $M > 0$, $|\!|\!|a|\!|\!| \leq M^2 \cdot \|a\| + \frac{\sqrt{3}}{2M} \cdot |\!|\!|a|\!|\!|$; hence

$$|\!|\!|a|\!|\!| \leq \frac{2M^3}{2M - \sqrt{3}} \cdot \|a\|$$

whenever $M > \sqrt{3}/2$. Elementary calculus reveals that the optimal choice of M is $3 \cdot \sqrt{3}/4$, and thus $|\!|\!|a|\!|\!| \leq (81/16) \cdot \|a\|$. QED

Now that we have Grothendieck's Inequality in hand, we turn to the proof of his theorem.

Proof of Grothendieck's Theorem. There is no loss in assuming that $\|u\| \leq 1$ and in restricting to real scalars.

Let (x_n) be an unconditionally summable sequence in ℓ_1. Since (x_n) is sign summable, $\sum_n \varepsilon_n x_n$ converges in ℓ_1 for any sequence of signs ε_n, and we have

$$\Big\|\sum_n \varepsilon_n x_n\Big\| \leq \sup_{x^* \in B_{\ell_\infty}} \sum_n |\langle x^*, x_n\rangle| = \|v\|$$

where v is the operator from $\ell_\infty = \ell_1^*$ to ℓ_1 referred to in the Omnibus Theorem.

We need to show that $\sum_n \|ux_n\|_2 < \infty$. The first step is to reduce to finite dimensions so that we can apply Grothendieck's Inequality.

Let $m \in \mathbf{N}$ and $\delta > 0$ be given. Choose $n \geq m$ and vectors $y_1,\dots,y_m$ in $\ell_1^n \subset \ell_1$ so that $\|x_i - y_i\| \leq \delta/2^i$ for $1 \leq i \leq m$. If n happens to be strictly greater than m, set $y_{m+1} = \dots = y_n = 0$ as well. For each i, write $y_i = \sum_{j=1}^n a_{ij}e_j$ for the expansion of y_i with respect to the unit coordinate vectors in ℓ_1^n. This gives us a matrix $a = (a_{ij})$ for use in Grothendieck's Inequality.

First think in terms of absolute summability:

$$\sum_{i=1}^n \|uy_i\|_2 = \sum_{i=1}^n \Big\|\sum_{j=1}^n a_{ij}ue_j\Big\|_2 = \sum_{i,j} a_{ij}(z_i|ue_j)$$

for appropriate $z_1,\dots,z_n \in B_{\ell_2^n}$. Note that this is well-suited for insertion into the left hand side of Grothendieck's Inequality.

Now switch to unconditional summability, interpreted as sign summability. Given $\varepsilon_1,\dots,\varepsilon_n = \pm 1$,

$$\begin{aligned}\Big\|\sum_{i=1}^n \varepsilon_i y_i\Big\|_1 &= \Big\|\sum_{j=1}^n\Big(\sum_{i=1}^n a_{ij}\varepsilon_i\Big)e_j\Big\|_1 = \sum_j \Big|\sum_i a_{ij}\varepsilon_i\Big| \\ &= \max\Big\{\Big|\sum_{i,j} a_{ij}\varepsilon_i\varepsilon_j'\Big| : \varepsilon_j' = \pm 1\Big\}.\end{aligned}$$

This is well-suited for insertion into the right hand side of Grothendieck's Inequality (see (G')).

As all this is in terms of the y_i's, we must arrange to recover the x_i's.

$$\begin{aligned}\sum_{i\leq n}\|ux_i\|_2 &\leq \sum_{i\leq n}\|uy_i\|_2 + \delta \leq \|a\| + \delta \leq \kappa_G\cdot\|a\| + \delta \\ &= \kappa_G\cdot\max\Big\{\Big\|\sum_{i\leq n}\varepsilon_i y_i\Big\|_1 : \varepsilon_i = \pm 1\Big\} + \delta \\ &\leq \kappa_G\cdot\max\Big\{\Big\|\sum_{i\leq n}\varepsilon_i x_i\Big\|_1 : \varepsilon_i = \pm 1\Big\} + (1+\kappa_G)\cdot\delta \leq \kappa_G\cdot\|v\| + (1+\kappa_G)\cdot\delta.\end{aligned}$$

A simple limit argument completes the proof. QED

The choice of δ has no relevance to the proof as it stands. However, we are free to let $\delta \to 0$, and then we get

$$\sum \|ux_i\|_2 \leq \kappa_G\cdot\|u\|\cdot \sup_{\varepsilon_i=\pm 1} \Big\|\sum \varepsilon_i x_i\Big\|_1$$

for any choice of (x_i) in ℓ_1. The next chapter will teach us how to appreciate inequalities like this.

It will turn out to be very convenient to have the following finite dimensional version of Grothendieck's Theorem at our disposal:

1.15 Corollary: *Let n and N be positive integers, and let $u : \ell_1^n \to \ell_2^N$ be any operator. Then, for any choice of $x_1,\ldots,x_m \in \ell_1^n$,*

$$\sum_{i=1}^{m} \|ux_i\|_{\ell_2^N} \leq \kappa_G \cdot \|u\| \cdot \sup_{\varepsilon_i = \pm 1} \Big\| \sum_{i=1}^{m} \varepsilon_i x_i \Big\|_{\ell_1^n}.$$

To derive this from 1.14, just embed ℓ_1^n and ℓ_2^N in the most obvious manner into ℓ_1 and ℓ_2, respectively.

Grothendieck's Theorem gives us good reason to believe what we mentioned after Theorem 1.12, namely that no subspace H of $L_1[0,1]$ which is isomorphic to ℓ_2 could ever be complemented in $L_1[0,1]$. All that is needed is a slight extension of Grothendieck's Theorem: every operator $L_1(\mu) \to \ell_2$ is absolutely summing. See Chapter 3 for details. Were H to be complemented in $L_1[0,1]$, it would follow that the identity of ℓ_2 would be absolutely summing – much to our distress.

Notes and Remarks

The starting point and central theme of this chapter is the pursuit of the first alternative encountered in the following dichotomy.

Theorem: *Let $\sum_n t_n$ be a convergent series of real numbers. Then precisely one of the following holds:*

(UC) *For every permutation π of $\mathbf{N}$, $\sum_n t_{\pi(n)}$ converges and the resulting sum is $\sum_n t_n$. Moreover, this occurs if and only if $\sum_n |t_n|$ converges.*

(CC) *There are permutations π of $\mathbf{N}$ for which $\sum_n t_{\pi(n)}$ is not convergent; in this case, for each $t \in \mathbf{R}$ there is a permutation π of $\mathbf{N}$ such that $t = \sum_n t_{\pi(n)}$. We can also find a permutation for which the resulting permuted sum diverges to $+\infty$ or $-\infty$.*

(UC) is due to J.P.G.L. Dirichlet [1837, p. 48], while (CC) made its first appearance in the work of B. Riemann [1854]. It was not until after Riemann's death that a small gap in his reasoning was discovered and closed by U. Dini [1868].

Proposition 1.1 takes note of the unconditional summability of absolutely summable sequences in a Banach space; this is one of the first consequences of completeness observed by S. Banach in his [1922] thesis. That the (unconditional) summability of an absolutely summable sequence in a normed linear space is tantamount to completeness seems to go back to early proofs of the completeness of L_p-spaces.

The question of unconditional summability implying absolute summability in infinite dimensional Banach spaces was raised in S. Banach's [1932] monograph (p. 240) and deemed interesting enough to be repeated as Problem 122 in the Scottish Book (see R.D. Mauldin [1981]). Though easy counterexamples could be found in many classical spaces, it was M.S. Macphail [1947] who first proved the existence of an unconditionally summable sequence in ℓ_1 that is not absolutely summable. Further, by making essential use of the Rademacher-Khinchin circle of ideas, he was the first to make explicit the basically Hilbertian character of the problem.

This connection was clarified by A. Dvoretzky and C.A. Rogers in their seminal paper [1950]. So provocative were the insights offered by this paper that A. Grothendieck [1953b] was moved to refer to the Dvoretzky - Rogers Lemma 1.3 (and its extension to be proved in 14.2) as the unique decisive result in the finer metric theory of general Banach spaces known in his day. He soon provided an alternative approach to this theorem in his [1955] thesis, a proof that relies on ideas which were to inspire A. Pietsch a decade later to introduce the notion of a p-summing operator. We return to these matters in Chapter 2.

The fact that finite dimensionality is necessary for unconditionally summable sequences to be absolutely summable is special to normed linear spaces. In his thesis, Grothendieck isolated the class of nuclear spaces, in part to address this issue for non-normable locally convex spaces. The standard source of information about general nuclear spaces, spaces in which the absolute summability of unconditionally summable sequences is a pivotal issue, is A. Pietsch's [1972a] monograph; for recent additions along these lines, see H. Jarchow [1981], [1984b] and H. Jarchow - K. John [1994].

It is curious that the second alternative, (CC), in the dichotomy opening this section was neglected for so long in Banach space theory. But, as P. Rosenthal [1987] has noted, "surprisingly few mathematicians know" even the complex scalar variant of the second alternative of (CC). It says that *given a convergent series of complex numbers, the set of sums of convergent rearrangements of the series is either a singleton, a line in* $\mathbf{C}$, *or the whole complex plane.* P. Lévy is responsible for the discovery of this. His [1905] proof is highly non-trivial. He offered his proof for the set of sums of rearrangements in $\mathbf{R}^k$ for $k \geq 3$, too, but it was flawed. E. Steinitz [1913] found a proper formulation and a correct proof; the result is as follows:

Lévy - Steinitz Theorem: *For any convergent series in* $\mathbf{R}^k$, *the set of sums of its convergent rearrangements is an affine set, that is, the translate of a linear subspace.*

Rosenthal's beautiful [1987] exposition of this theorem contains an intriguing ingredient which he credits to W. Gross [1917]:

- *for each* $n \in \mathbf{N}$ *there is a* $C_n > 0$ *such that whenever* $v_1, \dots, v_m$ *are in the Euclidean unit ball of* $\mathbf{R}^n$ *and satisfy* $v_1 + \dots + v_m = 0$, *we can find a permutation* π *of* $\{1, \dots, m\}$ *such that for each* $k \leq m$, $\sum_{j \leq k} v_{\pi(j)} \leq C_n$.

The best constant for the case $n = 2$ is $C_2 = \sqrt{5}/2$; this was established by

W. Banaszczyk [1987]. Generally, $C_n \leq \left((5/4) + \sum_{k=3}^{n}(\sqrt{k+2} - (1/2))^2\right)^{1/2}$ seems to be the best information available; the interested reader is referred to W. Banaszczyk [1990] for this and much more.

The problem of extending the Lévy - Steinitz Theorem to infinitely many dimensions was posed as Problem 106 in the Scottish Book (R.D. Mauldin [1981]). J. Marcinkiewicz apparently settled the question almost 'on the spot' but his exposition suffered from a case of scrambled subscripts. Here though is basically what Marcinkiewicz had to say. The setting is $L_2[0, 1)$. To begin, let $f_0 = 1$. Next, take $f_1 = 1_{[0,1/2)}$ and $f_2 = 1_{[1/2,1)}$, then $f_3 = 1_{[0,1/4)}, \ldots$, $f_6 = 1_{[3/4,1)}$, and so on. The series to consider is $\sum g_n$ where $g_{2n} = f_n$ and $g_{2n+1} = -f_n$. Although $\sum g_n = 0$, we clearly have

$$f_0 + f_1 + f_2 - f_0 + f_3 + f_4 - f_1 + f_5 + f_6 - f_2 + + - \ldots = 1.$$

Any convergent rearrangement of $\sum_n g_n$ can only sum to an integer-valued member of $L_2[0, 1)$; in particular, the set of sums is not even convex.

This example of Marcinkiewicz was rediscovered by P.A. Kornilov [1980]; see E.M. Nikishin [1971] as well.

Subsequently, V.M. Kadets [1986] used Dvoretzky's Theorem (a topic of our final chapter) to graft a version of the Marcinkiewicz - Kornilov example into general infinite dimensional Banach spaces. The result was the stunning achievement that

- *in any infinite dimensional Banach space there is a convergent series which has the property that the set of sums of convergent rearrangements of the series fails to be convex.*

But this was only the beginning of the story:

Theorem: *Any infinite dimensional Banach space contains a convergent series such that each of its convergent rearrangements has one of two (!) sums.*

This was discovered by P. Enflo (unpublished), M.I. Kadets and K. Woźniakowski [1989] and P.A. Kornilov [1988a,b].

Many intriguing questions regarding sums of rearrangements of convergent series remain open. The interested reader should indulge in the nice little monograph by V.M. Kadets and M.I. Kadets [1991].

As with alternative (UC), (CC) also holds reign in some locally convex spaces. A. Wald [1933a] showed the validity of (CC) in $\mathbf{R}^{\mathbf{N}}$ and discussed its validity for topological groups in [1933b]. Wald's work was overlooked for some time, only to be rediscovered by S.L. Troyanski [1967] and Y. Katznelson and O.C. McGehee [1974]. Remarkably, W. Banaszczyk [1990], [1993] has shown that

- *a Fréchet space E is nuclear if and only if the Lévy - Steinitz Theorem holds in E.*

Again, in infinite dimensional Banach spaces it is possible for a convergent series to have just one possible sum for all its convergent rearrangements yet still not be unconditionally convergent. This phenomenon, which was uncovered in ℓ_2 by H. Hadwiger [1941], was later shown to be common to all infinite dimensional Banach spaces by C.W. McArthur [1956].

The equivalence of (i), (iii) and (iv) in Theorem 1.5 is due to W. Orlicz [1933a], while (ii)'s inclusion is the work of T.H. Hildebrandt [1940]. For locally convex spaces, it is (ii) that opens the gate to the study of unconditional summability; see again A. Pietsch's monograph [1972a].

The Bounded Multiplier Test 1.6 is essential to the development of a Lebesgue type integral for vector-valued functions and so it is not surprising that I.M. Gelfand [1938] and N. Dunford [1938] each discovered forms of 1.6 in their developments of integration theory. The test is not shackled to locally convex spaces and is of interest, for example, in general complete metrizable topological vector spaces. More on these developments can be found in the monographs of J. Diestel and J.J. Uhl, Jr. [1977] and S. Rolewicz [1972].

Sometimes a Banach space sequence $(x_n)_n$ enjoys the property that $\sum_n \lambda_n x_n$ converges whenever $(\lambda_n) \in c_0$; such a series is called *weakly unconditionally Cauchy* since the condition means that $\sum_n |\langle x^*, x_n \rangle| < \infty$ for each $x^* \in X^*$. We will encounter such sequences again and again in these deliberations. It is of more than passing interest to observe that although unconditionally summable sequences are weakly unconditionally Cauchy, the converse is not true: in c_0, $\sum_n e_n$ is weakly unconditionally Cauchy but not convergent. In a precise sense, this is the only example of this phenomenon; we are indebted to C. Bessaga and A. Pełczyński [1958] for making this elegant and important observation.

Theorem: *If a Banach space X contains a sequence (x_n) which is weakly unconditionally summable but not convergent, then a sequence of 'blocks' of (x_n) must be equivalent to the standard basis of c_0.*

Consequently, X contains no copy of c_0 if and only if convergence of $\sum_n |\langle x^*, x_n \rangle|$ for each $x^* \in X^*$ always implies unconditional convergence of $\sum_n x_n$.

Schur's Theorem 1.7 was part of J. Schur's [1920] study of matrix summability methods. Banach reformulated the result in terms of the equivalence of weak and norm sequential convergence in ℓ_1. Our category argument follows J. Conway [1969]; N. Dunford and J.T. Schwartz [1958] also used category arguments to prove this result.

It is a consequence of H.P. Rosenthal's startling characterization [1974] of ℓ_1's presence in a Banach space that any infinite dimensional Banach space which has the 'Schur property' contains an isomorphic copy of ℓ_1. So, in principle, to prove that a space has the Schur property, the argument must be at least as delicate as Schur's, or that given in the text.

The Orlicz - Pettis Theorem 1.8 is due to W. Orlicz [1929b] and B.J. Pettis [1938]. The curious date differential has given rise to several interesting anecdotes; rather than pursue them here we prefer to recommend that the interested reader consult the monograph of J. Diestel and J.J. Uhl, Jr. [1977] or the essay of W. Filter and I. Labuda [1990]. Anyone who enjoys the Orlicz - Pettis Theorem ought to read N.J. Kalton's [1980] survey.

Parts (ix), (x) and (xi) of the Omnibus Theorem 1.9 were first discovered by I.M. Gelfand [1938].

H. Rademacher [1922] proved that the scalar sequence (t_n) satisfies $\sum_n |t_n|^2 < \infty$ provided that $\sum_n \pm t_n$ converges for almost all choices of signs. A. Khinchin showed the converse in his famous [1925] paper with A.N. Kolmogorov. All the ingredients needed for the standard proof of Khinchin's Inequality (as given in A. Zygmund [1935] or J. Lindenstrauss - L. Tzafriri [1977]) already appear in A. Khinchin [1923]; however, the precise relationship between the square summability of a sequence of scalars and the L_p-convergence of its Rademacher averages awaited exposure by R.E.A.C. Paley and A. Zygmund [1930].

Khinchin's Inequality, as it is known today, appears for the first time in A. Zygmund's treatise [1935]. Our proof is taken from E.M. Stein [1970].

S.J. Szarek [1976] showed that the best value of A_1 is $2^{-1/2}$. U. Haagerup ([1978], [1982]) found the precise values of A_p and B_p in general: for $p > 2$,

$$B_p = \sqrt{2} \cdot \Big(\frac{\Gamma\big((p+1)/2\big)}{\sqrt{\pi}}\Big)^{1/p}$$

while for $0 < p < 2$

$$A_p = \begin{cases} \sqrt{2} \cdot \Big(\frac{\Gamma\big((p+1)/2\big)}{\sqrt{\pi}}\Big)^{1/p} & p_0 \leq p < 2 \\ 2^{(1/2)-(1/p)} & 0 < p < p_0, \end{cases}$$

where $p_0 \approx 1.8474$ is determined by $\Gamma\big((p_0+1)/2\big) = \sqrt{\pi}/2$ and $1 < p_0 < 2$.

B. Tomaszewski [1987] offers a different path to Szarek's result. See R.M.G. Young [1976] as well. J. Sawa [1985] considered complex analogues to Khinchin's Inequality for $p = 1$ and found the best constant.

The 'trick' we use at the end of the proof of Khinchin's Inequality to deduce the L_1-case from the L_4-case and Hölder's Inequality is due to J.E. Littlewood [1930].

Theorem 1.11 is from W. Orlicz [1933a] where it is demonstrated that

- *if* $1 \leq p \leq 2$ *and* (f_n) *is unconditionally summable in* $L_p[0,1]$, *then* $\sum_n \|f_n\|_p^2 < \infty$.

Shortly afterwards, Orlicz [1933b] proved that

- *if* $q > 2$ *and* (f_n) *is unconditionally summable in* $L_q[0,1]$, *then* $\sum_n \|f_n\|_q^q < \infty$.

The only rôle of the unit interval here is to give the functions a domain; both theorems hold over any measure spaces. We shall derive them in 3.12 and 10.7, and generalize them within the theory of type and cotype. Orlicz used his results to show that if $p, q > 1$ and $p \neq q$ then $L_p[0,1]$ and $L_q[0,1]$ are not isomorphic.

In [1956], M.I. Kadets took another approach altogether and was able to give an abstract proof of Orlicz's results, at least in case $p > 1$. Recall that a Banach space X is *uniformly convex* if given $\varepsilon > 0$ there is a $\delta = \delta(\varepsilon) > 0$ such that $\| \frac{x+y}{2} \| \leq 1 - \varepsilon$ whenever $x, y \in S_X$ and $\|x - y\| \geq \delta$; the function $\varepsilon \mapsto \delta(\varepsilon)$ that best suits these purposes is called the *modulus of convexity* of X and is denoted by δ_X. Kadets proved the following:

Theorem: *If (x_n) is unconditionally summable in the uniformly convex Banach space X, then $\sum_n \delta_X(\|x_n\|) < \infty$.*

O. Hanner [1956] gave the following tight estimates for the moduli of convexity of $L_p[0,1]$: as $\varepsilon \to 0$,

$$\delta_{L_p[0,1]} \sim \begin{cases} \varepsilon^2 & \text{if } 1 < p \leq 2 \\ \varepsilon^p & \text{if } 2 < p < \infty. \end{cases}$$

Again $L_p[0,1]$ serves as a perfectly good model for general $L_p(\mu)$'s. Thus the results of Kadets and Hanner imply Orlicz's theorems for $p > 1$.

The moduli of convexity have been computed for a number of spaces. In the light of Kadets' Theorem, such computations are of interest in the context of this first chapter.

R.P. Maleev and S.L. Troyanski [1975] have dealt with uniformly convex Orlicz spaces, as have T. Figiel [1976a] and, more recently, H. Hudzik ([1987], [1991]). The [1976a] paper of Figiel is a basic contribution to the study of uniformly convex spaces and their duals, complementing the earlier fundamental contributions of M.I. Kadets [1956] and J. Lindenstrauss [1963].

Z. Altshuler [1980] has considered Lorentz sequence spaces while N. Carothers [1981], [1982], [1987] is a good source for Lorentz function spaces. C.A. McCarthy [1967] and N. Tomczak-Jaegermann [1974] have shown that the Schatten - von Neumann classes $\mathcal{S}_p$ share asymptotic estimates for their moduli of convexity with the L_p-spaces of the same index. This is of particular interest since, if $1 \leq p < \infty$ and $p \neq 2$, L_p is not isomorphic to a subspace of $\mathcal{S}_p$ and $\mathcal{S}_p$ is not isomorphic to a subspace of L_p.

Since $L_1[0,1]$ is not uniformly convex, there is no way to derive Orlicz's Theorem 1.11 from that of Kadets in case $p = 1$. On the other hand, $L_1[0,1]$'s failure to be uniformly convex is a problem of *real* scalars. There is a natural notion of uniform convexity for *complex* Banach spaces which is due to J. Globevnik [1975] and was refined by W.J. Davis, D.J.H. Garling and N. Tomczak-Jaegermann [1984], and which includes complex L_1-spaces as examples. For a complex Banach space X, we introduce the *moduli of complex convexity* H_p^X by setting, for each $\varepsilon > 0$,

$$H_p^X(\varepsilon) := \inf\left\{\left(\int_{-\pi}^{\pi} \|x+e^{it}y\|^p \frac{dt}{2\pi}\right)^{1/p} - 1 : \|x\|=1,\ \|y\|=\varepsilon\right\}$$

when $1 \leq p < \infty$, and

$$H_\infty^X(\varepsilon) := \inf\left\{\sup\left\{\|x+e^{it}y\| - 1 : t \in [-\pi,\pi]\right\} : \|x\|=1,\ \|y\|=\varepsilon\right\}.$$

Davis, Garling and Tomczak-Jaegermann showed that for finite p, the moduli $H_p^X(\varepsilon)$ are all equivalent as $\varepsilon \to 0$: if $H_p^X(\varepsilon) > 0$ for all $\varepsilon > 0$ holds for one p it holds for all p. This is also true when we include $p = \infty$. It is easy to see that $H_1^X(\varepsilon) \leq H_\infty^X(\varepsilon)$, and S.J. Dilworth [1986] showed that there is a constant c such that $c \cdot H_\infty^X(\varepsilon)^2 \leq H_1^X(\varepsilon)$ for all complex Banach spaces X and all $0 < \varepsilon \leq 1$. X is called *complex uniformly convex* if $H_1^X(\varepsilon) > 0$ for each $\varepsilon > 0$.

S.J. Dilworth [1986] gave a beautiful companion to Kadets' Theorem:

Theorem: *If (x_n) is an unconditionally summable sequence in the complex uniformly convex Banach space X, then $\sum_n H_\infty^X(\|x_n\|) < \infty$.*

Davis, Garling and Tomczak-Jaegermann computed the moduli H_p for complex L_p-spaces: $H_p^{L_p} \sim \varepsilon^p$ if $p \geq 2$ and $H_p^{L_p} \sim \varepsilon^2$ if $1 \leq p \leq 2$. So the complex uniform convexity of $L_1[0,1]$, which goes back to J. Globevnik [1975], also allows to rederive Orlicz's Theorem in the complex case.

D.J.H. Garling and N. Tomczak-Jaegermann [1983] and U. Haagerup and G. Pisier [1989] are other excellent sources for further information regarding these and related ideas, in particular regarding non-commutative generalizations of the results mentioned above.

Theorem 1.12(b) seems to be 'known to all'; it is part of the motivation behind the construction of H.P. Rosenthal's ℓ_1 Theorem [1974]. Its possible rôle was suggested by A. Pełczyński's [1968] work on ℓ_1-subspaces of general Banach spaces and reinforced by J. Hagler [1973]. The [1975a] note of L.E. Dor should also be inspected in this context.

Though 1.12(c) might have been known earlier, it is first formulated in A. Pełczyński's classical paper [1960].

M.I. Kadets and A. Pełczyński [1962] showed that

- *if $p \geq 2$ and X is a closed subspace of $L_p[0,1]$ which is isomorphic to Hilbert space, then X is complemented in $L_p[0,1]$.*

In fact, they proved that the projection can be taken to be an averaging projection with respect to some p-lacunary basic sequence. In 12.22 we shall derive a much more general result due to B. Maurey [1974b].

For $1 < p < 2$ the situation is more complicated. The Rademacher functions form a uniformly bounded sequence of independent, identically distributed random variables of mean zero and variance one. However, not all Hilbertian subspaces of $L_p[0,1]$ are generated by such a well-behaved sequence.

Indeed, G. Bennett, L.E. Dor, V. Goodman, W.B. Johnson and C.M. Newman [1977] showed that

- *each $L_p[0,1]$, $1 < p < 2$, contains an uncomplemented isomorphic copy of Hilbert space.*

Earlier H.P. Rosenthal [1966] had used harmonic analytic arguments to prove this for $1 < p < 4/3$. Rosenthal's argument will actually work for the whole range $1 < p < 2$, but at the time he lacked a vital ingredient which was only supplied by J. Bourgain [1989] when he settled a long-standing problem about $\Lambda(p)$-sets.

By the way, Bennett et al. [1977] also succeeded in completing an earlier circle of results of H.P. Rosenthal [1970] to conclude that

- *if $1 < p < \infty$, $p \neq 2$, then $L_p[0,1]$ contains an uncomplemented copy of ℓ_p.*

For $p = 1$, J. Bourgain [1981a] called on tools from harmonic analysis to do the same.

L.E. Dor and T. Starbird [1979] showed that any copy of ℓ_p ($1 \leq p < \infty$) spanned by independent random variables in $L_p[0,1]$ *is* complemented.

The isomorphic character of the span of the Rademacher functions also plays an important rôle in the classification of rearrangement invariant function spaces; the [1979] monograph of W.B. Johnson, B. Maurey, G. Schechtman and L. Tzafriri is replete with examples where the Rademacher system plays its part in uncovering some fine flaw in the structure of a space.

The original formulation of Grothendieck's Inequality is found in Théorème 1, §4.2, of his [1953a] Résumé. It is chiselled in the language of tensor norms.

J. Lindenstrauss and A. Pełczyński [1968] bypassed the theory of tensor products, in favour of what we now call operator ideals, to arrive at the formulation of Grothendieck's Inequality which appears as 1.14 in the text. Their proof, which owed much to that of Grothendieck, rested on:

Grothendieck's Identity: *Let S be the unit sphere of ℓ_2^n and let λ be the normalized rotation invariant measure on its Borel sets. For $x, y \in S$, denote the angle (in $[0,\pi]$) between them by $\theta(x,y)$, so $\cos\theta(x,y) = (x|y)$. Then*

$$\int_S \operatorname{sign}(x|s) \cdot \operatorname{sign}(y|s)\, d\lambda(s) = 1 - \frac{2}{\pi} \cdot \theta(x,y)$$

The proof is an easy though messy exercise in integration with respect to spherical coordinates.

Using this identity, Lindenstrauss and Pełczyński gave a perfectly elementary proof of Grothendieck's Inequality, arriving at the estimate $\kappa_G \leq \sinh(\pi/2) \approx 2.301$ for the real Grothendieck constant.

R.E. Rietz [1974] followed the same basic strategy but managed to improve the estimate for κ_G by using a modification of Grothendieck's Identity based on averaging over the whole of $\mathbf{R}^n$ with normalized Gaussian measure. His approach had the added benefit of showing that for positive definite matrices $\pi/2$ can replace κ_G. Grothendieck himself had observed that $\kappa_G \geq \pi/2$, even for positive definite matrices. However, the best known estimate for the real case is $\kappa_G \leq \dfrac{\pi}{2 \cdot \sinh^{-1} 1} \approx 1.782$. J.L. Krivine [1977a],[1979] was led to this through his work on Banach lattices. His proof was scrutinized by many – G.J.O. Jameson [1987] and A. Pełczyński amongst them – with an eye to further simplifications. Here is the gist of Krivine's proof à la Jameson - Pełczyński. It too relies on Grothendieck's Identity but uses the available formalities in a more sophisticated manner than Grothendieck's original proof as exposed by Lindenstrauss and Pełczyński. The key is the following lemma.

Lemma: *Choose $c > 0$. Then for each (real) Hilbert space H there exist a bigger Hilbert space $\mathcal{H}$ and (non-linear) mappings $u, v : H \to \mathcal{H}$ such that, for any $x, y \in H$,*

$$c \cdot (x|y) = \arcsin (ux|vy)$$

and

$$\|ux\|^2 = \sinh (c \cdot \|x\|^2) \quad , \quad \|vy\|^2 = \sinh (c \cdot \|y\|^2).$$

The construction of $\mathcal{H}$ involves taking a direct ℓ_2-sum of tensor products of H with itself. We establish some notation before embarking on the proof.

View H as the set of all $x = (x_i)_{i \in I} \in \mathbf{R}^I$ such that $\|x\|^2 = \sum_i |x_i|^2 < \infty$. The k-fold tensor product of H with itself consists of all vectors $\xi = (\xi_{i_1,..,i_k})_{i_1,..,i_k \in I}$ such that $\|\xi\|^2 = \sum_{i_1,..,i_k \in I} |\xi_{i_1,..,i_k}|^2 < \infty$. This is a Hilbert space with respect to the inner product $(\xi|\eta) = \sum_{i_1,..,i_k \in I} \xi_{i_1,..,i_k} \eta_{i_1,..,i_k}$, and we denote it by $H^{(k)}$.

For us it will be important to work with special vectors in $H^{(k)}$. If $x \in H$ we write $x^{(k)} = x \otimes \cdots \otimes x = (x_{i_1} x_{i_2} \cdots x_{i_k})_{i_1,..,i_k \in I} \in H^{(k)}$. It is plain that $\|x^{(k)}\| = \|x\|^k$ and that $\big(x^{(k)} \,\big|\, y^{(k)}\big) = (x|y)^k$ for $x, y \in H$.

Proof of the Lemma. $\mathcal{H}$ is just the Hilbert space direct sum $\big(\oplus_{k=0}^{\infty} H^{(2k+1)}\big)_2$. The construction of u and v emerges naturally from the Taylor expansion of the sine function. Note that

$$\sin c \cdot (x|y) = \sum_{k=0}^{\infty} (-1)^k \frac{c^{2k+1}}{(2k+1)!} (x|y)^{2k+1} = \sum_{k=0}^{\infty} (-1)^k \frac{c^{2k+1}}{(2k+1)!} \big(x^{(2k+1)} \big| y^{(2k+1)}\big).$$

Now set $ux = \Big((-1)^k \big(\frac{c^{2k+1}}{(2k+1)!}\big)^{1/2} x^{(2k+1)}\Big)_k$ and $vy = \Big(\big(\frac{c^{2k+1}}{(2k+1)!}\big)^{1/2} y^{(2k+1)}\Big)_k$.
The norm computations are immediate. QED

We can now make short shrift of Grothendieck's Inequality. Suppose that $\|(a_{ij})\| \leq 1$ and let $x_1, ..., x_n, y_1, ..., y_n$ be in S_H. Apply the lemma with $c = \sinh^{-1} 1$, and for convenience let $ux_i = u_i$ and $vy_j = v_j$. Then

$$c \cdot \sum_{i,j \leq n} a_{ij} (x_i|y_j) = \sum_{i,j \leq n} a_{ij} \arcsin(u_i|v_j)$$

and $u_1, ..., u_n, v_1, ..., v_n$ are unit vectors. Consider these as elements of ℓ_2^{2n} and let λ be the unique rotation invariant Borel probability on $S = S_{\ell_2^{2n}}$. Appeal to Grothendieck's Identity:

$$c \cdot \sum_{i,j \le n} a_{ij}(x_i|y_j) = \frac{\pi}{2} \cdot \int_S \sum_{i,j \le n} a_{ij} \cdot \mathrm{sign}\,(u_i|w) \cdot \mathrm{sign}\,(v_j|w)\, d\lambda(w).$$

It follows that

$$\Big|\sum_{i,j} a_{ij}(x_i|y_j)\Big| \le \frac{\pi}{2 \cdot c} \le \frac{\pi}{\sinh^{-1} 1}\,.$$

Over the years, Grothendieck's Inequality has attracted a great deal of attention and has inspired several new proofs. Among the most exciting proofs of its equivalent counterpart 1.13 – where κ_G is the gauge by which we compare the operator norm with the absolutely summing norm – is that of A. Pełczyński and P. Wojtaszczyk (see A. Pełczyński [1976b] or P. Wojtaszczyk [1991]). The kernel of their proof is a still wonderful inequality of R.E.A.C. Paley [1933] connected with lacunary sequences.

There are other approaches to Grothendieck's Inequality which rely on integral representations of the inner product different from Grothendieck's Identity or Rietz's adaptation. J.J. Fournier [1979] noticed that if x and y are vectors in an $(N+1)$-dimensional complex Hilbert space, each with norm $\le 2^{-1/2}$, then there are functions f and g in the unit ball ball of $C(\mathbf{T}^n)$ such that $(x|y) = \int_{\mathbf{T}^n} f(\omega)\overline{g(\omega)}\, d\lambda(\omega)$, where λ is Haar measure on $\mathbf{T}^n$. This gives 2 as an upper bound for κ_G in the case of complex scalars. Fournier relies on the so-called 'Schur algorithm' and his techniques also yield a proof of Paley's Inequality with the best possible constant.

An integral representation of the inner product was also the basis for the work of R.C. Blei [1977]. Working again with complex scalars, he managed to show that there exist a compact abelian group G, a constant $K > 0$ and a non-linear map $\phi : \ell_2 \to L_\infty(G)$ such that $\|\phi(x)\|_\infty \le K \cdot \|x\|_2$ and $(x|y) = \big(\phi(x) * \phi(\overline{y})\big)(0)$ for all $x, y \in \ell_2$. Here $*$ denotes convolution and $\overline{y}$ is the conjugate vector $(\overline{y_n})$ to $y = (y_n) \in \ell_2$. The construction of ϕ used Riesz products, yet another standard tool from harmonic analysis. Once the inner product has been represented as indicated, Grothendieck's Inequality follows at once.

Blei quickly transferred his approach to other settings. In [1980] he blithely ignored the fact that Lindenstrauss and Pełczyński [1968] had shown that a Banach space X is isomorphic to a Hilbert space if and only if there is a constant C such that for all finite collections $x_1, \dots, x_n$ from B_X and $x_1^*, \dots, x_n^*$ from B_{X^*} we have $\big|\sum_{i,j} a_{ij}\langle x_i, x_j^*\rangle\big| \le C \cdot \|a\|$ (see 7.5). He showed that for $X = L_p(G)$ $(1 \le p < 2)$ a uniform inequality does prevail, but only when $\|a\|$ is replaced by $\sup\big\{\big|\sum_{i,j} a_{ij}(f_i * \mu_j)(0)\big| : \|f_i\|_\infty \le 1,\ \|\hat{f}_i\|_p \le 1,\ \|\mu_j\|_{C(G)^*} \le 1,\ \|\hat{\mu}_j\|_{p^*} \le 1\big\}$.

A.M. Tonge [1987] also studied the extent to which Grothendieck's Inequality fails in L_p-spaces.

Another profitable direction taken by R.C. Blei ([1979], [1980], [1988]) was to look into the possibility of multilinear extensions of Grothendieck's Inequality. This theme was also taken up by T.K. Carne [1980], J.J. Fournier

[1981] and A.M. Tonge ([1978], [1987]). They found that such extensions are only possible for a restricted class of multilinear forms. It turns out that there are close relations between this topic and the failure of natural many-variable extensions of the von Neumann - Andô Inequalities for commuting systems of contractions on complex Hilbert spaces (see Chapter 18).

Incidentally, the failure of the von Neumann - Andô Inequalities, as proved by N.Th. Varopoulos [1974], [1976], seems to be the only context in which the fact that $\kappa_G > 1$ has ever been of use. We discuss this in 18.12.

To be sure, when speaking of the Grothendieck constant, we ought to be careful about whether we are in the case of real or complex scalars. This in mind, the notations $\kappa_G^{\mathbf{R}}$ and $\kappa_G^{\mathbf{C}}$ are self-explanatory. Considerable effort has been expended on the search for the precise values of $\kappa_G^{\mathbf{R}}$ and $\kappa_G^{\mathbf{C}}$, as well as on their analogues $\kappa_G^{\mathbf{R}}(n)$ and $\kappa_G^{\mathbf{C}}(n)$ when the matrices are restricted to $n \times n$ format.

Grothendieck himself announced that $\pi/2 \leq \kappa_G^{\mathbf{R}} \leq \sinh(\pi/2)$ and his calculations gave $4/\pi \leq \kappa_G^{\mathbf{C}} \leq 2 \cdot \sinh(\pi/2)$. It is easily checked that $\kappa_G^{\mathbf{R}} \leq \kappa_G^{\mathbf{R}}(2) \cdot \kappa_G^{\mathbf{C}}$. J.L. Krivine [1977b] established that $\kappa_G^{\mathbf{R}}(2) = \sqrt{2}$ and also provided much information about $\kappa_G^{\mathbf{R}}(n)$ for $n > 2$. Later on, H. König [1990] came up with very tight estimates for the numbers $\kappa_G^{\mathbf{C}}(n)$.

Nevertheless, much mystery remains about $\kappa_G^{\mathbf{R}}$ and $\kappa_G^{\mathbf{C}}$. The most recent information is that

$$\frac{\pi}{2} \leq \kappa_G^{\mathbf{R}} \leq \frac{\pi}{2 \cdot \sinh^{-1} 1} \approx 1.782 \quad \text{and} \quad 1.338 < \kappa_G^{\mathbf{C}} \leq 1.405.$$

This may be gleaned from U. Haagerup [1987] and A.M. Davie (unpublished). H. König's [1990] estimates for the $\kappa_G^{\mathbf{C}}(n)$'s also lead to $\kappa_G^{\mathbf{C}} \leq 1.405$. Krivine has unpublished work showing that $\kappa_G^{\mathbf{R}} > \pi/2$.

G. Pisier [1979] was the first to observe that $\kappa_G^{\mathbf{C}} < \pi/2$ and so that $\kappa_G^{\mathbf{C}} < \kappa_G^{\mathbf{R}}$. He presented a proof of Grothendieck's Inequality for complex scalars which was based on the use of Gaussian variables and interpolation theory. He found that $\kappa_G^{\mathbf{C}} \leq e^{1-\gamma} < 1.527 < \pi/2$, where γ is the Euler - Mascheroni constant.

Pisier's proof actually appears in an appendix to [1978a]; the main body of his work is devoted to conjectures of A. Grothendieck [1953a] and J. Ringrose [1976] on bilinear forms on C^*-algebras. He proved a non-commutative version of Grothendieck's Inequality for 'approximable' operators. Soon after, U. Haagerup [1985] removed the approximability hypothesis. The work of Haagerup and Pisier inspired S. Kaijser [1983] to produce notable simplifications. Kaijser's ideas, specialized to the classical commutative case, are what we used for the proof of Grothendieck's Inequality in the text. Informative expositions can be found in G. Pisier's [1985] lecture notes and the paper of S. Kaijser and A.M. Sinclair [1984]. The beautiful expository essay of E. Christensen and A.M. Sinclair [1989] offers valuable insight into how this inequality has affected the study of operator ideals and operator algebras.

The inequality has also been extended to JB^*-triples by T. Barton and Y. Friedman [1987].

The original Grothendieck Inequality has found a home in Banach lattices through the work of T.K. Carne [1980] and J.L. Krivine [1973/74]; more on this in the notes and remarks to Chapter 17.

Here are some suggestions for further reading.

Banach Algebras: A.M. Mantero and A.M. Tonge ([1979],[1980]), A.M. Sinclair [1984], A.M. Tonge [1976], A. Ülger [1988];

Harmonic Analysis: R.C. Blei ([1977],[1980],[1983]), M. Cowling [1978], J.J. Fournier [1979], J.E. Gilbert ([1977],[1979]), J.E. Gilbert, T. Ito and B.M. Schreiber [1985], C. Graham and B.M. Schreiber [1984], G.J.O. Jameson [1985], S. Kaijser [1975/76], S. Kwapień and A. Pełczyński ([1970],[1978],[1980]), A. Pełczyński ([1969],[1974],[1976b],[1985]), G. Racher [1983], K. Ylinen [1984];

Operator Algebras and C^*-Algebras: E. Christensen and A.M. Sinclair [1989], E.G. Effros [1988], U. Haagerup ([1983a],[1983b],[1985]), A.M. Sinclair [1984], R.R. Smith [1988], K. Ylinen ([1984],[1988a]);

Operator Theory: M. Cotlar and S. Sadosky [1991], H. Niemi and A. Weron [1981], V.I. Ovčinnikov [1976], V.V. Peller [1982], A.M. Tonge [1978], N.Th. Varopoulos ([1974],[1976]).

Probability and Measure Theory: G. Bennett [1976], R.C. Blei ([1988],[1989]), D. Dehay [1987], C. Houdré ([1990a],[1990b]), S. Karni and E. Merzbach [1990], S. Kwapień and A. Pełczyński [1970], A. Makagon and H. Salehi [1987], B. Maurey and A. Nahoum [1973], H. Niemi [1984], P. Ørno [1976], M. Rosenberg [1982], K. Ylinen [1984].

2. Fundamentals of p-Summing Operators

Definition

It is high time for us to introduce the main topic of this book: p-summing operators, known in some quarters as absolutely p-summing operators. They may be defined in a variety of ways; let us begin with the most common.

Suppose that $1 \leq p < \infty$ and that $u : X \to Y$ is a linear operator between Banach spaces. We say that u is *p-summing* if there is a constant $c \geq 0$ such that regardless of the natural number m and regardless of the choice of $x_1, \ldots, x_m$ in X we have

$$(\Pi_p) \qquad \Big(\sum_{i=1}^{m} \|ux_i\|^p\Big)^{1/p} \leq c \cdot \sup\Big\{\Big(\sum_{i=1}^{m} |\langle x^*, x_i\rangle|^p\Big)^{1/p} : x^* \in B_{X^*}\Big\}.$$

The least c for which inequality (Π_p) always holds is denoted by

$$\pi_p(u).$$

We shall write

$$\Pi_p(X,Y)$$

for the set of all p-summing operators from X into Y.

- It is a simple matter to check that $\Pi_p(X,Y)$ is a *linear subspace* of $\mathcal{L}(X,Y)$, the space of all bounded linear operators from X into Y, and that π_p defines a *norm* on $\Pi_p(X,Y)$ with

  $$\|u\| \leq \pi_p(u)$$

  for all $u \in \Pi_p(X,Y)$.

A notable characteristic of this definition is its finitary nature; (Π_p) is tested by using finitely many vectors at a time. This feature ensures that local properties of Banach spaces, that is, properties of their finite dimensional subspaces, lie at the heart of many of the most penetrating applications of p-summing operators.

Why is one interested in estimates of the ilk of (Π_p)? They certainly turn out to be convenient for computational purposes, but there is a slightly more abstract formulation which exposes the class of p-summing operators as a very natural object of study. It also facilitates the development of some of the properties of these operators and has the advantage of tying in neatly with the considerations of Chapter 1.

VECTOR-VALUED SEQUENCE SPACES

As a prelude we introduce some vector-valued sequence spaces. We shall work with an index $1 \leq p < \infty$ and a Banach space X. The vector sequence (x_n) in X is *strongly p-summable* (alternatively, a *strong ℓ_p sequence*) if the corresponding scalar sequence $(\|x_n\|)$ is in ℓ_p. We denote by

$$\ell_p^{strong}(X)$$

the set of all such sequences in X. This is clearly a vector space under pointwise operations, and a natural norm is given by

$$\|(x_n)\|_p^{strong} := \big(\sum_n \|x_n\|^p\big)^{1/p}.$$

An effortless adaptation of the usual proof that ℓ_p is a Banach space rapidly leads to the conclusion that $\ell_p^{strong}(X)$ is a Banach space. In fact, $\ell_p^{strong}(X)$ is nothing but the ℓ_p direct sum of countably many copies of the Banach space X.

The easiest examples of members of $\ell_p^{strong}(X)$ are the finite sequences $(x_1,..., x_m)$ – identified with $(x_1,..., x_m, 0, 0,...)$. These are readily seen to form a dense subspace.

Strong p-summability makes reference to the strong (or norm) topology on X. What about the natural analogue for the weak topology? The vector sequence (x_n) in X is *weakly p-summable* (alternatively, a *weak ℓ_p sequence*) if the scalar sequences $(\langle x^*, x_n\rangle)$ are in ℓ_p for every $x^* \in X^*$. We denote by

$$\ell_p^{weak}(X)$$

the set of all such sequences in X. Once again, this carries an obvious linear structure. We would, of course, like $\ell_p^{weak}(X)$ to be a Banach space. It takes a little work to see that a suitable norm is given by

$$\|(x_n)\|_p^{weak} := \sup\big\{\big(\sum_n |\langle x^*, x_n\rangle|^p\big)^{1/p} : x^* \in B_{X^*}\big\}.$$

The first step is to show that this quantity is finite, and for this we call on the Closed Graph Theorem. Take (x_n) in $\ell_p^{weak}(X)$ and associate with it the map $u : X^* \to \ell_p$ given by $u(x^*) = (\langle x^*, x_n\rangle)$. Certainly, u is well-defined and linear. Moreover, if (x_k^*) converges to x_0^* in X^*, then for each n the scalar sequence $(\langle x_k^*, x_n\rangle)_k$ converges to $\langle x_0^*, x_n\rangle$. As a consequence, u has a closed graph and so is bounded; in other words

$$\|u\| = \sup\big\{\big(\sum_n |\langle x^*, x_n\rangle|^p\big)^{1/p} : x^* \in B_{X^*}\big\} < \infty,$$

which is what we wanted.

With the finiteness in hand, it only takes a moment to check that $\|\cdot\|_p^{weak}$ is indeed a norm on $\ell_p^{weak}(X)$. The completeness requires some thought. Here we use a direct argument; a little later (see 2.2), we shall indicate a more devious route. Take a Cauchy sequence of members $x^{(k)} = (x_n^{(k)})_n$ of $\ell_p^{weak}(X)$. Our first aim is to locate a candidate for the limit of $(x^{(k)})_k$.

Given any $\varepsilon > 0$, there is a natural number N such that for every $k, k' \geq N$ we have

$$(*) \qquad \sum_n \big| \langle x^*, x_n^{(k)} \rangle - \langle x^*, x_n^{(k')} \rangle \big|^p \leq \varepsilon^p$$

for each $x^* \in B_{X^*}$. Each term in this series is dominated by ε^p, so for every n

$$\big\| x_n^{(k)} - x_n^{(k')} \big\| = \sup \big\{ \big| \langle x^*, x_n^{(k)} \rangle - \langle x^*, x_n^{(k')} \rangle \big| : x^* \in B_{X^*} \big\} \leq \varepsilon.$$

This tells us that the sequences $(x_n^{(k)})_k$ are all Cauchy in the Banach space X, and so converge to x_n, say; thus is born a candidate $x = (x_n)$ for the limit of $(x^{(k)})_k$. We must show that x really is such a limit in $\ell_p^{weak}(X)$. Return to the Cauchy condition $(*)$ and, with all necessary care, let k' tend to infinity. What emerges is that whenever $k \geq N$ we have $(\sum_n |\langle x^*, x_n - x_n^{(k)} \rangle|^p)^{1/p} \leq \varepsilon$ for every $x^* \in B_{X^*}$. Reinterpreted, this says that $x - x^{(k)}$, and hence x, belong to $\ell_p^{weak}(X)$, with $\|x - x^{(k)}\|_p^{weak} \leq \varepsilon$ for $k \geq N$. Since $\varepsilon > 0$ was arbitrary, $(x^{(k)})_k$ converges to x in the norm $\|\cdot\|_p^{weak}$.

So far, our discussion has excluded the case $p = \infty$. There is a simple reason for this: if (x_n) is a bounded sequence in the Banach space X, then

$$\sup_n \|x_n\| = \sup_{x^* \in B_{X^*}} \sup_n |\langle x^*, x_n \rangle|.$$

In other words, if we use natural definitions, then the spaces $\ell_\infty^{strong}(X)$ and $\ell_\infty^{weak}(X)$ are identical and $\|(x_n)\|_\infty^{strong} = \|(x_n)\|_\infty^{weak}$. Henceforth we shall refer them simply as

$$\ell_\infty(X)$$

and use $\|(x_n)\|_\infty$ for the norm.

Things heat up a little when we turn to the corresponding spaces of null sequences. We write

$$c_0^{weak}(X)$$

for the closed subspace of $\ell_\infty(X)$ consisting of all sequences (x_n) in X with $\lim_{n\to\infty} \langle x^*, x_n \rangle = 0$ for all $x^* \in X^*$. This in turn has as a closed subspace the collection

$$c_0^{strong}(X)$$

of all sequences (x_n) in X with $\lim_{n\to\infty} \|x_n\| = 0$. The members of $c_0^{weak}(X)$ and $c_0^{strong}(X)$ are called respectively the *weak null sequences* and the *strong null sequences* in the Banach space X.

We shall see later (2.18) that when $1 \leq p < \infty$, we have $\ell_p^{weak}(X) = \ell_p^{strong}(X)$ if and only if X is finite dimensional. It is noteworthy that Schur's ℓ_1 Theorem 1.7 shows that it is possible to have $c_0^{weak}(X) = c_0^{strong}(X)$ for infinite dimensional spaces X – in particular $X = \ell_1$. However, this phenomenon is by no means typical: the unit vector basis in ℓ_p $(1 < p < \infty)$ is weakly null but certainly not strongly null.

The sequence spaces ℓ_p^{weak} and ℓ_p^{strong} are now defined; let us put them to use. First note that $\ell_p^{strong}(X)$ is a linear subspace of $\ell_p^{weak}(X)$ and that the inclusion is continuous with norm one. We just mentioned that the inclusion is strict, unless X is finite dimensional. In certain cases this is evident; for example, the standard unit vector basis in ℓ_{p^*} (resp. c_0 if $p = 1$) is always a weak ℓ_p sequence in ℓ_{p^*} (resp. c_0). Actually, this remark has more than passing significance: it will serve us soon to generate all weak ℓ_p sequences.

Next, if $u : X \to Y$ is a bounded linear operator between Banach spaces, the correspondence

$$\hat{u} : (x_n)_n \mapsto (ux_n)_n$$

always induces a bounded linear operator $\ell_p^{weak}(X) \to \ell_p^{weak}(Y)$, as well as a bounded linear operator $\ell_p^{strong}(X) \to \ell_p^{strong}(Y)$. In both cases, the norm is clearly $\|u\|$. *Sometimes,* this process even produces a linear operator from $\ell_p^{weak}(X)$ to $\ell_p^{strong}(Y)$; such is the case precisely when u is p-summing.

2.1 Proposition: *u is p-summing if and only if $\hat{u}(\ell_p^{weak}(X))$ is contained in $\ell_p^{strong}(Y)$. In this case, $\|\hat{u} : \ell_p^{weak}(X) \to \ell_p^{strong}(Y)\| = \pi_p(u)$.*

Proof. Suppose first that u is p-summing. Then, for any finite collection of vectors $x_1,\dots,x_m$ in X, we have

$$\Big(\sum_{n\le m} \|ux_n\|^p\Big)^{1/p} \le \pi_p(u) \cdot \sup\Big\{\Big(\sum_{n\le m} |\langle x^*, x_n\rangle|^p\Big)^{1/p} : x^* \in B_{X^*}\Big\}.$$

Let $(x_n) \in \ell_p^{weak}(X)$. Then

$$\begin{aligned}\|\hat{u}((x_n)_n)\|_p^{strong} &= \sup_m\Big(\sum_{n\le m}\|ux_n\|^p\Big)^{1/p} \le \pi_p(u)\cdot\sup_m \sup_{x^*\in B_{X^*}}\Big(\sum_{n\le m}|\langle x^*,x_n\rangle|^p\Big)^{1/p} \\ &= \pi_p(u)\cdot \sup_{x^*\in B_{X^*}}\sup_m\Big(\sum_{n\le m}|\langle x^*,x_n\rangle|^p\Big)^{1/p} = \pi_p(u)\cdot\|(x_n)\|_p^{weak}.\end{aligned}$$

Consequently, $\hat{u}$ maps $\ell_p^{weak}(X)$ continuously into $\ell_p^{strong}(Y)$, and $\|\hat{u}\| \le \pi_p(u)$.

Suppose conversely that $\hat{u}(\ell_p^{weak}(X)) \subset \ell_p^{strong}(Y)$. The resulting operator $\ell_p^{weak}(X) \to \ell_p^{strong}(Y)$ has a closed graph, and so is bounded; after all, $\hat{u} : \ell_p^{weak}(X) \to \ell_p^{weak}(Y)$ is continuous and the $\ell_p^{strong}(Y)$ norm dominates the $\ell_p^{weak}(Y)$ norm. Our conclusion now follows immediately: for finite sequences, $\|(ux_n)\|_p^{strong} \le \|\hat{u}\| \cdot \|(x_n)\|_p^{weak}$, and so u is p-summing with $\pi_p(u) \le \|\hat{u}\|$.

QED

When $p = 1$, a bit more can be said. Recall that in Chapter 1 we introduced the notion of absolutely summing operators between Banach spaces, that is, (bounded) linear operators which take unconditionally summable sequences to absolutely summable sequences. Clearly, 1-summing operators are absolutely summing; happily, the converse holds as well. To prove this, we choose once again to invoke the Closed Graph Theorem, together with the Omnibus Theorem 1.9 on unconditionally summable sequences. Suppose then that $u : X \to Y$ is absolutely summing. What we need from the Omnibus Theorem 1.9 is that the sequence (x_n) in X is unconditionally summable if and only if the linear map $w : c_0 \to X$, defined by $w(e_n) = x_n$ for all n, is compact. This will allow us to create a linear operator

$$\mathcal{K}(c_0, X) \longrightarrow \ell_1^{strong}(Y) : w \mapsto (ux_n)_n$$

with closed graph. To understand why this is so, first observe that the issue is not the existence of the operator; this follows from the fact that u is absolutely summing. What does require justification is that its graph is closed. The key is that $\ell_1^{strong}(Y)$ embeds continuously into $\mathcal{K}(c_0, Y)$ under the mapping $(y_n) \mapsto (e_n \mapsto y_n)$. To establish the compactness of the map $c_0 \to Y : e_n \mapsto y_n$, recall that absolute summability implies unconditional summability, and then apply the Omnibus Theorem once more. The continuity of the composite map $\mathcal{K}(c_0, X) \to \ell_1^{strong}(Y) \hookrightarrow \mathcal{K}(c_0, Y) : w \mapsto (ux_n) \mapsto uw$ now forces our original map $\mathcal{K}(c_0, X) \to \ell_1^{strong}(Y)$ to have a closed graph.

The upshot of all this is that there must be a constant $c \geq 0$ such that regardless of the finite (and so unconditionally summable) sequence (x_n),

$$\begin{aligned} \sum_n \|ux_n\| \leq c\cdot\|w\|_{\mathcal{K}(c_0,X)} &= c\cdot\sup\left\{\left\|\sum_n b_n x_n\right\| : (b_n)\in B_{c_0}\right\} \\ &= c\cdot\sup\left\{\left|\sum_n b_n\langle x^*, x_n\rangle\right| : (b_n)\in B_{c_0},\ x^*\in B_{X^*}\right\} \\ &= c\cdot\sup\left\{\sum_n |\langle x^*, x_n\rangle| : x^*\in B_{X^*}\right\} = c\cdot\|(x_n)\|_1^{weak}. \end{aligned}$$

Thus u is 1-summing.

As it is going to be extremely useful to have alternative methods for computing the weak ℓ_p norm of a sequence, we investigate our options carefully.

Let $1 \leq p < \infty$ and write p^* for the conjugate index. Take $x_1, \ldots, x_n$ in the Banach space X and set $K = B_{X^*}$. The following interchange of suprema argument exploits the usual duality between ℓ_p^m and $\ell_{p^*}^m$:

$$(*) \qquad \begin{aligned} \sup_{x^*\in K}\Big(\sum_{i\leq m}|\langle x^*, x_i\rangle|^p\Big)^{1/p} &= \sup_{x^*\in K}\sup_{a\in B_{\ell_{p^*}^m}}\Big|\sum_{i\leq m} a_i\langle x^*, x_i\rangle\Big| \\ &= \sup_{a\in B_{\ell_{p^*}^m}}\sup_{x^*\in K}\Big|\Big\langle x^*, \sum_{i\leq m} a_i x_i\Big\rangle\Big| = \sup_{a\in B_{\ell_{p^*}^m}}\Big\|\sum_{i\leq m} a_i x_i\Big\|. \end{aligned}$$

We have restricted our attention to finite sums, but only for technical reasons; we have to be sure that $\sum a_i x_i$ exists. This is no problem as soon as (x_n) is in $\ell_p^{weak}(X)$ and (a_n) belongs to ℓ_{p^*}; then we see that for $n > m$

$$\begin{aligned} \Big\|\sum_{k=m}^n a_k x_k\Big\| &= \sup_{x^*\in B_{X^*}}\Big|\sum_{k=m}^n a_k\cdot\langle x^*, x_k\rangle\Big| \\ &\leq \Big(\sum_{k=m}^n |a_k|^{p^*}\Big)^{1/p^*}\cdot\sup_{x^*\in B_{X^*}}\Big(\sum_k |\langle x^*, x_k\rangle|^p\Big)^{1/p}. \end{aligned}$$

Let us step back and reflect on the preceding chain $(*)$ of equalities.

- The first pleasant fact to emerge is that knowledge of the dual space X^* is not critical for the computation of the weak ℓ_p norm, at least for sequences with only finitely many non-zero terms.
- Notice also the rôle of $K = B_{X^*}$ in the third equality. We could have replaced this by any *norming subset*, that is, by any set $K \subset X^*$ with the property that $\|x\| = \sup\{|\langle x^*, x\rangle| : x^* \in K\}$. To be specific: if K is a norming subset of X^* and (x_n) is in $\ell_p^{weak}(X)$, then arguing as before we get

$$\begin{aligned}\|(x_n)\|_p^{weak} &= \sup_m \sup_{a\in B_{\ell_{p^*}^m}} \sup_{x^*\in B_{X^*}} \Big|\Big\langle x^*, \sum_{n\le m} a_n x_n\Big\rangle\Big| = \sup_m \sup_{a\in B_{\ell_{p^*}^m}} \Big\|\sum_{n\le m} a_n x_n\Big\| \\ &= \sup_m \sup_{a\in B_{\ell_{p^*}^m}} \sup_{x^*\in K} \Big|\Big\langle x^*, \sum_{n\le m} a_n x_n\Big\rangle\Big| = \sup_{x^*\in K} \Big(\sum_n |\langle x^*, x_n\rangle|^p\Big)^{1/p}.\end{aligned}$$

A particularly useful example of such a K is the set of all extreme points of B_{X^*}.

- Most significant of all is the fact that the quantities we are manipulating are norms of linear operators: if $x_1,\ldots,x_m$ are in X consider the operator

$$u : \ell_{p^*}^m \longrightarrow X : (a_i) \mapsto \sum_{i\le m} a_i x_i$$

and its adjoint

$$u^* : X^* \longrightarrow \ell_p^m : x^* \mapsto (\langle x^*, x_i\rangle).$$

It is apparent that $\sup\{\|\sum_{i\le m} a_i x_i\| : a \in B_{\ell_{p^*}^m}\}$ is just the norm of u, whereas $\sup\{(\sum_{i\le m} |\langle x^*, x_i\rangle|^p)^{1/p} : x^* \in B_{X^*}\}$ is the norm of u^*. The identity we established above amounts to nothing more than the familiar fact that u and u^* have the same norm.

The operator formulation nudges us gently in the direction of an elegant identification of the Banach spaces $\ell_p^{weak}(X)$.

2.2 Proposition: *The correspondence $u \mapsto (ue_n)_n$ provides an isometric isomorphism of $\mathcal{L}(\ell_p^*, X)$ onto $\ell_p^{weak}(X)$ when $1 < p < \infty$. For $p = 1$, the isometric isomorphism is from $\mathcal{L}(c_0, X)$ onto $\ell_1^{weak}(X)$.*

Recall that we already made the trivial observation that the unit vector basis of ℓ_{p^*} (c_0 when $p = 1$) is a weak ℓ_p sequence in ℓ_{p^*} (c_0 when $p = 1$); it is now plain that we have identified the prototype for all weak ℓ_p sequences.

Constructions of p-Summing Operators

Before we immerse ourselves too deeply in the theory of p-summing operators we had better make sure that we have at our disposition a sufficient supply of examples. Fortunately, we don't have to look far.

Let X and Y be Banach spaces. An operator $u \in \mathcal{L}(X,Y)$ is said to have *rank one* if $u(X)$ is one-dimensional. It is apparent that an operator has rank one if and only if it has the form

$$u = x^* \otimes y : x \mapsto \langle x^*, x\rangle \cdot y$$

for some non-zero $x^* \in X$ and $y \in Y$.

It is not hard to establish that $u = x^* \otimes y$ is in $\Pi_p(X,Y)$, with $\pi_p(u) = \|x^*\| \cdot \|y\|$. Clearly $\|x^*\| \cdot \|y\| = \|u\| \le \pi_p(u)$, so we only need to check that $\pi_p(u) \le \|x^*\| \cdot \|y\|$. But this follows from

$$\Big(\sum_{k=1}^m \|ux_k\|^p\Big)^{1/p} = \|x^*\| \cdot \|y\| \cdot \Big(\sum_{k=1}^m |\langle \frac{x^*}{\|x^*\|}, x_k\rangle|^p\Big)^{1/p} \le \|x^*\| \cdot \|y\| \cdot \|(x_k)_1^m\|_p^{weak},$$

which is valid for all choices of finitely many vectors $x_1, ..., x_m$ from X.

With this trivial example in hand, we can produce many more. An operator $u \in \mathcal{L}(X,Y)$ is said to have *finite rank* if $u(X)$ is finite dimensional.

2.3 Proposition: *Let $u \in \mathcal{L}(X,Y)$ have finite rank. Then u is p-summing for every $1 \le p < \infty$.*

To see why this is so, take $y_1, ..., y_n$ to be a basis for $u(X)$. Then we can find $x_1^*, ..., x_n^*$ in X^* with $ux = \sum_{k=1}^n \langle x_k^*, x\rangle y_k$ for all $x \in X$. This exhibits u as a sum of rank one operators and so as a member of the vector space $\Pi_p(X,Y)$.

Starting with the most basic p-summing operators – those of rank one – we have created new p-summing operators – those of finite rank – by the simple device of addition. Composition is another useful device for the creation of new p-summing operators from old.

2.4 Ideal Property of p-Summing Operators: *Let $1 \le p < \infty$ and let $v \in \Pi_p(X,Y)$. Then the composition of v with any bounded linear operator is p-summing. More specifically, if X_0 and Y_0 are Banach spaces then, regardless of how we choose $u \in \mathcal{L}(Y, Y_0)$ and $w \in \mathcal{L}(X_0, X)$, we always have $uvw \in \Pi_p(X_0, Y_0)$ with $\pi_p(uvw) \le \|u\| \cdot \pi_p(v) \cdot \|w\|$.*

Proof. It would be simple enough to prove this directly from the definition of a p-summing operator, but we prefer a slicker approach. The operators u, v, w give rise to canonical operators $\hat{u} : \ell_p^{strong}(Y) \to \ell_p^{strong}(Y_0)$, $\hat{v} : \ell_p^{weak}(X) \to \ell_p^{strong}(Y)$ and $\hat{w} : \ell_p^{weak}(X_0) \to \ell_p^{weak}(X)$ with $\|u\| = \|\hat{u}\|$, $\pi_p(v) = \|\hat{v}\|$ and $\|w\| = \|\hat{w}\|$. The operator $\widehat{uvw}$ corresponding to $uvw : X_0 \to Y_0$ coincides with $\hat{u}\hat{v}\hat{w}$ and thus induces a map $\ell_p^{weak}(X_0) \to \ell_p^{strong}(Y_0)$. Hence uvw is p-summing, with

$$\pi_p(uvw) = \|\widehat{uvw}\| \le \|\hat{u}\| \cdot \|\hat{v}\| \cdot \|\hat{w}\| = \|u\| \cdot \pi_p(v) \cdot \|w\|. \qquad \text{QED}$$

It is worth isolating easy but important special cases:

- If X_0 is a subspace of X and $v : X \to Y$ is p-summing, then the restriction map $v|_{X_0} : X_0 \to Y$ is also p-summing, with $\pi_p(v|_{X_0}) \le \pi_p(v)$.

This follows from the ideal property when we take $u : X_0 \to X$ to be the inclusion map and set w to be the identity operator on Y.

- If Y is a subspace of Y_0 and $v : X \to Y$ is p-summing, then v is also p-summing when considered as a map $X \to Y_0$.

This follows again from the ideal property: this time take u to be the identity operator on X and set w to be the inclusion map $Y \to Y_0$.

However, in this situation more can be said:

2.5 *Injectivity of* Π_p: *If $i : Y \to Y_0$ is isometric, then $v \in \Pi_p(X, Y)$ if and only if $iv \in \Pi_p(X, Y_0)$. In this case, we even have $\pi_p(iv) = \pi_p(v)$.*

Indeed, just observe that $\|(iy_n)_n\|_p^{strong} = \|(y_n)\|_p^{strong}$ holds for all (y_n) in $\ell_p^{strong}(Y)$.

The operation of taking limits is yet another basic technique for constructing new p-summing operators from old.

2.6 *Proposition:* *$\Pi_p(X, Y)$ is a Banach space under the norm π_p.*

Proof. Only completeness requires an argument. We could work directly with the definition of p-summing operators and dive into straightforward but unpleasant calculations, but we prefer an approach which is useful in a variety of other situations.

Let (u_n) be a π_p-Cauchy sequence in $\Pi_p(X, Y)$. Since $\|\cdot\| \leq \pi_p(\cdot)$, (u_n) is also Cauchy in $\mathcal{L}(X, Y)$ and so converges to some $u \in \mathcal{L}(X, Y)$ in the uniform norm. Now u gives rise to the operator $\hat{u} : \ell_p^{weak}(X) \to \ell_p^{weak}(Y) : (x_k) \mapsto (ux_k)$. For each n, $\hat{u}_n : \ell_p^{weak}(X) \to \ell_p^{strong}(Y) : (x_k) \mapsto (u_n x_k)$ is well-defined, and our hypothesis can be reformulated to assert that $(\hat{u}_n)$ is a Cauchy sequence in $\mathcal{L}(\ell_p^{weak}(X), \ell_p^{strong}(Y))$. This is a Banach space, thanks to the completeness of $\ell_p^{strong}(Y)$, and so $(\hat{u}_n)$ converges. Its limit can be regarded as a map with values in $\ell_p^{weak}(Y)$ and so must be $\hat{u}$. Consequently, u is p-summing, and of course we have $\lim_{n\to\infty} \pi_p(u - u_n) = 0$. QED

We remark that the various properties of p-summing operators developed in this section are shared by many other classes of operators on Banach spaces. 2.4 and 2.6 make it clear why such classes are called *Banach (operator) ideals.*

Recall that every uniform limit of finite rank operators is compact. Since the uniform norm is dominated by the p-summing norm, π_p-limits of finite rank operators from $\mathcal{L}(X, Y)$ are compact. Simple examples, such as $\ell_2 \to \ell_2 : (x_n)_n \mapsto \big(x_n / \log(n+1)\big)_n$, show that there are compact operators which fail to be p-summing for all p : just think about the effect of the operator on the unit vector basis. Also, we do not have to look too far to find examples of p-summing operators which fail to be compact. The unit vector basis has no convergent subsequence in any ℓ_p; in particular, the formal identity $\ell_1 \hookrightarrow \ell_2$ is not compact. However, it is 1-summing by Grothendieck's Theorem (1.13).

In fact, the full force of Grothendieck's Theorem is not really needed to show that this map is 1-summing. We could appeal directly to Khinchin's Inequality (1.10) and argue as follows.

Take any finite sequence $(x^{(k)})_{k=1}^m$ of vectors in ℓ_1. Then, by Khinchin's Inequality and since $(r_n(t)) \in B_{\ell_\infty}$ for each $t \in [0,1]$,

$$\sum_{k=1}^m \|x^{(k)}\|_{\ell_2} = \sum_{k=1}^m \Big(\sum_{n=1}^\infty |x_n^{(k)}|^2\Big)^{1/2} \le A_1^{-1} \cdot \int_0^1 \sum_{k=1}^m \Big|\sum_{n=1}^\infty r_n(t) x_n^{(k)}\Big|\, dt$$

$$\le A_1^{-1} \cdot \int_0^1 \sup_{x^* \in B_{\ell_\infty}} \sum_{k=1}^m |\langle x^*, x^{(k)}\rangle|\, dt = A_1^{-1} \cdot \|(x^{(k)})_1^m\|_2^{weak}.$$

We continue with a simple, but extremely useful, consequence of Proposition 2.2.

2.7 Proposition: Let $1 \le p < \infty$. *The following statements about an operator $u : X \to Y$ are equivalent:*

(i) *u is p-summing.*

(ii) *For each $v \in \mathcal{L}(\ell_{p^*}, X)$ ($v \in \mathcal{L}(c_0, X)$ if $p = 1$) uv is p-summing.*

(iii) *There is a constant c such that $\pi_p(uv) \le c \cdot \|v\|$ for each $n \in \mathbf{N}$ and each $v \in \mathcal{L}(\ell_{p^*}^n, X)$.*

Let $A := \sup \pi_p(uv)$ where the supremum is taken over all v as in (ii) *with norm at most one, and let B be the smallest of all numbers c permitted in* (iii). *Then $A = B = \pi_p(u)$.*

Proof. Both (i)⇒(ii) and (ii)⇒(iii) are immediate consequences of the ideal property, and we get $B \le A \le \pi_p(u)$. To prove (iii)⇒(i) along with the missing inequality fix $(x_n) \in \ell_p^{weak}(X)$ and let $v : \ell_{p^*} \to X$ ($v : c_0 \to X$ if $p = 1$) be the associated operator à la 2.2. For each $N \in \mathbf{N}$, denote the canonical restriction of v to $\ell_{p^*}^N$ by v_N. We get

$$\Big(\sum_{k=1}^N \|ux_k\|^p\Big)^{1/p} = \Big(\sum_{k=1}^N \|uv_N e_k\|^p\Big)^{1/p} \le \pi_p(uv_N)$$

$$\le B \cdot \|v_N\| \le B \cdot \|v\| = B \cdot \|(x_k)\|_p^{weak}.$$

Consequently, u is p-summing with $\pi_p(u) \le B$. QED

Once we know that a map is 1-summing we can conclude that it is p-summing for any $1 < p < \infty$. In fact, we can use Hölder's Inequality to establish useful inclusion relations.

2.8 Inclusion Theorem: *If $1 \le p < q < \infty$, then $\Pi_p(X,Y) \subset \Pi_q(X,Y)$. Moreover, for $u \in \Pi_p(X,Y)$ we have $\pi_q(u) \le \pi_p(u)$.*

Proof. Take $x_1, \ldots, x_n \in X$ and observe that if $\lambda_k = \|ux_k\|^{(q/p)-1}$ then $\|ux_k\|^q = \|u(\lambda_k x_k)\|^p$. If u is p-summing, then

$$\Big(\sum_{k=1}^n \|ux_k\|^q\Big)^{1/p} = \Big(\sum_{k=1}^n \|u(\lambda_k x_k)\|^p\Big)^{1/p} \le \pi_p(u) \cdot \sup_{x^* \in B_{X^*}} \Big(\sum_{k=1}^n \lambda_k^p |\langle x^*, x_k\rangle|^p\Big)^{1/p}.$$

Now, since $q > p$, we can apply Hölder's Inequality with the conjugate indices q/p and $q/(q-p)$ to get

$$\begin{aligned}\Big(\sum_{k=1}^{n} \|ux_k\|^q\Big)^{1/p} &\leq \pi_p(u)\cdot\Big(\sum_{k=1}^{n} \lambda_k^{qp/(q-p)}\Big)^{(q-p)/(qp)}\cdot \sup_{x^*\in B_{X^*}} \Big(\sum_{k=1}^{n} |\langle x^*, x_k\rangle|^q\Big)^{1/q} \\ &= \pi_p(u)\cdot\Big(\sum_{k=1}^{n} \|ux_k\|^q\Big)^{(1/p)-(1/q)}\cdot \|(x_k)_{k=1}^n\|_q^{weak}.\end{aligned}$$

Rearrange to obtain

$$\Big(\sum_{k=1}^{n} \|ux_k\|^q\Big)^{1/q} \leq \pi_p(u)\cdot\|(x_k)_1^n\|_q^{weak}. \qquad \text{QED}$$

A 'conceptual' proof of the Inclusion Theorem will be available later in this chapter when we have developed a little more theory.

Basic Examples

The first few examples are of archetypal character and will show up in all of our coming discussions.

2.9 Examples:

(a) Multiplication Operators From Continuous Function Spaces

Let K be a compact Hausdorff space, let μ be a positive regular Borel measure on K, and let $1 \leq p < \infty$. Each $\varphi \in L_p(\mu)$ induces a 'multiplication operator'

$$M_\varphi : C(K) \longrightarrow L_p(\mu) : f \mapsto f\cdot\varphi.$$

This map is p-summing with $\pi_p(M_\varphi) = \|\varphi\|_p$.

(b) Formal Inclusion Operators

Let K be a compact Hausdorff space, let μ be a positive regular Borel measure on K, and let $1 \leq p < \infty$. The canonical map

$$j_p : C(K) \longrightarrow L_p(\mu)$$

is p-summing with $\pi_p(j_p) = \mu(K)^{1/p}$.

(c) Multiplication Operators From $L_\infty(\mu)$ Spaces

Let (Ω, Σ, μ) be a measure space, and let $1 \leq p < \infty$. Each $\varphi \in L_p(\mu)$ induces a 'multiplication operator'

$$M_\varphi : L_\infty(\mu) \longrightarrow L_p(\mu) : f \mapsto f\cdot\varphi.$$

This map is p-summing with $\pi_p(M_\varphi) = \|\varphi\|_p$.

(d) Formal Inclusion Operators

Let (Ω, Σ, μ) be a finite measure space and let $1 \leq p < \infty$. The formal inclusion map

$$i_p : L_\infty(\mu) \longrightarrow L_p(\mu)$$

is p-summing, with $\pi_p(i_p) = \mu(\Omega)^{1/p}$.

(e) DIAGONAL OPERATORS

Let $1 \leq p < \infty$. Any member (λ_n) of ℓ_p induces a 'diagonal operator'

$$D_\lambda : \ell_\infty \longrightarrow \ell_p : (a_n) \mapsto (\lambda_n a_n),$$

which is p-summing with $\pi_p(D_\lambda) = \|\lambda\|_p$.

Proof. (a) First we find a convenient form for $\|(f_i)_1^m\|_p^{weak}$ when $f_1,\dots,f_m$ belong to $\mathcal{C}(K)$. To each $\omega \in K$ there corresponds a 'point mass' $\delta_\omega \in \mathcal{C}(K)^*$ given by $\langle \delta_\omega, f\rangle = f(\omega)$. The point masses evidently form a norming subset of $\mathcal{C}(K)^*$, so

$$\begin{aligned}\|(f_i)_1^m\|_p^{weak} &= \sup_{\omega\in K} \big(\sum_{i\leq m} |\langle \delta_\omega, f_i\rangle|^p\big)^{1/p} \\ &= \sup_{\omega\in K} \big(\sum_{i\leq m} |f_i(\omega)|^p\big)^{1/p} = \big\| \big(\sum_{i\leq m} |f_i|^p\big)^{1/p} \big\|_\infty .\end{aligned}$$

With this, the result is within easy reach:

$$\begin{aligned}\big(\sum_{i\leq m} \|M_\varphi f_i\|_p^p\big)^{1/p} &= \big(\sum_{i\leq m} \int |f_i\varphi|^p d\mu\big)^{1/p} = \Big(\int |\varphi(\omega)|^p \cdot \big(\sum_{i\leq m} |f_i(\omega)|^p\big) d\mu(\omega)\Big)^{1/p} \\ &\leq \Big(\int |\varphi(\omega)|^p d\mu(\omega)\Big)^{1/p} \cdot \sup_{\omega\in K} \big(\sum_{i\leq m} |f_i(\omega)|^p\big)^{1/p} \\ &= \|\varphi\|_p \cdot \big\| \big(\sum_{i\leq m} |f_i|^p\big)^{1/p} \big\|_\infty = \|\varphi\|_p \cdot \|(f_i)_1^m\|_p^{weak}.\end{aligned}$$

We have shown that M_φ is p-summing with $\pi_p(M_\varphi) \leq \|\varphi\|_p$. It is straightforward to get equality of norms since $\pi_p(M_\varphi) \geq \|M_\varphi\| \geq \|M_\varphi 1\|_p = \|\varphi\|_p$.

(b) Plainly, the action of the operator j_p is the same as that of multiplication by the function which is constantly one. This function has L_p norm equal to the p-th root of the measure of the ambient space.

(c) For those who are aware that every $L_\infty(\mu)$ space may be represented as a $\mathcal{C}(K)$ space this example should not come as a surprise. But let us give a complete argument.

All that prevents us from copying the proof of (a) word for word is a minor irritation with the essential supremum, which makes it slightly trickier to justify the identity

$$\|(f_i)_1^m\|_p^{weak} = \big\| \big(\sum_{i\leq m} |f_i|^p\big)^{1/p} \big\|_\infty$$

for any choice of $f_1,\dots,f_m \in L_\infty(\mu)$. Let $u : \ell_{p^*}^m \to L_\infty(\mu)$ be the operator given by $e_i \mapsto f_i$; see 2.2. For each μ-null set $N \in \Sigma$ we have

$$\begin{aligned}\|(f_i)_1^m\|_p^{weak} &= \|u\| = \sup\big\{\big\| \sum_{i\leq m} a_i f_i \big\|_\infty : a \in B_{\ell_{p^*}^m}\big\} \\ &\leq \sup\big\{\big| \sum_{i\leq m} a_i f_i(\omega) \big| : a \in B_{\ell_{p^*}^m},\, \omega \in \Omega \setminus N\big\},\end{aligned}$$

and hence

$$\|(f_i)_1^m\|_p^{weak} \leq \big\| \big(\sum_{i\leq m} |f_i(\omega)|^p\big)^{1/p} \big\|_\infty .$$

Separability of $\ell_{p^*}^m$ enables us to establish the reverse inequality. For any $a = (a_1, ..., a_m) \in B_{\ell_{p^*}^m}$ we have

$$\big|\sum_{i\le m} a_i f_i(\omega)\big| \le \big\|\sum_{i\le m} a_i f_i\big\|_\infty$$

except perhaps for ω in a set of measure zero. Applying this to each member of a countable dense subset D of $B_{\ell_{p^*}^m}$, we soon find a μ-null set $N \in \Sigma$ with the property that the above inequality is satisfied for every $a \in D$ and every $\omega \in \Omega$ outside N. Consequently,

$$\begin{aligned}\big\|\big(\sum_{i\le m}|f_i|^p\big)^{1/p}\big\|_\infty &\le \sup_{\omega\in\Omega\setminus N}\big(\sum_{i\le m}|f_i(\omega)|^p\big)^{1/p} = \sup_{a\in D}\sup_{\omega\in\Omega\setminus N}\big|\sum_{i\le m} a_i f_i(\omega)\big| \\ &\le \sup_{a\in D}\big\|\sum_{i\le m} a_i f_i\big\|_\infty = \|u\| = \|(f_i)_1^m\|_p^{weak},\end{aligned}$$

which was what we wanted.

(d) follows from (c) as (b) from (a), and (e) is obtained by applying (c) to the case when μ is the counting measure on $\mathbf{N}$: then $L_\infty(\mu)$ is ℓ_∞, $L_p(\mu)$ is ℓ_p, and the multiplication operator induced by the function $\lambda \in \ell_p = L_p(\mu)$ is D_λ. We are done. QED

Whereas the formal inclusion operator i_p in (d) is always injective, the corresponding operator j_p in (b) is injective only when no non-empty open subset V of K satisfies $\mu(V) = 0$.

Formal inclusion operators merit particular attention; they will recur throughout this text. For example, they provide a convenient tool to show that the relationships expressed in the Inclusion Theorem 2.8 are proper. We leave it to the reader to verify that if $1 < q < \infty$, then the formal identity i_q from $L_\infty[0,1]$ to $L_q[0,1]$ (Lebesgue measure) cannot be p-summing for any $1 \le p < q$.

A few more simple examples of p-summing operators are worth mentioning. For the first of these, let Ω be a bounded open domain in $\mathbf{R}^n$, let $\ell \in \mathbf{N}$ and denote by

$$C^\ell(\bar\Omega)$$

the space of functions $f : \Omega \to \mathbf{R}$ whose derivatives $D^\alpha f$ exist for all $\alpha = (m_1, ..., m_n) \in \mathbf{N}^n$ satisfying $|\alpha| = m_1 + ... + m_n \le \ell$ and which extend continuously to $\bar\Omega$. A natural norm is given by

$$\|f\|_{C^\ell(\bar\Omega)} = \sup_{|\alpha|\le\ell}\|D^\alpha f\|_\infty.$$

If we complete $C^\ell(\bar\Omega)$ in the norm

$$\|f\|_{W_p^\ell(\Omega)} := \Big(\int_\Omega\sum_{\alpha\le\ell}|D^\alpha f(t)|^p\,dt\Big)^{1/p},$$

we get the *Sobolev space*

$$W_p^\ell(\Omega).$$

2.10 Example: *The inclusion map $C^\ell(\bar\Omega) \to W_p^\ell(\Omega)$ is p-summing.*

The proof is a straightforward modification of the proof in 2.9(a). If $f_1,\dots,f_k$ are in $\mathbf{C}^\ell(\bar\Omega)$, then

$$\sum_{j\le k}\|f_j\|_{W_p^\ell}^p = \sum_{j\le k}\int_\Omega \sum_{|\alpha|\le\ell} |D^\alpha f_j(t)|^p\,dt = \int_\Omega \sum_{|\alpha|\le\ell}\sum_{j\le k} |\langle \delta_t D^\alpha, f_j\rangle|^p\,dt$$
$$\le N\cdot M\cdot \sup\big\{\sum_{j\le k}|\langle x^*, f_j\rangle|^p : x^* \in B_{C^\ell(\bar\Omega)^*}\big\},$$

where N is the number of α's such that $|\alpha|\le\ell$ and M is the Lebesgue measure of Ω.

Next we look at operators defined by integrable functions. We work with $L_p(\mu, X)$, the space of all *Bochner p-integrable* X-valued 'functions' on Ω, that is, of all (μ-a.e. equivalence classes of) strongly measurable functions $f:\Omega\to X$ such that $\|f\|_p = (\int_\Omega \|f(\omega)\|^p d\mu(\omega))^{1/p} < \infty$. Of course, if $x^* \in X^*$, then $\langle x^*, f(\cdot)\rangle$ is in $L_p(\mu)$.

2.11 Example: *Let (Ω,Σ,μ) be any measure space and X any Banach space. For each $f \in L_p(\mu, X)$, the operator $u_f : X^* \to L_p(\mu) : x^* \mapsto x^* f$ is p-summing, with $\pi_p(u_f) \le \|f\|_{L_p(\mu,X)}$.*

For the simple proof, take $x_1^*,\dots,x_n^* \in X^*$ and let E be the support of f. Then

$$\big(\sum_{k\le n}\|u_f x_k^*\|_p^p\big)^{1/p} = \big(\sum_{k\le n}\int_E |\langle x_k^*, f(\omega)\rangle|^p d\mu(\omega)\big)^{1/p} \le \|(x_k^*)_{k\le n}\|_p^{weak}\cdot\|f\|_{L_p(\mu,X)}.$$

We highlight an immediate consequence.

- *Let (Ω,Σ,μ) be any measure space and let $1 < p < \infty$. Suppose that $k : \Omega\times\Omega \to \mathbf{K}$ is $\mu\times\mu$-measurable and*
$$c_p = \Big(\int_\Omega \big(\int_\Omega |k(\omega,\omega')|^{p^*} d\mu(\omega')\big)^{p/p^*} d\mu(\omega)\Big)^{1/p}$$
is finite. Then the 'kernel operator'
$$K : L_p(\mu) \longrightarrow L_p(\mu) : f \mapsto \int_\Omega k(\cdot,\omega') f(\omega')\,d\mu(\omega')$$
is p-summing with $\pi_p(K) \le c_p$.

Indeed, if $X = L_{p^*}(\mu) = L_p(\mu)^*$ and if we define $f \in L_p(\mu, X)$ by $f(\omega)(\omega') := k(\omega,\omega')$ $(\omega,\omega' \in \Omega)$ then $K = u_f$ as in Example 2.11.

Operators of this kind are called *Hille - Tamarkin operators.*

Domination and Factorization

We have seen (2.9(d)) that the canonical operators $j_p : C(K) \to L_p(\mu)$ are simple examples of p-summing operators, and we have asserted that they are especially important examples. This is the hidden content of a fundamental inequality:

2.12 Pietsch Domination Theorem: *Suppose that $1 \leq p < \infty$, that $u : X \to Y$ is a Banach space operator, and that K is a weak* compact norming subset of B_{X^*}. Then u is p-summing if and only if there exist a constant C and a regular probability measure μ on K such that for each $x \in X$*

$$\|ux\| \leq C\cdot\Big(\int_K |\langle x^*, x\rangle|^p d\mu(x^*)\Big)^{1/p}.$$

In such a case, $\pi_p(u)$ is the least of all the constants C for which such a measure exists.

Proof. One direction is easy. If such an inequality holds then

$$\sum_{k\leq n} \|ux_k\|^p \leq C^p\cdot\sum_{k\leq n}\int_K |\langle x^*, x_k\rangle|^p d\mu(x^*) \leq C^p \cdot \sup_{x^*\in B_{X^*}} \sum_{k\leq n} |\langle x^*, x_k\rangle|^p$$

for any finite collection of vectors $x_1,\dots, x_n$ in X. It follows that u is p-summing and $\pi_p(u) \leq C$.

The converse lies deeper and relies on the geometric version of the Hahn-Banach Theorem in tandem with the representation of $\mathcal{C}(K)^*$ as the space of regular Borel measures on K.

To start, we consider any finite subset M of X and define $g_M : K \to \mathbf{R}$ by

$$g_M(x^*) := \sum_{x\in M} \big(\|ux\|^p - \pi_p(u)^p\cdot|\langle x^*, x\rangle|^p\big).$$

Let Q be the set of all such g_M's; this is plainly a subset of $\mathcal{C}(K, \mathbf{R})$. Also, Q is a convex set: indeed, if M and M' are finite subsets of X and $0 < \lambda < 1$, then $\lambda \cdot g_M + (1 - \lambda)\cdot g_{M'} = g_{M''}$, where M'' is the union of $\{\lambda^{1/p}x : x \in M\}$ and $\{(1 - \lambda)^{1/p}x' : x' \in M'\}$.

A quick reference to the definition of $\pi_p(u)$ shows that Q is disjoint from the positive cone $P = \{f \in \mathcal{C}(K, \mathbf{R}) : f(x^*) > 0 \ \forall x^* \in K\}$. Now P is clearly open and convex, and so the geometric version of the Hahn-Banach Theorem provides us with a $\mu \in \mathcal{C}(K, \mathbf{R})^*$ such that, for some c,

$$\langle\mu, g\rangle \leq c < \langle\mu, f\rangle$$

for all $g \in Q$ and $f \in P$. Observe that on the one hand $c \geq 0$ since $0 \in Q$, and on the other hand, $c \leq 0$ since every positive constant function belongs to P. Since μ is a continuous linear form, $\langle\mu, f\rangle \geq 0$ follows for all $f \geq 0$ in $\mathcal{C}(K, \mathbf{R})$. Therefore we can assume that μ is a positive regular Borel measure on K, and the above inequality reads $\int_K g\, d\mu \leq 0 < \int_K f\, d\mu$ for all $g \in Q$ and $f \in P$. This inequality is impervious to positive scalings of μ, so we may as well assume that μ is a probability measure. If we now test against functions in Q of the form $g_{\{x\}}$ for $x \in X$, we see that

$$\int_K \big(\|ux\|^p - \pi_p(u)^p\cdot|\langle x^*, x\rangle|^p\big)\, d\mu(x^*) \leq 0$$

or, since μ is a probability measure,

$$\|ux\|^p \leq \pi_p(u)^p\cdot\int_K |\langle x^*, x\rangle|^p\, d\mu(x^*).$$

QED

When we referred to the hidden content of the Pietsch Domination Theorem, we had in mind the pivotal rôle of the canonical p-summing operators $j_p : C(K) \to L_p(\mu)$. To make this explicit we require some preliminaries.

Recall that any Banach space X can be viewed as a subspace of ℓ_∞^K for an appropriate set K. Indeed, for any norming subset K of B_{X^*}, the map

$$i_X : X \longrightarrow \ell_\infty^K \,, \quad i_X(x)(x^*) = \langle x^*, x\rangle$$

is plainly linear and isometric. If K is weak∗compact and norming, then i_X takes its values in the subspace $C(K)$; we still denote the resulting isometric embedding by

$$i_X : X \longrightarrow C(K).$$

Recall that a Banach space Z is *injective* if, whenever W_0 is a subspace of a Banach space W, any $v \in \mathcal{L}(W_0, Z)$ has an extension $\tilde{v} \in \mathcal{L}(W, Z)$ with $\|v\| = \|\tilde{v}\|$:

$$\begin{array}{ccc} W & \xrightarrow{\tilde{v}} & Z \\ \cup & \nearrow_v & \\ W_0 & & \end{array}$$

The spaces ℓ_∞^K are always injective: working coordinate by coordinate, this is an easy corollary to the Hahn - Banach Theorem. Injective Banach spaces are characterized by the property of being complemented by a norm one projection in any bigger Banach space. Indeed, if Z is injective, the identity $id_Z : Z \to Z$ can always be extended to a norm one projection. Conversely, if Z is a Banach space having the indicated complementation property, let i_Z be the embedding of Z into ℓ_∞^B where $B = B_{Z^*}$, and let p be a norm one projection of ℓ_∞^B onto Z. Now take a subspace W_0 of a Banach space W and any $v \in \mathcal{L}(W_0, Z)$. As ℓ_∞^B is injective, $i_Z v$ has an extension $u \in \mathcal{L}(W, \ell_\infty^B)$ such that $\|u\| = \|i_Z v\| = \|v\|$. Summarize by a diagram:

$$\begin{array}{ccc} W & \xrightarrow{\quad u \quad} & \ell_\infty^B \\ \cup & & {\scriptstyle p}\downarrow\uparrow{\scriptstyle i_Z} \\ W_0 & \xrightarrow{\quad v \quad} & Z\,. \end{array}$$

Clearly $\tilde{v} := pu : W \to Z$ is an extension of v with $\|\tilde{v}\| = \|v\|$.

2.13 Pietsch Factorization Theorem: *Let $1 \le p < \infty$, let X and Y be Banach spaces, let K be a weak∗compact norming subset of B_{X^*}, and let B be B_{Y^*} (or a norming subset thereof). For every operator $u : X \to Y$, the following are equivalent:*

(i) *u is p-summing.*

(ii) *There exist a regular Borel probability measure μ on K, a (closed) subspace X_p of $L_p(\mu)$ and an operator $\hat{u} : X_p \to Y$ such that*
(a) $j_p i_X(X) \subset X_p$ and (b) $\hat{u} j_p i_X(x) = ux$ for all $x \in X$.
In other words, if j_p^X is the map $i_X(X) \to X_p$ induced by j_p, then the following diagram commutes:

$$\begin{array}{ccc} X & \xrightarrow{u} & Y \\ i_X \downarrow & & \uparrow \hat{u} \\ i_X(X) & \xrightarrow{j_p^X} & X_p \\ \cap & & \cap \\ C(K) & \xrightarrow{j_p} & L_p(\mu)\ . \end{array}$$

(iii) *There exist a regular Borel probability measure μ on K and an operator $\tilde{u} : L_p(\mu) \to \ell_\infty^B$ such that the following diagram commutes:*

$$\begin{array}{ccc} X & \xrightarrow{u} & Y \overset{i_Y}{\searrow} \\ \downarrow i_X & & \ell_\infty^B \\ C(K) & \xrightarrow{j_p} & L_p(\mu) \overset{\tilde{u}}{\nearrow} . \end{array}$$

(iv) *There exist a probability space (Ω, Σ, μ) and operators $\tilde{u} : L_p(\mu) \to \ell_\infty^B$ and $v : X \to L_\infty(\mu)$ such that the following diagram commutes:*

$$\begin{array}{ccc} X & \xrightarrow{u} & Y \overset{i_Y}{\searrow} \\ \downarrow v & & \ell_\infty^B \\ L_\infty(\mu) & \xrightarrow{i_p} & L_p(\mu) \overset{\tilde{u}}{\nearrow} . \end{array}$$

In addition, we may choose μ and $\hat{u}$ in (ii) *or μ and $\tilde{u}$ in* (iii) *so that $\|\hat{u}\| = \|\tilde{u}\| = \pi_p(u)$; in* (iv) *we may arrange that $\|v\| = 1$ and $\|\tilde{u}\| = \pi_p(u)$.*

Proof. (i)$\Rightarrow$(ii): The essential work has already been done. If u is p-summing, the Pietsch Domination Theorem 2.12 provides a regular Borel probability measure μ on K for which

$$\|ux\| \le \pi_p(u) \cdot \Big(\int_K |\langle x^*, x\rangle|^p \, d\mu(x^*) \Big)^{1/p} \qquad \text{for all } x \in X.$$

This informs us that if we denote the range of $j_p i_X$ by S and consider it to be a normed subspace of $L_p(\mu)$, the map $S \to Y : j_p i_X(x) \mapsto ux$ is a well-defined operator with norm at most $\pi_p(u)$. Let X_p be the closure of S in $L_p(\mu)$. Then the natural extension of our map to X_p is the operator $\hat{u}$ we are looking for. It certainly satisfies $\hat{u} j_p^X i_X = u$ and $\|\hat{u}\| \le \pi_p(u)$. For the reverse inequality bring the ideal property 2.4 and the injectivity property 2.5 to bear: $\pi_p(u) = \pi_p(\hat{u} j_p^X i_X) \le \|\hat{u}\| \cdot \pi_p(j_p^X) \cdot \|i_X\| \le \|\hat{u}\| \cdot \pi_p(j_p) = \|\hat{u}\|$.

(ii)$\Rightarrow$(iii): The key is the injectivity of ℓ_∞^B. Starting from the diagram in (ii), we see that $i_Y \hat{u} : X_p \to \ell_\infty^B$ has an extension $\tilde{u} : L_p(\mu) \to \ell_\infty^B$ with $\|\tilde{u}\| = \|i_Y \hat{u}\| = \|\hat{u}\| = \pi_p(u)$. This is obviously the operator we are looking for.

(iii)$\Rightarrow$(iv): The only thing we have to observe is that we can factor j_p using canonical mappings: $j_p : C(K) \xrightarrow{j_\infty} L_\infty(\mu) \xrightarrow{i_p} L_p(\mu)$. Inserting this into the commutative diagram in (iii) we see that $v = j_\infty i_X$ is the required operator.

(iv)$\Rightarrow$(i): Since we know from our basic example 2.9(d) that i_p is p-summing, an appeal to the ideal property 2.4 establishes that $i_Y u$ is p-summing. The injectivity property 2.5 is enough to settle the issue. QED

There are many further variants of the Pietsch Factorization Theorem. For example, a variant of (ii) may be constructed using $L_\infty(\mu)$-spaces as in (iv). We leave the details, as well as the invention of other variants, to the reader's imagination.

On its way to $L_p(\mu)$ the canonical map $j_p : C(K) \to L_p(\mu)$ meets all the spaces $L_q(\mu)$ for $p < q < \infty$, that is, it factorizes $C(K) \xrightarrow{j_q} L_q(\mu) \xrightarrow{i} L_p(\mu)$, i being the formal identity. In conjunction with the Pietsch Factorization Theorem, this provides an alternative proof of the Inclusion Theorem 2.8.

The reason for introducing ℓ_∞^B in case (iii) of the Factorization Theorem was that its injectivity allowed us to extend maps from X_p. If Y is itself injective this device is no longer necessary and factorization takes a sleeker form.

2.14 Corollary: *If Y is injective then $u : X \to Y$ is p-summing if and only if there exist a regular Borel probability measure μ on K and a map $\tilde{u} \in \mathcal{L}(L_p(\mu), Y)$ such that $u = \tilde{u} j_p i_X$:*

$$\begin{array}{ccc} X & \xrightarrow{u} & Y \\ \downarrow{\scriptstyle i_X} & & \uparrow{\scriptstyle \tilde{u}} \\ C(K) & \xrightarrow{j_p} & L_p(\mu) \, . \end{array}$$

Again μ and $\tilde{u}$ can be chosen so that $\|\tilde{u}\| = \pi_p(u)$.

There is of course a corresponding $L_\infty(\mu)$-version of this corollary, modelled on case (iv) of the Pietsch Factorization Theorem:

- If Y is injective, then $u : X \to Y$ is p-summing if and only if there exist a probability space (Ω, Σ, μ) along with operators $\tilde{u} : L_p(\mu) \to Y$ and $v : X \to L_\infty(\mu)$ such that $u = \tilde{u} i_p v$:

$$\begin{array}{ccc} X & \xrightarrow{u} & Y \\ \downarrow{\scriptstyle v} & & \uparrow{\scriptstyle \tilde{u}} \\ L_\infty(\mu) & \xrightarrow{i_p} & L_p(\mu) \, . \end{array}$$

Again, we may arrange that $\|v\| = 1$ and $\|\tilde{u}\| = \pi_p(u)$.

It is inescapable that such pleasant forms of the Factorization Theorem are not available for general range spaces (see 5.11 and 5.13 for examples). However, there are further remarkable situations where the theorem has an uncluttered appearance: the presence of an injective range is not always essential for the existence of 'good' extensions. The first example also serves to illustrate why we insisted on having the freedom to choose weak$*$compact norming sets other than the dual unit ball.

2.15 Corollary: *Let K be a compact Hausdorff space. An operator $u : \mathcal{C}(K) \to Y$ is p-summing if and only if there exist a regular Borel probability measure μ on K and a map $\tilde{u} \in \mathcal{L}(L_p(\mu), Y)$ such that $\tilde{u} j_p = u$:*

$$\begin{array}{ccc} \mathcal{C}(K) & \xrightarrow{u} & Y \\ & {}_{j_p}\searrow \quad \nearrow_{\tilde{u}} & \\ & L_p(\mu) . & \end{array}$$

Moreover, we may arrange that $\|\tilde{u}\| = \pi_p(u)$.

Proof. The only thing which requires proof is that u admits such a factorization if it is p-summing. Let $\delta_\omega \in \mathcal{C}(K)^*$ be the point mass associated with $\omega \in K$ (see the proof of 2.9(a)). Then $\omega \mapsto \delta_\omega$ identifies K with a weak$*$compact norming subset of $B_{\mathcal{C}(K)^*}$. Consequently, when $X = \mathcal{C}(K)$, the corresponding map i_X is nothing but the identity. As $j_p : \mathcal{C}(K) \to L_p(\mu)$ has dense range, the diagram of case (ii) of the Pietsch Factorization Theorem collapses to the one we are looking for. QED

Simplicity also rules when $p = 2$, thanks this time to the fact that every closed subspace of a Hilbert space is nicely complemented.

2.16 Corollary: *An operator $u : X \to Y$ is 2-summing if and only if there exist a regular probability measure μ on K and a map $\tilde{u} \in \mathcal{L}(L_2(\mu), Y)$ such that the following diagram commutes:*

$$\begin{array}{ccc} X & \xrightarrow{u} & Y \\ \downarrow{\scriptstyle i_X} & & \uparrow{\scriptstyle \tilde{u}} \\ \mathcal{C}(K) & \xrightarrow{j_2} & L_2(\mu) . \end{array}$$

Moreover, we may arrange that $\|\tilde{u}\| = \pi_2(u)$.

Proof. Once again, only one direction requires proof. Assume that u is 2-summing and refer back to case (ii) of the Pietsch Factorization Theorem: we have a regular Borel probability measure μ on K, a closed subspace X_2 of $L_2(\mu)$, and an operator $\hat{u} : X_2 \to Y$ such that $\hat{u} j_2 i_X(x) = ux$ for all $x \in X$ and $\|\hat{u}\| = \pi_2(u)$. Let P be the orthogonal projection from $L_2(\mu)$ onto X_2. Then $\tilde{u} := \hat{u}P$ satisfies $\tilde{u} j_2 i_X = u$ (draw a diagram!). Also $\|\tilde{u}\| \leq \|\hat{u}\| = \pi_2(u)$, and since $\tilde{u}$ extends $\hat{u}$, we even have $\|\tilde{u}\| = \pi_2(u)$. QED

It is worth observing for later use that, taking case (iv) of the Factorization Theorem as a starting point, we may characterize 2-summing operators $u : X \to Y$ as those which admit a factorization $X \xrightarrow{v} L_\infty(\mu) \xrightarrow{i_2} L_2(\mu) \xrightarrow{\tilde{u}} Y$, where μ is a probability measure, and v and $\tilde{u}$ satisfy $\|v\| = 1$ and $\|\tilde{u}\| = \pi_2(u)$.

Some Consequences

In discussing the notion of p-summing operators we found that a linear map $u : X \to Y$ is p-summing precisely when it takes weakly p-summable sequences in X to strongly p-summable sequences in Y. Thus a natural companion to p-summing operators is the class of linear maps $u : X \to Y$ which take weakly null sequences in X to norm null sequences in Y. These operators are said to be *completely continuous.*

Taking limits into account, a linear map $u : X \to Y$ is completely continuous if and only if it takes weakly convergent sequences to norm convergent sequences; thanks to the Eberlein - Šmulian Theorem this happens exactly when u takes weakly compact sets into norm compact sets.

Another frequently useful feature is that if $u : X \to Y$ is a completely continuous operator then in fact u takes each weakly Cauchy sequence into a norm Cauchy (hence convergent) sequence. The reason is simple enough:

If (x_n) is a sequence in X, then (x_n) is weakly Cauchy (respectively, norm Cauchy) if and only if, given strictly increasing sequences (j_n) and (k_n) of positive integers, the sequence $(x_{k_n} - x_{j_n})$ is weakly null (respectively, norm null). Our assertion is now readily verified.

The completely continuous operators from X to Y form a closed linear subspace of $\mathcal{L}(X, Y)$ which we denote by

$$\mathcal{V}(X, Y).$$

It contains the space

$$\mathcal{K}(X, Y)$$

of compact operators, but the containment may be proper: if a Banach space X has the Schur property (see after 1.7), then the identity id_X is completely continuous – but is not compact unless X is finite dimensional.

Next, we recall that $u \in \mathcal{L}(X, Y)$ is *weakly compact* if $u(B_X)$ is relatively weakly compact in Y. The Eberlein - Šmulian Theorem alerts us to the fact that $u \in \mathcal{L}(X, Y)$ is weakly compact if and only if, given any bounded sequence (x_n) in X, (ux_n) has a weakly convergent subsequence in Y. The weakly compact operators from X to Y form a closed linear subspace of $\mathcal{L}(X, Y)$ which we will denote by

$$\mathcal{W}(X, Y).$$

Again, every compact operator is weakly compact.

Trivially, each of the classes $\mathcal{K}$, $\mathcal{V}$, $\mathcal{W}$ has the *ideal property:* if X_0, X, Y and Y_0 are Banach spaces and $u : Y \to Y_0$, $v : X \to Y$, $w : X_0 \to X$ are bounded linear operators, then $v \in \mathcal{A}(X, Y)$ implies $uvw \in \mathcal{A}(X_0, Y_0)$, where $\mathcal{A}$ is $\mathcal{K}$, $\mathcal{V}$, or $\mathcal{W}$. These classes also enjoy the injectivity property: if

$i : Y \to Y_0$ is an isometric embedding, then $u : X \to Y$ is in $\mathcal{A}(X,Y)$ if and only if $iu \in \mathcal{A}(X,Y_0)$.

2.17 Theorem: *Regardless of* $1 \le p < \infty$, *every p-summing operator between Banach spaces is weakly compact and completely continuous.*

Proof. By virtue of the Inclusion Theorem 2.8, it suffices to deal with the case $1 < p < \infty$.

The ideal property and the injectivity of Π_p, $\mathcal{V}$ and $\mathcal{W}$, taken in combination with the Pietsch Factorization Theorem 2.13, allow considerable reduction. Think about it: they indicate that we merely need to show that if μ is a positive regular Borel measure on a compact Hausdorff space K, then the canonical map $j_p : C(K) \to L_p(\mu)$ is both weakly compact and completely continuous.

Since $L_p(\mu)$ is reflexive, bounded subsets are relatively weakly compact, and so j_p is weakly compact. To see that it is completely continuous, let (f_n) be any weak null sequence in $C(K)$. Then (f_n) is bounded and converges pointwise to zero. Application of Lebesgue's Dominated Convergence Theorem shows that $(j_p f_n)$ is a norm null sequence in $L_p(\mu)$. QED

- It is plain and easy to see that if $v \in \mathcal{W}(X,Y)$ and $u \in \mathcal{V}(Y,Z)$, then uv is compact.
- In particular, the composition of a p-summing operator with a q-summing operator is compact, no matter how we choose $1 \le p, q < \infty$.

This leads to the following weak version of the Dvoretzky - Rogers Theorem 1.2:

2.18 Weak Dvoretzky - Rogers Theorem: *Let* $1 \le p < \infty$. *Every infinite dimensional Banach space* X *contains a weakly p-summable sequence which fails to be strongly p-summable.*

Proof. If not, id_X would be p-summing. But $id_X = (id_X)^2$, so id_X would be compact, which is only possible if X is finite dimensional. QED

In view of Proposition 2.7, we could also have stated this corollary as follows: if a Banach space X satisfies $\mathcal{L}(\ell_{p^*}, X) = \Pi_p(\ell_{p^*}, X)$ $(1 < p < \infty)$ or $\mathcal{L}(c_0, X) = \Pi_1(c_0, X)$, then X must be finite dimensional. In fact, the condition just stated is equivalent to id_X being p-summing. It will follow from 3.2 that ℓ_{p^*} (c_0 if $p = 1$) can be replaced by any infinite dimensional space $L_{p^*}(\mu)$ ($C(K)$ if $p = 1$). In 10.5 we shall even extend these results and present a proof of 2.18 which is much closer to the one of the original Dvoretzky - Rogers Theorem 1.2.

Weak compactness of p-summing operators is the key to the following result:

2.19 Proposition: *Let* $1 \le p < \infty$. *A Banach space operator* $u : X \to Y$ *is p-summing if and only if its second adjoint* $u^{**} : X^{**} \to Y^{**}$ *is p-summing. In this case,* $\pi_p(u) = \pi_p(u^{**})$.

Proof. The map u^{**} extends u. By the ideal property 2.4 and by the injectivity property 2.5, if $u^{**} : X^{**} \to Y^{**}$ is p-summing so is u, with $\pi_p(u) \leq \pi_p(u^{**})$.

Suppose conversely that u is p-summing. The Pietsch Factorization Theorem 2.13 furnishes a regular Borel probability measure μ on the weak$*$compact set B_{X^*} and an operator $\tilde{u} : L_p(\mu) \to \ell_\infty^{B_{Y^*}}$ with $\|\tilde{u}\| = \pi_p(u)$, having the property that $i_Y u$ factors as $X \xrightarrow{v} L_\infty(\mu) \xrightarrow{i_p} L_p(\mu) \xrightarrow{\tilde{u}} \ell_\infty^{B_{Y^*}}$, where $v = j_\infty i_X$. Note that i_p^{**} can be viewed as i_p composed with the canonical projection $P : L_\infty(\mu)^{**} \to L_\infty(\mu)$, which is simply the adjoint of the usual embedding $L_1(\mu) \to L_1(\mu)^{**}$. By weak compactness, we may and do consider u^{**} as a map from X^{**} to Y. Since $i_Y u^{**} = \tilde{u} i_p P v^{**}$, we may again invoke the ideal property and injectivity to infer that u is p-summing and

$$\pi_p(u^{**}) = \pi_p(i_Y u^{**}) \leq \|\tilde{u}\| \cdot \pi_p(i_p) \cdot \|P\| \cdot \|v^{**}\| = \|\tilde{u}\| = \pi_p(u). \qquad \text{QED}$$

2.20 Remark: We take this occasion to point out that in general the adjoint of a p-summing operator need not be p-summing. Further, there are even operators with a p-summing adjoint which are not themselves p-summing.

For an easy example, take the formal identity $i : \ell_2 \to c_0$. Then $i^* : \ell_1 \to \ell_2$ and $i^{**} : \ell_2 \to \ell_\infty$ are just the corresponding formal identities. But i^* is 1-summing and so p-summing for all $1 \leq p < \infty$, whereas i and i^{**} even fail to be completely continuous.

On the other hand, if our Banach spaces are such that the adjoint of a q-summing operator is always p-summing for some p, then we are in a very interesting situation. Here we discuss one such case; for more on this see 4.19 and 9.12.

2.21 Theorem: *Let X be a Banach space and H a Hilbert space. If $u \in \mathcal{L}(X, H)$ is such that u^* is q-summing for some $1 \leq q < \infty$, then u is 1-summing and $\pi_1(u) \leq A_1^{-1} \cdot B_q \cdot \pi_q(u^*)$.*

Here A_1 and B_q are the constants from Khinchin's Inequality 1.10.

Proof. Consider first the case of an operator $u : X \to \ell_2^n$ ($n \in \mathbf{N}$). Given $x_1, \ldots, x_m \in X$, we can use Khinchin's Inequality to get

$$\begin{aligned}
\sum_{j \leq m} \|u x_j\| &= \sum_{j \leq m} \Big(\sum_{k \leq n} |(u x_j | e_k)|^2 \Big)^{1/2} = \sum_{j \leq m} \Big(\int_0^1 \Big| \sum_{k \leq n} (u x_j | e_k) r_k(t) \Big|^2 dt \Big)^{1/2} \\
(1) \qquad &\leq A_1^{-1} \cdot \int_0^1 \sum_{j \leq m} \Big| \Big\langle x_j, \sum_{k \leq n} r_k(t) u^* e_k \Big\rangle \Big| \, dt \\
&\leq A_1^{-1} \cdot \Big(\int_0^1 \Big\| \sum_{k \leq n} r_k(t) u^* e_k \Big\| \, dt \Big) \cdot \|(x_j)_1^m\|_1^{weak}.
\end{aligned}$$

We must, then, get a good bound on the integral.

Consider u^* as a member of $\Pi_q(\ell_2^n, X^*)$. By Pietsch's Domination Theorem 2.12 there is a regular Borel probability measure μ on the compact Hausdorff space $K = B_{\ell_2^n}$ such that, for each $x \in \ell_2^n$,

$$\|u^*x\| \leq \pi_q(u^*)\cdot\Big(\int_K |(y\,|\,x)|^q\, d\mu(y)\Big)^{1/q}.$$

Apply Fubini's Theorem and Khinchin's Inequality 1.10 and the fact that μ is a probability measure to get

$$\begin{aligned}
\Big(\int_0^1 \Big\|\sum_{k\leq n} r_k(t)u^*e_k\Big\|^q\, dt\Big)^{1/q} &= \Big(\int_0^1 \Big\|u^*\Big(\sum_{k\leq n} r_k(t)e_k\Big)\Big\|^q\, dt\Big)^{1/q} \\
&\leq \pi_q(u^*)\cdot\Big(\int_K\int_0^1 \Big|\Big(y\,\Big|\,\sum_{k\leq n} r_k(t)e_k\Big)\Big|^q\, dt\, d\mu(y)\Big)^{1/q} \qquad (2)\\
&\leq B_q\cdot\pi_q(u^*)\cdot\Big(\int_K\Big(\sum_{k\leq n}|(y\,|\,e_k)|^2\Big)^{q/2} d\mu(y)\Big)^{1/q} \leq B_q\cdot\pi_q(u^*).
\end{aligned}$$

The important work is done. To conclude, take any operator $u : X \to H$ with q-summing adjoint and fix $x_1,\ldots,x_m \in X$. Identify the span of the ux_j's with ℓ_2^n for the appropriate n, let $p \in \mathcal{L}(H)$ be the orthogonal projection onto this span, and use (1) and (2) to obtain

$$\begin{aligned}
A_1\cdot\pi_1(pu) &\leq \int_0^1 \Big\|\sum_{k\leq n} r_k(t)(pu)^*e_k\Big\|\, dt \leq \Big(\int_0^1 \Big\|\sum_{k\leq n} r_k(t)u^*pe_k\Big\|^q\, dt\Big)^{1/q} \\
&= \Big(\int_0^1 \Big\|\sum_{k\leq n} r_k(t)u^*e_k\Big\|^q\, dt\Big)^{1/q} \leq B_q\cdot\pi_q(u^*).
\end{aligned}$$

It follows that

$$\begin{aligned}
\sum_{j\leq m}\|ux_j\| &= \sum_{j\leq m}\|pux_j\| \leq \pi_1(pu)\cdot\|(x_j)_1^m\|_1^{weak} \\
&\leq A_1^{-1}\cdot B_q\cdot\pi_q(u^*)\cdot\|(x_j)_1^m\|_1^{weak}.
\end{aligned}$$

QED

COMPOSITION

We end this chapter with an important application of factorization and domination. We have already seen that if $v : X \to Y$ is p-summing and $u : Y \to Z$ is q-summing, then uv will be compact. By the ideal property, it will also be r-summing for $r = \min\{p, q\}$. Much more can be said:

2.22 Composition Theorem: *Let $u \in \Pi_p(Y, Z)$ and $v \in \Pi_q(X, Y)$ with $1 \leq p, q < \infty$. Define $1 \leq r < \infty$ by $1/r := \min\{1,\ (1/p) + (1/q)\}$. Then uv is r-summing, and $\pi_r(uv) \leq \pi_p(u)\cdot\pi_q(v)$.*

We require a technical lemma.

2.23 Lemma: *Suppose $1 \le p,q,r < \infty$ are such that $1/r = (1/p) + (1/q)$. Let X and Y be arbitrary Banach spaces, and let $v : X \to Y$ be any q-summing operator. Then, given any sequence $(x_n) \in \ell_r^{weak}(X)$, there are sequences $(\sigma_n) \in \ell_q$ and $(y_n) \in \ell_p^{weak}(Y)$ such that, if $\gamma := \|(x_n)\|_r^{weak}$,*

(a) $\quad vx_n = \sigma_n \cdot y_n \quad \forall n \in \mathbf{N}$,

(b) $\quad \|(\sigma_n)\|_q \le \gamma^{r/q}$,

(c) $\quad \|(y_n)\|_p^{weak} \le \gamma^{r/p} \cdot \pi_q(v)$.

Proof. By the Pietsch Factorization Theorem 2.13(ii), we have the commutative diagram indicated below:

$$\begin{array}{ccc} X & \xrightarrow{\ v\ } & Y \\ \downarrow i_X & & \uparrow \hat{v} \\ i_X(X) & \xrightarrow{\ j_q^X\ } & X_q \\ \cap & & \cap \\ C(B_{X^*}) & \xrightarrow{\ j_q\ } & L_q(\mu)\ . \end{array}$$

As usual, μ is a regular Borel probability measure on the weak$*$compact set B_{X^*}. We may assume $\|\hat{v}\| = \pi_q(u)$. For convenience, set $j = j_q^X i_X$.

For each $y^* \in Y^*$, consider $t_{y^*} := \hat{v}^*(y^*) \in X_q^*$. Since for all $x \in X$,

$$|\langle t_{y^*}, jx\rangle| = |\langle \hat{v}^* y^*, jx\rangle| \le \|\hat{v}^*\|\cdot\|y^*\|\cdot\|jx\|_{L_q} = \pi_q(v)\cdot\|y^*\|\cdot\|jx\|_{L_q}$$

it follows that

$$\|t_{y^*}\| \le \pi_q(v)\cdot\|y^*\|.$$

Let $f_{y^*} \in L_{q^*}(\mu)$ be a norm preserving Hahn-Banach extension of t_{y^*}, so that

$$\|f_{y^*}\|_{L_q^*} \le \pi_q(v)\cdot\|y^*\|. \tag{1}$$

We may now write

$$\langle y^*, vx\rangle = \langle t_{y^*}, jx\rangle = \int_K \langle x^*, x\rangle \cdot f_{y^*}(x^*)\, d\mu(x^*) \quad \forall\ (x,y^*) \in X \times Y^*.$$

Set $\sigma := \left(\int_K |\langle x^*, x\rangle|^r d\mu(x^*)\right)^{1/q}$ and apply Hölder's Inequality twice, using $(r/q) + (r/p) = (q^*/p) + (q^*/r^*) = 1$:

$$\begin{aligned} |\langle y^*, vx\rangle| &\le \int_K |\langle x^*, x\rangle|^{r/q}\cdot|\langle x^*, x\rangle|^{r/p}\cdot|f_{y^*}(x^*)|^{q^*/p}\cdot|f_{y^*}(x^*)|^{q^*/r^*}\, d\mu(x^*) \\ &\le \sigma\cdot\Big(\int_K \big(|\langle x^*, x\rangle|^r\cdot|f_{y^*}(x^*)|^{q^*}\big)^{q^*/p}\cdot\big(|f_{y^*}(x^*)|^{q^*}\big)^{q^*/r^*}\, d\mu(x^*)\Big)^{1/q^*} \qquad (2) \\ &\le \sigma\cdot\Big(\int_K |\langle x^*, x\rangle|^r\cdot|f_{y^*}(x^*)|^{q^*}\, d\mu(x^*)\Big)^{1/p}\cdot\|f_{y^*}\|_{q^*}^{q^*/r^*}. \end{aligned}$$

If we now set $\sigma_n := \left(\int_K |\langle x^*, x_n\rangle|^r d\mu(x^*)\right)^{1/q}$ $(n \in \mathbf{N})$, then

$$\sum_n |\sigma_n|^q = \int_K \sum_n |\langle x^*, x_n\rangle|^r d\mu(x^*) \le \gamma^r.$$

Hence $(\sigma_n) \in \ell_q$, and (b) obtains.

The next step is of course to define $y_n := \sigma_n^{-1} \cdot vx_n$ whenever $\sigma_n \neq 0$ and $y_n := 0$ otherwise. This ensures that (a) comes for free once (c) is established. From (2) we obtain

$$|\langle y^*, y_n\rangle|^p \le \left(\int_K |\langle x^*, x_n\rangle|^r \cdot |f_{y^*}(x^*)|^{q^*} d\mu(x^*)\right) \cdot \|f_{y^*}\|_{q^*}^{pq^*/r^*}$$

for all $y^* \in Y^*$ and $n \in \mathbf{N}$. Hence

$$\left(\sum_{n\le k} |\langle y^*, y_n\rangle|^p\right)^{1/p} \le \left(\int_K \sum_{n\le k} |\langle x^*, x_n\rangle|^r \cdot |f_{y^*}(x^*)|^{q^*} d\mu(x^*)\right)^{1/p} \cdot \|f_{y^*}\|_{q^*}^{q^*/r^*}$$

$$\le \left(\int_K \gamma^r \cdot |f_{y^*}(x^*)|^{q^*} d\mu(x^*)\right)^{1/p} \cdot \|f_{y^*}\|_{q^*}^{q^*/r^*} = \gamma^{r/p} \cdot \|f_{y^*}\|_{q^*}$$

for all $y^* \in Y^*$ and $k \in \mathbf{N}$. Combining this with (1), we obtain $(y_n) \in \ell_p^{weak}(Y)$ together with $\|(y_n)\|_p^{weak} \le \gamma^{r/p} \cdot \pi_q(v)$. QED

We now have sufficient ammunition to dispose of the Composition Theorem.

Proof of 2.22. Suppose first that $1 \ge (1/p) + (1/q)$, so that $1/r = (1/p) + (1/q)$. Let $(x_n) \in \ell_r^{weak}(X)$ be given; without loss of generality we may assume that $\|(x_n)\|_r^{weak} = 1$. Apply the preceding lemma to produce sequences (σ_n) in ℓ_q and (y_n) in $\ell_p^{weak}(Y)$ such that $\|(\sigma_n)\|_q \le 1$, $\|(y_n)\|_p^{weak} \le \pi_q(v)$, and $vx_n = \sigma_n y_n$ for all n. Then $(uy_n) \in \ell_p^{strong}(Z)$ since $u \in \Pi_p(Y, Z)$, and $\|(uy_n)\|_p^{strong} \le \pi_p(u) \cdot \|(y_n)\|_p^{weak} \le \pi_p(u) \pi_q(v)$. Now Hölder's Inequality, using $(r/p) + (r/q) = 1$, yields

$$\left(\sum_n \|uvx_n\|^r\right)^{1/r} \le \|(\sigma_n)\|_q \cdot \left(\sum_n \|uy_n\|^p\right)^{1/p} \le \pi_p(u) \cdot \pi_q(v),$$

which was what we wanted.

The case $1 < (1/p) + (1/q)$ is taken care of by the Inclusion Theorem 2.8. QED

The index r in the Composition Theorem cannot generally be improved. For an easy counterexample observe that if $D_\lambda : c_0 \to c_0$ is the diagonal operator induced by $\lambda \in \ell_\infty$, then D_λ is p-summing if and only if $\lambda \in \ell_p$. Then recall that the pointwise product of elements in ℓ_p and ℓ_q lies in ℓ_r whenever $1/r = (1/p) + (1/q)$ and that this is best possible.

NOTES AND REMARKS

The classes of p-summing operators were introduced by A. Pietsch [1967] and virtually everything in this chapter can be found in his remarkable paper.

Actually, Π_1 and Π_2 were studied before in A. Grothendieck's Résumé [1953a]: Π_1 is the class of 'applications préintégrales droites', and Π_2 is the class of so-called 'opérateurs \H-intégrales'. Whereas Grothendieck was in possession of the norm π_1 in the first case, the norm which he attributed to the Π_2-operators is only equivalent to π_2. This difference has important ramifications, thanks to Pietsch's Domination Theorem 2.12. Indeed, the depth of Pietsch's [1967] contribution comes in large part from his isolation of the simple finitary defining inequality governing an operator's inclusion in Π_p. The discovery of the Domination Theorem and the deduction of the factorizations that flow from it make his paper a classic. Such schemes in combination with the finitary description offer new computational flexibility in studying delicate analytic phenomena. The presence of a formal identity mapping $C(K) \to L_p(\mu)$ in every p-summing operator has profound consequences, just some of which are to be pursued in this text.

The late sixties saw a general revival of interest in the theory of Banach spaces. With few exceptions, the subject had been in a state of abeyance for more than twenty years. Pietsch's paper was but one of several significant signals. Already in the very next issue of Studia Mathematica, J. Lindenstrauss and A. Pełczyński published their seminal [1968] paper. Pietsch's p-summing operators played a central part in the Lindenstrauss - Pełczyński scheme most especially because they permitted the circumvention of certain very technical aspects of Grothendieck's powerful yet difficult theory of tensor norms.

Things developed quickly thereafter. Whereas Lindenstrauss and Pełczyński needed more or less only Π_1 and Π_2, H.P. Rosenthal made use of the full scale of the ideals Π_p in his penetrating [1973] study of subspaces of $L_1[0,1]$. Since then, p-summing operators have proved their importance in manifold applications and it is the exposition of these that will occupy much of our time and effort.

The central result of this chapter, indeed of the entire basic theory of absolutely summing operators, is the remarkable connection with measure theory, as expressed in Pietsch's [1967] twin Theorems 2.12 and 2.13. The proof presented in the text takes into account certain technical improvements suggested by J. Lindenstrauss and A. Pełczyński [1968] and by B. Maurey's fundamental study [1974a] of factorization of L_p-valued operators.

The somewhat unexpected measure theoretic connections which are hidden in the finitary definition of p-summing operators lead to delightful global conclusions of the sort expressed in Theorem 2.17. Its corollary, the Weak Dvoretzky-Rogers Theorem (2.18), was known to A. Grothendieck [1953b] though his derivation is not nearly so crisp as Pietsch's.

The p-summing operators improve summability properties of sequences, and so it is not surprising that they also improve the integrability of vector-valued functions. Indeed, a Banach space operator $u : X \to Y$ is p-summing if and only if, given any probability space (Ω, Σ, μ) and any strongly measurable function $f : \Omega \to X$ which is weakly p-integrable (that is, $x^* \circ f \in L_p(\mu)$ for each $x^* \in X$), $u \circ f$ is Bochner p-integrable (that is, $u \circ f$ is strongly measurable and $\|u \circ f(\cdot)\| \in L_p(\mu)$). For $p = 1$, this observation was made by J. Diestel [1972]; his proof is easily modified to cover the case $p > 1$. More generally, A. Bélanger and P. Dowling [1988] proved that if $f : \Omega \to X$ is weakly 1-integrable and $u : X \to Y$ is 1-summing, then there is a Bochner 1-integrable function $g : \Omega \to Y$ such that $y^* \circ g = y^* \circ u \circ f$ μ-almost everywhere for each $y^* \in Y^*$.

Proposition 2.19 is due to Pietsch and reflects the tensorial nature of the class Π_p; this property is common to the α-integral operators for any of Grothendieck's tensor norms α. Theorem 2.21 is due to S. Kwapień [1970b]. The important Composition Theorem 2.22 and its proof is due to A. Pietsch [1967].

For a Banach space operator $u : X \to Y$ define $\pi_p^{(n)}(u)$ in a similar way to $\pi_p(u)$, but using no more than n vectors from X. A famous result of N. Tomczak-Jaegermann [1979] asserts that $\pi_2(u) \leq \sqrt{2} \cdot \pi_2^{(n)}(u)$ holds whenever u is a rank n operator. Generalizations to the p-summing case and further references can be found in W.B. Johnson and G. Schechtman [1994].

The probability measure arising in the Pietsch Domination Theorem is sometimes called a 'Pietsch measure'. In general, its existence is accessible only by transfinite means. When we are given additional structural information about the operator, it may happen that we can directly detect such a measure which then naturally reflects properties of the operator. Here is a simple example.

Theorem: *Let G be a compact topological group. If F is a closed translation invariant subspace of $\mathcal{C}(G)$, X is a Banach space, and $u : F \to X$ is a translation invariant p-summing operator $(1 \leq p < \infty)$, then the normalized Haar measure on G is a Pietsch measure for u.*

Proof. Consider G as a norming set for F, and let μ be a Pietsch measure on G for u; so $\mu \in \mathcal{C}(G)^*$ is a probability measure such that, for all $f \in F$,

$$\|uf\|^p \leq \pi_p(u)^p \cdot \int_G |f(g)|^p d\mu(g).$$

For each $h \in G$ and $f \in \mathcal{C}(G)$ define the translate $f_h \in \mathcal{C}(G)$ by $f_h(g) := f(hg)$. Next, define $\mu_h \in \mathcal{C}(G)^*$ by $\langle \mu_h, f \rangle := \langle \mu, f_h \rangle$. Again, μ_h is a probability measure; what is more, it is a Pietsch measure for u too: if $f \in F$ then

$$\|uf\|^p = \|u(f_h)\|^p \leq \pi_p(u)^p \cdot \int_G |f_h(g)|\, d\mu(g) = \pi_p(u)^p \cdot \int_G |f(g)|\, d\mu_h(g).$$

Now the map $G \to C(G)^* : h \to \mu_h$ is continuous when we endow $C(G)^*$ with its weak∗topology, so if σ_G denotes the normalized Haar measure on G the Gelfand integral $\int_G \mu_h \, d\sigma_G(h)$ exists. The Gelfand integral is just a measure $\nu \in C(G)^*$ such that for every $f \in C(G)$

$$\langle \nu, f \rangle = \int_G \langle \mu_h, f \rangle \, d\sigma_G(h).$$

The ν so described is quickly seen to be a non-negative regular Borel measure which, since

$$\nu(1_G) = \int_G \mu_h(1_G) \, d\sigma_G(h) = \int_G 1_G \, d\sigma_G = 1,$$

is a probability. Also, the Gelfand integral averages well, allowing us to use the fact that for each $h \in G$ and for any $f \in F$,

$$\pi_p(u)^{-p} \cdot \|uf\|^p \leq \langle \mu_h, |f|^p \rangle.$$

This ensures that for $f \in F$

$$\frac{\|uf\|^p}{\pi_p(u)^p} \leq \int_G \langle \mu_h, |f|^p \rangle \, d\sigma_G(h) = \langle \nu, |f|^p \rangle$$

showing that ν is a Pietsch measure for u, too.

To complete the proof, we only need to observe that ν is translation invariant, and so must coincide with G's Haar measure σ_G. In fact, regardless of how we select $h_0 \in G$ and $f \in C(G)$, we have

$$\begin{aligned} \langle \nu, f_{h_0} \rangle &= \int_G \langle \mu_h, f_{h_0} \rangle \, d\sigma_G(h) = \int_G \int_G f_{h_0}(g) \, d\mu_h(g) \, d\sigma_G(h) \\ &= \int_G \int_G f(h_0 hg) \, d\mu(g) \, d\sigma_G(h) = \int_G \int_G f(h_0 hg) \, d\sigma_G(h) \, d\mu(g) \\ &= \int_G \int_G f(hg) \, d\sigma_G(h) \, d\mu(g) = \ldots. = \langle \nu, f \rangle. \end{aligned}$$

QED

The above proof was inspired by Y. Gordon [1969] and A. Pełczyński [1969]. S. Kwapień and A. Pełczyński [1978] used the theorem just proved in case G is the circle group $\mathbf{T} = \{z \in \mathbf{C} : |z| = 1\}$ to show that

- *the space of translation invariant operators from the Hardy spaces H^p to H^2 is naturally identifiable (isomorphically and isometrically) with the space of translation invariant p-summing operators from the disk algebra A to H^2.*

Key to their proof was the realization that for a translation invariant p-summing operator from A, the Pietsch measure can be chosen to be normalized Lebesgue measure on $\mathbf{T}$. Thus the Pietsch factorization can be taken through $L^p(\mathbf{T})$ and, since analytic functions are involved, actually through H^p via the canonical inclusion $A \hookrightarrow H^p$.

Considerable effort has been expended on the precise determination of the p-summing nature of certain frequently encountered operators in analysis. A nearly complete characterization of when a diagonal map from ℓ_u to ℓ_v is p-summing was given by D.J.H. Garling [1974], [1975].

G. Bennett [1977] related 'Schur multipliers' to the theory of summing operators and went on to use that to classify multipliers between classical sequence spaces.

B. Beauzamy and B. Maurey [1973] and B. Beauzamy [1976] have studied convolution operators on Lebesgue spaces modelled on compact abelian groups and have given a nearly complete description of p-summing convolvers. Using the Fourier transform, they were able to call on Garling's work on diagonal operators and order bounded operators to treat several bothersome cases.

The topic of the p-summing character of composition operators between Hardy spaces was first taken up by J.H. Shapiro and P.D. Taylor [1973] and later on continued by H. Hunziker and H. Jarchow [1991], H. Jarchow [1993] and R. Riedl [1994].

H. König [1975] gave conditions for certain weakly singular operators to be p-summing and used this to develop and apply a determinant theory for such operators. The complete determination of when weakly singular operators are p-summing remains open.

Both H. König [1986] and A. Pietsch [1987] discussed the p-summing character of classical kernel operators, like Hille - Tamarkin operators and Carleman operators, and certain Sobolev embeddings.

E.D. Gluskin [1978] gave estimates for the p-summing norms of inclusion maps of Sobolev spaces into L_p-spaces and of certain Besov spaces into others.

A. Pełczyński and M. Wojciechowski [1992b] have studied p-summing surjections whose domains are Sobolev spaces with a view to extending R.E.A.C. Paley's [1933] work and classifying the isomorphic types of Sobolev spaces.

In the same direction, A. Pełczyński [1992] has built on earlier work of M. Wojciechowski [1991] and M. Wojciechowski and himself [1992] to show that the disk algebra does not contain a complemented subspace isomorphic to the space $\mathcal{C}^k(\mathbf{T}^d)$ of k-times continuously differentiable functions on the d-dimensional torus $\mathbf{T}^d$ for $d \geq 2$. His proof relies on the fact that there is no 1-summing surjection from $\mathcal{C}^2(\mathbf{T}^2)$ onto Hilbert space.

Because they are outside our present interests, we have ignored the classes of p-summing operators when $0 \leq p < 1$. For $0 < p < 1$, the definition is the same as for $p \geq 1$, while a Banach space operator $u : X \to Y$ is called 0-*summing* if for each $\varepsilon > 0$ there is a $\delta > 0$ so that if $x_1, \ldots, x_n \in X$ and

$$\sup_{x^* \in B_{X^*}} \frac{1}{n} \sum_{k \leq n} \min \{1, |\langle x^*, x_i \rangle|\} < \delta,$$

then

$$\frac{1}{n} \sum_{k \leq n} \min \{1, \|u x_i\|\} < \varepsilon.$$

This definition (due to S. Kwapień [1970b]) is a viable alternative to that of L. Schwartz [1969a,b,c], [1970] in the context of p-summing operators.

B. Maurey [1974a] proved the remarkable fact, already conjectured by A. Pietsch, that if $0 < p < 1$, then the p-summing operators and the 0-summing ones are the same.

It is noteworthy that for $0 < p < 1$ the class of p-summing operators is well-defined for operators with values in a homogeneous linear metric space, and it is in that context that many of the most basic applications are found. B. Maurey's [1974a] monograph is *the* place to see for this.

B. Maurey [1974a] showed that all operators from an $\mathcal{L}_1$-space to a Hilbert space are 0-summing. L. Schwartz [1970b] characterized the 0-summing diagonal operators from ℓ_p to ℓ_{p^*}, and B. Maurey and A. Nahoum [1973] showed that a diagonal operator D_λ from ℓ_1 to the space of convergent series is 0-summing if and only if $(\lambda_n \cdot \log n)$ is a bounded sequence; they used this to prove their version of the Bennett - Maurey - Nahoum Theorem, a topic to be discussed in 12.32.

3. Summing Operators on $\mathcal{L}_p$-Spaces

The definition of a summing operator involves only finitely many vectors at a time. Our knowledge of these operators will thus be greatly enhanced if we have a good understanding of the finite dimensional subspaces of the Banach spaces we work with. This is possible in many 'classical' situations: $L_p(\mu)$-spaces have the convenient property that their finite dimensional subspaces are always contained in slightly distorted copies of ℓ_p^n's. It turns out that this property is just what is required to 'localize' arguments about certain operators on $L_p(\mu)$-spaces to arguments about operators on ℓ_p^n's. This reduction process is so fundamental that we start by isolating a general class of Banach spaces in which it is always applicable.

$\mathcal{L}_p$-Spaces

Let $1 \leq p \leq \infty$ and let $\lambda > 1$. The Banach space X is said to be an *$\mathcal{L}_{p,\lambda}$-space* if every finite dimensional subspace E of X is contained in a finite dimensional subspace F of X for which there is an isomorphism $v : F \to \ell_p^{\dim F}$ with $\|v\| \cdot \|v^{-1}\| < \lambda$. We say that X is an *$\mathcal{L}_p$-space* if it is an $\mathcal{L}_{p,\lambda}$-space for some $\lambda > 1$.

We immediately seize the opportunity to illustrate the 'localization procedure' for which $\mathcal{L}_p$-spaces are so suitable. Here is an extension of Grothendieck's Theorem (1.13). Similar arguments recur frequently, so the proof merits close attention.

3.1 Theorem: *If X is an $\mathcal{L}_{1,\lambda}$-space and Y is an $\mathcal{L}_{2,\lambda'}$-space, then every operator $u : X \to Y$ is 1-summing with $\pi_1(u) \leq \kappa_G \cdot \lambda \cdot \lambda' \cdot \|u\|$.*

Proof. Choose $x_1,..., x_m$ in X. Since X is an $\mathcal{L}_{1,\lambda}$-space, $\{x_1,..., x_m\}$ (and hence its span) sits inside a finite dimensional subspace E of X for which we can find an isomorphism $v : E \to \ell_1^n$, where $n = \dim E$, with $\|v\| \cdot \|v^{-1}\| < \lambda$. On the other hand, Y is an $\mathcal{L}_{2,\lambda'}$-space, so $u(E)$ is contained in a finite dimensional subspace F of Y which can be chosen so that there is an isomorphism $w : F \to \ell_2^N$, where $N = \dim F$, with $\|w\| \cdot \|w^{-1}\| < \lambda'$.

Now u induces an operator $u_0 : E \to F$ and so we can create an operator $w u_0 v^{-1} : \ell_1^n \to \ell_2^N$. With the aim of applying Corollary 1.15 to Grothendieck's Theorem, we embark on a factorization game in which the main feature is the ideal property of Π_1. Decomposing u_0 as $E \xrightarrow{v} \ell_1^n \xrightarrow{v^{-1}} E \xrightarrow{u_0} F \xrightarrow{w} \ell_2^N \xrightarrow{w^{-1}} F$, we get

$$\begin{aligned}\sum_{k=1}^m \|ux_k\| &= \sum_{k=1}^m \|w^{-1}(wu_0v^{-1})vx_k\| \le \|w^{-1}\|\cdot\sum_{k=1}^m \|(wu_0v^{-1})(vx_k)\| \\ &\le \|w^{-1}\|\cdot\kappa_G\cdot\|wu_0v^{-1}\|\cdot\|(vx_k)_1^m\|_1^{weak} \\ &\le \kappa_G\cdot\|w^{-1}\|\cdot\|w\|\cdot\|u_0\|\cdot\|v^{-1}\|\cdot\|v\|\cdot\|(x_k)_1^m\|_1^{weak} \\ &\le \kappa_G\cdot\lambda\cdot\lambda'\cdot\|u\|\cdot\|(x_k)_1^m\|_1^{weak},\end{aligned}$$

which was what we wanted. QED

Of course, we ought to make sure that we have indeed improved upon our old version 1.13 of Grothendieck's Theorem. The next result takes care of this: it reveals that the best-known classical function and sequence spaces can be found lurking among the $\mathcal{L}_p$-spaces.

3.2 Theorem:

(I) *If (Ω, Σ, μ) is any measure space and $1 \le p \le \infty$, then $L_p(\mu)$ is an $\mathcal{L}_{p,\lambda}$-space for all $\lambda > 1$.*

(II) *If K is a compact Hausdorff space, then $C(K)$ is an $\mathcal{L}_{\infty,\lambda}$-space for all $\lambda > 1$.*

Note that (II) subsumes the case $p = \infty$ in (I).

We start the proof of this important theorem with an equally important lemma.

3.3 Lemma: *Suppose X is an $L_p(\mu)$-space $(1 \le p \le \infty)$ or a $C(K)$-space. Assume M is a non-empty compact subset of X and let $\varepsilon > 0$. Then there exists a finite rank projection $P \in \mathcal{L}(X)$ such that*

(a) $\|P\| = 1$,

(b) $\|Pf - f\| \le \varepsilon$ *for all $f \in M$,*

(c) *$P(X)$ is isometrically isomorphic to ℓ_p^n, where $n = \dim P(X)$. (In case $X = C(K)$ we let $p = \infty$.)*

Proof. We first consider the case $X = L_p(\mu)$, $1 \le p \le \infty$.

Let $\{f_1, \ldots, f_k\}$ be an $(\varepsilon/2)$-net for M consisting of simple functions, and let $A_1, \ldots, A_n \in \Sigma$ be pairwise disjoint sets of finite positive μ-measure such that each f_j is constant on each A_i and vanishes outside $\bigcup_{i=1}^n A_i$.

Define $P : L_p(\mu) \to L_p(\mu)$ by

$$Pf := \sum_{i=1}^n \frac{1}{\mu(A_i)} \cdot \Big(\int_{A_i} f \, d\mu\Big) \cdot 1_{A_i}.$$

P is plainly linear, and since $P(1_{A_i}) = 1_{A_i}$ for all i, it is clear that P is a projection. As the A_i's are disjoint, $\|P\| \le 1$ when $p = \infty$. If $1 \le p < \infty$, an application of Hölder's Inequality gives

$$\|Pf\|^p = \sum_{i=1}^n \Big|\int_{A_i} f \, d\mu\Big|^p \cdot \mu(A_i)^{1-p} \le \sum_{i=1}^n \int_{A_i} |f|^p \, d\mu \le \|f\|^p,$$

and so once again $\|P\| \leq 1$. But P is a non-trivial projection since $M \neq \emptyset$; hence $\|P\| = 1$. Moreover, if $f \in M$, then $\|f - f_i\| \leq \varepsilon/2$ for some $1 \leq i \leq k$, and as $Pf_i = f_i$, it follows that $\|Pf - f\| \leq \|P(f - f_i)\| + \|Pf_i - f\| \leq 2\cdot\|f - f_i\| \leq \varepsilon$. Finally, the operator $\ell_p^n \to P(X)$ generated by $e_i \mapsto \mu(A_i)^{-1}\cdot 1_{A_i}$ is easily seen to be an isometric isomorphism.

We pass now to the case $X = \mathcal{C}(K)$ and start by choosing an $(\varepsilon/4)$-net for M which we label $\{f_1,..., f_n\}$. By compactness, there is an open cover $\Omega_1,..., \Omega_n$ of K such that, as long as ω and ω' belong to the same Ω_i, $|f_j(\omega) - f_j(\omega')| \leq \varepsilon/2$ for all $1 \leq j \leq n$. Without loss of generality we may assume $\Omega_i \setminus \bigcup_{i' \neq i} \Omega_{i'} \neq \emptyset$ for each i and then select an element ω_i in the corresponding set.

Now let $(\varphi_1,..., \varphi_n)$ be a partition of unity subordinate to $(\Omega_1,..., \Omega_n)$: each φ_i belongs to $\mathcal{C}(K)$, has values in $[0,1]$ and vanishes off Ω_i, and we have $\sum_{i=1}^n \varphi_i(\omega) = 1$ for all $\omega \in K$. As $\varphi_i(\omega_j) = \delta_{ij}$, we obtain a projection $P \in \mathcal{L}(\mathcal{C}(K))$ by setting

$$Pf := \sum_{i=1}^n f(\omega_i)\cdot\varphi_i.$$

The range of P is the span of the φ_i's and so is n-dimensional. Also, $\|P\| = 1$ since P is non-trivial and satisfies $\|Pf\| \leq \max_{i \leq n} |f(\omega_i)| \leq \|f\|$ for each $f \in \mathcal{C}(K)$. If $f \in M$ and $\omega \in K$ then, for some $1 \leq j \leq n$, $\|f - f_j\| < \varepsilon/4$, and so

$$\begin{aligned}
|Pf(\omega) - f(\omega)| &= \Big|\sum_{i=1}^n \big(f(\omega_i) - f(\omega)\big)\cdot\varphi_i(\omega)\Big| \\
&\leq \sum_{i=1}^n \big(|f(\omega_i) - f_j(\omega_i)| + |f_j(\omega_i) - f_j(\omega)| + |f_j(\omega) - f(\omega)|\big)\cdot\varphi_i(\omega) \\
&\leq \sum_{i=1}^n |f_j(\omega_i) - f_j(\omega)|\cdot\varphi_i(\omega) + (\varepsilon/2) \\
&= \sum_{\{i:\, \omega \in \Omega_i\}} |f_j(\omega_i) - f_j(\omega)|\cdot\varphi_i(\omega) + (\varepsilon/2) \leq \varepsilon.
\end{aligned}$$

Therefore $\|Pf - f\| \leq \varepsilon$. Finally, $(a_i)_1^n \mapsto \sum_{i=1}^n a_i\varphi_i$ defines an isometric isomorphism $\ell_\infty^n \to P(\mathcal{C}(K))$. To see this, notice that for each $\omega \in K$, $(\varphi_i(\omega))_1^n$ is a unit vector in ℓ_1^n and that since $(a_i)_1^n \in \ell_\infty^n$, there is some $1 \leq i_0 \leq n$ such that $|a_{i_0}| = \|(a_i)_1^n\|_\infty$. Then apply our construction to reveal

$$\begin{aligned}
\Big\|\sum_{i=1}^n a_i\varphi_i\Big\| &= \sup_{\omega \in K} \Big|\sum_{i=1}^n a_i\varphi_i(\omega)\Big| \leq \|(a_i)_1^n\|_\infty \\
&= |a_{i_0}| = \Big|\sum_{i=1}^n a_i\varphi_i(\omega_{i_0})\Big| \leq \Big\|\sum_{i=1}^n a_i\varphi_i\Big\|.
\end{aligned}$$

QED

Preparations are complete; we return to the business at hand.

Proof of Theorem 3.2. We cover cases (I) and (II) simultaneously.

Let E be any finite dimensional subspace of X, where X is $L_p(\mu)$ $(1 \leq p \leq \infty)$ or $\mathcal{C}(K)$. Let $Q \in \mathcal{L}(X)$ be any projection onto E, and choose

$\delta > 0$ so that $\delta \cdot \|Q\| < 1$. Since B_E is compact, Lemma 3.3 produces a finite rank projection $P \in \mathcal{L}(X)$ of norm one such that $\|Px - x\| \leq \delta$ for all $x \in B_E$ and $P(X)$ is isometrically isomorphic to ℓ_p^n, where $n = \dim P(X)$.

Set $u := id_X + PQ - Q$. Then $\|u - id_X\| = \|PQ - Q\| \leq \delta \cdot \|Q\| < 1$, so u^{-1} exists in $\mathcal{L}(X)$ (and is in fact $\sum_{k\geq 0}(id_X - u)^k$). Of course, $\|u\| \leq 1 + \delta \cdot \|Q\|$ and $\|u^{-1}\| \leq (1 - \delta \cdot \|Q\|)^{-1}$. Note that $u^{-1}Pu$ is a finite rank projection in $\mathcal{L}(X)$. Its range is an n-dimensional subspace F of X. It contains E since if $x \in E$, then $ux = x + PQx - Qx = Px = PPx = Pux$; in other words, $x = u^{-1}Pux$. Restricting u to F gives an isomorphism $u_0 : F \to P(X)$, and as $P(X)$ is isometrically isomorphic to ℓ_p^n, we may conclude that $\|u_0\| \cdot \|u_0^{-1}\| \leq (1 + \delta\|Q\|) \cdot (1 - \delta\|Q\|)^{-1}$. This can be made smaller than any $\lambda > 1$ by a judicious choice of δ. QED

The two preceding theorems can be combined fruitfully to obtain a significant corollary:

3.4 Theorem: *Regardless of the measures μ and ν, every operator $u : L_1(\mu) \to L_2(\nu)$ is 1-summing with $\pi_1(u) \leq \kappa_G \cdot \|u\|$.*

We stress that this theorem is a consequence of the finite dimensional structure of the spaces involved and of the finitary nature of p-summing operators. A process like the reduction of Theorem 3.1, which is infinite dimensional in character, to the entirely finite dimensional statement of Corollary 1.15 is what is understood by *'localization'* in Banach space theory. We shall encounter various other examples in the course of this book.

We can use Theorem 3.4 to corroborate an announcement at the end of Chapter 1. When taken in combination with the Weak Dvoretzky - Rogers Theorem 2.18 it shows that no $\mathcal{L}_1$-space X can contain a complemented subspace which is isomorphic to an infinite dimensional Hilbert space H: after all, if p is a 1-summing projection of X onto H, then $p^2 = p$ is compact and so cannot have an infinite dimensional range.

It is worth remarking here that duals of $C(K)$-spaces are $L_1(\mu)$-spaces, though the measures μ that arise in this context are stupendously non-σ-finite. Be not of faint heart; it is relatively simple to use localization to show that $C(K)^*$ is an $\mathcal{L}_{1,\lambda}$-space for each $\lambda > 1$.

We recognize $C(K)^*$ as the space of regular Borel measures defined on K. So, if $\mu_1, ..., \mu_n \in C(K)^*$ are given, each is absolutely continuous with respect to $\mu := |\mu_1| + ... + |\mu_n|$, where $|\mu_i| \in C(K)^*$ denotes the variation of μ_i. It follows that the Radon - Nikodým derivatives $f_i = \dfrac{d\mu_i}{d\mu}$ exist and belong to $L_1(\mu)$ and that, regardless of the scalars $a_1, ..., a_n$, $\|a_1\mu_1 + ... + a_n\mu_n\|_{C(K)^*} = \|a_1 f_1 + ... + a_n f_n\|_{L_1(\mu)}$. Now look inside $L_1(\mu)$ for the appropriate finite dimensional subspace F containing $f_1,\ldots, f_n$ for which there is an isomorphism $u : F \to \ell_1^{\dim F}$ with $\|u\|\cdot\|u^{-1}\| < \lambda$. Observe that the same F can be viewed as a subspace of $\mathcal{C}(K)^*$ without changing its isometric structure one iota.

As a result we see that every operator from $\mathcal{C}(K)^*$ to an $\mathcal{L}_2$-space is 1-summing.

Operators on $\mathcal{L}_\infty$-Spaces

The following is a companion to Theorem 3.4, and its proof involves yet another application of Grothendieck's Inequality.

3.5 Theorem: *Let K be a compact Hausdorff space and let μ be any measure. If $1 \le p \le 2$ then any operator $u : \mathcal{C}(K) \to L_p(\mu)$ is 2-summing with $\pi_2(u) \le \kappa_G \cdot \|u\|$.*

We follow the same strategy as before. A finite dimensional lemma lies at the heart of the matter:

3.6 Lemma: *Let $1 \le p \le 2$ and let n, N be positive integers. Then every operator $u : \ell_\infty^n \to \ell_p^N$ satisfies $\pi_2(u) \le \kappa_G \cdot \|u\|$.*

Once this has been established, the reader may apply localization techniques, as in the proof of 3.1, to obtain the next result, which then readily gives 3.5 as a corollary.

3.7 Theorem: *Let $1 \le p \le 2$, let X be an $\mathcal{L}_{\infty,\lambda}$-space and let Y be an $\mathcal{L}_{p,\lambda'}$-space. Then every operator $u : X \to Y$ is 2-summing, with $\pi_2(u) \le \kappa_G \cdot \lambda \cdot \lambda' \cdot \|u\|$.*

Thus, the only place where we really have to work is in the proof of the lemma.

Proof of 3.6. We wish to show that for any choice of finitely many vectors $x_1,\ldots, x_m$ in ℓ_∞^n we have

$$\Big(\sum_{k=1}^m \|ux_k\|^2\Big)^{1/2} \le \kappa_G \cdot \|u\| \cdot \sup\Big\{\Big(\sum_{k=1}^m |\langle x^*, x_k\rangle|^2\Big)^{1/2} : x^* \in B_{\ell_1^n}\Big\}.$$

We may assume $m = n$: if $m < n$, augment $x_1,\ldots, x_m$ by $n - m$ zero vectors; if $m > n$, just think of u as an operator on ℓ_∞^m mapping the additional basis vectors to zero. Arguing similarly, we may make a further reduction and assume $m = n = N$ for the rest of the proof. For even more convenience, we may also assume that $\|(x_k)_1^n\|_2^{weak} = 1$.

To set up use of Grothendieck's Inequality we let (u_{ij}) be the $n \times n$ matrix determined by $ue_i = \sum_{j=1}^n u_{ij}e_j$, $1 \le i \le n$. Let $s = (s_i)_1^n$ and $t = (t_j)_1^n$ be arbitrary members of $B_{\ell_\infty^n}$. Our aim is to bound norms in ℓ_p^n, and duality will help in this regard; so let $y = (y_j)_1^n$ be an element of $B_{\ell_{p^*}^n}$. Notice that $y_t = (t_j y_j)_1^n \in B_{\ell_{p^*}^n}$, and so

$$\Big|\sum_{i,j=1}^n u_{ij} y_j s_i t_j\Big| = |\langle y_t, u(s)\rangle| \le \|u(s)\|_p \le \|u\|.$$

Applying Grothendieck's Inequality 1.14 to the matrix $(u_{ij} y_j)$ gives us

$$\Big|\sum_{i,j} u_{ij} y_j (w_i|z_j)\Big| \le \kappa_G \cdot \|u\| \tag{1}$$

for all $w_i, z_j \in B_{\ell_2^n}$, $1 \le i,j \le n$. Judicious choice of the w_i's and z_j's will pave our way.

For each $1 \le k \le n$, write $x_k = (x_{ki})_{i=1}^n = \sum_{i=1}^n x_{ki} e_i$ and notice that

$$\Big(\sum_{k=1}^n |x_{ki}|^2\Big)^{1/2} = \Big(\sum_{k=1}^n |\langle e_i, x_k\rangle|^2\Big)^{1/2} \le \|(x_k)_1^n\|_2^{weak} = 1.$$

Then $w_i := (x_{ki})_{k=1}^n$ is in $B_{\ell_2^n}$ for $1 \le i \le n$.

Next we choose $z_j \in B_{\ell_2^n}$, $1 \le j \le n$, so that

$$\Big(\sum_{i=1}^n u_{ij} y_j w_i \,\Big|\, z_j\Big) = \Big\|\sum_{i=1}^n u_{ij} y_j w_i\Big\|_2 .$$

From (1) we infer

$$\sum_{j=1}^n \Big\|\sum_{i=1}^n u_{ij} y_j w_i\Big\|_2 \le \kappa_G \cdot \|u\|;$$

in other words, passing to coordinates,

$$\sum_{j=1}^n |y_j| \cdot \Big(\sum_{k=1}^n \Big|\sum_{i=1}^n u_{ij} x_{ki}\Big|^2\Big)^{1/2} \le \kappa_G \cdot \|u\|.$$

Since this holds for any $y \in B_{\ell_{p^*}^n}$, the vector $\big((\sum_{k=1}^n |\sum_{i=1}^n u_{ij} x_{ki}|^2)^{1/2}\big)_{j=1}^n$ has ℓ_p^n norm at most $\kappa_G \cdot \|u\|$, that is

$$\Big(\sum_{j=1}^n \Big(\sum_{k=1}^n \Big|\sum_{i=1}^n u_{ij} x_{ki}\Big|^2\Big)^{p/2}\Big)^{1/p} \le \kappa_G \cdot \|u\|. \tag{2}$$

It is only now that the hypothesis $1 \le p \le 2$ comes into play. It makes the triangle inequality available in $\ell_{2/p}^n$, and together with (2), this validates the following chain of events:

$$\begin{aligned}
\Big(\sum_{k=1}^n \|u x_k\|^2\Big)^{1/2} &= \Big(\sum_{k=1}^n \Big(\sum_{j=1}^n \Big|\sum_{i=1}^n u_{ij} x_{ki}\Big|^p\Big)^{2/p}\Big)^{1/2} \\
&= \Big(\Big\|\sum_{j=1}^n \Big(\Big|\sum_{i=1}^n u_{ij} x_{ki}\Big|^p\Big)_{k=1}^n\Big\|_{2/p}\Big)^{1/p} \le \Big(\sum_{j=1}^n \Big\|\Big(\Big|\sum_{i=1}^n u_{ij} x_{ki}\Big|^p\Big)_{k=1}^n\Big\|_{2/p}\Big)^{1/p} \\
&= \Big(\sum_{j=1}^n \Big(\sum_{k=1}^n \Big|\sum_{i=1}^n u_{ij} x_{ki}\Big|^2\Big)^{p/2}\Big)^{1/p} \le \kappa_G \cdot \|u\|.
\end{aligned}$$

QED

3.8 Remarks: (a) Later on, in 10.6, we shall exploit the meaning of inequality (2) above for $2 < p < \infty$. Also, in 10.5, we shall see that when the spaces in 3.5 and 3.7 are infinite dimensional, it is impossible to replace Π_2 by Π_r for $r < 2$.

(b) Theorem 3.5 remains valid for operators $u : X \to L_p(\mu)$, $1 \le p \le 2$, when X^{**} is a $C(K)$-space. This can easily be seen by passage to second adjoints (and using that $L_1(\mu)^{**}$ is again an $L_1(\nu)$-space if $p = 1$). Another way would be to take advantage of 2.14 after showing that X is an $\mathcal{L}_{\infty,\lambda}$-space if and only if X^{**} is; this will follow from 8.14.

This remark applies in particular if $X = c_0^\Gamma$ for any set Γ. It is however easy to see that c_0^Γ is an $\mathcal{L}_{\infty,\lambda}$-space for any $\lambda > 1$: adjust the proof of (I) in Theorem 3.2 and the relevant part of Lemma 3.3 by choosing the net to consist of finitely supported members of c_0^Γ.

(c) It would have been sufficient to deal with $p = 1$ in 3.5 – 3.7. In fact, one can show that every $\mathcal{L}_p$-space ($1 < p \le 2$) is isomorphic to a subspace of an $\mathcal{L}_1$-space. For $p = 2$ this is a consequence of Khinchin's Inequality 1.10, as we observe in Remark 3.10 below.

(d) In contrast with the situation when $p = 1$ or ∞, if we suppose that X is an infinite dimensional $\mathcal{L}_p$-space for $1 < p < \infty$, then $\Pi_r(X, \ell_2) \neq \mathcal{L}(X, \ell_2)$, however we choose $1 \le r < \infty$.

To understand why this is so, first take $X = L_p[0,1]$ and recall Theorem 1.12 which assures us that ℓ_2 is complemented in X. So the equality of $\Pi_r(X, \ell_2)$ and $\mathcal{L}(X, \ell_2)$ would imply that id_{ℓ_2} is r-summing, and this impossible (see 2.18).

The result for general $\mathcal{L}_p$-spaces follows from a local analysis of the argument we have just presented; we omit the technicalities.

Some Applications

Recall from basic functional analysis that every Banach space is isometrically isomorphic to a subspace of a $C(K)$-space and to a quotient of an $L_1(\mu)$-space. What can we say about a Banach space which is simultaneously a quotient of a $C(K)$-space and a subspace of an $L_1(\mu)$-space? The results of this chapter lead to a surprisingly clear-cut answer.

3.9 Theorem: *If a Banach space X is simultaneously isomorphic to a quotient of a $C(K)$-space and to a subspace of an $L_1(\mu)$-space, then X is isomorphic to a Hilbert space.*

Proof. Let q be an operator from $C(K)$ onto X. Since X is a subspace of $L_1(\mu)$, Theorem 3.5 conspires with the injectivity property (2.5) to ensure that q is 2-summing, and so, by Corollary 2.15, q factors through a Hilbert space:

$$q : C(K) \xrightarrow{v} H \xrightarrow{u} X.$$

As q is onto, so is u. If in addition u is injective, it must be an isomorphism on account of the Open Mapping Theorem. However, injectivity of u can always be arranged by replacing H by the orthogonal complement of $\ker u$. QED

3.10 Remark: The converse is also true: every Hilbert space H is isomorphic to a subspace of an $L_1(\mu)$-space and to a quotient of a $\mathcal{C}(K)$-space.

To see this, write $H = \ell_2^I$ where I is an appropriate set. Of course, the only case of interest is when I is infinite. If I is countable, then H is isomorphic to a subspace of $L_1[0,1]$ on account of Khinchin's Inequality 1.10. By duality, H is then also isomorphic to a quotient of $L_\infty[0,1]$. We shall reduce the case when I is uncountable to this simple situation.

To this end, let D be the set of all countable (finite or infinite) subsets of I. For each $J \in D$, fix a copy Ω_J of [0,1]; let Σ_J be the Borel field of Ω_J and let μ_J be the corresponding Lebesgue measure on Σ_J. Write Ω for the disjoint union of the Ω_J's, denote by Σ the smallest σ-field on Ω which contains all the Σ_J's, and define μ to be the unique measure on Σ whose restriction to each Σ_J is μ_J.

Given $x = (x_i)_{i\in I}$ in ℓ_2^I, the set $J_x := \{i \in I : x_i \neq 0\}$ belongs to D, and so we may consider x as an element of $L_1(\mu_J)$, using the Khinchin embedding as above. But in the canonical fashion, $L_1(\mu_J)$ is a subspace of $L_1(\mu)$; so we have created a map $\ell_2^I \to L_1(\mu)$. This is clearly an isomorphic embedding. By duality, ℓ_2^I is then also isomorphic to a quotient of $L_1(\mu)^*$ which is certainly a $\mathcal{C}(K)$-space.

We have already seen after 2.9 that if $p \neq q$, then there are Banach spaces X, Y such that $\Pi_p(X,Y) \neq \Pi_q(X,Y)$. On the other hand, Theorem 3.1 and Theorem 3.7 describe situations where $\Pi_p(X,Y) = \Pi_q(X,Y)$ holds for a wide range of p and q. There are more special occasions of coincidence; here is another:

3.11 Theorem: *Let $1 \le p \le 2$. If X is a subspace of an $\mathcal{L}_{p,\lambda}$-space and Y is a Banach space, all 2-summing operators $u : X \to Y$ are 1-summing with $\pi_1(u) \le \kappa_G \cdot \lambda \cdot \pi_2(u)$.*

Proof. Let $x_1,\dots,x_n \in X$, and let $v : \ell_\infty^n \to X$ be the operator given by $ve_k = x_k$, $1 \le k \le n$. Thanks to Theorem 3.7 and Proposition 2.2, $\pi_2(v) \le \kappa_G \cdot \lambda \cdot \|v\| = \kappa_G \cdot \lambda \cdot \|(x_k)_1^n\|_1^{weak}$, and by the Composition Theorem 2.22, $\pi_1(uv) \le \pi_2(u) \cdot \pi_2(v)$. Consequently,

$$\sum_{k=1}^n \|ux_k\| = \sum_{k=1}^n \|uve_k\| \le \pi_1(uv) \le \kappa_G \cdot \lambda \cdot \pi_2(u) \cdot \|(x_k)_1^n\|_1^{weak},$$

as asserted. QED

The theorem is no longer true when $p > 2$. Indeed, we shall prove much more in 14.5. Other related information can be found in 10.11.

The same strategy leads us to a generalization of Orlicz's Theorem 1.11:

3.12 Theorem: *Let X be (a subspace of) an $\mathcal{L}_p$-space, where $1 \le p \le 2$. Then every weak ℓ_1 sequence in X is a strong ℓ_2 sequence.*

Proof. For a change, let us work in infinite dimensions. Take a sequence $(x_n) \in \ell_1^{weak}(X)$, and let $v \in \mathcal{L}(c_0, X)$ be given by $ve_n = x_n$ for all n, so that $\|v\| = \|(x_n)\|_1^{weak}$ by 2.2. By 3.7 (and 3.8(b)), v is 2-summing. But as (e_n) is a weak ℓ_1 sequence in c_0, it is certainly a weak ℓ_2 sequence. We deduce that $(x_n) = (ve_n)$ is a strong ℓ_2 sequence in X. QED

As we shall see later on (6.20), 3.12 actually holds for any Banach space X such that $\Pi_1(X,Y) = \Pi_2(X,Y)$ regardless of the Banach space Y.

Recall that a basis (x_n) for a Banach space X is *unconditional* if there is a constant C such that

$$\text{(UC)} \qquad \Big\|\sum_{k=1}^n \varepsilon_k a_k x_k\Big\| \le C \cdot \Big\|\sum_{k=1}^n a_k x_k\Big\|$$

no matter how we choose finite sets $\{a_1,\dots,a_n\}$ of scalars and $\{\varepsilon_1,\dots,\varepsilon_n\}$ of ± 1's. It is immediate that if (x_n^*) is the corresponding biorthogonal sequence in X^*, then $\sum_n \langle x_n^*, x\rangle x_n$ converges unconditionally for each $x \in X$, and the limit is x.

The theorems of Orlicz and Grothendieck are handy tools for investigating whether certain spaces have an unconditional basis.

3.13 Theorem: *If an infinite dimensional complemented subspace X of an $\mathcal{L}_1$-space Z has an unconditional basis (x_n), then (x_n) is equivalent to the unit vector basis of ℓ_1 and so X must be isomorphic to ℓ_1.*

Proof. It does no harm to normalize so that $\|x_n\| = 1$ for each n. Our aim is to show that $(a_n) \mapsto \sum_n a_n x_n$ defines an isomorphism from ℓ_1 onto X.

We exploit the fact that X has the Orlicz property to construct an operator $u : X \to \ell_2 : x \mapsto \big(\langle x_n^*, x\rangle\big)_n$ where (x_n^*) is the biorthogonal sequence of the preamble: the unconditional convergence of $\sum_n \langle x_n^*, x\rangle x_n$ is essential here. Note that $ux_k = e_k$ for each k.

Now, as X is complemented in Z, we can consider u to be the restriction of an operator $\tilde{u} : Z \to \ell_2$. Since Z is an $\mathcal{L}_1$-space, Grothendieck's Theorem 3.1 steps in to ensure that $\tilde{u}$, and so u, is 1-summing. Consequently, if $\{a_1,\dots,a_n\}$ is any finite set of scalars and C is the constant from (UC), then

$$\begin{aligned}\sum_{k=1}^n |a_k| &= \sum_{k=1}^n \|a_k e_k\|_{\ell_2} = \sum_{k=1}^n \|u(a_k x_k)\|_{\ell_2} \\ &\le \pi_1(u) \cdot \sup\Big\{\Big\|\sum_{k=1}^n \varepsilon_k a_k x_k\Big\| : \varepsilon_1,\dots,\varepsilon_n = \pm 1\Big\} \\ &\le C \cdot \pi_1(u) \cdot \Big\|\sum_{k=1}^n a_k x_k\Big\| = C \cdot \pi_1(u) \cdot \sum_{k=1}^n |a_k|,\end{aligned}$$

and all is as it should be. QED

There is a stunning consequence. It follows from the fact that

- $L_1[0,1]$ *is not isomorphic to* ℓ_1*:*

indeed, Khinchin's Inequality 1.10 implies that $L_1[0,1]$ contains an isomorphic copy of ℓ_2 while Schur's Theorem 1.7 plainly rules out isomorphs of ℓ_2 among the subspaces of ℓ_1.

3.14 Corollary: *$L_1[0,1]$ has no unconditional basis.*

Now for another situation where different p's and q's produce coincidental p-summing and q-summing operators. This will require recourse to Khinchin's Inequality.

3.15 Theorem: *Let $1 \le p \le 2$ and $2 < q < \infty$. If X is a subspace of an $\mathcal{L}_{p,\lambda}$-space and Y is a Banach space, then every q-summing operator $u : Y \to X$ is 2-summing, with $\pi_2(u) \le A_p^{-1} \cdot B_q \cdot \lambda \cdot \pi_q(u)$.*

Here A_p and B_q are the constants from Khinchin's Inequality 1.10.

Proof. First we deal with $X = \ell_p^n$ for some $n \in \mathbf{N}$. Let $y_1, \dots, y_m \in Y$. Since $p \le 2$ the triangle inequality is available in $\ell_{2/p}^m$, and we get

$$\Big(\sum_{k=1}^m \|uy_k\|_X^2\Big)^{1/2} = \Big(\sum_{k=1}^m \Big(\sum_{j=1}^n |\langle uy_k, e_j\rangle|^p\Big)^{2/p}\Big)^{1/2} = \Big\| \Big(\sum_{j=1}^n |\langle uy_k, e_j\rangle|^p\Big)_{k=1}^m \Big\|_{2/p}^{1/p}$$

$$\le \Big(\sum_{j=1}^n \big\| \big(|\langle uy_k, e_j\rangle|^p\big)_{k=1}^m \big\|_{2/p}\Big)^{1/p} = \Big(\sum_{j=1}^n \Big(\sum_{k=1}^m |\langle uy_k, e_j\rangle|^2\Big)^{p/2}\Big)^{1/p}$$

Application of Khinchin's Inequality 1.10 leads to

$$\Big(\sum_{k=1}^m \|uy_k\|_X^2\Big)^{1/2} \le A_p^{-1} \cdot \Big(\sum_{j=1}^n \int_0^1 \Big|\sum_{k=1}^m r_k(t) \cdot \langle uy_k, e_j\rangle\Big|^p \, dt\Big)^{1/p}$$

$$= A_p^{-1} \cdot \Big(\int_0^1 \sum_{j=1}^n \Big|\Big\langle u\Big(\sum_{k=1}^m r_k(t) y_k\Big), e_j\Big\rangle\Big|^p \, dt\Big)^{1/p}$$

$$= A_p^{-1} \cdot \Big(\int_0^1 \Big\| u\Big(\sum_{k=1}^m r_k(t) y_k\Big)\Big\|_X^p \, dt\Big)^{1/p}.$$

But, as $p < q$,

$$\Big(\sum_{k=1}^m \|uy_k\|_X^2\Big)^{1/2} \le A_p^{-1} \cdot \Big(\int_0^1 \Big\| u\Big(\sum_{k=1}^m r_k(t) y_k\Big)\Big\|_X^q \, dt\Big)\Big)^{1/q}.$$

Up to this point the map u is completely irrelevant; what matters is that the vectors uy_k belong to ℓ_p^n. But now we apply the Pietsch Domination Theorem 2.12: there is a regular Borel probability measure μ on the weak$*$compact set B_{Y^*} such that

$$\|uy\| \le \pi_q(u) \cdot \Big(\int_{B_{Y^*}} |\langle y^*, y\rangle|^q d\mu(y^*)\Big)^{1/q} \quad \text{for all } \ y \in Y.$$

Inserting this into the above inequality and appealing to Fubini's Theorem and again to Khinchin's Inequality, we arrive at

$$\begin{aligned}\Big(\sum_{k=1}^m \|uy_k\|_X^2\Big)^{1/2} &\le A_p^{-1}\cdot\pi_q(u)\cdot\Big(\int_0^1\int_{B_{Y^*}}\Big|\sum_{k=1}^m r_k(t)\cdot\langle y^*,y_k\rangle\Big|^q\,d\mu(y^*)\,dt\Big)^{1/q}\\ &= A_p^{-1}\cdot\pi_q(u)\cdot\Big(\int_{B_{Y^*}}\int_0^1\Big|\sum_{k=1}^m r_k(t)\cdot\langle y^*,y_k\rangle\Big|^q\,dt\,d\mu(y^*)\Big)^{1/q}\\ &\le A_p^{-1}\cdot B_q\cdot\pi_q(u)\cdot\Big(\int_{B_{Y^*}}\Big(\sum_{k=1}^m|\langle y^*,y_k\rangle|^2\Big)^{q/2}d\mu(y^*)\Big)^{1/q}\\ &\le A_p^{-1}\cdot B_q\cdot\pi_q(u)\cdot\|(y_k)_1^m\|_2^{weak}.\end{aligned}$$

The theorem works for $X=\ell_p^n$.

To settle the case where X is any $\mathcal{L}_{p,\lambda}$-space we proceed by localization – of course! Given $y_1,\dots,y_m\in Y$, locate a finite dimensional subspace F of X which contains $uy_1,\dots,uy_m$ and for which there is an isomorphism $v:F\to\ell_p^{\dim F}$ such that $\|v\|\cdot\|v^{-1}\|<\lambda$. Then

$$\begin{aligned}\Big(\sum_{k=1}^m\|uy_k\|^2\Big)^{1/2} = \Big(\sum_{k=1}^m\|v^{-1}vuy_k\|^2\Big)^{1/2} &\le \|v^{-1}\|\cdot\Big(\sum_{k=1}^m\|vuy_n\|^2\Big)^{1/2}\\ &\le \|v^{-1}\|\cdot A_p^{-1}\cdot B_q\cdot\pi_q(vu)\cdot\|(y_k)_1^m\|_2^{weak}\\ &\le A_p^{-1}\cdot B_q\cdot\|v\|\cdot\|v^{-1}\|\cdot\pi_q(u)\cdot\|(y_k)_1^m\|_2^{weak}.\end{aligned}$$

It follows that u is 2-summing, with $\pi_2(u)\le A_p^{-1}\cdot B_q\cdot\lambda\cdot\pi_q(u)$. QED

When considered in tandem, 3.11 and 3.15 lead to a remarkable conclusion:

3.16 Corollary: *Let $1\le p,\ q\le 2$. If X is a subspace of an $\mathcal{L}_p$-space and Y is a subspace of an $\mathcal{L}_q$-space, then for all $1\le r<\infty$,*

$$\Pi_1(X,Y)=\Pi_r(X,Y).$$

This holds in particular if X and Y are (isomorphic to) Hilbert spaces.

Generalizations of the preceding results will be derived in 11.16 when we discuss the theory of type and cotype.

To round off our discussion of the coincidence of p-summing and r-summing operators, we present another beautiful consequence of domination and factorization.

3.17 Extrapolation Theorem: *Let $1<r<p<\infty$, and let X be a Banach space such that*

$$(1)\qquad \Pi_p(X,\ell_p)=\Pi_r(X,\ell_p).$$

Then, for every Banach space Y,

$$(2)\qquad \Pi_p(X,Y)=\Pi_1(X,Y).$$

Some comments on the hypothesis are in order, since it is possible to express it in a variety of different forms. First, the Closed Graph Theorem shows that (1) is equivalent to the existence of a constant c such that

$$\pi_r(v) \leq c \cdot \pi_p(v) \tag{3}$$

for all $v \in \Pi_p(X, \ell_p)$.

Next, localization may be brought to bear: (1) is equivalent to postulating that (3) holds for all $v \in \mathcal{L}(X, \ell_p^n)$ for all $n \in \mathbf{N}$. The same idea reveals that we could substitute any infinite dimensional $L_p(\mu)$-space for ℓ_p in the statement of the theorem (or any infinite dimensional $\mathcal{L}_p$-space if we are willing to accept a change in the constant).

To see this, recall that the r-summing characteristics of $v \in \Pi_p(X, L_p(\mu))$ are determined by restrictions to spans of finitely many elements of X. But since $L_p(\mu)$ is an $\mathcal{L}_{p,\lambda}$-space for each $\lambda > 1$, we can assert that for each $x_1, \ldots, x_n \in X$, the subspace of $L_p(\mu)$ generated by $\{vx_1, \ldots, vx_n\}$ embeds λ-isomorphically into ℓ_p. Consequently, the ideal property ensures that $\pi_r(v) \leq c \cdot \lambda \cdot \pi_p(v)$ for each $\lambda > 1$, whence $\pi_r(v) \leq c \cdot \pi_p(v)$.

Proof. In the same spirit as our first comment, we aim to prove that, regardless of the Banach space Y, there is a constant C such that for each $u \in \Pi_p(X, Y)$,

$$\pi_1(u) \leq C \cdot \pi_p(u). \tag{4}$$

Our conclusion will follow at once.

As is customary, we regard $K := B_{X^*}$ as a compact Hausdorff space under the weak$*$topology of X^*. It will be convenient to denote by $P(K)$ the collection of all regular Borel probability measures on K.

Since we have available the natural isometric embedding $X \to C(K)$, each $\mu \in P(K)$ gives rise to an operator

$$j_\mu : X \longrightarrow L_p(\mu)$$

by restricting the canonical map $C(K) \longrightarrow L_p(\mu)$. Consequently, j_μ has p-summing norm at most one. In view of the remarks preceding the proof, it follows from our hypothesis that $\pi_r(j_\mu) \leq c$. We may use this in conjunction with Pietsch's Domination Theorem 2.12 to produce a measure $\hat{\mu} \in P(K)$ such that

$$\|j_\mu x\|_{L_p(\mu)} \leq c \cdot \|j_{\hat{\mu}} x\|_{L_r(\hat{\mu})} \tag{5}$$

for all $x \in X$. The correspondence $\mu \mapsto \hat{\mu}$ will be fruitful.

To reach our conclusion, we must consider an arbitrary p-summing operator $u : X \to Y$, where Y is any Banach space. The Domination Theorem provides us with a $\mu_0 \in P(K)$ for which

$$\|ux\| \leq \pi_p(u) \cdot \|j_{\mu_0} x\|_{L_p(\mu_0)}$$

for all $x \in X$. We shall construct a measure $\lambda \in P(K)$ together with a constant C – depending only on X – such that

$$\|j_{\mu_0} x\|_{L_p(\mu_0)} \leq C \cdot \|j_\lambda x\|_{L_1(\lambda)} \tag{6}$$

for all $x \in X$. Once this has been done, we can deduce (4), thanks to the Domination Theorem again, and the proof will be complete.

Our approach is 'heavy-hatted'. Starting with μ_0, we define a sequence $(\mu_n)_{n=0}^{\infty}$ in $P(K)$ by setting $\mu_{n+1} := \hat{\mu}_n$ for $n = 0, 1, 2, \ldots$. It is simple enough to check that $\lambda := \sum_{n=0}^{\infty} 2^{-n-1}\mu_n$ exists as a member of $P(K)$. What requires proof is that λ satisfies (6).

Since $1 < r < p$, we can produce $0 < \theta < 1$ with $1/r = \theta + (1-\theta)/p$. Use Hölder's Inequality to find that

$$\|j_{\mu_n} x\|_{L_r(\mu_n)} \le \|j_{\mu_n} x\|_{L_1(\mu_n)}^{\theta} \cdot \|j_{\mu_n} x\|_{L_p(\mu_n)}^{1-\theta} \tag{7}$$

for each $x \in X$ and each $n = 0, 1, 2, \ldots$. From (5), (7) and another application of Hölder's Inequality

$$\begin{aligned}
\sum_{n=0}^{\infty} 2^{-n-1} \|j_{\mu_n} x\|_{L_p(\mu_n)} &\le c \cdot \sum_{n=0}^{\infty} 2^{-n-1} \|j_{\mu_{n+1}} x\|_{L_r(\mu_{n+1})} \\
&\le c \cdot \sum_{n=0}^{\infty} 2^{-n-1} \|j_{\mu_{n+1}} x\|_{L_1(\mu_{n+1})}^{\theta} \|j_{\mu_{n+1}} x\|_{L_p(\mu_{n+1})}^{1-\theta} \\
&\le c \cdot \Big(\sum_{n=0}^{\infty} 2^{-n-1} \|j_{\mu_{n+1}} x\|_{L_1(\mu_{n+1})}\Big)^{\theta} \cdot \Big(\sum_{n=0}^{\infty} 2^{-n-1} \|j_{\mu_{n+1}} x\|_{L_p(\mu_{n+1})}\Big)^{1-\theta} \\
&\le c \cdot \Big(\sum_{n=0}^{\infty} 2^{-n-1} \|j_{\mu_{n+1}} x\|_{L_1(\mu_{n+1})}\Big)^{\theta} \cdot \Big(2 \cdot \sum_{n=0}^{\infty} 2^{-n-1} \|j_{\mu_n} x\|_{L_p(\mu_n)}\Big)^{1-\theta}
\end{aligned}$$

for each $x \in X$. Simplification gives, for all $x \in X$,

$$\begin{aligned}
\sum_{n=0}^{\infty} 2^{-n-1} \cdot \|j_{\mu_n} x\|_{L_p(\mu_n)} &\le c^{1/\theta} \cdot 2^{(1-\theta)/\theta} \cdot \sum_{n=0}^{\infty} 2^{-n-1} \cdot \|j_{\mu_{n+1}} x\|_{L_1(\mu_{n+1})} \\
&\le (2c)^{1/\theta} \cdot \sum_{n=0}^{\infty} 2^{-n-1} \cdot \|j_{\mu_n} x\|_{L_1(\mu_n)} = (2c)^{1/\theta} \cdot \|j_\lambda x\|_{L_1(\lambda)}
\end{aligned}$$

and so, discarding terms on the left hand side,

$$\|j_{\mu_0} x\|_{L_p(\mu_0)} \le 2 \cdot (2c)^{1/\theta} \cdot \|j_\lambda x\|_{L_1(\lambda)}.$$

This is exactly what we wanted. QED

The power of the Extrapolation Theorem can be seen at once. We use it to give an alternative proof of Grothendieck's Theorem 1.13.

- For this we shall make use of the fact that Banach spaces ℓ_1^I enjoy the *'lifting property'*: if $q : X \to Y$ is a surjective Banach space operator then, given $\varepsilon > 0$, each $u \in \mathcal{L}(\ell_1^I, Y)$ is of the form $u = q\,\hat{u}$ where $\hat{u} \in \mathcal{L}(\ell_1^I, X)$ satisfies $\|\hat{u}\| \le (1+\varepsilon) \cdot \|q\| \cdot \|u\|$; the operator $\hat{u}$ is called a *'lifting'* of u relative to q.

The argument is straightforward. First observe that thanks to the Open Mapping Theorem we may assume that q is a quotient map. Now let

e_i, $i \in I$, be the standard unit basis vectors in ℓ_1^I. Look at the definition of a quotient norm to find, for each $i \in I$, an $x_i \in X$ such that $qx_i = ue_i$ and $\|x_i\| \leq (1+\varepsilon) \cdot \|ue_i\|$. The correspondence $e_i \mapsto x_i$ defines an operator $\hat{u} : \ell_1^I \to X$ with the announced properties.

We shall also employ two facts established earlier: first that, regardless of $1 \leq p < \infty$, the formal identity $i_p : L_\infty[0,1] \to L_p[0,1]$ is p-summing (see 2.9), and second that the Rademacher functions generate an isomorphic embedding $R_p : \ell_2 \to L_p[0,1] : (a_n) \mapsto \sum_n a_n r_n$ (see 1.12).

Fix $1 < r \leq 2$. Note that i_r is nothing but the adjoint of the formal identity $k_{r^*} : L_{r^*}[0,1] \to L_1[0,1]$, and that $R_1 = k_{r^*} R_{r^*}$. This gives a factorization $R_1^* : L_\infty[0,1] \xrightarrow{i_r} L_r[0,1] \xrightarrow{R_{r^*}^*} \ell_2$ and shows that R_1^* shares i_r's r-summing nature.

On the other hand, as R_1 is an isomorphic embedding, R_1^* is onto. Consequently, since ℓ_1 enjoys the lifting property, each $u \in \mathcal{L}(\ell_1, \ell_2)$ can be written $u = R_1^* \hat{u}$ for some $\hat{u} \in \mathcal{L}(\ell_1, L_\infty[0,1])$, and so must be r-summing.

What this establishes is that $\mathcal{L}(\ell_1, \ell_2) = \Pi_2(\ell_1, \ell_2) = \Pi_r(\ell_1, \ell_2)$ for each $1 < r \leq 2$. The Extrapolation Theorem now steps in to reveal that $\mathcal{L}(\ell_1, \ell_2)$ and $\Pi_1(\ell_1, \ell_2)$ coincide.

Notes and Remarks

This chapter is an exposition of some of the results in the classic [1968] paper by J. Lindenstrauss and A. Pełczyński. They introduced the $\mathcal{L}_p$-spaces and, with Pietsch's p-summing operators as a powerful tool, showed how to attack many fundamental questions about $C(K)$ and L_p-spaces. Inspired by long-ignored ideas from Grothendieck's Résumé, able to divest these ideas of their tensor product trappings and to combine them successfully with what was known in abstract analysis in the late sixties, they heralded a new period in the study of Banach spaces. During this period, many of the classical problems were solved and new ones, often with close links to more established areas in mathematics, arose.

Theorem 3.1 is a natural generalization of Corollaire 1 in §4.2 of Grothendieck's [1953a] Résumé; the corollary itself appears here as 3.4. The new-found $\mathcal{L}_p$-form of 3.1 is directly from Lindenstrauss - Pełczyński [1968]. The crucial results 3.2 and 3.3 are gleaned from this paper, and from J. Lindenstrauss - H.P. Rosenthal [1969] and J. Lindenstrauss [1964]. Similarly, 3.5 - 3.7 have close relatives in Grothendieck's Résumé (Cor. 4 in §3.5 and Cor. 3 in §4.2); however, as in the case of 2-summing operators, constants differ and the language is that of tensor norms. In all these affairs we have followed the Lindenstrauss - Pełczyński lead at a somewhat pedantic pace.

Theorem 3.9 also was lost for more than a decade in the labyrinths of the Résumé (Prop. 5 of §4.4) until Lindenstrauss and Pełczyński recovered it and gave it a full and clear exposition. Appropriately enough it appears as a natural application of the theory of $\mathcal{L}_p$-spaces.

Theorems 3.11 through 3.13 are right out of Lindenstrauss - Pełczyński [1968]. This paper also contains a predual companion to 3.13:

- *If an infinite dimensional complemented subspace of an $\mathcal{L}_\infty$-space has an unconditional basis (x_n), then (x_n) is equivalent to the unit vector basis of c_0.*

Lindenstrauss - Pełczyński [1968] noted that, just like ℓ_2, the spaces c_0 and ℓ_1 have a unique unconditional basis, and they pondered the converse. In a remarkable turnabout, J. Lindenstrauss and M. Zippin ([1969a], [1969b]) showed that

- *c_0, ℓ_1 and ℓ_2 are the only infinite dimensional Banach spaces with a unique unconditional basis.*

Corollary 3.14 has a longer history. It was first proved by A.Pełczyński [1960] who derived it from R.C. James' [1950] famous result that a weakly sequentially complete Banach space with an unconditional basis must be isomorphic to a dual. That $L_1[0,1]$ is not a dual space had been known since I.M. Gelfand [1938]. In fact, $L_1[0,1]$ is not even a subspace of a separable dual, or of a space with unconditional basis; see A. Pełczyński [1961], F.G. Artjunjan [1962] and V.D. Milman [1970a].

Theorems 3.15 and 3.16 are due to S. Kwapień [1970a]. The particular case of 3.16 in which the domain and range spaces are both Hilbertian is due to A. Pietsch [1967] and A. Pełczyński [1967]. D.J.H. Garling [1970] has studied this problem as well.

The Extrapolation Theorem 3.17 is due to B. Maurey [1974a].

Grothendieck's Inequality and its consequences for operators between $\mathcal{L}_p$-spaces (most particularly 3.1 and 3.7) were exploited by S. Kwapień and S.J. Szarek [1979] and S.J. Szarek [1980] in their study of bases and biorthogonal systems in $L_1[0,1]$ and $C[0,1]$. They used p-summing operators to uncover close ties between inequalities of S.V. Bočkariev [1975] and A.M. Olevskiĭ [1975] and drew from this several poignant conclusions:

- *no order bounded basis for $L_1[0,1]$ exists, and if (f_n) is a normalized basis for $C[0,1]$ with coefficient functionals μ_n, then (μ_n) is not order bounded.*

Recall that a basis in a lattice is order bounded if some positive element dominates the absolute value of each member of the basis.

In another direction, S.J. Szarek [1980] showed that

- *there is no normalized basis (g_n) for $C[0,1]$ such that convergence of $\sum_n t_n g_n$ in $C[0,1]$ implies that (t_n) belongs to ℓ_2*

and that

- *no normalized basis (f_n) for $L_1[0,1]$ can have the property that $\sum_n s_n f_n$ converges in $L_1[0,1]$ for all $(s_n) \in \ell_2$.*

All these results, and much more, were subsumed by S.J. Szarek's [1979] remarkable proof that

- *no normalized basis for* $L_1[0,1]$ *is uniformly integrable*

and, its dual partner, that

- *if* (g_n) *is a normalized basis for* $C[0,1]$ *then there is a strictly increasing sequence of indices* (n_k) *so that the map* $\sum_n t_n g_n \mapsto (t_{n_k})$ *is a surjection of* $C[0,1]$ *onto* c_0.

Another noteworthy use of Grothendieck's Inequality that was to inspire later applications of p-summing operators was S.V. Kisliakov's [1975] proof that if $k \geq 2$ and $\ell \geq 1$ are integers, then the Banach space $C^\ell(I^k)$ of ℓ-times continuously differentiable functions on the k-cube is not isomorphic to any $C(K)$-space. The proof proceeds by constructing operators from $C^\ell(I^k)^*$ to ℓ_2 which are not 1-summing, an impossibility if $C^\ell(I^k)$ were to be isomorphic to a $C(K)$-space. Insofar as these spaces share many of the linear topological properties of $C(K)$-spaces (see J. Bourgain [1984d]) this is surprising and demonstrates the strength of the tools developed in this chapter.

Other than J. Hagler's unpublished notes from a Berkeley course by H.P. Rosenthal in the early seventies, no detailed exposition of the basic theory of the $\mathcal{L}_p$-spaces exists.

Perhaps the success of the theory in uncovering many of the mysteries of the classical spaces was outweighed by the impetus it provided to the investigation of the local theory of Banach spaces. In any case, since the Lindenstrauss-Pełczyński classic, the subject has evidenced many beautiful papers besides those already mentioned. Here are a few more choice morsels to digest at leisure: D. Alspach, P. Enflo and E. Odell [1977], J. Bourgain ([1980], [1981b]), J. Bourgain and F. Delbaen [1980], J. Bourgain, H.P. Rosenthal and G. Schechtman [1981], J. Bourgain and G. Pisier [1983], Y. Gordon, D.R. Lewis and J.R. Retherford [1972], W.B. Johnson, H.P. Rosenthal and M. Zippin [1971], D.R. Lewis and C. Stegall [1971], J. Lindenstrauss [1964], J. Lindenstrauss and H.P. Rosenthal [1969], J.R. Retherford [1972], G. Schechtman [1975], C. Stegall and J.R. Retherford [1972].

4. OPERATORS ON HILBERT SPACES AND SUMMING OPERATORS

We know from Chapter 2 that p-summing operators with a reflexive domain must be compact. In particular we are dealing with compact operators when we consider summing operators between Hilbert spaces. There are strong decomposition properties for such operators and, for the sake of completeness, we discuss some of these in detail.

In what follows, $H_1, H_2, \ldots$ will always be Hilbert spaces and we shall use $(\cdot|\cdot)$ to denote scalar products.

Throughout this text, whenever an operator u is acting between Hilbert spaces, the symbol u^* shall denote not only the Banach space adjoint but also the Hilbert space adjoint of u, so that $(ux|y) = (x|u^*y)$. This might seem a bit inconvenient but a few words of clarification should prevent undue confusion.

It is basic to Hilbert space theory that if H is a Hilbert space, then so is H^*; moreover, we identify H with H^* by a conjugate linear isometric surjection j that takes $x \in H$ to the functional $(\cdot|x) \in H^*$. Now if H_1 and H_2 are Hilbert spaces and $u: H_1 \to H_2$ is a bounded linear operator, then the Hilbert space adjoint $H_2 \to H_1$ of u can be obtained from the Banach space adjoint $u^*: H_2^* \to H_1^*$ quite directly: it is just $j_1^{-1} \circ u^* \circ j_2$ where $j_k : H_k \to H_k^*$ is the aforementioned identification. It is easy to see from this that most 'ideal' properties enjoyed by one adjoint are enjoyed by both; for example, being compact or p-summing is something one adjoint cannot be without the other following suit. For such reasons, we do not distinguish in notation between the two adjoints.

COMPACT HILBERT SPACE OPERATORS

Compact operators between Hilbert spaces admit a particular representation involving orthonormal sequences, a result which is known as the Spectral Theorem. It reads as follows:

4.1 Spectral Theorem: *An operator $u : H_1 \to H_2$ is compact if and only if it has a representation of the form*

$$(*) \qquad u = \sum_n \tau_n(\cdot|e_n)f_n,$$

where (e_n) is an orthonormal sequence in H_1, (f_n) is an orthonormal sequence in H_2, and (τ_n) is a null sequence of scalars.

If u is a finite rank operator, $(*)$ is a finite sum. Otherwise, $(*)$ represents a series which converges with respect to the operator norm $\|\cdot\|$ of $\mathcal{L}(H_1, H_2)$.

It will be convenient to consider finite sequences as sequences of infinite length by adding zeros.

We shall refer to $(*)$ as an *orthonormal representation – ON-representation* for short – of the compact operator u. Note that $(*)$ immediately gives rise to the ON-representation $u^* = \sum_n \tau_n(\,\cdot\,|f_n)e_n$ of the Hilbert space adjoint of u.

The proof of the Spectral Theorem will be obtained by induction, using the following lemma.

4.2 Lemma: *Let $u : H_1 \to H_2$ be a compact operator. Then there is a unit vector $x_0 \in H_1$ such that $u^*ux_0 = \|u\|^2 \cdot x_0$.*

In other words, the operator u^*u has eigenvalue $\|u\|^2$.

Proof. Naturally, we assume that $u \neq 0$. Since u is compact, we can find a sequence $(x_n)_n$ in B_{H_1} such that (ux_n) converges to some z_0 in H and such that $(\|ux_n\|)_n$ converges to $\|u\|$. For each $n \in \mathbf{N}$, we have

$$\begin{aligned}\big\|\, u^*ux_n - \|u\|^2 \cdot x_n \,\big\|^2 &= \|u^*ux_n\|^2 - 2\cdot\|u\|^2\cdot\|ux_n\|^2 + \|u\|^4\cdot\|x_n\|^2 \\ &\leq \|u\|^4 - \|u\|^2\cdot\|ux_n\|^2,\end{aligned}$$

and so $\lim_n(u^*ux_n - \|u\|^2 x_n) = 0$. It follows that $(x_n)_n$ converges to $x_0 := \|u\|^{-2}\cdot u^*(z_0)$. This limit is the vector we are looking for. Certainly, $\|x_0\| \leq 1$, and $u^*ux_0 = \lim_{n\to\infty} u^*ux_n = \|u\|^2 \cdot x_0$. But now

$$\|u\| = \lim_{n\to\infty} \|ux_n\| = \|ux_0\| \leq \|u\|\cdot\|x_0\|,$$

and so we see that actually $\|x_0\| = 1$. We are done. QED

It is handy to observe that, for the purpose of the theorem, it is good enough to work with one-to-one operators between *separable* Hilbert spaces. To understand why, let $u : H_1 \to H_2$ be any compact operator, let $K := u^{-1}(0)$ be its kernel, and let $R := \overline{u(H_1)}$ be the closure of its range. In the natural fashion, u induces an operator $u_0 : K^\perp \to R$ which is clearly injective with dense range; the same is true for its adjoint $u_0^* : R \to K^\perp$. Noticing that u_0 and u_0^*, like u, are compact and that a compact metric space is separable, we see that R and $K^\perp$ are separable. Moreover, if $p : H_1 \to K^\perp$ is the canonical orthogonal projection and $j : R \to H_2$ is the natural embedding, the factorization $u : H_1 \xrightarrow{p} K^\perp \xrightarrow{u_0} R \xrightarrow{j} H_2$ shows that we can make the announced reduction.

With this in mind, we can now proceed to our goal.

Proof of 4.1. If u has a representation $(*)$, then it is approximable by the finite rank operators $\sum_{n=1}^m \tau_n(\,\cdot\,|e_n)f_n$ and so is compact.

Assume conversely that u is compact. Clearly, we may suppose that $u \neq 0$ and, by the preceding argument, we may take H_1 to be separable and u to be injective. If (e_n) is some orthonormal basis for H_1, then $ux = \sum_n (x|e_n)ue_n$

for each $x \in H_1$. Since u is injective, $ue_n \neq 0$ for every n, and so

$$ux = \sum_n \|ue_n\| \cdot (x|e_n) \cdot \frac{ue_n}{\|ue_n\|}.$$

Notice that (e_n) is weakly null and that u, being compact, is completely continuous. It follows that $\lim_n \|ue_n\| = 0$. Consequently, if we choose our basis (e_n) in such a way that (ue_n) is an orthogonal sequence in H_2, then we can set $f_n = ue_n/\|ue_n\|$ and $\tau_n = \|ue_n\|$ for each n to reach our conclusion. Actually, we can do better and even arrange for (τ_n) to be a decreasing sequence.

The construction of suitable e_n's will be recursive. To begin with, we shall assume that H_1 has finite dimension, N say. By the lemma, there is a unit vector $e_1 \in H_1$ with $u^*ue_1 = \|u\|^2 e_1$, and so $\|u\| = \|ue_1\|$.

Now let $1 \leq n < N$ and suppose that we have found orthonormal vectors $e_1, \ldots, e_n$ in H_1 with the properties that

(1) $$u^*ue_k = \|ue_k\|^2 e_k \quad \text{for each} \quad 1 \leq k \leq n,$$

and setting $u_0 := u$ and $u_k := u\big|_{\{e_1,\ldots,e_k\}^\perp}$ $(1 \leq k \leq n)$,

(2) $$\|u_k\| = \|ue_{k+1}\| \quad \text{for each} \quad 0 \leq k \leq n-1.$$

Note that (2) implies that $\|ue_1\| \geq \|ue_2\| \geq \ldots \geq \|ue_n\|$.

Observe that u_n maps $\{e_1, \ldots, e_n\}^\perp \neq \{0\}$ into $\{ue_1, \ldots, ue_n\}^\perp$: if x is in $\{e_1, \ldots, e_n\}^\perp$ and $1 \leq k \leq n$, then $(ux|ue_k) = (x|u^*ue_k) = \|ue_k\|^2(x|e_k) = 0$. Since u_n is a compact operator, we can apply the lemma again to come up with a unit vector $e_{n+1} \in \{e_1, \ldots, e_n\}^\perp$ satisfying $u_n^* u_n e_{n+1} = \|u_n\|^2 e_{n+1}$, and so $\|ue_{n+1}\| = \|u_n\| \leq \|u_{n-1}\| = \|ue_n\|$.

In this way, we can construct an orthonormal basis $e_1, \ldots, e_N$ for H_1 with $\|ue_1\| \geq \ldots \geq \|ue_N\|$. Also, if $(ue_k|ue_m) = (u^*ue_k|e_m) = \|ue_k\|^2(e_k|e_m)$ for all $1 \leq k, m \leq n$, and so the vectors $ue_1, \ldots, ue_N$ are orthogonal in H_2.

This completes the proof in case H_1 is finite dimensional. If H_1 is infinite dimensional (and separable), we can still apply the recursive process just described, arriving at an infinite orthonormal sequence (e_n) such that (ue_n) is an orthogonal sequence and $(\|ue_n\|)$ is a decreasing null sequence. We just need to check that (e_n) is in fact an orthonormal basis for H_1.

If it were not, we could use the arguments above to apply the lemma once more to $\hat{u} := u\big|_{\{e_n : n \in \mathbf{N}\}^\perp} : \{e_n : n \in \mathbf{N}\}^\perp \to \{ue_n : n \in \mathbf{N}\}^\perp$, a compact operator which is not identically zero. But then $0 < \|\hat{u}\| \leq \|u_n\| = \|ue_n\|$ for all $n \in \mathbf{N}$. This prevents $(\|ue_n\|)$ from being a null sequence: contradiction.

QED

The proof of 4.1 shows that every compact operator $u : H_1 \to H_2$ admits an ON-representation

$$u = \sum_n \tau_n(\cdot|e_n) f_n$$

where the τ_n's satisfy

(**) $$0 \leq \tau_{n+1} \leq \tau_n$$

for all admissible indices.

Our next goal is to show that, with this extra condition, the numbers τ_n are uniquely determined by u. To see why, let us introduce, for an arbitrary operator $u : H_1 \to H_2$ and each $n \in \mathbf{N}$, the *n-th approximation number* (or *singular number*)

$$a_n(u) := \inf\{\|v-u\| : v \in \mathcal{L}(H_1, H_2),\ \dim\, v(H_1) < n\}.$$

Geometrically, $a_n(u)$ is just u's distance from the operators of rank $< n$.

4.3 Lemma: *Let the compact operator $u : H_1 \to H_2$ be represented à la (*) with the τ_n's satisfying (**). Then*

$$\tau_n = a_n(u)$$

for all $n \in \mathbf{N}$.

Proof. Fix $n \in \mathbf{N}$. Since $v := \sum_{k=1}^{n-1} \tau_k(\cdot\,|e_k)f_k \in \mathcal{L}(H_1, H_2)$ has rank $<n$,

$$a_n(u) \leq \|u-v\| = \sup\Big\{\Big\|\sum_{k\geq n} \tau_k(x|e_k)f_k\Big\| : x \in B_{H_1}\Big\}$$

$$= \sup\Big\{\Big(\sum_{k\geq n} \tau_k^2|(x|e_k)|^2\Big)^{1/2} : x \in B_{H_1}\Big\} \leq \tau_n.$$

To prove the reverse inequality, take any operator $v \in \mathcal{L}(H_1, H_2)$ of rank $< n$, and let $x_0 = \sum_{k=1}^{n} \xi_k e_k$ be a unit vector in v's kernel. Note that since $ux_0 = \sum_{k=1}^{n} \tau_k \xi_k f_k$ we have, thanks to our hypothesis,

$$\|u-v\| \geq \|ux_0 - vx_0\| = \|ux_0\| = \Big(\sum_{k=1}^{n} \tau_k^2|\xi_k|^2\Big)^{1/2} \geq \tau_n.$$

It follows that $a_n(u) \geq \tau_n$. QED

Trivially, $\|u\| = a_1(u) \geq a_2(u) \geq a_3(u) \geq \ldots$ for any $u \in \mathcal{L}(H_1, H_2)$. Moreover:

4.4 Lemma: *Let H_1, H_2 and H_3 be Hilbert spaces and let m and n be natural numbers.*

(a) *For any $u, v \in \mathcal{L}(H_1, H_2)$,*

$$a_{m+n-1}(u+v) \leq a_m(u) + a_n(v).$$

(b) *For any $u \in \mathcal{L}(H_2, H_3)$ and $v \in \mathcal{L}(H_1, H_2)$,*

$$a_{m+n-1}(uv) \leq a_m(u)\cdot a_n(v) \quad \text{for all } m, n \in \mathbf{N}.$$

Proof. (a) Take any u_0, $v_0 \in \mathcal{L}(H_1, H_2)$ with rank $u_0 < m$ and rank $v_0 < n$. Then rank $(u_0 + v_0) < m + n - 1$, and so

$$a_{m+n-1}(u+v) \leq \|u+v-(u_0+v_0)\| \leq \|u-u_0\| + \|v-v_0\|.$$

Passing to the appropriate infima, we get what we wanted.

(b) If $u_0 \in \mathcal{L}(H_2, H_3)$ has rank $< n$ and $v_0 \in \mathcal{L}(H_1, H_2)$ has rank $< m$, then $w := u_0 v + (u-u_0)v_0 \in \mathcal{L}(H_1, H_3)$ has rank $< m + n - 1$, and so

$$a_{m+n-1}(uv) \leq \|uv - w\| = \|(u-u_0)(v-v_0)\| \leq \|u-u_0\|\cdot\|v-v_0\|.$$

The proof is completed as before. QED

Schatten-von Neumann Classes

For $1 \leq p < \infty$, the *p-th Schatten-von Neumann class*

$$\mathcal{S}_p(H_1, H_2)$$

consists of all compact operators $u : H_1 \to H_2$ which admit an ON-representation (*) with

$$(\tau_n) \in \ell_p.$$

It follows from the preceding discussion that if some ON-representation has this property then any ON-representation must have it, and that in fact

$$\sigma_p(u) := \big(\sum_n |\tau_n|^p\big)^{1/p}$$

does not depend on the specific representation. In particular,

$$\sigma_p(u) = \big(\sum_n a_n(u)^p\big)^{1/p}.$$

Note that $\sigma_p((\cdot\,|x)y) = \|x\|\cdot\|y\|$ holds for all $x \in H_1$ and $y \in H_2$.

It is true, but by no means routine to show, that

$$[\mathcal{S}_p(H_1, H_2), \sigma_p]$$

is a Banach space. We do not stop to give a direct argument; rather, we prefer derive the result as a by-product of work performed later in this book.

For the time being, we only take note that, given $u, v \in \mathcal{S}_p(H_1, H_2)$, we can use 4.4(a) to conclude that

$$\begin{aligned}\sigma_p(u+v) &= \big(\sum_n a_n(u+v)^p\big)^{1/p} = \big(\sum_n a_{2n-1}(u+v)^p + \sum_n a_{2n}(u+v)^p\big)^{1/p}\\ &\leq 2^{1/p}\cdot\big(\sum_n a_{2n-1}(u+v)^p\big)^{1/p} \leq 2^{1/p}\cdot\big(\sum_n (a_n(u)+a_n(v))^p\big)^{1/p}\\ &\leq 2^{1/p}\cdot\big(\sigma_p(u)+\sigma_p(v)\big).\end{aligned}$$

This entails that $\mathcal{S}_p(H_1, H_2)$ is a linear space and that σ_p is a quasinorm on $\mathcal{S}_p(H_1, H_2)$.

Later, in 6.4, it will become clear that $[\mathcal{S}_p(H_1, H_2), \sigma_p]$ is isometrically isomorphic to the dual of $[\mathcal{S}_{p^*}(H_2, H_1), \sigma_{p^*}]$ if $1 < p < \infty$, and to the dual of $[\mathcal{K}(H_2, H_1), \|\cdot\|]$ if $p = 1$. Therefore σ_p is indeed a Banach space norm. For $p = 2$ we will actually see this in 4.10; for $p = 1$, compare also with 5.30, and for $2 \leq p < \infty$ with 10.3

We start by listing a number of immediate consequences of the definition.

4.5 Proposition: (a) *For each $1 \leq p \leq \infty$, the finite rank operators form a dense linear subspace of $[\mathcal{S}_p(H_1, H_2), \sigma_p]$.*

(b) *When $1 \leq p < q \leq \infty$, $\mathcal{S}_p(H_1, H_2)$ is a dense linear subspace of $[\mathcal{S}_q(H_1, H_2), \sigma_q]$, and $\sigma_q(u) \leq \sigma_p(u)$ for all $u \in \mathcal{S}_p(H_1, H_2)$.*

(c) *Let $1 \le p \le \infty$ and let $u \in \mathcal{L}(H_1, H_2)$. Then $u \in \mathcal{S}_p(H_1, H_2)$ if and only if $u^* \in \mathcal{S}_p(H_2, H_1)$, and in this case, $\sigma_p(u) = \sigma_p(u^*)$.*

(d) *Let $1 \le p \le \infty$ and let $w : H_0 \to H_1$, $v : H_1 \to H_2$ and $u : H_2 \to H_3$ be Hilbert space operators. If v is in $\mathcal{S}_p(H_1, H_2)$, then uvw is in $\mathcal{S}_p(H_0, H_3)$, and $\sigma_p(uvw) \le \|u\| \cdot \sigma_p(v) \cdot \|w\|$.*

It is simple to check for membership of the Schatten - von Neumann classes by using the following useful criterion.

4.6 Theorem: (a) *An operator $u : H_1 \to H_2$ is compact if and only if, no matter how we choose orthonormal sequences (e_n) in H_1 and (f_n) in H_2, the sequence $\big((ue_n|f_n)\big)_n$ belongs to c_0.*

(b) *Let $1 \le p < \infty$. An operator $u : H_1 \to H_2$ belongs to $\mathcal{S}_p(H_1, H_2)$ if and only if, however we select orthonormal sequences (e_n) in H_1 and (f_n) in H_2, the sequence $\big((ue_n|f_n)\big)_n$ belongs to ℓ_p.*

Proof. To avoid trivialities, assume that $u \neq 0$.

(a) Suppose first that u is compact and that we are given orthonormal sequences (e_n) in H_1 and (f_n) in H_2. Since (e_n) is a weak null sequence and compact operators are completely continuous, $\lim_n \|ue_n\| = 0$. By the Cauchy - Schwarz Inequality, it follows that $\lim_n (ue_n|f_n) = 0$.

To prove the converse we will show that for any $0 < \varepsilon < \|u\|$ there is a finite rank operator $v : H_1 \to H_2$ such that $\|u - v\| \le \varepsilon$.

We begin with a general observation. Let $u \in \mathcal{L}(H_1, H_2)$ be any non-zero operator, and let $0 < \varepsilon < \|u\|$. Then there are unit vectors $x \in H_1$ and $y \in H_2$ such that $|(ux|y)| > \varepsilon$. Consequently, the collection Φ of all pairs $\big((x_i)_{i \in I}, (y_i)_{i \in I}\big)$ of orthonormal families $(x_i)_{i \in I}$ in H_1 and $(y_i)_{i \in I}$ in H_2 which satisfy $|(ux_i|y_i)| > \varepsilon$ for all $i \in I$ is non-empty. We partially order Φ by writing

$$\big((x_i)_{i \in I}, (y_i)_{i \in I}\big) \le \big((x'_j)_{j \in J}, (y'_j)_{j \in J}\big)$$

whenever $I \subset J$ and $x_i = x'_i$ and $y_i = y'_i$ for each $i \in I$. It is easy to see that Zorn's Lemma applies: it provides us with a maximal element of Φ; we label it $\big((e_i)_{i \in I_\varepsilon}, (f_i)_{i \in I_\varepsilon}\big)$.

Now, if u satisfies the convergence condition of (a), then I_ε must be finite: $I_\varepsilon = \{1, ..., N\}$. Define orthogonal projections $p = \sum_{i=1}^N (\,\cdot\,|e_i)e_i \in \mathcal{L}(H_1)$ and $q = \sum_{i=1}^N (\,\cdot\,|f_i)f_i \in \mathcal{L}(H_2)$: each has rank N, and so $v := up + qu - qup$ is a finite rank operator in $\mathcal{L}(H_1, H_2)$. We claim that $\|u - v\| \le \varepsilon$.

Assume to the contrary that $\|u - v\| > \varepsilon$. Then there are unit vectors $x \in H_1$ and $y \in H_2$ such that $\big|\big((u - v)x \mid y\big)\big| > \varepsilon$. Put $e_0 := x - px$ and $f_0 := y - qy$ and observe that $e_0 \in \{e_i : i \in I_\varepsilon\}^\perp$ and $f_0 \in \{f_i : i \in I_\varepsilon\}^\perp$. Both have norm ≤ 1. Since $(id_{H_2} - q)u(id_{H_1} - p) = u - v$,

$$\begin{aligned} |(ue_0|f_0)| &= \big|\big(u(id_{H_1} - p)x \mid (id_{H_2} - q)y\big)\big| = \big|\big((id_{H_2} - q)u(id_{H_1} - p)x \mid y\big)\big| \\ &= \big|\big((u - v)x|y)\big)\big| > \varepsilon \ge \varepsilon \cdot \|e_0\| \cdot \|f_0\|. \end{aligned}$$

Accordingly, e_0 and f_0 are non-zero vectors, and since $\big(u(e_0/\|e_0\|)\,|f_0/\|f_0\|\big) > \varepsilon$, trouble ensues with the maximality of I_ε.

(b) Suppose that $1 \le p < \infty$ and $u : H_1 \to H_2$ is an operator such that, regardless of the orthonormal sequences (e_n) in H_1 and (f_n) in H_2, the sequence $(ue_n|f_n)$ belongs to ℓ_p. By (a) we know that u must be compact, and hence it admits an ON-representation $u = \sum_n \tau_n(\,\cdot\,|x_n)y_n$. Note that $\tau_n = (ux_n|y_n)$ for each n to see that our hypothesis gives $(\tau_n) \in \ell_p$ and conclude that u belongs to $\mathcal{S}_p(H_1, H_2)$.

Conversely, if u is a member of $\mathcal{S}_p(H_1,H_2)$ and if $u = \sum_m \tau_m(\,\cdot\,|x_m)y_m$ is an ON-representation with $\tau_m \ge 0$ for each m, then for any orthonormal sequence (e_n) in H_1 and each $n \in \mathbf{N}$,

$$\sum_m \tau_m|(e_n|x_m)|^2 = \sum_m \tau_m|(e_n|x_m)|^{2/p}|(e_n|x_m)|^{2/p^*}$$
$$\le \Big(\sum_m \tau_m^p|(e_n|x_m)|^2\Big)^{1/p}\cdot\Big(\sum_m |(e_n|x_m)|^2\Big)^{1/p^*} \le \Big(\sum_m \tau_m^p|(e_n|x_m)|^2\Big)^{1/p}.$$

Similarly, for any orthonormal sequence (f_n) in H_2 and any $n \in \mathbf{N}$,

$$\sum_m \tau_m|(y_m|f_n)|^2 \le \Big(\sum_m \tau_m^p|(y_m|f_n)|^2\Big)^{1/p}.$$

(Actually, as written, this argument covers the case $1 < p < \infty$. The case $p = 1$ requires no argument.) But for all n,

$$|(ue_n|f_n)| \le \sum_m \tau_m^{1/2}|(e_n|x_m)|\tau_m^{1/2}|(y_m|f_n)|$$
$$\le \Big(\sum_m \tau_m|(e_n|x_m)|^2\Big)^{1/2}\cdot\Big(\sum_m \tau_m|(y_m|f_n)|^2)\Big)^{1/2},$$

and so we finally arrive at

$$\sum_n |(ue_n|f_n)|^p \le \sum_n\Big(\sum_m \tau_m^p|(e_n|x_m)|^2\Big)^{1/2}\cdot\Big(\sum_m \tau_m^p|(y_m|f_n)|^2\Big)^{1/2}$$
$$\le \Big(\sum_{m,n} \tau_m^p|(e_n|x_m)|^2\Big)^{1/2}\cdot\Big(\sum_{m,n} \tau_m^p|(y_m|f_n)|^2\Big)^{1/2} \le \sum_m \tau_m^p.$$

This was what we wanted. QED

Now that we have a satisfactory test for membership in $\mathcal{S}_p(H_1,H_2)$, we turn our attention to estimates for $\sigma_p(u)$.

4.7 Theorem: *Let $u \in \mathcal{L}(H_1,H_2)$ and let $(g_i)_{i\in I}$ be any orthonormal basis for H_1.*

(a) *If $1 \le p \le 2$ and $\big(\|ug_i\|\big)_{i\in I} \in \ell_p^I$, then $u \in \mathcal{S}_p(H_1,H_2)$ and*

$$\sigma_p(u) \le \Big(\sum_{i\in I} \|ug_i\|^p\Big)^{1/p}.$$

(b) *If $2 \le p < \infty$ and $u \in \mathcal{S}_p(H_1,H_2)$, then $\big(\|ug_i\|\big)_{i\in I} \in \ell_p^I$ and*

$$\Big(\sum_{i\in I} \|ug_i\|^p\Big)^{1/p} \le \sigma_p(u).$$

Proof. (a) We begin by showing that our hypothesis implies the compactness of u. To achieve this, we fix $\varepsilon > 0$ and seek a finite rank operator $v \in \mathcal{L}(H_1, H_2)$ with $\|u - v\| \leq \varepsilon$.

Choose a finite set $J \subset I$ such that $(\sum_{i \in I \setminus J} \|ug_i\|^p)^{1/p} \leq \varepsilon$. Then $v := \sum_{i \in J}(\,\cdot\,|g_i)ug_i$ is certainly a finite rank operator in $\mathcal{L}(H_1, H_2)$, and it satisfies $\|u - v\| \leq \varepsilon$ because

$$\begin{aligned}\|(u-v)x\| &= \Big\| \sum_{i \in I \setminus J} (x|g_i)ug_i \Big\| \leq \Big(\sum_{i \in I} |(x|g_i)|^2\Big)^{1/2} \cdot \Big(\sum_{i \in I \setminus J} \|ug_i\|^2\Big)^{1/2} \\ &\leq \|x\| \cdot \Big(\sum_{i \in I \setminus J} \|ug_i\|^p\Big)^{1/p} \leq \varepsilon \cdot \|x\|.\end{aligned}$$

As u is compact, it admits an ON-representation $u = \sum_n \tau_n(\,\cdot\,|e_n)f_n$; as usual, we can assume that $\tau_n \geq 0$ for each n. For each $x \in H_1$, we have

$$\|ux\|^2 = \sum_n \tau_n^2 |(x|e_n)|^2,$$

and so, by application of Hölder's Inequality with the indices $2/p$ and $2/(2-p)$,

$$\begin{aligned}\sum_n \tau_n^p &= \sum_n \tau_n^p \sum_{i \in I} |(g_i|e_n)|^2 = \sum_{i \in I} \sum_n \tau_n^p |(g_i|e_n)|^p |(g_i|e_n)|^{2-p} \\ &\leq \sum_{i \in I} \Big(\sum_n \tau_n^2 |(g_i|e_n)|^2\Big)^{p/2} \cdot \Big(\sum_n |(g_i|e_n)|^2\Big)^{(2-p)/2} \\ &\leq \sum_{i \in I} \Big(\sum_n \tau_n^2 |(g_i|e_n)|^2\Big)^{p/2} = \sum_{i \in I} \|ug_i\|^p.\end{aligned}$$

This shows that u belongs to $\mathcal{S}_p(H_1, H_2)$, with $\sigma_p(u) \leq \big(\sum_{i \in I} \|ug_i\|^p\big)^{1/p}$.

(b) Take any $u \in \mathcal{S}_p(H_1, H_2)$, where now $2 \leq p < \infty$. Let $u = \sum_n \tau_n(\,\cdot\,|e_n)f_n$ be any ON-representation with $\tau_n \geq 0$ for each n. This time, apply Hölder's Inequality with indices $p/2$ and $p/(p-2)$ to get, for each $i \in I$,

$$\begin{aligned}\|ug_i\|^2 &= \sum_n \tau_n^2 \cdot |(g_i|e_n)|^{4/p} \cdot |(g_i|e_n)|^{2(p-2)/p} \\ &\leq \Big(\sum_n \tau_n^p \cdot |(g_i|e_n)|^2\Big)^{2/p} \cdot \Big(\sum_n |(g_i|e_n)|^2\Big)^{(p-2)/p} \leq \Big(\sum_n \tau_n^p \cdot |(g_i|e_n)|^2\Big)^{2/p}.\end{aligned}$$

It follows that

$$\sum_{i \in I} \|ug_i\|^p \leq \sum_{i \in I} \sum_n \tau_n^p \cdot |(g_i|e_n)|^2 = \sum_n \tau_n^p = \sigma_p(u)^p,$$

and the proof is complete. QED

The case $p = 2$ of the preceding theorem is of fundamental importance:

4.8 Corollary: *An operator $u : H_1 \to H_2$ belongs to $\mathcal{S}_2(H_1, H_2)$ if and only if there is an orthonormal basis $(e_i)_{i \in I}$ in H_1 such that $\sum_{i \in I} \|ue_i\|^2 < \infty$. In this case, the quantity $\sum_{i \in I} \|ue_i\|^2$ is independent of the choice of orthonormal basis $(e_i)_{i \in I}$; in fact, for any orthonormal basis $(e_i)_{i \in I}$ of H_1*

$$\sigma_2(u) = \Big(\sum_{i \in I} \|ue_i\|^2\Big)^{1/2}.$$

The elements of $\mathcal{S}_2(H_1, H_2)$ are called *Hilbert - Schmidt operators.* We already know that $[\mathcal{S}_2(H_1, H_2), \sigma_2]$ is a Banach space. Even more is true.

4.9 Proposition: *$[\mathcal{S}_2(H_1, H_2), \sigma_2]$ is a Hilbert space.*

The norm σ_2 is induced by the inner product

$$(u \mid v) = \sum_{i \in I} (ue_i \mid ve_i)$$

where $(e_i)_{i \in I}$ is an orthonormal basis for H_1. By polarization, the fact that σ_2 is independent of the choice of orthonormal basis implies that the inner product also has this property.

In this chapter, we concentrate on the Hilbert - Schmidt class. We will return to general Schatten - von Neumann classes later on in Chapters 6 and 10.

Hilbert - Schmidt Operators and Summing Operators

We are now ready to link up with the theory of summing operators.

4.10 Theorem: *An operator $u : H_1 \to H_2$ is Hilbert - Schmidt if and only if it is 2-summing. In this case $\sigma_2(u) = \pi_2(u)$.*

Proof. Take $u \in \Pi_2(H_1, H_2)$ and any orthonormal basis $(e_i)_{i \in I}$ for H_1. Then, given any finite subset J of I, we have

$$\Big(\sum_{i \in J} \|ue_i\|^2\Big)^{1/2} \leq \pi_2(u) \cdot \|(e_i)_{i \in J}\|_2^{weak} = \pi_2(u).$$

It follows that u is a Hilbert - Schmidt operator and $\sigma_2(u) \leq \pi_2(u)$.

Conversely, take $u \in \mathcal{S}_2(H_1, H_2)$ and let $(x_n) \in \ell_2^{weak}(H_1)$. Define v in $\mathcal{L}(\ell_2, H_1)$ by $ve_n = x_n$ for each $n \in \mathbf{N}$; we saw in 2.2 that $\|v\| = \|(x_n)\|_2^{weak}$. By the ideal property of $\mathcal{S}_2$ we know that uv is Hilbert - Schmidt and since

$$\Big(\sum_n \|ux_n\|^2\Big)^{1/2} = \Big(\sum_n \|uve_n\|^2\Big)^{1/2} = \sigma_2(uv) \leq \sigma_2(u) \cdot \|v\| = \sigma_2(u) \cdot \|(x_n)\|_2^{weak},$$

it follows that $u \in \Pi_2(H_1, H_2)$ and $\pi_2(u) \leq \sigma_2(u)$. QED

As a by-product of this theorem we obtain again that $[\mathcal{S}_2(H_1, H_2), \sigma_2]$ is a Banach space.

The identification we made in Theorem 4.10 allows us to spot several far from obvious properties of the Hilbert - Schmidt class that are, however, fairly immediate consequences of the theory of 2-summing operators. The first such corollary shows that the range of a Hilbert space operator may constrain it to be Hilbert - Schmidt.

4.11 Corollary: (a) *Suppose that μ is a probability measure and that the operator $u : L_2(\mu) \to L_2(\mu)$ has its range inside $L_\infty(\mu)$. Then u is a Hilbert - Schmidt operator.*

(b) *Suppose that the operator $u : \ell_2 \to \ell_2$ has its range inside ℓ_1. Then u is a Hilbert - Schmidt operator.*

Proof. (a) The Closed Graph Theorem can be relied upon to show that u, viewed as a map into $L_\infty(\mu)$, is a bounded linear operator. But this is tantamount to u having $i_2 : L_\infty(\mu) \to L_2(\mu)$ as a factor. As we know i_2 is 2-summing, u is also 2-summing, hence Hilbert - Schmidt.

(b) Proceeding as above, the operator u will this time have the natural 2-summing inclusion $\ell_1 \hookrightarrow \ell_2$ as a factor. QED

Summoning Grothendieck's Theorem we can go further.

4.12 Corollary: *When u is a Hilbert space operator, the following are equivalent:*

(i) *u is a Hilbert - Schmidt operator.*

(ii) *u factors through an $\mathcal{L}_\infty$-space.*

(iii) *u factors through an $\mathcal{L}_1$-space.*

Proof. (i)⇔(ii): On the one hand, if u is Hilbert - Schmidt, then u is 2-summing, so Corollary 2.16 informs us that it actually factors through a $\mathcal{C}(K)$-space.

On the other hand, if u factors through an $\mathcal{L}_\infty$-space, then Grothendieck's Theorem 3.7 steps in to assure that u is 2-summing and so Hilbert - Schmidt.

(i)⇔(iii): This is a dual version of the previous equivalence. We know from a remark after 3.4 that the dual of a $\mathcal{C}(K)$-space is an $\mathcal{L}_1$-space. If $u : H_1 \to H_2$ is Hilbert - Schmidt, then so is u^* and any factorization $u^* : H_2 \to \mathcal{C}(K) \to H_1$ leads to a factorization $H_1 \to \mathcal{L}_1 \to H_2$ of $u = u^{**}$.

Conversely, if $u : H_1 \to H_2$ factors through an $\mathcal{L}_1$-space, we can use Grothendieck's Theorem 3.1 to see that it is 2-summing and so Hilbert - Schmidt. QED

We shall prove a far-reaching extension of this result in 19.2. It will turn out that the $\mathcal{L}_\infty$- and $\mathcal{L}_1$-spaces can be replaced by any infinite dimensional Banach space.

From Corollary 3.16 we know that $\Pi_2(H_1, H_2)$ and $\Pi_p(H_1, H_2)$ coincide for any $1 \leq p < \infty$. A reformulation in the light of Theorem 4.10 reveals a mapping property of Hilbert - Schmidt operators.

4.13 Corollary: *If $1 \leq p < \infty$, then the operator $u : H_1 \to H_2$ is Hilbert - Schmidt if and only if it takes weakly p-summable sequences in H_1 to strongly p-summable sequences in H_2.*

Extension Property

The 2-summing operators between general Banach spaces enjoy a striking extension property due to the particular form of their factorization through an $L_\infty(\mu)$-space, where μ is a suitable probability measure. To establish this property we first isolate an important feature of such spaces.

4.14 Theorem: *For any probability measure μ, $L_\infty(\mu)$ is an injective Banach space.*

Proof. Choose Banach spaces X and Z, with X a subspace of Z, and let $u \in \mathcal{L}(X, L_\infty(\mu))$. Our task is to extend u to an operator $\tilde{u} : Z \to L_\infty(\mu)$ with $\|\tilde{u}\| = \|u\|$. We are already familiar with this result when $L_\infty(\mu) = \ell_\infty^n$ (see the comments before 2.13). Our strategy hinges on a localization argument which allows us to exploit this case.

By Theorem 3.2, $L_\infty(\mu)$ is an $\mathcal{L}_{\infty,\lambda}$-space for each $\lambda > 1$. Accordingly, if we fix $\varepsilon > 0$, then given a finite dimensional subspace E of X, we can arrange to put $u(E)$ inside a finite dimensional subspace F of $L_\infty(\mu)$ which admits an isomorphism $v : F \to \ell_\infty^{\dim F}$ such that $\|v\|\cdot\|v^{-1}\| \leq 1+\varepsilon$. Let $u_E : E \to F$ be the operator induced by u. Now $vu_E : E \to \ell_\infty^{\dim F}$ admits an extension $\tilde{u}_E : Z \to \ell_\infty^{\dim F}$ with $\|\tilde{u}_E\| = \|vu_E\|$. Clearly, $u_E^\varepsilon := v^{-1}\tilde{u}_E$, considered as a map $Z \to L_\infty(\mu)$, extends $u\,\big|_E$ and is constrained by $\|u_E^\varepsilon\| \leq \|v^{-1}\|\cdot\|vu_E\| \leq (1+\varepsilon)\cdot\|u\|$.

Let $\mathcal{F}_X$ be the collection of all finite dimensional subspaces of X. To each $E_0 \in \mathcal{F}_X$ and $\varepsilon_0 > 0$ associate $\{(E,\varepsilon) \in \mathcal{F}_X \times \mathbf{R}^+ : E_0 \subset E,\ 0 < \varepsilon < \varepsilon_0\}$. As these sets form a filter basis, they are members of some ultrafilter $\mathcal{U}$ on $\mathcal{F}_X \times \mathbf{R}^+$. By design, for each $z \in Z$, the map $\mathcal{F}_X \times \mathbf{R}^+ \to L_\infty(\mu) = L_1(\mu)^* : (E,\varepsilon) \mapsto u_E^\varepsilon z$ is bounded. In $L_\infty(\mu)$, bounded sets are relatively weak$*$ compact, so as $\mathcal{U}$ is an ultrafilter, the weak$*$limit of its image under this map exists: we label it $\tilde{u}z$.

It is plain that we have constructed a map $\tilde{u} : Z \to L_\infty(\mu)$ which is linear and bounded, with $\|\tilde{u}\| \leq \|u\|$. Naturally, $\tilde{u}$ extends u, and because of this, $\|u\| \leq \|\tilde{u}\|$. QED

The hypothesis on μ was only needed to allow a weak$*$limit procedure. The theorem actually holds any time $L_\infty(\mu)$ is a dual space.

Theorem 4.14 will be an invaluable aid on many occasions. To begin, we use it in the study of Π_2.

4.15 Π_2-Extension Theorem: *Let X, Y, Z be Banach spaces with X a subspace of Z. Each 2-summing operator $u : X \to Y$ admits a 2-summing extension $\tilde{u} : Z \to Y$ with $\pi_2(\tilde{u}) = \pi_2(u)$.*

Proof. In view of the comments made following Corollary 2.16, any 2-summing operator $u : X \to Y$ has a factorization

$$u : X \xrightarrow{v} L_\infty(\mu) \xrightarrow{i_2} L_2(\mu) \xrightarrow{w} Y,$$

where μ is a probability measure, $\|v\| = 1$ and $\|w\| = \pi_2(u)$. By $L_\infty(\mu)$'s injectivity, v has an extension $\tilde{v} \in \mathcal{L}(Z, L_\infty(\mu))$ with $\|\tilde{v}\| = 1$. As a consequence, the operator

$$\tilde{u} : Z \xrightarrow{\tilde{v}} L_\infty(\mu) \xrightarrow{i_2} L_2(\mu) \xrightarrow{w} Y$$

is 2-summing, satisfies $\pi_2(\tilde{u}) \leq \|w\|\cdot\pi_2(i_2)\cdot\|\tilde{v}\| = \pi_2(u)$ and, since it extends u, also $\pi_2(u) \leq \pi_2(\tilde{u})$. QED

Here is an appealing Banach space consequence of Theorem 4.15.

4.16 Corollary: *Suppose that X is a Banach space with a subspace isomorphic to ℓ_1. Then X admits a quotient space isomorphic to ℓ_2 where the quotient map is 2-summing.*

Proof. Thanks to the Open Mapping Theorem, we just have to produce a 2-summing surjective operator $X \to \ell_2$. Like all separable Banach spaces, ℓ_2 is a quotient of ℓ_1, so if X has a subspace Y which is isomorphic to ℓ_1, we are guaranteed a surjective operator u in $\mathcal{L}(Y, \ell_2)$. On account of Grothendieck's Theorem 1.13, u is 2-summing. The proof is readily completed by applying the previous theorem. QED

The factorization scheme 2.16 supplied much of what was needed to prove Theorem 4.15. It also features in the proof of another stunning result.

4.17 Theorem: *If E is any n-dimensional normed space, then*

$$\pi_2(id_E) = \sqrt{n} .$$

Proof. First we show that $\pi_2(id_E) \leq \sqrt{n}$. To this end, we take $x_1, \ldots, x_m$ in E and aim to prove that $(\sum \|x_k\|^2)^{1/2} \leq \sqrt{n} \cdot \|(x_k)\|_2^{weak}$. Use the x_k's to define $u \in \mathcal{L}(\ell_2^m, E)$ via $u(e_k) = x_k$. We know from 2.2 that $\|u\| = \|(x_k)\|_2^{weak}$, and so $(\sum \|ue_k\|^2)^{1/2} \leq \pi_2(u)$. Hence it is adequate to show that $\pi_2(u) \leq \sqrt{n} \cdot \|u\|$.

Now u admits a factorization $\ell_2^m \xrightarrow{q} \ell_2^d \xrightarrow{v} E$, where q is the quotient map onto $\ell_2^d = \ell_2^m / \ker u$ and v is injective with $\|v\| = \|u\|$. Hence

$$\pi_2(u) = \pi_2(v\, id_{\ell_2^d}\, q) \leq \|v\| \cdot \pi_2(id_{\ell_2^d}) \cdot \|q\| = \|u\| \cdot \sigma_2(id_{\ell_2^d})$$

through the good graces of Theorem 4.10. But $\sigma_2(id_{\ell_2^d})$ is obviously $\sqrt{d}$, so $\pi_2(u) \leq \sqrt{d} \cdot \|u\|$ follows.

For the reverse inequality, we apply Corollary 2.16 to id_E. There exists a regular Borel probability measure μ on the compact (!) space B_{E^*}, so that id_E has a factorization $E \xrightarrow{v} L_2(\mu) \xrightarrow{w} E$ with $\pi_2(v) \leq 1$ and $\|w\| = \pi_2(id_E)$. As v is injective, $H = v(E) \subset L_2(\mu)$ may be identified with ℓ_2^n, and since $x = vwx$ for each $x \in H$, we obtain

$$\sqrt{n} = \sigma_2(id_H) = \pi_2(id_H) \leq \pi_2(v) \cdot \|w\| \leq \pi_2(id_E). \qquad \text{QED}$$

In tandem with the Π_2-Extension Theorem 4.15, the preceding theorem provides the material needed to prove a famous result.

4.18 Kadets - Snobar Theorem: *If E is an n-dimensional subspace of the Banach space X, then E is the range of a projection $P \in \mathcal{L}(X)$ with $\|P\| \leq \sqrt{n}$.*

Proof. The previous theorem tells us that $\pi_2(id_E) = \sqrt{n}$, whereas Theorem 4.15 provides us with a 2-summing extension $P : X \to E$ of id_E such that $\pi_2(P) = \sqrt{n}$. When considered as a map from X to X, P is a projection with range E and norm $\|P\| \leq \pi_2(P) = \sqrt{n}$. QED

ADJOINTS OF 2-SUMMING OPERATORS

We have seen (4.5) that an operator between Hilbert spaces is 2-summing (alias Hilbert - Schmidt) precisely when its adjoint is. It is only natural to ask about the 2-summing nature of adjoints of 2-summing operators between general Banach spaces.

We present two contrasting results which highlight important classes of Banach spaces.

4.19 Theorem: *Let X be a Banach space. The following statements about X are equivalent:*

(i) *Every 2-summing operator with domain X has a 2-summing adjoint.*

(ii) *Every 2-summing operator from X to a Hilbert space has a 2-summing adjoint.*

(iii) *X is isomorphic to a Hilbert space.*

Proof. (i)$\Rightarrow$(ii) is trivial.

(ii)$\Rightarrow$(iii): Let K be B_{X^*} with its weak$*$topology, and let $i_X : X \to \mathcal{C}(K)$ be the natural embedding. Let (μ_n) be a weak ℓ_2 sequence in $\mathcal{C}(K)^*$ with $\|(\mu_n)\|_2^{weak} \leq 1$. Using Proposition 2.2 we see that $w : \mathcal{C}(K) \to \ell_2 : f \mapsto (\langle \mu_n, f \rangle)_n$ satisfies $\|w\| = \|(\mu_n)\|_2^{weak} \leq 1$. Thanks to Theorem 3.5, w is 2-summing, and this forces $w i_X : X \to \ell_2$ to be 2-summing. By hypothesis, $i_X^* w^* : \ell_2 \to X$ is also 2-summing. Even more: i_X^* is 2-summing since

$$\left(\sum \|i_X^*(\mu_n)\|^2\right)^{1/2} = \left(\sum \|i_X^* w^* e_n\|^2\right)^{1/2} \leq \pi_2(i_X^* w^*) < \infty$$

and since $(\mu)_n$ was taken arbitrarily from the unit ball of $\ell_2^{weak}(\mathcal{C}(K)^*)$.

Appealing again to Corollary 2.16, we can create a factorization $i_X^* : \mathcal{C}(K)^* \xrightarrow{b} H \xrightarrow{a} X^*$ where H is a Hilbert space. The surjectivity of i_X^* transfers to a, so the Open Mapping Theorem tells us that X^* is isomorphic to a Hilbert space; X follows suit.

(iii)$\Rightarrow$(i): Without loss of generality, we may assume that X is a Hilbert space. If $u : X \to Y$ is 2-summing, we may once more invoke Corollary 2.16 to obtain a factorization $u : X \xrightarrow{v} L_2(\mu) \xrightarrow{w} Y$ where the Hilbert space operator v is 2-summing, hence Hilbert - Schmidt. As we know, this implies that v^* is also Hilbert - Schmidt, so 2-summing. Consequently, $u^* = v^* w^*$ is 2-summing as well. QED

There are significant changes when X is considered as a range space rather than a domain space.

4.20 Theorem: *The following statements about the Banach space X are equivalent.*

(i) *Every 2-summing operator with range in X has a 2-summing adjoint.*

(ii) *Every operator from X to a Hilbert space is 2-summing.*

(iii) *Every operator with domain X having a 2-summing adjoint must itself be 2-summing.*

(iv) *Every operator from X^* to a Hilbert space is 2-summing.*

Proof. (i)⇒(ii): Let H be a Hilbert space. Our aim is to show that every operator $u : X \to H$ is 2-summing, and for this we make double use of Proposition 2.7.

It will be enough to show that $uv : \ell_2 \to H$ is 2-summing for every operator $v : \ell_2 \to X$; so let us fix $v \in \mathcal{L}(\ell_2, X)$. What we shall actually prove is that uv is 1-summing. (This is not overkill; see 3.11.) To do this, we shall utilize Proposition 2.7 one more time and show that $uvw : c_0 \to H$ is 1-summing for every operator $w : c_0 \to \ell_2$. So let us also fix $w \in \mathcal{L}(c_0, \ell_2)$.

We had better use our hypothesis. Fortunately, $vw : c_0 \to X$ is 2-summing thanks to Grothendieck's Theorem (3.5); so we can assert that $(vw)^*$ is also 2-summing and thus has a factorization

$$(vw)^* : X^* \xrightarrow{a} L_\infty(\mu) \xrightarrow{i_2} L_2(\mu) \xrightarrow{b} \ell_1,$$

with μ a probability measure. Predually,

$$\begin{array}{ccccccc}
c_0 & \xrightarrow{w} & \ell_2 & \xrightarrow{v} & X & \xrightarrow{u} & H \\
 & & & & & k_X \searrow \quad \nearrow u^{**} & \\
b^*|_{c_0} \downarrow & & & & & X^{**} & \\
 & & & & & \nearrow a^* & \\
L_2(\mu) & & \xrightarrow{i_2^*} & & L_1(\mu)^{**} . & &
\end{array}$$

A tiny bit of thought reveals that i_2^* is nothing but the formal identity $L_2(\mu) \to L_1(\mu)$ regarded as a map into $L_1(\mu)^{**}$. We thus have a factorization of the Hilbert space operator $u^{**}a^*i_2^*$ through $L_1(\mu)$: it is Hilbert - Schmidt by 4.12, hence 1-summing by 4.13. Consequently, uvw is 1-summing – which was what we wanted.

(ii)⇒(iii): Assume that the operator $u : X \to Y$ has 2-summing adjoint $u^* : Y^* \to X^*$. By the Pietsch Factorization Theorem, $u^* : Y^* \xrightarrow{b} H \xrightarrow{a} X^*$ where H is a Hilbert space. To show that u is 2-summing, it suffices, by 2.5, to show that $k_Y u = u^{**}k_X$ is 2-summing. However, dualization brings us to $k_Y u : X \xrightarrow{k_X} X^{**} \xrightarrow{a^*} H \xrightarrow{b^*} Y^{**}$. The map a^*k_X is 2-summing by hypothesis; we are done.

(iii)⇒(iv): The argumentation of (i)⇒(ii) applies again. Let H be a Hilbert space and take $u \in \mathcal{L}(X^*, H)$. Everything will be as we want it, provided we can show that for all operators $w : c_0 \to \ell_2$ and $v : \ell_2 \to X^*$ the composition $uvw : c_0 \to H$ is 1-summing.

Since w is 2-summing, so is uvw. To go further we need our hypothesis. It will show that vw, and hence uvw, factors through an $L_1(\mu)$-space. Grothendieck's Theorem (3.4) takes care of the rest.

If we are to use our hypothesis, we must find an operator with domain X whose adjoint is 2-summing. The operator $w^*v^*k_X : X \to \ell_1$ fits the bill; its adjoint has $w^{**} : \ell_\infty \to \ell_2$ as a 2-summing factor.

The upshot is that $w^*v^*k_X$ is a 2-summing operator and factors like $X \xrightarrow{a} L_\infty(\mu) \xrightarrow{i_2} L_2(\mu) \xrightarrow{b} \ell_1$ where μ is some probability measure. Take adjoints to get a factorization of $vw : c_0 \to X^*$:

$$\begin{array}{ccccccc} c_0 & \hookrightarrow & \ell_\infty & \xrightarrow{w^{**}} & \ell_2 & \xrightarrow{v} & X^* \\ & & \downarrow{\scriptstyle b^*} & & & & \uparrow{\scriptstyle a^*} \\ & & L_2(\mu) & & \xrightarrow{i_2^*} & & L_1(\mu)^{**} \end{array}.$$

Since i_2^* takes its values in $L_1(\mu)$, we have done what we set out to do.

(iv)$\Rightarrow$(i): Assume the operator $u : Z \to X$ is 2-summing. Then u factors through a Hilbert space H, say. Dually, $u^* : X^* \to Z^*$ factors through H and so is 2-summing, by hypothesis. QED

Many spaces satisfy the conditions we have laboured to prove equivalent. For example, $\mathcal{L}_1$-spaces and $\mathcal{L}_\infty$-spaces belong to this class, by virtue of Theorems 3.1 and 3.7. For further examples, we refer to 15.16.

It is also worth recording that, thanks to 2.7 and 4.10, a Hilbert space operator is Hilbert - Schmidt if and only if it factors through a Banach space which satisfies the equivalent conditions of 4.20. In 19.2 we shall see that for factorization of a Hilbert - Schmidt operator through a Banach space X there is no need to place any restriction on X, other than infinite dimensionality.

Notes and Remarks

The results on compact Hilbert space operators are standard and are found in many books on operator theory. Among many excellent treatises, we cite R. Schatten [1960], I.C. Gohberg and M.G. Krein [1969] and B.Simon [1979]. The crucial representation in 4.1 is often referred to as 'Schmidt representation of compact operators'. It bears plain resemblence to the Gram - Schmidt orthonormalization procedure.

What we now call Schatten - von Neumann classes were introduced in a series of papers by R. Schatten and J. von Neumannn [1946], [1948]. Hilbert - Schmidt operators, however, are much older. They trace their history back to D. Hilbert [1904], [1906a,b], [1909], [1910] and his student, E. Schmidt [1907a,b], [1908]. Favoured by the existence of the Lebesgue integral, which allowed the production of complete function spaces, and inspired by the ideas of Fredholm, Hilbert and Schmidt launched a program on integral equations whose success continues to be felt to this very day.

Lemma 4.3 is due to D.E.Allahverdiev [1957]. In the light of this result, 4.5(c) is a close relative of the finite dimensional theorem that a complex number λ belongs to the spectrum $\sigma(A)$ of a matrix A if and only $\overline{\lambda}$ belongs to $\sigma(A^*)$. Proposition 4.5, along with several other basic results on the classes $\mathcal{S}_p$, appears in R. Schatten - J. von Neumann [1948].

The results 4.5, 4.6 and 4.7(a) can be extended directly to include all p's in the interval $(0, 1)$. However, as we want to stay with Banach spaces, we do not treat this case.

Theorem 4.10 is due to A. Pietsch [1967]. Again, A. Grothendieck proved a similar result in [1953a] but his norm doesn't match up exactly with σ_2. Corollary 4.11, however, is due to Grothendieck [1953a], whereas 4.12 is from Lindenstrauss - Pelczyński [1968]; extensions will be available once 19.2 is proved. Corollary 4.13 is due to A. Pietsch [1967] and A. Pełczyński [1967]. Here it appears as a consequence of 3.16, but historically, its position is much more central: it initiated much of the later work on the classification of Banach spaces, in particular the theory of type and cotype. It is one of the first places where Khinchin's Inequality is used in abstract functional analysis; till now no non-probabilistic proof of (the Pełczyński part of) 4.13 seems to be known. The range of p's can be extended: it was shown by S. Kwapień [1970b] that the 0-summing operators (see Notes and Remarks to Chapter 2) between Hilbert spaces coincide with the corresponding Hilbert - Schmidt operators.

Theorem 4.14 can also be proved by mimicking the standard proof of the scalar-valued Hahn - Banach Extension Theorem. The key ingredient is the fact that $L_\infty(\mu)$ is an *order complete* Banach lattice whose closed unit ball has a biggest element, 1; if one now proceeds through the proof of the real Hahn - Banach Theorem with an operator $u : X \to L_\infty(\mu)$ instead of a functional defined on the subspace X of the Banach space Y, the end result will be Theorem 4.14. In the complex case, the classical Bohnenblust - Sobczyk trick still works.

Injectivity for infinite dimensional Banach spaces is still a mystery as far as the isomorphic theory is concerned. The isometric theory has been completely understood for decades, thanks to the work of L. Nachbin [1950], D. Goodner [1950] and J.L. Kelley [1952] in the real case, and M. Hasumi [1958] in the complex case. For real Banach spaces, the result reads as follows:

Theorem: *The following are equivalent statements about the real Banach space* X:

(i) *For any subspace* Y *of any normed space* Z, *each* $u \in \mathcal{L}(Y, X)$ *is the restriction of some* $\tilde{u} \in \mathcal{L}(Z, X)$ *such that* $\|\tilde{u}\| = \|u\|$.

(ii) *If X is a closed subspace of a Banach space W, then there is a norm one projection $p \in \mathcal{L}(W)$ such that $p(W) = X$.*

(iii) *X has the 'binary intersection property for balls', that is, given any family $\mathcal{B}$ of closed balls in X such that any two members of $\mathcal{B}$ have non-empty intersection, it is true that $\bigcap \mathcal{B} \neq \emptyset$.*

(iv) *X is isometrically isomorphic to the space $C(K)$ of all continuous real-valued functions on some extremally disconnected compact Hausdorff space K.*

Recall that K is extremally disconnected if the closure of each open subset of K is again open.

Nachbin and Goodner provided the details for most of the above theorem but needed to assume the existence of extreme points at a crucial juncture; Kelley was able to dispose of this. Following a necessarily different path, Hasumi extended the equivalence of (i) and (iv) to complex spaces.

In the light of this, it is of interest to note that H. Nakano [1941] and M.H. Stone [1949] showed that a $C(K)$-space is order complete if and only if K is extremally disconnected. It is then not hard to see, arguing once again as in the Hahn-Banach Theorem, that $C(K)$-spaces are injective whenever K is extremally disconnected.

The Π_2-Extension Theorem 4.15 became part of public knowledge once the injectivity of $L_\infty(\mu)$-spaces became known. It can be spotted in Grothendieck's Résumé [1953a], and it is cited in Theorem 6.3 of Lindenstrauss-Pełczyński [1968] in the case where the domain is a Hilbert space.

Corollary 4.16 is due to R.I. Ovsepian and A. Pełczyński [1975]. It has a converse, but this requires H.P. Rosenthal's [1974] celebrated ℓ_1 theorem: a Banach space X doesn't contain an isomorphic copy of ℓ_1 if and only if every bounded sequence in X has a weakly Cauchy subsequence. The converse of 4.16 is this:

- *if the Banach space X admits a 2-summing surjection $q \in \mathcal{L}(X, \ell_2)$, then X must contain an isomorphic copy of ℓ_1.*

If not, we would be able to pick from any given bounded sequence (x_n) in X a weakly Cauchy subsequence, say (x_{n_k}), and thanks to the complete continuity of q (see 2.17), (qx_{n_k}) would be norm convergent in ℓ_2. Thus q would be compact: contradiction.

Theorem 4.17 was discovered independently by M.G. Snobar [1972] and by D.J.H. Garling and Y. Gordon [1971]; the proof presented here is apparently due to S. Kwapień. Snobar derived 4.17 from Theorem 4.18 which was proved shortly before by M.I. Kadets and M.G. Snobar [1971].

H. König, D.R. Lewis and P.K. Lin [1983] showed that for an n-dimensional subspace X of a k-dimensional normed space Y one can find a projection of Y onto X with norm no more than $\sqrt{n}\cdot\left(1 - (\sqrt{n} - 1)^2/2k\right)$. Later, H. König and D.R. Lewis [1987] showed that if E is a finite dimensional subspace of any Banach space X and $n := \dim E \geq 2$, then there is a projection of X onto E with norm $< \sqrt{n}$. Their proof relies on estimating the π_1-norm of the identity and does not give a concrete estimate for the projection constant. D.R. Lewis [1988] showed that for spaces of finite dimension $n \geq 2$ there is a projection of norm $\leq \sqrt{n} \cdot (1 - 5^{-2n-11} n^{-2})$; Lewis' proof also computes the 1-summing norm of the identity in a very elegant way by using the Pietsch Factorization Theorem and his earlier [1978] study of finite dimensional subspaces of L_p. Finally, verifying a conjecture of König and Lewis [1987], H. König and N. Tomczak-Jaegermann [1990] showed the existence of

- *a universal constant $c > 0$ such that if $n \geq 2$ and E is an n-dimensional subspace of a Banach space X, then there is a linear projection of X onto E with norm $\leq \sqrt{n} - (c/\sqrt{n})$.*

H. König and N. Tomczak-Jaegermann [1995], using trace duality and calling on some deep results in spherical harmonics, have continued this work and have computed best values of c for low dimensional cases. For two dimensional spaces, the best constants are $4/3$ in the real case and $(1 + \sqrt{3})/2$ in the complex case. For three dimensional spaces the values are $(1 + \sqrt{5})/2$ in the real case and $5/3$ in the complex case. In the real case these numbers are attained when the ball is a regular hexagon and a regular dodecahedron, respectively. H. König [1995] has used similar techniques to improve the general bounds in case symmetry assumptions are made on the range of the projection.

Theorem 4.19 was S. Kwapień's [1970c] response to a question of J.S. Cohen [1970] who had previously established an isometric version. A close look at our proof of (iii)⇒(i) shows that if u is 2-summing with a Hilbert space domain, then u^* is 2-summing with $\pi_2(u^*) \leq \pi_2(u)$. Cohen showed that this property of operators characterizes Hilbert spaces isometrically by proving the validity of the parallelogram law in two dimensional subspaces. The essence of the proof we present is a fact to be proved in 7.11: the ideal Γ_2 is isometrically the trace dual of $\Delta_2 = \Pi_2^d \circ \Pi_2$.

Theorem 4.20 is part of the folklore of trace duality and may first have been formulated by H. Jarchow [1982]. It is a highly non-trivial fact, to be proved in 19.2, that a Banach space X enjoys the properties isolated in 4.20 precisely when Hilbert space operators factoring through X are exactly the Hilbert - Schmidt operators.

A further spectacular demonstration of the power of the ideas we are developing emerges in results involving eigenvalue distribution. As we saw in 2.17, summing operators are weakly compact and completely continuous. Consequently, if X is a complex Banach space and u is in $\Pi_p(X)$ for some $1 \leq p < \infty$, then u^2 is compact and so there is a (possibly finite) sequence $(\lambda_n(u))_n$ of non-zero complex numbers of decreasing moduli (repetitions allowing for multiplicities) with $\lim_{n\to\infty} \lambda_n(u) = 0$ such that $\{\lambda_n(u) : n \in \mathbf{N}\}$ is $\sigma(u) \setminus \{0\}$ where $\sigma(u)$ is the spectrum of u. Off hand, the $\lambda_n(u)$'s could tend to zero as slowly as one might wish if u were just compact; but $u \in \Pi_p(X)$ forces $\sum_n |\lambda_n(u)|^r < \infty$, where $r = \max\{p, 2\}$. The proof for the case $1 \le p \le 2$ is due to A. Pietsch [1963], [1967]. The case $2 < p < \infty$ turned out to be much harder; it was settled by W.B. Johnson, H. König, B. Maurey and J.R. Retherford [1979] using very involved arguments. Fortunately, A. Pietsch [1986] found a much more accessible proof based on tensor product methods. The monographs of H. König [1986] and A. Pietsch [1987] are recommended for complete expositions of all this and much more.

If $1 \le p \le 2$ and $u \in \Pi_p(X)$, then one cannot in general expect to have more than $(\lambda_n(u)) \in \ell_2$. This was first stated explicitly in A. Pietsch [1963], but in fact goes back to A. Grothendieck [1955] whose proof even works when we assume nuclearity of our operators (see Chapter 5). Even worse, R.J. Kaiser and J.R. Retherford showed [1983] that given $(\lambda_n) \in \ell_2$ with all non-zero entries, one can find a Banach space X and a nuclear operator u in $\mathcal{L}(X)$ whose eigenvalues are precisely the λ_n's. The spaces X they use for this purpose are those constructed by G. Pisier [1983]; they have the property that the projective and injective tensor norms coincide on $X \otimes X$.

As a consequence of the work of Johnson, König, Maurey and Retherford [1979] and of H. König, J.R. Retherford and N. Tomczak-Jaegermann [1980], the best possible p such that for a given Banach space X every nuclear operator $u : X \to X$ satisfies $(\lambda_n(u)) \in \ell_p$ depends heavily on the geometry of X. Whereas it is known from the work of H. Weyl [1949] that $p = 1$ whenever X is a Hilbert space, the fact that this characterizes Hilbert spaces isomorphically is a deep result from the paper by Johnson, König, Maurey and Retherford quoted above.

5. p-Integral Operators

Definition and Elementary Properties

We shall say that a linear mapping $u : X \to Y$ between Banach spaces is a *p-integral operator* $(1 \le p \le \infty)$ if there are a probability measure μ and (bounded linear) operators $a : L_p(\mu) \to Y^{**}$ and $b : X \to L_\infty(\mu)$ giving rise to the commutative diagram

$$\begin{array}{ccccc} X & \xrightarrow{u} & Y & \xrightarrow{k_Y} & Y^{**} \\ \downarrow{\scriptstyle b} & & & & \uparrow{\scriptstyle a} \\ L_\infty(\mu) & & \xrightarrow{i_p} & & L_p(\mu) \ . \end{array}$$

As usual, $i_p : L_\infty(\mu) \to L_p(\mu)$ is the formal identity, and $k_Y : Y \to Y^{**}$ is the canonical isometric embedding. The collection of all p-integral operators from X to Y is denoted by

$$\mathcal{I}_p(X, Y).$$

With each $u \in \mathcal{I}_p(X, Y)$ we associate its *p-integral norm,*

$$\iota_p(u) = \inf \|a\| \cdot \|b\|,$$

where the infimum is extended over all measures μ and operators a and b as above. We have taken something of a liberty by calling ι_p a norm. Before we justify this, let us take note of the following immediate consequence of the definition:

5.1 Proposition: *Let $1 \le p < q \le \infty$. Then $\mathcal{I}_p(X, Y) \subset \mathcal{I}_q(X, Y)$ with $\iota_q(u) \le \iota_p(u)$ for each $u \in \mathcal{I}_p(X, Y)$.*

Next we pass to some fundamental properties of the newly defined operators.

5.2 Theorem: *Let $1 \le p \le \infty$, and let X_0, X, Y and Y_0 be Banach spaces.*

(a) *$\mathcal{I}_p(X, Y)$ is a linear subspace of $\mathcal{L}(X, Y)$ containing all the finite rank members of $\mathcal{L}(X, Y)$. Further, ι_p is a Banach space norm on $\mathcal{I}_p(X, Y)$, and $\|u\| \le \iota_p(u)$ for all $u \in \mathcal{I}_p(X, Y)$.*

(b) *(Ideal Property) The composition of a p-integral operator with any operator is p-integral. More specifically, if $w \in \mathcal{L}(X_0, X)$, $v \in \mathcal{I}_p(X, Y)$ and $u \in \mathcal{L}(Y, Y_0)$, then $uvw \in \mathcal{I}_p(X_0, Y_0)$ and $\iota_p(uvw) \le \|u\| \cdot \iota_p(v) \cdot \|w\|$.*

Proof. (a) The very definition of $\iota_p(u)$ ensures that $\|u\| \leq \iota_p(u)$. Since the normed space axioms are plain and simple to verify, we confine ourselves to the proof of completeness. The astute reader will appreciate, by the way, that this proof subsumes the additivity of $\mathcal{I}_p(X,Y)$ and the attendant triangle inequality.

It is enough to verify that absolutely convergent series converge (see 1.1); accordingly, let (u_n) be a sequence in $\mathcal{I}_p(X,Y)$ for which $\sum_n \iota_p(u_n) < \infty$. Immediately, $\sum_n \|u_n\| < \infty$, so $\sum_n u_n$ converges, say to u, in $\mathcal{L}(X,Y)$. Our expectation is that $u \in \mathcal{I}_p(X,Y)$ with $\iota_p(u) \leq \sum_n \iota_p(u_n)$; this will then enable us to show that u is the ι_p-limit of $(\sum_{k\leq n} u_k)_n$.

Here is how to proceed. Start with the ever-present $\varepsilon > 0$, and for each n find a probability space $(\Omega_n, \Sigma_n, \mu_n)$ and operators $a_n : L_p(\mu_n) \to Y^{**}$ and $b_n : X \to L_\infty(\mu_n)$ such that $k_Y u_n$ factors as

$$X \xrightarrow{b_n} L_\infty(\mu_n) \xrightarrow{i_p} L_p(\mu_n) \xrightarrow{a_n} Y^{**}$$

with $\|b_n\| = 1$ and $\|a_n\| \leq \iota_p(u_n) + \varepsilon/2^n$. Let (Ω, Σ) be the direct sum measurable space of the (Ω_n, Σ_n), that is, assuming, as we may, that the Ω_n's are pairwise disjoint, let $\Omega := \bigcup_n \Omega_n$ and $\Sigma := \{S \subset \Omega : S \cap \Omega_n \in \Sigma_n \text{ for all } n\}$. Define a probability measure μ on Σ by specifying that for each m and for each $S_m \in \Sigma_m$,

$$\mu(S_m) = \mu_m(S_m) \cdot \frac{\|a_m\|}{\sum_n \|a_n\|}.$$

Define $a : L_p(\mu) \to Y^{**}$ by $a(f) = \sum_n a_n(f|_{\Omega_n})$ and $b : X \to L_\infty(\mu)$ by $b(x) = \sum_n b_n(x) \cdot 1_{\Omega_n}$. Plainly, a and b are linear, and $\|b\| \leq 1$. When $p = \infty$, it is clear that $\|a\| \leq \sum_n \|a_n\|$. For $1 \leq p < \infty$, this still holds: when $f \in L_p(\mu)$,

$$\|af\| \leq \sum_n \|a_n\| \cdot \|f|_{\Omega_n}\|_{L_p(\mu_n)} = \Big(\sum_n \|a_n\|\Big)^{1/p} \cdot \sum_n \|a_n\|^{1/p^*} \cdot \|f|_{\Omega_n}\|_{L_p(\mu)},$$

and so, by Hölder's Inequality,

$$\begin{aligned} \|af\| &\leq \Big(\sum_n \|a_n\|\Big)^{1/p} \cdot \Big(\sum_n \|a_n\|\Big)^{1/p^*} \cdot \Big(\sum_n \|f|_{\Omega_n}\|^p_{L_p(\mu)}\Big)^{1/p} \\ &= \Big(\sum_n \|a_n\|\Big) \cdot \|f\|_{L_p(\mu)}. \end{aligned}$$

In any case, $\|a\| \leq \sum_n \|a_n\| \leq \sum_n \iota_p(u_n) + \varepsilon$.

It is easy to see that $k_Y u = a i_p b$, and so u is p-integral with

$$\iota_p(u) \leq \|a\| \cdot \|b\| \leq \sum_n \iota_p(u_n) + \varepsilon.$$

The arbitrariness of $\varepsilon > 0$ ensures that $\iota_p(u) \leq \sum_n \iota_p(u_n)$.

The end is nigh. In the above situation, let $t_n : L_p(\mu) \to Y^{**}$ be given by $t_n(f) = \sum_{k>n} a_k(f|_{\Omega_k})$. When $p = \infty$, $\|t_n\| = \sum_{k>n} \|a_k\|$, whereas when $1 \leq p < \infty$, the argument used to estimate $\|af\|$ leads to $\|t_n f\| \leq (\sum_k \|a_k\|)^{1/p} \cdot (\sum_{k>n} \|a_k\|)^{1/p^*} \cdot \|f\|_{L_p(\mu)}$. Anyway, $\lim_{n\to\infty} \|t_n\| = 0$, and since $\iota_p(u - \sum_{k\leq n} u_k) \leq \|a\| \cdot \|t_n\|$, we have $\lim_{n\to\infty} \iota_p(u - \sum_{k\leq n} u_k) = 0$.

We turn to the ideal property (b), and consider a composition of operators $X_0 \xrightarrow{w} X \xrightarrow{v} Y \xrightarrow{u} Y_0$. If v is p-integral, then $k_Y v$ admits a typical factorization $X \xrightarrow{b} L_\infty(\mu) \xrightarrow{i_p} L_p(\mu) \xrightarrow{a} Y^{**}$, and so, taking note of the fact that $k_{Y_0} u = u^{**} k_Y$, we arrive at a factorization for $k_{Y_0} uvw$:

$$X_0 \xrightarrow{bw} L_\infty(\mu) \xrightarrow{i_p} L_p(\mu) \xrightarrow{u^{**}a} Y_0^{**}.$$

Hence $uvw \in \mathcal{I}_p(X_0, Y_0)$ and $\iota_p(uvw) \leq \|bw\| \cdot \|u^{**}a\| \leq \|u\| \cdot \|a\| \cdot \|b\| \cdot \|w\|$. The factorization $k_Y u = a i_p b$ was arbitrary, so $\iota_p(uvw) \leq \|u\| \cdot \iota_p(v) \cdot \|w\|$.

QED

It is not unreasonable to ask why we insist on factoring p-integral operators into a bidual. The reason will become clear in the next chapter when we investigate a powerful duality relationship between summing and integral operators. However, the usefulness will already make itself felt when we discuss adjoints of integral operators (5.14 and 5.15).

There are, none the less, important cases where the passage to the bidual is superfluous. We shall say that a Banach space operator $u : X \to Y$ is *strictly p-integral* if it admits a bounded linear factorization

$$(*) \qquad \begin{array}{ccc} X & \xrightarrow{u} & Y \\ \downarrow b & & \uparrow a \\ L_\infty(\mu) & \xrightarrow{i_p} & L_p(\mu) \end{array}$$

where μ is a probability measure.

5.3 Proposition: *When Y is norm one complemented in Y^{**}, every p-integral operator $u : X \to Y$ is strictly p-integral. Moreover, $\iota_p(u) = \inf \|a\| \cdot \|b\|$, where the infimum is extended over all factorizations of the form $(*)$.*

A quick diagram chase should make this apparent.

Since the adjoint of $k_Z : Z \to Z^{**}$ provides a norm one projection of Z^{***} onto Z^*, there is an important special case which is worth singling out.

5.4 Corollary: *When $Y = Z^*$ is a dual space, every p-integral operator is strictly p-integral.*

From now on in this chapter we shall mainly restrict our attention to the range $1 \leq p < \infty$. It will not be until the end of Chapter 6 that the importance of $\mathcal{I}_\infty(X, Y)$ becomes apparent.

Relations to p-Summing Operators

The p-integral operators come well within our proposed area of investigation.

5.5 Proposition: *Let $1 \leq p < \infty$. If $u : X \to Y$ is p-integral, then it is p-summing with $\pi_p(u) \leq \iota_p(u)$.*

Proof. Start from a typical p-integral factorization

$$k_Y u : X \xrightarrow{b} L_\infty(\mu) \xrightarrow{i_p} L_p(\mu) \xrightarrow{a} Y^{**}$$

and use the fact that i_p is p-summing with $\pi_p(i_p) = 1$ (2.9(d)). The ideal property (2.4) informs us that $k_Y u$ is p-summing, and the the injectivity property (2.5) reveals that u is too. The relation $\pi_p(u) \leq \iota_p(u)$ follows readily from $\pi_p(u) = \pi_p(k_Y u) = \pi_p(a i_p b) \leq \|a\| \cdot \|b\|$. QED

With this connection between summing and integral operators in mind, it is only natural to expect a parallel development of their factorization schemes.

5.6 Theorem: *Let $1 \leq p \leq \infty$. An operator $u : X \to Y$ is p-integral if and only if every time we take a weak$*$compact norming subset K of B_{X^*} we can find a regular Borel probability measure ν on K and an operator $\tilde{u} : L_p(\nu) \to Y^{**}$ such that the following diagram commutes:*

$$\begin{array}{ccccc} X & \xrightarrow{u} & Y & \xrightarrow{k_Y} & Y^{**} \\ \downarrow i_X & & & & \uparrow \tilde{u} \\ C(K) & & \xrightarrow{j_p} & & L_p(\nu) \end{array} .$$

As usual, j_p denotes the canonical map, and i_X is the natural isometric embedding defined by $i_X(x)(x^*) = \langle x^*, x\rangle$.

In this case

$$\iota_p(u) = \inf \|\tilde{u}\|,$$

where the infimum is taken over all possible ν's and $\tilde{u}$'s.

Proof. The injectivity of $L_\infty(\mu)$ (see 4.14) makes the result straightforward when $p = \infty$. Accordingly, we assume $1 \leq p < \infty$ and write Δ for the proposed infimum.

One direction is easy: if $k_Y u$ factors in the way described, then we can introduce the obvious factorization $j_p : C(K) \xrightarrow{j_\infty} L_\infty(\nu) \xrightarrow{i_p} L_p(\nu)$ to obtain

$$k_Y u : X \xrightarrow{j_\infty i_X} L_\infty(\nu) \xrightarrow{i_p} L_p(\nu) \xrightarrow{\tilde{u}} Y^{**}$$

and to deduce that u is p-integral with $\iota_p(u) \leq \Delta$.

Conversely, if $u \in \mathcal{I}_p(X, Y)$, select a typical factorization

$$k_Y u : X \xrightarrow{b} L_\infty(\mu) \xrightarrow{i_p} L_p(\mu) \xrightarrow{a} Y^{**}.$$

Bring the injectivity (see 4.14) of $L_\infty(\mu)$ into play to display b in the form $b : X \xrightarrow{i_X} C(K) \xrightarrow{\tilde{b}} L_\infty(\mu)$ with $\|\tilde{b}\| = \|b\|$. We know that i_p is p-summing, so $i_p \tilde{b} : C(K) \to L_p(\mu)$ is p-summing as well (see again 2.9(d) and 2.4), and we may summon a form of the Pietsch Factorization Theorem (2.15) to decompose $i_p \tilde{b}$ as $i_p \tilde{b} : C(K) \xrightarrow{j_p} L_p(\nu) \xrightarrow{w} L_p(\mu)$, where ν is a regular Borel probability measure on K and $\|w\| = \pi_p(i_p \tilde{b})$. Assembling this information we find

$$k_Y u : X \xrightarrow{i_X} C(K) \xrightarrow{j_p} L_p(\nu) \xrightarrow{aw} Y^{**}.$$

This is a diagram of the desired form; it shows that

$$\Delta \leq \|aw\| \leq \pi_p(i_p\tilde{b})\cdot\|a\| \leq \pi_p(i_p)\cdot\|\tilde{b}\|\cdot\|a\| = \|a\|\cdot\|b\|.$$

Passing to the infimum we arrive at $\Delta \leq \iota_p(u)$. QED

A corresponding theorem can of course also be established for strictly p-integral operators as long as $1 \leq p < \infty$.

Our new formulation of the p-integral factorization scheme allows us to reap the benefits of the groundwork of Chapter 2. Glance at Corollary 2.14:

5.7 Corollary: *Let $1 \leq p < \infty$. If Y is an injective Banach space, then every p-summing operator $u : X \to Y$ is p-integral, and $\pi_p(u) = \iota_p(u)$.*

Take a peek at Corollary 2.15:

5.8 Corollary: *Let $1 \leq p < \infty$. If K is a compact Hausdorff space, then every p-summing operator $u : C(K) \to Y$ is p-integral, and $\pi_p(u) = \iota_p(u)$.*

Turn expectantly to Corollary 2.16:

5.9 Corollary: *The 2-summing and 2-integral operators are the same. More precisely, if X and Y are Banach spaces, then*

$$\Pi_2(X,Y) = \mathcal{I}_2(X,Y)$$

with equality of norms.

5.10 Corollary: *If Y is a subspace of an $\mathcal{L}_p$-space, $1 \leq p \leq 2$, then regardless of the Banach space X, we have*

$$\Pi_q(X,Y) = \mathcal{I}_q(X,Y) = \mathcal{I}_2(X,Y),$$

for all $2 \leq q < \infty$.

Proof. For $2 \leq q < \infty$, merely observe that the combination of Theorem 3.15 with Corollary 5.9 and Propositions 5.1 and 5.5 brings to light

$$\Pi_q(X,Y) = \Pi_2(X,Y) = \mathcal{I}_2(X,Y) \subset \mathcal{I}_q(X,Y) \subset \Pi_q(X,Y). \qquad \text{QED}$$

As soon as we know more about $\mathcal{L}_p$-spaces (see 8.14) it will be apparent that Grothendieck's Theorem allows us to incorporate $\mathcal{I}_\infty(X,Y)$ into this chain of equalities.

p-Summing Operators Failing to be p-Integral

Before we delve deeper into the properties of integral operators, it may be useful to display side by side the revealing factorization schemes for p-summing and p-integral operators ($1 \leq p < \infty$). There is a noticeable resemblance, but the difference is very important.

$$\begin{array}{ccc} X & \xrightarrow{u} Y \xrightarrow{i_Y} & \ell_\infty^{B_{Y^*}} \\ \downarrow{\scriptstyle i_X} & & \uparrow{\scriptstyle \tilde{u}} \\ C(K) & \xrightarrow{\quad j_p \quad} & L_p(\mu) \\ & (\Pi_p) & \end{array} \qquad \begin{array}{ccc} X & \xrightarrow{u} Y \xrightarrow{k_Y} & Y^{**} \\ \downarrow{\scriptstyle i_X} & & \uparrow{\scriptstyle \tilde{u}} \\ C(K) & \xrightarrow{\quad j_p \quad} & L_p(\mu)\,. \\ & (\mathcal{I}_p) & \end{array}$$

In both cases, K is an arbitrary weak$*$compact norming subset of B_{X^*} and μ is a regular probability measure on K; i_X, i_Y, k_Y and j_p are the usual natural maps.

In both cases we are able to factor u through $j_p : C(K) \to L_p(\mu)$, but at the expense of ending up in a space which is potentially much vaster than Y. The Π_p-factorization scheme is an extension of the scheme for $\mathcal{I}_p$: $j(y^{**}) = \big(\langle y^{**}, y^* \rangle\big)_{y^* \in B_{Y^*}}$ gives an isometric embedding $j : Y^{**} \to \ell_\infty^{B_{Y^*}}$ such that $j\,k_Y = i_Y$.

These schemes also grant us good heuristic insight into one major difference between p-integral and p-summing operators. The injectivity property of p-summing operators (2.5) is manifestly related to the injectivity of $\ell_\infty^{B_{Y^*}}$. In general, there is no reason to ascribe to Y^{**} any such pleasant extension properties; in fact, so long as $p \neq 2$, examples of p-summing operators that are not p-integral are often the harbingers of deep analytic phenomena. The easiest case is when $p = 1$.

5.11 Example: *The natural inclusion map $i : \ell_1 \to \ell_2$ is 1-summing but not 1-integral.*

That i is 1-summing needs no more explanation. Were it to be 1-integral it would have to have a factorization

$$i : \ell_1 \xrightarrow{\ b\ } L_\infty(\mu) \xrightarrow{\ i_1\ } L_1(\mu) \xrightarrow{\ a\ } \ell_2 .$$

But $i_1 b$ and a are both 1-summing, the first because i_1 is, and the second thanks to Grothendieck's Theorem (3.4). By a remark following 2.17, $i = a i_1 b$ would manifest itself as a compact operator, a plain affront to the fact that $\|e_m - e_n\|_2 = \sqrt{2}$ for $m \neq n$.

The next proposition shows that such a simple example is insufficient to distinguish p-integral and p-summing operators for $p > 1$.

5.12 Proposition: *If $1 < p < \infty$, then $\mathcal{L}(\ell_1, \ell_2) = \mathcal{I}_p(\ell_1, \ell_2)$.*

Proof. We shall use standard properties of the Rademacher functions and the lifting property of ℓ_1 (see after 3.17).

Let $R_p : \ell_2 \longrightarrow L_p[0,1] : (a_n) \mapsto \sum_n a_n r_n$ be the usual Rademacher embedding (Theorem 1.12), and write $i_p : L_\infty[0,1] \to L_p[0,1]$ and $h_p : L_p[0,1] \to L_1[0,1]$ for the natural maps. As we have already observed on several occasions, $i_p = h_{p^*}^*$. Consequently, when we dualize the factorization

$$R_1 : \ell_2 \xrightarrow{R_{p^*}} L_{p^*}[0,1] \xrightarrow{h_{p^*}} L_1[0,1],$$

we obtain

$$R_1^* : L_\infty[0,1] \xrightarrow{i_p} L_p[0,1] \xrightarrow{R_{p^*}^*} \ell_2.$$

Now this map R_1^* is onto since R_1 is an isomorphic embedding, and is p-integral since i_p is p-integral. So, if we apply the lifting property of ℓ_1, we see that any operator $u : \ell_1 \to \ell_2$ factors as $u : \ell_1 \to L_\infty[0,1] \overset{R_1^*}{\to} \ell_2$ and hence must be p-integral. QED

Coincidence results of this sort are a recurring theme throughout this text.

If we are to distinguish p-summing from p-integral operators $1 < p < \infty$, $p \neq 2$, we must flex our intellectual muscles in a more serious way.

5.13 Example: Our example will be based on a familiar p-summing operator: the canonical map $\mathcal{C}(K) \to L_p(\mu)$. We assume all scalars to be complex.

We take K to be the discrete set $\{1,..., N\}$ and take $\mu(\{k\}) = 1/N$ for each $1 \leq k \leq N$. Of course, $\mathcal{C}(K)$ and $L_p(\mu)$ are the same vector space and the canonical map above is the identity.

It is important to observe that the N-dimensional vector space $\mathcal{C}(K)$ has a basis consisting of functions e_m defined by $e_m(k) = \exp(2\pi ikm/N)$ where $1 \leq m \leq N$.

To see this, it is enough to establish linear independence. Observe that

$$\sum_k \Big|\sum_m c_m e_m(k)\Big|^2 = \sum_{k,m,n} c_m \bar{c}_n e_m(k)\overline{e_n(k)}$$
$$= \sum_{m,n} c_m \bar{c}_n \Big(\sum_k \exp(2\pi i(m-n)k/N)\Big).$$

Now if w is any N-th root of unity other than 1, the relation $w^N - 1 = 0$ yields that $\sum_k w^k = 0$. Consequently

$$\sum_k \Big|\sum_m c_m e_m(k)\Big|^2 = \sum_{m,n} c_m \cdot \bar{c}_n \cdot N \cdot \delta_{m,n} = N \cdot \sum_m |c_m|^2.$$

Were $\sum_m c_m e_m$ to be the zero function, we would thus have $\sum_m |c_m|^2 = 0$, and so each c_m would be zero.

The point of all this is that to each $f \in \mathcal{C}(K)$ we can associate unique scalars $\hat{f}(m)$ such that

$$f = \sum_m \hat{f}(m) e_m.$$

For $\Lambda \subset \{1,..., N\}$ we now introduce subspaces

$$\mathcal{C}_\Lambda := \{f \in \mathcal{C}(K) : f = \sum_{m\in\Lambda} \hat{f}(m)e_m\}$$

of $\mathcal{C}(K)$, and

$$L_{p,\Lambda} := \{f \in L_p(\mu) : f = \sum_{m\in\Lambda} \hat{f}(m)e_m\}$$

of $L_p(\mu)$. Let $u_\Lambda : \mathcal{C}_\Lambda \to L_{p,\Lambda}$ be the identity. We know that $\pi_p(u_\Lambda) \le 1$ since u_Λ is derived by restriction from the canonical map $\mathcal{C}(K) \to L_p(\mu)$. Our aim is to show that $\iota_p(u_\Lambda)$ may be as large as we please, provided we are given the freedom to choose suitable Λ and N.

To begin, the finite rank operator u_Λ admits a p-integral factorization

$$\mathcal{C}_\Lambda \xrightarrow{\ i\ } \mathcal{C}(K) \xrightarrow{\ j_p\ } L_p(\nu) \xrightarrow{\ \tilde{u}\ } L_{p,\Lambda}$$

where ν is an appropriate probability measure on $K(!)$, i and j_p are the natural maps, and $\iota_p(u) = \|\tilde{u}\|$. It turns out that the original measure μ can be expressed in terms of ν.

To understand why, it is necessary to introduce the *translates* f_k of functions f in $\mathcal{C}(K)$. For $k \in \mathbf{Z}$, we define

$$f_k(n) := f(k+n),$$

where the addition $k+n$ is performed modulo N. By the nature of the e_m's, it is clear that $f_k = \sum_m \hat{f}(m) \cdot \exp(2\pi i k m/N) \cdot e_m$, so the map $f \mapsto f_k$ is an isometric isomorphism on any of the spaces $\mathcal{C}(K)$, $\mathcal{C}_\Lambda$, $L_p(\mu)$ and $L_{p,\Lambda}$.

We use the translates of functions to define translates ν_k of the measure ν:

$$\langle \nu_k, f \rangle := \langle \nu, f_k \rangle.$$

These measures are evidently probability measures. Also

$$\mu = \frac{1}{N} \cdot \sum_{k=1}^{N} \nu_k,$$

since

$$\begin{aligned} \mu(\{m\}) &= \frac{1}{N} = \frac{1}{N} \cdot \nu(K) = \frac{1}{N} \cdot \sum_{k=1}^{N} \nu(\{k\}) \\ &= \frac{1}{N} \cdot \sum_{k=1}^{N} \nu(\{k+m\}) = \frac{1}{N} \cdot \sum_{k=1}^{N} \nu_k(\{m\}). \end{aligned}$$

Our new formulation of μ will enable us to show that the natural projection $L_p(\mu) \to L_{p,\Lambda}$ has norm $\iota_p(u_\Lambda)$, and this will lead via Orlicz's Theorem 3.12 to the desired conclusion.

To this end, consider the map

$$w : \mathcal{C}(K) \longrightarrow L_{p,\Lambda} : f \mapsto \frac{1}{N} \cdot \sum_{k=1}^{N} (v f_k)_{-k}$$

where $v := \tilde{u}\, j_p$, with $\tilde{u}$ and j_p from u_Λ's factorization. By our representation for μ,

$$\|f\|_{L_p(\mu)}^p = \langle \mu, |f|^p \rangle = \frac{1}{N} \cdot \sum_{k=1}^{N} \langle \nu_k, |f|^p \rangle.$$

However, as $\|g_k\|_{L_p(\mu)} = \|g\|_{L_p(\mu)}$ for every $g \in L_p(\mu)$ and every $1 \le k \le N$, we discover that

$$\begin{aligned}\|wf\|_{L_{p,\Lambda}} &\leq \frac{1}{N}\cdot\sum_{k=1}^{N}\|vf_k\|_{L_{p,\Lambda}} \leq \|\tilde{u}\|\cdot\frac{1}{N}\cdot\sum_{k=1}^{N}\|j_p f_k\|_{L_p(\nu)}\\ &= \iota_p(u_\Lambda)\cdot\frac{1}{N}\cdot\sum_{k=1}^{N}\langle\nu,|f_k|^p\rangle^{1/p} = \iota_p(\mu_\Lambda)\cdot\frac{1}{N}\cdot\sum_{k=1}^{N}\langle\nu_k,|f|^p\rangle^{1/p}\\ &\leq \iota_p(u_\Lambda)\cdot\Big(\frac{1}{N}\cdot\sum_{k=1}^{N}\langle\nu_k,|f|^p\rangle\Big)^{1/p} = \iota_p(u_\Lambda)\cdot\|f\|_{L_p(\mu)}.\end{aligned}$$

What this tells us is that we may consider w to be an operator $L_p(\mu) \to L_{p,\Lambda}$ with norm at most $\iota_p(u_\Lambda)$. Actually, w is the natural projection.

Identify C_Λ with $L_{p,\Lambda}$ in the natural algebraic way. If $f \in L_{p,\Lambda}$, then

$$wf = \frac{1}{N}\cdot\sum_{k=1}^{N}(vf_k)_{-k} = \frac{1}{N}\cdot\sum_{k=1}^{N}(f_k)_{-k} = f.$$

It remains to show that $we_n = 0$ for $n \notin \Lambda$.

For this, we remark that $(e_n)_k = \exp(2\pi ikn/N)e_n = e_n(k)e_n$ and

$$w(f_k) = \frac{1}{N}\cdot\sum_{n}(vf_{k+n})_{-n} = \frac{1}{N}\cdot\sum_{n}(vf_n)_{-n+k} = (wf)_k.$$

All this reveals that

$$(we_n)(k) = (we_n)(N+k) = (we_n)_k(N) = w((e_n)_k)(N) = (we_n)(N)\cdot e_n(k),$$

and so we_n is a multiple of e_n for each n. In particular, if $n \notin \Lambda$, then $we_n \in L_{p,\Lambda}$ must be zero.

Now that we have seen that the natural projection $L_p(\mu) \to L_{p,\Lambda}$ has norm at most $\iota_p(u_\Lambda)$, we are in a position to find lower bounds for $\iota_p(u_\Lambda)$.

First, assume $1 \leq p < 2$. Select $f = \sum_n \hat{f}(n)e_n \in L_p(\mu)$ and signs $\varepsilon_n = \pm 1$ $(1 \leq n \leq N)$. Writing

$$\Lambda_\varepsilon = \{n : \varepsilon_n = -1\},$$

we find

$$\begin{aligned}\Big\|\sum_{n=1}^{N}\varepsilon_n\hat{f}(n)e_n\Big\|_{L_p(\mu)} &= \Big\|f - 2\cdot\sum_{n\in\Lambda_\varepsilon}\hat{f}(n)e_n\Big\|_{L_p(\mu)}\\ &\leq \|f\|_{L_p(\mu)} + 2\cdot\Big\|\sum_{n\in\Lambda_\varepsilon}\hat{f}(n)e_n\Big\|_{L_p(\mu)} \leq \big(1+2\cdot\iota_p(u_{\Lambda_\varepsilon})\big)\cdot\|f\|_{L_p(\mu)}.\end{aligned}$$

It follows that $(1+2\cdot\iota_p(u_{\Lambda_\varepsilon}))\cdot\|f\|_{L_p(\mu)}$ is an upper bound for $\|(\hat{f}(n)e_n)_{n=1}^N\|_1^{weak}$. Since

$$\big\|\big(\hat{f}(n)e_n\big)_{n=1}^{N}\big\|_1^{weak} = \sup_{\varepsilon_n=\pm1}\Big\|\sum_{n=1}^{N}\varepsilon_n\hat{f}(n)e_n\Big\|_{L_p(\mu)},$$

Orlicz's Theorem 3.12 leads to

$$\begin{aligned}\|f\|_{L_2(\mu)} &= \Big\|\sum_{n=1}^{N} \hat{f}(n)e_n\Big\|_{L_2(\mu)} = \Big(\sum_{n=1}^{N} \|\hat{f}(n)e_n\|^2_{L_2(\mu)}\Big)^{1/2}\\ &= \Big(\sum_{n=1}^{N} \|\hat{f}(n)e_n\|^2_{L_p(\mu)}\Big)^{1/2} \le K_p \cdot \sup_{\varepsilon_n=\pm 1} \Big\|\sum_{n=1}^{N} \varepsilon_n \hat{f}(n)e_n\Big\|_{L_p(\mu)}\\ &\le K_p \cdot \sup_{\Lambda} \big(1+2\cdot\iota_p(u_\Lambda)\big)\cdot\|f\|_{L_p(\mu)}\end{aligned}$$

for some absolute constant K_p, depending only on p.

Taking $f(n) = N^{1/p}\cdot\delta_{n,1}$ $(1 \le n \le N)$ we get $\|f\|_{L_p(\mu)} = 1$, but $\|f\|_{L_2(\mu)} = N^{(1/p)-(1/2)}$, from which we derive

$$\sup_{\Lambda} \iota_p(u_\Lambda) \ge K'_p \cdot N^{(1/p)-(1/2)}$$

for some absolute constant K'_p, depending of course on K_p.

To obtain information when $p > 2$, notice that $\iota_p(u_\Lambda)$ is an upper bound for the norm of the natural projection $L_p(\mu) \to L_{p,\Lambda}$. However, the adjoint of this map is just the natural projection $L_{p^*}(\mu) \to L_{p^*,\Lambda}$, and since $|(1/p)-(1/2)| = |(1/p^*)-(1/2)|$ we find that, regardless of $1 \le p \le \infty$,

$$\sup_{\Lambda} \iota_p(u_\Lambda) \ge K'_p \cdot N^{|(1/p)-(1/2)|}.$$

To sum up: when $N \ge 2$, we can find a set $\Lambda_N \subset \{1,..., N\}$ with the property that $u_N := u_{\Lambda_N}$ satisfies $\pi_p(u_N) \le 1$ but $\iota_p(u_N) \ge K'_p \cdot N^{|(1/p)-(1/2)|}$.

When $p \ne 2$, the construction of infinite dimensional Banach spaces X and Y for which $\Pi_p(X,Y) \ne \mathcal{I}_p(X,Y)$ is now straightforward. Just set $X := (\oplus_N \mathcal{C}_{\Lambda_N})_2$ and $Y := (\oplus_N L_{p,\Lambda_N})_2$. For either space, P_N will denote the natural projection onto the N-th coordinate and J_N will be the corresponding injection. If $\bar{u}_N : X \overset{P_N}{\to} \mathcal{C}_{\Lambda_N} \overset{u_N}{\to} L_{p,\Lambda_N} \overset{J_N}{\to} Y$, then $u_N : \mathcal{C}_{\Lambda_N} \overset{J_N}{\to} X \overset{\bar{u}_N}{\to} Y \overset{P_N}{\to} L_{p,\Lambda_N}$, so $\pi_p(\bar{u}_N) = \pi_p(u_N)$ and $\iota_p(\bar{u}_N) = \iota_p(u_N)$. We have constructed a sequence of operators $\bar{u}_N : X \to Y$ with $\pi_p(\bar{u}_N) \le 1$ and $\iota_p(\bar{u}_N) \to \infty$, and the Closed Graph Theorem intervenes to forbid equality of $\Pi_p(X,Y)$ and $\mathcal{I}_p(X,Y)$.

Further Structural Results

Let us continue by developing some structure. Both similarities and differences crop up when we compare p-integral operators with the p-summing ones. First, a similarity.

5.14 Theorem: *Let $1 \le p \le \infty$. An operator $u : X \to Y$ is p-integral precisely when its second adjoint is. In fact, the following are equivalent:*

(i) $u : X \to Y$ *is p-integral.*

(ii) $u^{**} : X^{**} \to Y^{**}$ *is p-integral.*

(iii) $k_Y u : X \to Y^{**}$ *is p-integral.*

Moreover, $\iota_p(u) = \iota_p(u^{**}) = \iota_p(k_Y u)$.

Proof. We shall only carry out the proof for $1 \leq p < \infty$, but we urge the reader to make the simple natural adjustments necessary to treat the case $p = \infty$.

We show that (i)⇒(ii)⇒(iii)⇒(i), establishing along the way that $\iota_p(u) \geq \iota_p(u^{**}) \geq \iota_p(k_Y u) \geq \iota_p(u)$.

(i)⇒(ii): Let $u \in \mathcal{I}_p(X,Y)$ and let $k_Y u : X \xrightarrow{b} L_\infty(\mu) \xrightarrow{i_p} L_p(\mu) \xrightarrow{a} Y^{**}$ be a typical p-integral factorization. We take second adjoints of this composition. As in 2.19, ${i_p}^{**}$ can and will be regarded as the composition of $i_p : L_\infty(\mu) \to L_p(\mu)$ and the natural norm one projection $P : L_\infty(\mu)^{**} \to L_\infty(\mu)$. Then

$$u^{**} : X^{**} \xrightarrow{Pb^{**}} L_\infty(\mu) \xrightarrow{i_p} L_p(\mu) \xrightarrow{a} Y^{**}$$

identifies u^{**} as p-integral and discloses that $\iota_p(u^{**}) \leq \|Pb^{**}\| \cdot \|a\| \leq \|a\| \cdot \|b\|$, from which $\iota_p(u^{**}) \leq \iota_p(u)$.

(ii)⇒(iii): This is an easy consequence of the ideal property. Just note that $k_Y u = u^{**} k_X$ to see that $\iota_p(k_Y u) \leq \iota_p(u^{**})$.

(iii)⇒(i): First recollect that if $k_Y u$ is p-integral it is strictly p-integral, by the nature of its range. This allows us, for each $\varepsilon > 0$, a factorization $k_Y u : X \xrightarrow{b} L_\infty(\mu) \xrightarrow{i_p} L_p(\mu) \xrightarrow{a} Y^{**}$, with μ a probability measure and $\|a\| \cdot \|b\| \leq \iota_p(k_Y u) + \varepsilon$. So, u is visibly p-integral, with $\iota_p(u) \leq \iota_p(k_Y u) + \varepsilon$. We are done, after passing with ε to zero. QED

For $1 < p < \infty$, the adjoint of a p-integral operator need not be p-integral. In fact, we know from Remark 2.20 that it need not even be r-summing for any r. The exclusion of the case $p = 1$ is conspicuous and essential:

5.15 Theorem: *A Banach space operator* $u : X \to Y$ *is* 1*-integral if and only if its adjoint* $u^* : Y^* \to X^*$ *is; in this case,* $\iota_1(u^*) = \iota_1(u)$.

Proof. Thanks to the theorem we just proved it is enough to show that when $u \in \mathcal{I}_1(X,Y)$, we have $u^* \in \mathcal{I}_1(Y^*, X^*)$ and $\iota_1(u^*) \leq \iota_1(u)$.

For $u \in \mathcal{I}_1(X,Y)$, take a typical 1-integral factorization

$$k_Y u : X \xrightarrow{b} L_\infty(\mu) \xrightarrow{i_1} L_1(\mu) \xrightarrow{a} Y^{**}.$$

Note that $i_1^* = k_{L_1(\mu)} i_1$ and $id_{Y^*} = (k_Y)^* k_{Y^*}$ to understand that there is a decomposition

$$u^* : Y^* \xrightarrow{k_{Y^*}} Y^{***} \xrightarrow{a^*} L_\infty(\mu) \xrightarrow{i_1} L_1(\mu) \xrightarrow{k_{L_1(\mu)}} L_1(\mu)^{**} \xrightarrow{b^*} X^*.$$

This contains the information that u^* is 1-integral with

$$\iota_1(u^*) \leq \|a^* k_{Y^*}\| \cdot \|b^* k_{L_1(\mu)}\| \leq \|a\| \cdot \|b\|.$$

It follows that $\iota_1(u^*) \leq \iota_1(u)$. QED

Building on the Composition Theorem (2.22) for summing operators, we derive a similar, but stronger, result using our new ingredients.

5.16 Theorem: *Let $1 \leq p, q, r \leq \infty$ satisfy $1/r = (1/p) + (1/q)$, and consider operators $v : X \to Y$ and $u : Y \to Z$ between Banach spaces.*

(a) *If u is p-summing and v is q-integral, then uv is r-integral, with $\iota_r(uv) \leq \pi_p(u) \cdot \iota_q(v)$.*

(b) *If u is p-integral and v is q-summing, then uv is r-integral, with $\iota_r(uv) \leq \iota_p(u) \cdot \pi_q(v)$.*

As always, we interpret $\pi_\infty(\cdot)$ as $\|\cdot\|$.

Proof. (a) We piece together a typical q-integral factorization of $k_Y v$, provided by Theorem 5.6, and the by now familiar identity $u^{**}k_Y = k_Z u$. Here is the resulting (commutative) picture:

$$\begin{array}{ccccccc} X & \xrightarrow{\quad v \quad} & & & Y & \xrightarrow{u} & Z \\ \downarrow{\scriptstyle i_X} & & & & \downarrow{\scriptstyle k_Y} & & \downarrow{\scriptstyle k_Z} \\ C(K) & \xrightarrow{j_q} & L_q(\mu) & \xrightarrow{\tilde{v}} & Y^{**} & \xrightarrow{u^{**}} & Z^{**} . \end{array}$$

By Theorem 5.14, we only need to prove that $k_Z uv$ is r-integral and that $\iota_r(k_Z uv) \leq \pi_p(u) \cdot \iota_q(v)$. Recall Proposition 2.19: u^{**} is p-summing with $\pi_p(u^{**}) = \pi_p(u)$. We also know that j_q has q-summing norm one; thus the Composition Theorem 2.22 for summing operators ensures that $u^{**}\tilde{v}j_q : C(K) \to Z^{**}$ is r-summing, and so, by Corollary 5.8, r-integral with

$$\iota_r(u^{**}\tilde{v}j_q) = \pi_r(u^{**}\tilde{v}j_q) \leq \pi_p(u^{**}) \cdot \pi_q(\tilde{v}j_q) \leq \pi_p(u) \cdot \|\tilde{v}\|.$$

Hence $k_Z uv$ is r-integral and $\iota_r(k_Z uv) \leq \pi_p(u) \cdot \|\tilde{v}\|$. The result follows at once.

(b) Examine again a typical factorization

$$\begin{array}{ccccccc} X & \xrightarrow{v} & Y & \xrightarrow{u} & Z & \xrightarrow{k_Z} & Z^{**} \\ & & \downarrow{\scriptstyle b} & & & & \uparrow{\scriptstyle a} \\ & & L_\infty(\mu) & & \xrightarrow{\quad i_p \quad} & & L_p(\mu) \end{array}$$

and notice that $bv : X \to L_\infty(\mu)$ is q-integral, since v is q-summing and $L_\infty(\mu)$ is injective (see 5.7). Now ai_p is p-summing (since i_p is), so we can revert to case (a) to see that $k_Z uv = ai_p bv$ is r-integral and

$$\iota_r(k_Z uv) \leq \pi_p(ai_p) \cdot \iota_q(bv) \leq \|a\| \cdot \|b\| \cdot \pi_q(v).$$

Once more, the result is immediate. QED

In part (a) it is even true that uv is strictly r-integral, since, as u factors through a reflexive space, u^{**} maps Y^{**} into Z.

Moreover, if u is strictly p-integral and v is q-summing, then uv is strictly r-integral, as the proof of (b) clearly indicates.

The composition theorem we have just proved will be especially important when $r = 1$, that is, when p and q are conjugate indices.

ORDER BOUNDEDNESS

The $L_p(\mu)$-spaces enjoy a fundamental structure that we have largely ignored so far: they are Banach lattices. We now use this extra structure to throw more light on the nature of certain p-summing and p-integral operators.

To enter the discussion, we once again compare and contrast $\ell_p^{weak}(X)$ and $\ell_p^{strong}(X)$. Recall from 2.2 that $\ell_p^{weak}(X)$, the space of all weakly p-summable sequences in the Banach space X, is isometrically isomorphic to $\mathcal{L}(\ell_{p^*}, X)$ when $1 < p < \infty$, and to $\mathcal{L}(c_0, X)$ when $p = 1$. The isomorphism is obtained by associating with $(x_n) \in \ell_p^{weak}(X)$ the operator v from ℓ_{p^*} (or c_0) to X given by $ve_n = x_n$ for all n. The relevant consequence for our present purposes is a triviality: $v^*(B_{X^*})$ is norm bounded in ℓ_p; so there is an $A > 0$ for which $\|v^*x^*\| = (\sum |\langle x^*, x_n\rangle|^p)^{1/p} \leq A$ whenever $x^* \in B_{X^*}$. What more can we say if $(x_n) \in \ell_p^{strong}(X)$?

5.17 Proposition: *Let $1 \leq p \leq \infty$. Then (x_n) belongs to $\ell_p^{strong}(X)$ if and only if there exists an $a = (a_n) \in \ell_p$ such that $|\langle x^*, x_n\rangle| \leq a_n$ for each $x^* \in B_{X^*}$ and each $n \in \mathbf{N}$.*

In other terms, we require $|v^*x^*| \leq a$ to hold pointwise.

Proof. All we need to do is to show that $(\sup_{x^* \in B_{X^*}} |\langle v^*x^*, e_n\rangle|)_{n=1}^{\infty}$ is in ℓ_p if and only if (x_n) is in $\ell_p^{strong}(X)$. But this is very simple: for each n, $\sup_{x^* \in B_{X^*}} |\langle v^*x^*, e_n\rangle| = \sup_{x^* \in B_{X^*}} |\langle x^*, x_n\rangle| = \|x_n\|$. QED

We translate this into lattice language. Let (Ω, Σ, μ) be any measure space, and let $1 \leq p \leq \infty$. The non-empty subset M of $L_p(\mu)$ is *order bounded* if there is a non-negative $h \in L_p(\mu)$ such that $|f| \leq h$ μ-almost everywhere for each $f \in M$. Two properties are worth singling out.

- When $M \subset L_p(\mu)$ is order bounded, the set $\{|f| : f \in M\}$ has a supremum in the lattice $L_p(\mu)$.
- If the non-empty subset M of $L_p(\mu)$ consists of non-negative functions, and if it is directed upwards and norm bounded, then it must be order bounded. Moreover, $\|\sup_M f\|_p = \sup_M \|f\|_p$.

We focus our attention on an operator version of this. A Banach space operator $u : X \to L_p(\mu)$ is *order bounded* if $u(B_X)$ is an order bounded subset of $L_p(\mu)$. Of course, this definition makes sense with $L_p(\mu)$ replaced by any Banach lattice but we will only use this in Chapter 16. In our new terminology, Proposition 5.17 says that

- *(x_n) belongs to $\ell_p^{strong}(X)$ if and only if $X^* \to \ell_p : x^* \mapsto (\langle x^*, x_n\rangle)_n$ is an order bounded operator.*

It is easy to see that this operator is then also p-integral. This reveals a special case of a general phenomenon.

5.18 Proposition: *Let $1 \le p \le \infty$. Then any order bounded operator $u : X \to L_p(\mu)$ is p-integral, and $\iota_p(u) \le \| \sup_{x\in B_X} |ux| \|_p$.*

Proof. Disregarding the uninteresting case $u = 0$, we can assume that $h := \sup_{x\in B_X} |ux|$ exists as a non-zero element in $L_p(\mu)$. To obtain a p-integral factorization scheme for u, we introduce a probability measure ν on (Ω, Σ) by $d\nu := (h/\|h\|_p)^p d\mu$. Now $w : X \to L_\infty(\nu) : x \mapsto u(x)/h$ and $v : L_p(\nu) \to L_p(\mu) : f \mapsto fh$ are well-defined operators with $\|w\| \le 1$ and $\|v\| \le \|h\|_p$. Since the operator u is nothing other than the composition $X \stackrel{w}{\to} L_\infty(\nu) \stackrel{i_p}{\to} L_p(\nu) \stackrel{v}{\to} L_p(\mu)$, it has to be p-integral and must satisfy $\iota_p(u) \le \|v\| \cdot \|w\| \le \| \sup_{x\in B_X} |ux| \|_p$. QED

Have we stumbled on a new characterization of p-integral operators $X \to L_p(\mu)$? In case $1 < p < \infty$, the answer is negative. We shall see in 5.20 that if this happened, every p-integral operator $u : X \to L_p(\mu)$ would have a p-summing adjoint, and this is easily seen to be false. For an example, consider the formal identity $i_p : L_\infty[0,1] \to L_p[0,1]$. Its adjoint is the formal identity $L_{p^*}[0,1] \to L_1[0,1]$, regarded as a map with values in $L_1[0,1]^{**}$. This map cannot be p-summing since then its composition with i_{p^*} would render the formal identity $L_\infty[0,1] \to L_1[0,1]$ compact.

When $p = 1$, the situation changes.

5.19 Theorem: *An operator $u : X \to L_1(\mu)$ is 1-integral if and only if it is order bounded, and then $\iota_1(u) = \| \sup_{x\in B_X} |ux| \|_1$.*

Proof. Thanks to the last proposition, we only have to prove that if u is 1-integral, then it is order bounded with $\| \sup_{x\in B_X} |ux| \|_1 \le \iota_1(u)$.

Recall that $L_1(\mu)$ is norm one complemented in its bidual. Hence any $u \in \mathcal{I}_1(X, L_1(\mu))$ is even strictly integral and so admits a factorization $u : X \stackrel{w}{\to} L_\infty(\nu) \stackrel{i_1}{\to} L_1(\nu) \stackrel{v}{\to} L_1(\mu)$ where ν is a probability measure; moreover, $\iota_1(u)$ is the infimum of $\|v\| \cdot \|w\|$ taken over all such factorizations.

In order to show that $u(B_X)$ is order bounded, we examine the set M of all functions $f \in L_1(\mu)$ which have a representation as a finite sum $f = \sum_{k\le m} |ux_k| \cdot 1_{A_k}$ where $x_1,\dots, x_m$ are taken from B_X and $A_1,\dots, A_m$ are pairwise disjoint members of Σ having finite μ-measure. It is clearly enough to show that M is order bounded, and to do this, we shall prove that M is directed upwards and norm bounded.

Directedness is straightforward. Take two typical members of M, say

$$f = \sum_{k\le m} |ux_k| \cdot 1_{A_k} \quad \text{and} \quad g = \sum_{\ell\le n} |uy_\ell| \cdot 1_{B_\ell}.$$

Now define $C_{k\ell} := \{|ux_k| > |uy_\ell|\} \cap A_k \cap B_\ell$ and $D_{k\ell} := (A_k \cap B_\ell) \setminus C_{k\ell}$ for each $1 \leq k \leq m$ and $1 \leq \ell \leq n$. Manifestly, $\sum_{k,\ell}(|ux_k| \cdot 1_{C_{k\ell}} + |uy_\ell| \cdot 1_{D_{k\ell}})$ belongs to M and dominates both f and g.

Norm boundedness is a bit more delicate; we are going to prove that $\iota_1(u)$ is a suitable bound. To this end, we fix a typical 1-integral factorization $u : X \overset{w}{\to} L_\infty(\nu) \overset{i_1}{\to} L_1(\nu) \overset{v}{\to} L_1(\mu)$ where ν is a probability measure and show that $\|v\| \cdot \|w\|$ is a bound for the norm of every function in M.

Fix $\sum_{k\leq m} |ux_k| \cdot 1_{A_k}$ in M, and for each $1 \leq k \leq m$ define g_k in $L_\infty(\mu)$ by $g_k := (\operatorname{sign} ux_k) \cdot 1_{A_k}$. For any $a = (a_k)_1^m \in B_{\ell_\infty^m}$, we have ν-almost everywhere

$$
\begin{aligned}
\Big|\sum_{k\leq m} a_k v^* g_k\Big| &= \Big|v^*\Big(\sum_{k\leq m} a_k g_k\Big)\Big| \leq \Big\|v^*\Big(\sum_{k\leq m} a_k g_k\Big)\Big\|_{L_\infty(\nu)} \\
&\leq \|v^*\| \cdot \Big\|\sum_{k\leq m} a_k g_k\Big\|_{L_\infty(\mu)} \leq \|v\|.
\end{aligned}
$$

We can make this pointwise estimate a uniform estimate by allowing a to range over a countable dense subset D of $B_{\ell_\infty^m}$. Then, except on a set of ν-measure zero, $|\sum_{k\leq m} a_k v^* g_k| \leq \|v\|$ for every $a \in D$, which implies

$$
\Big\|\sum_{k\leq m} |v^* g_k|\Big\|_{L_\infty(\nu)} \leq \|v\|.
$$

A few computations settle the issue:

$$
\begin{aligned}
\Big\|\sum_{k\leq m} |ux_k| \cdot 1_{A_k}\Big\|_{L_1(\mu)} &= \sum_{k\leq m} \int u(x_k) g_k \, d\mu = \sum_{k\leq m} \langle g_k, ux_k\rangle \\
&= \sum_{k\leq m} \langle g_k, v i_1 w x_k\rangle = \sum_{k\leq m} \langle v^* g_k, i_1 w x_k\rangle \\
&= \sum_{k\leq m} \int v^*(g_k) w(x_k) \, d\nu \leq \sum_{k\leq m} \Big(\int |v^*(g_k)| \, d\nu\Big) \cdot \|w(x_k)\|_{L_\infty(\nu)} \\
&\leq \|w\| \cdot \int \sum_{k\leq m} |v^*(g_k)| \, d\nu \leq \|v\| \cdot \|w\|.
\end{aligned}
$$

QED

Order boundedness yields an important characterization of Banach space operators whose duals are p-summing.

5.20 Theorem: *Let $1 \leq p < \infty$. An operator $u : X \to Y$ between Banach spaces has p-summing adjoint if and only if, however we choose a measure μ and an operator $w : Y \to L_p(\mu)$, the composition $wu : X \to L_p(\mu)$ is order bounded.*

Proof. Necessity will be an immediate consequence of the ideal property of p-summing operators, provided we can show that any operator $u : X \to L_p(\mu)$ with a p-summing adjoint is order bounded.

If $p = 1$ this follows from 5.8, 5.15 and 5.19.

In case $1 < p < \infty$, the proof runs along much the same lines as that of Theorem 5.19. We aim to show that the set M of all members of $L_p(\mu)$ of the form $\sum_{k\le m} |ux_k|\cdot 1_{A_k}$, with $x_1,\ldots, x_m$ chosen from B_X and $A_1,\ldots, A_m$ pairwise disjoint sets of finite μ-measure, is directed upwards and norm bounded.

The directedness is obtained by a mere repetition of the corresponding arguments from 5.19. The norm boundedness results from the p-summing nature of u^*. To compute the norm of a typical member $\sum_{k\le m} |ux_k|\cdot 1_{A_k}$ of M, first choose, for each $1 \le k \le m$, a function $g_k \in L_{p^*}(\mu)$ with $\|g_k\|_{p^*} = 1$ so that

$$\|u(x_k)1_{A_k}\|_p = \int_{A_k} u(x_k)g_k\,d\mu = \langle ux_k, 1_{A_k}g_k\rangle = \langle x_k, u^*(1_{A_k}g_k)\rangle.$$

By the disjointness of the A_k's, we get

$$\Big\|\sum_{k\le m} |ux_k|\cdot 1_{A_k}\Big\|_p = \Big(\sum_{k\le m} \|u(x_k)\cdot 1_{A_k}\|_p^p\Big)^{1/p} = \Big(\sum_{k\le m} |\langle x_k, u^*(1_{A_k}g_k)\rangle|^p\Big)^{1/p}$$
$$\le \Big(\sum_{k\le m} \|u^*(1_{A_k}g_k)\|^p\Big)^{1/p} \le \pi_p(u^*)\cdot\|(1_{A_k}g_k)_{k\le m}\|_p^{weak} \le \pi_p(u^*).$$

For the last inequality we used

$$\|(1_{A_k}g_k)_{k\le m}\|_p^{weak} = \sup\Big\{\Big\|\sum_{k\le m} \lambda_k 1_{A_k}g_k\Big\|_{p^*} : (\lambda_k) \in B_{\ell_{p^*}^m}\Big\}$$
$$= \sup\Big\{\Big(\sum_{k\le m} |\lambda_k|^{p^*}\cdot\int |1_{A_k}g_k|^{p^*}\,d\mu\Big)^{1/p^*} : (\lambda_k) \in B_{\ell_{p^*}^m}\Big\} \le 1.$$

Now we proceed to the proof of the other implication. Assume that $wu : X \to L_p(\mu)$ is order bounded for any choice of the measure μ and of the operator $w : Y \to L_p(\mu)$. We wish to show that $u^* : Y^* \to X^*$ is p-summing.

Accordingly, we fix a sequence (y_n^*) in $\ell_p^{weak}(Y^*)$ and, in the standard way, we let $v : \ell_{p^*} \to Y^*$ (or $v : c_0 \to Y^*$ if $p = 1$) be the operator defined by $ve_n = y_n^*$ for each n. By hypothesis, $v^*k_Y u = v^*u^{**}k_X : X \to \ell_p$ is order bounded, and so there must be an element $a = (a_n) \in \ell_p$ such that

$$|\langle v^*u^{**}x, e_n\rangle| \le a_n$$

for each n and for every $x \in B_X$. However, each $x^{**} \in B_{X^{**}}$ is the weak$*$limit of a net (x_α) from B_X and consequently satisfies

$$|\langle v^*u^{**}(x^{**}), e_n\rangle| = |\langle x^{**}, u^*ve_n\rangle| = \lim_\alpha |\langle x_\alpha, u^*ve_n\rangle|$$
$$= \lim_\alpha |\langle v^*u^{**}x_\alpha, e_n\rangle| \le a_n$$

for each n. It follows that $v^*u^{**} = (u^*v)^*$ is order bounded and so $(u^*y_n^*) = (u^*ve_n)$ belongs to $\ell_p^{strong}(X^*)$, by Proposition 5.17. QED

From the proof of the first part of the preceding theorem, we can obtain a special result deserving special mention.

5.21 Corollary: *If the operator $u : X \to L_p(\mu)$ $(1 \leq p < \infty)$ has a p-summing adjoint, then u must be order bounded, and $\pi_p(u^*)$ is an order bound for $u(B_X)$.*

In particular, incorporating 5.18, we find that u is p-integral with $\iota_p(u) \leq \pi_p(u^*)$. In case of a favourable duality situation, the converse is also true.

5.22 Corollary: *Let $1 \leq p < \infty$, let μ and ν be measures, and let $u : L_{p^*}(\mu) \to L_p(\nu)$ be any operator. The following are equivalent statements:*

(i) *u is p-integral.*

(ii) *u is p-summing.*

(iii) *u^* is p-integral.*

(iv) *u^* is p-summing.*

It is also easy to see that $\iota_p(u) = \pi_p(u) = \iota_p(u^*) = \pi_p(u^*)$ in such a case.

p-Nuclear Operators

We close this chapter with a somewhat detailed study of the discrete analogue of strictly p-integral operators, $1 \leq p < \infty$. Recall our basic example: if (Ω, Σ, μ) is a measure space (not necessarily finite) and if $\lambda \in L_p(\mu)$, then the multiplication operator $M_\lambda : L_\infty(\mu) \to L_p(\mu) : f \mapsto f \cdot \lambda$ is p-summing. Thanks to 5.8, a Banach space operator $u : X \to Y$ is strictly p-integral if and only if there are a measure space (Ω, Σ, μ), operators $b \in \mathcal{L}(L_p(\mu), Y)$ and $a \in \mathcal{L}(X, L_\infty(\mu))$, and a function $\lambda \in L_p(\mu)$ such that $u = bM_\lambda a$.

We shall be interested in the particular case when μ is counting measure on $\mathbf{N}$. Then $L_\infty(\mu)$ and $L_p(\mu)$ are the standard sequence spaces ℓ_∞ and ℓ_p, respectively, and

$$M_\lambda : \ell_\infty \longrightarrow \ell_p : (\xi_n)_n \mapsto (\lambda_n \xi_n)_n$$

is the *diagonal operator* induced by $\lambda = (\lambda_n)_n \in \ell_p$. Use 2.9(e) to see that

$$\|M_\lambda\| = \pi_p(M_\lambda) = \iota_p(M_\lambda) = \left(\sum_n |\lambda_n|^p\right)^{1/p} = \|\lambda\|_p.$$

A Banach space operator $u : X \to Y$ is called *p-nuclear* $(1 \leq p < \infty)$ if there are operators $a \in \mathcal{L}(\ell_p, Y)$, $b \in \mathcal{L}(X, \ell_\infty)$ and a sequence $\lambda \in \ell_p$ such that the following diagram commutes:

$$(*) \qquad \begin{array}{ccc} X & \xrightarrow{u} & Y \\ b \downarrow & & \uparrow a \\ \ell_\infty & \xrightarrow{M_\lambda} & \ell_p . \end{array}$$

We set

$$\nu_p(u) := \inf \|a\| \cdot \|M_\lambda\| \cdot \|b\|,$$

the infimum being extended over all factorizations as above. We denote by

$$\mathcal{N}_p(X,Y)$$

the collection of all p-nuclear operators from X into Y.

Frequently, 1-nuclear operators are just referred to as *nuclear operators.*

Before we prove that we have in fact encountered another scale of Banach ideals, let us derive some alternate descriptions which will it make more convenient to handle our newly defined objects.

5.23 Proposition: *Let $1 \leq p < \infty$. The following statements about the operator $u : X \to Y$ are equivalent:*

(i) *u is p-nuclear.*

(ii) *u admits a factorization $u : X \xrightarrow{b} c_0 \xrightarrow{M_\tau} \ell_p \xrightarrow{a} Y$, where a and b are compact and M_τ is the diagonal operator induced by $\tau \in \ell_p$.*

(iii) *The same as (ii), but with a and b only bounded.*

(iv) *There are sequences $(\lambda_n) \in \ell_p$, (x_n^*) in B_{X^*} and $(y_n) \in \ell_{p^*}^{weak}(Y)$ such that*

$$u = \sum_{n=1}^{\infty} \lambda_n x_n^* \otimes y_n,$$

and the series converges in $\mathcal{L}(X,Y)$.

(v) *There are sequences $(x_n^*) \in \ell_p^{strong}(X^*)$ and $(y_n) \in \ell_{p^*}^{weak}(Y)$ such that*

$$u = \sum_{n=1}^{\infty} x_n^* \otimes y_n,$$

and again the series converges in $\mathcal{L}(X,Y)$.

When u is p-nuclear,

$$\nu_p(u) = \inf \|a\| \cdot \|M_\tau\| \cdot \|b\|$$

where the infimum extends over all factorizations as in (ii) or (iii). Further

$$\nu_p(u) = \inf \{ (\sum |\lambda_n|^p)^{1/p} \cdot \|(y_n)\|_{p^*}^{weak} \},$$

with the infimum taken over all representations of u as in (iv), or, if you prefer,

$$\nu_p(u) = \inf \|(x_n^*)\|_p^{strong} \cdot \|(y_n)\|_{p^*}^{weak}$$

with all representations of u as in (v) accounted for when taking the infimum.

Proof. To see why (i) implies (ii) just recall that given any sequence $\lambda \in \ell_p$ we can find sequences $\alpha \in c_0$ and $\tau \in \ell_p$ such that $\lambda_n = \alpha_n \tau_n$ for each $n \in \mathbf{N}$; in addition, we may choose $\alpha_n \geq 0$ for each n and, for a given $\varepsilon > 0$, arrange to have $\|\tau\|_p \leq (1+\varepsilon) \cdot \|\lambda\|_p$. Note that $(\alpha_n^{1/2})_n \in c_0$, so this sequence can be used to define compact diagonal operators $b_0 : \ell_\infty \to c_0$ and $a_0 : \ell_p \to \ell_p$, and our diagram $(*)$ can be put in the form

$$u : X \xrightarrow{b} \ell_\infty \xrightarrow{b_0} c_0 \xrightarrow{M_\tau} \ell_p \xrightarrow{a_0} \ell_p \xrightarrow{a} Y.$$

Since $b_0 b$ and aa_0 are both compact we have the factorization we wanted. Moreover,

$$\nu_p(u) \leq \|aa_0\| \cdot \|\tau\|_p \cdot \|b_0 b\| \leq (1+\varepsilon) \cdot \|a\| \cdot \|\tau\|_p \cdot \|b\|.$$

This ensures that $\nu_p(u) \leq \inf \|a\| \cdot \|M_\tau\| \cdot \|b\|$, where a, τ and b are as in (ii).

Thanks to 2.2, the remaining implications are straightforward, and so are the assertions about $\nu_p(u)$. QED

5.24 Corollary: *Regardless of how we choose the Banach spaces X and Y and indices $1 \leq p < q < \infty$,*

(a) *every operator in $\mathcal{N}_p(X,Y)$ is the ν_p-limit of finite rank operators, and so is compact;*

(b) *$\mathcal{N}_p(X,Y) \subset \mathcal{N}_q(X,Y)$, and $\nu_q(u) \leq \nu_p(u)$ for each $u \in \mathcal{N}_p(X,Y)$;*

(c) *$\mathcal{N}_p(X,Y) \subset \mathcal{I}_p(X,Y)$, and $\iota_p(u) \leq \nu_p(u)$ for each $u \in \mathcal{N}_p(X,Y)$.*

Proof. Appropriate amputations of elements of c_0 make (a) follow immediately from (ii) of the preceding proposition.

For the proof of (b) the essential thing to realize is that a diagonal operator $M_\lambda : \ell_\infty \to \ell_p$ can be factored as $M_\lambda : \ell_\infty \overset{M_\beta}{\to} \ell_q \overset{M_\alpha}{\to} \ell_p$ where M_α and M_β are diagonal operators which are given by $\alpha_n = |\lambda_n|^{1-(p/q)}$ and $\beta_n = (\operatorname{sign} \lambda_n) \cdot |\lambda_n|^{p/q}$ $(n \in \mathbf{N})$.

Finally, (c) is clear from the introductory remarks. QED

Formal identities $L_\infty(\mu) \hookrightarrow L_p(\mu)$ are examples of non-compact p-integral operators, so the inclusion in (c) may be proper. Nevertheless, ι_p and ν_p are much more closely related than they might appear at first sight. We shall return to this soon, but first of all we must do some basic book-keeping.

5.25 Theorem: *For each $1 \leq p < \infty$ and for any two Banach spaces X and Y, $[\mathcal{N}_p(X,Y), \nu_p]$ is a Banach space. Moreover, the spaces $[\mathcal{N}_p(X,Y), \nu_p]$ enjoy the ideal property: if X_0 and Y_0 are Banach spaces then, regardless of how we choose the operators $u \in \mathcal{L}(Y, Y_0)$, $v \in \mathcal{N}_p(X,Y)$ and $w \in \mathcal{L}(X_0, X)$, we always have $uvw \in \mathcal{N}_p(X_0, Y_0)$ with $\nu_p(uvw) \leq \|u\| \cdot \nu_p(v) \cdot \|w\|$.*

Proof. First, let us show that $[\mathcal{N}_p(X,Y), \nu_p]$ is always a normed linear space. Whereas this is obvious for $p = 1$, the statement that $u + v \in \mathcal{N}_p(X,Y)$ and $\nu_p(u+v) \leq \nu_p(u) + \nu_p(v)$ whenever $u, v \in \mathcal{N}_p(X,Y)$ requires proof if $1 < p < \infty$.

Accordingly, fix $\varepsilon > 0$ and choose representations $u = \sum_n x_{2n-1}^* \otimes y_{2n-1}$ and $v = \sum_n x_{2n}^* \otimes y_{2n}$ such that $\|(x_{2n-1}^*)_n\|_p^{strong} \leq \nu_p(u) + \varepsilon$, $\|(x_{2n}^*)_n\|_p^{strong} \leq \nu_p(v) + \varepsilon$ and $\|(y_{2n-1})_n\|_{p^*}^{weak} = \|(y_{2n})_n\|_{p^*}^{weak} = 1$. Fix $r, s > 0$ arbitrarily and define $(\tilde{x}_n^*)_n \in \ell_p^{strong}(X^*)$ by $\tilde{x}_{2n-1}^* := r \cdot x_{2n-1}^*$ and $\tilde{x}_{2n}^* := s \cdot x_{2n}^*$, $n \in \mathbf{N}$. Similarly, define $(\tilde{y}_n)_n \in \ell_{p^*}^{weak}(Y)$ by $\tilde{y}_{2n-1} := r^{-1} \cdot y_{2n-1}$ and $\tilde{y}_{2n} := s^{-1} \cdot y_{2n}$, $n \in \mathbf{N}$. Note that

$$\left(\|(\tilde{x}_n^*)_n\|_p^{strong}\right)^p = \left(r\cdot\|(x_{2n-1}^*)_n\|_p^{strong}\right)^p + \left(s\cdot\|(x_{2n}^*)_n\|_p^{strong}\right)^p$$
$$\leq r^p\cdot(\nu_p(u)+\varepsilon)^p + s^p\cdot(\nu_p(v)+\varepsilon)^p$$

and

$$\left(\|(\tilde{y})_n\|_{p^*}^{weak}\right)^{p^*} \leq \left(\frac{1}{r}\cdot\|(y_{2n-1})_n\|_{p^*}^{weak}\right)^{p^*} + \left(\frac{1}{s}\cdot\|(y_{2n})_n\|_{p^*}^{weak}\right)^{p^*}$$
$$= r^{-p^*} + s^{-p^*}.$$

Writing $u+v=\sum_n \tilde{x}_n^*\otimes\tilde{y}_n^*$, we see that $u+v$ is p-nuclear with

$$\nu_p(u+v) \leq \|(\tilde{x}_n^*)_n\|_p^{strong}\cdot\|(\tilde{y}_n)_n\|_{p^*}^{weak} \leq \frac{1}{p}\left(\|(\tilde{x}_n^*)_n\|_p^{strong}\right)^p + \frac{1}{p^*}\left(\|(\tilde{y}_n\|_{p^*}^{weak}\right)^{p^*}$$
$$\leq \frac{r^p}{p}\left(\nu_p(u)+\varepsilon\right)^p + \frac{r^{-p^*}}{p^*} + \frac{s^p}{p}\left(\nu_p(v)+\varepsilon\right)^p + \frac{s^{-p^*}}{p^*}\,.$$

With $r=(\nu_p(u)+\varepsilon)^{-1/p^*}$ and $s=(\nu_p(v)+\varepsilon)^{-1/p^*}$ we obtain

$$\nu_p(u+v)\leq\nu_p(u)+\nu_p(v)+2\varepsilon.$$

Since $\varepsilon>0$ was arbitrary, we have achieved our first goal.

All but one of the remaining properties of a Banach ideal are readily verified; it is completeness which requires some work.

Let (u_k) be a sequence in $[\mathcal{N}_p(X,Y),\nu_p]$ such that $\sum_k \nu_p(u_k)<\infty$. Then $\sum_k\|u_k\|<\infty$ too, and so $u=\sum_k u_k$ exists in $\mathcal{L}(X,Y)$. We shall show that $u\in\mathcal{N}_p(X,Y)$ and $\nu_p(u)\leq\sum_k\nu_p(u_k)$.

Let $\varepsilon>0$. For each natural number k, choose $(x_{k,n}^*)_n$ in $\ell_p^{strong}(X^*)$ and $(y_{k,n})_n$ in $\ell_{p^*}^{weak}(Y)$ such that $u_k=\sum_n x_{k,n}^*\otimes y_{k,n}$ and

$$\|(x_{k,n}^*)_n\|_p^{strong}\leq\left(\nu_p(u_k)+\frac{\varepsilon}{2^k}\right)^{1/p}\;,\quad \|(y_{k,n})_n\|_{p^*}^{weak}\leq\left(\nu_p(u_k)+\frac{\varepsilon}{2^k}\right)^{1/p^*}.$$

Notice that $(x_{k,n}^*)_{k,n}\in\ell_p^{strong}(X^*)$ and $\|(x_{k,n}^*)_{k,n}\|_p^{strong}\leq\left(\sum_k\nu_p(u_k)+\varepsilon\right)^{1/p}$; also, $(y_{k,n})_{k,n}\in\ell_{p^*}^{weak}(Y)$ and $\|(y_{k,n})_{k,n}\|_{p^*}^{weak}\leq\left(\sum_k\nu_p(u_k)+\varepsilon\right)^{1/p^*}$. Since $u=\sum_{k,n}x_{k,n}^*\otimes y_{k,n}$, it is in $\mathcal{N}_p(X,Y)$ with $\nu_p(u)\leq\sum_k\nu_p(u_k)$.

Naturally we can apply the above construction to $u-\sum_{k\leq n}u_k$ to get

$$\nu_p\left(u-\sum_{k\leq n}u_k\right)\leq\sum_{k\leq n}\nu_p(u)$$

for each $n\in\mathbf{N}$ and $u=\sum_k u_k$ in $\mathcal{N}_p(X,Y)$, too. In view of Proposition 1.1, we are done. QED

Relations to Integral Operators

The deep relationships of p-integral and p-nuclear operators are laid bare by our next result. To make the proof palatable we call on a fundamental result, the Principle of Local Reflexivity, which will be proved in Chapter 8;

the form of this principle that we use goes as follows: whenever G is a finite dimensional subspace of Y^{**} and ε is a positive number, there is a linear map $w : G \to Y$ of norm $\leq 1+\varepsilon$ which is the identity on $G \cap Y$.

The only reason we need Local Reflexivity is to cope with the bidual. We could avoid it if we chose to work with strictly p-integral operators.

5.26 Theorem: *If one of the Banach spaces X and Y is finite dimensional then, for any $1 \leq p < \infty$ and all $u \in \mathcal{L}(X,Y)$,*

$$\nu_p(u) = \iota_p(u).$$

Proof. We consider first of all the case where, for some finite set $K = \{1,..., N\}$, our Banach space X is $C(K) = \ell_\infty^N$. Let $u : C(K) \to Y$ be any operator. By 2.15 and 5.8, there exists a probability measure $\mu \in C(K)^*$ together with an operator $\tilde{u} : L_p(\mu) \to Y$ such that $u : C(K) \xrightarrow{j_p} L_p(\mu) \xrightarrow{\tilde{u}} Y$ and $\|\tilde{u}\| = \iota_p(u)$. For each $1 \leq k \leq N$, define $\lambda_k := \mu(\{k\})$, so that $\mu = \sum_{k\leq N} \lambda_k \delta_k$. Set $\varrho_k := \lambda_k^{-1}$ for $\lambda_k > 0$ and $\varrho_k := 0$ otherwise. The operator $v : \ell_p^N \to L_p(\mu) : (f_k) \mapsto (\varrho_k f_k)f$ has norm ≤ 1, and $j_p = vM_\lambda$ where $M_\lambda : \ell_\infty^N \to \ell_p^N$ is the diagonal operator induced by $\lambda = (\lambda_k)_{k\leq N}$. Since $\|M_\lambda\| \leq 1$, it follows that $\nu_p(u) \leq \|\tilde{u}\| \cdot \|v\| \cdot \|M_\lambda\| \leq \|\tilde{u}\| = \iota_p(u)$.

Let now X be any finite dimensional normed space. Given $u \in \mathcal{L}(X,Y)$, consider a typical p-integral factorization $k_Y u : X \xrightarrow{b} L_\infty(\mu) \xrightarrow{i_p} L_p(\mu) \xrightarrow{a} Y^{**}$. Fix $\varepsilon > 0$ and apply 3.2 to find a finite dimensional subspace E of $L_\infty(\mu)$, of dimension N say, together with an isomorphism $v : E \to \ell_\infty^N$ such that $\|v\| \cdot \|v^{-1}\| \leq 1+\varepsilon$ and $b(X) \subset E$. Thanks to the remark following the proof of 3.2, $i_p(E)$ is contained in a $(1+\varepsilon)$-complemented subspace F of $L_p(\mu)$. Let $J : F \to L_p(\mu)$ be the corresponding canonical embedding and $P : L_p(\mu) \to F$ the corresponding projection with $\|P\| \leq 1+\varepsilon$.

The Principle of Local Reflexivity (8.16) guarantees the existence of an operator $w : a(F) \to Y$ such that $\|w\| \leq 1+\varepsilon$ and $wk_Y ux = ux$ for all $x \in X$. Compose the restriction of Π_p to E with v^{-1} to obtain an operator $\tilde{u} : \ell_\infty^N \to F$. It follows from the special case we considered first that

$$\nu_p(\tilde{u}) = \iota_p(\tilde{u}) \leq \|P\| \cdot \iota_p(i_p) \cdot \|v^{-1}\| \leq (1+\varepsilon) \cdot \|v^{-1}\|.$$

If we look at b as an operator $X \to E$ and at aJ as a map $F \to a(F)$, then we get

$$\begin{aligned}\nu_p(u) &= \nu_p(waJ\tilde{u}vb) \leq \|waJ\| \cdot \nu_p(\tilde{u}) \cdot \|vb\| \\ &\leq (1+\varepsilon)^2 \cdot \|a\| \cdot \|v^{-1}\| \cdot \|v\| \cdot \|b\| \leq (1+\varepsilon)^3 \cdot \|a\| \cdot \|b\|.\end{aligned}$$

Take the infimum over all factorizations as above and over all $\varepsilon > 0$ to see that $\nu_p(u) \leq \iota_p(u)$. Equality follows from 5.24.

Now let Y be finite dimensional, and let $u : X \to Y$ be any operator. Let $\varepsilon > 0$ be given and set $\varrho = (1+\varepsilon)^{1/2}$. There is a probability measure μ and a factorization

$$u : X \xrightarrow{b} L_\infty(\mu) \xrightarrow{i_p} L_p(\mu) \xrightarrow{a} Y$$

such that $\|b\| = 1$ and $\|a\| \leq \varrho \cdot \iota_p(u)$.

Denote the kernel of ai_p by M. As $E := L_\infty(\mu)/M$ is finite dimensional, it is the dual of some finite dimensional subspace F of $L_1(\mu)$. By 3.2, F is contained in a subspace G of $L_1(\mu)$ which has finite dimension, say N, and for which there exists an isomorphism $v : G \to \ell_1^N$ such that $\|v\| \leq \varrho$ and $\|v^{-1}\| \leq 1$. As above, G can be chosen $(1+\varepsilon)$-complemented in $L_1(\mu)$. Let $P : L_1(\mu) \to G$ be the corresponding projection with $\|P\| \leq 1+\varepsilon$, and let $J : G \to L_1(\mu)$ be the natural embedding. The operator $w := ai_pP^* : G^* \to Y$ satisfies $wJ^* = ai_p$.

By the first part, $wv^* : \ell_\infty^N \to Y$ has a representation

$$wv^* = \sum_{k=1}^N \lambda_k z_k \otimes y_k,$$

with $z_k \in B_{\ell_1^N}$ for $1 \leq k \leq N$, $\|(y_k)_k\|_{p^*}^{weak} \leq 1$ and $(\sum_k |\lambda_k|^p)^{1/p} \leq \iota_p(wv^*)$. But

$$\iota_p(wv^*) = \iota_p(ai_pP^*v^*) \leq \|a\|\iota_p(i_p)\|P^*\|\|v^*\| \leq (1+\varepsilon)^2\iota_p(u)$$

and $u = wv^*(v^{-1})^*J^*b$, and so

$$u = \sum_{k=1}^N \lambda_k x_k^* \otimes y_k,$$

where $x_k^* := b^*Jv^{-1}z_k \in B_{X^*}$ for each $1 \leq k \leq N$. This completes the proof.

QED

This leads us to an important composition/decomposition theorem.

5.27 Theorem: *Let $1 \leq p < \infty$. A Banach space operator $u : X \to Y$ is p-nuclear if and only if there exist a Banach space Z, a compact operator $v : Z \to Y$ and a p-integral operator $w : X \to Z$ such that $u = vw$. In this case,*

$$\nu_p(u) = \inf \|v\| \cdot \iota_p(w),$$

where the infimum is extended over all such factorizations.

Proof. It follows quickly from 5.23 that the condition is necessary (with $Z = L_p$) and that $\nu_p(u)$ dominates the infimum under discussion.

To prove sufficiency, suppose next that $u : X \xrightarrow{w} Z \xrightarrow{v} Y$ is a factorization of the form described. We aim to express u as a limit in $[\mathcal{N}_p(X,Y), \nu_p]$ of a sequence of finite rank operators.

Fix any p-integral factorization $k_Zw : X \xrightarrow{b} L_\infty(\mu) \xrightarrow{i_p} L_p(\mu) \xrightarrow{a} Z^{**}$. Since v is compact, so is $a^*k_{Z^*}v^* : Y^* \to L_{p^*}(\mu)$. An appeal to 3.3 reveals that there are finite rank norm one projections $P_n \in \mathcal{L}(L_{p^*}(\mu))$ such that $\|P_na^*k_{Z^*}v^* - a^*k_{Z^*}v^*\| \leq 1/n$ for each n. The operators $P_na^*k_{Z^*}v^*$ are adjoints of finite rank operators $v_n : L_p(\mu) \to Y$; so, using compactness to view v^{**} as an operator $Z^{**} \to Y$, we obtain $\|v^{**}a - v_n\| \leq 1/n$ for each n, and so (v_n) is a $\|\cdot\|$-Cauchy sequence.

We shall now show that $(v_n i_p b)$ converges to u in $[\mathcal{N}_p(X,Y), \nu_p]$.

For natural numbers k and m, let Z be the kernel of $v_k - v_m$ and let $v_k - v_m : L_p(\mu) \xrightarrow{q} L_p(\mu)/Z \xrightarrow{t} Y$ be the corresponding canonical factorization. As $\dim L_p(\mu)/Z < \infty$, 5.26 tells us that $\nu_p(qi_pb) = \iota_p(qi_pb)$, and so

$$\nu_p(v_k i_p b - v_m i_p b) \leq \|t\| \cdot \nu_p(qi_pb) = \|v_k - v_m\| \cdot \iota_p(qi_pb) \leq \|v_k - v_m\| \cdot \|b\|.$$

We conclude that $(v_n i_p b)$ is a Cauchy sequence in $[\mathcal{N}_p(X,Y), \nu_p]$; by 5.25, it converges. The limit must be u since for each $x \in X$ and $y^* \in Y^*$,

$$\begin{aligned} \langle ux, y^* \rangle &= \langle v^{**} k_Z wx, y^* \rangle = \langle i_p bx, a^* k_{Z^*} v^* y^* \rangle \\ &= \lim_{n\to\infty} \langle i_p bx, P_n a^* k_{Z^*} v^* y^* \rangle = \lim_{n\to\infty} \langle v_n i_p bx, y^* \rangle. \end{aligned}$$

Moreover,

$$\nu_p(u) = \lim_{n\to\infty} \nu_p(v_n i_p b) \leq \|v^{**} a\| \cdot \|b\| \leq \|v\| \cdot \|a\| \cdot \|b\|.$$

Since we were working with an arbitrary p-integral factorization for w, $\nu_p(u) \leq \|v\| \cdot \iota_p(w)$. QED

The same type of argument leads to a companion result:

5.28 Theorem: *Let $1 \leq p < \infty$. The Banach space operator $u : X \to Y$ is p-nuclear if and only if there exist a Banach space Z, a strictly p-integral operator $v : Z \to Y$ and a compact operator $w : X \to Z$ such that $u = vw$.*

Proof. Necessity is again straightforward by 5.23. Suppose conversely that u factors as announced: there exist a Banach space Z, a strictly p-integral operator $v : Z \to Y$ and a compact operator $w : X \to Z$ such that $u = vw$. Look at a strictly p-integral factorization $v : Z \xrightarrow{b} L_\infty(\nu) \xrightarrow{i_p} L_p(\nu) \xrightarrow{a} Y$. Apply 3.3 to generate finite rank operators $w_n : X \to L_\infty(\nu)$ such that $\lim_{n\to\infty} \|bw - w_n\| = 0$. By 5.26, $\nu_p(ai_pw_m - ai_pw_n) = \iota_p(ai_p(w_m - w_n)) \leq \|a\| \cdot \|w_m - w_n\|$, so (ai_pw_n) is Cauchy in $\mathcal{N}_p(X,Y)$; clearly, its limit is $vw = u$. QED

As another consequence of 5.27, we obtain the following composition theorem for p-nuclear operators:

5.29 Theorem: *Suppose that $1 \leq p, q, r < \infty$ satisfy $\dfrac{1}{r} = \dfrac{1}{p} + \dfrac{1}{q}$. Then, regardless of how we choose Banach spaces X, Y, Z and operators $v \in \mathcal{L}(X,Y), u \in \mathcal{L}(Y,Z)$,*

(a) *if v is q-summing and u is p-nuclear, then uv is r-nuclear with*

$$\nu_r(uv) \leq \nu_p(u) \cdot \pi_q(v);$$

(b) *if v is q-nuclear and u is p-summing, then uv is r-nuclear with*

$$\nu_r(uv) \leq \pi_p(u) \cdot \nu_q(v).$$

Proof. (a) Let $\varepsilon > 0$. As u is p-nuclear, 5.27 tells us that $u = u_1u_2$, where u_1 is compact and u_2 is p-integral, and that we may arrange for $\|u_1\| \cdot \iota_p(u_2) \leq (1+\varepsilon) \cdot \nu_p(u)$. Since v is p-summing, u_2v is r-integral by 5.16, and

$\iota_r(uv) \leq \iota_p(u_2)\cdot\pi_q(v)$. It takes another appeal to 5.27 to see that $uv = u_1(u_2v)$ is r-nuclear, and $\nu_r(uv) \leq \|u_1\|\cdot\iota_r(u_2v) \leq (1+\varepsilon)\cdot\nu_p(u)\cdot\pi_q(v)$. This is good enough.

(b) Use 5.23 to get a factorization $v : X \xrightarrow{v_2} c_0 \xrightarrow{v_1} Y$ where v_1 is q-nuclear and v_2 is compact. By 5.24 and the comments after 5.16, uv_1 is strictly r-integral, and so 5.28 shows that $uv = uv_1v_2$ is r-nuclear. – We omit the straightforward estimates to get the statement about the norms. QED

Since $[\mathcal{N}_s, \nu_s] \subset [\mathcal{I}_s, \iota_s] \subset [\Pi_s, \pi_s]$, the above result is still true with Π_q replaced by $\mathcal{I}_q$ or $\mathcal{N}_q$ in part (a), and with Π_p replaced by $\mathcal{I}_p$ or $\mathcal{N}_p$ in part (b).

Hilbert Space Operators

We now characterize p-integral and p-nuclear operators on Hilbert spaces.

5.30 Theorem: *Let H_1 and H_2 be Hilbert spaces.*

(a) *If $1 < p < \infty$, then $\mathcal{I}_p(H_1,H_2) = \mathcal{N}_p(H_1,H_2) = \mathcal{S}_2(H_1,H_2)$ isomorphically, and even isometrically if $p = 2$.*

(b) *$\mathcal{I}_1(H_1,H_2) = \mathcal{N}_1(H_1,H_2) = \mathcal{S}_1(H_1,H_2)$ isometrically.*

Proof. (a) We start with the case $p = 2$. Using 4.10 and 5.9 we get

$$\mathcal{N}_2(H_1,H_2) \subset I_2(H_1,H_2) = \Pi_2(H_1,H_2) = \mathcal{S}_2(H_1,H_2),$$

with $\sigma_2(u) = \pi_2(u) = \iota_2(u) \leq \nu_2(u)$ for all $u \in \mathcal{N}_2(H_1,H_2)$. To show that we have equality everywhere, take $u \in \mathcal{S}_2(H_1,H_2)$, and let $u = \sum_n \tau_n(\cdot|e_n)f_n$ be any ON-representation. Then $u = a\Delta b$ where $b : H_1 \to \ell_\infty : x \mapsto \big((x|e_n)\big)_n$, $a : \ell_2 \to H_2 : (\xi_n) \mapsto \sum_n \xi_n f_n$, and $\Delta : \ell_\infty \to \ell_2$ is the diagonal operator induced by $(\tau_n) \in \ell_2$. It follows that u is 2-nuclear with $\nu_2(u) \leq \|a\|\cdot\|\Delta\|\cdot\|b\| \leq \big(\sum_n |\tau_n|^2\big)^{1/2} = \sigma_2(u)$.

Now let $1 < p < \infty$ be arbitrary. Using 4.13 we see that $\mathcal{N}_p(H_1,H_2) \subset \mathcal{I}_p(H_1,H_2) \subset \Pi_p(H_1,H_2) = \mathcal{S}_2(H_1,H_2)$. On the other hand, any $u \in \mathcal{S}_2(H_1,H_2)$ can be written in the form $u = u_1u_2$ where $u_1 : \ell_2 \to H_2$ is Hilbert-Schmidt and $u_2 : H_1 \to \ell_2$ is compact: as in the proof of 5.23, just write the sequence of u's approximation numbers as the coordinate-wise product of a null sequence and another sequence from ℓ_2. Let $u_1 = \sum_n \sigma_n(\cdot|g_n)h_n$ be any ON-representation. By setting $v_1(\xi) := \sum_n \xi_n h_n$ for $\xi \in \ell_1$ and $v_2(\xi) := \sum_n \sigma_n(\xi|g_n)e_n$ for $\xi \in \ell_2$, we obtain a factorization $u_1 : \ell_2 \xrightarrow{v_2} \ell_1 \xrightarrow{v_1} H_2$. By 5.12, v_1, and hence u_1, is (strictly) p-integral. By 5.28, $u = u_1u_2$ is p-nuclear.

(b) Let $u : H_1 \to H_2$ be 1-integral. Look at a typical 1-integral factorization $u : H_1 \xrightarrow{b} L_\infty(\mu) \xhookrightarrow{i_1} L_1(\mu) \xrightarrow{a} H_2$. Note that $i_1 = i\,i^*$ where $i : L_2(\mu) \to L_1(\mu)$ is the formal identity. By 4.12, $v_1 := ai : L_2(\mu) \to H_2$ and $v_2 := i^*b : H_1 \to L_2(\mu)$ are Hilbert-Schmidt operators. If $u = \sum_n \tau_n(\cdot|e_n)f_n$ is an ON-representation then, since v^* and w are Hilbert-Schmidt,

$$\sum_n |\tau_n| = \sum_n |(vwe_n|f_n)| = \sum_n |(we_n|v^*f_n)|$$
$$\leq \big(\sum_n \|ve_n\|^2\big)^{1/2}\cdot\big(\sum_n \|w^*f_n\|^2\big)^{1/2} < \infty,$$

and so $u \in \mathcal{S}_1(H_1, H_2)$. Also,

$$\sigma_1(u) \leq \sigma_2(v)\cdot\sigma_2(w^*) = \sigma_2(v^*)\cdot\sigma_2(w) = \pi_2(v^*)\cdot\pi_2(w) \leq \|a\|\cdot\|b\|.$$

Taking infima into account, $\sigma_1(u) \leq \iota_1(u)$. QED

In 6.3, we shall return to the topic of composition and factorization for operators in the Schatten classes.

We conclude this section by a closer look at compositions of 2-summing operators.

5.31 Theorem: *A Banach space operator $u : X \to Y$ can be represented as a composition of two 2-summing operators if and only if there exist a Hilbert space H, a nuclear operator $w : X \to H$ and a bounded operator $v : H \to Y$ such that $u = vw$.*

In particular, the composition of two 2-summing operators is nuclear.

Proof. Suppose first that u is the composition of two 2-summing operators. By 2.16, there are probability measures μ and ν such that u admits a factorization

$$X \xrightarrow{b} L_\infty(\mu) \xrightarrow{j} L_2(\mu) \xrightarrow{a} L_\infty(\nu) \xrightarrow{i} L_2(\nu) \xrightarrow{v} Y,$$

where i, j are formal identities and v, a, b are suitable operators. By 4.12, ia is a Hilbert - Schmidt operator, hence 2-nuclear by what we have just seen. Clearly, jb is 2-summing. By 5.29, $iajb$ is nuclear and so u is too.

Next suppose that u admits a factorization $X \xrightarrow{w} H \xrightarrow{v} Y$ where H is a Hilbert space and w is nuclear. Make the decomposition $w : X \xrightarrow{b} \ell_\infty \xrightarrow{\Delta} \ell_1 \xrightarrow{a} H$ where Δ is a diagonal operator. Δb is nuclear, hence 1-summing, and a is 1-summing by Grothendieck's Theorem 1.13. The conclusion follows. QED

An examination of the proof of 5.31 shows that Π_2 can be replaced by many other ideals, for example by Π_1.

Notes and Remarks

The 1-integral (or, simply, integral) operators are the creation of A. Grothendieck, and are central to his theory of tensor products, both for general locally convex spaces (as in his Memoir [1955]) and for Banach spaces (as in his Résumé [1953a]). Grothendieck introduced the integral operators while investigating duals of tensor products.

For $1 < p \leq \infty$, it was A. Pietsch [1971] who first discussed the p-integral operators. However, A. Persson and A. Pietsch [1969] had previously studied the strictly p-integral operators and they had already made some accommodation for the need to use the bidual of the range of a p-integral operator in its canonical factorization. The main object in these two papers was trace duality.

To describe the work of Grothendieck that led to his invention of integral operators we need to say a few words about tensor products. If X and Y are vector spaces, their algebraic tensor product $X \otimes Y$ has a notable 'universal mapping property': when Z is a vector space, the assignation

$$(1) \qquad \beta(x,y) = u_\beta(x \otimes y) \quad \text{for any } x \in X \text{ and } y \in Y$$

provides a natural isomorphism $\beta \mapsto u_\beta$ between the vector space of bilinear maps $\beta : X \times Y \to Z$ and the vector space of linear maps $X \otimes Y \mapsto Z$.

Now there are many ways to equip the algebraic tensor product $X \otimes Y$ of two Banach spaces with a 'reasonable' norm. But if the universal mapping property is to be successfully transferred to this new situation, only one norm will do the trick – the 'projective' norm π. For $t \in X \otimes Y$, this is given by

$$\pi(t) := \inf \sum_{k=1}^{n} \|x_k\| \cdot \|y_n\|,$$

where the infimum is taken over all possible representations of t of the form $t = \sum_{k=1}^{n} x_k \otimes y_k$, $x_k \in X$, $y_k \in Y$. The completion of $[X \otimes Y, \pi]$ is the *projective tensor product* $X \hat{\otimes} Y$ of X and Y.

Recall that for Banach spaces X, Y and Z, the collection $\mathcal{B}(X,Y;Z)$ of all continuous bilinear mappings $\beta : X \times Y \to Z$ is also a Banach space under the norm

$$\|\beta\| := \sup \{ \|\beta(x,y)\| : x \in B_X, \, y \in B_Y \}.$$

The 'universal mapping property' in this context includes the extra fact that the assignment (1) induces an *isometric isomorphism* between $\mathcal{B}(X,Y;Z)$ and $\mathcal{L}(X \hat{\otimes} Y, Z)$.

As a consequence, we have an alternative description of the norm π on $X \otimes Y$, namely, whenever $t = \sum_{k \le n} x_k \otimes y_k$ is in $X \otimes Y$,

$$\pi(t) = \sup \{ \big| \sum_{k \le n} \beta(x_n, y_n) \big| : \, \beta \in \mathcal{B}(X,Y;\mathbf{K}), \, \|\beta\| \le 1 \}.$$

Among the other norms of interest on $X \otimes Y$, one of the most useful is the 'injective' norm ε. If $t = \sum_{k \le n} x_k \otimes y_k \in X \otimes Y$, then

$$\varepsilon(t) := \sup \{ \big| \sum_{k \le n} \langle x^*, x_k \rangle \cdot \langle y^*, y_k \rangle \big| : \, x^* \in B_{X^*}, \, y^* \in B_{Y^*} \}.$$

The completion $X \check{\otimes} Y$ of $[X \otimes Y, \varepsilon]$ is the *injective tensor product* of X and Y.

Note that π calls on every available continuous bilinear form while ε asks for the bare minimum. It is part of the theory that among 'reasonable' tensor norms, π is the greatest and ε is the least. It is also part of the theory that $X \check{\otimes} Y$ occurs naturally as a closed subspace of the function space $C(K)$ where $K := B_{X^*} \times B_{Y^*}$ and B_{X^*} and B_{Y^*} are both taken in their respective compact weak$*$topologies. With the Hahn-Banach Theorem and the Riesz Representation Theorem in hand, Grothendieck recognized that the elements in $(X\check{\otimes}Y)^*$ correspond exactly to the bilinear forms $\beta : X \times Y \to \mathbf{K}$ such that, for all $x \in X$ and $y \in Y$,

$$\beta(x,y) = \int_K \langle x^*, x\rangle\langle y^*, y\rangle \, d\mu(x^*, y^*)$$

for some measure $\mu \in \mathcal{C}(K)^*$, bilinear forms which he called *integral* for obvious reasons.

A Banach space operator $u : X \to Y$ is called *integral* if the induced bilinear form $\beta_u : X \times Y^* \to \mathbf{K} : (x, y^*) \mapsto \langle ux, y^*\rangle$ is integral: there is a measure $\mu \in \mathcal{C}(B_{X^*} \times B_{Y^{**}})^*$ such that, for all $x \in X$ and $y^* \in Y^*$,

$$\langle ux, y^*\rangle = \int_{B_{X^*} \times B_{Y^{**}}} \langle x^*, x\rangle\langle y^{**}, y^*\rangle \, d\mu(x^*, y^{**}).$$

Grothendieck unravelled this formula to obtain a factorization $k_Y u : X \overset{b}{\to} L_\infty(\mu) \overset{i_1}{\to} L_1(\mu) \overset{a}{\to} Y^{**}$: integral operators are what we called 1-integral. Here Y^{**} enters the picture simply because μ is forced to live on $B_{X^*} \times B_{Y^{**}}$!

The elementary results 5.1 - 5.9 are all to be found in one form or another in A. Persson - A. Pietsch [1969].

T. Figiel and W.B. Johnson [1973] showed the existence of an integral operator that is not strictly integral. Their construction depends on the existence of a Banach space with the approximation property but without the bounded approximation property and so ultimately on P. Enflo's famous [1973] example of a Banach space without the approximation property. O. Reinov [1982] has found examples of p-integral operators that are not strictly p-integral for $p \neq 2$. It seems to be the case that any examples of this kind require delicate approximation considerations.

Grothendieck knew (though under other guises) that absolutely summing operators from or to $\mathcal{C}(K)$-spaces are integral, with equality of the π_1- and ι_1-norms.

A. Pietsch ([1965], [1969]) gave a description of absolutely p-summing operators on $\mathcal{C}(K)$-spaces involving representing vector measures. This shows that such operators are (strictly) p-integral with coincidence of the π_p- and ι_p-norms for all $p \geq 1$.

Proposition 5.12 is due to P. Saphar [1972].

Example 5.13 is a special case of a more general result by A. Pełczyński [1969]; he even showed that

- *if $\Lambda \subset \mathbf{Z}$ and if $\mathcal{C}_\Lambda(\mathbf{T})$ ($L_{p,\Lambda}(\mathbf{T})$) denotes the span of the monomials z^n for $n \in \Lambda$ in $\mathcal{C}(\mathbf{T})$ ($L_p(\mathbf{T})$ for $1 \leq p < \infty$), then the p-summing operator $\mathcal{C}_\Lambda(\mathbf{T}) \hookrightarrow L_{p,\Lambda}(\mathbf{T})$ is p-integral if and only if $L_{p,\Lambda}(\mathbf{T})$ is complemented in $L_p(\mathbf{T})$.*

B. Carl [1976] provided further examples of p-summing operators which fail to be p-integral by showing that if $1<p<t<2<s<p^*$ or $1<p^*<s<2<t<p$, then $I_p(\ell_s, \ell_t) \neq \Pi_p(\ell_s, \ell_t)$; in fact, there are even p-summing diagonal maps from ℓ_s to ℓ_t which are not p-integral. Carl's proof is existential in nature; he

shows that $\sup\{\iota_p(u): u\in\mathcal{L}(\ell_s^n, \ell_t^n),\ \pi_p(u)\leq 1\}$ – the '*(ι_p/π_p)-ratio*' – tends to ∞ as n does, thereby precluding the existence of a constant $K>0$ such that $\iota_p(u) \leq K\cdot\pi_p(u)$ for all $u \in \Pi_p(\ell_s, \ell_t)$. Carl's work relies in part on that of D.J.H. Garling [1974] and A. Pietsch [1970/71]. Garling's paper, by the way, gives a complete classification of p-integral diagonal operators from ℓ_s to ℓ_t if $1\leq p, s, t<\infty$.

On the other side of the ledger, B. Mitiagin and A. Pełczyński (compare A. Pełczyński [1976b]) discovered that

- *if $1 < p < \infty$, then any p-summing operator whose domain is the disk algebra $A(D)$, where $D := \{z \in \mathbf{C} : |z| < 1\}$ is the open unit disk, is strictly p-integral.*

The key ingredients of the proof are outer functions, M.Riesz's Projection Theorem, the Rudin - Carleson Extrapolation Theorem, and the Lebesgue Decomposition Theorem mixed with the Pietsch Factorization Theorem. Later, S.V. Kisliakov [1981b] gave another proof of this theorem using weighted norm inequalities; he also proved an analogous result to be true when H^∞ replaces $A(D)$. Again the result is stated in terms of the (ι_p/π_p)-ratio of a Banach space X, that is, the quantity

$$(\iota_p/\pi_p)(X) := \sup \iota_p(u)$$

where the supremum extends over all Banach spaces Z and $u \in \Pi_p(X, Z)$ with $\pi_p(u) \leq 1$. The Mitiagin - Pełczyński Theorem is the statement that $(\iota_p/\pi_p)(A(D))$ is finite for all $1 < p < \infty$, and Kisliakov's extension is that $(\iota_p/\pi_p)(H^\infty) < \infty$ for the same p's. On the other hand, in his investigations into isomorphic invariants of spaces of analytic functions in higher dimensions, J. Bourgain [1984a,b], [1985] found that the polyball and polydisk algebras as well as the H^∞-spaces modelled on polyballs or polydisks also have infinite (ι_p/π_p)-ratios for any $1 \leq p < \infty$, $p \neq 2$.

Theorem 5.14 was proved in a more general context by P. Saphar [1970], 5.15 is due to A. Grothendieck [1955], and 5.16 is due to A. Persson and A. Pietsch [1969].

The simple change-of-densities argument encountered in 5.17 seems to have first been exploited by A. Grothendieck in his Memoir [1955]. It is, of course, one motivation behind his remarkable Theorem 5.19.

Operators from a Banach space X into an order complete Banach lattice L that take norm bounded sets into order bounded ones are very special and possessed of many peculiar yet attractive properties. For instance, they enjoy a Hahn - Banach - like property: if $u : X \to L$ is such an operator and $|ux| \leq h$ for all $x \in B_X$ and some $h \in L$, then given any Banach space Y which contains X as a subspace, there is an extension $\tilde{u} \in \mathcal{L}(Y, L)$ such that $|\tilde{u}y| \leq h$ for all $y \in B_Y$. Grothendieck uses 5.19 throughout his Memoir [1955] and his Résumé [1953a]. It provides, for example, an important link between the theory of differentiation of vector measures and the representability of $(X\check{\otimes}Y)^*$ as $X^*\hat{\otimes}Y^*$.

Theorem 5.20 and its corollary 5.21 are due to S. Kwapień [1970b] and L. Schwartz [1970a], whereas 5.22 is a special case of results proved by S. Kwapień [1972a]. D.J.H. Garling [1975] gave a succinct account of 5.20 - 5.22 and much more. Further details can also be found in B.M. Makarov's [1991] survey. The relevance of order boundedness for composition operators on Hardy spaces and other spaces of analytic functions was investigated in a series of papers by H. Hunziker [1989], H. Hunziker and H. Jarchow [1991], H. Hunziker, V. Mascioni and H. Jarchow [1990], H. Jarchow ([1992],[1993],[1995]), H. Jarchow and R. Riedl [1995] and R. Riedl [1994].

Nuclear operators were introduced and studied by A. Grothendieck ([1953a],[1955]) though their Hilbert space counterparts, the trace class operators, go back considerably further. The p-nuclear operators were first investigated, with trace duality in mind, by A. Persson and A. Pietsch [1969]. In our discussion of their general properties (in particular, 5.23 through 5.30) we follow Persson and Pietsch.

That the product of two 2-summing operators is nuclear is due to A. Grothendieck [1953a] who proved it using tensor products; an alternative path was found by A. Pietsch [1967]. A third path, following vector measure arguments, can be found in J. Diestel and J.J. Uhl, Jr. [1977]; this approach is, in fact, much in the spirit of Grothendieck's original proof, differing only in execution. It seems however that Theorem 5.31 as it stands was first formulated by H. Jarchow and R. Ott [1982].

It follows from 5.31 and 2.22 that if $N \in \mathbf{N}$ is even, then the composition of N N-summing operators is nuclear. It seems to be unknown what happens if $N \geq 3$ is an odd integer.

In case $p = 1$, Theorems 5.27 and 5.28 are open to considerable generalization. It is an old and remarkable theorem of A. Grothendieck [1956] that

- *if $v: X \to Y$ is integral and $u: Y \to Z$ is weakly compact, then $uv: X \to Z$ is nuclear;*

on the other hand,

- *if v is weakly compact and u is strictly integral, then uv is nuclear.*

In particular, *integral operators to or from reflexive spaces are nuclear.* The anomaly arising in the second statement is unavoidable: if u is only presumed integral then the best that can be concluded is that $(uv)^*$ is nuclear. The culprit is the bounded approximation property; see T. Figiel and W.B. Johnson [1973] for explanations.

The study of differentiability of vector-valued measures is close to the root of the 'integral = nuclear' question. Grothendieck's proof of the aforementioned results rely on by-now classical results of N. Dunford and B.J. Pettis [1940] and R.S. Phillips [1940] on the kernel representation of operators from L_1 to any Banach space. Proceeding carefully along the direction marked by Grothendieck one can show:

- *if one of the operators $v \in \mathcal{L}(X,Y)$ and $u \in \mathcal{L}(Y,Z)$ is integral and the other 'Asplund', then $(uv)^*$ is nuclear.*

An Asplund operator is one which factors through a Banach space each of whose separable subspaces has a separable dual; C. Stegall [1981] is *the* source on Asplund operators. A discussion of why Asplund operators play such a crucial rôle in the 'integral = nuclear' question can be found in the monograph of J. Diestel and J.J. Uhl, Jr. [1977].

Grothendieck was aware of the close ties that exist between the differentiability of certain vector measures and nuclearity of integral operators. In this vein he established a crisp companion to Theorem 5.19.

Theorem: *A bounded linear operator $u : X \to L_1(\mu)$ is nuclear if and only if $u(B_X)$ is order bounded and equimeasurable.*

Recall that a subset K of $L_1(\mu)$ is *equimeasurable* if, given $\varepsilon > 0$, there is a measurable set Ω_ε such that $\|f\,1_{\Omega_\varepsilon^c}\| \leq \varepsilon$ for all $f \in K$ and $\{f\,1_{\Omega_\varepsilon} : f \in K\}$ is relatively compact in $L_\infty(\mu)$.

This characterization of nuclear operators into $L_1(\mu)$ was overlooked for some time until the work of C. Stegall [1981] brought it to the fore again. Soon after, W. Schachermayer [1981a,b] used equimeasurability in his attack on problems of P.R. Halmos and V.S. Sunder [1978]. Later, the notion of equimeasurability surfaced in the monograph of N. Ghoussoub, G. Godefroy, B. Maurey and W. Schachermayer [1987]

For $p > 1$ the 'p-integral = p-nuclear' question has some expected responses and some unexpected ones. On the one hand, A. Persson [1969] showed that if X is reflexive or if X^* is separable, then strictly p-integral operators with domain X are p-nuclear. This result was extended by C.S. Cardassi and J. Diestel:

- *if $v : X \to Y$ is an Asplund operator and $u : Y \to Z$ is strictly p-integral $(1 \leq p < \infty)$, then uv is p-nuclear.*

Cardassi went on to show that if, for some $1 < p < \infty$, all strictly p-integral operators with domain X are p-nuclear, then X does not contain an isomorphic copy of ℓ_1. On the other hand, we have the surprising result of Cardassi stating that *if $v : X \to Y$ is strictly p-integral and $u : Y \to Z$ is completely continuous, then uv is p-nuclear* and that, surprisingly enough, p-integral operators $(1 \leq p < \infty)$ into Tsirelson space are p-nuclear. The source for these developments is C.S. Cardassi [1989].

The question of when 0-summing operators (see Notes and Remarks to Chapter 2) are nuclear turned out to be important for Banach spaces from harmonic analysis. S.V. Kisliakov proved in [1981b] that every 0-summing operator from a space $C(K)$ to any Banach space is nuclear. It was shown by J. Bourgain [1984c] that this property of $C(K)$ is shared with the disk algebra. Other examples of Banach spaces X such that every 0-summing operator with domain X is nuclear would be welcomed additions.

6. TRACE DUALITY

TRACE

When $u \in \mathcal{F}(X)$, the space of finite rank operators on the Banach space X, it may be represented in the form

$$(*) \qquad u = \sum_{i=1}^{N} x_i^* \otimes y_i$$

with $y_1,\ldots,y_N \in X$ and $x_1^*,\ldots,x_N^* \in X^*$. We define the *trace* of u by

$$\operatorname{tr}(u) := \sum_{i=1}^{N} \langle x_i^*, y_i \rangle.$$

This is certainly consistent with the usual definition of the trace of a matrix: if X is finite dimensional, if u has a matrix representation (u_{ij}) with respect to the basis $y_1,\ldots,y_n$ of X, and if $y_1^*,\ldots,y_n^*$ is the dual basis of X^* (so that $\langle y_i^*, y_j \rangle = \delta_{ij}$ for all $1 \le i,j \le n$), then $u = \sum_{j=1}^n (\sum_{i=1}^n u_{ij} y_i^*) \otimes y_j$, and so

$$\operatorname{tr}(u) = \sum_{j=1}^{n} \Big\langle \sum_{i=1}^{n} u_{ij} y_i^*, y_j \Big\rangle = \sum_{i,j=1}^{n} u_{ij}\delta_{ij} = \sum_{i=1}^{n} u_{ii}.$$

Of course, for true consistency, we are obliged to demonstrate that the trace is independent of the operator's representation $(*)$. For this it is adequate to check that if $u = \sum_{i=1}^N x_i^* \otimes y_i$ is the zero operator, then $\sum_{i=1}^N \langle x_i^*, y_i \rangle = 0$.

To this end, choose a basis $z_1,\ldots,z_k$ of $\operatorname{span}\{y_1,\ldots,y_N\}$, and let $z_1^*,\ldots,z_k^*$ in X^* satisfy $\langle z_i^*, z_j \rangle = \delta_{ij}$. Writing $\eta_{ij} = \langle y_i, z_j^* \rangle$, we have $y_i = \sum_{j=1}^k \eta_{ij} z_j$ for all $1 \le i \le N$, and

$$\sum_{i=1}^{N} \langle x_i^*, y_i \rangle = \sum_{j=1}^{k} \Big\langle \sum_{i=1}^{N} \eta_{ij} x_i^*, z_j \Big\rangle = \sum_{j=1}^{k} \Big\langle \sum_{i=1}^{N} \langle x_i^*, z_j \rangle y_i, z_j^* \Big\rangle = \sum_{j=1}^{k} \langle u z_j, z_j^* \rangle = 0.$$

As another immediate consequence of the definition we get

$$\operatorname{tr}(u) = \operatorname{tr}(u^*)$$

for all $u \in \mathcal{F}(X)$.

Manifestly, the trace acts linearly on the vector space $\mathcal{F}(X)$. An elementary but very useful feature of the trace is found in the next lemma.

6.1 Lemma: *If $u \in \mathcal{L}(X,Y)$, $v \in \mathcal{L}(Y,X)$, and one of these operators has finite rank, then*

$$\operatorname{tr}(uv) = \operatorname{tr}(vu).$$

Proof. Of course, $uv \in \mathcal{F}(Y)$ and $vu \in \mathcal{F}(X)$, so that both traces exist. Suppose u has finite rank and is represented in the form $u = \sum_{i=1}^N x_i^* \otimes y_i$. Then $uv = \sum_{i=1}^N v^* x_i^* \otimes y_i$ and $vu = \sum_{i=1}^N x_i^* \otimes vy_i$, and so

$$\operatorname{tr}(uv) = \sum_{i=1}^N \langle v^* x_i^*, y_i \rangle = \sum_{i=1}^N \langle x_i^*, vy_i \rangle = \operatorname{tr}(vu). \qquad \text{QED}$$

Our interest in the trace stems from the fact that it permits us to set up a canonical duality, called *trace duality*, between $\mathcal{L}(E,F)$ and $\mathcal{L}(F,E)$, where E and F are finite dimensional vector spaces. With any $v \in \mathcal{L}(F,E)$ we associate a linear functional φ_v on $\mathcal{L}(E,F)$:

$$\langle \varphi_v, u \rangle := \operatorname{tr}(uv).$$

It is plain and easy to see that if $\operatorname{tr}(uv) = 0$ for all $u \in \mathcal{L}(E,F)$, then $v = 0$; in other words, the map

$$\mathcal{L}(F,E) \longrightarrow \mathcal{L}(E,F)^* : v \mapsto \varphi_v$$

is injective. Linearity is evident, so dimension considerations constrain this map to be an isomorphism. Naturally, a similar process identifies $\mathcal{L}(E,F)$ as the dual of $\mathcal{L}(F,E)$.

For finite dimensional spaces E and F, we have so far discussed trace duality between $\mathcal{L}(E,F)$ and $\mathcal{L}(F,E)$ in purely algebraic terms. It behoves us to introduce topological considerations into the proceedings. Given any norm α on $\mathcal{L}(E,F)$, we can define the *adjoint norm* α^* on $\mathcal{L}(F,E)$ by

$$\alpha^*(v) := \sup\big\{|\operatorname{tr}(uv)| : u \in \mathcal{L}(E,F),\ \alpha(u) \le 1\big\}.$$

In this way, $[\mathcal{L}(F,E), \alpha^*]$ becomes the Banach space dual of $[\mathcal{L}(E,F), \alpha]$. Iteration ensures that $\alpha^{**} := (\alpha^*)^* = \alpha$.

The goal of this chapter is to find an appropriate extension of trace duality to the infinite dimensional setting. For the Schatten classes that we introduced in Chapter 4, this can be accomplished in a way that mimics the finite dimensional theory. However, when we turn our attention to the summing operators, things do not run so smoothly. Only after technical modifications are satisfactorily developed can we expose a type of duality relationship between summing operators and their cousins, the integral operators.

Duality for Schatten-von Neumann Classes

Let H_1 and H_2 be Hilbert spaces. We know that the Banach space $[\mathcal{S}_2(H_1,H_2), \sigma_2]$ of all Hilbert-Schmidt operators is actually a Hilbert space: no matter how we choose an orthonormal basis $(e_i)_{i\in I}$ in H_1, the quantity

$$(u|v) = \sum_{i\in I} (ue_i|ve_i)$$

doesn't depend on this choice, is an inner product on $\mathcal{S}_2(H_1,H_2)$, and defines the norm σ_2 (see 4.9). Moreover, however we choose orthonormal families

$(e_i)_{i\in I}$ in H_1 and $(f_j)_{j\in J}$ in H_2, $(e_i \otimes f_j)_{(i,j)\in I\times J}$ is easily seen to be orthonormal in $\mathcal{S}_2(H_1, H_2)$, and it is an orthonormal basis whenever $(e_i)_{i\in I}$ and $(f_j)_{j\in J}$ are. From now on, we also write $x \otimes y$ in place of $(\cdot\,|x)y$.

We are going to exploit the Hilbert space structure of $\mathcal{S}_2(H_1, H_2)$ to develop a satisfactory trace duality. To this end, we associate with each $v \in \mathcal{L}(H_1, H_2)$ the map

$$T_v : \mathcal{S}_2(H_1, H_1) \to \mathcal{S}_2(H_2, H_2) : u \mapsto vuv^*.$$

T_v is linear and bounded, with

$$\|T_v\| = \|v\|^2.$$

Clearly, $\|T_v\| \le \|v\|^2$. As for the reverse inequality, fix $\lambda > 1$ and choose a unit vector $x \in H_1$ such that $\|v\| \le \lambda\|vx\|$. Then $\sigma_2(x \otimes x) = 1$ and

$$\begin{aligned}\|v\|^2 \le \lambda^2\|vx\|^2 &= \lambda^2\sigma_2(vx \otimes vx)\\ &= \lambda^2\sigma_2(v^*(x\otimes x)v) = \lambda^2\sigma_2(T_v(x\otimes x)) \le \lambda^2\|T_v\|.\end{aligned}$$

Since $\lambda > 1$ was arbitrary, we are done.

More generally, we can prove:

6.2 Theorem: *No matter how we choose $1 \le r < \infty$, v belongs to $\mathcal{S}_r$ if and only if T_v does, and if this is the case*

$$\sigma_r(T_v) = \sigma_r(v)^2.$$

Proof. Suppose first that v is in $\mathcal{S}_r$ and let $v = \sum_n \alpha_n e_n \otimes f_n$ be an ON-representation with $\alpha_n \ge 0$ for all n. For each $u \in \mathcal{S}_2(H_1, H_1)$ we find

$$T_v(u) = \sum_{n,m} \alpha_n\alpha_m(ue_m|e_n)(\cdot\,|f_m)f_n.$$

This can be written in terms of the inner product of $\mathcal{S}_2(H_1, H_1)$ as

$$T_v(u) = \sum_{n,m} \alpha_n\alpha_m\big(u\,\big|\,e_m \otimes e_n\big)\, f_m \otimes f_n,$$

and this is an ON-representation for T_v! Since

$$\sigma_r(T_v)^r = \sum_{n,m} \alpha_n^r\alpha_m^r = \Big(\sum_n \alpha_n^r\Big)^2 = \sigma_r(v)^{2r},$$

T_v belongs to $\mathcal{S}_r$, and we have the asserted equality of norms.

For the converse, we can assume $\|v\| = 1$. First select $x_0 \in H_1$ with $\|x_0\| = 1$ and $\|vx_0\| = 1$. This is possible because of the reflexivity of H_1 and the compactness of v. Now it is straightforward to check that $x \mapsto x_0 \otimes x$ defines an isometric embedding $\varphi : H_1 \to \mathcal{S}_2(H_1, H_2)$, and that $y \mapsto (vx_0) \otimes y$ likewise gives an isometric embedding $\psi : H_2 \to \mathcal{S}_2(H_2, H_2)$. Since we know from 4.9 that $\mathcal{S}_2(H_2, H_2)$ is a Hilbert space, there is an orthogonal projection P from $\mathcal{S}_2(H_2, H_2)$ onto $\psi(H_2)$.

Observe that for any $x \in H_1$, $T_v\varphi(x) = T_v(x_0 \otimes v) = vx_0 \otimes vx = \psi v(x)$. Therefore it is possible to write $v = (\psi^{-1}P)\, T_v\, \varphi$ and to use 4.5(d) to reach our conclusion. QED

The path is now clear to an elegant factorization result.

6.3 Theorem: *Suppose that $1 \le p, q, r < \infty$ are such that $\frac{1}{p} + \frac{1}{q} = \frac{1}{r}$. An operator $u \in \mathcal{L}(H_1, H_2)$ belongs to $\mathcal{S}_r(H_1, H_2)$ if and only if there exist a Hilbert space H and operators $v \in \mathcal{S}_p(H, H_2)$ and $w \in \mathcal{S}_q(H_1, H)$ such that $u = vw$. In such a case,*

$$\sigma_r(u) = \inf \sigma_p(v)\cdot\sigma_q(w),$$

where the infimum is extended over all such factorizations. The infimum is actually a minimum.

Proof. If $u = vw$ with $v \in \mathcal{S}_p(H, H_2)$ and $w \in \mathcal{S}_q(H_1, H)$ for some Hilbert space H, then, by Lemma 4.4 (and the accompanying discussion) and Hölder's Inequality,

$$\begin{aligned}\sigma_r(u) &= \Big(\sum_n a_n(u)^r\Big)^{1/r} = \Big(\sum_n a_{2n}(u)^r + a_{2n-1}(u)^r\Big)^{1/r} \\ &\le 2^{1/r}\cdot\Big(\sum_n a_{2n-1}(u)^r\Big)^{1/r} = 2^{1/r}\cdot\Big(\sum_n a_{2n-1}(vw)^r\Big)^{1/r} \\ &\le 2^{1/r}\cdot\Big(\sum_n a_n(v)^r a_n(w)^r\Big)^{1/r} \le 2^{1/r}\cdot\sigma_p(v)\cdot\sigma_q(w).\end{aligned}$$

It follows that u belongs to $\mathcal{S}_r(H_1, H_2)$. We claim that this estimate can be upgraded to

$$\sigma_r(u) \le \sigma_p(v)\cdot\sigma_q(w).$$

To see why this is so, consider the operators $T_u : \mathcal{S}_2(H_1, H_1) \to \mathcal{S}_2(H_2, H_2)$, $T_v : \mathcal{S}_2(H, H) \to \mathcal{S}_2(H_2, H_2)$ and $T_w : \mathcal{S}_2(H_1, H_1) \to \mathcal{S}_2(H, H)$ which correspond to u, v, and w. As we know, $T_u \in \mathcal{S}_r$, $T_v \in \mathcal{S}_p$ and $T_w \in \mathcal{S}_q$, with $\sigma_r(T_u) = \sigma_r(u)^2$, $\sigma_p(T_v) = \sigma_p(v)^2$ and $\sigma_q(T_w) = \sigma_q(w)^2$. Obviously, $T_vT_w = T_{vw} = T_u$, and so we can use 6.2 to conclude that

$$\sigma_r(u) = \sigma_r(T_u)^{1/2} \le 2^{1/(2r)}\cdot\sigma_p(T_v)^{1/2}\cdot\sigma_q(T_w)^{1/2} = 2^{1/(2r)}\cdot\sigma_p(v)\cdot\sigma_q(w).$$

Iterate to get the desired estimate.

For the other implication, consider $u \in \mathcal{S}_r(H_1, H_2)$ having ON-representation $u = \sum_n \tau_n e_n \otimes f_n$ with $\tau_n \ge 0$ for all n. Notice that $w : H_1 \to \ell_2 : x \mapsto (\tau_n^{r/q}(x|e_n))_n$ belongs to $\mathcal{S}_q(H_1, \ell_2)$, and $v : \ell_2 \to H_2 : (\xi_n) \mapsto \sum_n \tau_n^{r/p}\xi_n f_n$ belongs to $\mathcal{S}_p(\ell_2, H_2)$. Since $u = vw$ and $\sigma_p(v)\cdot\sigma_q(w) = \big(\sum_n \tau_n^r\big)^{1/p}\cdot\big(\sum_n \tau_n^r\big)^{1/q}$ $= \sigma_r(u)$, the theorem is proved. QED

Let H be a Hilbert space, and let $(e_i)_{i\in I}$ be any orthonormal basis in H. Given $u \in \mathcal{S}_1(H, H)$, we know from 4.6(b) that the family $(ue_i|e_i)_{i\in I}$ belongs to ℓ_1^I. Use 6.3 to write $u = vw$ with $v \in \mathcal{S}_2(H_0, H)$ and $w \in \mathcal{S}_2(H, H_0)$ for some Hilbert space H_0. Cast an eye over the proof, and notice that we may take $H_0 = H$. Refer to 4.5(c) and 4.7 to infer

$$\sum_k |(ue_k|e_k)| = \sum_k |(we_k|v^*e_k)| \le \big(\sum_k \|we_k\|^2\big)^{1/2} \cdot \big(\sum_k \|v^*e_k\|^2\big)^{1/2}$$
$$= \sigma_2(v)\cdot\sigma_2(w).$$

By another application of 6.3, $\sum_k |(ue_k|e_k)| \le \sigma_1(u)$, and so

$$(\circ) \qquad \tau : \mathcal{S}_1(H,H) \to \mathbf{K} : u \mapsto \sum_k (ue_k|e_k)$$

is a well-defined bounded linear form.

If $u \in \mathcal{L}(H,H)$ is a finite rank operator, $u = \sum_{j=1}^n x_j^* \otimes y_j$, then $\tau(u)$ is nothing but the trace of u discussed before: $\tau(u) = \sum_{j=1}^n \langle x_j^*, y_j\rangle$. Since the finite rank operators are dense in $[\mathcal{S}_1(H,H), \sigma_1]$, τ is the only continuous extension of the trace functional tr to $[\mathcal{S}_1(H,H), \sigma_1]$; it therefore deserves to be denoted by tr as well; occasionally we may write

$$\mathrm{tr}_H,$$

for finer distinction. Note that

$$|\mathrm{tr}_H(u)| \le \sigma_1(u) \quad \text{for all} \quad u \in \mathcal{S}_1(H,H).$$

It follows from our discussion that the sum in $(\circ)$ depends only on u and not on the special choice of the orthonormal basis $(e_i)_{i\in I}$.

When combined with 6.3, our considerations lead to the following extension of 6.1:

- Let H_1 and H_2 be Hilbert spaces. If $u \in \mathcal{S}_p(H_1,H_2)$ and $v \in \mathcal{S}_{p^*}(H_2,H_1)$ $(1 \le p < \infty)$, then $\mathrm{tr}_{H_1}(vu) = \mathrm{tr}_{H_2}(uv)$.

It is clear that for all $x, y \in H$, $\mathrm{tr}\,(x \otimes y) = (y|x)$, and that if u is in $\mathcal{S}_1(H,H)$, then $\mathrm{tr}\,(u^*) = \overline{\mathrm{tr}\,(u)}$ holds for the *Hilbert space adjoint* of u.

All this comes in handy to prove:

6.4 Theorem (Trace duality for $\mathcal{S}_p$):

(a) *If $1 < p < \infty$, then an isometric isomorphism*

$$\phi : \mathcal{S}_{p^*}(H_2,H_1) \longrightarrow \mathcal{S}_p(H_1,H_2)^*$$

is obtained by setting

$$\phi(v)(u) := \mathrm{tr}_{H_2}(uv)$$

for all $v \in \mathcal{S}_{p^*}(H_2,H_1)$ and $u \in \mathcal{S}_p(H_1,H_2)$.

(b) *In the same way, $\mathcal{S}_1(H_2,H_1)$ is isometrically isomorphic to $\mathcal{S}_\infty(H_1,H_2)^*$, and $\mathcal{L}(H_2,H_1)$ is isometrically isomorphic to $\mathcal{S}_1(H_1,H_2)^*$.*

Proof. (a) By 6.3, if $v \in \mathcal{S}_{p^*}(H_2,H_1)$, then $uv \in \mathcal{S}_1(H_2,H_2)$ for all $u \in \mathcal{S}_p(H_1,H_2)$, and $\sigma_1(uv) \le \sigma_p(u)\cdot\sigma_{p^*}(v)$. Consequently, $\phi(v)$ exists as a bounded linear form on $\mathcal{S}_p(H_1,H_2)$. The resulting map ϕ is of course linear, and, again by 6.3, it is bounded: for each $v \in \mathcal{S}_{p^*}(H_2,H_1)$ we have

$$\begin{aligned} \|\phi(v)\| &= \sup\{|\mathrm{tr}\,(uv)| : u \in B_{\mathcal{S}_p(H_1,H_2)}\} \\ &\leq \sup\{\sigma_1(uv) : u \in B_{\mathcal{S}_p(H_1,H_2)}\} \;=\; \sigma_{p^*}(v). \end{aligned}$$

But if we now take any ON-representation $v = \sum_n \alpha_n(\cdot|f_n)e_n$ of our operator $v \in \mathcal{S}_{p^*}(H_2, H_1)$, then

$$\begin{aligned} \sigma_{p^*}(v) &= \Big(\sum_n |\alpha_n|^{p^*}\Big)^{1/p^*} = \sup\Big\{\Big|\sum_n \alpha_n\tau_n\Big| : \sum_n |\tau_n|^p \leq 1\Big\} \\ &= \sup\Big\{|\mathrm{tr}\,(uv)| : u = \sum_n \tau_n(\cdot|e_n)f_n, \sum_n |\tau_n|^p \leq 1\Big\} \\ &= \sup\big\{|\mathrm{tr}\,(uv)| : u \in B_{\mathcal{S}_p(H_1,H_2)}\big\} \;=\; \|\phi(v)\|. \end{aligned}$$

Thus ϕ is even isometric.

To see that ϕ is surjective, fix $\varphi \in \mathcal{S}_p(H_1, H_2)^*$. Then $\beta : H_1 \times H_2 \to \mathbf{K} : (x, y) \mapsto \overline{\langle\varphi, x \otimes y\rangle}$ is sesquilinear, that is, it is linear in the first variable and conjugate linear in the second variable. Now β is continuous since $|\beta(x, y)| \leq \|\varphi\|\cdot\sigma_p(x \otimes y) = \|\varphi\|\cdot\|x\|\cdot\|y\|$ for all $x \in H_1$, $y \in H_2$. Each $\beta(\cdot, y)$ $(y \in H_2)$ is a member of H_1^* and so has the form $\beta(\cdot, y) = (\cdot|x_y)$ where $x_y \in H_1$ is uniquely determined. The map $v : H_2 \to H_1 : y \mapsto x_y$ is clearly linear and bounded, with $\beta(x, y) = (x|vy)$ for all $(x, y) \in H_1 \times H_2$.

We claim that v belongs to $\mathcal{S}_{p^*}(H_2, H_1)$. To prove this, fix orthonormal sequences (e_n) in H_1 and (f_n) in H_2. Each $(\tau_n) \in \ell_p$ gives rise to the operator $u = \sum_n \tau_n e_n \otimes f_n \in \mathcal{S}_p(H_1, H_2)$, and $\sum_n \tau_n(e_n|vf_n) = \sum_n \tau_n \overline{\varphi(e_n \otimes f_n)} = \overline{\langle\varphi, u\rangle}$ exists in $\mathbf{K}$. Since $(\tau_n) \in \ell_p$ was arbitrary, $((e_n|f_n))_n$ belongs to ℓ_{p^*}. But (e_n) and (f_n) were arbitrary orthonormal sequences, and so 4.6 tells us that $v \in \mathcal{S}_{p^*}(H_2, H_1)$.

Finally, if $w = \sum_n \sigma_n g_n \otimes h_n$ is an ON-representation of any operator in $\mathcal{S}_p(H_1, H_2)$, then

$$\begin{aligned} \phi(v)(w) &= \mathrm{tr}\,(vw) = \mathrm{tr}\,\Big(\sum_n \sigma_n g_n \otimes vh_n\Big) = \sum_n \sigma_n(vh_n|g_n) \\ &= \sum_n \sigma_n \overline{(g_n|vh_n)} = \sum_n \sigma_n\langle\varphi, g_n \otimes h_n\rangle = \langle\varphi, w\rangle\;. \end{aligned}$$

We are done with part (a). Part (b) is obtained in the same manner, by using the standard duality relations between c_0, ℓ_1 and ℓ_∞. QED

In particular, each $\sigma_p (1 \leq p \leq \infty)$ is a Banach space norm.

Banach Ideals

It may be disappointing to learn that there is little hope of generalizing a result like Theorem 6.4 to general Banach spaces. The failure is due to a lack of approximation properties, a topic which will not be covered in this text. However, if we are working with a finitely determined norm on operators between Banach spaces, like the p-summing norm π_p, we still can aspire to gain information by investigating local behaviour. Though we have the p-summing operators and their relatives in mind, there is an abstract framework, that of Banach ideals, which is particularly appropriate for our development. In addition to affording us considerable flexibility it also provides valuable insight into the most essential inner workings of trace duality.

An *operator ideal* $\mathcal{A}$ is a method of ascribing to each pair (X, Y) of Banach spaces a linear subspace $\mathcal{A}(X, Y)$ of $\mathcal{L}(X, Y)$ such that the following two conditions are satisfied:

(I 1) *$x^* \otimes y \in \mathcal{A}(X, Y)$ for any $x^* \in X^*$, $y \in Y$,*

(I 2) *If X_0 and Y_0 are Banach spaces, then $uvw \in \mathcal{A}(X_0, Y_0)$ whenever $u \in \mathcal{L}(Y, Y_0)$, $v \in \mathcal{A}(X, Y)$, $w \in \mathcal{L}(X_0, X)$.*

Further, if for each (X, Y), $\mathcal{A}(X, Y)$ is supplied with a norm α in such a way that

(N 1) *$\alpha(x^* \otimes y) = \|x^*\| \cdot \|y\|$ for all $x^* \in X^*$ and $y \in Y$,*

(N 2) *$\alpha(uvw) \leq \|u\| \cdot \alpha(v) \cdot \|w\|$ whenever X_0 and Y_0 are Banach spaces and $u \in \mathcal{L}(Y, Y_0)$, $v \in \mathcal{A}(X, Y)$ and $w \in \mathcal{L}(X_0, X)$,*

(N 3) *$[\mathcal{A}(X, Y), \alpha]$ is a Banach space,*

then $[\mathcal{A}, \alpha]$ is called a *Banach operator ideal,* or *Banach ideal,* for short.

We shall say that the Banach ideal $[\mathcal{A}, \alpha]$ is *injective* if $\alpha(uv) = \alpha(v)$ whenever X, Y, Y_0 are Banach spaces, $u \in \mathcal{L}(Y, Y_0)$ is isometric and $v \in \mathcal{A}(X, Y)$.

Two basic remarks are in order:

- Regardless of the operator ideal $\mathcal{A}$, we have $\mathcal{F}(X, Y) \subset \mathcal{A}(X, Y)$ for all Banach spaces X, Y. This is an immediate consequence of (I 1) and the linear structure of $\mathcal{A}(X, Y)$.
- Regardless of the Banach ideal $[\mathcal{A}, \alpha]$, we have $\|u\| \leq \alpha(u)$ for any u in $\mathcal{A}(X, Y)$.

  To see this, begin by noting that

$$\|u\| = \|u^*\| = \sup\{\|u^*(y^*)\| : y^* \in B_{Y^*}\}.$$

But each $u^*(y^*) = y^*u \in X^*$ is unwittingly an operator of rank at most one and so, according to (N 1) and (N 2), $\|y^*u\| = \alpha(y^*u) \leq \|y^*\| \cdot \alpha(u)$. At once, $\|u\| \leq \alpha(u)$. Note that this information is independent of the completeness axiom (N 3).

Familiar examples of Banach ideals abound: the ideals $[\mathcal{L}, \|\cdot\|]$ of all (bounded linear) operators, $[\mathcal{K}, \|\cdot\|]$ of all compact operators, $[\mathcal{V}, \|\cdot\|]$ of all completely continuous operators, $[\mathcal{W}, \|\cdot\|]$ of all weakly compact operators, and so on, where $\|\cdot\|$ is the usual operator norm. Further, for any fixed $1 \leq p < \infty$, there are the Banach ideals $[\Pi_p, \pi_p]$ of all p-summing operators and $[\mathcal{I}_p, \iota_p]$ of all p-integral operators. However, there is no way to make $\mathcal{F}$, the ideal of finite rank operators, a Banach ideal: we shall see later in this chapter that the nuclear operators form the smallest Banach ideal.

The topological structure of a Banach ideal is uniquely determined:

6.5 Proposition: *Suppose $[\mathcal{A}, \alpha]$ and $[\mathcal{B}, \beta]$ are Banach ideals where $\mathcal{A}(X,Y) \subset \mathcal{B}(X,Y)$ for all Banach spaces X and Y. Then there is a constant C such that, regardless of the Banach spaces X and Y,*

$$\beta(u) \leq C \cdot \alpha(u)$$

for every $u \in \mathcal{A}(X,Y)$.

Proof. Suppose that no such C exists. Then, for each positive integer n, there are Banach spaces X_n, Y_n and an operator $u_n \in \mathcal{A}(X_n, Y_n)$ such that $\beta(u_n) \geq n$ even though $\alpha(u_n) \leq 2^{-n}$. Let $X = (\oplus X_n)_2$. Introduce the natural projection $p_k^X : X \to X_k : (x_n)_n \mapsto x_k$ and the natural injection $j_k^X : X_k \to X : x \mapsto (x_n)_n$ with $x_k = x$ and $x_n = 0$ if $n \neq k$. Each of these maps has norm one. Similarly, we work with $Y = (\oplus Y_n)_2$ and the attendant maps $p_k^Y : Y \to Y_k$ and $j_k^Y : Y_k \to Y$. Evidently,

$$\alpha\Big(\sum_{k=m}^{m+n} j_k^Y u_k p_k^X\Big) \leq \sum_{k=m}^{m+n} \|j_k^Y\| \cdot \alpha(u_k) \cdot \|p_k^X\| \leq 2^{-m+1}$$

for all $m, n \in \mathbf{N}$, so that $(\sum_{k \leq n} j_k^Y u_k p_k^X)_n$ is a Cauchy sequence in $[\mathcal{A}(X,Y), \alpha]$. By completeness, $u := \sum_{k=1}^{\infty} j_k^Y u_k p_k^X$ exists in $[\mathcal{A}(X,Y), \alpha]$. But then u belongs to $\mathcal{B}(X,Y)$, and so $\beta(u) < \infty$. This cannot be reconciled with the following consequence of axiom (N 2): for each n,

$$n \leq \beta(u_n) = \beta(p_n^Y u j_n^X) \leq \|p_n^Y\| \cdot \beta(u) \cdot \|j_n^X\| = \beta(u). \qquad \text{QED}$$

6.6 Corollary: *If $[\mathcal{A}, \alpha]$ and $[\mathcal{A}, \beta]$ are Banach ideals, then there is a universal constant $C > 0$ such that $(1/C) \cdot \beta \leq \alpha \leq C \cdot \beta$.*

Adjoint Ideals

We are now primed for an extension of trace duality to Banach ideals. Take a Banach ideal $[\mathcal{A}, \alpha]$, and let X and Y be Banach spaces. We say that the operator $v : X \to Y$ belongs to

$$\mathcal{A}^*(X,Y)$$

provided there is a constant $c \geq 0$ such that regardless of the finite dimensional normed spaces E and F and operators $w \in \mathcal{L}(E,X)$, $u \in \mathcal{L}(Y,F)$ and $t \in \mathcal{L}(F,E)$, the composition

$$E \xrightarrow{w} X \xrightarrow{v} Y \xrightarrow{u} F \xrightarrow{t} E$$

satisfies

$$|\mathrm{tr}\,(tuvw)| \leq c \cdot \alpha(t) \cdot \|u\| \cdot \|w\|.$$

The collection of all such c has an infimum – readily seen to be a minimum – which is denoted by

$$\alpha^*(v).$$

We ought to take care that we are employing this symbol in a way that is consistent with its earlier use within their common setting; let us distinguish temporarily between the two by writing α^*_{new} and α^*_{old}. To show they are the same, consider the composition of linear maps

$$E \xrightarrow{w} E \xrightarrow{v} F \xrightarrow{u} F \xrightarrow{t} E$$

where E and F are finite dimensional. Since

$$|\operatorname{tr}(tuvw)| = |\operatorname{tr}(wtuv)| \le \alpha^*_{old}(v)\cdot\|w\|\cdot\alpha(t)\cdot\|u\|,$$

we certainly have $\alpha^*_{new}(v) \le \alpha^*_{old}(v)$. On the other hand, if we exercise our freedom of choice and take u and w to be identity maps we find that

$$|\operatorname{tr}(tv)| \le \alpha^*_{new}(v)\cdot\alpha(t).$$

Take the supremum over all $\alpha(t) \le 1$ to obtain $\alpha^*_{old}(v) \le \alpha^*_{new}(v)$.

With any apprehensions about the notation $\alpha^*(v)$ laid to rest, we prove that we have created a desirable object.

6.7 Theorem: *$[\mathcal{A}^*, \alpha^*]$ is a Banach ideal.*

Proof. Routinely $[\mathcal{A}^*(X,Y), \alpha^*]$ is a normed space for any Banach spaces X and Y.

Of some note is the expected fact that $\|v\| \le \alpha^*(v)$ for any $v \in \mathcal{A}^*(X,Y)$. Indeed, given $y^* \in Y^*$ and $x \in X$, if we define $m_x \in \mathcal{L}(\mathbf{K}, X)$ by $m_x(\lambda) = \lambda x$, then it is plain that

$$\begin{aligned} |\langle v(x), y^*\rangle| &= |\operatorname{tr}(id_{\mathbf{K}} y^* v m_x)| \\ &\le \alpha^*(v)\cdot\alpha(id_{\mathbf{K}})\cdot\|y^*\|\cdot\|m_x\| = \alpha^*(v)\cdot\|y^*\|\cdot\|x\|, \end{aligned}$$

and so $\|v\| \le \alpha^*(v)$.

In particular, for any $x^* \in X^*$ and $y \in Y$,

$$\|x^*\|\cdot\|y\| = \|x^* \otimes y\| \le \alpha^*(x^* \otimes y).$$

On the other hand, if E and F are finite dimensional and $w: E \to X$, $u: Y \to F$ and $t: F \to E$ are any operators, then

$$\begin{aligned} \big|\operatorname{tr}\big(tu(x^* \otimes y)w\big)\big| &= |\langle x^* w, tuy\rangle| \le \|x^*\|\cdot\|y\|\cdot\|t\|\cdot\|u\|\cdot\|w\| \\ &\le \|x^*\|\cdot\|y\|\cdot\alpha(t)\cdot\|u\|\cdot\|w\|. \end{aligned}$$

Combining this with the above estimate, we see that $\alpha^*(x^* \otimes y) = \|x^*\|\cdot\|y\|$.

As the ideal property (axioms (I 2) and (N 2)) is routine, we finish by proving the completeness of $[\mathcal{A}^*(X,Y), \alpha^*]$ for arbitrary Banach spaces X and Y. We do this by showing that absolutely convergent series converge (see 1.1).

So let $v_n \in \mathcal{A}^*(X,Y)$ satisfy $\sum_n \alpha^*(v_n) < \infty$. Immediately $\sum_n \|v_n\| < \infty$ and so $\sum_n v_n$ converges to v, say, in the Banach space $[\mathcal{L}(X,Y), \|\cdot\|]$. For finite dimensional Banach spaces E and F, consider the composition of operators

$$E \xrightarrow{w} X \xrightarrow{v} Y \xrightarrow{u} F \xrightarrow{t} E.$$

Since tr is a member of $\mathcal{L}(E,E)^*$ we have

$$\begin{aligned} |\mathrm{tr}\,(tuvw)| &= \big|\,\mathrm{tr}\,\big(tu(\sum_n v_n)w\big)\,\big| \leq \sum_n |\mathrm{tr}\,(tuv_nw)| \\ &\leq \sum_n \alpha^*(v_n)\cdot\alpha(t)\cdot\|u\|\cdot\|w\|. \end{aligned}$$

It follows that $v \in \mathcal{A}^*(X,Y)$, with $\alpha^*(v) \leq \sum_n \alpha^*(v_n)$. From this it is a simple matter to complete the proof. QED

The Banach ideal $[\mathcal{A}^*, \alpha^*]$ is called the *adjoint ideal* of $[\mathcal{A}, \alpha]$.

The promised relations between summing and integral operators will come to fruition only after we have investigated the *biadjoint ideal*

$$[\mathcal{A}^{**}, \alpha^{**}]$$

of $[\mathcal{A}, \alpha]$, that is, the adjoint ideal of $[\mathcal{A}^*, \alpha^*]$.

Some notation is called for. As before, we write

$$\mathcal{F}_X$$

for the collection of all finite dimensional subspaces of X, and if $E \in \mathcal{F}_X$, we label the canonical inclusion as

$$i_E : E \longrightarrow X.$$

The symbol

$$\mathcal{C}_X$$

is reserved for the collection of all (closed) finite codimensional subspaces of X; for $Z \in \mathcal{C}_X$,

$$q_Z : X \longrightarrow X/Z$$

is the natural quotient map.

6.8 Biadjoint Criterion: *Suppose that $[\mathcal{A}, \alpha]$ is a Banach ideal and that X and Y are Banach spaces. An operator $v \in \mathcal{L}(X,Y)$ belongs to $\mathcal{A}^{**}(X,Y)$ precisely when*

$$c := \sup\big\{\alpha(q_Z v i_E) : E \in \mathcal{F}_X,\ Z \in \mathcal{C}_Y\big\} < \infty,$$

and in this case, $\alpha^{**}(v) = c$.

Proof. Since we have already noted that α and α^{**} coincide when the spaces are finite dimensional, we certainly know that if $v \in \mathcal{A}^{**}(X,Y)$, then

$$c = \sup \alpha^{**}(q_Z v i_E) \leq \alpha^{**}(v) < \infty.$$

Conversely, suppose c is finite. Let E_0 and F_0 be finite dimensional Banach spaces, and consider the composition of operators t, u, v and w

$$\begin{array}{ccccccccc} E_0 & \xrightarrow{w} & X & \xrightarrow{v} & Y & \xrightarrow{u} & F_0 & \xrightarrow{t} & E_0 \\ & {\scriptstyle w_0}\searrow & \nearrow{\scriptstyle i_E} & & & {\scriptstyle q_Z}\searrow & \nearrow{\scriptstyle u_0} & & \\ & & E & & & & Y/Z & & \end{array}$$

augmented by the natural factorizations of w through $E = w(E_0)$ and of u through Y/Z where $Z=\ker u$. Then

$$\begin{aligned} |\mathrm{tr}\,(tuvw)| &= |\mathrm{tr}\,(tu_0(q_Z v i_E)w_0)| \le \alpha^{**}(q_Z v i_E)\cdot\alpha^*(t)\cdot\|u_0\|\cdot\|w_0\| \\ &= \alpha(q_Z v i_E)\cdot\alpha^*(t)\cdot\|u\|\cdot\|w\| \le c\cdot\alpha^*(t)\cdot\|u\|\cdot\|w\|. \end{aligned}$$

It follows that $v \in \mathcal{A}^{**}(X,Y)$, with $\alpha^{**}(v) \le c$. QED

Maximal Ideals

The family of Banach ideals has a natural partial order. Given two Banach ideals $[\mathcal{A},\alpha]$ and $[\mathcal{B},\beta]$, we shall write

$$[\mathcal{A},\alpha] \subset [\mathcal{B},\beta]$$

if, regardless of the Banach spaces X and Y, we have $\mathcal{A}(X,Y) \subset \mathcal{B}(X,Y)$, and $\beta(v) \le \alpha(v)$ for all $v \in \mathcal{A}(X,Y)$. Naturally,

$$[\mathcal{A},\alpha] = [\mathcal{B},\beta]$$

means that both relations $[\mathcal{A},\alpha] \subset [\mathcal{B},\beta]$ and $[\mathcal{B},\beta] \subset [\mathcal{A},\alpha]$ hold simultaneously.

It is worth noting that if $[\mathcal{A},\alpha]\subset[\mathcal{B},\beta]$, then $[\mathcal{B}^*,\beta^*]\subset[\mathcal{A}^*,\alpha^*]$. Moreover:

6.9 Corollary: *If $[\mathcal{A},\alpha]$ is any Banach ideal, then $[\mathcal{A},\alpha] \subset [\mathcal{A}^{**},\alpha^{**}]$.*

In fact, a glance at the Biadjoint Criterion will reveal more:

- *With respect to $\subset$, $[\mathcal{A}^{**},\alpha^{**}]$ is the largest of the Banach ideals $[\mathcal{B},\beta]$ which satisfy $\alpha(u) = \beta(u)$ for all $u \in \mathcal{L}(E,F)$ whenever E and F are finite dimensional.*

Naturally enough, we say that a Banach ideal $[\mathcal{A},\alpha]$ is *maximal* if

$$[\mathcal{A},\alpha] = [\mathcal{A}^{**},\alpha^{**}].$$

Maximal ideals are ubiquitous:

6.10 Theorem: *A Banach ideal is maximal if and only if it is the adjoint of a Banach ideal.*

Proof. Only one implication requires proof: if $[\mathcal{A},\alpha]$ is a Banach ideal, we must show that $[\mathcal{A}^*,\alpha^*]$ is maximal.

Let $v \in \mathcal{A}^{***}(X,Y)$ where X and Y are Banach spaces. By our characterization 6.8 of biadjoints, we have

$$\alpha^{***}(v) = \sup\{\alpha^*(q_Z v i_E) : E \in \mathcal{F}_X,\ Z \in \mathcal{C}_Y\}.$$

If E_0 and F_0 are finite dimensional we can, as in the proof of 6.8, consider the composition of bounded linear maps t, u, v and w

$$\begin{array}{ccccccccc} E_0 & \xrightarrow{w} & X & \xrightarrow{v} & Y & \xrightarrow{u} & F_0 & \xrightarrow{t} & E_0 \\ & {\scriptstyle w_0}\searrow & \nearrow{\scriptstyle i_E} & & {\scriptstyle q_Z}\searrow & \nearrow{\scriptstyle u_0} & & & \\ & E & & & Y/Z & & & & \end{array}$$

where we have tacked on the natural factorizations of w through $E = w(E_0)$ and of u through Y/Z with $Z = \ker u$. Then

$$\begin{aligned} |\operatorname{tr}(tuvw)| &= |\operatorname{tr}(tu_0(q_Z v i_E)w_0)| \leq \alpha^*(q_Z v i_E)\cdot\alpha(t)\cdot\|u_0\|\cdot\|w_0\| \\ &\leq \alpha^{***}(v)\cdot\alpha(t)\cdot\|u\|\cdot\|w\|, \end{aligned}$$

and so $v \in \mathcal{A}^*(X,Y)$ with $\alpha^*(v) \leq \alpha^{***}(v)$. QED

This proliferation of maximal ideals would go undetected amongst the *classical Banach ideals,* that is, those with the operator norm.

6.11 Proposition: *$[\mathcal{L}, \|\cdot\|]$ is maximal and is the only classical Banach ideal that is maximal.*

$[\mathcal{L}, \|\cdot\|]$ is evidently the largest of the maximal ideals. It turns out that the smallest is $[\mathcal{I}_1, \iota_1]$, but this requires some solid effort. As a prelude to proving the maximality of $[\mathcal{I}_p, \iota_p]$ for any $p \geq 1$, we establish an extension theorem for p-integral operators.

6.12 Proposition: *Suppose that $1 \leq p < \infty$, that X, Y and Z are Banach spaces with X a subspace of Z. If $u \in \mathcal{I}_p(X,Y)$, there is some $\tilde{u} \in \mathcal{I}_p(Z, Y^{**})$, with $\iota_p(\tilde{u}) = \iota_p(u)$, such that the diagram*

$$\begin{array}{ccc} Z & \xrightarrow{\tilde{u}} & Y^{**} \\ \cup & & \uparrow {\scriptstyle k_Y} \\ X & \xrightarrow{u} & Y \end{array}$$

commutes.

Consequently, if Y is 1-complemented in Y^{**}, every p-integral operator $X \to Y$ can be extended to a p-integral operator $Z \to Y$ without changing the p-integral norm.

Proof. Consider a typical p-integral factorization

$$k_Y u : X \xrightarrow{b} L_\infty(\mu) \xrightarrow{i_p} L_p(\mu) \xrightarrow{a} Y^{**}$$

for $u \in \mathcal{I}_p(X,Y)$. Since $L_\infty(\mu)$ is injective (4.14), b admits an extension $\tilde{b} \in \mathcal{L}(Z, L_\infty(\mu))$ with $\|\tilde{b}\| = \|b\|$. This creates a p-integral extension $\tilde{u} : Z \to Y^{**}$ of $k_Y u$ having a factorization

$$\tilde{u} : Z \xrightarrow{\tilde{b}} L_\infty(\mu) \xrightarrow{i_p} L_p(\mu) \xrightarrow{a} Y^{**}.$$

As $\iota_p(\tilde{u}) \leq \|a\|\cdot\|\tilde{b}\| = \|a\|\cdot\|b\|$, we find that $\iota_p(\tilde{u}) \leq \iota_p(u)$. The reverse inequality is trivial. QED

6.13 Theorem: *The ideals $[\mathcal{I}_p, \iota_p]$ of p-integral operators, $1 \leq p \leq \infty$, are maximal.*

Proof. We shall only carry out the proof for $1 \le p < \infty$. The simple adjustments necessary to treat the case $p = \infty$ are left to the reader.

Take $v \in \mathcal{I}_p^{**}(X, Y)$, where X and Y are any Banach spaces. What is required is to show that $v \in \mathcal{I}_p(X, Y)$ with $\iota_p(v) \le \iota_p^{**}(v)$. We shall make full use of the development to date.

Let $\mathcal{P}$ denote the family of all pairs (E, Z) with $E \in \mathcal{F}_X$ and $Z \in \mathcal{C}_Y$. Given $(E, Z) \in \mathcal{P}$, let

$$(E, Z)^{\#} := \{(\tilde{E}, \tilde{Z}) \in \mathcal{P} : E \subset \tilde{E},\ \tilde{Z} \subset Z\}.$$

As the sets $(E, Z)^{\#}$ form a filter basis of subsets of $\mathcal{P}$, they are contained in some ultrafilter $\mathcal{U}$ on $\mathcal{P}$.

According to the Biadjoint Criterion 6.8,

$$\iota_p^{**}(v) = \sup\{\iota_p(q_Z v i_E) : (E, Z) \in \mathcal{P}\}.$$

Let K be the unit ball of X^* endowed with the weak$*$topology, and regard X as a subspace of $\mathcal{C}(K)$. The preceding proposition allows us to extend each $q_Z v i_E : E \to Y/Z$ to an operator

$$v_{(E,Z)} : \mathcal{C}(K) \longrightarrow Y/Z$$

with

$$\iota_p(v_{(E,Z)}) = \iota_p(q_Z v i_E) \le \iota_p^{**}(v).$$

Call the Pietsch Domination Theorem 2.12 into play, identifying K with the weak$*$compact norming subset $\{\delta_{x^*} : x^* \in K\}$ of $B_{\mathcal{C}(K)^*}$: it furnishes a probability measure $\mu_{(E,Z)} \in \mathcal{C}(K)^*$ such that

$$\begin{aligned} \|v_{(E,Z)} f\| &\le \pi_p(v_{(E,Z)}) \cdot \Big(\int_K |f(x^*)|^p d\mu_{(E,Z)}(x^*) \Big)^{1/p} \\ &\le \iota_p^{**}(v) \cdot \Big(\int_K |f(x^*)|^p d\mu_{(E,Z)}(x^*) \Big)^{1/p} \end{aligned}$$

for each $f \in \mathcal{C}(K)$.

The strategy now is to take limits along $\mathcal{U}$. We shall see that this will enable us to provide an operator $\tilde{v} : \mathcal{C}(K) \to Y^{**}$ with $\tilde{v}\,|_X = k_Y v$ and a probability measure $\mu \in \mathcal{C}(K)^*$ such that

$$(\bullet) \qquad \|\tilde{v} f\| \le \iota_p^{**}(v) \cdot \Big(\int_K |f(x^*)|^p \, d\mu(x^*) \Big)^{1/p}$$

for every $f \in \mathcal{C}(K)$. Deciphering this again requires the Domination Theorem 2.12: we find that $\tilde{v}$ is p-summing with $\pi_p(\tilde{v}) \le \iota_p^{**}(v)$. Take note of the domain of $\tilde{v}$ and summon up Corollary 5.8: $\tilde{v}$ is p-integral, with $\iota_p(\tilde{v}) = \pi_p(\tilde{v})$. All in all, v is p-integral, and $\iota_p(v) \le \iota_p(\tilde{v}) \le \iota_p^{**}(v)$.

To complete the proof, then, we must supply the technical details which culminate in $(\bullet)$. We begin with the construction of μ. All the $\mu_{(E,Z)}$ are probability measures, so for each $g \in \mathcal{C}(K)$, the map

$$\mathcal{P} \longrightarrow \mathbf{K} : (E, Z) \mapsto \int_K g(x^*) \, d\mu_{(E,Z)}(x^*)$$

is certainly bounded. As $\mathcal{U}$ is an ultrafilter, the limit $\lim_{\mathcal{U}} \int_K g(x^*) \, d\mu_{(E,Z)}(x^*)$ of $\mathcal{U}$'s image under the above map exists, and scarcely a moment's thought is needed to see that

$$\mu : \mathcal{C}(K) \longrightarrow \mathbf{K} : g \mapsto \lim_{\mathcal{U}} \int_K g(x^*)\, d\mu_{(E,Z)}(x^*)$$

is linear, positive and obliges by taking the constantly one function to 1. In short, μ is a regular Borel probability measure on K satisfying

$$\int_K g(x^*)\, d\mu(x^*) = \lim_{\mathcal{U}} \int_K g(x^*)\, d\mu_{(E,Z)}(x^*)$$

for all $g \in \mathcal{C}(K)$.

We finally turn our attention to the construction of $\tilde{v}$. This must in some sense represent a limit of the $v_{(E,Z)}$'s. Proceed pointwise: fix $y^* \in Y^*$ and $f \in \mathcal{C}(K)$, and for $(E, Z) \in \mathcal{P}$ define

$$s_{y^*,f} : \mathcal{P} \longrightarrow \mathbf{K} : (E, Z) \mapsto \begin{cases} \langle y^*, v_{(E,Z)} f\rangle & \text{if } y^* \in Z^\perp = (Y/Z)^* \\ 0 & \text{if not .} \end{cases}$$

Evidently, $s_{y^*,f}$ is a bounded function, so $\lim_{\mathcal{U}} s_{y^*,f}(E, Z)$ exists. Routine manipulations show that

$$s_{y^*} : \mathcal{C}(K) \longrightarrow \mathbf{K} : f \mapsto \lim_{\mathcal{U}} s_{y^*,f}(E, Z)$$

is linear. It is bounded since $\|v_{(E,Z)}\| \leq \iota_p^{**}(v)$ for all $(E, Z) \in \mathcal{P}$, and so

$$|s_{y^*}(f)| \leq \iota_p^{**}(v) \cdot \|y^*\| \cdot \|f\|.$$

More: we have linearity in y^*, so the map $Y^* \to \mathcal{C}(K)^* : y^* \mapsto s_{y^*}$ is linear and bounded. Its adjoint takes $\mathcal{C}(K)^{**}$ to Y^{**}, and so restriction creates a bounded linear map

$$\tilde{v} : \mathcal{C}(K) \longrightarrow Y^{**}.$$

That $\tilde{v}\,|_X = k_Y v$ is straightforward: simply observe that

$$\langle y^*, \tilde{v}x\rangle = s_{y^*}(x) = \lim_{\mathcal{U}} s_{y^*,x}(E, Z)$$

for all $x \in X$ and $y^* \in Y^*$, whereas, if $y^* \in Z^\perp$ and $x \in E$, as they will be eventually, then

$$s_{y^*,x}(E, Z) = \langle y^*, q_Z v i_E x\rangle = \langle y^*, vx\rangle.$$

Property ($\bullet$) follows easily, since for all $f \in \mathcal{C}(K)$,

$$\begin{aligned} \|\tilde{v}f\| &= \sup_{y^* \in B_{Y^*}} |\lim_{\mathcal{U}} s_{y^*,f}(E, Z)| \leq \lim_{\mathcal{U}} \sup_{y^* \in B_{Y^*}} |s_{y^*,f}(E, Z)| \\ &\leq \lim_{\mathcal{U}} \|v_{(E,Z)} f\| \leq \iota_p^{**}(v) \cdot \lim_{\mathcal{U}} \Big(\int_K |f(x^*)|^p d\mu_{(E,Z)}(x^*)\Big)^{1/p} \\ &= \iota_p^{**}(v) \cdot \Big(\int_K |f(x^*)|^p d\mu(x^*)\Big)^{1/p}. \end{aligned}$$

All is as it should be. QED

We have already mentioned the extreme position of the Banach ideal of 1-integral operators. The next lemma is the key to understanding this.

6.14 Lemma: *For any finite dimensional Banach space E and any operator $u \in \mathcal{L}(E)$,*

$$|\mathrm{tr}\,(u)| \leq \iota_1(u).$$

Proof. Let $\varepsilon > 0$ be given, and let

$$u : E \xrightarrow{b} L_\infty(\mu) \xrightarrow{i_1} L_1(\mu) \xrightarrow{a} E$$

be a 1-integral factorization of u, where μ is a probability measure, $\|b\| \le 1$ and $\|a\| \le \sqrt{1+\varepsilon}\cdot\iota_1(u)$. Since $b(E)$ is a finite dimensional subspace of $L_\infty(\mu)$, a known $\mathcal{L}_{\infty,\lambda}$-space for each $\lambda > 1$, it follows that there are a finite, say k, dimensional subspace F of $L_\infty(\mu)$ containing $u(E)$, and bounded linear operators $v_1 : \ell_\infty^k \to F$, $v_2 : F \to \ell_\infty^k$ satisfying $v_1 v_2 = id_F$, $\|v_1\| \le \sqrt{1+\varepsilon}$, $\|v_2\| \le 1$. The factorization now looks like this:

$$u : E \xrightarrow{w} \ell_\infty^k \xrightarrow{v} E.$$

Here $w = v_2\, b$ has norm at most one, and $v = a i_1 v_1$ satisfies $\iota_1(v) \le (1+\varepsilon)\iota_1(u)$.

By 2.15, v has a bounded linear factorization

$$v : \ell_\infty^k \xrightarrow{j_1} L_1(\nu) \xrightarrow{\tilde{v}} E,$$

where ν is a regular Borel probability measure on $\{1,\ldots,k\}$, $\|\tilde{v}\| = \pi_1(v)$, and j_1 is the canonical map of $\mathcal{C}(\{1,\ldots,k\}) = \ell_\infty^k$ into $L_1(\nu)$. We may as well assume $\nu_i = \nu(\{i\}) > 0$ for all $i \in \{1,\ldots,k\}$, so, using point masses, we can express $\nu = \sum_{i\le k} \nu_i \delta_i$. Consequently, $L_1(\nu)$ appears as the space of all k-tuples $\zeta = (\zeta_1,\ldots,\zeta_k)$ of scalars with norm $\|\zeta\| = \sum_{i\le k} \nu_i |\zeta_i|$, and $j_1 : \ell_\infty^k \to L_1(\nu)$ is nothing other than the formal identity.

We are in a position to compute. With respect to the usual unit coordinate vectors $e_i = (\delta_{ij})_{j\le k}$ we can express $w\tilde{v}j_1 : \ell_\infty^k \to \ell_\infty^k$ as $w\tilde{v}j_1 = \sum_{i\le k} e_i \otimes w\tilde{v}j_1 e_i$. Consequently,

$$\begin{aligned} |\mathrm{tr}\,(u)| &= |\mathrm{tr}\,(\tilde{v}j_1 w)| = |\mathrm{tr}\,(w\tilde{v}j_1)| = \Big|\sum_{i\le k} \langle e_i, w\tilde{v}j_1 e_i\rangle\Big| \\ &\le \|w\|\cdot\|\tilde{v}\|\cdot\sum_{i\le k} \|j_1 e_i\| \le \pi_1(v)\cdot\sum_{i\le k} \nu_i \le (1+\varepsilon)\cdot\iota_1(u). \end{aligned}$$

Let ε tend to zero to complete the proof. QED

6.15 Corollary: *$[\mathcal{I}_1, \iota_1]$ is the smallest of all the maximal Banach ideals.*

Proof. In view of Theorem 6.10 all we need to show is that $[\mathcal{I}_1, \iota_1] \subset [\mathcal{A}^*, \alpha^*]$ holds, regardless of the Banach ideal $[\mathcal{A}, \alpha]$. But if E and F are finite dimensional normed spaces then, no matter how we choose operators $w : E \to X$, $u : Y \to F$ and $t : F \to E$, we know that whenever $v : X \to Y$ is integral,

$$\begin{aligned} |\mathrm{tr}\,(tuvw)| &\le \iota_1(tuvw) \le \iota_1(v)\cdot\|t\|\cdot\|u\|\cdot\|w\| \\ &\le \iota_1(v)\cdot\|u\|\cdot\|w\|\cdot\alpha(t). \end{aligned}$$

QED

Now for the long-promised computations identifying the adjoint ideals of our main examples.

6.16 Theorem: (a) *$[\mathcal{I}_1{}^*, \iota_1{}^*] = [\mathcal{L}, \|\cdot\|]$, and $[\mathcal{L}^*, \|\cdot\|^*] = [\mathcal{I}_1, \iota_1]$.*

(b) *If $1 < p \le \infty$, then $[\mathcal{I}_p{}^*, \iota_p{}^*] = [\Pi_{p^*}, \pi_{p^*}]$ and $[\Pi_{p^*}^*, \pi_{p^*}^*] = [\mathcal{I}_p, \iota_p]$.*

Proof. (a) Corollary 6.15 assures us that $[\mathcal{I}_1, \iota_1] \subset [\mathcal{L}^*, \|\cdot\|^*]$, and so $[\mathcal{L}^{**}, \|\cdot\|^{**}] \subset [\mathcal{I}_1{}^*, \iota_1{}^*] \subset [\mathcal{L}, \|\cdot\|] \subset [\mathcal{L}^{**}, \|\cdot\|^{**}]$. Thus $[\mathcal{L}, \|\cdot\|] = [\mathcal{I}_1{}^*, \iota_1{}^*]$.

Given this and the fact that the ideal $[\mathcal{I}_1, \iota_1]$ is maximal, we see that $[\mathcal{L}^*, \|\cdot\|^*] = [\mathcal{I}_1{}^{**}, \iota_1{}^{**}] = [\mathcal{I}_1, \iota_1]$.

(b) We assume $1 < p < \infty$. The case $p = \infty$ requires minimal adjustments which we leave to the reader.

Let $v \in \Pi_{p^*}(X, Y)$, suppose E and F are finite dimensional, and let $w \in \mathcal{L}(E, X)$, $u \in \mathcal{L}(Y, F)$, $t \in \mathcal{L}(F, E)$. From Lemma 6.14 we get $|\mathrm{tr}\,(tuvw)| \leq \iota_1(tuvw)$ which thanks to the composition theorem 5.16 leads to

$$|\mathrm{tr}\,(tuvw)| \leq \pi_{p^*}(vw) \cdot \iota_p(tu) \leq \pi_{p^*}(v) \cdot \|w\| \cdot \|u\| \cdot \iota_p(t).$$

It follows that $v \in \mathcal{I}_p{}^*(X, Y)$ and $\iota_p{}^*(v) \leq \pi_{p^*}(v)$.

Suppose on the other hand that $v \in \mathcal{I}_p{}^*(X, Y)$ and let $x_1, \ldots, x_n \in X$ be given. Choose scalars $\lambda_1, \ldots, \lambda_n \geq 0$ such that $\sum_{k \leq n} \lambda_k^p = 1$ and $\sum_{k \leq n} \lambda_k \|vx_k\| = (\sum_{k \leq n} \|vx_k\|^{p^*})^{1/p^*}$. Next choose $y_1^*, \ldots, y_n^* \in B_{Y^*}$ such that $\langle y_k^*, vx_k \rangle = \|vx_k\|$ for $k = 1, \ldots, n$. Define operators w, u, t as follows:

$$w : \ell_p^n \longrightarrow X : (\zeta_k) \mapsto \textstyle\sum_{k \leq n} \zeta_k x_k \quad , \quad u : Y \longrightarrow \ell_\infty^n : y \mapsto (\langle y_k^*, y \rangle)_{k \leq n}$$

and

$$t : \ell_\infty^n \longrightarrow \ell_p^n : (\zeta_k)_{k \leq n} \mapsto (\lambda_k \zeta_k)_{k \leq n}.$$

Notice that $\|w\| = \|(x_k)\|_{p^*}^{weak}$ (see 2.2), $\|u\| \leq 1$, and $\iota_p(t) = \pi_p(t) = 1$ (thanks to 5.8 and example 2.9(e)). It follows that

$$\Big(\sum_{k=1}^n \|vx_k\|^{p^*}\Big)^{1/p^*} = \sum_{k=1}^n \lambda_k \langle y_k^*, vx_k \rangle = \sum_{k=1}^n \lambda_k \langle u^* e_k, vwe_k \rangle = \sum_{k=1}^n \langle e_k, tuvwe_k \rangle$$
$$= \mathrm{tr}\,(tuvw) \leq \iota_p(t) \cdot \|u\| \cdot \iota_p{}^*(v) \cdot \|w\| \leq \iota_p{}^*(v) \cdot \|(x_k)\|_{p^*}^{weak}.$$

Lo and behold: $v \in \Pi_{p^*}(X, Y)$ and $\pi_{p^*}(v) \leq \iota_p{}^*(v)$.

Now that we have $[\Pi_{p^*}, \pi_{p^*}] = [\mathcal{I}_p{}^*, \iota_p{}^*]$, we can use the maximality of $[\mathcal{I}_p, \iota_p]$ to conclude that $[\Pi_{p^*}^*, \pi_{p^*}^*] = [\mathcal{I}_p, \iota_p]$. QED

As we are already aware (5.9) that $[\Pi_2, \pi_2] = [\mathcal{I}_2, \iota_2]$, we have an interesting instance of 'self-adjointness':

6.17 Corollary: $[\Pi_2{}^*, \pi_2{}^*] = [\Pi_2, \pi_2]$.

6.18 Remark: The maximality of the ideal $[\Pi_p, \pi_p]$ $(1 \leq p < \infty)$ can easily be established directly. We present the argument, since it may serve as a model in many other instances.

Suppose $v \in \Pi_p{}^{**}(X, Y)$, and let $x_1, \ldots, x_n \in X$ be arbitrary. Let E be the span of $\{x_1, \ldots, x_n\}$. Choose $y_1^*, \ldots, y_n^* \in B_{Y^*}$ such that $\|vx_k\| = \langle y_k^*, vx_k \rangle$ for $1 \leq k \leq n$. Then $Z := \bigcap_{k \leq n} \ker(y_k^*)$ is a finite codimensional subspace of Y, and $\langle q_Z y, y_k^* \rangle = \langle y, y_k^* \rangle$ holds for all $y \in Y$ and $1 \leq k \leq n$. Our assertion now follows from

$$\Big(\sum_{k=1}^{n}\|vx_k\|^p\Big)^{1/p} = \Big(\sum_{k\le n}\langle y_k^*, vx_k\rangle^p\Big)^{1/p} = \Big(\sum_{k\le n}\langle y_k^*, q_Z v i_E x_k\rangle^p\Big)^{1/p}$$
$$\le \Big(\sum_{k\le n}\|q_Z v i_E x_k\|^p\Big)^{1/p}$$
$$\le \pi_p(q_Z v i_E)\cdot\|(x_k)_k\|_p^{weak} \le \pi_p{}^{**}(v)\cdot\|(x_k)_k\|_p^{weak}.$$

Using our new-found 'duality', we can quickly obtain some cases of coincidence, thereby throwing new light on 3.12.

6.19 Corollary: (a) *If $1 \le p \le 2$, if X is an $\mathcal{L}_p$-space and if Y is any Banach space, then for each $1 < r < 2$,*

$$\mathcal{I}_r(X,Y) = \Pi_r(X,Y) = \Pi_2(X,Y).$$

(b) *If $1 \le p,\ q \le 2$, if X is an $\mathcal{L}_p$-space and if Y is an $\mathcal{L}_q$-space, then for each $1 < r \le \infty$,*

$$\mathcal{I}_r(X,Y) = \mathcal{I}_2(X,Y).$$

Proof. (a) The basic inclusion relations between summing operators and integral operators (2.8, 5.1 and 5.5) mean that we only have to show that $v \in \mathcal{I}_2(X,Y) = \Pi_2(X,Y)$ implies that $v \in \mathcal{I}_r(X,Y)$. Accordingly, consider a composition of operators $E \xrightarrow{w} X \xrightarrow{v} Y \xrightarrow{u} F \xrightarrow{t} E$ with E and F finite dimensional. Using first 6.14 and 5.16 and then 3.15 (since $2 \le r^*$),

$$|\mathrm{tr}\,(tuvw)| \le \iota_1(tuvw) \le \pi_2(tu)\cdot\pi_2(vw) \le C\cdot\pi_{r^*}(tu)\cdot\pi_2(vw)$$
$$\le C\cdot\|w\|\cdot\pi_{r^*}(t)\cdot\|u\|\cdot\pi_2(v),$$

where C is a constant depending on X and r. So we end up with

$$\iota_r(v) = \pi_{r^*}^*(v) \le C\cdot\pi_2(v) = C\cdot\iota_2(v),$$

as asserted.

(b) follows from (a) and 3.7. QED

Orlicz's Theorem 3.12 can be generalized as follows.

6.20 Proposition: *The following statements about a Banach space X are equivalent:*

(i) $\Pi_1(X,Y) = \Pi_2(X,Y)$ *for every Banach space Y.*

(ii) $\mathcal{I}_\infty(Y,X) = \Pi_2(Y,X)$ *for every Banach space Y.*

In such a case, every weak ℓ_1 sequence in X is a strong ℓ_2 sequence.

Proof. (i)⇒(ii): By localization, it is good enough to show that there is a constant $c > 0$ such that for each $n \in \mathbf{N}$ and every $v \in \mathcal{L}(\ell_\infty^n, X)$ we have $\pi_2(v) \le c\cdot\|v\|$.

For this, look at a composition $E \xrightarrow{w} \ell_\infty^n \xrightarrow{v} X \xrightarrow{u} F \xrightarrow{t} E$ of operators with E and F finite dimensional. Then, thanks to 6.14, 5.8 and (i)

$$|\mathrm{tr}\,(tuvw)| = \iota_1(tuvw) \le \iota_1(tuv)\cdot\|w\| = \pi_1(tuv)\cdot\|w\|$$
$$\le c\cdot\pi_2(tuv)\cdot\|w\| \le c\cdot\|u\|\cdot\|v\|\cdot\|w\|\cdot\pi_2(t).$$

Using 6.17, it follows that $\pi_2(v) = \pi_2^*(v) \le c\cdot\|v\|$.

(ii)⇒(i): Select $u \in \Pi_2(X,Y)$; we need to prove that $u \in \Pi_1(X,Y)$. But if $v \in \mathcal{L}(c_0, X)$, the hypothesis (ii) tells us that $v \in \Pi_2(c_0, X)$ and so, by the Composition Theorem 2.22, we obtain $uv \in \Pi_1(c_0, Y)$. Applying 2.7, we draw our conclusion.

The argument used in the last step also yields the final statement. QED

Nuclear operators fit into the picture as follows.

6.21 Corollary: *For any* $1 \leq p < \infty$,

$$[\mathcal{N}_p^{**}, \nu_p^{**}] = [\mathcal{I}_p, \iota_p] \quad \text{and} \quad [\mathcal{N}_p^{*}, \nu_p^{*}] = [\Pi_{p^*}, \pi_{p^*}].$$

Here we put $[\Pi_\infty, \pi_\infty] := [\mathcal{L}, \|\cdot\|]$.

The proof is a simple combination of 6.8, 5.26, 6.13 and 6.16.

We can express this differently by referring to the remark after 6.9. With respect to $\subset$, $[\mathcal{I}_p, \iota_p]$ is the largest of the Banach ideals $[\mathcal{B}, \beta]$ which satisfy $\nu_p(u) = \beta(u)$ for all $u \in \mathcal{L}(E, F)$ whenever E and F are finite dimensional. There is an important counterpart in which the rôles of $\mathcal{I}_p$ and $\mathcal{N}_p$ are reversed.

6.22 Theorem: *For each* $1 \leq p < \infty$, $[\mathcal{N}_p, \nu_p]$ *is the smallest of all Banach ideals* $[\mathcal{B}, \beta]$ *such that* $\iota_p(u) = \beta(u)$ *for all* $u \in \mathcal{L}(E, F)$ *whenever* E *and* F *are finite dimensional normed spaces.*

Proof. Take any u in $\mathcal{N}_p(X, Y)$ and fix $\varepsilon > 0$. Consider a factorization $u : X \xrightarrow{b} c_0 \xrightarrow{M_\lambda} \ell_p \xrightarrow{a} Y$ with a and b compact and $\|a\| \cdot \|M_\lambda\| \cdot \|b\| \leq (1+\varepsilon) \cdot \nu_p(u)$. Composition with projections given by 3.3 relative to the sets $a^*(B_{Y^*})$ and $b(B_X)$ will yield finite rank operators $a_n : \ell_p \to Y$ and $b_n : X \to c_0$ such that $\lim_n \|a - a_n\| = \lim_n \|b - b_n\| = 0$. Given any Banach space operator $w : Z \to W$, let $Z \xrightarrow{w_1} Z/\ker(w) \xrightarrow{w_0} W$ be the canonical factorization where, naturally, $\|w_1\| = 1$ and $\|w_0\| = \|w\|$. For any m and n,

$$\begin{aligned} \beta(a_m M_\lambda b_m - a_n M_\lambda b_n) &\leq \beta\big((a_m - a_n) M_\lambda b_m\big) + \beta\big(a_n M_\lambda (b_m - b_n)\big) \\ &\leq \beta\big((a_m - a_n)_1 M_\lambda (b_m)_0\big) \cdot \|a_m - a_n\| + \beta\big((a_n)_1 M_\lambda (b_m - b_n)_0\big) \cdot \|a_n\|. \end{aligned}$$

In view of our hypothesis, this gives, with the help of 5.26,

$$\begin{aligned} &\beta(a_m M_\lambda b_m - a_n M_\lambda b_n) \\ &\qquad \leq \iota_p\big((a_m - a_n)_1 M_\lambda (b_m)_0\big) \cdot \|a_m - a_n\| + \iota_p\big((a_n)_1 M_\lambda (b_m - b_n)_0\big) \cdot \|a_n\| \\ &\qquad \leq \|a_m - a_n\| \cdot \|M_\lambda\| \cdot \|b_m\| + \|a_n\| \cdot \|M_\lambda\| \cdot \|b_m - b_n\|. \end{aligned}$$

Consequently, $(a_n M_\lambda b_n)$ is a Cauchy sequence in $[\mathcal{B}(X, Y), \beta]$. By completeness, its β-limit exists, and it must coincide with $u = a M_\lambda b$. In addition,

$$\begin{aligned} \beta(u) &= \lim_{n \to \infty} \beta(a_n M_\lambda b_n) \leq \lim_{n \to \infty} \beta((a_n)_1 M_\lambda (b_n)_0) \cdot \|a_n\| \\ &= \lim_{n \to \infty} \iota_p((a_n)_1 M_\lambda (b_n)_0) \cdot \|a_n\| \leq \lim_{n \to \infty} \|a_n\| \cdot \|M_\lambda\| \cdot \|b_n\| \\ &= \|a\| \cdot \|M_\lambda\| \cdot \|b\| \leq (1 + \varepsilon) \cdot \nu_p(u). \end{aligned}$$

The conclusion follows. QED

If we combine this with 6.15, then we get that

- *$[\mathcal{N}_1, \nu_1]$ is the smallest of all Banach ideals,*

but such an approach is overkill. A direct simple argument runs as follows. Let $[\mathcal{A}, \alpha]$ be any Banach ideal. Fix Banach spaces X and Y and let $u : X \to Y$ be a nuclear operator. Represent $u = \sum_n x_n^* \otimes y_n$ with $\sum_n \|x_n^*\| < \infty$ and $\|y_n\| = 1$ for each n. Then $\sum_n \alpha(x_n^* \otimes y_n) = \sum_n \|x_n^*\| < \infty$. By completeness $u = \sum x_n^* \otimes y_n$ is in $\mathcal{A}(X, Y)$ and $\alpha(u) \le \sum_n \alpha(x_n^* \otimes y_n) = \sum_n \|x_n^*\|$. It follows that $\alpha(u) \le \nu_1(u)$, and we are done.

We continue with a characterization of $\mathcal{I}_\infty = \Pi_1{}^*$ which shows that the original definition could have been phrased more loosely.

6.23 Proposition: *A Banach space operator $u : X \to Y$ belongs to $\mathcal{I}_\infty(X, Y)$ if and only if for each $\lambda > 1$ there exists a factorization*

$$(*) \qquad k_Y u : X \xrightarrow{b} W \xrightarrow{a} Y^{**}$$

where W is an $\mathcal{L}_{\infty,\lambda}$-space. Moreover, in this case

$$\iota_\infty(u) = \inf \|a\| \cdot \|b\|,$$

where the infimum extends over all factorizations $(*)$ as above.

Proof. Write Δ for the above infimum. If $u \in \mathcal{I}_\infty(X, Y)$, then $k_Y u$ factors through $L_\infty(\mu)$ for some probability measure μ, and we get $\Delta \le \iota_\infty(u)$.

Conversely, suppose $k_Y u$ admits, for each $\lambda > 1$, a factorization $(*)$ with W an $\mathcal{L}_{\infty,\lambda}$-space. Let $E \in \mathcal{F}_X$ and $Z \in \mathcal{C}_Y$ be given, and let $G \subset W$ be a finite dimensional subspace containing $b(E)$ having the property that there is an isomorphism $v : G \to \ell_\infty^n$, $n = \dim G$, with $\|v\| \cdot \|v^{-1}\| \le \lambda$. Writing b_0 for the map $E \to G$ induced by b, and a_0 for a's restriction to G, we end up with the commutative diagram

$$\begin{array}{ccccccc} X & & \xrightarrow{\quad u \quad} & & Y & & \\ \uparrow{\scriptstyle i_E} & & & & \downarrow{\scriptstyle k_Y} & \searrow^{q_Z} & \\ & & & & & & Y/Z \;, \\ & & & & & \nearrow_{q_Z^{**}} & \\ E & \xrightarrow{v b_0} & \ell_\infty^n & \xrightarrow{a_0 v^{-1}} & Y^{**} & & \end{array}$$

showing that

$$\iota_\infty(q_Z u i_E) \le \|q_Z^{**} a_0 v^{-1}\| \cdot \|v b_0\| \le \|a_0\| \cdot \|v^{-1}\| \cdot \|v\| \cdot \|b_0\| \le \lambda \cdot \|a\| \cdot \|b\|.$$

It follows that

$$\iota_\infty^{**}(u) = \sup_{E,Z} \iota_\infty(q_Z u i_E) \le \Delta,$$

and we are done by maximality of $[\mathcal{I}_\infty, \iota_\infty]$. QED

Trace duality enables us to include the following characterizations of $\mathcal{L}_\infty$-spaces.

6.24 Corollary: *A Banach space X is an $\mathcal{L}_\infty$-space if and only if $\Pi_1(X, Y) = \mathcal{I}_1(X, Y)$ for every Banach space Y if and only if $\Pi_1(Z, X) = \mathcal{I}_1(Z, X)$ for every Banach space Z.*

Proof. First notice that if $\Pi_1(X, \cdot) = \mathcal{I}_1(X, \cdot)$, then $\mathcal{I}_\infty(\cdot, X) = \mathcal{L}(\cdot, X)$ by 6.16. In particular, $id_X \in \mathcal{I}_\infty$ and X is an $\mathcal{L}_\infty$-space by 6.23. Conversely, if X is an $\mathcal{L}_\infty$-space, then $\Pi_1(X, \cdot) = \mathcal{I}_1(X, \cdot)$ follows from 5.8 by localization.

The other part is proved in an analogous manner, using 5.7 instead of 5.8.

QED

In Chapters 7 and 9, we will characterize those operators that factor through $\mathcal{L}_p$-spaces for $1 \leq p < \infty$.

APPLICATIONS

We conclude this chapter with a profound geometrical consequence of trace duality.

6.25 Lewis' Theorem: *Let E and F be n-dimensional normed spaces, and let α be any norm on $\mathcal{L}(E, F)$. Denote by α^* the trace dual norm on $\mathcal{L}(F, E)$.*

(a) *There is always an isomorphism $u_0 \in \mathcal{L}(E, F)$ with $\alpha(u_0) = 1$ and $\alpha^*(u_0^{-1}) = n$.*

(b) *If α is an ideal norm, if E (or F) is ℓ_2^n, and if $u_0, v_0 \in \mathcal{L}(E, F)$ are isomorphisms as described in* (a), *then $v_0^{-1}u_0 : E \to E$ (or $v_0 u_0^{-1} : F \to F$) is an isometry.*

Proof. (a) Given bases $(e_i)_1^n$ of E and $(f_j^*)_1^n$ of F^*, define $D : \mathcal{L}(E, F) \to \mathbf{K}$ by

$$D(u) = \det\left(\langle ue_i, f_j^*\rangle\right)_{i,j=1}^n.$$

Thanks to finite dimensionality, the continuous map D achieves its supremum on the unit ball of $[\mathcal{L}(E, F), \alpha]$. More specifically, we are assured of the existence of $u_0 \in \mathcal{L}(E, F)$ with

$$\alpha(u_0) = 1$$

and

$$|D(u_0)| = \sup\left\{|D(u)| : u \in \mathcal{L}(E, F), \alpha(u) \leq 1\right\}.$$

By the nature of determinants, basis changes will only change D by a constant factor, and so will not affect this extremal property of u_0.

Naturally, u_0 is the map we are seeking. We need to establish that u_0 is invertible and satisfies $\alpha^*(u_0^{-1}) = n$.

Invertibility presents no problem. Since E and F have the same dimension, there is an isomorphism $u : E \to F$. This must satisfy $D(u) \neq 0$ and may be normalized so that $\alpha(u) = 1$. But then it must be true, by definition of u_0, that $|D(u_0)| \geq |D(u)| > 0$, and so u_0, too, is invertible.

By trace duality,

$$n = \operatorname{tr}(id_E) = \operatorname{tr}(u_0^{-1}u_0) \leq \alpha(u_0)\cdot\alpha^*(u_0^{-1}) = \alpha^*(u_0^{-1}),$$

so we only need to show that

$$\alpha^*(u_0^{-1}) \leq n.$$

This will take some doing, but we can proceed by a perturbation argument along the lines of Lemma 1.3, which was the key to the proof of the Dvoretzky-Rogers Theorem. We take advantage of our freedom to choose bases and select $(f_j^*)_1^n$ with $f_j^* := (u_0^{-1})^* e_j^*$ $(1 \le j \le n)$ where $(e_j^*)_1^n$ is the basis dual to $(e_i)_1^n$ (and so satisfies $\langle e_i, e_j^* \rangle = \delta_{ij}$ for $1 \le i, j \le n$). In particular, it follows that $D(u_0) = 1$. By finite dimensionality again, we can fix $v \in \mathcal{L}(E, F)$ with

$$\alpha(v) = 1 \quad \text{and} \quad \alpha^*(u_0^{-1}) = \operatorname{tr}(u_0^{-1} v).$$

But then, if t is a scalar and id is the $(n \times n)$ identity matrix,

$$D(u_0 + tv) = \det\big(\langle u_0 e_i, f_j^* \rangle + t\langle v e_i, f_j^* \rangle\big) = \det\big(id + t\langle v e_i, f_j^* \rangle\big).$$

Notice that the determinant can be expanded as a polynomial in t to obtain

$$D(u_0 + tv) = 1 + c_1 t + \dots + c_n t^n$$

where

$$c_1 = \sum_{i=1}^n \langle v e_i, f_i^* \rangle = \sum_{i=1}^n \langle u_0^{-1} v e_i, e_i^* \rangle = \operatorname{tr}(u_0^{-1} v) = \alpha^*(u_0^{-1}).$$

However, by the extremal property of u_0,

$$\begin{aligned} |D(u_0 + tv)| &= \alpha(u_0 + tv)^n \cdot \Big| D\Big(\frac{u_0 + tv}{\alpha(u_0 + tv)}\Big)\Big| \\ &\le \alpha(u_0 + tv)^n \cdot |D(u_0)| \le (1 + |t|)^n. \end{aligned}$$

Thus, for small positive t,

$$1 + \alpha^*(u_0^{-1}) t + O(t^2) \le 1 + nt + O(t^2),$$

and it follows that

$$\alpha^*(u_0^{-1}) \le n.$$

(b) We treat only one case, that where $E = \ell_2^n$. The other case is analogous. So, if u_0, v_0 have the form described in (a), we must show that the Hilbert space operator $v_0^{-1} u_0$ is an isometry. As is well-known, we may express $v_0^{-1} u_0$ as vw, where v is a positive operator on E, and w is an isometry on E. All we need, then, is to show that $v = id_E$.

As v is certainly invertible, we can select an orthonormal basis $(e_i)_1^n$ of E consisting of eigenvectors of v. To achieve our goal, we must show that if λ_i is the eigenvalue corresponding to e_i, then $\lambda_i = 1$ $(1 \le i \le n)$. Note that each λ_i is positive, so the desired conclusion will follow at once if we can establish

$$(*) \qquad \sum_{i=1}^n \lambda_i = \sum_{i=1}^n \lambda_i^{-1} = n.$$

To prove $(*)$ we first invoke the ideal norm property of α, and, of course, the defining properties of u_0 and v_0. We have

$$\sum_i \lambda_i = \operatorname{tr}(v) = \operatorname{tr}(v_0^{-1} u_0 w^{-1}) \le \alpha^*(v_0^{-1}) \cdot \alpha(u_0) \cdot \|w^{-1}\| \le n$$

and

$$\sum_i \lambda_i^{-1} = \operatorname{tr}(v^{-1}) = \operatorname{tr}(w u_0^{-1} v_0) \le \|w\| \cdot \alpha^*(u_0^{-1}) \cdot \alpha(v_0) \le n;$$

hence $\sum_{i,j} \lambda_i \lambda_j^{-1} \le n^2$. Now as $\dfrac{\lambda_i}{\lambda_j} + \dfrac{\lambda_j}{\lambda_i} \ge 2$ for all $1 \le i, j \le n$, we must have $\sum_{i,j} \lambda_i \lambda_j^{-1} \ge n^2$. All these inequalities must therefore be equalities. But then $\dfrac{\lambda_i}{\lambda_j} + \dfrac{\lambda_j}{\lambda_i} = 2$ for all i, j and so all the λ_i's are equal; the common value is clearly 1. QED

By specializing to the case where E is ℓ_1^n and α is the usual operator norm, we can present a high-tech proof of a classical result which is known as Auerbach's Lemma. In fact, the idea of maximizing determinant functions in Lewis' Theorem goes back to the original proof of this lemma.

6.26 Auerbach's Lemma: *Let E be an n-dimensional normed space. Then there are unit vectors $x_1,\dots,x_n \in E$ and unit vectors $x_1^*,\dots,x_n^* \in E^*$ such that*

$$\langle x_i^*, x_j \rangle = \delta_{ij} \qquad (1 \le i, j \le n).$$

Evidently, the vectors $x_1,\dots,x_n$ form a basis for E, and the vectors $x_1^*,\dots,x_n^*$ constitute the corresponding dual basis for E^*. The striking feature of Auerbach's Lemma is that it allows us to choose a normalized basis whose dual basis is also normalized. Each such system $(x_i, x_i^*)_{i=1}^n$ is usually referred to as an *Auerbach basis* for E.

Proof. We apply Lewis' Theorem with α equal to the usual operator norm (so by 6.16 $\alpha^* = \iota_1$). We find an isomorphism $u_0 \in \mathcal{L}(\ell_1^n, E)$ such that $\|u_0\| = 1$ and $\iota_1(u_0^{-1}) = n$.

Set $x_i = u_0 e_i$ and $x_i^* = (u_0^{-1})^* e_i$, $1 \le i \le n$. It is plain that $\|x_i\| \le 1$. Also, $\langle x_j^*, x_i \rangle = \langle e_i, e_j \rangle = \delta_{ij}$, from which it follows that $\|x_j^*\| \ge 1$. But, summoning 5.15 and noting that $(u_0^{-1})^* = \sum_{j \le n} e_j \otimes x_j^*$, we discover that

$$n = \iota_1(u_0^{-1}) = \iota_1\big((u_0^{-1})^*\big) \le \sum_{j \le n} \|x_j^*\| \le n,$$

and this forces the conclusion that $\|x_j^*\| = 1$ for each $1 \le j \le n$.

All that remains is to observe that $\|x_j\| \ge |\langle x_j, x_j^* \rangle| = 1$, and so that $\|x_j\| = 1$ for each j. QED

We seize this opportunity to prove a companion to Lemma 6.14:

6.27 Corollary: *If E is an n-dimensional normed space, then $\iota_1(id_E) = n$.*

Proof. According to 6.14, $n = \operatorname{tr}(id_E) \le \iota_1(id_E)$. For the converse, pick an Auerbach basis $(x_i, x_i^*)_1^n$ for E. Then, as $id_E = \sum_{i \le n} x_i^* \otimes x_i$, we have $\iota_1(id_E) \le \sum_{i \le n} \|x_i^*\| \cdot \|x_i\| = n$. QED

We could have avoided the use of Auerbach's lemma by appealing, instead, to Theorem 4.17 and the composition theorem 5.16:

$$n \le \iota_1(id_E) \le \pi_2(id_E)^2 = n.$$

At this point it is also good to think back to the Kadets-Snobar Theorem 4.18: every n-dimensional subspace E of a Banach space X is the range of a projection in $\mathcal{L}(X)$ of norm $\leq \sqrt{n}$. If we select an Auerbach basis $(x_i, x_i^*)_1^n$ for E and produce a norm-one extension $\tilde{x}_i^* \in X^*$ of each x_i^*, then even though $P = \sum_i \tilde{x}_i^* \otimes x_i$ is a projection in $\mathcal{L}(X)$ with range E, we only get $\|P\| \leq \iota_1(P) \leq n$.

We *can* recover the full Kadets - Snobar Theorem from the approach developed here, but to do so we need to exploit a fundamental geometrical construct found in the fabric of Lewis' Theorem.

Any isomorphism u taking ℓ_2^n onto an n-dimensional normed space E displays Hilbertian structure on the space E with $u(B_{\ell_2^n})$ as the unit ball. Geometric tradition compels us to call such balls *ellipsoids* in E.

If we consider the operator norm $\|\cdot\|$ on $\mathcal{L}(\ell_2^n, E)$, then the dual norm is ι_1, and Lewis' Theorem gives us an isomorphism $u_0 \in \mathcal{L}(\ell_2^n, E)$ satisfying $\|u_0\| = 1$ and $\iota_1(u_0^{-1}) = n$. The ellipsoid $u_0(B_{\ell_2^n})$ has several special features, the first being that it is unique. In fact, if $v_0 \in \mathcal{L}(\ell_2^n, E)$ satisfies $\|v_0\| = 1$ and $\iota_1(v_0^{-1}) = n$ then, by 6.25, $v_0^{-1}u_0$ is an isometry on ℓ_2^n, and so $v_0(B_{\ell_2^n}) = u_0(B_{\ell_2^n})$. We may thus legitimately introduce the notation

$$\mathcal{E}_E := u_0(B_{\ell_2^n}).$$

$\mathcal{E}_E$ is known as the *John ellipsoid* of the normed space E.

Since $\|u_0\| = 1$, $\mathcal{E}_E$ must be contained in B_E. The next thing to observe is that no other ellipsoid inside B_E occupies more 'space' than $\mathcal{E}_E$.

Specifically, if $u : \ell_2^n \to E$ is any isomorphism and λ is Lebesgue measure on ℓ_2^n, then $\lambda(u^{-1}(\cdot))$ is a measure on the Borel subsets A of E, and it is natural enough to think of

$$\text{vol } A := \lambda(u^{-1}(A))$$

as representing the 'volume' of A. Although this quantity changes with u, for any two Borel subsets A and B of E, the ratio $\dfrac{\text{vol } A}{\text{vol } B}$ is certainly independent of u. Now we can make the vague statement about the size of $\mathcal{E}_E$ precise: of all ellipsoids $\mathcal{E}$ in B_E, the John ellipsoid maximizes $\dfrac{\text{vol } \mathcal{E}}{\text{vol } B_E}$.

The important thing to notice is that for any isomorphism $u \in \mathcal{L}(\ell_2^n, E)$ with $\|u\| \leq 1$ we have

$$\lambda(u^{-1}(B_E)) \geq \lambda(u_0^{-1}(B_E)).$$

This inequality stems from the definition of u_0 in terms of the determinant. Recall that

$$|D(u_0)| = \sup\{|D(u)| : u \in \mathcal{L}(\ell_2^n, E),\ \|u\| = 1\}$$

regardless of how we define D in terms of bases of ℓ_2^n and E^*. Given any isomorphism $u \in \mathcal{L}(\ell_2^n, E)$, let us work with the standard basis $(e_i)_1^n$ of ℓ_2^n and the basis $\big((u^{-1})^* e_i\big)_1^n$ of E^*. Then

$$1 = \det\big(\langle ue_i, (u^{-1})^* e_j\rangle\big) = D(u) \leq |D(u_0)|,$$

and hence

$$\begin{aligned}\lambda\big(u^{-1}(B_E)\big) &= |\det(u^{-1}u_0)|\cdot\lambda\big(u_0^{-1}(B_E)\big)\\ &= \big|\det\big(\langle u_0e_i,(u^{-1})^*e_j\rangle\big)\big|\cdot\lambda\big(u_0^{-1}(B_E)\big)\\ &= |D(u_0)|\cdot\lambda\big(u_0^{-1}(B_E)\big) \geq \lambda\big(u_0^{-1}(B_E)\big).\end{aligned}$$

Consequently

$$\begin{aligned}\frac{\mathrm{vol}\,u(B_{\ell_2^n})}{\mathrm{vol}\,B_E} &= \frac{\lambda(u^{-1}u(B_{\ell_2^n}))}{\lambda(u^{-1}(B_E))} \leq \frac{\lambda(B_{\ell_2^n})}{\lambda(u^{-1}(B_E))}\\ &\leq \frac{\lambda(B_{\ell_2^n})}{\lambda(u_0^{-1}(B_E))} = \frac{\lambda(u_0^{-1}u_0(B_{\ell_2^n}))}{\lambda(u_0^{-1}(B_E))} = \frac{\mathrm{vol}\,\mathcal{E}_E}{\mathrm{vol}\,B_E}\,.\end{aligned}$$

This makes it clear why the John ellipsoid is also referred to as the *ellipsoid of maximal volume* inside B_E.

It should come as no surprise that a construct as natural as the John ellipsoid enjoys various alternative descriptions. To gain access to these, we need to demonstrate that in finite dimensions the various infima in the descriptions 5.23 of a p-nuclear norm are attained. Of course, the finite dimensional setting is conducive to reaching a minimum, but will be to no avail unless we can bound the length of the sum representing the operator. Another classical result comes to our rescue.

6.28 Carathéodory's Theorem: *Let E be an n-dimensional real vector space and let M be a subset of E. Then each point in the convex hull of M can be written as a convex combination of at most $n+1$ elements of M.*

Note that if we had an n-dimensional complex vector space, it could be regarded as a $2n$-dimensional real linear space, and so the convex hull of any subset could be constructed by convexly combining at most $2n+1$ points from the set.

Proof. To fix notation, let $x \in M$ be represented as $x = \sum_{i=1}^m \alpha_i x_i$, with $x_1,\dots,x_m \in M$, $\alpha_1,\dots,\alpha_m > 0$, $\sum_{i=1}^m \alpha_i = 1$. If $m \leq n+1$, we are where we want to be; so assume that the contrary, $m > n+1$, holds. We shall show how to express x as a convex combination of $m-1$ points from M.

As the vectors $x_m - x_i$ $(1 \leq i \leq m-1)$ are linearly dependent, there are scalars $\beta_1,\dots,\beta_{m-1}$, not all zero, such that $\sum_{i=1}^{m-1}\beta_i(x_m - x_i) = 0$. With $\beta_m := -\sum_{i=1}^{m-1}\beta_i$, we have $\sum_{i=1}^m \beta_i = 0$ and $\sum_{i=1}^m \beta_i x_i = 0$.

Consider $\gamma := \max_{1\leq i\leq m}\beta_i/\alpha_i$. It is no loss to assume that $\gamma = \beta_1/\alpha_1$, and it is clear that $\gamma > 0$. But now $\alpha_1 x_1 = (\alpha_1/\beta_1)\beta_1 x_1 = -\sum_{i=2}^m(\beta_i/\gamma)x_i$, so that $x = \sum_{i=2}^m\big(\alpha_i - (\beta_i/\gamma)\big)x_i$. This is a convex combination of the $m-1$ vectors $x_2,\dots,x_m$ from M since, by construction, $\alpha_i - (\beta_i/\gamma) \geq 0$ $(2 \leq i \leq m)$ and

$$\sum_{i=2}^m \alpha_i - \frac{\beta_i}{\gamma} = \sum_{i=1}^m \alpha_i - \frac{1}{\gamma}\cdot\sum_{i=1}^m \beta_i = 1.$$

The reduction game can only continue as long as we are guaranteed linearly dependent vectors, and so we must abandon the process as soon as m reaches $n+1$.

QED

What we are going to use is the following consequence of Carathéodory's Theorem.

6.29 Corollary: *Let E and F be finite dimensional normed spaces, and write $m = \dim E$ and $n = \dim F$. Then, if $u \in \mathcal{L}(E,F)$, we can find vectors $x_1^*,\dots,x_N^* \in E^*$ and $y_1,\dots,y_N \in F$ such that*

$$u = \sum_{i=1}^N x_i^* \otimes y_i \quad \text{and} \quad \iota_1(u) = \sum_{i=1}^N \|x_i^*\| \cdot \|y_i\|,$$

where $N \le mn+1$ in the real case and $N \le 2mn+1$ in the complex case.

Proof. Denote $[\mathcal{L}(F,E), \|\cdot\|]$ by L and write L^* for its trace dual $[\mathcal{L}(E,F), \iota_1]$.

If $K := \{x^* \otimes y : x^* \in E^*, y \in F, \|x^*\| = \|y\| = 1\}$, then K is obviously a symmetric subset of B_{L^*}; it is norming for $\|\cdot\|$. Since K is compact, its convex hull, $\operatorname{conv}(K)$, is also compact, and so the Bipolar Theorem tells us that $B_{L^*} = \operatorname{conv}(K)$.

Now for Carathéodory's Theorem: with N as above, $\operatorname{conv}(K)$ consists of all $\sum_{i=1}^N \lambda_i x_i^* \otimes y_i$, where $x_i^* \otimes y_i \in K$ and $\lambda_i \ge 0$ satisfy $\sum_{i=1}^N \lambda_i = 1$. Consequently, if $u \in \mathcal{L}(E,F)$ and $\iota_1(u) = 1$, then u has a representation $u = \sum_{i=1}^N z_i^* \otimes y_i$ where $z_i^* \in E^*$ and $y_i \in F$ are such that $\sum_{i=1}^N \|z_i^*\| \cdot \|y_i\| = 1$. Homogeneity takes care of the rest. QED

With this technical point out of the way, we can now proceed to the promised characterizations of the John ellipsoid.

6.30 Theorem: *Let E be an n-dimensional normed space, and let $u \in \mathcal{L}(\ell_2^n, E)$. The following statements are equivalent:*

(i) $u(B_{\ell_2^n}) = \mathcal{E}_E$.

(ii) $\|u\| = 1$ *and* $\iota_1(u^{-1}) = n$.

(iii) $\|u\| = 1$ *and* $\pi_2(u^{-1}) = n^{1/2}$.

(iv) $\pi_2(u) = n^{1/2} = \pi_2(u^{-1})$.

Proof. (i)⇒(ii): As we know, $\mathcal{E}_E$ is generated by an isomorphism $u_0 : \ell_2^n \to E$ such that $\|u_0\| = 1$ and $\iota_1(u_0^{-1}) = n$; see the first part of 6.25. By hypothesis, then, $u_0(B_{\ell_2^n}) = \mathcal{E}_E = u(B_{\ell_2^n})$. But now, by the second part of 6.25, both $u_0^{-1}u$ and $u^{-1}u_0$ are isometries. From here, we derive $\iota_1(u^{-1}) = \iota_1(u_0^{-1}uu^{-1}) = \iota_1(u_0^{-1}) = n$ and $\|u\| = \|uu^{-1}u_0\| = \|u_0\| = 1$.

(ii)⇒(iii): By 6.29, if we fix $u \in \mathcal{L}(\ell_2^n, E)$ with $\|u\| = 1$ and $\iota_1(u^{-1}) = n$, then we can find vectors $x_1^*,\dots,x_N^*$ in E^* and $y_1,\dots,y_N$ in ℓ_2^n such that $u^{-1} = \sum_{i=1}^N x_i^* \otimes y_i$ and $\sum_{i=1}^N \|x_i^*\| \cdot \|y_i\| = \iota_1(u^{-1}) = n$. In order to show that $\pi_2(u^{-1}) = n^{1/2}$, it is convenient to tinker with the x_i^*'s and y_i's to arrange for $\sum_{i=1}^N \|x_i^*\|^2 = n$ and $\sum_{i=1}^N \|y_i\|^2 = n$. Our plan is to create a factorization $u^{-1} = vw$ in such a way that v and wu are Hilbert space operators and then to apply our knowledge of the Hilbert-Schmidt class to get where we want to be.

To begin, define $a : \ell_2^N \to \ell_2^n$ by $a\big((\xi_i)_1^N\big) := \sum_{i=1}^N \xi_i y_i$, and $b : E \to \ell_2^N$ by $bx := \big(\langle x_i^*, x\rangle\big)_1^N$. Plainly, $u^{-1} = ab$ and, by 4.8,

$$\sigma_2(a) = \Big(\sum_{i\le N} \|ae_i\|^2\Big)^{1/2} = \Big(\sum_{i\le N} \|y_i\|^2\Big)^{1/2} = n^{1/2}.$$

Moreover, $b^*e_i = x_i^*$ for each i. So, using 2.7 and 4.10, and 4.8 once again,

$$\begin{aligned}\pi_2(b) &= \sup\big\{\sigma_2(bt) : t \in B_{\mathcal{L}(\ell_2,E)}\big\} = \sup\Big\{\Big(\sum_k \|bte_k\|^2\Big)^{1/2} : t \in B_{\mathcal{L}(\ell_2,E)}\Big\} \\ &= \sup\Big\{\Big(\sum_k \sum_{i\le N} |\langle te_k, x_i^*\rangle|^2\Big)^{1/2} : t \in B_{\mathcal{L}(\ell_2,E)}\Big\} \\ &= \sup\Big\{\Big(\sum_{i\le N} \|t^*x_i^*\|^2\Big)^{1/2} : t \in B_{\mathcal{L}(\ell_2,E)}\Big\} \le \Big(\sum_{i\le N} \|x_i^*\|^2\Big)^{1/2} = n^{1/2}.\end{aligned}$$

We are now in a position to define v and w. Notice that $N \ge n$, since otherwise u^{-1} would pull insufficient rank to be invertible. We can therefore consider the orthogonal projection $P : \ell_2^N \to b(E)$ and the natural embedding $j : b(E) \to \ell_2^N$, and write $u^{-1} = vw$, where $v = aj$ and $w = Pb$. Simple algebra shows that v and w are isomorphisms. What is more, our estimates above yield

$$\sigma_2(v) \le \sigma_2(a)\cdot\|j\| \le n^{1/2} \quad \text{and} \quad \pi_2(w) \le \|P\|\cdot\pi_2(b) \le n^{1/2}.$$

Looking at both v and wu as Hilbert-Schmidt operators on ℓ_2^n and exploiting trace duality, we will show next that $v^* = v^{-1} = wu$.

First notice that

$$\sigma_2(v)\cdot\sigma_2(wu) \le n = \operatorname{tr}(u^{-1}u) = \operatorname{tr}(vwu) \le \pi_2(v)\cdot\pi_2(wu) = \sigma_2(v)\cdot\sigma_2(wu).$$

All are equal. We conclude from the Cauchy-Schwarz *equality* that $wu = cv^*$ for some $c > 0$. Another consequence is that, in fact, $\sigma_2(v) = n^{1/2} = \sigma_2(wu)$. But now, $n^{1/2} = \sigma_2(wu) = c\cdot\sigma_2(v^*) = c\cdot\sigma_2(v) = c\cdot n^{1/2}$, from which we get $c = 1$ and $wu = v^*$. As $id_{\ell_2^n} = u^{-1}u = vwu = vv^*$, it must be further the case that $wu = v^* = v^{-1}$.

The upshot of all this is that, conveniently, v is an isometry, so that $\pi_2(u^{-1}) = \pi_2(vw) = \pi_2(w)$. But $\pi_2(w) \le n^{1/2} = \pi_2(wu) \le \pi_2(w)$. Our long struggle is at an end: $\pi_2(u^{-1}) = \pi_2(w) = n^{1/2}$.

(iii)$\Rightarrow$(iv): As we are assuming $\pi_2(u^{-1}) = n^{1/2}$, we may deduce that $n = \operatorname{tr}(u^{-1}u) \le \pi_2(u^{-1})\cdot\pi_2(u) = n^{1/2}\cdot\pi_2(u)$. Consequently, $\pi_2(u) \ge n^{1/2}$. The reverse inequality also holds, since $\pi_2(u) = \pi_2(u \circ id_{\ell_2^n}) \le \|u\|\cdot\sigma_2(id_{\ell_2^n}) = n^{1/2}$.

(iv)$\Rightarrow$(i): Select an isomorphism $u_0 : \ell_2^n \to E$ such that $u_0(B_{\ell_2^n}) = \mathcal{E}_E$. Since we already know that (i) implies (iv), we have not only $\pi_2(u) = \pi_2(u^{-1}) = n^{1/2}$, but also $\pi_2(u_0) = \pi_2(u_0^{-1}) = n^{1/2}$.

Make a temporary adjustment: $n^{-1/2}\cdot u_0$ and $n^{-1/2}\cdot u$ both satisfy part (a) of Lewis' Theorem with $\alpha = \alpha^* = \pi_2$. This allows us to implement part (b) of that theorem and conclude that $u_0^{-1}u$ $\big($alias $(n^{-1/2}u_0)^{-1}(n^{-1/2}u)\big)$ is an isometry of ℓ_2^n, whence $u(B_{\ell_2^n}) = u_0(B_{\ell_2^n}) = \mathcal{E}_E$. QED

Some consequences are worth mentioning. Let E be a n-dimensional normed space, and let $u : \ell_2^n \to E$ be an isomorphism which generates $\mathcal{E}_E$.

- Since $\|u^{-1}\| \leq \pi_2(u^{-1}) = n^{1/2}$, we have $\mathcal{E}_E \subset B_E \subset n^{1/2}\mathcal{E}_E$. We certainly cannot improve on $n^{1/2}$ if we take $E = \ell_\infty^n$ or $E = \ell_1^n$.
- As $\pi_2(id_E) = \pi_2(uu^{-1}) \leq \|u\| \cdot \pi_2(u^{-1}) = n^{1/2}$ and $n = \mathrm{tr}\,(id_E) \leq \pi_2(id_E)^2$, we have a new proof of Theorem 4.17: $\pi_2(id_E) = n^{1/2}$.
- When E is a subspace of some Banach space X, Theorem 4.15 provides us with an extension $v \in \mathcal{L}(X, \ell_2^n)$ of u^{-1} such that $\pi_2(v) = \pi_2(u^{-1}) = n^{1/2}$. Let $j : E \to X$ be the canonical injection, and note that $vju = id_{\ell_2^n}$. As juv is a projection on X with range E and $\|juv\| \leq \|u\| \cdot \|v\| \leq \pi_2(v) = n^{1/2}$, we have reproved the Kadets - Snobar Theorem 4.18.

Notes and Remarks

The notion of trace stems of course from the early days of matrix theory; it was a natural object of study as it is one of the symmetric functions of the characteristic polynomial of a matrix. For instance, 6.1 goes back to J.J. Sylvester [1883], though the first published proof seems to be due to H.S. Thurston [1931]. The idea of using the trace for purposes of duality was already present in J. von Neumann's work when he showed in [1937] that $|\mathrm{tr}\, AB| \leq |(\sigma A \mid \sigma B)|$ in $\mathbf{C}^n$'s scalar product where σA is the sequence of singular numbers of an $n \times n$ matrix A arranged so that their moduli are in decreasing order.

The first explicit use of trace to describe duality in spaces of operators is in R. Schatten and J. von Neumann [1948]. They introduced the $\mathcal{S}_p$-classes and developed their basic theory. Notably, 6.2, 6.3 and 6.4 are be found there.

The proofs of 6.2 and 6.3 can be extended without essential changes to include all the exponents in the range $0 < p < \infty$; we have only shied away from this to stay in the comfortable realm of Banach spaces. The case $0 < p < 1$ is important but we do not use it and so we do not treat it.

Again, this material, and much more, can be found in the treatises of R. Schatten [1960], I.C. Gohberg and M.G. Krein [1969] and B. Simon [1979]. The proof of 6.2 found in the text seems new; in a sense, it is a dual version of an argument of B. Carl and A. Defant [1992].

An attempt at using the trace to describe the duality of spaces of Banach space operators was made by A. Grothendieck [1953a], [1955]; he realized that the approximation property is intimately tied to the viability of such an attempt.

The notion of an operator ideal, as formalized in this chapter, is due to A. Pietsch [1971]. It is a powerful notion, an elegant abstraction. There are again many traces of this idea in A. Grothendieck's Résumé [1953a]. The work of Grothendieck and Pietsch not only gave a far broader base to the by then well-established Schatten-von Neumann theory, but also opened the gate to completely new insights into the theory of Banach spaces and their operators. Pietsch's program, however, is on the mark in that it subsumes much of Grothendieck's work, often clarifying and occasionally eliminating, approximation assumptions. This is not to say that Grothendieck's theory of tensor norms no longer needs to be studied; rather Pietsch's approach is a bit more accessible. The recent monograph of A. Defant and K. Floret [1993] is an up to date account of what is presently known about relations between both theories (and much more).

Lemma 6.14 is due to A. Grothendieck [1953a].

The duality relationships of 6.16 are found in A. Persson and A. Pietsch [1969] and A. Pietsch [1971]. Y. Gordon, D.R. Lewis and J.R. Retherford [1972] gave a fascinating guide to the duality of operator ideals. Corollary 6.19 is due to S. Kwapień [1972b], but a close look at D.R. Lewis [1973] would not be amiss in this regard.

Corollary 6.24 is due to C. Stegall and J.R. Retherford [1972] who also discovered a number of other interesting operator ideal relationships present only when one of the spaces is an $\mathcal{L}_1$- or $\mathcal{L}_\infty$-space.

The geometric application of trace duality in the text follows D.R. Lewis' inspired [1979] treatment of the ellipsoid of minimal volume containing a given convex body (introduced by F. John [1948]) and the attendant inequalities:

Theorem: *Let $E = [E, \|\cdot\|]$ be an n-dimensional real Banach space, and let $\|\cdot\|_2$ be the Euclidean norm on E which has the property that $B_{[E,\|\cdot\|_2]}$ is the ellipsoid of least volume containing B_E. Then there exist $s \in \mathbf{N}$, $x_1,\ldots,x_s \in E$ and $\lambda_1,\ldots,\lambda_s > 0$ such that*

(a) $n \le s \le n(n+1)/2$,

(b) $1 = \|x_i\|_E = \|x_i\|_2$ *for all* $1 \le i \le s$,

(c) $x = \sum_{i\le s} \lambda_i (x|y_i) y_i$ *for all* $x \in E$.

It should be mentioned that the existence and uniqueness of the ellipsoid is a consequence of 6.24: it is just the 'polar body' $\{x \in E : |(x|y)| \le 1 \; \forall y \in \mathcal{E}_E\}$ of the John ellipsoid $\mathcal{E}_E$.

Our presentation of Lewis' Theorem was much influenced by A. Pełczyński's [1980] lecture notes. Far-reaching extensions of the material given here can be found in the books of V.D. Milman - G. Schechtman [1986], G. Pisier [1989] and N. Tomczak-Jaegermann [1989].

Auerbach's Lemma 6.26 enjoys a proud history. In Banach's [1932] book, H. Auerbach is credited with having noted this lemma. In [1947], A.E. Taylor gave a geometric proof (without reference to Auerbach); M.M. Day [1947] also pursued lines of thought similar to Taylor but without explicitly aiming at Auerbach's Lemma. The proof in the text is in much the same spirit as A.F.

Ruston's [1962]; he used the lemma to show the existence of a projection of integral norm $\leq n$ onto an n-dimensional subspace of a Banach space.

S. Banach [1932] asked whether, for any separable Banach space X, there is a biorthogonal sequence (x_n, x_n^*) such that the span of the x_n's is dense in X (that is, (x_n) is *'fundamental'*), the span of the x_n^*'s is weak$*$dense in X^* (that is, (x_n^*) is *'total'*), and $\sup \|x_n\| \cdot \|x_n^*\| < \infty$ The positive response to Banach's question was provided by R.I. Ovsepian and A. Pełczyński [1975]; A. Pełczyński [1976a] soon succeeded in showing that for any given $\varepsilon > 0$ one can even arrange to have $\sup_n \|x_n\| \cdot \|x_n^*\| \leq 1 + \varepsilon$. It remains open whether $\varepsilon = 0$ is admissible, that is, whether there is an infinite Auerbach system for any separable Banach space.

As mentioned, 6.27 was noted, more or less, by A.F. Ruston [1962]. It is explicitly cited and generalized by D.J.H. Garling and Y. Gordon [1971].

Theorem 6.28 appears in C. Carathéodory [1907], [1911]. It is a bit of historical marginalia that this theorem was but a stepping stone to the solution of a problem in Fourier series: the characterization of the Fourier coefficients of a non-negative member of $L_1[-\pi, \pi]$. The proof in the text is essentially that given by H.G. Eggleston [1958]. A thorough discussion of the theorem and its close relatives can be found in L. Danzer, B. Grünbaum and V.L. Klee, Jr. [1963].

7. 2-Factorable Operators

Generalities

The defining property of a p-integral operator degenerates to a particularly simple form when $p = \infty$ – a form which calls for generalization. We will be led to a new family of operator ideals which turns out to be highly relevant to the study of summing operators.

Let $1 \leq p \leq \infty$. A Banach space operator $u : X \to Y$ is labelled as *p-factorable* if there exists a measure space (Ω, Σ, μ) and operators $a : L_p(\mu) \to Y^{**}$, $b : X \to L_p(\mu)$ such that

$$(*) \qquad k_Y u : X \xrightarrow{b} L_p(\mu) \xrightarrow{a} Y^{**}.$$

We write

$$\gamma_p(u) := \inf \|a\| \cdot \|b\|,$$

where the infimum extends over all conceivable factorizations of the form we have indicated. The collection of all p-factorable operators from X into Y is denoted by

$$\Gamma_p(X, Y).$$

Our efforts in this chapter will focus on the 2-factorable operators. Their theory is more elegant and more easily accessible than that of their p-factorable cousins for $p \neq 2$; these we shall investigate in Chapter 9.

However, it does behove us to start with some simple, but useful general results. The first is hardly surprising.

7.1 Theorem: *For each $1 \leq p \leq \infty$, $[\Gamma_p, \gamma_p]$ is a Banach ideal.*

The proof lends itself to the now standard format, and requires only minor adjustments to the arguments used to prove Theorem 5.2. We leave the details to the dutiful reader.

The appearance of the second dual in our definition of a p-factorable operator, distressing as it may be, is necessary if we are to develop a satisfactory theory. It is needed, for example, to prove that p-factorable operators behave as well as we could hope under the process of taking adjoints.

7.2 Proposition: *Let p^* be the index conjugate to $1 \leq p \leq \infty$. The following are equivalent statements about a Banach space operator $u : X \to Y$:*

(i) $u \in \Gamma_p(X, Y)$,

(ii) $u^* \in \Gamma_{p^*}(Y^*, X^*)$,

(iii) $u^{**} \in \Gamma_p(X^{**}, Y^{**})$,
(iv) $k_Y u \in \Gamma_p(X, Y^{**})$.

In this case, $\gamma_p(u) = \gamma_{p^*}(u^*) = \gamma_p(u^{**}) = \gamma_p(k_Y u)$.

Proof. The observation that a dual space is norm one complemented in its bidual renders trivial the equivalence of (i) and (iv), and the equality of $\gamma_p(u)$ and $\gamma_p(k_Y u)$. The remaining implications are hardly more difficult.

(i)⇒(ii): Let $k_Y u : X \xrightarrow{b} L_p(\mu) \xrightarrow{a} Y^{**}$ be a typical factorization of $u \in \Gamma_p(X, Y)$. We can use the identity $k_Y^* k_{Y^*} = id_{Y^*}$ to obtain the factorization

$$u^* : Y^* \xrightarrow{k_{Y^*}} Y^{***} \xrightarrow{a^*} L_p(\mu)^* \xrightarrow{b^*} X^*.$$

Since $L_p(\mu)^*$ is an $L_{p^*}(\nu)$-space, u^* is a p^*-factorable operator, with $\gamma_{p^*}(u^*) \leq \|b^*\| \cdot \|a^* k_{Y^*}\| = \|a\| \cdot \|b\|$. The inequality $\gamma_{p^*}(u^*) \leq \gamma_p(u)$ is immediate.

(ii)⇒(iii): A repetition of the argument above, starting with $u^* \in \Gamma_{p^*}(Y^*, X^*)$, gives $u^{**} \in \Gamma_p(X^{**}, Y^{**})$ and $\gamma_p(u^{**}) \leq \gamma_{p^*}(u^*)$.

(iii)⇒(iv): Observe that $k_Y u = u^{**} k_X$. If $u^{**} \in \Gamma_p(X^{**}, Y^{**})$, the ideal property makes it clear that $k_Y u \in \Gamma_p(X, Y^{**})$, and $\gamma_p(k_Y u) \leq \gamma_p(u^{**})$. QED

Again as with the p-integral operators, there are several instances where we can do without the bidual in the definition of a p-factorable operator $u : X \to Y$. One example is when Y is norm one complemented in Y^{**}, as is the case whenever Y is a dual space; this was just applied in the proof of 7.2. What is of greater interest is that the bidual is never required when $p = 2$: since every closed subspace of a Hilbert space is norm one complemented,

- *an operator* $u : X \to Y$ *belongs to* $\Gamma_2(X, Y)$ *if and only if it has a factorization* $u : X \xrightarrow{b} H \xrightarrow{a} Y$*, where* H *is a Hilbert space. In this case,* $\gamma_2(u) = \inf \|a\| \cdot \|b\|$*, where the infimum is taken over all such factorizations.*

  Even more:

7.3 Proposition: *Let* X_0, X, Y, Y_0 *be Banach spaces, let* $q \in \mathcal{L}(X_0, X)$ *be a quotient map, and let* $j \in \mathcal{L}(Y, Y_0)$ *be an isometric embedding. An operator* $u : X \to Y$ *is 2-factorable if and only if* $juq : X_0 \to Y_0$ *is, and in this case* $\gamma_2(u) = \gamma_2(juq)$.

Proof. One half of the assertion and the inequality $\gamma_2(juq) \leq \gamma_2(u)$ follow from the ideal property of $[\Gamma_2, \gamma_2]$. For the other part, start with a factorization $X_0 \xrightarrow{b} H \xrightarrow{a} Y_0$ of juq through a Hilbert space H. Consider the closed subspace $H_0 := \ker(a)^{\perp} \cap \{h \in H : ah \in j(Y)\}$, and let $a_0 : H_0 \to Y$ be the operator such that ja_0 is a's restriction to H_0. Notice that a_0 is injective with $\|a_0\| \leq \|a\|$. Let p be the orthogonal projection of H onto H_0 and set $b_1 := pb$. Then $\|b_1\| \leq \|b\|$ and uq has the factorization $uq : X_0 \xrightarrow{b_1} H_0 \xrightarrow{a_0} Y$. The injectivity of a_0 implies the existence of a $b_0 \in \mathcal{L}(X, H_0)$ such that $b_0 q = b_1$; since q is a quotient map, we have $\|b_0\| = \|b_1\|$. Now we are done: since $u : X \xrightarrow{b_0} H_0 \xrightarrow{a_0} Y$, we see that $u \in \Gamma_2(X,Y)$ and $\gamma_2(u) \le \|a_0\| \cdot \|b_0\| \le \|a\| \cdot \|b\|$. Since we started with an arbitrary Γ_2-factorization of juq, $\gamma_2(u) \le \gamma_2(juq)$ follows. QED

Maximality of $[\Gamma_2, \gamma_2]$

We shall eventually see, thanks to some powerful ultrafilter techniques which we develop in the next chapter, that all the ideals $[\Gamma_p, \gamma_p]$ are maximal. For the moment, we restrict ourselves to the case $p = 2$; the argument proceeds along a familiar path.

7.4 Theorem: *The ideal $[\Gamma_2, \gamma_2]$ is maximal.*

Proof. Let $u \in \Gamma_2^{**}(X,Y)$. The Biadjoint Criterion 6.8 will give us enough leverage to show that $u \in \Gamma_2(X,Y)$.

Let $\mathcal{P}$ denote the family of all pairs (E,Z), where E is a finite dimensional subspace of X and Z is a closed finite codimensional subspace of Y. Given $(E,Z) \in \mathcal{P}$, let

$$(E,Z)^{\#} := \{(\tilde{E},\tilde{Z}) \in \mathcal{P} : E \subset \tilde{E},\ \tilde{Z} \subset Z\}.$$

As the $(E,Z)^{\#}$ form a filter basis of subsets of $\mathcal{P}$, they are contained in some ultrafilter $\mathcal{U}$ on $\mathcal{P}$.

We can read off from 6.8 that there is some positive constant C such that

$$\sup_{(E,Z)\in\mathcal{P}} \gamma_2(q_Z u i_E) < C^2;$$

here $i_E : E \to X$ is the natural inclusion, and $q_Z : Y \to Y/Z$ is the quotient map. For each $(E,Z) \in \mathcal{P}$, we may thus select a factorization

$$q_Z u i_E : E \xrightarrow{b_{E,Z}} H_{E,Z} \xrightarrow{a_{E,Z}} Y/Z,$$

where $H_{E,Z}$ is a Hilbert space, $\|a_{E,Z}\| \le C$ and $\|b_{E,Z}\| \le C$. We can immediately lighten our notational burden and assume that all the $H_{E,Z}$ are the same Hilbert space H: otherwise just replace each $H_{E,Z}$ by the ℓ_2 direct sum $(\oplus_{\mathcal{P}} H_{E,Z})_2$. We are going to use this Hilbert space to produce a factorization $u : X \xrightarrow{b} H \xrightarrow{a} Y$. Natural maps which demand consideration are

$$\tilde{b}_{E,Z} : X \to H : x \mapsto \begin{cases} b_{E,Z}x & \text{if } x \in E, \\ 0 & \text{if } x \notin E. \end{cases}$$

These are certainly far from linear; but not to worry – happily we can create linearity by taking limits along the ultrafilter $\mathcal{U}$. Noting that for each $x \in X$ the map $\mathcal{P} \to H : (E,Z) \mapsto \tilde{b}_{E,Z}(x)$ is bounded and that bounded sets are relatively weakly compact in H, we find that $b(x) := \lim_{\mathcal{U}} \tilde{b}_{E,Z}(x)$ exists in the weak topology. The linearity of the resulting map $b : X \to H$ is ensured by the fact that X is covered by its finite dimensional subspaces; boundedness, with $\|b\| \le C$, is by construction.

The rest is now straightforward: we follow the same strategy with the operators $a^*_{E,Z}$ to produce an operator $v : Y^* \to H$ with norm at most C, such that $vy^* = \lim_{\mathcal{U}}(a^*_{E,Z})^{\sim}(y^*)$ for each $y^* \in Y^*$. By the way things were set up, we have $k_Y u = v^*b$, and so $k_Y u \in \Gamma_2(X, Y^{**})$ with $\gamma_2(k_Y u) \leq C^2$. An appeal to Proposition 7.2 completes the proof. QED

Relations with Grothendieck's Inequality

Before we can proceed to the actual determination of $[\Gamma_2^*, \gamma_2^*]$, we change perspective and take up yet again the topic of Grothendieck's Inequality. As Theorem 4.19 shows that Hilbert spaces can be characterized in terms of 2-summing operators, it should come as no great surprise that Grothendieck-like inequalities are relevant in the present context as well.

We begin with an isomorphic characterization of 2-factorable operators. A modification will then lead us to a powerful isometric characterization (7.8). In view of the Biadjoint Criterion 6.8, the next result may also be seen as an alternate approach to the maximality of $[\Gamma_2, \gamma_2]$.

7.5 Theorem: *The following statements about an operator $u : X \to Y$ are equivalent:*

(i) $u \in \Gamma_2(X, Y)$,

(ii) *There is a $C \geq 0$ such that*

$$\Big|\sum_{i,j=1}^{n} a_{ij}\langle y_i^*, ux_j\rangle\Big| \leq C \cdot \sup\Big\{\Big|\sum_{i,j=1}^{n} a_{ij}s_it_j\Big| : (s_i), (t_j) \in B_{\ell_\infty^n}\Big\}$$

regardless of the choice of $n \in \mathbf{N}$, the $n \times n$ scalar matrix (a_{ij}), and the vectors $x_1,\ldots, x_n \in B_X$ and $y_1^*,\ldots, y_n^* \in B_{Y^*}$.

In case (i) and (ii) hold, we may take $C = \kappa_G \cdot \gamma_2(u)$ where κ_G is Grothendieck's constant.

Proof. (i)⇒(ii): Suppose that $u : X \xrightarrow{w} H \xrightarrow{v} Y$ is a typical factorization through a Hilbert space H. Let $n \in \mathbf{N}$, and take any scalar matrix $(a_{ij})_{i,j=1}^n$ and vectors $x_1,\ldots, x_n \in B_X$ and $y_1^*,\ldots, y_n^* \in B_{Y^*}$. Apply Grothendieck's Inequality (1.14):

$$\begin{aligned}\Big|\sum_{i,j=1}^{n} a_{ij}\langle y_i^*, ux_j\rangle\Big| &= \Big|\sum_{i,j=1}^{n} a_{ij}\langle v^*y_i^*, wx_j\rangle\Big| \\ &\leq \kappa_G \cdot \sup_{i\leq n}\|v^*y_i^*\| \cdot \sup_{j\leq n}\|wx_j\| \cdot \sup\Big\{\Big|\sum_{i,j=1}^{n} a_{ij}s_it_j\Big| : (s_i), (t_j) \in B_{\ell_\infty^n}\Big\} \\ &\leq \kappa_G \cdot \|v\| \cdot \|w\| \cdot \sup\Big\{\Big|\sum_{i,j=1}^{n} a_{ij}s_it_j\Big| : (s_i), (t_j) \in B_{\ell_\infty^n}\Big\}.\end{aligned}$$

The result is (ii) with $C = \kappa_G \cdot \gamma_2(u)$.

(ii)⇒(i): Since every Banach space is a quotient of an ℓ_1^I-space, Proposition 7.3 allows an immediate reduction to the case where $X = \ell_1^I$. Our strategy will be to show that our operator $u : \ell_1^I \to Y$ is actually 1-summing once it satisfies the condition in (ii). Corollary 2.16 will then take care of the rest.

Accordingly, fix $x_1,\dots,x_n \in \ell_1^I$ with $\|(x_i)\|_1^{weak} = 1$. We intend to prove that $\sum_i \|ux_i\| \le C$ where C is the constant in condition (ii).

Given $\varepsilon > 0$, there are an integer N and vectors $y_1,\dots,y_n \in \ell_1^N$ such that $\sum_i \|x_i - y_i\| \le \varepsilon\cdot(\|u\|+1)^{-1}$. We may clearly assume that $N \ge n$, and through the device of setting $x_{n+1} = \dots = x_N = 0$ and $y_{n+1} = \dots = y_N = 0$, we may also assume that $N = n$.

Write $y_i = \sum_{j=1}^n a_{ij}e_j$ for each $1 \le i \le n$. Then, for any $s, t \in B_{\ell_\infty^n}$, our hypothesis on $x_1,\dots,x_n$ gives the estimate

$$\begin{aligned}\Big|\sum_{i,j} a_{ij}s_i t_j\Big| &\le \sum_i \Big|\sum_j a_{ij}t_j\Big| = \sum_i |\langle y_i, t\rangle| \\ &\le \sum_i |\langle x_i, t\rangle| + \sum_i |\langle y_i - x_i, t\rangle| \le 1 + \varepsilon.\end{aligned}$$

But now we can choose $y_1^*,\dots,y_n^* \in B_{Y^*}$ so that $\sum_i \|uy_i\| = \sum_i \langle y_i^*, uy_i\rangle = \sum_{i,j} a_{ij}\langle y_i^*, ue_j\rangle$ and use condition (ii) to arrive at

$$\sum_j \|ux_j\| \le \sum_j \|uy_j\| + \varepsilon \le (1+\varepsilon)\cdot C + \varepsilon. \qquad \text{QED}$$

The difficulty with this description of 2-factorable operators is of course the presence of Grothendieck's constant: the matrix inequality does not allow us to read off the precise value of $\gamma_2(u)$. Nevertheless, the idea of a matricial formulation of u's membership in $\Gamma_2(X,Y)$ is a good one, and it will become effective when we shift the balance of the argument. We replace Grothendieck's Inequality with a much easier estimate and regain lost ground by introducing a delicate separation argument. The weak version of Grothendieck's Inequality is this:

7.6 Lemma: *Given any Hilbert space H, any positive integer n, any $n \times n$ matrix $a = (a_{ij})$ of scalars and any vectors $x_1,\dots,x_n, y_1,\dots,y_n$ in H satisfying $\sum_i \|x_i\|^2 \le 1$ and $\sum_j \|y_j\|^2 \le 1$, we have*

$$\Big|\sum_{i,j} a_{ij}(x_i|y_j)\Big| \le \max\Big\{\Big|\sum_{i,j} a_{ij}s_i t_j\Big| : (s_i), (t_j) \in B_{\ell_2^n}\Big\}.$$

For convenience, we abbreviate the maximum on the right hand side by

$$|\!|\!| a |\!|\!|.$$

After all, this expression is a norm – that of the matrix a as an operator on ℓ_2^n.

Proof. We may assume that H is finite dimensional by replacing it if necessary by the span of the x_i's and the y_j's. Then, representing each vector in terms of a standard Hilbert space basis, we have $\sum_{i,j} a_{ij}(x_i|y_j) = \sum_{i,j,n} a_{ij}x_{in}\overline{y_{jn}}$ and are in the fortunate situation of manipulating finite sums. The condition imposed on the x_i's and the y_j's simply asserts that $\sum_{i,n} |x_{in}|^2 \leq 1$ and $\sum_{j,n} |y_{j,n}|^2 \leq 1$. Now, putting it all together, we find

$$\begin{aligned}\Big|\sum_{i,j} a_{i,j}(x_i|y_j)\Big| &\leq \sum_n \Big|\sum_{i,j} a_{ij}x_{in}\overline{y_{jn}}\Big| \leq \|a\|\cdot\sum_n \Big(\sum_i |x_{in}|^2\Big)^{1/2}\cdot\Big(\sum_j |y_{jn}|^2\Big)^{1/2} \\ &\leq \|a\|\cdot\Big(\sum_{i,n} |x_{in}|^2\Big)^{1/2}\cdot\Big(\sum_{j,n} |y_{jn}|^2\Big)^{1/2} \leq \|a\|.\end{aligned}$$

QED

Despite the superficiality of this lemma, the absence of constants is attractive. Indeed, arguing as in (i)⇒(ii) of Theorem 7.5, we immediately find that for any $u \in \Gamma_2(X,Y)$ with $\gamma_2(u) \leq C$ we have

$$\Big|\sum_{i,j=1}^{n} a_{ij}\langle y_i^*, ux_j\rangle\Big| \leq C\cdot\|a\|,$$

no matter how we choose n, the $n \times n$ scalar matrix $a = (a_{ij})$ and the vectors $x_1,\ldots,x_n \in X$ and $y_1^*,\ldots,y_n^* \in Y^*$ satisfying $\sum_j \|x_j\|^2 \leq 1$ and $\sum_i \|y_i^*\|^2 \leq 1$.

What is remarkable and important is that the converse is also true, but to establish this we must undertake some more serious work. We shall be rewarded by obtaining a much stronger result than the one we were originally intending to prove. We isolate the crucial part of the argument as a lemma.

7.7 Lemma: *Let X and Y be Banach spaces, let Z be a subspace of X, and let $1 \leq p < \infty$. Suppose that an operator $u : Z \to Y$ has the property that there is a constant $C > 0$ for which*

$$\sum_{i=1}^{m} \|uz_i\|^p \leq C^p\cdot\sum_{j=1}^{n} \|x_j\|^p \tag{1}$$

whenever $(z_i)_{i=1}^m$ and $(x_j)_{j=1}^n$ are finite families in Z and X, respectively, satisfying

$$\sum_{i=1}^{m} |\langle x^*, z_i\rangle|^p \leq \sum_{j=1}^{n} |\langle x^*, x_j\rangle|^p \quad \text{for all } x^* \in X^*. \tag{2}$$

Then there are a measure μ and an operator $w : X \to L_p(\mu)$, of norm at most C, such that

$$\|uz\| \leq \|wz\| \quad \text{for all } z \in Z. \tag{3}$$

Proof. We ignore the trivial case $u = 0$ and proceed by localization. Accordingly, let E be a finite dimensional subspace of X, and write K for the norm compact unit sphere of E^*. Denote by

$$\Phi$$

the collection of all functions $\varphi : K \to \mathbf{R}$ which are defined by

$$\varphi(x^*) = \sum_{i=1}^{m} |\langle x^*, z_i\rangle|^p - \sum_{j=1}^{n} |\langle x^*, x_j\rangle|^p, \quad x^* \in K,$$

where $(z_i)_{i=1}^m$ and $(x_j)_{j=1}^n$ are finite families in $E \cap Z$ and E, respectively, such that

$$\sum_{i=1}^{m} \|uz_i\|^p > C^p \cdot \sum_{j=1}^{n} \|x_j\|^p. \tag{4}$$

Note that for each $\varphi \in \Phi$ we have $\sup\{\varphi(x^*) : x^* \in K\} > 0$; if not, (2) would hold, and this would imply (1) which is difficult to reconcile with (4).

It is simple enough to check that Φ is a convex cone in the Banach space $\mathcal{C}(K, \mathbf{R})$, and we have just observed that it is disjoint from the open convex cone

$$S := \{f \in \mathcal{C}(K, \mathbf{R}) : f(x^*) < 0 \text{ for all } x^* \in K\}.$$

The Hahn-Banach Separation Theorem furnishes us with a measure μ_E in $\mathcal{C}(K, \mathbf{R})^*$ and a real number α such that

$$\langle \mu_E, f \rangle < \alpha \leq \langle \mu_E, \varphi \rangle \quad \text{for all } f \in S \text{ and } \varphi \in \Phi.$$

Moreover, the properties of S and Φ entail that $\alpha = 0$, so that μ_E is a non-trivial positive measure. Our choice of K guarantees that

$$0 < \sup_{x \in B_E} \int_K |\langle x^*, x \rangle|^p d\mu_E(x^*) \leq \mu_E(K),$$

and this allows us to scale μ_E so that

$$C = \sup_{x \in B_E} \Big(\int_K |\langle x^*, x \rangle|^p d\mu_E(x^*) \Big)^{1/p}.$$

It follows that

$$w_E : E \longrightarrow L_p(\mu_E) : x \mapsto \langle \cdot , x \rangle$$

is a well-defined operator with $\|w_E\| \leq C$.

Our normalization will also ensure that for $z \in E \cap Z$

$$\|w_E z\| \geq C \quad \text{whenever} \quad \|uz\| > C. \tag{5}$$

To see this, note that if $\|uz\| > C$ and $x \in B_E$, the singletons $\{x\}$ and $\{z\}$ give rise to an element φ of Φ with $\varphi(x^*) = |\langle x^*, z \rangle|^p - |\langle x^*, x \rangle|^p$ for all $x^* \in K$. As $\langle \mu_E, \varphi \rangle \geq 0$, we obtain $\|w_E z\|^p \geq \int_K |\langle x^*, x \rangle|^p d\mu_E(x^*)$, and a passage to the supremum over all $x \in B_E$ leads to (5).

Putting (5) another way, we arrive at the conclusion that

$$\|uz\| \leq \|w_E z\| \quad \text{for all} \quad z \in E \cap Z. \tag{6}$$

An ultrafilter argument will allow us to create the operator w that we are looking for. To this end, we first take as $L_p(\mu)$ the space which is the ℓ_p direct sum of all $L_p(\mu_E)$, $E \in \mathcal{F}_X$. As always, $\mathcal{F}_X$ denotes the collection of all finite dimensional subspaces of X. Next, each w_E induces a map $\tilde{w}_E : X \to L_p(\mu)$ such that, after natural identifications, $\tilde{w}_E(x) = w_E(x)$ if $x \in E$ and $\tilde{w}_E(x) = 0$ if not. Fix an ultrafilter $\mathcal{U}$ on $\mathcal{F}_X$ which contains, for each $E \in \mathcal{F}_X$, the set $\{\tilde{E} \in \mathcal{F}_X : E \subset \tilde{E}\}$. Then, for each $x \in X$, the image of $\mathcal{U}$ under $E \mapsto \tilde{w}_E x$ is an ultrafilter on a bounded set in $L_p(\mu)$. In the reflexive case $1 < p < \infty$, it must converge weakly to some $wx \in L_p(\mu)$. It is a simple matter to check that we have produced a bounded linear operator $w : X \to L_p(\mu)$ with norm at most C. Equation (6) guarantees that (3) is satisfied.

If $p = 1$, we use weak$*$compactness to arrive at an operator $X \to L_1(\mu)^{**}$ in the same manner. The proof can be terminated either by noting that $L_1(\mu)^{**}$ is isometrically isomorphic to $L_1(\nu)$ for an appropriate ν, or by using the fact that $L_1(\mu)$ is norm one complemented in $L_1(\mu)^{**}$. QED

Let us put this lemma to work to obtain the announced characterization of 2-factorable operators.

7.8 Theorem: *Let X and Y be Banach spaces. The operator $u : X \to Y$ belongs to $\Gamma_2(X,Y)$ with $\gamma_2(u) \le C$ if and only if, for any positive integer n, any $n \times n$ matrix $a = (a_{ij})$ of scalars and any vectors $x_1,\dots,x_n \in X$, $y_1^*,\dots,y_n^* \in Y^*$ we have*

$$(1) \qquad \Big|\sum_{i,j=1}^{n} a_{ij}\langle y_i^*, ux_j\rangle\Big| \le C\cdot\|a\|\cdot\Big(\sum_{j=1}^{n}\|x_j\|^2\Big)^{1/2}\cdot\Big(\sum_{i=1}^{n}\|y_i^*\|^2\Big)^{1/2}.$$

Proof. We have already observed that each $u \in \Gamma_2(X,Y)$ satisfies condition (1), with $C = \gamma_2(u)$.

For the converse, rephrase inequality (1) as

$$(2) \qquad \sum_{i=1}^{n}\Big\|\sum_{j=1}^{n} a_{ij}ux_j\Big\|^2 \le C^2\cdot\|a\|^2\cdot\sum_{j=1}^{n}\|x_j\|^2;$$

we leave the verification of the equivalence to the reader.

Our aim is to use the lemma in the special case $Z = X$. Accordingly, suppose we are given sequences $(x_j)_{j=1}^n$ and $(z_i)_{i=1}^m$ in X with

$$(3) \qquad \sum_{i=1}^{m}|\langle x^*, z_i\rangle|^2 \le \sum_{j=1}^{n}|\langle x^*, x_j\rangle|^2 \quad \text{for all } x^* \in X^*.$$

By adding zeros if necessary we may as well assume that $m = n$. Let s and t be the operators in $\mathcal{L}(\ell_2^n, X)$ which are defined via $e_i \mapsto z_i$ and $e_i \mapsto x_i$, respectively $(1 \le i \le n)$. Note that (3) can be restated as

$$(4) \qquad \|s^*x^*\| \le \|t^*x^*\| \quad \text{for all } x^* \in X^*.$$

Since t^*'s range is a Hilbert space, this signifies that there exists an operator $a \in \mathcal{L}(\ell_2^n)$, of norm at most one, such that $s^* = a^*t^*$, that is, $s = ta$. If (a_{ij}) is the matrix of a then, with suitably chosen $y_1^*,\dots,y_n^* \in B_{Y^*}$, we get

$$\begin{aligned}\sum_{i=1}^{n}\|uz_i\|^2 &= \sum_{i=1}^{n}|\langle y_i^*, use_i\rangle|^2 = \sum_{i=1}^{n}|\langle y_i^*, utae_i\rangle|^2\\ &= \sum_{i=1}^{n}\Big|\Big\langle y_i^*, \sum_{j=1}^{n} a_{ij}ux_j\Big\rangle\Big|^2 \le \sum_{i=1}^{n}\Big\|\sum_{j=1}^{n} a_{ij}ux_j\Big\|^2 \le C^2\cdot\sum_{j=1}^{n}\|x_j\|^2,\end{aligned}$$

by (2) and since $\|a\| \le 1$.

The lemma now provides us with a Hilbert space H and an operator $w : X \to H$ such that

$$\|ux\| \le \|wx\| \le C\cdot\|x\| \quad \text{for all } x \in X.$$

We may of course assume that w has dense range. Call on the first of these inequalities to see that $wx \mapsto ux$ provides us with a well-defined map $v_0 : w(X) \to Y$; it is clearly linear and bounded with $\|v_0\| \le 1$. If $v \in \mathcal{L}(H,Y)$ is its extension, then $u = vw$ by construction, so that $u \in \Gamma_2(X,Y)$. Moreover, $\gamma_2(u) \le \|v\|\cdot\|w\| \le C$. QED

7.9 Remark: Suppose that we are given an operator $u : Z \to Y$ which is only defined on a subspace of our Banach space X. If we know that

$$\sum_{i=1}^{m} \|uz_i\|^p \le C^p\cdot\sum_{j=1}^{n} \|x_j\|^p$$

holds for all finite families $(z_i)_{i=1}^m$ in Z and $(x_j)_{j=1}^n$ in X which satisfy

$$\sum_{i=1}^{m} |\langle x^*, z_i\rangle|^p \le \sum_{j=1}^{n} |\langle x^*, x_j\rangle|^p$$

for all $x^* \in X$, and if $w : X \to L_p(\mu)$ is the operator with $\|w\| \le C$ and $\|uz\| \le \|wz\|$ for all $z \in Z$ whose existence is guaranteed by Lemma 7.7, then we may, as before, construct an operator $v \in \mathcal{L}(\overline{w(X)}, Y)$ such that $\|v\| \le 1$ and $uz = vwz$ for all $z \in Z$. Thus $vw : X \to Y$ is an extension of u which factors through a closed subspace of $L_p(\mu)$.

We shall come back to this topic in 9.13.

In the special case $p = 2$, $vw \in \Gamma_2(X,Y)$ extends u, and $\gamma_2(vw) \le C$. Important geometric conditions on X and Y are available to ensure that the above hypotheses are automatically satisfied. They will be discussed in 12.22.

2-Dominated Operators

Our next goal is to give a precise identification of $[\Gamma_2^*, \gamma_2^*]$, so that we can bring trace duality to bear on our uses of $[\Gamma_2, \gamma_2]$. Once again a factorization scheme lies at the heart of the matter; this time it involves 2-summing operators.

In preparation for our search for a likely candidate for $[\Gamma_2^*, \gamma_2^*]$ we recall '(iii)⇒(i)' in Theorem 4.19: if $u : H \to X$ is a 2-summing operator from the Hilbert space H to the Banach space X, then $u^* : X^* \to H$ is also 2-summing. A close look at the proof shows that $\pi_2(u^*) \le \pi_2(u)$.

Now, if $u : E \to F$ is an operator between finite dimensional normed spaces, then

$$\gamma_2^*(u) = \sup\{|\mathrm{tr}\,(vu)| : v \in \mathcal{L}(F,E),\ \gamma_2(v) \le 1\}.$$

Fix $v \in \mathcal{L}(F,E)$ with $\gamma_2(v) \le 1$ and settle on a factorization $F \overset{v_2}{\to} \ell_2^n \overset{v_1}{\to} E$ of v. Let Z be a Banach space and suppose that u has a factorization $u : E \overset{u_2}{\to} Z \overset{u_1}{\to} F$. Then, by 6.14 and 5.16

$$\begin{aligned} |\mathrm{tr}\,(uv)| &= |\mathrm{tr}\,(v^*u^*)| = |\mathrm{tr}\,(v_2^*v_1^*u_2^*u_1^*)| = |\mathrm{tr}\,((u_2v_1)^*u_1^*v_2^*)| \\ &\le \iota_1((u_2v_1)^*u_1^*v_2^*) \le \pi_2\big((u_2v_1)^*\big)\cdot\pi_2(u_1^*v_2^*) \\ &\le \pi_2(u_2v_1)\cdot\pi_2(u_1^*v_2^*) \le \pi_2(u_1^*)\cdot\pi_2(u_2)\cdot\|v_1\|\cdot\|v_2\|. \end{aligned}$$

If we now vary v, we conclude that

$$(*) \qquad \gamma_2^*(u) \le \inf \pi_2(u_1^*)\cdot\pi_2(u_2)$$

where the infimum is taken over all factorizations $E \overset{u_2}{\to} Z \overset{u_1}{\to} F$ of u through any Banach space.

One of our principal objectives is to show that equality holds in $(*)$. With this as a perspective, we introduce a new class of Banach space operators.

We say that a Banach space operator $u : X \to Y$ is 2-*dominated,* and write

$$u \in \Delta_2(X,Y),$$

if there exist a Banach space Z and a factorization $u : X \overset{u_2}{\to} Z \overset{u_1}{\to} Y$ such that both u_1^* and u_2 are 2-summing. We then set

$$\delta_2(u) := \inf \pi_2(u_1^*)\cdot\pi_2(u)$$

where the infimum extends over all factorizations of this kind.

Since the factorization scheme 2.16 is available for 2-summing operators, the Banach space Z in the definition never needs to be more complicated than Hilbert space. Likewise, the infimum defining $\delta_2(u)$ only has to be extended over all such special factorizations. This helps, for example, to shorten the dreary path to the conclusion that $[\Delta_2, \delta_2]$ is a Banach ideal. We skip the proof of this in pursuance of a more interesting goal:

7.10 Theorem: *The Banach ideal* $[\Delta_2, \delta_2]$ *is maximal.*

Proof. By the Biadjoint Criterion 6.8 we have to show that if u belongs to $\Delta_2^{**}(X,Y)$, then $u \in \Delta_2(X,Y)$ and $\delta_2(u) \le \delta_2^{**}(u)$.

As in the proof of Theorem 7.4, we let $\mathcal{P}$ denote the family of all pairs (E,Z), where E is a finite dimensional subspace of X and Z is a closed finite codimensional subspace of Y. Again let $\mathcal{U}$ be an ultrafilter on $\mathcal{P}$ which contains all sets $(E,Z)^\# = \{(\tilde{E},\tilde{Z}) \in \mathcal{P} : E \subset \tilde{E}, \tilde{Z} \subset Z\}$, $(E,Z) \in \mathcal{P}$. Keeping the notation from 7.4, we start by fixing a constant C such that $\sup_{\mathcal{P}} \delta_2(q_Z u i_E) < C^2$, and we select, for each $(E,Z) \in \mathcal{P}$, a factorization

$$q_Z u i_E : E \overset{b_{E,Z}}{\longrightarrow} H_{E,Z} \overset{a_{E,Z}}{\longrightarrow} Y/Z$$

where $H_{E,Z}$ is a Hilbert space and $\pi_2(b_{E,Z}) \le C$, $\pi_2(a_{E,Z}^*) \le C$. As in 7.4, we may and shall assume that all the $H_{E,Z}$ are equal to a fixed Hilbert space H.

Our aim is to produce a Δ_2-factorizaton $u : X \overset{b}{\to} H \overset{a}{\to} Y_0$.

As X can be considered to be a subspace of $\mathcal{C}(B_{X^*})$, we may rely on the Π_2-Extension Theorem 4.15 to view each map $b_{E,Z}$ as the restriction of some $\tilde{b}_{E,Z} \in \Pi_2(\mathcal{C}(B_{X^*}), H)$ with $\pi_2(\tilde{b}_{E,Z}) = \pi_2(b_{E,Z})$. According to Pietsch's Domination Theorem 2.12 there is, for each $(E, Z) \in \mathcal{P}$, a probability measure $\mu_{E,Z} \in \mathcal{C}(B_{X^*})^*$ such that

$$\|\tilde{b}_{E,Z}f\| \leq C\cdot\Big(\int_{B_{X^*}} |f(x^*)|^2 d\mu_{E,Z}(x^*)\Big)^{1/2} \quad \text{for all } f \in \mathcal{C}(B_{X^*}).$$

By weak$*$compactness of the unit ball of $\mathcal{C}(B_{X^*})^*$, $\mu := \lim_{\mathcal{U}} \mu_{E,Z}$ exists in the weak$*$topology: this is clearly a probability measure. In particular,

$$\int_{B_{X^*}} |f(x^*)|^2 d\mu(x^*) = \lim_{\mathcal{U}} \int_{B_{X^*}} |f(x^*)|^2 d\mu_{E,Z}(x^*)$$

holds for each $f \in \mathcal{C}(B_{X^*})$.

Next, weak compactness of the closed bounded subsets in H guarantees the existence of the weak limit

$$\tilde{b}(f) := \lim_{\mathcal{U}} \tilde{b}_{E,Z}(f)$$

for each $f \in \mathcal{C}(B_{X^*})$. A routine check divulges that we have defined a bounded linear operator $\tilde{b} : \mathcal{C}(B_{X^*}) \to H$. Even more,

$$\begin{aligned}\|\tilde{b}(f)\| &= \lim_{\mathcal{U}} \|\tilde{b}_{E,Z}(f)\| \leq C\cdot\lim_{\mathcal{U}} \Big(\int_{B_{X^*}} |f(x^*)|^2 d\mu_{E,Z}(x^*)\Big)^{1/2} \\ &= C\cdot\Big(\int_{B_{X^*}} |f(x^*)|^2 d\mu(x^*)\Big)^{1/2}\end{aligned}$$

for all $f \in \mathcal{C}(B_{X^*})$. Consequently, $\tilde{b}$ is 2-summing with $\pi_2(\tilde{b}) \leq C$.

The sought-after $b : X \to H$ is just $\tilde{b}|_X$. We will actually think of b as having values in $H_0 = \overline{b(X)}$. Regardless, we still have $\pi_2(b) \leq C$.

The same strategy can be applied to the operators $a^*_{E,Z}$; the result is another 2-summing operator, v, this time acting from Y^* to H, such that

$$v(y^*) = \lim_{\mathcal{U}} (a^*_{E,Z})^{\sim}(y^*) \quad \text{for all } y^* \in Y^*$$

in the weak topology of H, and $\pi_2(v) \leq C$. By design, $k_Y u = v^* b$ and $v^*(H_0) \subset Y$. Hence, if $a : H_0 \to Y$ is the operator induced by v^*, then $u = ab$. Since $a^* = pv^{**}$ where $p : H \to H_0$ is the canonical projection, a^* is 2-summing and $\pi_2(a^*) \leq C$.

We have shown that u belongs to $\Delta_2(X, Y)$ with $\Delta_2(u) \leq C^2$. QED

We are now ready for the announced identification of $[\Gamma_2^*, \gamma_2^*]$.

7.11 Theorem: $[\Gamma_2^*, \gamma_2^*] = [\Delta_2, \Delta_2]$ *and* $[\Delta_2^*, \Delta_2^*] = [\Gamma_2, \gamma_2]$.

Proof. We only need to prove one identity, since the other will then follow from the maximality result.

Our first step is to show that $[\Delta_2(X, Y), \Delta_2] \subset [\Gamma_2^*(X, Y), \gamma_2^*]$. To this end, we fix $v \in \Delta_2(X, Y)$ and we introduce finite dimensional normed spaces E and F and the usual operator composition for computing $\gamma_2^*(v)$:

$$\begin{array}{ccccccccc} E & \xrightarrow{w} & X & \xrightarrow{v} & Y & \xrightarrow{u} & F & \xrightarrow{t} & E \\ & & & {}_{v_2}\searrow & & \nearrow_{v_1} & & {}_{t_2}\searrow & & \nearrow_{t_1} \\ & & & & H & & & & H_0\,. \end{array}$$

We have included a factorization $v = v_1 v_2$ of the 2-dominated operator through a Hilbert space H with v_1^* and v_2 both 2-summing. Since $\gamma_2(t)$ is required, we have also adjoined a factorization $t = t_1 t_2$ through some other Hilbert space H_0.

Now we can estimate $\operatorname{tr}(tuvw)$ by applying (in this order) 6.14, 5.30, 6.3, 4.5 and 4.10:

$$\begin{aligned} |\operatorname{tr}(tuvw)| &= |\operatorname{tr}(t_2 u v_1 v_2 w t_1)| \le \iota_1(t_2 u v_1 v_2 w t_1) = \sigma_1(t_2 u v_1 v_2 w t_1) \\ &\le \sigma_2(t_2 u v_1)\cdot\sigma_2(v_2 w t_1) = \sigma_2((t_2 u v_1)^*)\cdot\sigma_2(v_2 w t_1) \\ &= \pi_2(v_1^* u^* t_2^*)\cdot\pi_2(v_2 w t_1) \le \pi_2(v_1^*)\cdot\pi_2(v_2)\cdot\|t_1\|\cdot\|t_2\|\cdot\|u\|\cdot\|w\|. \end{aligned}$$

Taking the necessary infima we obtain $|\operatorname{tr}(tuvw)| \le \Delta_2(v)\cdot\gamma_2(t)\cdot\|u\|\cdot\|w\|$. Consequently, $v \in \Gamma_2^*(X,Y)$ and $\gamma_2^*(v) \le \Delta_2(v)$, as required.

Since $[\Delta_2, \Delta_2] \subset [\Gamma_2^*, \gamma_2^*]$, we also have $[\Gamma_2^{**}, \gamma_2^{**}] \subset [\Delta_2^*, \Delta_2^*]$. We prove that the reverse inclusion holds by showing that $\gamma_2(v) \le \Delta_2^*(v)$ whenever $v : E \to F$ is an operator between finite dimensional normed spaces. We are going to apply 7.8.

So, given $v \in \Delta_2^*(E,F)$, fix $n \in \mathbf{N}$, $x_1,\dots,x_n$ in E and $y_1^*,\dots,y_n^*$ in F^* with $\sum \|x_j\|^2 \le 1$ and $\sum \|y_i^*\|^2 \le 1$, and a $n \times n$ scalar matrix $a = (a_{ij})_{i,j=1}^n$. Define $u \in \mathcal{L}(F, \ell_2^n)$ by $uy = (\langle y_i^*, y\rangle)_{i=1}^n$ and $w \in \mathcal{L}(\ell_2^n, E)$ by $we_i = x_i$ $(1 \le i \le n)$. No matter how we choose $y_1,\dots,y_N \in F$, we have

$$\begin{aligned} \sum_{k=1}^N \|uy_k\|^2 &= \sum_{i,k} |\langle y_i^*, y_k\rangle|^2 = \sup\Big\{\Big|\sum_{i,k}\langle y_i^*, b_{ik}y_k\rangle\Big|^2 : \sum_{i,k}|b_{ik}|^2 \le 1\Big\} \\ &\le \sup\Big\{\Big(\sum_i \|y_i^*\|^2\Big)\cdot\Big(\sum_i \Big\|\sum_k b_{ik}y_k\Big\|^2\Big) : \sum_{i,k}|b_{ik}|^2 \le 1\Big\} \\ &\le \sup\Big\{\sum_i \Big(\sup_{y^*\in B_{Y^*}} \sum_k |b_{ik}\langle y^*, y_k\rangle|\Big)^2 : \sum_{i,k}|b_{ik}|^2 \le 1\Big\} \\ &\le \sup_{y^*\in B_{Y^*}} \sum_k |\langle y^*, y_k\rangle|^2, \end{aligned}$$

so that $\pi_2(u) \le 1$. Likewise, $\pi_2(w^*) \le 1$.

Considering a as an operator $\ell_2^n \to \ell_2^n$ and using our hypothesis we may write

$$\begin{aligned} \Big|\sum_{i,j=1}^n a_{ij}\langle y_j^*, vx_i\rangle\Big| &= |\operatorname{tr}(wauv)| \le \Delta_2(wau)\cdot\Delta_2^*(v) \\ &\le \pi_2(w^*)\cdot\|a\|\cdot\pi_2(u)\cdot\Delta_2^*(v) \le \|a\|\cdot\Delta_2^*(v). \end{aligned}$$

It follows that $\gamma_2(v) \le \Delta_2^*(v)$. But 7.8, in conjunction with the technique of 6.21, tells us even more: the ideal $[\Gamma_2, \gamma_2]$ is maximal. Therefore $[\Delta_2^*, \Delta_2^*] \subset [\Gamma_2, \gamma_2]$, and we are done. QED

We conclude this chapter by returning to our starting point and establishing an attractive link with the ideals Γ_1 and Γ_∞:

7.12 Theorem: *A Banach space operator $u : X \to Y$ is 2-dominated if and only if there is a Banach space Z such that $u : X \xrightarrow{b} Z \xrightarrow{a} Y$ with $a \in \Gamma_1(Z, Y)$ and $b \in \Gamma_\infty(X, Z)$.*

A self-explanatory shorthand notation for this is $\Delta_2 = \Gamma_1 \circ \Gamma_\infty$.

This theorem throws the 1-integral and 2-dominated operators into sharp contrast. On the one hand, an operator $u : X \to Y$ is 1-integral if and only if there is a probability measure μ such that the formal identity $L_\infty(\mu) \hookrightarrow L_1(\mu)$ is a factor of $k_Y u : X \to Y^{**}$. On the other hand, the operator u is 2-dominated precisely when some operator from an $L_\infty(\mu)$-space to an $L_1(\nu)$-space is a factor of $k_Y u$.

Proof. By Corollary 2.16, every 2-summing operator is ∞-factorable. Summoning 7.2, we see that necessity is an immediate consequence of the definition of Δ_2.

As for sufficiency, assume that $u \in \mathcal{L}(X, Y)$ admits a factorization $u : X \xrightarrow{b} Z \xrightarrow{a} Y$ with $a \in \Gamma_1(Z, Y)$ and $b \in \Gamma_\infty(X, Z)$. The identity $a^{**} k_Z b = k_Y u$ in conjunction with 7.2 gives rise to a factorization $k_Y u : X \xrightarrow{b_2} L_\infty(\mu) \xrightarrow{b_1} Z^{**} \xrightarrow{a_2} L_1(\nu) \xrightarrow{a_1} Y^{**}$ for appropriately chosen measures μ and ν. Appealing to 3.7 and 2.16 we see that $a_2 b_1$ is 2-summing and so can be factored as $a_2 b_1 : L_\infty(\mu) \xrightarrow{w} H \xrightarrow{v} L_1(\nu)$ for a suitable Hilbert space H, and that both v^* and w are 2-summing. Consequently $a_2 b_1$ and hence $k_Y u$ are 2-dominated, and $u \in \Delta_2(X, Y)$ follows when we repeat, for example, the argument at the end of proof of 7.10. QED

If we let $\|u\| := \inf \gamma_1(a) \cdot \gamma_\infty(b)$, the infimum being extended over all factorizations of u as in the statement of 7.12, then we obtain an (ideal) norm on $\Delta_2(X, Y)$ which is equivalent to $\Delta_2(\cdot)$. Equality is excluded due to the unavoidable influence of Theorem 3.7 and the attendant Grothendieck constant in the proof.

We shall derive further characterizations of $[\Delta_2, \Delta_2]$ in a broader context in Chapter 9.

Notes and Remarks

Operators that factor through a Hilbert space arise naturally in A. Grothendieck's Résumé [1953a]: they are the h-integral operators. Their study was revived and energized by J. Lindenstrauss and A. Pełczyński [1968]. Among other things, they established 7.5.

Most of the basic results in this chapter come from S. Kwapień [1972b]. Theorem 7.8, however, appears in slightly altered form in G. Pisier's [1985] lectures which are certainly one of the most informative sources on this topic.

Although the duality in 7.11 is from S. Kwapień [1972b], it was J.S. Cohen [1973] who first isolated the ideal Δ_2. It has a natural extremal property which underlines its importance.

Proposition: *$[\Delta_2, \delta_2]$ is the largest Banach ideal such that $[\Delta_2(H_1, H_2), \delta_2] = [\mathcal{S}_1(H_1, H_2), \sigma_1]$ whenever H_1 and H_2 are Hilbert spaces.*

Proof. First recall from 6.3 that if $u \in \mathcal{S}_1(H_1, H_2)$ then $u = u_1 u_2$ where $u_1 \in \mathcal{S}_2(\ell_2, H_2)$ and $u_2 \in \mathcal{S}_2(H_1, \ell_2)$; moreover we can arrange to have $\sigma_1(u) = \sigma_2(u_1)\cdot\sigma_2(u_2)$. But $\sigma_2(u_1) = \sigma_2(u_1^*) = \pi_2(u_1^*)$ and $\sigma_2(u_2) = \pi_2(u_2)$, so that $u \in \Delta_2(H_1, H_2)$ and $\delta_2(u) \le \sigma_1(u)$.

The inclusion $\Delta_2(H_1, H_2) \subset \mathcal{S}_1(H_1, H_2)$, along with the appropriate norm estimate, is proved analogously; the point of departure is a decompostion $u = u_1 u_2$ where u_1^* and u_2 are 2-summing.

Next, let $[\mathcal{A}, \alpha]$ be any Banach ideal such that, whenever H_1 and H_2 are Hilbert spaces, $\mathcal{A}(H_1, H_2) = \mathcal{S}_1(H_1, H_2)$ isometrically. Let X and Y be Banach spaces, and let $u \in \mathcal{A}(X, Y)$. Take finitely many vectors $x_1, ..., x_n \in X$ and $y_1^*, ..., y_n^* \in Y$ and define $a \in \mathcal{L}(Y, \ell_2^n)$ by $ay := (\langle y_k^*, y\rangle)_{k=1}^n$ and $b \in \mathcal{L}(\ell_2^n, X)$ via $be_k := x_k$ $(1 \le k \le n)$. Note that

$$\begin{aligned} \Big|\sum_{k\le n}\langle ux_k, y_k^*\rangle\Big| &= |\mathrm{tr}\,(aub)| \le \sigma_1(aub) = \alpha(aub) \le \alpha(u)\cdot\|a\|\cdot\|b\| \\ &= \alpha(u)\cdot\|(x_k)\|_2^{weak}\cdot\|(y_k^*)\|_2^{weak}. \end{aligned}$$

Since $\|(x_k)\|_2^{weak}$ is not affected by changing the signs of the x_k's, we might as well assume that $\langle ux_k, y_k^*\rangle \ge 0$ for all $1 \le k \le n$.

As we shall see in 9.7, we have enough to conclude $u \in \Delta_2(X, Y)$ and $\delta_2(u) \le \alpha(u)$. QED

A great deal of effort has been expended on deciding when $\mathcal{L}(X, Y) = \Gamma_2(X, Y)$. In Chapter 12 we shall study general classes of pairs of Banach spaces for which this coincidence occurs. One case of special note is when X is a quotient space of a C^*-algebra and Y is a subspace of the dual of a C^*-algebra. Results of this nature rely on a C^*-algebra version of Grothendieck's Inequality, due to G. Pisier [1978a] and U. Haagerup [1985], and can be found in G. Pisier [1985] and W.J. Davis, D.J.H. Garling and N. Tomczak-Jaegermann [1984].

In the study of a Banach algebra X it is of interest to know whether certain special operators from X to X^* belong to Γ_2. More precisely, each $x^* \in X^*$ gives rise to an element $\widetilde{x^*}$ of $\mathcal{L}(X, X^*)$ defined by $\langle \widetilde{x^*(x)}, y\rangle := \langle x^*, yx\rangle$ for all $x, y \in X$. Banach algebras X with the property that $\widetilde{x^*} \in \Gamma_2(X, X^*)$ for all $x^* \in X^*$ have been studied by P. Charpentier [1974], R.M. Redheffer and

P. Volkmann [1983], S. Kaijser [1976] and A.M. Tonge [1986], among others. It only requires the classical Grothendieck Inequality to verify that quotient algebras and subalgebras of C^*-algebras fall into this class. It was a delicate problem to decide whether or not these are the only examples; the negative solution was given by T.K. Carne [1979]. Related topics will form the subject matter of Chapter 18.

At times, factorization through Hilbert space occurs in a somewhat unexpected manner. For example, B. Carl and A. Defant [1992] proved the following result.

Theorem: *Let $1 \le p, q < \infty$ satisfy $1/q \ge |(1/p) - (1/2)|$. If X is an $\mathcal{L}_p$-space and Y is any Banach space, then $\Pi_q(X, Y) \subset \Gamma_2(X, Y)$.*

In particular, 4-summing operators are 2-factorable whenever their domain is an $\mathcal{L}_4$-space.

In proving their theorem, Carl and Defant relied on certain tensor product techniques; interpolation arguments are also possible. When $p > 2$ there is an alternative route, and the argument runs as follows.

It suffices to consider the case $1/q = (1/2) - (1/p)$. Localization and injectivity allow us to assume that X is ℓ_p^n and Y is finite dimensional. We exploit the fact that $\Gamma_2(X, Y)$ is the trace dual of $\Delta_2(Y, X)$.

Let $u \in \mathcal{L}(X, Y)$ be given. We are going to show that, for any $v \in \mathcal{L}(Y, X)$,

$$|\mathrm{tr}\,(vu)| \le \Delta_2(v) \cdot \pi_q(u).$$

For this we make the decomposition $v : Y \xrightarrow{v_2} Z \xrightarrow{v_1} X$ where Z is any Banach space (a Hilbert space would do!). By the composition theorem 5.16 and our choice of p and q,

$$\iota_{p^*}(v_2 u) \le \iota_2(v_2) \cdot \pi_q(u) = \pi_2(v_2) \cdot \pi_q(u).$$

Since $X = \ell_p^n$ any p^*-integral factorization of $v_2 u$ allows us to apply 5.22 to obtain that $\iota_{p^*}((v_2 u)^*) \le \iota_{p^*}(v_2 u)$. Another appeal to 5.16 in combination with 5.15 and 6.14 now reveals that

$$\begin{aligned} |\mathrm{tr}\,(vu)| &\le \iota_1(vu) = \iota_1((v_2 u)^* v_1^*) \le \iota_{p^*}((v_2 u)^*) \cdot \pi_2(v_1^*) \\ &\le \iota_{p^*}(v_2 u) \cdot \pi_2(v_1) \le \pi_q(u) \cdot \pi_2(v_1^*) \cdot \pi_2(v_2). \end{aligned}$$

Since our factorization $v = v_1 v_2$ was arbitrary, we get $|\mathrm{tr}\,(vu)| \le \pi_q(u) \cdot \Delta_2(v)$, and $\gamma_2(u) \le \pi_q(u)$ follows.

Unfortunately, we do not know of any equally simple argument to cover the case $1 < p < 2$.

8. Ultraproducts and Local Reflexivity

We have already witnessed several situations in which theorems were proved by exploiting the existence and properties of ultrafilters. The point has been reached where it is desirable to take time out to develop a more systematic approach to such arguments. The setup will be based on a few elementary observations:

- A topological space Ω is compact if and only if every ultrafilter on Ω converges to some point in Ω.
- If $f : \Omega \to \Omega'$ is a map between sets and if $\mathcal{U}$ is an ultrafilter on Ω, then $f(\mathcal{U})$ is an ultrafilter on Ω'. Here we have followed common usage and written $f(\mathcal{U})$ for the filter on Ω' which is generated by all sets $f(U)$ with $U \in \mathcal{U}$.
- Let I be a set and let f be a bounded scalar-valued map on I. Then, no matter how we choose an ultrafilter $\mathcal{U}$ on I, the filter $f(\mathcal{U})$ will converge. The limit will be denoted by

$$\lim_{\mathcal{U}} f(i).$$

The last observation, which of course follows from the first two, is the basis for our main construction.

Generalities on Ultraproducts

Given a family $(X_i)_{i\in I}$ of Banach spaces, let

$$\ell_\infty(X_i; I)$$

be the ℓ_∞ direct sum of the X_i's, that is, the collection of all $(x_i)_{i\in I} \in \prod_{i\in I} X_i$ such that $(\|x_i\|)_{i\in I}$ is bounded. Recall that the norm of this Banach space is given by

$$\|(x_i)_{i\in I}\|_\infty := \sup_{i\in I} \|x_i\|.$$

Now choose an ultrafilter $\mathcal{U}$ on I, and select $(x_i)_{i\in I} \in \ell_\infty(X_i; I)$. The boundedness of the map $I \to \mathbf{R} : i \mapsto \|x_i\|$ ensures that $\lim_{\mathcal{U}} \|x_i\|$ exists in $\mathbf{R}$. Evidently,

$$N_{\mathcal{U}} := \left\{(x_i) \in \ell_\infty(X_i; I) : \lim_{\mathcal{U}} \|x_i\| = 0\right\}$$

is a closed linear subspace of $\ell_\infty(X_i; I)$, and so we can create the Banach space $\ell_\infty(X_i; I)/N_{\mathcal{U}}$ equipped with the usual quotient norm. This quotient space is the *ultraproduct* of the family $(X_i)_{i\in I}$ with respect to the ultrafilter $\mathcal{U}$. We shall usually denote it by

$$(\prod_{i\in I} X_i)_{\mathcal{U}} \quad \text{or even} \quad (\prod X_i)_{\mathcal{U}}.$$

The element of $(\prod_{i\in I} X_i)_{\mathcal{U}}$ which is generated by $(x_i)_{i\in I} \in \ell_\infty(X_i; I)$ will be written

$$(x_i)_{\mathcal{U}},$$

and its norm is given by

$$\|(x_i)_{\mathcal{U}}\| = \lim_{\mathcal{U}} \|x_i\|.$$

An important special case merits its own notation. If $X_i = X$ for each $i \in I$, we prefer to write

$$X^{\mathcal{U}}$$

instead of $(\prod_{i\in I} X_i)_{\mathcal{U}}$, and we shall refer to $X^{\mathcal{U}}$ as the *ultrapower* of X with respect to the ultrafilter $\mathcal{U}$.

The close interaction between Banach spaces and their operators compels us to introduce the idea of ultraproducts of operators. Suppose then that we are given two families of Banach spaces, $(X_i)_{i\in I}$ and $(Y_i)_{i\in I}$. If, for each $i \in I$, we have an operator $u_i \in \mathcal{L}(X_i, Y_i)$, and if $C := \sup_{i\in I} \|u_i\|$ is finite, then we can construct a bounded linear operator $\ell_\infty(X_i; I) \to \ell_\infty(Y_i; I) : (x_i)_{i\in I} \mapsto (ux_i)_{i\in I}$ of norm C. Now, if $\mathcal{U}$ is any ultrafilter on I and if $\lim_{\mathcal{U}} \|x_i\| = 0$, then clearly $\lim_{\mathcal{U}} \|u_i x_i\| = 0$. Consequently,

$$u : (\prod_{i\in I} X_i)_{\mathcal{U}} \longrightarrow (\prod_{i\in I} Y_i)_{\mathcal{U}} : (x_i)_{\mathcal{U}} \mapsto (u_i x_i)_{\mathcal{U}}$$

is well-defined; it is a bounded linear operator with norm at most C.

On the other hand, our hypothesis assures us of the existence of $(u_i)_{\mathcal{U}}$ in $\big(\prod_{i\in I} \mathcal{L}(X_i, Y_i)\big)_{\mathcal{U}}$, and only a little book-keeping is required to obtain:

8.1 Proposition: *The correspondence $(u_i)_{\mathcal{U}} \mapsto u$ gives a linear isometric embedding of $\big(\prod_{i\in I} \mathcal{L}(X_i, Y_i)\big)_{\mathcal{U}}$ into $\mathcal{L}\big((\prod_{i\in I} X_i)_{\mathcal{U}}, (\prod_{i\in I} Y_i)_{\mathcal{U}}\big)$.*

Consequently, there is no reason to distinguish between u and $(u_i)_{\mathcal{U}}$; either will be referred to as the *ultraproduct of the operators u_i* with respect to $\mathcal{U}$.

A particularly important case of the proposition occurs when all the u_i's are linear forms. To deal with this, we need our first concrete example.

8.2 Lemma: *If $\mathcal{U}$ is an ultrafilter on I and $\mathbf{K}$ is the scalar field, then $\mathbf{K}^{\mathcal{U}}$ is isometrically isomorphic to $\mathbf{K}$.*

Proof. By definition, $\mathbf{K}^{\mathcal{U}} = \ell_\infty^I / N_{\mathcal{U}}$, where $N_{\mathcal{U}} = \{(\xi_i)_{i\in I} : \lim_{\mathcal{U}} \xi_i = 0\}$ is the kernel of the bounded linear form $\ell_\infty^I \to \mathbf{K} : (\xi_i) \mapsto \lim_{\mathcal{U}} \xi_i$, and so $\mathbf{K}^{\mathcal{U}}$ is one-dimensional. To see that the isomorphism $\mathbf{K}^{\mathcal{U}} \to \mathbf{K} : (\xi_i)_{\mathcal{U}} \mapsto \lim_{\mathcal{U}} \xi_i$ is actually isometric, it is enough to observe that if $\xi_i = 1$ for all $i \in I$, then $\|(\xi_i)_{\mathcal{U}}\| = 1$. QED

8.3 Corollary: *If $(X_i)_{i\in I}$ is a family of Banach spaces and $\mathcal{U}$ is an ultrafilter on I, then $(\prod X_i^*)_{\mathcal{U}}$ embeds isometrically into $(\prod X_i)_{\mathcal{U}}^*$.*

In fact, according to 8.1 we may write

$$\big\langle (x_i^*)_{\mathcal{U}}, (x_i)_{\mathcal{U}} \big\rangle = \lim_{\mathcal{U}} \langle x_i^*, x_i \rangle$$

for all $(x_i^*)_{\mathcal{U}}$ in $(\prod X_i^*)_{\mathcal{U}}$ and $(x_i)_{\mathcal{U}}$ in $(\prod X_i)_{\mathcal{U}}$.

The next concrete example is now within reach.

8.4 Proposition: *Let $n \in \mathbf{N}$, and let $(E_i)_{i\in I}$ be a family of n-dimensional Banach spaces. Regardless of how we choose an ultrafilter $\mathcal{U}$ on I, the ultraproduct $(\prod_{i\in I} E_i)_{\mathcal{U}}$ is n-dimensional.*

Proof. Select an Auerbach basis $(x_{i,k}, x^*_{i,k})^n_{k=1}$ for each E_i (see 6.26). Note that $x^*_k := (x^*_{i,k})_{\mathcal{U}}$ exists in $(\prod E^*_i)_{\mathcal{U}} \subset (\prod E_i)^*_{\mathcal{U}}$ for each $1 \le k \le n$. We shall show that $(x^*_k)^n_{k=1}$ is a basis for $(\prod E_i)^*_{\mathcal{U}}$.

The independence follows from the fact that if $x_m := (x_{i,m})_{\mathcal{U}} \in (\prod E_i)_{\mathcal{U}}$, then $\langle x^*_k, x_m\rangle = \lim_{\mathcal{U}}\langle x^*_{i,k}, x_{i,m}\rangle = \delta_{k,m}$ for all $1 \le k, m \le n$.

To see that we have a spanning set, we take $x = (x_i)_{\mathcal{U}}$ in $(\prod X_i)_{\mathcal{U}}$ with $\langle x^*_k, x\rangle = 0$ for all $k \in \mathbf{N}$ and show that $x = 0$. But $\langle x^*_k, x\rangle = 0$ means that $\lim_{\mathcal{U}}\langle x^*_{i,k}, x_i\rangle = 0$ $(1 \le k \le n)$, and so by the nature of Auerbach bases,

$$\|x\| = \lim_{\mathcal{U}} \|x_i\| = \lim_{\mathcal{U}} \Big\|\sum_{k=1}^{n}\langle x^*_{i,k}, x_i\rangle x_{i,k}\Big\| \le \lim_{\mathcal{U}} \sum_{k=1}^{n} |\langle x^*_{i,k}, x_i\rangle| = 0.$$

QED

We shall see next that we cannot continue this progression any further. As soon as dimensional constraints à la 8.4 are lifted, ultraproducts of Banach spaces have a tendency to be 'enormous'.

To be more precise, recall that an ultrafilter can only be detected by transfinite means, such as Zorn's Lemma, unless it is *trivial* in the sense that it consists of all sets which contain a fixed element. If $\mathcal{U}$ is the trivial ultrafilter $\{S \subset I : i_0 \in S\}$ on I generated by $i_0 \in I$ and if $(X_i)_{i\in I}$ is any family of Banach spaces, then $\left(\prod X_i\right)_{\mathcal{U}}$ is isometrically isomorphic to X_{i_0}. For this reason, *we only consider non-trivial ultrafilters.*

8.5 Proposition: *Suppose that E_n is an n-dimensional Banach space for every $n \in \mathbf{N}$ and that $\mathcal{U}$ is a (non-trivial) ultrafilter on $\mathbf{N}$. Then $(\prod_{n\in\mathbf{N}} E_n)_{\mathcal{U}}$ is non-separable.*

Proof. We shall work with the collection ϕ of all functions $\varphi : \mathbf{N} \to \mathbf{N}$ which satisfy $\varphi(n) \le n$ for each $n \in \mathbf{N}$.

Consider sets $S \subset \phi$ with the property that whenever $\varphi, \psi \in S$, either $\varphi = \psi$ or we only have $\varphi(n) = \psi(n)$ for finitely many $n \in \mathbf{N}$. The collection Φ of all such sets is partially ordered by inclusion, and all chains have upper bounds. So, by Zorn's Lemma, there is a maximal element, S_0 say, in Φ.

We claim that S_0 is uncountable. Suppose, to the contrary, that S_0 consists of countably many functions $\varphi_k : \mathbf{N} \to \mathbf{N}$. Define $\varphi : \mathbf{N} \to \mathbf{N}$ recursively by setting $\varphi(1) = 1$ and by choosing $\varphi(k)$ in $\{\varphi_1(k),..., \varphi_{k-1}(k)\}$'s complement in $\{1,..., k\}$ for each $k > 1$. Notice that $\varphi \in \phi$ and $S_0 \cup \{\varphi\} \in \Phi$. This conflicts with the maximality of S_0.

We can now proceed to the situation under discussion. For each $n \in \mathbf{N}$, select an Auerbach basis $(x_{i,n})^n_{i=1}$ in E_n, and use the φ's in S_0 to produce unit vectors $x_\varphi := (x_{\varphi(n),n})_\mathcal{U}$ in $(\prod E_n)_\mathcal{U}$. Since $\mathcal{U}$ is a non-trivial ultrafilter, it cannot contain any finite subset of $\mathbf{N}$. Therefore, if $\varphi, \psi \in S_0$ are different, we conclude that

$$\|x_\varphi - x_\psi\| = \lim_{\mathcal{U}} \|x_{\varphi(n),n} - x_{\psi(n),n}\| \geq 1.$$

The uncountability of S_0 now entails that $(\prod E_n)_\mathcal{U}$ must be non-separable.

QED

In particular, when we use non-trivial ultrafilters, ultraproducts of infinite dimensional Banach spaces can never be separable. Furthermore, we shall see in 8.8 that every Banach space is contained in an ultraproduct of finite dimensional spaces.

Some Stability Properties

Certain important classes of Banach spaces are stable under the formation of ultraproducts. We shall be particularly interested in Banach lattices.

8.6 Proposition: *Let $(X_i)_{i \in I}$ be a family of Banach lattices. If $\mathcal{U}$ is an ultrafilter on I, then $(\prod_{i \in I} X_i)_\mathcal{U}$ has a natural Banach lattice structure.*

In fact, given (x_i) and (y_i) in $\ell_\infty(X_i; I)$, define

$$(x_i)_\mathcal{U} \leq (y_i)_\mathcal{U}$$

whenever there is an element (z_i) of $\ell_\infty(X_i; I)$ with $\lim_\mathcal{U} \|z_i\| = 0$ such that $x_i + z_i \leq y_i$ for each $i \in I$. It is straightforward to check that we have defined a partial order under which $(\prod X_i)_\mathcal{U}$ becomes a Banach lattice. It turns out that, for any $(x_i), (y_i) \in \ell_\infty(X_i; I)$,

$$(x_i)_\mathcal{U} \vee (y_i)_\mathcal{U} = (x_i \vee y_i)_\mathcal{U}\,,\ (x_i)_\mathcal{U} \wedge (y_i)_\mathcal{U} = (x_i \wedge y_i)_\mathcal{U}\,,\ \text{and}\ \left|(x_i)_\mathcal{U}\right| = (|x_i|)_\mathcal{U}.$$

8.7 Theorem: (a) *Let $1 \leq p < \infty$. Ultraproducts of $L_p(\mu)$-spaces are isometrically isomorphic (as Banach lattices) to $L_p(\mu)$-spaces.*

(b) *Ultraproducts of $\mathcal{C}(K)$-spaces are isometrically isomorphic (as Banach lattices) to $\mathcal{C}(K)$-spaces.*

Proof. We rely on some standard results from Banach lattice theory.

(a) Recall that, for $1 \leq p < \infty$, a Banach lattice X is isometrically isomorphic (as a Banach lattice) to an $L_p(\mu)$-space if and only if $\|x + y\|^p = \|x\|^p + \|y\|^p$ for every $x, y \in X$ satisfying $x \wedge y = 0$. Consequently, if $X_i = L_p(\mu_i)$ $(i \in I)$, if $\mathcal{U}$ is an ultrafilter on I, and if $(x_i)_\mathcal{U}$, $(y_i)_\mathcal{U}$ in $(\prod_{i \in I} X_i)_\mathcal{U}$ satisfy $(x_i)_\mathcal{U} \wedge (y_i)_\mathcal{U} = (x_i \wedge y_i)_\mathcal{U} = 0$, then

$$\begin{aligned}\|(x_i)_\mathcal{U} + (y_i)_\mathcal{U}\|^p &= \lim_{\mathcal{U}} \|x_i + y_i\|^p = \lim_{\mathcal{U}} \|x_i - (x_i \wedge y_i) + y_i - (x_i \wedge y_i)\|^p \\ &= \lim_{\mathcal{U}} \|x_i - (x_i \wedge y_i)\|^p + \lim_{\mathcal{U}} \|y_i - (x_i \wedge y_i)\|^p = \|(x_i)\|_\mathcal{U}^p + \|(y_i)\|_\mathcal{U}^p,\end{aligned}$$

and our conclusion follows.

(b) Recall that a Banach lattice X with order unit is isometrically isomorphic (as a Banach lattice) to a $\mathcal{C}(K)$-space if and only if $\|x \vee y\| = \|x\| \vee \|y\|$ for every $x \geq 0$, $y \geq 0$ in X. It is simple to see that an ultraproduct of Banach lattices with order unit also has an order unit, and the criterion can be applied along the same lines as the proof of (a). QED

The result we called on to prove part (b) is known as the Kakutani Representation Theorem. In Chapter 16, we shall again find its power useful when we develop 'functional calculus' for Banach lattices; see 16.1 and the subsequent discussion.

There is another way to prove part (b) which is more algebraic in nature. It depends on several simple observations:

- If each X_i $(i \in I)$ is a (commutative) Banach algebra, then any ultraproduct $(\prod_{i \in I} X_i)_{\mathcal{U}}$ will also be a (commutative) Banach algebra with multiplication $(x_i)_{\mathcal{U}} \cdot (y_i)_{\mathcal{U}} := (x_i y_i)_{\mathcal{U}}$;
- If each of the Banach algebras X_i has a unit e_i, then $(e_i)_{\mathcal{U}}$ is a unit for the Banach algebra $(\prod X_i)_{\mathcal{U}}$;
- If the scalar field is complex, if each X_i is a C^*-algebra, and if we set $(x_i)_{\mathcal{U}}^* := (x_i^*)_{\mathcal{U}}$, then $(\prod X_i)_{\mathcal{U}}$ becomes a C^*-algebra.

For complex scalars, 8.7(b) now follows from the Gelfand Representation Theorem. The obvious decomposition into real and imaginary parts allows us to settle the real case as well.

Ultraproducts and Finite Dimensional Structure

In order for ultraproducts to serve us well, the ambient ultrafilter must be adapted to specific situations. Usually we may select an appropriate filter and then apply Zorn's Lemma to refine it by an ultrafilter.

A typical and particularly important case occurs when the index set I is directed by a partial order $\leq$. The *order filter* on I, which is determined by the filter basis $\big\{\{j \in I : i \leq j\} : i \in I\big\}$, is then the appropriate choice for refinement. We apply this in the proof of the following result:

8.8 Theorem: *Every Banach space X is isometrically isomorphic to a subspace of an ultraproduct of its finite dimensional subspaces.*

Proof. The set $\mathcal{F}_X$ of all finite dimensional subspaces of X is directed by inclusion. We consider an ultrafilter $\mathcal{U}$ on $\mathcal{F}_X$ which refines the corresponding order filter and show that X is isometrically isomorphic to a subspace of $(\prod_{E \in \mathcal{F}_X} E)_{\mathcal{U}}$.

Given $x \in X$ and $E \in \mathcal{F}_X$, we create an element x_E of E by setting $x_E := x$ if $x \in E$ and $x_E := 0$ otherwise. Since $\|x_E\| \le \|x\|$ for all $E \in \mathcal{F}_X$,

$$J_X : X \longrightarrow \left(\prod_{E\in\mathcal{F}_X} E\right)_{\mathcal{U}} : x \mapsto (x_E)_{\mathcal{U}}$$

is a well-defined map. It is clearly linear, and it is isometric since for each $x \in X$, x_E eventually equals x. QED

There is a dual version of this theorem.

8.9 Theorem: *The bidual of any Banach space X is isometrically isomorphic to a quotient of the finite dimensional quotient spaces of X.*

Proof. The collection $\mathcal{C}_X$ of all closed finite codimensional subspaces of X is a directed set under reverse inclusion. This allows us to consider an ultrafilter $\mathcal{U}$ on $\mathcal{C}_X$ refining the corresponding order filter. However, as $\mathcal{F}_{X^*}$ is just $\{Z^\perp : Z \in \mathcal{C}_X\}$ and $Z_1^\perp \subset Z_2^\perp$ precisely when $Z_2 \subset Z_1$, the previous theorem provides us with a canonical isometric embedding $J_{X^*} : X^* \to \left(\prod_{Z\in\mathcal{C}_X} Z^\perp\right)_{\mathcal{U}}$.

As we have seen before (8.3), each element $(x_Z^*)_{\mathcal{U}}$ of $\left(\prod_{Z\in\mathcal{C}_X} Z^\perp\right)_{\mathcal{U}} = \left(\prod_{Z\in\mathcal{C}_X} (X/Z)^*\right)_{\mathcal{U}}$ may be considered as a member of $\left(\prod_{Z\in\mathcal{C}_X} X/Z\right)_{\mathcal{U}}^*$ by setting

$$\left\langle (x_Z^*)_{\mathcal{U}}, (x_Z)_{\mathcal{U}} \right\rangle = \lim_{\mathcal{U}} \langle x_Z^*, x_Z \rangle;$$

the norm satisfies $\|(x_Z^*)_{\mathcal{U}}\| = \lim_{\mathcal{U}} \|x_Z^*\|$. Consequently, we may view J_{X^*} as an isometric embedding $X^* \to \left(\prod_{Z\in\mathcal{C}_X} X/Z\right)_{\mathcal{U}}^*$. Taking adjoints, we find that $(J_{X^*})^*$ induces a map

$$Q_X : \left(\prod_{Z\in\mathcal{C}_X} X/Z\right)_{\mathcal{U}} \longrightarrow X^{**},$$

and our task will be completed once we show that Q_X is onto.

Fix $x^{**} \in X^{**}$. Then we need to find $(x_Z)_{\mathcal{U}} \in \left(\prod_{Z\in\mathcal{C}_X} X/Z\right)_{\mathcal{U}}$ such that $Q_X((x_Z)_{\mathcal{U}}) = x^{**}$. The surjectivity of the mappings making up $(J_{X^*})^*$ ensures that there is some $w \in \left(\prod_{Z\in\mathcal{C}_X} Z^\perp\right)_{\mathcal{U}}^*$ with $(J_{X^*})^*(w) = x^{**}$. We proceed to our goal by constructing, for each $Z_0 \in \mathcal{C}_X$, an appropriate operator $v_{Z_0} : Z_0^\perp \to \left(\prod_{Z\in\mathcal{C}_X} Z^\perp\right)_{\mathcal{U}}$ and then defining $x_{Z_0} := w v_{Z_0}$. Since X/Z_0 is finite dimensional, $x_{Z_0} \in (Z_0^\perp)^* = (X/Z_0)^{**} = X/Z_0$.

The operators v_{Z_0} must be constructed to mesh appropriately with the definition of J_{X^*} from the previous theorem. We set $v_{Z_0}(x^*) := (x_Z^*)_{\mathcal{U}}$ where x_Z^* is x^* if $Z_0 \supset Z$ and zero if not; this clearly guarantees that $Q_X((x_Z)_{\mathcal{U}}) = x^{**}$.
QED

There is a certain amount of flexibility in the preceding constructions. Instead of working with the order filters on $\mathcal{F}_X$ and $\mathcal{C}_X$, we could have used the order filters on $\mathcal{F}_X \times I$ and $J \times \mathcal{C}_X$ (product order) where I and J are conveniently chosen directed sets.

This type of observation is extremely useful when we work with operators: it allows us to reconstruct an operator $u : X \to Y$ from its action on finite dimensional subspaces of X evaluated on finite dimensional subspaces of Y^*.

To be precise, consider the set $\mathcal{P} := \mathcal{F}_X \times \mathcal{C}_Y$ directed by the natural product order and let $\mathcal{U}$ be an ultrafilter on $\mathcal{P}$ which refines the corresponding order filter. Just as we remarked above, we can construct a natural isometric embedding $J_X : X \to \left(\prod_{(E,Z)\in\mathcal{P}} E\right)_{\mathcal{U}}$ and a natural quotient map $Q_Y : (\prod_{(E,Z)\in\mathcal{P}} Y/Z)_{\mathcal{U}} \to Y^{**}$. Also, given $u \in \mathcal{L}(X,Y)$ and $(E,Z) \in \mathcal{P}$, we can look at the familiar maps $u_{E,Z} \in \mathcal{L}(E, Y/Z)$ given by $u_{E,Z} := q_Z u i_E$, where $i_E : E \to X$ is the natural embedding and where $q_Z : Y \to Y/Z$ is the natural quotient map. As it is clear that $\|u_{E,Z}\| \le \|u\|$ for all $(E,Z) \in \mathcal{P}$, the ultraproduct

$$(u_{E,Z})_{\mathcal{U}} : \left(\prod\nolimits_{\mathcal{P}} E\right)_{\mathcal{U}} \longrightarrow \left(\prod\nolimits_{\mathcal{P}} Y/Z\right)_{\mathcal{P}}$$

is well-defined.

8.10 Theorem: *With the notation above,*

$$k_Y u = Q_Y (u_{E,Z})_{\mathcal{U}} J_X$$

for each $u \in \mathcal{L}(X,Y)$.

Proof. Choose $x \in X$ and $y^* \in Y^*$. Then we have $(x, y^*) \in E \times Z^\perp$ for eventually all $(E,Z) \in \mathcal{P}$. As our construction yields

$$\langle ux, y^* \rangle = \lim_{\mathcal{U}} \langle u_{E,Z} x, y^* \rangle = \big\langle (u_{E,Z})_{\mathcal{U}} J_X x\,,\, J_{Y^*} y^* \big\rangle = \big\langle (Q_Y (u_{E,Z})_{\mathcal{U}} J_X) x\,,\, y^* \big\rangle,$$

we have what we wanted. QED

This last result is of particular interest when applied to the identity operator of a Banach space X. It gives a factorization $k_X : X \xrightarrow{b} W \xrightarrow{a} X^{**}$ where W may be taken either as an ultraproduct of the finite dimensional subspaces of X or as an ultraproduct of the finite dimensional quotients of X. In addition, both a and b have norm at most one.

If we take biadjoints and recall the identity $(k_{X^*})^* k_{X^{**}} = id_{X^{**}}$, we end up with a striking result:

8.11 Corollary: *X^{**} is always isometrically isomorphic to a norm one complemented subspace of both $(\prod_{(E,Z)\in\mathcal{P}} E)^{**}_{\mathcal{U}}$ and $(\prod_{(E,Z)\in\mathcal{P}} X/Z)^{**}_{\mathcal{U}}$.*

Finite Representability

We now introduce a notion which has wide ramifications for general Banach spaces and which will also give additional insight into the structure of $\mathcal{L}_p$-spaces.

Let X and Y be Banach spaces, and let $\lambda \ge 1$. We say that Y is *λ-representable* in X if, no matter how we choose $\varepsilon > 0$, each finite dimensional subspace of Y is $(\lambda+\varepsilon)$-isomorphic to a subspace of X. More precisely: given $\varepsilon > 0$ and $F \in \mathcal{F}_Y$ we can find $E \in \mathcal{F}_X$ and an isomorphism $u : F \to E$ such that $\|u\|\cdot\|u^{-1}\| \le \lambda + \varepsilon$. The case when $\lambda = 1$ is so important that it has acquired special terminology; we shall write *finitely representable* in place of 1-representable.

The $\mathcal{L}_{p,\lambda}$-spaces are a motivating example for the notion of λ-representability: a Banach space X is a $\mathcal{L}_{p,\lambda}$-space precisely when it is λ-representable in ℓ_p. Alternatively, X is a $\mathcal{L}_{p,\lambda}$-space provided it is λ-representable in any infinite dimensional $L_p(\mu)$-space.

The next result uncovers the key relationship between λ-representability and ultraproducts.

8.12 Theorem: *A Banach space Y is λ-representable in a Banach space X if and only if Y is λ-isomorphic to a subspace of some ultrapower of X.*

We isolate a crucial component of the proof.

8.13 Theorem: *Let X be a Banach space. For any index set I and any ultrafilter $\mathcal{U}$ on I, $X^{\mathcal{U}}$ is finitely representable in X.*

Proof. Fix $\varepsilon > 0$, choose a finite dimensional subspace E of $X^{\mathcal{U}}$, and select a basis for E consisting of unit vectors, $z_1,\ldots, z_n$, say. Then, for each $1 \leq j \leq n$, there is an element $x^{(j)} = (x_i^{(j)})_{i\in I}$ of $\ell_\infty(X_i; I)$ with $z_j = (x_i^{(j)})_{\mathcal{U}}$. For each $i \in I$, we can define a linear map $u_i : E \to X$ such that $u_i(z_j) = x_i^{(j)}$, $1 \leq j \leq n$. Our strategy is to show that there is at least one $i \in I$ such that

$$(1+\varepsilon)^{-1}\cdot\|z\| \leq \|u_i z\| \leq (1+\varepsilon)\cdot\|z\|$$

for every $z \in E$.

The problem is to get the uniform estimate. A pointwise estimate comes easily, since if $z = \sum_{j=1}^n \lambda_j z_j \in E$, then

$$\lim_{\mathcal{U}} \|u_i z\| = \lim_{\mathcal{U}} \Big\|\sum_{j=1}^n \lambda_j x_i^{(j)}\Big\| = \Big\|\sum_{j=1}^n \lambda_j (x_i^{(j)})_{\mathcal{U}}\Big\| = \|z\|.$$

Consequently, there is a set $J_z \in \mathcal{U}$ such that, for each $i \in J_z$,

$$(1+\varepsilon/2)^{-1}\cdot\|z\| \leq \|u_i z\| \leq (1+\varepsilon/2)\cdot\|z\|.$$

To obtain the uniform estimate, we first demonstrate that the u_i's are uniformly bounded. Since E is certainly isomorphic to ℓ_1^n, there is constant $C > 0$ such that for all $z = \sum_{j=1}^n \lambda_j z_j$ in E,

$$\|u_i z\| = \Big\|\sum_{j=1}^n \lambda_j x_i^{(j)}\Big\| \leq \max_{1\leq j\leq n} \|x_i^{(j)}\|\cdot\sum_{j=1}^n |\lambda_j| \leq C\cdot M\cdot\|z\|,$$

where $M := \max_{1\leq j\leq n} \|(x_i^{(j)})_{i\in I}\|_\infty$.

Now, E's finite dimensionality alerts us to the fact that it has compact unit sphere S_E. We set $\delta := \varepsilon/\big(C\cdot M\cdot(1+\varepsilon)\cdot(2+\varepsilon)\big)$, fix a δ-net $y_1,\ldots, y_s$ in S_E, and select any $i \in \bigcap_{k=1}^s J_{y_k}$ (!). No matter how we choose $z \in S_E$, we can find $1 \leq k \leq s$ for which $\|z - y_k\| \leq \delta$, and so

$$\|u_i z\| \leq \|u_i(z-y_k)\| + \|u_i y_k\| \leq C\cdot M\cdot\delta + (1+\varepsilon/2) \leq 1+\varepsilon$$

and

$$\|u_i z\| \geq \|u_i y_k\| - \|u_i(z-y_k)\| \geq (1+\varepsilon/2)^{-1} - C\cdot M\cdot\delta = (1+\varepsilon)^{-1}.$$

We have achieved our aim. QED

Proof of 8.12. Assume first that Y is λ-isomorphic to a subspace of an ultraproduct $X^{\mathcal{U}}$ of X. Theorem 8.13 assures us that $X^{\mathcal{U}}$ is finitely representable in X, so a simple composition argument will give the desired conclusion.

For the converse, the trick is to identify an appropriate ultrafilter $\mathcal{U}$. We begin by considering the collection I of all pairs (F, ε) with $F \in \mathcal{F}_Y$ and $\varepsilon > 0$. Now I is a directed set with respect to the partial order given by $(F_1, \varepsilon_1) \leq (F_2, \varepsilon_2)$ if and only if $F_1 \subset F_2$ and $\varepsilon_1 \geq \varepsilon_2$. We take $\mathcal{U}$ to be an ultrafilter on I which refines the order filter.

Our assumption that Y is λ-representable in X allows us to find, for each $(F, \varepsilon) \in I$, an isomorphic embedding $u_{F,\varepsilon} : F \to X$ with $\|u_{F,\varepsilon}\| \leq \lambda + \varepsilon$ and $\|u_{F,\varepsilon}^{-1}\| \leq 1$. (Of course, we consider $u_{F,\varepsilon}^{-1}$ as a map from $u_{F,\varepsilon}(F)$ to F.) This, in turn, allows us to define $j_{F,\varepsilon} : Y \to X$ by $j_{F,\varepsilon}(y) := u_{F,\varepsilon}(y)$ if $y \in F$ and $j_{F,\varepsilon}(y) := 0$ if not. Since $\|j_{F,\varepsilon}(y)\| \leq (\lambda + \varepsilon) \cdot \|y\|$ for each $y \in Y$, we can properly define $j : Y \to X^{\mathcal{U}}$ by $j(y) := \big(j_{F,\varepsilon}(y)\big)_{\mathcal{U}}$. From the construction it follows at once that j is linear and that $\|y\| \leq \|jy\| \leq \lambda \cdot \|y\|$ for all $y \in Y$, and this is precisely what we are looking for. QED

There is an immediate corollary concerning standard spaces:

8.14 Corollary: (a) *A Banach space is λ-representable in an $L_p(\nu)$-space $(1 \leq p \leq \infty)$ if and only if it is λ-isomorphic to a subspace of an $L_p(\mu)$-space.*

(b) *An $\mathcal{L}_{p,\lambda}$-space is λ-isomorphic to a norm one complemented subspace of some $L_p(\mu)$ for $1 < p < \infty$. For $p = 1$ or ∞, this is still true of the bidual.*

In particular, the $\mathcal{L}_2$-spaces are precisely the isomorphic copies of Hilbert spaces.

Proof. Part (a) is obvious when $p = \infty$, and follows from the theorem and the stability of $L_p(\mu)$-spaces under ultraproducts when $p < \infty$.

For part (b), one needs to use the theorem in conjunction with 8.10.

QED

Local Reflexivity

Recall a fundamental fact about biduals of Banach spaces.

8.15 Helly's Lemma: *If X is a Banach space and G is a finite dimensional subspace of X^*, then for each $x^{**} \in X^{**}$ and each $\delta > 0$ there is an $x \in X$ such that $\|x\| \leq (1 + \delta) \cdot \|x^{**}\|$ and $\langle x^*, x \rangle = \langle x^*, x^{**} \rangle$ for all $x^* \in G$.*

This result yields, for example, *Goldstine's Theorem:*

- *The unit ball of X is weak$*$dense in the unit ball of X^{**}.*

Here, we dig deeper into the relationship between a Banach space and its bidual, and we unearth a cornerstone of modern functional analysis: every Banach space bidual X^{**} is finitely representable in X. We shall actually prove a much more precise statement.

8.16 Principle of Local Reflexivity: *Let X be a Banach space, and let E and F be finite dimensional subspaces of X^{**} and X^*, respectively. Then, for each $\varepsilon > 0$, there is an injective operator $u : E \to X$ with the following properties:*

(a) $ux = x$ *for all* $x \in E \cap X$,

(b) $\|u\| \cdot \|u^{-1}\| \leq 1 + \varepsilon$,

(c) $\langle ux^{**}, x^* \rangle = \langle x^{**}, x^* \rangle$ *for all* $x^{**} \in E$ *and* $x^* \in F$.

A note on imprecisions is in order: X has been identified with its canonical image $k_X(X)$ in X^{**}, and u^{-1} is considered as a map from $u(E)$ to E.

The key both to the proof and to the understanding of the Principle is a variant of Helly's Lemma which will appear very natural when we adopt the appropriate viewpoint. To set things up, observe that we can naturally identify $\mathcal{L}(\mathbf{K}, X)$ with X and $\mathcal{L}(\mathbf{K}, X^{**})$ with X^{**}. In other words, if E is a 1-dimensional Banach space, $\mathcal{L}(E, X)^{**}$ is isometrically isomorphic to $\mathcal{L}(E, X^{**})$. The main issue is to show that this is true for any finite dimensional space E. This generalization will require some effort, but once the effort has been made, the path to the Principle of Local Reflexivity will be clear.

8.17 Theorem: *Let E and X be Banach spaces, with E finite dimensional. Then there is a natural isometric isomorphism of $\mathcal{L}(E, X)^{**}$ onto $\mathcal{L}(E, X^{**})$.*

Proof. To begin with, we shall establish this result when $E = \ell_1^N$. In fact, we can work quite formally. Write as usual $\ell_1^N(X)$, resp. $\ell_\infty^N(X)$, for the N-fold product X^N, endowed with the norm $\|(x_k)_1^N\|_1 = \sum_{k=1}^N \|x_k\|$, resp. $\|(x_k)_1^N\|_\infty = \max_{1 \leq k \leq N} \|x_k\|$. Recall that there are natural identifications between $\ell_1^N(X)^*$ and $\ell_\infty^N(X^*)$ and between $\ell_\infty^N(X)^*$ and $\ell_1^N(X^*)$; in each case the duality is given by $\langle (x_k^*), (x_k) \rangle = \sum_{k=1}^N \langle x_k^*, x_k \rangle$. From this we obtain another natural identification:

$$\ell_\infty^N(X)^{**} = \ell_\infty^N(X^{**})$$

isometrically and isomorphically. But we know from Proposition 2.2 that the map $\mathcal{L}(\ell_1^N, X) \to \ell_\infty^N(X) : u \mapsto (ue_k)_1^N$ is an isometric isomorphism, and so we may reformulate to reach the conclusion that there is a linear isometry

$$\psi_N : \mathcal{L}(\ell_1^N, X)^{**} \longrightarrow \mathcal{L}(\ell_1^N, X^{**}).$$

More precisely, let $u \in \mathcal{L}(\ell_1^N, X)^{**}$ be given. By Goldstine's Theorem, u is the weak$*$limit of a net $(u_\gamma)_{\gamma \in \Gamma}$ in $\mathcal{L}(\ell_1^N, X)$ satisfying $\|u_\gamma\| \leq \|u\|$ for all $\gamma \in \Gamma$. For each $1 \leq k \leq N$, the weak$*$limit z_k of the net $(u_\gamma e_k)_{\gamma \in \Gamma}$ exists in X^{**}, and $\psi_N(u) \in \mathcal{L}(\ell_1^N, X^{**})$ is the operator defined by $\psi_N(u)(e_k) = z_k$ for all $1 \leq k \leq N$.

We use this special case to establish the general result. First we extend the previous remark to discover a good candidate for the isomorphism between $\mathcal{L}(E, X)^{**}$ and $\mathcal{L}(E, X^{**})$. Let $u \in \mathcal{L}(E, X)^{**}$ be given. Again, Goldstine's Theorem reveals that u is the weak$*$limit of a net $(u_\gamma)_{\gamma \in \Gamma}$ in $\mathcal{L}(E, X)$ which satifies $\|u_\gamma\| \le \|u\|$ for all $\gamma \in \Gamma$. So if $e \in E$ and $x^* \in X^*$, we can define $\varphi_{e,x^*} \in \mathcal{L}(E,X)^*$ by $\langle \varphi_{e,x^*}, v\rangle = \langle ve, x^*\rangle$ for all $v \in \mathcal{L}(E,X)$ and deduce that

$$\langle \varphi_{e,x^*}, u\rangle = \lim_{\gamma\in\Gamma} \langle \varphi_{e,x^*}, u_\gamma\rangle = \lim_{\gamma\in\Gamma} \langle u_\gamma e, x^*\rangle.$$

This informs us that for each $e \in E$ there is a weak$*$limit

$$\hat{u}e := \lim_{\gamma\in\Gamma} u_\gamma e$$

in X^{**}. Little effort is required to see that $\hat{u}e$ does not depend on the particular net $(u_\gamma)_{\gamma\in\Gamma}$ that was chosen, and so we have a well-defined map $\hat{u} : E \to X^{**}$. Moreover, it is clear that $\hat{u} \in \mathcal{L}(E, X^{**})$.

Now consider the map

$$\psi_E : \mathcal{L}(E,X)^{**} \longrightarrow \mathcal{L}(E,X^{**}) : u \mapsto \hat{u}.$$

By construction, ψ_E is linear and bounded with norm at most one. It will be shown to be the isometric isomorphism we are seeking.

It is clear from the preceding remarks that $\psi_{\ell_1^N}$ is nothing but the isometric isomorphism ψ_N considered above. We are now going to reduce the general case to this special one.

The first step is to show that ψ_E is isometric. To this end, we fix $0 < \varepsilon < 1$ and choose an ε-net $\{x_1,\ldots,x_N\}$ for the unit sphere of E. The map $j : E^* \to \ell_\infty^N : x^* \mapsto \big(\langle x^*, x_k\rangle\big)_{k=1}^N$ is then an isomorphic embedding satisfying

$$(1-\varepsilon)\cdot\|x^*\| \le \|jx^*\| \le \|x^*\| \quad \text{for all } x^* \in E^*.$$

If Y is any Banach space, it follows that

$$J_Y : \mathcal{L}(E,Y) \longrightarrow \mathcal{L}(\ell_1^N, Y) : u \mapsto uj^*$$

is an isomorphic embedding satisfying

$$(*) \qquad (1-\varepsilon)\cdot\|u\| \le \|J_Y u\| \le \|u\| \quad \text{for all } u \in \mathcal{L}(E,Y).$$

The upper bound is evident, and the lower bound is derived by noting that

$$\begin{aligned}\|J_Y u\| &= \sup\big\{|\langle (J_Y u)z, y^*\rangle| : z \in B_{\ell_1^N},\ y^* \in B_{Y^*}\big\}\\ &= \sup\big\{|\langle z, ju^*y^*\rangle| : z \in B_{\ell_1^N},\ y^* \in B_{Y^*}\big\}\\ &= \sup\big\{\|ju^*y^*\| : y^* \in B_{Y^*}\big\} \ge (1-\varepsilon)\cdot\|u\|.\end{aligned}$$

We shall be particularly interested in $J_{X^{**}}$ and $J_X{}^{**}$. Note that $J_{X^{**}}$ is continuous when $\mathcal{L}(E,X^{**})$ is given the topology of pointwise convergence and X^{**} carries its weak$*$topology. Using this and Goldstine's Theorem, we see that the diagram

$$\begin{array}{ccc} \mathcal{L}(E,X)^{**} & \xrightarrow{(J_X)^{**}} & \mathcal{L}(\ell_1^N,X)^{**} \\ \downarrow \psi_E & & \psi_N \downarrow \\ \mathcal{L}(E,X^{**}) & \xrightarrow{J_{X^{**}}} & \mathcal{L}(\ell_1^N,X^{**}) \end{array}$$

commutes, and we may conclude that ψ_E must be an isometric embedding: after all, (*) yields that $(1-\varepsilon)\cdot\|u\| \leq \|\psi_E u\| \leq \|u\|$ holds for all $u \in \mathcal{L}(E,X)^{**}$ and all $\varepsilon > 0$.

It remains to show that ψ_E is onto. Accordingly, let $v \in \mathcal{L}(E, X^{**})$ be given. Fix a basis $(z_k)_1^m$ in E, and let $(z_k^*)_1^m$ be the corresponding dual basis in E^*. We may as well assume that $\|z_k^*\| = 1$ for all $1 \leq k \leq m$. Writing $x_k^{**} := vz_k$, $1 \leq k \leq m$, we obtain $v = \sum_{k=1}^m z_k^* \otimes x_k^{**}$. Each x_k^{**} is the weak$*$ limit of a net $\big(x_k^{(\gamma)}\big)_{\gamma\in\Gamma}$ in X. (We may in fact use the same directed index set Γ for all $1 \leq k \leq m$: just approximate $(x_k^{**})_{k=1}^m$ in the weak$*$topology of $\prod_{k=1}^m X_k^{**} = \big(\prod_{k=1}^m X_k\big)^{**}$ by a net in $\prod_{k=1}^m X_k$.) For each $\gamma \in \Gamma$, define v_γ in $\mathcal{L}(E,X)$ by $v_\gamma := \sum_{k=1}^m z_k^* \otimes x_k^{(\gamma)}$, and then identify $v_\gamma j^* \in \mathcal{L}(\ell_1^N, X) \subset \mathcal{L}(\ell_1^N, X^{**})$ with $\big(v_\gamma(j^*e_i)\big)_1^N \in \ell_\infty^N(X^{**})$. In the same manner, identify $vj^* \in \mathcal{L}(\ell_1^N, X^{**})$ with $\big(v(j^*e_i)\big)_1^N \in \ell_\infty^N(X^{**})$.

Given $(x_i^*)_1^N$ in the predual $\ell_1^N(X^*)$ of $\ell_\infty^N(X^{**})$ and $\delta > 0$, choose $\gamma_\delta \in \Gamma$ astutely so that $|\langle x_i^*, x_k^{**} - x_k^{(\gamma)}\rangle| \leq \delta/(mN)$ for all $1 \leq i \leq N$, $1 \leq k \leq m$ and $\gamma \geq \gamma_\delta$. It follows that

$$\big|\big\langle \big((v - v_\gamma)(j^*e_i)\big)_1^N, (x_i^*)_1^N\big\rangle\big| = \Big|\sum_{i=1}^N \langle x_i^*, (v - v_\gamma)j^*e_i\rangle\Big|$$
$$\leq \sum_{i=1}^N\sum_{k=1}^m |\langle x_i^*, x_k^{**} - x_k^{(\gamma)}\rangle|\cdot|\langle z_k^*, j^*e_i\rangle| \leq \delta$$

whenever $\gamma \geq \gamma_\delta$. We conclude that $(J_X{}^{**}v_\gamma)_{\gamma\in\Gamma}$ converges to $\psi_N^{-1}(J_{X^{**}}(v))$ in the weak$*$topology of $\mathcal{L}(\ell_1^N, X)^{**}$. Since $J_X{}^{**}$ is a homeomorphic embedding for the weak$*$topologies, $(v_\gamma)_{\gamma\in\Gamma}$ must have a weak$*$limit, say u, in $\mathcal{L}(E,X)^{**}$. But the identity $(\psi_N J_X{}^{**})(u) = J_{X^{**}}(v)$ forces $v = \psi_E(u)$. QED

We use this theorem to give the promised variant of Helly's Lemma:

8.18 Corollary: *Let E be a finite dimensional Banach space. If X is a Banach space and F is a finite dimensional subspace of X^*, then for each v in $\mathcal{L}(E, X^{**})$ and each $\delta > 0$, there is a $u \in \mathcal{L}(E,X)$ such that $\|u\| \leq (1+\delta)\cdot\|v\|$ and $\langle vz, x^*\rangle = \langle uz, x^*\rangle$ for all $z \in E$ and $x^* \in F$.*

Proof. First, with each $z \in E$ and $x^* \in F$ we can associate as before an element φ_{z,x^*} of $\mathcal{L}(E,X)^*$ given by $\langle \varphi_{z,x^*}, w\rangle = \langle wz, x^*\rangle$ for all $w \in \mathcal{L}(E,X)$. Note that the φ_{z,x^*} ($z \in E$, $x^* \in F$) span a finite dimensional subspace G of $\mathcal{L}(E,X)^*$: if $z_1,\dots,z_m$ is a basis for E and $x_1^*,\dots,x_n^*$ is a basis for F, then the φ_{z_i,x_j^*} clearly span G.

Since, by the theorem, $\mathcal{L}(E,X^{**}) = \mathcal{L}(E,X)^{**}$, our claim follows by applying Helly's Lemma 8.15 to $\mathcal{L}(E,X)$ and the finite dimensional subspace G of $\mathcal{L}(E,X)^*$. QED

This corollary is just what we need to achieve our goal.

Proof of the Principle of Local Reflexivity. We are going to apply 8.18 to the operator $v : E \to X^{**}$ which is the restriction of the identity operator. This will immediately yield property (c). However, in order to achieve properties (a) and (b), we must enlarge the space F.

We choose $0 < \delta < 1/3$ with $1+\delta \leq (1+\varepsilon)\cdot(1-3\delta)$ and let $\{x_1^{**},\dots,x_n^{**}\}$ be a δ-net for the unit sphere of E. Then, for each $1 \leq i \leq n$, we can select x_i^* in the unit sphere of X^* such that $|\langle x_i^{**}, x_i^*\rangle| \geq 1-\delta$. Now, let F_1 be a finite dimensional subspace of X^* which contains F and all the x_i^* $(1 \leq i \leq n)$.

At this stage, we may apply the above corollary to X, E, F_1, δ and the canonical embedding $v{:}E \to X^{**}$. This provides us with an operator $u{:}E \to X$ which satisfies $\|u\| \leq 1+\delta$ and $\langle ux^{**}, x^*\rangle = \langle x^{**}, x^*\rangle$ for all $x^{**} \in F_1$. Property (c) holds since $F \subset F_1$. Next, if $x \in E \cap X$, then $\langle ux - x, x^*\rangle = 0$ for all $x^* \in F_1$, and we claim that this implies $ux = x$.

If not, we would have $\left\| \dfrac{ux-x}{\|ux-x\|} - x_k^{**} \right\| \leq \delta$ for some $1 \leq k \leq n$, and this would lead to

$$1-\delta \leq |\langle x_k^*, x_k^{**}\rangle| = \left|\left\langle x_k^*, \frac{ux-x}{\|ux-x\|} - x_k^{**} \right\rangle\right| \leq \delta,$$

contradicting our choice of δ. Thus property (a) holds as well.

To finish, we establish property (b). We know already that $\|u\| \leq 1+\delta \leq 1+\varepsilon$. Fix a norm one vector x^{**} in E and choose $1 \leq i \leq n$ for which $\|x^{**} - x_i^{**}\| \leq \delta$. Then

$$\|ux_i^{**}\| \leq \|ux^{**}\| + \|u(x_i^{**} - x^{**})\| \leq \|ux^{**}\| + (1+\delta)\cdot\delta \leq \|ux^{**}\| + 2\delta.$$

On the other hand,

$$\|ux_i^{**}\| \geq |\langle ux_i^{**}, x_i^*\rangle| = |\langle x_i^{**}, x_i^*\rangle| \geq 1-\delta,$$

so that

$$\|ux^{**}\| \geq \|ux_i^{**}\| - 2\delta \geq 1-3\delta.$$

It follows that u is injective and that $\|u^{-1}\| \leq (1-3\delta)^{-1}$. Thanks to our choice of δ, we obtain $\|u\|\cdot\|u^{-1}\| \leq 1+\varepsilon$, and so the proof is complete. QED

If we combine 8.12 and 8.16, we see that, for any $\varepsilon > 0$, the bidual X^{**} of a Banach space X is always $(1+\varepsilon)$-isomorphic to a subspace of some ultrapower of X. The Principle of Local Reflexivity allows us to derive a much stronger conclusion.

8.19 Theorem: *Whenever X is a Banach space, its bidual X^{**} is isometrically isomorphic to a norm one complemented subspace of some ultrapower $X^{\mathcal{U}}$ of X.*

We trust our readers to expel $\{0\}$ automatically from the club of honest Banach spaces, at least for the time being.

Proof. The appropriate index set is $I := \mathcal{F}_{X^{**}} \times \mathcal{F}_{X^*} \times (0,\infty)$. It becomes a directed set when we define $(E,F,\varepsilon) \leq (\hat{E},\hat{F},\hat{\varepsilon})$ to mean $E \subset \hat{E}$, $F \subset \hat{F}$ and

$\hat{\varepsilon} \leq \varepsilon$. We shall work with a fixed ultrafilter $\mathcal{U}$ on I which refines the corresponding order filter.

Given $i = (E_i, F_i, \varepsilon_i)$ in I, we use local reflexivity to find a $(1+\varepsilon_i)$-isomorphic embedding $u_i : E_i \to X$ which satisfies $u_i x = x$ if $x \in E_i \cap X$ together with $\langle u_i x^{**}, x^* \rangle = \langle x^{**}, x^* \rangle$ for all $x^{**} \in E_i$ and $x^* \in F_i$.

We repeat arguments from 8.12. Given $x^{**} \in X^{**}$, set $x_i := u_i x^{**}$ if x^{**} is in E_i and $x_i := 0$ if not. Then $(x_i)_{\mathcal{U}} \in X^{\mathcal{U}}$ is well-defined, and

$$j : X^{**} \to X^{\mathcal{U}} : x^{**} \mapsto (x_i)_{\mathcal{U}}$$

is a linear and isometric embedding.

To see that j's range is norm one complemented, we use the fact that bounded sets in X^{**} are relatively weak$*$compact. In particular, $\lim_{\mathcal{U}} x_i$ exists in the weak$*$topology of X^{**} whenever $(x_i)_{i \in I}$ is bounded in X. The corresponding map $\ell_\infty(X; I) \to X^{**} : (x_i)_{i\in I} \mapsto \lim_{\mathcal{U}} x_i$ is clearly linear, and since $\lim_{\mathcal{U}} x_i$ is obviously zero whenever $\lim_{\mathcal{U}} \|x_i\| = 0$, we obtain a linear map

$$p : X^{\mathcal{U}} \to X^{**} : (x_i)_{\mathcal{U}} \mapsto \lim_{\mathcal{U}} x_i;$$

it is easy to see that $\|p\| \leq 1$. By construction, if $(x_i)_{\mathcal{U}} = jx^{**}$ with $x^{**} \in X^{**}$, then

$$\langle pjx^{**}, x^* \rangle = \langle \lim_{\mathcal{U}} x_i, x^* \rangle = \lim_{\mathcal{U}} \langle x_i, x^* \rangle = \langle x^{**}, x^* \rangle$$

for all $x^* \in X^*$, so that $pjx^{**} = x^{**}$. It follows that jp is the sought-after projection. QED

Notes and Remarks

We owe to D. Dacunha-Castelle and J.L. Krivine [1972] the introduction of ultraproduct techniques to the study of Banach spaces. Some years later, S. Heinrich produced some beautifully executed notes [1980] on the topic; these have greatly influenced our treatment.

There is another approach to many of the results in this chapter – through non-standard hulls. The work of C.W. Henson ([1974], [1975], [1976]), C.W. Henson and L.C. Moore, Jr. ([1972], [1974a], [1974b]) and W.A.J. Luxemburg [1969] should be consulted for this alternative, yet equivalent, viewpoint.

Proposition 8.1 is due to S. Heinrich [1980], as is 8.3. Proposition 8.6 and Theorem 8.7 have their roots in the work of D. Dacunha-Castelle and J.L. Krivine [1972] and are explicitly formulated in Heinrich's notes. These notes also contain an ultraproduct rendition of a curious feature of ℓ_∞: $(\ell_\infty)_{\mathcal{U}}$ is *not* complemented in its bidual. This result had first been proved in the context of non-standard hulls by C.W. Henson and L.C. Moore, Jr. [1974a].

Finite representability is a product of the inventive mind of R.C. James [1972]. Theorems 8.12 and 8.13 are due independently to C.W. Henson and L.C. Moore, Jr. [1974b] and to J. Stern [1978].

Helly's Lemma first arose, in a somewhat different formulation, in the work of E. Helly [1921]. The identity of 8.17 was first discovered by A. Grothendieck [1953a], and he used it in tandem with Helly's Lemma for his theory of tensor norms. Although it seems likely that Grothendieck was aware that there had to be a Principle of Local Reflexivity, he never came close to formulating it in print. The first statement, proof, and application of the Principle are the achievement of J. Lindenstrauss and H.P. Rosenthal [1969]. It allowed them to show the still beguiling duality of $\mathcal{L}_p$-spaces: *if* $1 \leq p \leq \infty$, *then* X *is an* $\mathcal{L}_p$*-space if and only if* X^* *is an* $\mathcal{L}_{p^*}$*-space.* We also mention that 8.14(b) has a converse: *if* X *is isomorphic to a complemented subspace of an* $\mathcal{L}_p$*-space, then* X *must be an* $\mathcal{L}_r$*-space, where* $r = p$ *when* $p = 1, \infty$, *and* $r \in \{p, 2\}$ *when* $1 < p < \infty$.

The version of the Principle of Local Reflexivity stated in the text is an upgrade taken from W.B. Johnson, H.P. Rosenthal and M. Zippin [1971]. With this, they were able to carry out a penetrating analysis of finite dimensional structure, in which they dug up many gems. Here are just two examples: *if* X^* *has a basis, so does* X, and *every separable* $\mathcal{L}_p$*-space has a basis.*

Our proof of 8.16 goes back to D.W. Dean [1973]; we expand on the exposition given by J. Lindenstrauss and L. Tzafriri [1977] and H.P. Lotz [1973].

As might be expected, the Principle has attracted a variety of proofs and generalizations. An extremely general version was presented by E. Behrends [1991]. He called on ideas from D.W. Dean [1973], C. Stegall [1980] and E. Behrends [1986] and succeeded in bestowing considerable flexibility on the functional equations and approximations that arise in the Principle. Here is a modest sample of the kind of results derived by Behrends.

Theorem: *Let* $u : X \to Y$, $v : X \to Z$ *be Banach space operators. Suppose that* $x_0^{**} \in S_{X^{**}}$ *is weak∗continuous on the weak∗closure of* $u^*(Y^*)$ *and that* $y_0 := u^{**}(x_0^{**}) \in Y$. *Let* $C \subset Z$ *be a convex set whose weak∗closure in* Z^{**} *contains* $v^{**}(x_0^{**})$. *Then for any finite dimensional subspace* F *of* X^* *and for any* $\varepsilon > 0$ *it is possible to locate* $x_0 \in X$ *such that*

$$(1+\varepsilon)^{-1} \leq \|x_0\| \leq 1+\varepsilon, \quad \langle x_0^{**}, x^* \rangle = \langle x^*, x_0 \rangle \ \forall x^* \in F,$$

and such that $ux_0 = y_0$ *and* vx_0 *is within* ε *of* C.

Behrends pushed much further; he was able to generalize the above to the consideration of arbitrary finite dimensional subspaces E of X^{**} (instead of just the span of one vector) and to the approximation of arbitrary finite systems of operators on E. The identity $\mathcal{L}(E, X)^{**} = \mathcal{L}(E, X^{**})$ features prominently in these developments.

T. Barton and X.T. Yu [1995] took another noteworthy direction.

Theorem: *Let* E *be a finite dimensional subspace of* X^{**} *and let* F *be a reflexive subspace of* X^*. *Then given* $\varepsilon > 0$ *there is an injective operator* $u : E \to X$ *such that* $\|u\| \cdot \|u^{-1}\| < 1 + \varepsilon$, $ux = x$ *for all* $x \in E \cap X$ *and* $\langle x^*, ux^{**} \rangle = \langle x^{**}, x^* \rangle$ *for all* $x^{**} \in E$ *and* $x^* \in F$.

S.F. Bellenot [1984], S.J. Bernau [1980], P. Domański [1990], V.A. Geĭler and I.I. Chuchaev [1982] and K.D. Kürsten [1984] have presented further interesting variations of the Principle of Local Reflexivity, with both Bernau and Kürsten giving special consideration to ordered Banach spaces.

H. Pfitzner [1992] formulated a Principle of Local Reflexivity especially for C^*-algebras and applied it to show that unconditionally converging operators on a C^*-algebra are weakly compact.

The reader may have noticed that certain proofs in earlier chapters can be shortened and sweetened with the aid of ultraproducts. In later chapters we shall find many occasions to call on ultraproducts and the Principle of Local Reflexivity. These techniques are examples of abstract analysis at its best, and, together or separately, they find use in diverse situations that we shall not broach. As illustrations, by no means exhaustive, we cite the work on spreading models (B. Beauzamy and J.T. Lapresté [1984]), non-linear classification of Banach spaces (S. Heinrich and P. Mankiewicz [1982]), infinite dimensional holomorphy (S. Dineen [1986a], M. Lindström and R. Ryan [1991]), C^*-algebras (A.I. Singh and N. Mittal [1991]), triple systems (S. Dineen [1986a,b]) and fixed point theory (B. Maurey [1980/81]).

9. p-FACTORABLE OPERATORS

MAXIMALITY OF Γ_p

In this chapter, we return to the theme broached in Chapter 7, and see to what extent the properties of Γ_2 can be extended to general Γ_p. We shall be gratified to find much similarity in the theory, but chastened by the greater complexity of some of the proofs.

As we already know from 7.1 that $[\Gamma_p, \gamma_p]$ is a Banach ideal for any $1 \leq p \leq \infty$, it is only natural to try to extend 7.4 and show that all these ideals are maximal. Our first opportunity to use ultraproducts is at hand.

9.1 Theorem: *For every $1 \leq p \leq \infty$, the Banach ideal $[\Gamma_p, \gamma_p]$ is maximal.*

Proof. As usual, our strategy will be to use the Biadjoint Criterion 6.8 to show that whenever X and Y are Banach spaces and $u \in \Gamma_p^{**}(X,Y)$, then $u \in \Gamma_p(X,Y)$, with $\gamma_p(u) \leq \gamma_p^{**}(u)$.

We select an ultrafilter $\mathcal{U}$ on $\mathcal{P} = \mathcal{F}_X \times \mathcal{C}_Y$ which refines the usual order filter. For any $\varepsilon > 0$ and $(E,Z) \in \mathcal{P}$ there exist, thanks to the Biadjoint Criterion, a measure $\mu_{E,Z}$ and operators $a_{E,Z}$ in $\mathcal{L}(L_p(\mu_{E,Z}), Y/Z)$ and $b_{E,Z}$ in $\mathcal{L}(E, L_p(\mu_{E,Z}))$ such that $u_{E,Z} := q_Z u i_E = a_{E,Z} b_{E,Z}$, $\|b_{E,Z}\| \leq 1$ and $\|a_{E,Z}\| \leq \gamma_p(u_{E,Z}) + \varepsilon \leq \gamma_p^{**}(u) + \varepsilon$. This allows us to consider the ultraproducts $L := (\Pi_{\mathcal{P}} L_p(\mu_{E,Z}))_{\mathcal{U}}$, $a := (a_{E,Z})_{\mathcal{U}}$, $b := (b_{E,Z})_{\mathcal{U}}$ and $(u_{E,Z})_{\mathcal{U}}$. The stability results 8.7 tell us that L is an $L_p(\mu)$-space for $1 \leq p < \infty$, and a $C(K)$-space for $p = \infty$. Moreover, $a \in \mathcal{L}(L, (\Pi_{\mathcal{P}} Y/Z)_{\mathcal{U}})$ with $\|a\| \leq \gamma_p^{**}(u) + \varepsilon$, whereas $b \in \mathcal{L}((\Pi_{\mathcal{P}} E)_{\mathcal{U}}, L)$ with $\|b\| \leq 1$.

Since we made our construction so that $(u_{E,Z})_{\mathcal{U}} = (a_{E,Z})_{\mathcal{U}} (b_{E,Z})_{\mathcal{U}}$, we can deduce that $(u_{E,Z})_{\mathcal{U}} \in \Gamma_p\big((\Pi_{\mathcal{P}} E)_{\mathcal{U}}, (\Pi_{\mathcal{P}} Y/Z)_{\mathcal{U}}\big)$ and use 8.10 to come to the conclusion that $u \in \Gamma_p(X,Y)$ with $\gamma_p(u) \leq \gamma_p((u_{E,Z})_{\mathcal{U}}) \leq \gamma_p^{**}(u) + \varepsilon$. The ε may be chosen as small as we please, so we are done. QED

Here is an immediate application.

9.2 Corollary: *If $1 < p < \infty$, then $\Gamma_2(X,Y) \subset \Gamma_p(X,Y)$ for all Banach spaces X and Y.*

Proof. As usual, maximality allows us to make a reduction. We need only find a constant $c_p > 0$ such that $\gamma_p(u) \leq c_p \cdot \gamma_2(u)$ for all operators $u : E \to F$ between finite dimensional normed spaces.

Accordingly, let $u : E \xrightarrow{w} H \xrightarrow{v} F$ be any factorization through some Hilbert space H. We may as well assume that $H = \ell_2$. But thanks to the Rademacher projection (1.12), we have a complementation $\ell_2 \xrightarrow{b} L_p[0,1] \xrightarrow{a} \ell_2$; evidently we can take $c_p = \|a\| \cdot \|b\|$. QED

Dual Ideals

Theorem 9.1 immediately raises the question of the identity of the adjoint ideals $[\Gamma_p^*, \gamma_p^*]$. We have already seen in 7.11 that $[\Gamma_2^*, \gamma_2^*] = [\Delta_2, \Delta_2]$ and in 6.16 that $[\Gamma_\infty^*, \gamma_\infty^*] = [\Pi_1, \pi_1]$.

We turn first to the identification of $[\Gamma_1^*, \gamma_1^*]$. This will afford us an opportunity to use the Principle of Local Reflexivity; the context will be a new operation on Banach ideals.

Let $[\mathcal{A}, \alpha]$ be a Banach ideal. Given Banach spaces X and Y, introduce the notation

$$\mathcal{A}^d(X,Y)$$

for the set of all $u \in \mathcal{L}(X,Y)$ with $u^* \in \mathcal{A}(Y^*, X^*)$. For such u's we also stipulate that

$$\alpha^d(u) := \alpha(u^*).$$

Routine hackwork reveals that we have created a new Banach ideal,

$$[\mathcal{A}^d, \alpha^d],$$

and we choose to call it the *dual ideal* of $[\mathcal{A}, \alpha]$.

This concept is not entirely new to us. For example, part of Proposition 7.2 can be rephrased as $[\Gamma_p^d, \gamma_p^d] = [\Gamma_{p^*}, \gamma_{p^*}]$ $(1 \leq p \leq \infty)$, and Theorem 5.15 is just the statement that $[\mathcal{I}_1^d, \iota_1^d] = [\mathcal{I}_1, \iota_1]$ in disguise. Moreover, in 4.19 we discovered that a Banach space X is isomorphic to a Hilbert space if and only if $\Pi_2(X,Y) \subset \Pi_2^d(X,Y)$ for all Banach spaces Y – and that it even suffices to test with $Y = \ell_2$.

To identify $[\Gamma_1^*, \gamma_1^*]$, we only need one general result on dual ideals.

9.3 Proposition: *Let $[\mathcal{A}, \alpha]$ be a Banach ideal. Then*

$$[\mathcal{A}^{*d}, \alpha^{*d}] = [\mathcal{A}^{d*}, \alpha^{d*}].$$

Proof. One inclusion is a simple consequence of trace duality.

Consider a given $v \in \mathcal{A}^{*d}(X,Y)$ as part of a composition of operators $E \xrightarrow{w} X \xrightarrow{v} Y \xrightarrow{u} F \xrightarrow{t} E$, with E and F finite dimensional normed spaces. Turning the handle, we find that

$$\begin{aligned} |\mathrm{tr}\,(tuvw)| &= |\mathrm{tr}\,(w^*v^*u^*t^*)| = |\mathrm{tr}\,(t^*w^*v^*u^*)| \\ &\leq \alpha^*(v^*)\cdot\alpha(t^*)\cdot\|w^*\|\cdot\|u^*\| = \alpha^{*d}(v)\cdot\alpha^d(t)\cdot\|u\|\cdot\|w\| \end{aligned}$$

and notice that v belongs to $\mathcal{A}^{d*}(X,Y)$ with $\alpha^{d*}(v) \leq \alpha^{*d}(v)$.

The other inclusion also uses trace duality, but lies deeper. Suppose that $v \in \mathcal{A}^{d*}(X,Y)$, and consider a composition of operators $E \xrightarrow{w} Y^* \xrightarrow{v^*} X^* \xrightarrow{u} F \xrightarrow{t} E$, with E and F finite dimensional. Dualization leads to the composition $F^* \xrightarrow{u^*} X^{**} \xrightarrow{v^{**}} Y^{**} \xrightarrow{w^*} E^* \xrightarrow{t^*} F^*$. If X is reflexive, this is a very favourable situation, since then we can use

$$(\bullet)\qquad \begin{aligned}|\operatorname{tr}(tuv^*w)| &= |\operatorname{tr}(t^*w^*v^{**}u^*)| = |\operatorname{tr}(t^*w^*k_Y vu^*)| \\ &\leq \alpha^{d*}(v)\cdot\alpha^d(t^*)\cdot\|w^*k_Y\|\cdot\|u\| \leq \alpha^{d*}(v)\cdot\alpha(t)\cdot\|u\|\cdot\|w\|\end{aligned}$$

to obtain the desired conclusion that v^* is in $\mathcal{A}^*(Y^*, X^*)$ with $\alpha^*(v^*) \leq \alpha^{d*}(v)$.

If X is not reflexive, the Principle of Local Reflexivity saves the day, at the cost of a little epsilonics. Look at the finite dimensional subspaces $u^*(F^*)$ of X^{**} and $v^*w(E)$ of X^*. Given any $\varepsilon > 0$, 8.16 provides us with an operator $a : u^*(F^*) \to X$ with $\|a\| \leq 1+\varepsilon$ and $\langle au^*y^*, v^*wx\rangle = \langle u^*y^*, v^*wx\rangle$ for any $x \in E$ and $y^* \in F^*$. In other words, after making natural identifications, $w^*k_Y vau^* = w^*v^{**}u^*$. This allows us to modify $(\bullet)$ by substituting $k_Y va$ for $k_Y v$ and then to continue in the same vein until we find that v^* is in $\mathcal{A}^*(Y^*, X^*)$ with $\alpha^*(v^*) \leq (1+\varepsilon)\cdot\alpha^{d*}(v)$. There are traditional methods to eliminate the ε. QED

Iteration of the proposition shows that

$$[\mathcal{A}^{d**}, \alpha^{d**}] = [\mathcal{A}^{**d}, \alpha^{**d}]$$

holds for any Banach ideal $[\mathcal{A}, \alpha]$. In particular:

9.4 Corollary: *The Banach ideal $[\mathcal{A}, \alpha]$ is maximal if and only if $[\mathcal{A}^d, \alpha^d]$ is maximal.*

We need no more to make our first identification.

9.5 Corollary: $[\Gamma_1^*, \gamma_1^*] = [\Pi_1^d, \pi_1^d]$ *and* $[\Pi_1^{d*}, \pi_1^{d*}] = [\Gamma_1, \gamma_1]$.

Proof. It is enough to establish the first identity. To do this, draw on 7.2 which tells us that $[\Gamma_1, \gamma_1] = [\Gamma_\infty^d, \gamma_\infty^d]$. Then our knowledge of $[\Gamma_\infty, \gamma_\infty] = [\Pi_1^*, \pi_1^*]$ together with the maximality of $[\Pi_1, \pi_1]$ yields $[\Gamma_1^*, \gamma_1^*] = [\Pi_1^d, \pi_1^d]$. QED

q-Dominated Operators

The identification of $[\Gamma_p^*, \gamma_p^*]$ for $1 < p < \infty$ will require much more sustained effort. We already know that $[\Gamma_2^*, \gamma_2^*] = [\Delta_2, \Delta_2]$, and this is a useful model to follow. The selection of a good candidate for $[\Gamma_p^*, \gamma_p^*]$ can be motivated in much the same way as for the case $p = 2$. We need a little preliminary lemma.

9.6 Lemma: *Let X be any Banach space, and let $1 \leq p < \infty$. Then, for any measure μ, we have $\Pi_p^d(X, L_p(\mu)) \subset \mathcal{I}_p(X, L_p(\mu))$ with $\iota_p(u) \leq \pi_p^d(u)$ for all $u \in \Pi_p^d(X, L_p(\mu))$.*

Proof. By localization, it is good enough to restrict our attention to operators $u : X \to \ell_p^n$, where $n \in \mathbf{N}$. Setting $x_k^* := u^*e_k$ for each $1 \leq k \leq n$, we can produce a factorization $u : X \xrightarrow{w} \ell_\infty^n \xrightarrow{d} \ell_p^n$, where $d\big((a_k)_1^n\big) = \big(\|x_k^*\|\cdot a_k\big)_1^n$ and $w(x) = (\langle x, \|x_k^*\|^{-1}\cdot x_k^*\rangle)_1^n$ (modulo the usual precautions not to divide by zero).

We know by 5.8 that $\iota_p(d) = \pi_p(d)$. As we can use 2.9(e) to compute $\pi_p(d)$, we see that

$$\iota_p(u) \leq \iota_p(d) \cdot \|w\| = \Big(\sum_{k=1}^n \|x_k^*\|^p\Big)^{1/p} = \Big(\sum_{k=1}^n \|u^* e_k\|^p\Big)^{1/p} \leq \pi_p(u^*). \quad \text{QED}$$

We shall extend this lemma in 9.11 and 9.12.

Let $u : E \to F$ be an operator between finite dimensional normed spaces, and let $1 < p < \infty$. Given an operator $v : F \to E$, we can affirm, thanks to a simple localization argument, that $\gamma_p(v) = \inf \|v_1\| \cdot \|v_2\|$ where the infimum is extended over all factorizations $v : F \xrightarrow{v_2} \ell_p^n \xrightarrow{v_1} E$ with $n \in \mathbf{N}$. Fix such a factorization, and select any factorization $E \xrightarrow{u_2} Z \xrightarrow{u_1} F$ of u through a Banach space Z. Then, as $[\Pi_p, \pi_p]$ and $[\mathcal{I}_{p^*}, \iota_{p^*}]$ are in trace duality (6.16), we may apply the lemma above in combination with 6.14 and 5.16 to derive

$$\begin{aligned} |\mathrm{tr}\,(uv)| &= |\mathrm{tr}\,((u_2 v_1)^* u_1^* v_2^*)| \leq \iota_1((u_2 v_1)^* u_1^* v_2^*) \leq \iota_{p^*}((u_2 v_1)^*) \cdot \pi_p(u_1^* v_2^*) \\ &\leq \pi_{p^*}^d((u_2 v_1)^*) \cdot \pi_p(u_1^*) \cdot \|v_2\| = \pi_{p^*}(u_2 v_1) \cdot \pi_p^d(u_1) \cdot \|v_2\| \\ &\leq \pi_p^d(u_1) \cdot \pi_{p^*}(u_2) \cdot \|v_1\| \cdot \|v_2\|. \end{aligned}$$

We conclude that

$$(*) \qquad \gamma_p^*(u) \leq \inf \pi_p^d(u_1) \cdot \pi_{p^*}(u_2)$$

where the infimum is taken over all factorizations $E \xrightarrow{u_2} Z \xrightarrow{u_1} F$ of u through a Banach space.

So far, so good. Things are moving almost as smoothly as in the case $p = 2$. It is natural to press on and aim to establish equality in $(*)$. This turns out to be feasible but it will lead us into deeper water. To begin, with $(*)$ in mind, we introduce new classes of operators.

For $1 < q < \infty$ we say that a Banach space operator $u : X \to Y$ is *q-dominated* and write

$$u \in \Delta_q(X, Y)$$

if there exist a Banach space Z and operators $a \in \Pi_{q^*}^d(Z, Y)$ and $b \in \Pi_q(X, Z)$ with $u = ab$. We even set

$$\Delta_q(u) := \inf \pi_{q^*}^d(a) \cdot \pi_q(b)$$

with the infimum taken over all such factorizations.

We can extend our definition to include the cases $q = 1$ and $q = \infty$ by adopting the convention that $[\Pi_\infty, \pi_\infty] = [\mathcal{L}, \|\cdot\|]$. We then find $[\Delta_1, \Delta_1] = [\Pi_1, \pi_1] = [\Gamma_\infty^*, \gamma_\infty^*]$ and $[\Delta_\infty, \Delta_\infty] = [\Pi_1^d, \pi_1^d] = [\Gamma_1^*, \gamma_1^*]$. Our program should be clear: we strive to show that $[\Delta_q, \Delta_q]$ is a maximal Banach ideal whose adjoint is $[\Gamma_{q^*}, \gamma_{q^*}]$.

It is clear from the definition that any of the conditions $u \in \Delta_q(X, Y)$, $u^* \in \Delta_{q^*}(Y^*, X^*)$, $u^{**} \in \Delta_q(X^{**}, Y^{**})$ and $k_Y u \in \Delta_q(X, Y^{**})$ implies all the others, with $\Delta_q(u) = \Delta_{q^*}(u^*) = \Delta_q(u^{**}) = \Delta_q(k_Y u)$.

The fundamental characterizations of q-dominated operators are combined in the following theorem.

9.7 Theorem: *Let $1 < q < \infty$, and let X and Y be Banach spaces with K and L weak$*$compact norming subsets of B_{X^*} and $B_{Y^{**}}$, respectively. For every operator $u : X \to Y$ and every constant $c \geq 0$ the following assertions are equivalent:*

(i) $u \in \Delta_q(X,Y)$, *with* $\Delta_q(u) \leq c$.

(ii) *No matter how we choose finitely many vectors $x_1,...,x_n$ in X and $y_1^*,...,y_n^*$ in Y^*, we have*
$$\sum_{k=1}^{n} |\langle ux_k, y_k^*\rangle| \leq c\cdot \|(x_k)_1^n\|_q^{weak} \cdot \|(y_k^*)_1^n\|_{q^*}^{weak}.$$

(iii) *There exist probability measures $\mu \in \mathcal{C}(K)^*$ and $\nu \in \mathcal{C}(L)^*$ such that, for all $x \in X$ and $y^* \in Y^*$,*
$$|\langle ux, y^*\rangle| \leq c\cdot \Big(\int_K |\langle x^*, x\rangle|^q d\mu(x^*)\Big)^{1/q} \cdot \Big(\int_L |\langle y^{**}, y^*\rangle|^{q^*} d\nu(y^{**})\Big)^{1/q^*}.$$

The proof of the theorem is strenuous, so we prefer first to present the relatively easy passage to the duality result we are aiming for.

9.8 Theorem: *Let $1 \leq q \leq \infty$ and set $p = q^*$. Then $[\Delta_q, \Delta_q]$ is a maximal Banach ideal, and $[\Gamma_p^*, \gamma_p^*] = [\Delta_q, \Delta_q]$ and $[\Delta_q^*, \Delta_q^*] = [\Gamma_p, \gamma_p]$.*

Proof. We can of course assume that $1 < q < \infty$.

As the Banach ideal properties are routine consequences of part (ii) of the preceding theorem, we pass immediately to the proof of the identities.

Let $u \in \Delta_q(X,Y)$. Using the estimate $(*)$, we find that
$$\gamma_p^*(q_Z u i_E) \leq \Delta_q(q_Z u i_E) \leq \Delta_q(u)$$
for all $E \in \mathcal{F}_X$ and $Z \in \mathcal{C}_Y$. Since $[\Gamma_p^*, \gamma_p^*]$ is maximal, the Biadjoint Criterion tells us that $u \in \Gamma_p^*(X,Y)$ and that $\gamma_p^*(u) \leq \Delta_q(u)$.

Conversely, suppose that $u \in \Gamma_p^*(X,Y)$. We plan to invoke part (ii) of Theorem 9.7. Accordingly, we choose $x_1,...,x_n \in X$ and $y_1^*,...,y_n^* \in Y^*$ and define $v_1 \in \mathcal{L}(\ell_p^n, X)$ and $v_2 \in \mathcal{L}(Y, \ell_p^n)$ through $v_1 e_k = x_k$ and $v_2^* e_k = y_k^*$ $(1 \leq k \leq n)$. By 2.2, $\|v_1\| = \|(x_k)\|_q^{weak}$ and $\|v_2\| = \|(y_k^*)\|_p^{weak}$. But now
$$\Big|\sum_{k\leq n} \langle ux_k, y_k^*\rangle\Big| = \big|\operatorname{tr}(v_2 u v_1)\big| \leq \gamma_p^*(u)\cdot\|v_1\|\cdot\|v_2\|\cdot\gamma_p(id_{\ell_p^n})$$
$$= \gamma_p^*(u)\cdot\|(x_k)\|_q^{weak}\cdot\|(y_k^*)\|_p^{weak}.$$
We can apply Theorem 9.7 to obtain $u \in \Delta_q(X,Y)$ and $\Delta_q(u) \leq \gamma_p^*(u)$.

This has established $[\Delta_q, \Delta_q] = [\Gamma_p^*, \gamma_p^*]$; the other identity follows at once from the maximality of $[\Gamma_p, \gamma_p]$. QED

There is an immediate consequence of this theorem which is equivalent to 9.2.

9.9 Corollary: *If $1 < q < \infty$ and X, Y are any Banach spaces, then $\Delta_q(X,Y) \subset \Delta_2(X,Y)$.*

Proof of the Main Result

It is now time to concentrate on the proof of Theorem 9.7. A topological result is at the heart of the issue.

9.10 Ky Fan's Lemma: *Let E be a Hausdorff topological vector space, and let C be a compact convex subset of E. Let M be a set of functions on C with values in $(-\infty, \infty]$ having the following properties:*

(a) *Each $f \in M$ is convex and lower semicontinuous.*

(b) *If $g \in \operatorname{conv}(M)$, there is an $f \in M$ with $g(x) \leq f(x)$ for all $x \in C$.*

(c) *There is an $r \in \mathbf{R}$ such that each $f \in M$ has a value $\leq r$.*

Then there is an $x_0 \in C$ such that $f(x_0) \leq r$ for all $f \in M$.

Proof. Lower semicontinuity entails that each of the sets

$$S(f,\varepsilon) := \{x \in C : f(x) \leq r + \varepsilon\} \quad (f \in M,\, \varepsilon > 0)$$

is closed in C. By (c), each $S(f,\varepsilon)$ is non-empty. Our goal is to show that $\bigcap\{S(f,\varepsilon) : f \in M,\, \varepsilon > 0\}$ is non-empty, since any member of this intersection must satisfy our conclusion.

The compactness of C helps us out: we need only show that no matter how we choose $n \in \mathbf{N}$, $\varepsilon_k > 0$ and $f_k \in M$ $(1 \leq k \leq n)$, the set $\bigcap_{k=1}^n S(f_k, \varepsilon_k)$ is non-empty.

For this, we introduce two auxiliary subsets of $\mathbf{R}^n$. The first of these is $A := \prod_{k \leq n}(-\infty, r + \varepsilon_k)$: it is open and convex. The second is the convex hull B of $\{(f_k(x))_{k=1}^n : x \in C\} \cap \mathbf{R}^n$. Properties (b) and (c) ensure that $B \neq \emptyset$. We shall soon prove that $A \cap B \neq \emptyset$. Accepting this for now, we show how this implies our conclusion.

Choose $s = (s_1, \ldots, s_n)$ in $A \cap B$. By the definition of B, we can write $s_k = \sum_{j=1}^N \alpha_j f_k(x_j)$ $(1 \leq k \leq n)$ where the x_j's are in C and the α_j's are non-negative and sum to 1. However, C is convex, so $x = \sum_{j=1}^N \alpha_j x_j$ belongs to C, and each f_k is a convex function, so $f_k(x) \leq \sum_{j=1}^N \alpha_j f_k(x_j) = s_k$ $(1 \leq k \leq n)$. As $s \in A$, we deduce that $x \in S(f_k, \varepsilon_k)$ for each $1 \leq k \leq n$.

It only remains to show that $A \cap B \neq \emptyset$. We argue by contradiction and assume that $A \cap B = \emptyset$. The Hahn - Banach Separation Theorem (in $\mathbf{R}^n$) steps in to assure us that for some $\mu = (\mu_1, \ldots, \mu_n) \in \mathbf{R}^n$ and some $\lambda \in \mathbf{R}$ we have $\langle \mu, a \rangle < \lambda \leq \langle \mu, b \rangle$ for each $a \in A$ and $b \in B$. The strict inequality forces $\mu \neq 0$, and so we may, without loss of any generality, normalize so that $\sum_{k \leq n} |\mu_k| = 1$.

In fact, $\mu_k \geq 0$ for each k. To see why note that, no matter how we choose $\varrho \leq r, a = \sum_{j\neq k} re_j + \varrho e_k$ belongs to A and so $\langle \mu, a\rangle = r\cdot\sum_{j\neq k}\mu_j + \varrho\mu_k$. Now let ϱ tend to $-\infty$.

This allows us to use hypothesis (b) to find an $f \in M$ such that for $x \in C$, either $f(x) = \infty$ or

$$f(x) \geq \sum_{k\leq n} \mu_k f_k(x) = \langle \mu, (f_k(x))_1^n\rangle \geq \lambda > \langle \mu, (r + \tfrac{\varepsilon_k}{2})_1^n\rangle$$
$$= \sum_{k\leq n} \mu_k\cdot(r + \tfrac{\varepsilon_k}{2}) > r.$$

This contradicts hypothesis (c). QED

Our tools are all assembled.

Proof of Theorem 9.7.

(i)⇒(ii): This is straightforward. We select $u \in \Delta_q(X,Y)$ and exhibit a typical factorization $u = vw$ with $v \in \Pi^d_{q^*}(Z,Y)$ and $w \in \Pi_q(X,Z)$. Then, for any $x_1,\ldots,x_n \in X$ and $y_1^*,\ldots,y_n^* \in Y^*$, it follows that

$$\sum_{k\leq n}|\langle ux_k, y_k^*\rangle| = \sum_{k\leq n}|\langle wx_k, v^*y_k^*\rangle| \leq \Big(\sum_{k\leq n}\|wx_k\|^q\Big)^{1/q}\cdot\Big(\sum_{k\leq n}\|v^*y_k^*\|^{q^*}\Big)^{1/q^*}$$
$$\leq \pi_q(w)\cdot\pi^d_{q^*}(v)\cdot\|(x_k)_{k=1}^n\|_q^{weak}\cdot\|(y_k^*)_{k=1}^n\|_{q^*}^{weak}.$$

Take the infimum over all permissible factorizations to reach the desired conclusion.

(ii)⇒(iii): This is the main difficulty. We consider the sets $P(K)$ and $P(L)$ of probability measures in $\mathcal{C}(K)^*$ and $\mathcal{C}(L)^*$, respectively. These are convex sets which are compact when we endow $\mathcal{C}(K)^*$ and $\mathcal{C}(L)^*$ with their weak$*$ topologies. We are going to apply Ky Fan's Lemma with $E = \mathcal{C}(K)^* \times \mathcal{C}(L)^*$ and $C = P(K) \times P(L)$.

It behoves us to select an appropriate set M. Bearing in mind the conclusion we are heading for, and the traditional proof of Hölder's Inequality, we consider functions $f : C \to \mathbf{R}$ defined by

$$(*)\qquad f(\mu,\nu) := \sum_{k\leq n}\Big(|\langle ux_k, y_k^*\rangle| - \frac{c}{q}\cdot\int_K |\langle x_k, x^*\rangle|^q d\mu(x^*) - \frac{c}{q^*}\cdot\int_L |\langle y_k^*, y^{**}\rangle|^{q^*} d\nu(y^{**})\Big)$$

where $x_1,\ldots,x_n \in X$ and $y_1^*,\ldots,y_n^* \in Y^*$. These functions are clearly convex and continuous. We let M denote the collection of all such functions.

If we can show that M obeys conditions (b) and (c) of Ky Fan's Lemma with $r = 0$, we shall be able to deduce the existence of $(\mu,\nu) \in C$ such that $f(\mu,\nu) \leq 0$ for all $f \in M$. Then, if f is generated by the single elements $x \in X$ and $y^* \in Y^*$,

$$|\langle ux, y^*\rangle| \leq \frac{c}{q}\cdot\int_K |\langle x, x^*\rangle|^q d\mu(x^*) + \frac{c}{q^*}\cdot\int_L |\langle y^*, y^{**}\rangle|^{q^*} d\nu(y^{**}).$$

The usual trick transforms this into

$$|\langle ux, y^*\rangle| \le c\cdot\Big(\int_K |\langle x, x^*\rangle|^q d\mu(x^*)\Big)^{1/q}\cdot\Big(\int_L |\langle y^*, y^{**}\rangle|^{q^*}\, d\nu(y^{**})\Big)^{1/q^*}$$

which is just what we want.

All hangs on showing that M has the right properties. For property (b), it is good enough to work with a convex combination $\alpha f + (1-\alpha)g$ of $f, g \in M$. Suppose that f is defined in terms of $(x_k)_1^n$ in X and $(y_k^*)_1^n$ in Y^*, whereas g is defined in terms of $(x_k)_{n+1}^N$ in X and $(y_k^*)_{n+1}^N$ in Y^*. Squeezing and glueing, we create sequences $(\overline{x_k})_1^N$ in X and $(\overline{y_k}^*)_1^N$ in Y^* where $\overline{x_k} := \alpha^{1/q}x_k$ and $\overline{y_k}^* := \alpha^{1/q^*}y_k^*$ if $1 \le k \le n$, whereas $\overline{x_k} := (1-\alpha)^{1/q}x_k$ and $\overline{y_k}^* := (1-\alpha)^{1/q^*}y_k^*$ if $n < k \le N$. These new sequences give rise to another function $h \in M$, and it is a breeze to check that $\alpha f + (1-\alpha)g = h$. This gives us more than we need.

For property (c), we take f as in $(*)$ and select $x_0^* \in K$ and $y_0^{**} \in L$ such that $\xi := \|(x_k)_1^n\|_q^{weak} = (\sum_{k\le n}|\langle x_k, x_0^*\rangle|^q)^{1/q}$ and $\eta := \|(y_k^*)_1^n\|_{q^*}^{weak} = (\sum_{k\le n}|\langle y_k^*, y_0^{**}\rangle|^{q^*})^{1/q^*}$. Then

$$\begin{aligned} f(\delta_{x_0^*}, \delta_{y_0^{**}}) &= \sum_{k\le n}|\langle ux_k, y_k^*\rangle| - c\cdot\Big(\frac{1}{q}\cdot\xi^q + \frac{1}{q^*}\cdot\eta^{q^*}\Big) \\ &\le \sum_{k\le n}|\langle ux_k, y_k^*\rangle| - c\cdot\|(x_k)\|_q^{weak}\cdot\|(y_k^*)\|_{q^*}^{weak}. \end{aligned}$$

But f satisfies (ii), so we have completed our task: $f(\delta_{x_0^*}, \delta_{y_0^{**}}) \le 0$.

(iii)$\Rightarrow$(i): With our hypothesis in mind, we now consider the operator $w : X \to L_q(\mu)$ which is given by $x \mapsto \langle\,\cdot\,, x\rangle$ and notice that $\|ux\| \le c\cdot\|wx\|$ for all $x \in X$. Let Z be the closure in $L_q(\mu)$ of the range of w, and let $w_0 : X \to Z$ be the induced operator. Note that $w_0 \in \Pi_q(X, Z)$ with $\pi_q(w_0) \le 1$. By now it is routine to write $u = vw_0$ for some $v \in \mathcal{L}(Z, Y)$. We use the hypothesis again to show that if $y^* \in Y^*$ then

$$\begin{aligned} \|v^*y^*\| &= \sup\big\{|\langle w_0x, v^*y^*\rangle| : \|w_0x\| \le 1\big\} \\ &= \sup\Big\{|\langle ux, y^*\rangle| : \int_K |\langle x^*, x\rangle|^q d\mu(x^*) \le 1\Big\} \le c\cdot\Big(\int_L |\langle y^{**}, y^*\rangle|^{q^*} d\nu(y^{**})\Big)^{1/q^*}. \end{aligned}$$

Thanks to the Pietsch Domination Theorem 2.12, $v \in \Pi_{q^*}^d(Z, Y)$ with $\pi_{q^*}^d(v) \le c$.

Put everything together to see that $u \in \Delta_q(X, Y)$ with $\Delta_q(u) \le c$. QED

APPLICATIONS

Remember that when we were working towards the definition of q-dominated operators, we made use of the fact (9.6) that $\Pi_p^d(X, L_p(\mu)) \subset \mathcal{I}_p(X, L_p(\mu))$. This basic observation has a powerful extension.

9.11 Theorem: *Let $1 \le p \le \infty$, $c > 0$ and let $v : X \to Y$ be a Banach space operator. Then $v \in \Gamma_p(X, Y)$ with $\gamma_p(v) < c$ if and only if, regardless of the Banach space Z and the operator $w \in \Pi_p^d(Z, X)$, the composition $vw : Z \to Y$ is p-integral, with $\iota_p(vw) \le c\cdot\pi_p^d(w)$.*

Proof. If $v \in \Gamma_p(X,Y)$ with $\gamma_p(v) < c$, then there exists a factorization $k_Y v : X \xrightarrow{b} L_p(\mu) \xrightarrow{a} Y^{**}$ with $\|a\|\cdot\|b\| \le c$. When we make our choice of a Banach space Z and an operator $w \in \Pi_p^d(Z,X)$, we can use 9.6 to infer that bw is p-integral, with $\iota_p(bw) \le \|b\|\cdot\pi_p^d(w)$. Thus vw is also p-integral, with $\iota_p(vw) \le \|a\|\cdot\|b\|\cdot\pi_p^d(w) \le c\cdot\pi_p^d(w)$.

Suppose conversely that the condition is satisfied. The identity $[\Gamma_p, \gamma_p] = [\Delta_{p^*}^*, \Delta_{p^*}^*]$ means that to show $v \in \Gamma_p(X,Y)$ with $\gamma_p(v) \le c$, it suffices to take an arbitrary operator composition $E \xrightarrow{w} X \xrightarrow{v} Y \xrightarrow{u} F \xrightarrow{t} E$, with E and F finite dimensional, and to demonstrate that $|\mathrm{tr}\,(tuvw)| \le c\cdot\|u\|\cdot\|w\|\cdot\Delta_{p^*}(t)$.

We get at this by using trace duality between $[\mathcal{I}_p, \iota_p]$ and $[\Pi_{p^*}, \pi_{p^*}]$. If we factor t as $F \xrightarrow{t_2} Z \xrightarrow{t_1} E$ through some Banach space Z, we can use our hypothesis, again with 6.14 and 5.16, to obtain

$$\begin{aligned} |\mathrm{tr}\,(tuvw)| &= |\mathrm{tr}\,(uvwt)| \le \iota_1(uvwt_1t_2) \le \iota_p(uvwt_1)\cdot\pi_{p^*}(t_2) \\ &\le \|u\|\cdot\iota_p(vwt_1)\cdot\pi_{p^*}(t_2) \le c\cdot\|u\|\cdot\pi_p^d(wt_1)\cdot\pi_{p^*}(t_2) \\ &\le c\cdot\|u\|\cdot\|w\|\cdot\pi_p^d(t_1)\cdot\pi_{p^*}(t_2). \end{aligned}$$

Now the definition of $\Delta_{p^*}(t)$ enters; we conclude that

$$|\mathrm{tr}\,(tuvw)| \le c\cdot\|u\|\cdot\|w\|\cdot\Delta_{p^*}(t).$$ QED

This can be applied to derive various extensions of the Hilbert space characterizations obtained in 4.19.

Given ideals $\mathcal{A}$ and $\mathcal{B}$, let us use the notation $\mathcal{A}(\cdot,X) \subset \mathcal{B}(\cdot,X)$ to indicate that $\mathcal{A}(Z,X)$ is contained in $\mathcal{B}(Z,X)$ for every Banach space Z.

9.12 Corollary: *Let $1 \le p < \infty$, and let X be a Banach space.*

(a) *X^{**} is isomorphic to a complemented subspace of an $L_p(\mu)$-space if and only if $\Pi_p^d(\cdot,X) \subset \mathcal{I}_p(\cdot,X)$.*

(b) *X is isomorphic to a subspace of an $L_p(\mu)$-space if and only if $\Pi_p^d(\cdot,X) \subset \Pi_p(\cdot,X)$.*

(c) *X^{**} is isomorphic to a quotient of an $L_p(\mu)$-space if and only if $\mathcal{I}_p^d(\cdot,X) \subset \mathcal{I}_p(\cdot,X)$.*

(d) *X is isomorphic to a subspace of a quotient (equivalently: to a quotient of a subspace) of an $L_p(\mu)$-space if and only if $\mathcal{I}_p^d(\cdot,X) \subset \Pi_p(\cdot,X)$.*

Proof. We obtain (a) by applying the preceding theorem to the canonical embedding $k_X : X \to X^{**}$. Parts (b) and (c) follow in the same way by considering the canonical embedding $X \to \ell_\infty^{B_{X^*}}$ and the quotient map $\ell_1^{B_X} \to X$, respectively. Finally, (d) is obtained by combining (b) and (c). QED

There are immediate extensions of this corollary – which we leave to the reader's imagination – that describe operators which factor through subspaces, quotients, and so on, of $L_p(\mu)$-spaces. We prefer to take a different direction and derive a much deeper characterization of operators factoring through a subspace of a space $L_p(\mu)$, which completes the remarks we made in 7.9.

We take the occasion to note that the inclusion 9.6 allows another proof of Corollary 5.22:

- Let $1 \le p < \infty$ and let μ, ν be arbitrary measures. If $X = L_{p^*}(\mu)$ and $Y = L_p(\nu)$, then isometrically
$$\mathcal{I}_p(X,Y) = \Pi_p(X,Y) = \mathcal{I}_p^d(X,Y) = \Pi_p^d(X,Y).$$

Just take an operator $u : X \to Y$ and apply 5.5, 9.6 and 2.19 to obtain in fact
$$\pi_p(u^*) \le \iota_p(u^*) \le \pi_p(u^{**}) = \pi_p(u) \le \iota_p(u) \le \pi_p(u^*).$$

We apply this to prove the following result:

9.13 Theorem: *Let $1 \le p < \infty$, let $C \ge 0$, and let $u : X \to Y$ be a Banach space operator. The following are equivalent statements.*

(i) *There exist a subspace L of a space $L_p(\mu)$ and a factorization $u : X \xrightarrow{w} L \xrightarrow{v} Y$ with $\|v\|\cdot\|w\| \le C$.*

(ii) *Whenever the finite sequences $(z_i)_1^m$ and $(x_j)_1^n$ in X satisfy*
$$(*) \qquad \sum_{i=1}^m |\langle x^*, z_i\rangle|^p \le \sum_{j=1}^n |\langle x^*, x_j\rangle|^p$$
for all $x^* \in X^*$, we have
$$\sum_{i=1}^m \|uz_i\|^p \le C^p \cdot \sum_{j=1}^n \|x_j\|^p.$$

Proof. We only need to prove that (i) implies (ii); the argument for the converse was given in 7.9.

Let $u = vw$ be a factorization as described in (i), and let $(z_i)_1^m$ and $(x_j)_1^n$ be finite sequences in X satisfying $(*)$. We use these sequences to define operators $s \in \mathcal{L}(\ell_{p^*}^m, X)$ and $t \in \mathcal{L}(\ell_{p^*}^n, X)$ via $se_i = z_i$ $(1 \le i \le m)$ and $te_j = x_j$ $(1 \le j \le n)$. Our condition $(*)$ translates to read $\|s^*x^*\| \le \|t^*x^*\|$ for all $x^* \in X^*$, and this allows us to well-define an operator $a : t^*(X^*) \to \ell_p^m$ by $at^*x^* := s^*x^*$ for all $x^* \in X^*$. Plainly, $\|a\| \le 1$.

Thinking of w as an operator with values in $L_p(\mu)$, we can apply the above remark (or 5.22) to obtain
$$\begin{aligned}\Big(\sum_{i=1}^m \|uz_i\|^p\Big)^{1/p} &\le \|v\|\cdot\Big(\sum_{i=1}^m \|wse_i\|^p\Big)^{1/p} \le \|v\|\cdot\pi_p(ws) = \|v\|\cdot\pi_p(s^*w^*) \\ &= \|v\|\cdot\pi_p(at^*w^*) \le \|v\|\cdot\|w\|\cdot\pi_p(t^*) \le C\cdot\pi_p(t^*).\end{aligned}$$

To conclude, we need an appropriate upper bound for $\pi_p(t^*)$:

$$\begin{aligned}\pi_p(t^*) &= \sup\Big\{\big(\sum_k \|t^*x_k^*\|^p\big)^{1/p} : \|(x_k^*)\|_p^{weak} \le 1\Big\}\\ &= \sup\Big\{\big(\sum_k \sum_{j=1}^n |\langle x_k^*, te_j\rangle|^p\big)^{1/p} : \|(x_k^*)\|_p^{weak} \le 1\Big\}\\ &\le \big(\sum_{j=1}^n \|te_j\|^p\big)^{1/p} = \big(\sum_{j=1}^n \|x_j\|^p\big)^{1/p}.\end{aligned}$$ QED

NOTES AND REMARKS

The basic source for the ideals Γ_p is S. Kwapień [1972b]. As in the case $p = 2$, it was Kwapień who discovered the duality 9.8 between $[\Gamma_p, \gamma_p]$ and $[\Delta_{p^*}, \Delta_{p^*}]$. Theorem 9.7 is implicitly in Kwapień [1972b] and is made explicit in J.T. Lapresté's [1972/73] lectures.

Ky Fan's Lemma 9.10 is due to Ky Fan [1951]. We shall meet it at other critical junctures in our journey.

The characterizations of subspaces, quotients, and subspaces of quotients of L_p-spaces are again due to S. Kwapień [1972b]. The reader would be well-advised to look into D.R. Lewis' [1973] discussion of these affairs.

From the very definition, the q-dominated operators have close ties with summing operators. By stepping back a little, we briefly indicate how it is possible to incorporate both classes of operators into a much wider scheme.

To begin, observe that regardless of how we choose $1 \le p, q \le \infty$, any Banach space operator $u : X \to Y$ induces a bounded bilinear map

$$\check{u} : \ell_p^{weak}(X) \times \ell_q^{weak}(Y^*) \longrightarrow \ell_\infty : \big((x_j), (y_j^*)\big) \mapsto \big(\langle ux_j, y_j^*\rangle\big)$$

with $\|\check{u}\| = \|u\|$. Note that u is p-summing with $\pi_p(u) \le C$ if and only if $\big(\sum_{j=1}^n |\langle ux_j, y_j^*\rangle|^p\big)^{1/p} \le C\cdot\|(x_j)\|_p^{weak}\cdot\max_{j\le n}\|y_j^*\|$ no matter how we choose finite sets $\{x_1,\dots,x_n\} \subset X$ and $\{y_1^*,\dots,y_n^*\} \subset Y^*$. Thus Proposition 2.1 can be reformulated to say that u is p-summing if and only if $\check{u} : \ell_p^{weak}(X) \times \ell_\infty^{weak}(Y^*) \to \ell_\infty$ has range in ℓ_p, in which case $\pi_p(u)$ is just the norm of the resulting bilinear map $\ell_p^{weak}(X) \times \ell_\infty^{weak}(Y^*) \to \ell_p$.

Dually, u belongs to $\Pi_p^d(X, Y)$ if and only if $\big((x_j), (y_j^*)\big) \mapsto \big(\langle ux_j, y_j^*\rangle\big)$ defines a bounded bilinear map $\check{u} : \ell_\infty^{weak}(X) \times \ell_p^{weak}(Y^*) \to \ell_p$; moreover, $\pi_p^d(u) = \|\check{u}\|$.

Next, part (ii) of Theorem 9.7 tells us that for $u : X \to Y$ to be q-dominated it is necessary and sufficient that we have a bounded bilinear map $\check{u} : \ell_q^{weak}(X) \times \ell_{q^*}^{weak}(Y^*) \to \ell_1 : \big((x_j), (y_j^*)\big) \mapsto \big(\langle ux_j, y_j^*\rangle\big)$. Naturally, in such a case, $\|\check{u}\|$ is the q-dominated norm $\Delta_q(u)$.

To emphasize the generality of our observations, we begin by recalling that another way of stating that an operator $u : X \to Y$ satisfies Grothendieck's

Dunford - Pettis property is to say that it gives rise to a bounded bilinear map $c_0^{weak}(X) \times c_0^{weak}(Y^*) \to c_0 : ((x_j), (y_j^*)) \mapsto (\langle ux_j, y_j^* \rangle)$.

These patterns lead inexorably to a general multi-index definition. Given $1 \leq q, p, r \leq \infty$, we say that an operator $u : X \to Y$ is (q,p,r)*-summing* if it induces a bounded bilinear map $\check{u} : \ell_p^{weak}(X) \times \ell_r^{weak}(Y^*) \to \ell_q$. In this case, we write $u \in \Pi_{q,p,r}(X,Y)$, and $\pi_{q,p,r}(u) = \|\check{u}\|$.

Basic but boring manipulations reveal that u belongs to $\Pi_{q,p.r}(X,Y)$ with $\pi_{q,p,r}(u) \leq C$ if and only if

$$(*) \qquad \Big(\sum_{k=1}^{n} |\langle ux_k, y_k^* \rangle|^q\Big)^{1/q} \leq C \cdot \|(x_k)\|_p^{weak} \cdot \|(y_k^*)\|_r^{weak}$$

regardless of the choice of finite sets $\{x_1, ..., x_n\} \subset X$ and $\{y_1^*, ..., y_n^*\} \subset Y^*$.

Routinely, $[\Pi_{q,p,r}(X,Y), \pi_{q,p,r}]$ is a Banach space. However, when we examine inequality $(*)$ with $x_k = \lambda_k x$ and $y_k^* = \mu_k y^*$ $(1 \leq k \leq n)$ where $x \in X$ and $y^* \in Y^*$ are fixed and the λ_k's and the μ_k's are scalars, it soon becomes clear that $\Pi_{q,p,r}(X,Y) = \{0\}$ for $1/q > (1/p) + (1/r)$. Also, if $q = \infty$, then $\Pi_{q,p,r}(X,Y)$ is just $\mathcal{L}(X,Y)$ in disguise. Consequently, it is appropriate to assume $1 \leq q < \infty$, $1 \leq p, r \leq \infty$ and $1/q \leq (1/p) + (1/r)$ when dealing with (q,p,r)-summing operators. Then the standard processes show that

- $[\Pi_{q,p,r}, \pi_{q,p,r}]$ *is a maximal Banach ideal.*

For details, see J.T. Lapresté [1972/73] and A. Pietsch [1978].

To reiterate our introductory comments, we note that

$$[\Pi_p, \pi_p] = [\Pi_{p,p,\infty}, \pi_{p,p,\infty}] \quad , \quad [\Pi_p^d, \pi_p^d] = [\Pi_{p,\infty,p}, \pi_{p,\infty,p}]$$

and

$$[\Delta_q, \Delta_q] = [\Pi_{1,q,q^*}, \pi_{1,q,q^*}].$$

Besides these, there are only a few further significant groups of examples. One of these occurs when we choose $r = \infty$, and this is going to be the topic of the next chapter.

10. (q,p)-SUMMING OPERATORS

SOME BASIC PROPERTIES

Recall that a Banach space operator $u : X \to Y$ is p-summing if it induces an operator $\hat{u} : \ell_p^{weak}(X) \to \ell_p^{strong}(Y) : (x_n) \mapsto (ux_n)$ and that in this case $\pi_p(u) = \|\hat{u}\|$. It is fruitful to allow ourselves more room for manoeuvre.

We say that the operator $u : X \to Y$ is *(q,p)-summing* $(1 \leq p, q < \infty)$ if there is an induced operator

$$\hat{u} : \ell_p^{weak}(X) \longrightarrow \ell_q^{strong}(Y) : (x_n) \mapsto (ux_n).$$

These operators form a vector space,

$$\Pi_{q,p}(X,Y),$$

which becomes a Banach space under the norm

$$\pi_{q,p}(u) := \|\hat{u}\|.$$

The usual routine tells us that $u \in \mathcal{L}(X,Y)$ belongs to $\Pi_{q,p}(X,Y)$ if and only if there is some $K \geq 0$ for which

$$(*) \qquad \Big(\sum_{k=1}^{n} \|ux_k\|^q\Big)^{1/q} \leq K \cdot \sup\Big\{\Big(\sum_{k=1}^{n} |\langle x^*, x_k\rangle|^p\Big)^{1/p} : x^* \in B_{X^*}\Big\}$$

no matter how we choose a finite set $\{x_1,\ldots,x_n\}$ of vectors from X. Moreover, $\pi_{q,p}(u)$ is the least such K.

In Proposition 2.2 we exhibited the prototypical weak ℓ_p nature of the standard unit vector basis in ℓ_p^* (or c_0, if $p = 1$), and this turned out to be a useful tool in characterizing p-summing operators, for example in Proposition 2.7. Happily, all this extends with no fuss.

10.1 Proposition: *Let $1 \leq p, q < \infty$. For every Banach space operator $u : X \to Y$ and every constant $c > 0$, the following statements are equivalent:*

(i) *u is (q,p)-summing with $\pi_{q,p}(u) \leq c$.*

(ii) *For each $v \in \mathcal{L}(\ell_{p^*}, X)$ (or $v \in \mathcal{L}(c_0, X)$ if $p = 1$), uv is (q,p)-summing with $\pi_{q,p}(uv) \leq c \cdot \|v\|$.*

(iii) *However we choose $n \in \mathbf{N}$ and $v \in \mathcal{L}(\ell_{p^*}^n, X)$, we have $\pi_{q,p}(uv) \leq c \cdot \|v\|$.*

By using $(*)$ with $x_k = \lambda_k x$ where $\lambda_k \in \mathbf{K}$, we soon see that only the zero operator can be (q,p)-summing if $q < p$. Naturally enough then, the hypothesis

$$p \leq q$$

will be omnipresent in our considerations. It is now routine to prove:

10.2 Proposition: *If $1 \leq p \leq q < \infty$, then $[\Pi_{q,p}, \pi_{q,p}]$ is a Banach ideal; it is maximal and injective.*

If H_1 and H_2 are Hilbert spaces, $\Pi_{q,2}(H_1, H_2)$ turns out to be a familiar object. In fact, we have the following extension of 4.10:

10.3 Theorem: *If $2 \leq q < \infty$, then $\Pi_{q,2}(H_1, H_2)$ and the Schatten class $\mathcal{S}_q(H_1, H_2)$ coincide isometrically.*

Proof. Suppose first that u belongs to $\Pi_{q,2}(H_1, H_2)$ and let (e_n) and (f_n) be orthonormal sequences in H_1 and H_2, respectively. Since (e_n) belongs to $\ell_2^{weak}(H_1)$, we have $(ue_n) \in \ell_q^{strong}(H_2)$; in particular, $((ue_n|f_n)) \in \ell_q$. By 4.6(b), u is in $\mathcal{S}_q(H_1, H_2)$. To relate norms, consider an ON-representation $u = \sum_n \tau_n(\,\cdot\,|g_n)h_n$. Then

$$\sigma_q(u) = \Big(\sum_n |\tau_n|^q\Big)^{1/q} = \Big(\sum_n |(ug_n|h_n)|^q\Big)^{1/q} \leq \Big(\sum_n \|ug_n\|^q\Big)^{1/q} \leq \pi_{q,2}(u).$$

For the converse, take any $u \in \mathcal{S}_q(H_1, H_2)$ and fix $(x_n) \in \ell_2^{weak}(H_1)$. Follow the usual strategy and define $v \in \mathcal{L}(\ell_2, H_1)$ by $ve_n = x_n$ for each n. Then, by 4.7(b) and the ideal property 4.5(d),

$$\begin{aligned}\Big(\sum_n \|ux_n\|^q\Big)^{1/q} &= \Big(\sum_n \|uve_n\|^q\Big)^{1/q} \leq \sigma_q(uv) \\ &\leq \sigma_q(u)\cdot\|v\| = \sigma_q(u)\cdot\|(x_n)\|_2^{weak}.\end{aligned}$$

Hence $u \in \Pi_{q,2}(H_1, H_2)$ and $\pi_{q,2}(u) \leq \sigma_q(u)$. QED

If $1 \leq p_2 \leq p_1 \leq q_1 \leq q_2 < \infty$, the natural inclusions of vector-valued sequence spaces ensure that every (q_1, p_1)-summing operator $u : X \to Y$ is a fortiori (q_2, p_2)-summing with $\pi_{q_2,p_2}(u) \leq \pi_{q_1,p_1}(u)$. We may use Hölder's Inequality to derive inclusions for a much more interesting range of indices, thereby generalizing Theorem 2.8:

10.4 Inclusion Theorem: *Suppose that $1 \leq p_j \leq q_j < \infty$ $(j = 1, 2)$ satisfy $p_1 \leq p_2$, $q_1 \leq q_2$ and $\dfrac{1}{p_1} - \dfrac{1}{q_1} \leq \dfrac{1}{p_2} - \dfrac{1}{q_2}$. Then*

$$\Pi_{q_1,p_1}(X,Y) \subset \Pi_{q_2,p_2}(X,Y)$$

for all Banach spaces X and Y, and for each $u \in \Pi_{q_1,p_1}(X,Y)$ we have

$$\pi_{q_2,p_2}(u) \leq \pi_{q_1,p_1}(u).$$

Proof. We may as well assume that $p_1 < p_2$, since the case $p_1 = p_2$ is covered by the preamble. In this situation we must also have $q_1 < q_2$, and so we may define indices $1 < p, q < \infty$ by $1/p = (1/p_1) - (1/p_2)$ and $1/q = (1/q_1) - (1/q_2)$.

Now select an operator u in $\Pi_{q_1,p_1}(X,Y)$ and vectors $x_1, \ldots, x_n$ in X. Then, with $\lambda_k := \|ux_k\|^{q_2/q}$ $(1 \leq k \leq n)$, we have

$$\Big(\sum_{k=1}^n \|ux_k\|^{q_2}\Big)^{1/q_1} = \Big(\sum_{k=1}^n \|u(\lambda_k x_k)\|^{q_1}\Big)^{1/q_1} \leq \pi_{q_1,p_1}(u)\cdot\|(\lambda_k x_k)_1^n\|_{p_1}^{weak},$$

and an application of Hölder's Inequality reveals that this is

$$\leq \pi_{q_1,p_1}(u)\cdot\|(\lambda_k)_1^n\|_p\cdot\|(x_k)_1^n\|_{p_2}^{weak}.$$

Since $q \le p$, we end up with

$$\Big(\sum_{k=1}^{n} \|ux_k\|^{q_2}\Big)^{1/q_1} \le \pi_{q_1,p_1}(u)\cdot\Big(\sum_{k=1}^{n} \|ux_k\|^{q_2}\Big)^{1/q}\cdot\|(x_k)_1^n\|_{p_2}^{weak},$$

which can be rearranged to give the conclusion we wanted. QED

The (q,p)-summing operators provide us with an ideal setting for an extension of the Dvoretzky - Rogers Theorem (see 1.2 and 2.18).

10.5 Theorem: *Let $1 \le p \le q < \infty$. If $\frac{1}{p} - \frac{1}{q} < \frac{1}{2}$, then every infinite dimensional Banach space X contains a weakly p-summable sequence which fails to be strongly q-summable.*

Proof. We must show that for the values of p, q indicated, id_X is not (q,p)-summing. Our strategy will be to reduce to the situation $p = 2$ and then invoke the fundamental lemma 1.3.

If $p \ge 2$, it is trivially true that $\Pi_{q,p}$ is contained in $\Pi_{q,2}$, whereas, when $p < 2$, the Inclusion Theorem 10.4 tells us that $\Pi_{q,p}$ is contained in $\Pi_{r,2}$ where $r > 2$ is defined by $1/r = (1/2) - (1/p) + (1/q)$.

Thus we may suppose that $p = 2$. But then, for every $n \in \mathbf{N}$, Lemma 1.3 provides us with vectors $x_1,\dots,x_n \in X$, each of norm $\ge 1/2$, which satisfy $\|(x_k)_1^n\|_2^{weak} \le 1$. Since $\sum_{k=1}^n \|x_k\|^q \ge n\cdot 2^{-q}$ and n is arbitrary, this is all that is needed to see that id_X cannot have finite $\pi_{q,2}$-norm. QED

This extension of the Dvoretzky - Rogers Theorem is sharp. A rephrasing of Orlicz's Theorem 3.12 exhibits id_{ℓ_p} for $1 \le p \le 2$ as a (2,1)-summing operator and hence, by the above Inclusion Theorem, as a (q,r)-summing operator for any $1 \le r \le q$, $q \ge 2$, with $(1/r) - (1/q) \ge 1/2$. This explains part of Remark 3.8(a).

OPERATORS ON $\mathcal{L}_\infty$-SPACES

Clearly, in what we just noted, ℓ_p can be replaced by any infinite dimensional $\mathcal{L}_p$-space $(1 \le p \le 2)$; just recall that Orlicz's Theorem is a consequence of Theorem 3.7 which asserts that every operator from c_0 (or any other $\mathcal{L}_\infty$-space) to an $\mathcal{L}_p$-space is 2-summing. We note in passing that 3.7 cannot be improved: if Y is any infinite dimensional $\mathcal{L}_p$-space $(1 \le p \le 2)$ and if $1 \le q < 2$, then there is some operator from c_0 to Y which fails to be q-summing, or even $(q,1)$-summing. Otherwise the equivalences of Proposition 10.1 would tell us that id_Y would be $(q,1)$-summing, and this contradicts the generalized Dvoretzky - Rogers Theorem 10.5. A more elementary way to see this would be to observe that we can localize to choose $Y = \ell_p$ and then check with the standard unit vector basis.

Our new framework enables us to examine what can be said for operators $\mathcal{L}_\infty \to \mathcal{L}_p$ when $p > 2$. Familiar tools lead to a preliminary result:

10.6 Theorem: *Let X be an $\mathcal{L}_{\infty,\lambda}$-space and let Y be an $\mathcal{L}_{p,\lambda'}$-space with $2 < p < \infty$. Then all operators $u : X \to Y$ are $(p,2)$-summing and satisfy $\pi_{p,2}(u) \le \kappa_G \cdot \lambda \cdot \lambda' \cdot \|u\|$.*

Here κ_G is Grothendieck's constant.

Proof. By localization, it is good enough to show that if $u : \ell_\infty^n \to \ell_p^n$ is any operator, then $\pi_{p,2}(u) \le \kappa_G \cdot \|u\|$.

The work has already been done in Chapter 3. In fact, in the proof of Lemma 3.6 (see formula (2)) it was shown that if u has standard matrix representation (u_{ij}) and if $x_k = (x_{ki})_{i=1}^n$ $(1 \le k \le n)$ are vectors in ℓ_∞^n, then

$$\Big(\sum_{j=1}^n \Big(\sum_{k=1}^n \Big|\sum_{i=1}^n u_{ij}x_{ki}\Big|^2\Big)^{p/2}\Big)^{1/p} \le \kappa_G \cdot \|u\| \cdot \|(x_k)_1^n\|_2^{weak}.$$

But when $2 < p < \infty$,

$$\Big(\sum_{k=1}^n \|ux_k\|^p\Big)^{1/p} = \Big(\sum_{k=1}^n \sum_{j=1}^n \Big|\sum_{i=1}^n u_{ij}x_{ki}\Big|^p\Big)^{1/p} \le \Big(\sum_{j=1}^n \Big(\sum_{k=1}^n \Big|\sum_{i=1}^n u_{ij}x_{ki}\Big|^2\Big)^{p/2}\Big)^{1/p},$$

and so

$$\Big(\sum_{k=1}^n \|ux_k\|^p\Big)^{1/p} \le \kappa_G \cdot \|u\| \cdot \|(x_k)_1^n\|_2^{weak}.$$ QED

We can now retrace our previous steps to Orlicz's Theorem 3.12 and widen its range:

10.7 Corollary: *Let X be (a subspace of) an $\mathcal{L}_p$-space where $1 \le p < \infty$. Then every weak ℓ_1 sequence in X is a strong ℓ_r sequence where $r = \max\{p, 2\}$.*

Scrutinize, as before, the standard unit vector basis in ℓ_p^n $(n \in \mathbf{N})$ to realize that this cannot be improved.

It is remarkable that 10.6 does not tell the whole story. Our next objective is to show that if $2 < p < \infty$, then each $u \in \mathcal{L}(\mathcal{L}_\infty, \mathcal{L}_p)$ is not only $(p,2)$-summing, but is even (p,r)-summing for any $r < p$; moreover, u is q-summing for any $q > p$.

This requires substantial preparation which will culminate in a generalization of the familiar result 2.15: (q,p)-summing operators on $C(K)$ factor through certain Lorentz spaces.

The crucial tool that we apply to chisel out this information is provided by the following interpolation type inequality.

10.8 Theorem: *Let $1 \le p \le q < \infty$ and consider an operator $u : C(K) \to Y$, where K is a compact Hausdorff space and Y is a Banach space. Then u is (q,p)-summing if and only if there exist a Banach space Z and a p-summing operator $v : C(K) \to Z$ such that*

$$\|uf\| \le \|vf\|^{p/q} \cdot \|f\|^{1-(p/q)} \quad \text{for all} \quad f \in C(K).$$

Actually, we can choose Z to be $L_p(\mu)$ for an appropriate probability measure $\mu \in \mathcal{C}(K)^*$ and $v = q^{1/q} \cdot \pi_{q,p}(u) \cdot j_p$, a multiple of the canonical map $j_p : \mathcal{C}(K) \to L_p(\mu)$.

One direction in the proof of 10.8 is straightforward, but the other is intricate, and the complications are accentuated by having to use an approximation argument. Before giving the complete proof, we would like to isolate the essential ingredients of the tricky implication.

Assume that $u : \mathcal{C}(K) \to Y$ is (q,p)-summing. There is certainly no loss in assuming that $\pi_{q,p}(u) = 1$. Thus, however we choose $f_1, \ldots, f_n \in \mathcal{C}(K)$, we have

$$(1) \qquad \Big(\sum_{k\le n} \|uf_k\|^q\Big)^{1/q} \le \|(f_k)_1^n\|_p^{weak} = \Big\|\Big(\sum_{k\le n} |f_k|^p\Big)^{1/p}\Big\| :$$

in 2.9(a) we saw how to derive the last equality from the fact that K can be considered as a norming subset of $B_{\mathcal{C}(K)^*}$.

For the time being, let us make the (unjustified) hypothesis that even

$$(2) \qquad \Big(\sum_{k\le n} \|uf_k\|^q\Big)^{1/q} = \|(f_k)_1^n\|_p^{weak}$$

is true: this allows us to better illuminate the structure of the proof.

For convenience, we also normalize so that

$$(3) \qquad \Big(\sum_{k\le n} \|uf_k\|^q\Big)^{1/q} = 1.$$

Finally, we choose $b_1, \ldots, b_n \in Y^*$ so that

$$(4) \qquad \sum_{k\le n} \|b_k\|^{q^*} = 1 \quad \text{and} \quad \sum_{k\le n} \langle b_k, uf_k\rangle = \Big(\sum_{k\le n} \|uf_k\|^q\Big)^{1/q} = 1.$$

Note that this entails that $\langle b_k, uf_k\rangle \ge 0$ for each $1 \le k \le n$.

The linearization process that led to (4) allows us to define the measure that we are seeking:

$$\mu : \mathcal{C}(K) \longrightarrow \mathbf{K} : g \mapsto \sum_{k\le n} \langle b_k, u(gf_k)\rangle.$$

Thanks first to (4), then to (1), and finally to (2) and (3),

$$|\langle \mu, g\rangle| \le \Big(\sum_{k\le n} \|u(gf_k)\|^q\Big)^{1/q} \le \Big\|\Big(\sum_{k\le n} |gf_k|^p\Big)^{1/p}\Big\| \le \|g\|$$

for all $g \in \mathcal{C}(K)$, and this shows that $\|\mu\| \le 1$. It also follows from (4) that $\langle \mu, 1\rangle = 1$, and so μ is indeed a probability measure.

Now that we have our measure μ, we are going to prove that if $f \in \mathcal{C}(K)$ satisfies $\|f\| = 1$, then

$$(5) \qquad \|uf\|^q \le q \cdot \|f\|^p_{L_p(\mu)}.$$

By homogeneity, it will then follow that $\|uf\| \le q^{1/q} \cdot \|j_p f\|^{p/q} \cdot \|f\|^{1-(p/q)}$ for any $f \in \mathcal{C}(K)$, as we were hoping.

The trick to proving (5) is to incorporate the norm one function $f \in C(K)$ by considering the continuous functions

$$g_k := f_k \cdot (1 - |f|^p)^{1/p} \quad (1 \le k \le n) \quad \text{and} \quad g_{n+1} := f.$$

Then

$$\Big(\sum_{k \le n+1} \|ug_k\|^q\Big)^{1/q} \le \Big\|\Big(\sum_{k \le n+1} |g_k|^p\Big)^{1/p}\Big\| = \Big\|\Big(\sum_{k \le n} |f_k|^p\Big)(1 - |f|^p) + |f|^p\Big\|^{1/p}.$$

The normalization (3) guarantees that $\sum_{k \le n} |f_k|^p$ takes values in [0,1], and consequently $|f|^p \le (\sum_{k \le n} |f_k|^p)(1 - |f|^p) + |f|^p \le 1$. We deduce that $\sum_{k \le n+1} \|ug_k\|^q \le 1$, and hence

$$\begin{aligned}
\|uf\|^q &= \sum_{k \le n+1} \|ug_k\|^q - \sum_{k \le n} \|ug_k\|^q \le 1 - \sum_{k \le n} \|u(f_k(1 - |f|^p)^{1/p})\|^q \\
&\le 1 - \big|\langle \mu, (1 - |f|^p)^{1/p} \rangle\big|^q \le 1 - \big|\langle \mu, 1 - |f|^p \rangle\big|^q \\
&= 1 - \big|1 - \langle \mu, |f|^p \rangle\big|^q \le q \cdot \big|\langle \mu, |f|^p \rangle\big|,
\end{aligned}$$

which is just another way of writing (5).

For an accurate proof of the theorem, we need to bite the bullet and remove the unjustified assumption (2). Fortunately, an approximation argument is available to fill the gap.

Proof of 10.8. Let us start with the most difficult part and suppose that $u : C(K) \to Y$ is (q,p)-summing. We shall follow the course of the preamble as closely as honesty permits.

There is no problem in normalizing to achieve (1). However, when we restrict our sums to exactly n terms, it is possible that we might do better. Accordingly, we let C_n be the least constant for which, no matter how we choose $f_1, \dots, f_n$ in $C(K)$, we have

$$\text{(6)} \qquad \Big(\sum_{k \le n} \|uf_k\|^q\Big)^{1/q} \le C_n \cdot \Big\|\Big(\sum_{k \le n} |f_k|^p\Big)^{1/p}\Big\|;$$

we observe that $C_n \le 1$ and $\lim_{n \to \infty} C_n = 1$, but that there is no reason for equality to be achievable in (6). However, if we select $\delta_n = 1 + (1/n)$, then we can certainly make a choice of $f_1^{(n)}, \dots, f_n^{(n)}$ in $C(K)$ so that

$$\text{(7)} \qquad 1 = \Big(\sum_{k \le n} \|uf_k^{(n)}\|^q\Big)^{1/q} \ge C_n \cdot \delta_n^{-1} \cdot \Big\|\Big(\sum_{k \le n} |f_k^{(n)}|^p\Big)^{1/p}\Big\|.$$

Again, it is possible to find $b_1^{(n)}, \dots, b_n^{(n)}$ in Y^* so that

$$\text{(8)} \qquad \sum_{k \le n} \|b_k^{(n)}\|^{q^*} = 1 \quad \text{and} \quad \sum_{k \le n} \langle b_k^{(n)}, uf_k^{(n)} \rangle = \Big(\sum_{k \le n} \|uf_k^{(n)}\|^q\Big)^{1/q},$$

and this forces $\langle b_k^{(n)}, uf_k^{(n)} \rangle \ge 0$ for each $1 \le k \le n$.

Next we define measures

$$\mu_n : \mathcal{C}(K) \longrightarrow \mathbf{K} : g \mapsto \sum_{k\le n} \langle b_k^{(n)}, u(gf_k^{(n)})\rangle.$$

Just as in the previous discussion we can use (7) to find that $\|\mu_n\| \le \delta_n$ for each n. Consequently, (μ_n) is a bounded sequence in $\mathcal{C}(K)^*$; it must have a weak$*$accumulation point μ, say. Since $\lim_{n\to\infty} \delta_n = 1$, μ is constrained to have norm ≤ 1. By (8) and (7), $\langle \mu_n, 1\rangle = 1$ for each n, and so $\langle \mu, 1\rangle = 1$. Thus μ is a probability measure.

We are going to show that for any $f \in \mathcal{C}(K)$ with $\|f\| = 1$ we still have inequality (5). From there our conclusion will follow as before.

This time we incorporate the function f by considering, for any fixed n,

$$g_k := f_k^{(n)}\cdot(1-|f|^p)^{1/p}\ (1\le k\le n) \quad\text{and}\quad g_{n+1} := \delta_n\cdot C_n^{-1}\cdot f.$$

Arguing carefully as in the preamble, we discover that

$$\Big(\sum_{k\le n+1} \|ug_k\|^q\Big)^{1/q} \le \frac{C_{n+1}}{C_n}\cdot \delta_n,$$

from which it is a simple matter to derive

$$\begin{aligned}\|uf\|^q &= C_n^q\cdot\delta_n^{-q}\cdot\Big(\sum_{k\le n+1} \|ug_k\|^q - \sum_{k\le n} \|ug_k\|^q\Big)\\ &\le C_{n+1}^q - C_n^q\cdot\delta_n^{-q}\cdot\big|\langle \mu_n, (1-|f|^p)^{1/p}\rangle\big|.\end{aligned}$$

Since this is valid for all n, we may pass to the limit to obtain

$$\|uf\|^q \le 1-\big|\langle\mu, (1-|f|^p)^{1/p}\rangle\big|^q,$$

and (5) follows as before.

All that remains is the proof of the converse, but this is easy. Assume that we can associate with $u : \mathcal{C}(K) \to Y$ a p-summing operator $v : \mathcal{C}(K) \to Z$ satisfying $\|uf\| \le \|vf\|^{p/q}\cdot\|f\|^{1-(p/q)}$ for all $f \in \mathcal{C}(K)$. Select $f_1,\dots, f_n \in \mathcal{C}(K)$. As v is p-summing,

$$\begin{aligned}\Big(\sum_{k\le n}\|uf_k\|^q\Big)^{1/q} &\le \Big(\sum_{k\le n}\|vf_k\|^p\cdot\|f_k\|^{q-p}\Big)^{1/q} \le \Big(\sum_{k\le n}\|vf_k\|^p\Big)^{1/q}\cdot\max_{k\le n}\|f_k\|^{1-(p/q)}\\ &\le \pi_p(v)^{p/q}\cdot\big(\|(f_k)_1^n\|_p^{weak}\big)^{p/q}\cdot\max_{k\le n}\|f_k\|^{1-(p/q)} \le \pi_p(v)^{p/q}\cdot\|(f_k)_1^n\|_p^{weak}.\end{aligned}$$

We have shown that u is (q,p)-summing – and, incidentally, that $\pi_{q,p}(u) \le \pi_p(v)^{p/q}$. QED

We note in passing that the proof of the converse also applies when $\mathcal{C}(K)$ is replaced by an arbitrary Banach space.

We are now ready to prove the announced factorization theorem for (q,p)-summing operators on $\mathcal{C}(K)$, $1 \le p < q < \infty$. For this, we shall make use of the results we have just proved, together with some basic properties of the Lorentz space $L_{q,1}(\mu)$ where μ is a probability measure on K. Recall that this is the space which consists of all (μ-a.e. equivalence classes of) measurable functions $f : K \to \mathbf{K}$ whose decreasing rearrangement $f^* : [0,1] \to \mathbf{R} : t \mapsto \inf\{s > 0 : \mu(|f| > s) \le t\}$ satisfies

$$\|f\|_{q,1} := \int_0^1 t^{(1/q)-1} f^*(t)\, dt < \infty.$$

10.9 Theorem: *Let $1 \le p < q < \infty$, let K be a compact Hausdorff space, and let Y be a Banach space. The following statements about the operator $u : \mathcal{C}(K) \to Y$ are equivalent:*

(i) *u is $(q,1)$-summing.*

(ii) *There exist a probability measure $\mu \in \mathcal{C}(K)^*$ and a factorization $u : \mathcal{C}(K) \xrightarrow{j} L_{q,1}(\mu) \xrightarrow{\tilde{u}} Y$ where j is the canonical map.*

(iii) *u is (q,p)-summing.*

Proof. (i)⇒(ii): The preceding theorem combined with 2.15 tells us how to find a probability measure $\mu \in \mathcal{C}(K)^*$ and a constant c such that

$$\|uf\| \le c \cdot \|f\|_{L_1(\mu)}^{1/q} \cdot \|f\|_{\mathcal{C}(K)}^{1/q^*} \quad \text{for all} \quad f \in \mathcal{C}(K). \tag{1}$$

The opening move of our campaign is to 'extend' u to an operator $\hat{u} : L_\infty(\mu) \to Y$ with

$$\|\hat{u}f\| \le c \cdot \|f\|_{L_1(\mu)}^{1/q} \cdot \|f\|_{L_\infty(\mu)}^{1/q^*} \quad \text{for all} \quad f \in L_\infty(\mu). \tag{2}$$

For this we call on Lusin's Theorem, which informs us that for each $f \in L_\infty(\mu)$ we can find a sequence (g_n) in $\mathcal{C}(K)$ with $\|g_n\|_{\mathcal{C}(K)} \le \|f\|_{L_\infty(\mu)}$ for all n and $f = \lim_{n\to\infty} g_n$ μ-a.e. But now $\hat{u}f := \lim_{n\to\infty} ug_n$ exists in Y since (1), together with Lebesgue's Dominated Convergence Theorem, reveals that (ug_n) is a Cauchy sequence:

$$\begin{aligned}\|ug_m - ug_n\| &\le c \cdot \|g_m - g_n\|_{L_1(\mu)}^{1/q} \cdot \|g_m - g_n\|_{\mathcal{C}(K)}^{1/q^*} \\ &\le c \cdot \|g_m - g_n\|_{L_1(\mu)}^{1/q} \cdot \big(2 \cdot \|f\|_{L_\infty(\mu)}\big)^{1/q^*}\end{aligned}$$

for all m, n. As $\hat{u}f$ plainly does not depend on the particular choice of (g_n), a map $\hat{u} : L_\infty(\mu) \to Y$ is born. It is easily seen to be a linear and bounded 'extension' of u which satisfies (2).

Our next step is to show that (2) implies

$$\|\hat{u}f\| \le c \cdot q^{-1} \cdot \|f\|_{q,1} \tag{3}$$

for all simple functions f. Since these functions are dense in $L_{q,1}(\mu)$, we can then settle the issue by using $\hat{u}$ to construct the sought-after operator $\tilde{u} : L_{q,1}(\mu) \to Y$.

We establish proper contact with the decreasing rearrangement of a simple function $f : K \to \mathbf{K}$ with $\|f\|_{L_\infty(\mu)} = 1$ by choosing a representation

$$f = \sum_{k \le n} \alpha_k \cdot 1_{A_k}$$

using pairwise disjoint Borel sets A_k, enumerated so that the scalars satisfy

(4) $$1 = |\alpha_1| \geq |\alpha_2| \geq .... \geq |\alpha_n| > 0.$$

Then, if we set $t_0 = 0$ and $t_k = \sum_{i\leq k} \mu(A_i)$ $(1 \leq k \leq n)$, it is easy to check that for each $1 \leq k \leq n$, $f^*(t) = \alpha_k$ whenever $t_{k-1} < t < t_k$, whereas $f^*(t) = 0$ when $t > t_n$. Elementary integration shows that

(5) $$\|f\|_{q,1} = q\cdot \sum_{k\leq n} |\alpha_k|\cdot (t_k^{1/q} - t_{k-1}^{1/q}).$$

In order to estimate $\|\hat{u}f\|$, we apply (2) to particular functions whose moduli are characteristic functions: let $f_0 := 0$ and $f_k := \sum_{i\leq k}(\alpha_i/|\alpha_i|)\cdot 1_{A_i}$ for $1 \leq k \leq n$. Note that (2) gives $\|\hat{u}f_k\| \leq c\cdot t_k^{1/q}$ for all $0 \leq k \leq n$. Since

$$f = \sum_{k\leq n} |\alpha_k|(f_k - f_{k-1}) = \sum_{k\leq n-1} (|\alpha_k| - |\alpha_{k+1}|)f_k + |\alpha_n| f_n,$$

we can call on (4) to obtain

$$\|\hat{u}f\| \leq c\cdot \Big(\sum_{k\leq n-1} (|\alpha_k| - |\alpha_{k+1}|)t_k^{1/q} + |\alpha_n|\cdot t_n^{1/q}\Big) = c\cdot \sum_{k\leq n} |\alpha_k|\cdot (t_k^{1/q} - t_{k-1}^{1/q}).$$

Combining this with (5) we get (3).

(ii)$\Rightarrow$(iii): We shall again obtain our conclusion by way of 10.8. Since the canonical map $j_p : C(K) \to L_p(\mu)$ is p-summing, it will be enough to show that there is a constant c_1 such that

$$\|uf\| \leq c_1\cdot \|f\|_{L_p(\mu)}^{p/q}\cdot \|f\|_{C(K)}^{1-(p/q)} \quad \text{for each} \quad f \in C(K).$$

After all, if this is true then whenever we select $f_1,..., f_n \in C(K)$ with $\|(f_k)_{k=1}^n\|_p^{weak} = 1$, we have

$$\Big(\sum_{k\leq n} \|uf_k\|^q\Big)^{1/q} \leq c_1\cdot \Big(\sum_{k\leq n} \|f_k\|_{L_p(\mu)}^p\Big)^{1/q} \leq c_1\big(\|(f_k)\|_p^{weak}\big)^{p/q} \leq c_1.$$

Thanks to our hypothesis, it is adequate to demonstrate that for each continuous f,

(6) $$\|f\|_{q,1} \leq c_2\cdot \|f\|_{L_p(\mu)}^{p/q}\cdot \|f\|_{C(K)}^{1-(p/q)},$$

where c_2 is another constant.

Normalize so that $\|f\|_{C(K)} = 1$ and set $a := \|f\|_{L_p(\mu)}^p$. Then $f^*(t) \leq 1$ for all $0 \leq t \leq 1$, and

$$\begin{aligned}
\|f\|_{q,1} &= \int_0^1 t^{(1/q)-1} f^*(t)\,dt \leq \int_0^a t^{(1/q)-1}\,dt + \int_a^1 t^{-1/q^*} f^*(t)\,dt \\
&\leq q\cdot a^{1/q} + \Big(\int_a^\infty t^{-p^*/q^*}\,dt\Big)^{1/p^*}\cdot \|f^*\|_{L_p[0,1]} \\
&= q\cdot a^{1/q} + \Big(\frac{q^*}{p^* - q^*}\cdot a^{1-(p^*/q^*)}\Big)^{1/p^*}\cdot \|f\|_{L_p(\mu)} \\
&= q\cdot a^{1/q} + \Big(\frac{q^*}{p^* - q^*}\Big)^{1/p^*}\cdot a^{1/q} = c_2\cdot \|f\|_{L_p(\mu)}^{p/q}.
\end{aligned}$$

When we remove the normalization on f, we end up with (6).

As (iii)$\Rightarrow$(i) is a triviality, we are done. QED

We previously advertised a substantial improvement of 10.6. Here it is.

10.10 Corollary: *Let $2 < q < \infty$. If X is an $\mathcal{L}_\infty$-space and Y is an $\mathcal{L}_q$-space, then however we choose $1 \le p < q < r < \infty$, we have*

$$\mathcal{L}(X,Y) = \Pi_{q,p}(X,Y) = \Pi_r(X,Y).$$

Proof. Localize to reduce to the situation $X = C(K)$ and $Y = L_q(\nu)$, with K a compact Hausdorff space and ν some measure. We already know from 10.6 that each $u \in \mathcal{L}(X,Y)$ is $(q,2)$-summing, and Theorem 10.9 now assures us that it must be (q,p)-summing for any $p < q$.

It also tells us that there is a factorization $u : X = C(K) \xrightarrow{j} L_{q,1}(\mu) \xrightarrow{\tilde{u}} Y$ with $\mu \in C(K)^*$ a probability measure and j the formal identity. But $L_r(\mu)$ embeds canonically into $L_{q,1}(\mu)$ whenever $r > q$, and so u contains the natural map $j_r : C(K) \to L_r(\mu)$ as a factor; hence it is r-summing. QED

In 3.11 we discovered a coincidence theorem for classes of summing operators out of $\mathcal{L}_p$-spaces for $1 \le p \le 2$. We are now in a position to treat the case $2 < p < \infty$.

10.11 Corollary: *Let $2 < p < \infty$ and $1 < r < p^*$. If X is an $\mathcal{L}_p$-space and Y is any Banach space, then $\Pi_r(X,Y) = \Pi_1(X,Y)$.*

Proof. Choose any (x_k) in $\ell_1^{weak}(X)$. According to 2.2, we may identify (x_k) with the operator $v \in \mathcal{L}(c_0, X)$ which is given by $ve_k = x_k$, $k \in \mathbf{N}$. We have just seen that v is r^*-summing, so that if $u \in \Pi_r(X,Y)$, uv must be 1-summing by the Composition Theorem 2.22. In view of 2.7, this is what we wanted. QED

We conclude this section with a particularly beautiful application:

10.12 Corollary: *Let $1 < p < r < q < \infty$. If X is an $\mathcal{L}_q$-space and if Y is an $\mathcal{L}_p$-space, then $\mathcal{L}(X,Y) = \Gamma_r(X,Y)$.*

Proof. By localization, we may reduce to the case where $X = L_q(\mu)$ and $Y = L_p(\nu)$ for some measures μ, ν. Also, it suffices to consider the case $r \ge 2$; the case $r \le 2$ can then be settled by duality.

Take $u \in \mathcal{L}(X,Y)$. In view of 9.14, all we need to show is that no matter how we choose a Banach space Z and an operator $v \in \Pi_r^d(Z,X)$, the composition uv is r-integral.

Since $X = L_q(\mu)$, we certainly have $\Pi_r^d(Z,X) \subset \Pi_q^d(Z,X) \subset I_q(Z,X)$ thanks to 9.6, and this forces $uv \in \Gamma_\infty(Z,X)$. Now we can apply our knowledge of operators $\mathcal{L}_\infty \to \mathcal{L}_p$: using 3.7 when $p \le 2$ and 10.10 when $p > 2$ we find that uv has a factor which is r-summing on a $C(K)$-space. By 5.8 uv must be r-integral. QED

If $p < 2 < q$, then the last result remains valid when we make X a subspace of an $\mathcal{L}_q$-space and Y a quotient of an $\mathcal{L}_p$-space. But this is a different story, strongly tied, by the way, to Remark 7.9. See 12.25 for details.

NOTES AND REMARKS

The class $\Pi_{q,p}$ was introduced by B.S. Mitiagin and A. Pełczyński [1966] though the first in-depth study awaited S. Kwapień [1968]. Of course, using the notation of the Notes and Remarks of the last chapter,

$$[\Pi_{q,p}(X,Y), \pi_{q,p}] = [\Pi_{q,p,\infty}(X,Y), \pi_{q,p,\infty}].$$

But even at this restricted level of generality, new problems are abundant.

Considerable effort has been expended on identifying the (q,p)-summing operators on a Hilbert space H. Theorem 10.3 appears in S. Kwapień [1968] where it is attributed to B.S. Mitiagin. Following more or less classical lines, Kwapień also showed that

- *if* $p \leq 2$ *and* $1/r = (1/q) - (1/p) + (1/2) > 0$, *then* $\Pi_{q,p}(H) = \mathcal{S}_r(H)$

whereas

- *if* $(1/p) - (1/q) \geq 1/2$, *then every operator on* H *is* (q,p)*-summing.*

The remaining cases were settled by G. Bennett [1975/76] and G. Bennett, V. Goodman and C.M. Newman [1975] using powerful probabilistic tools. Here is the result:

Theorem: *If* $2 < p < q < \infty$, *then* $\Pi_{q,p}(H) = \mathcal{S}_{(2q/p),q}(H)$.

Here $\mathcal{S}_{(2q/p),q}(H)$ consists of all operators $u : H \to H$ such that the sequence $(a_n(u))_n$ of approximation numbers belongs to the Lorentz sequence space $\ell_{(2q/p),q}$, that is, it satisfies $\sum_n n^{(p/2)-1} a_n(u)^q < \infty$.

It follows that $\Pi_{q,p}(H)$ contains $\mathcal{S}_{2q/p}(H)$ and is contained in $\mathcal{S}_q(H)$ and that these are proper containments.

The inclusion relation 10.4 is essentially due to S. Kwapień [1968]. Several other results can be obtained by straightforward generalization of arguments applied in Chapter 2. For example, N. Tomczak-Jaegermann [1970] observed that 2.22 generalizes as follows. *Suppose that* $1 \leq p,q,r,s,t < \infty$ *satisfy* $s^{-1} = p^{-1} + r^{-1}$ *and* $t^{-1} = q^{-1} + r^{-1}$. *If* X, Y, Z *are Banach spaces and* $u \in \Pi_{q,p}(Y,Z)$ *and* $v \in \Pi_r(X,Y)$, *then* $uv \in \Pi_{t,s}(X,Z)$ *and* $\pi_{t,s}(uv) \leq \pi_{q,p}(u) \cdot \pi_r(v)$.

The variant 10.5 of the Dvoretzky-Rogers Theorem is related to more sophisticated composition theorems for (q,p)-summing operators. Using the technique of A. Brunel and L. Sucheston [1973] which we discuss in Chapter 14, B. Maurey and A. Pełczyński [1976] proved, for example:

Theorem: *If* $u_k : X_{k-1} \to X_k$ *are* (q_k, p_k)*-summing operators with* $(1/p_k) - (1/q_k) \leq 1/2$ $(1 \leq k \leq N)$, *then as soon as* $\sum_{k \leq N} \big((1/q_k) - (1/p_k) + (1/2)\big) > 1/2$, *the composition* $u_N u_{N-1} \cdots u_1$ *is compact.*

Using estimates for eigenvalues of $(q,2)$-summing operators, this result was soon improved by H. König, J.R. Retherford and N. Tomczak-Jaegermann [1980]:

Theorem: *Let $u_k : X_{k-1} \to X_k$ be $(q_k, 2)$-summing Banach space operators. If $1/q := \sum_{k\leq N}(1/q_k)$ satisfies $q < 2$, the composition $u_N \cdots u_1$ is 2-summing and compact.*

As in the proof of 10.5, $\Pi_{q_k,p_k} \subset \Pi_{r_k,2}$ if we define r_k via $1/r_k = (1/q_k) - (1/p_k) + (1/2)$. Therefore this theorem contains the Maurey - Pełczyński result.

For an elegant proof of the compostion theorem for $(q,2)$-summing operators using approximation numbers, see A. Pietsch [1980].

For more details about the topics just discussed, see the books of H. König [1986], A. Pietsch [1987] and N. Tomczak-Jaegermann [1989].

Theorem 10.6 is due to J. Lindenstrauss and A. Pełzcyński [1968]. The following result of S. Kwapień [1968] is of related interest.

Theorem: *If $1 \leq p < \infty$, then every operator from an $\mathcal{L}_1$-space to an $\mathcal{L}_p$-space is $(r,1)$-summing, where $1/r = 1 - \big|(1/p) - (1/2)\big|$.*

When $p = 1$, this is just 10.7, but the theorem does not follow from 10.7 when $p > 1$. The proof is based on Grothendieck's Inequality, and so on the case $p = 2$. It is actually an old result of J.E. Littlewood [1930] that the formal identity $\ell_1 \hookrightarrow \ell_{4/3}$ is $(4/3,1)$-summing: he had shown that there is a positive constant C such that $\sum_n \|ue_n\|_{\ell_{4/3}}^{4/3} \leq C \cdot \|u\|^{4/3}$ for all $u \in \mathcal{L}(c_0, \ell_1)$.

Note that the result of B. Carl and A. Defant [1992] cited in the Notes and Remarks to Chapter 7 improves upon Kwapień's theorem.

Theorems 10.8 and 10.9 are due to G. Pisier [1986b] but the equivalence of (i) and (iii) in 10.9 as well as Corollaries 10.11 and 10.12 are due to B. Maurey [1973]. Pisier even derived results of a similar nature when the operator's domain is an arbitrary C^*-algebra. S.V. Kisliakov [1991] adapted Pisier's argument to obtain a version of 10.8 when the domain is the disk algebra.

Corollary 10.10 is due to H.P. Rosenthal [1973]. Examples exist to show that 10.10 cannot be improved: if $q > 2$ then there are operators $\ell_\infty \to \ell_q$ which fail to be q-summing; see S. Kwapień [1970b}.

The idea of deriving sharp analytic information by interpolative inequalities is already present in G. Pisier's [1978b] work on new classes of Banach spaces satisfying Grothendieck's Theorem, a topic to be touched on in Chapter 15. A clear relationship exists between the [1986b] work of G. Pisier and a [1973] theorem of H.P. Rosenthal on reflexive subspaces of L_1. Building on U. Haagerup's [1985] C^*-algebra extension of Grothendieck's inequality and using H. Jarchow's [1986] result that a reflexive quotient of a C^*-algebra is superreflexive, Pisier established the following non-commutative formulation of Rosenthal's theorem:

Theorem: *Every reflexive subspace of the predual of a von Neumann algebra embeds isomorphically in a non-commutative L_p-space $(1 < p \leq 2)$ modelled on some normal state.*

We shall derive two proofs for the commutative analogue, one in 13.19, the other in 15.12; see also 11.18.

Theorems 10.8 and 10.9 are the results of efforts to find a Pietsch-like factorization scheme for $\Pi_{q,p}$. Aside from maximality and injectivity, Π_p and $\Pi_{q,p}$ do not have much in common when $p < q$. For example, in a remarkable exhibition of dexterity with Hölder's Inequality, G. Bennett [1973] and, independently, B. Carl [1974] established the following result.

Theorem: *Let* $1 \leq p \leq q \leq \infty$.

(a) *If* $q \leq 2$, *then* $\ell_p \hookrightarrow \ell_q$ *is* $(r, 1)$*-summing, where* $1/r = (1/p)+(1/2)-(1/q)$.

(b) *If* $2 \leq q$, *then* $\ell_p \hookrightarrow \ell_q$ *is* $(p, 1)$*-summing.*

(c) *Neither* (a) *nor* (b) *can be improved.*

It follows that for any $p > 1$ there are $(p, 1)$-summing operators which are *not* completely continuous. In fact, Bennett notes that there is a non-completely continuous operator that is $(p, 1)$-summing for all $p > 1$.

Bennett returned to the inclusion map $\ell_p \hookrightarrow \ell_q$ in his [1977] study of Schur multipliers. Schur multiplication of two square matrices $a = (a_{ij})_{i,j=1}^{\infty}$ and $b = (b_{ij})_{i,j=1}^{\infty}$ is a sophomore's delight: $a \bullet b := (a_{ij} \cdot b_{ij})_{i,j=1}^{\infty}$. Bennett considers *(Schur)* (p, q)*-multipliers:* matrices m such that $m \bullet a$ takes ℓ_p into ℓ_q whenever a does. One of his results is a remarkable characterization of $(2, 2)$-multipliers, one half of which is due to J. Schur [1911]:

Theorem: *The matrix* m *is a* $(2, 2)$*-multiplier if and only if* $m = a \bullet b$ *where* a *maps* ℓ_2 *to* ℓ_∞ *and* b *maps* ℓ_1 *to* ℓ_2.

With regard to our topic, the following theorem is particularly interesting. Some of this work was duplicated independently by R. Khalil [1980].

Theorem: *The matrix* m *is a* (p, q)*-multiplier if and only if for each* $d \in \ell_p$ *the operator* $m \circ D_d : \ell_{p^*} \to \ell_\infty$ *is* q*-summing, where* D_d *is the diagonal matrix with* d *for a diagonal.*

Bennett goes on to give a practically complete delineation of the (r, s)-summing behaviour of the inclusion maps $\ell_p \hookrightarrow \ell_q$ which includes the Bennett-Carl results as special cases.

Every p-summing operator is weakly compact, but such is not the case for (q, p)-summing operators when $p < q$. Already Orlicz's theorem for L_1 provides an example of a $(2, 1)$-summing operator that is not weakly compact. Again there is much more. S. Kwapień and A. Pełczyński [1970] showed that if $1 \leq p < q$, then

- *the sum operator* $\Sigma : \ell_1 \to \ell_\infty : (x_k) \mapsto \left(\sum_{k\le n} x_k\right)_n$ *is* (q,p)*-summing but not weakly compact.*

They derive this from a powerful inequality involving the space M of all scalar $\mathbf{N}\times\mathbf{N}$ matrices with only finitely many non-zero entries, equipped with an 'unconditional matrix norm', that is, a norm α on M which satisfies

(i) $\alpha(u)=1$ for each $u\in M$ which has one entry equal to 1 and all others equal to 0,

(ii) $\alpha(P_{m,n}a)\le\alpha(a)$ for all $a\in M$, where $P_{m,n}a$ is the matrix whose entries coincide with those of a in the north west $m\times n$ rectangle, but are zero elsewhere,

and

(iii) $\alpha(\widetilde{a}) = \alpha(a)$ for all $a \in M$, where $\widetilde{a}_{ij} = \sigma_{ij}a_{ij}$ with $|\sigma_{ij}| = 1$ for each $i,j \in \mathbf{N}$.

Here is the centrepiece of the Kwapień - Pełczyński paper:

Main Triangle Projection Theorem: *For each* $n \in \mathbf{N}$, *let* $T_n : M \to M$ *be given by*

$$T_n(a)_{ij} := \begin{cases} a_{ij} & \text{if } i+j \le n+1 \\ 0 & \text{otherwise} \end{cases} .$$

Then, for any unconditional matrix norm α,

$$\alpha(T_n(a)) \le \log_2(2n)\cdot\alpha(a) \quad \text{for each} \quad a \in M.$$

In addition to using the Main Triangle Projection Theorem to derive the (q,p)-summing nature of Σ for all $1 \le p < q$, Kwapień and Pełczyński anticipated the result of Bennett, Maurey and Nahoum which we treat in 12.32. They showed that for any unconditionally summable sequence (f_n) in $L_1[0,1]$, the series $\sum_n \dfrac{f_n}{\log(n+1)}$ converges in measure.

11. Type and Cotype: The Basics

Some of the most interesting results that we have seen on summing operators involved bounded linear maps from or to $\mathcal{L}_p$-spaces. Although there were occasions when we were able to obtain characterizations of all such spaces (9.12), it was much more usual to have hypotheses like $1 \le p \le 2$. More specifically, if X is an $\mathcal{L}_p$-space ($1 \le p \le 2$), then every operator from an $\mathcal{L}_\infty$-space into X is 2-summing (3.7), every 2-summing operator with domain X is 1-summing (3.11), and for $q > 2$ every q-summing operator into X is 2-summing (3.15). Further examples can be found in Chapters 3 and 10.

It is our purpose in this chapter to isolate common features of certain $\mathcal{L}_p$-spaces and to solidify their rôle in these matters.

The key issue at the heart of the results we have just mentioned is one of our strongest tools: Khinchin's Inequality (1.10). This inequality concludes that, for $0 < p < \infty$, the p-means $(\int_0^1 |\sum a_n r_n(t)|^p dt)^{1/p}$ are all equivalent in a way independent of the choice of the scalar sequence (a_n); in fact, these means are even uniformly equivalent to the ℓ_2 norm of (a_n):

$$A_p \cdot \big(\sum_{k\le n} |a_k|^2\big)^{1/2} \le \Big(\int_0^1 \big|\sum_{k\le n} r_k(t) a_k\big|^p \, dt\Big)^{1/p} \le B_p \cdot \big(\sum_{k\le n} |a_k|^2\big)^{1/2}$$

for all n, with absolute constants A_p and B_p.

In the study of summing operators, we needed to gauge the size of sums of the sort $\sum_n \varepsilon_n x_n$ where $\varepsilon_n = \pm 1$ and the x_n's are vectors in a Banach space. This suggests that we look closely at the means

$$\Big(\int_0^1 \big\|\sum_n r_n(t) x_n\big\|^p \, dt\Big)^{1/p}.$$

The idea is not new to us; for example, the term $\int_0^1 \|\sum r_n(t) f_n\|_{L_1} dt$ was a key player in the proof of Orlicz's Theorem (1.11). We shall see that in fact such averages behave remarkably well, and their study explains many of our $\mathcal{L}_p$-results in a more general, and almost optimal, context.

Our first goal is to prove

11.1 Kahane's Inequality: *If $0 < p, q < \infty$, then there is a constant $K_{p,q} > 0$ for which*

$$(*) \qquad \Big(\int_0^1 \big\|\sum_{k\le n} r_k(t) x_k\big\|^q \, dt\Big)^{1/q} \le K_{p,q} \cdot \Big(\int_0^1 \big\|\sum_{k\le n} r_k(t) x_k\big\|^p \, dt\Big)^{1/p}$$

regardless of the choice of a Banach space X and of finitely many vectors $x_1, \dots, x_n$ from X.

Beware: unlike the situation with Khinchin's Inequality, in general infinite dimensional Banach spaces none of these quantities can be compared with $\left(\sum_{k\leq n}\|x_k\|^2\right)^{1/2}$ in a uniform way.

The first clue to this comes from familiar spaces like ℓ_1 and c_0. Looking at the standard unit vector basis (e_k), we get $\left(\int_0^1 \|\sum_{k\leq n} r_k(t)e_k\|_{\ell_1}^p\, dt\right)^{1/p} = n$ and $\left(\int_0^1 \|\sum_{k\leq n} r_k(t)e_k\|_{c_0}^p\, dt\right)^{1/p} = 1$ for each n, whereas $\left(\sum_{k\leq n}\|e_k\|^2\right)^{1/2} = n^{1/2}$ for each n in both cases.

The initial disappointment of this realization will soon be allayed when we introduce very satisfying gradations of Banach spaces which allow us to measure the deviation from the classical Khinchin case. But first we set ourselves to the task of establishing Kahane's Inequality.

Randomized Sums

Our discussion is going to be couched in terms of *randomized sums*

$$\sum_{k=1}^{n} \chi_k x_k,$$

where the x_k's come from a Banach space X and the χ_k's are real-valued random variables on some probability space (Ω, Σ, P). *We shall always assume that the* χ_k*'s are independent and symmetric.*

To avoid any possible confusion, we recall some basic facts about real-valued random variables, alias measurable functions, on a probability space (Ω, Σ, P). The distribution of a random variable χ is a measure P_χ on the Borel sets $B \subset \mathbf{R}$ defined by $P_\chi(B) = P(\chi \in B)$. Here we are following a common probabilistic custom and writing $P(\chi \in B)$ in place of $P(\{\chi \in B\})$ where $\{\chi \in B\}$ is the event $\chi^{-1}(B) = \{\omega \in \Omega : \chi(\omega) \in B\}$.

The random variable χ is said to be *symmetric* if $P(\chi > a) = P(\chi < -a)$ for each $a \in \mathbf{R}$. This is equivalent to requiring that $P(\chi \in I) = P(-\chi \in I)$ for each interval I, and hence that P_χ equals $P_{-\chi}$.

Random variables $\chi_1, \ldots, \chi_n$ on (Ω, Σ, P) are labelled *independent* whenever $P\left(\bigcap_{k\leq n}\{\chi_k \in B_k\}\right) = \prod_{k\leq n} P(\chi_k \in B_k)$ holds for each n-tuple of Borel sets $B_k \subset \mathbf{R}$. This amounts to requiring that the measure P_n on the Borel sets B of $\mathbf{R}^n$, which is defined by $P_n(B) = P\big((\chi_1, \ldots, \chi_n) \in B\big)$, coincides with the product of the measures P_{χ_k}, $1 \leq k \leq n$.

From this approach one can deduce that if $(\chi_1, \ldots, \chi_n)$ and $(\chi_1^*, \ldots, \chi_n^*)$ are two n-tuples of independent random variables with $P_{\chi_k} = P_{\chi_k^*}$ for all $1 \leq k \leq n$ and if $f : \mathbf{R}^n \to \mathbf{R}$ is Borel measurable, then the random variables $f(\chi_1(\cdot), \ldots, \chi_n(\cdot))$ and $f(\chi_1^*(\cdot), \ldots, \chi_n^*(\cdot))$ have the same distribution.

Our interest in this observation will be a special case: if $\chi_1, \ldots, \chi_n$ are independent symmetric random variables, then $f(\chi_1(\cdot), \ldots, \chi_n(\cdot))$ and

$f(\varepsilon_1\chi_1(\cdot),\ldots,\varepsilon_n\chi_n(\cdot))$ have the same distribution, no matter how we assign signs $\varepsilon_i = \pm 1$.

It is also worth noting that if χ is a symmetric random variable on the probability space (Ω, Σ, μ), then it has the same distribution as $r_k(\cdot)\cdot|\chi(\cdot)|$ on the product space $[0,1]\times\Omega$; after all

$$P(r_k\cdot|\chi| > a) = P(|\chi| > a)/2 + P(|\chi| < -a)/2 = P(|\chi| > a)/2 = P(\chi > a).$$

Here is how we will apply these observations:

11.2 Lemma: *Let $\sum_{k\leq n}\chi_k x_k$ be a randomized sum in the Banach space X, where the underlying probability space is (Ω, Σ, P).*

(a) *For any choice of $\varepsilon_1,\ldots,\varepsilon_n = \pm 1$, $\|\sum_{k\leq n}\varepsilon_k\chi_k(\cdot)x_k\|$ and $\|\sum_{k\leq n}\chi_k(\cdot)x_k\|$ have the same distribution.*

(b) *Also, $\|\sum_{k\leq n}\chi_k(\cdot)x_k\|$ and $\|\sum_{k\leq n}r_k(\cdot)\cdot|\chi_k(\cdot)|\,x_k\|$ have the same distribution.*

In particular, for each $0 < p < \infty$,

$$\begin{aligned}\int_\Omega \Big\|\sum_{k\leq n}\chi_k(\omega)x_k\Big\|^p\,dP(\omega) &= \int_0^1\int_\Omega \Big\|\sum_{k\leq n} r_k(t)\chi_k(\omega)x_k\Big\|^p\,dP(\omega)\,dt \\ &= \int_0^1\int_\Omega \Big\|\sum_{k\leq n} r_k(t)\,|\chi_k(\omega)|\,x_k\Big\|^p\,dP(\omega)\,dt.\end{aligned}$$

Our first application is in the derivation of an important unconditionality result which is the key to our discussions.

11.3 Lévy's Inequality: *Let $\sum_{k\leq n}\chi_k x_k$ be a randomized sum in the Banach space X, acting on the probability space (Ω, Σ, P). Then, for each $a > 0$, we have*

$$P\big(\max_{k\leq n}\Big\|\sum_{j\leq k}\chi_j x_j\Big\| \geq a\big) \leq 2\cdot P\big(\Big\|\sum_{j\leq n}\chi_j x_j\Big\| \geq a\big).$$

Proof. With $n \in \mathbf{N}$ fixed, write $S_0 := 0$ and $S_k := \chi_1 x_1 + \ldots + \chi_k x_k$ for $1 \leq k \leq n$. Define

$$A := \{\max_{k\leq n}\|S_k\| \geq a\}\ ,\ B := \{\|S_n\| \geq a\},$$

and let A_k $(1 \leq k \leq n)$ be the event of 'first passage':

$$A_k := \bigcap_{j<k}\{\|S_j\| < a\} \cap \{\|S_k\| \geq a\}.$$

Note that the A_k's form a partition of A and that B is contained in A.

Symmetry and independence will be exploited by investigation of

$$\bar{S}_n := S_k - \chi_{k+1}x_{k+1} - \ldots - \chi_n x_n$$

for fixed $1 \leq k \leq n$. If $\omega \in A_k$, then $\|S_n(\omega)+\bar{S}_n(\omega)\| = \|2\cdot S_k(\omega)\| \geq 2a$, so that $\|S_n(\omega)\| \geq a$ or $\|\bar{S}_n(\omega)\| \geq a$. Since the χ_i's are symmetric and independent,

11.2 ensures that the events $U := A_k \cap \{\|S_n\| \geq a\}$ and $V := A_k \cap \{\|\bar{S}_n\| \geq a\}$ occur with equal probability. But $A_k = U \cup V$ and so $P(A_k) \leq P(U) + P(V) = 2P(U) = 2P(A_k \cap B)$. Summing over $1 \leq k \leq n$ we get

$$P(A) \leq 2P(A \cap B) = 2P(B),$$

as asserted. QED

Rademacher Sums

When our randomized sums are based on Rademacher functions, we call them *Rademacher sums.* Later on, they will occur as partial sums of suitable convergent series, and in this case the following lemma provides powerful estimates for the 'tail behaviour' of the sum.

Lebesgue measure on $[0,1]$ will be denoted by m.

11.4 Lemma: *Consider a Rademacher sum $\sum_{k\leq n} r_k x_k$ in the Banach space X. Then, for each $a > 0$,*

$$m(\|\sum_{k\leq n} r_k x_k\| \geq 2a) \leq 4\cdot\big(m(\|\sum_{k\leq n} r_k x_k\| \geq a)\big)^2.$$

Proof. For each $1 \leq k \leq n$, write $S_k := \sum_{j\leq k} r_j x_j$; in addition, put $S_0 := 0$. Given $a > 0$, consider

$$A := \{\max_{k\leq n} \|S_k\| \geq a\}\ ,\ B := \{\|S_n\| \geq a\}$$

and notice that

$$m(A) \leq 2\cdot m(B)$$

by Lévy's Inequality. We shall also need to work with

$$C := \{\|S_n\| \geq 2a\};$$

after all, our aim is to show $m(C) \leq 4\cdot m(B)^2$!

To gain access, we partition A as before into the events

$$A_k := \bigcap_{j<k} \{\|S_j\| < a\} \cap \{\|S_k\| \geq a\}\ ,\quad k \leq n.$$

Note that $A_k \cap C$ is plainly a subset of

$$C_k := \{\|S_n - S_{k-1}\| \geq a\},$$

so that

$$m(A_k \cap C) \leq m(A_k \cap C_k).$$

Since $C \subset B \subset A$, we find

$$m(C) = m(A \cap C) = m(\bigcup_k A_k \cap C) = \sum_k m(A_k \cap C) \leq \sum_k m(A_k \cap C_k).$$

We shall see next that A_k and C_k are independent. It is exactly here that we use the special nature of the Rademacher variables. In fact, since $r_k^2 = 1$ m-a.e., a little factoring reveals

$$C_k = \{\,\|x_k + \sum_{k+1\leq j\leq n} r_k r_j x_j\| \geq a\,\},$$

so that C_k depends on $\{r_k r_{k+1},\ldots, r_k r_n\}$, whereas A_k depends on $\{r_1,\ldots, r_k\}$. Notice that $r_1,\ldots, r_k, r_k r_{k+1},\ldots, r_k r_n$ are independent. This follows from the independence of $r_1,\ldots, r_n$, since

$$\begin{aligned}P(r_1 = \varepsilon_1,\ldots, r_k = \varepsilon_k, r_k r_{k+1} = \varepsilon_{k+1},\ldots, r_k r_n = \varepsilon_n)& \\ = P(r_1 = \varepsilon_1,\ldots, r_k = \varepsilon_k, r_{k+1} = \varepsilon_k\varepsilon_{k+1},\ldots, r_n = \varepsilon_k\varepsilon_n) = 2^{-n}& \\ = P(r_1 = \varepsilon_1)\cdot\ldots\cdot P(r_k = \varepsilon_k)\cdot P(r_k r_{k+1} = \varepsilon_{k+1}))\cdot\ldots\cdot P(r_k r_n = \varepsilon_n)&\end{aligned}$$

no matter how we choose $\varepsilon_i = \pm 1$, $1 \leq i \leq n$.

It follows that $m(A_k \cap C_k) = m(A_k)\cdot m(C_k)$. Thus, if we resume our previous chain of inequalities, we see that

$$\begin{aligned}m(C) &\leq \sum_{k\leq n} m(A_k \cap C_k) = \sum_{k\leq n} m(A_k)\cdot m(C_k) \leq \Big(\sum_{k\leq n} m(A_k)\Big)\cdot \max_{k\leq n} m(C_k) \\ &= m(A)\cdot \max_{k\leq n} m(C_k) \leq 2\cdot m(B)\cdot \max_{k\leq n} m(C_k).\end{aligned}$$

To go further, we have to break up the C_k's. For any choice of $\varepsilon_1,\ldots, \varepsilon_{k-1} = \pm 1$, we can define

$$C^{\pm}_{k,\varepsilon} := \Big(\bigcap_{j<k} \{r_j = \varepsilon_j\}\Big) \cap C_k \cap \big\{\|S_n - S_{k-1}) \pm S_{k-1}\| \geq a\big\}.$$

Of course, $C^{+}_{k,\varepsilon} = \big(\bigcap_{j<k}\{r_j = \varepsilon_j\}\big) \cap C_k \cap B$. Thanks the nature of the r_j's, the events $C^{+}_{k,\varepsilon}$ and $C^{-}_{k,\varepsilon}$ are equally probable, and

$$C_k \cap \bigcap_{j<k} \{r_j = \varepsilon_j\} = C^{+}_{k,\varepsilon} \cup C^{-}_{k,\varepsilon}$$

follows from the triangle inequality. So

$$\begin{aligned}m(C_k) &= \sum_{\varepsilon_1,\ldots,\varepsilon_{k-1}=\pm 1} m\Big(C_k \cap \bigcap_{j<k} \{r_j = \varepsilon_j\}\Big) = \sum_{\varepsilon_1,\ldots,\varepsilon_{k-1}=\pm 1} m(C^{+}_{k,\varepsilon} \cup C^{-}_{k,\varepsilon}) \\ &\leq 2 \cdot \sum_{\varepsilon_1,\ldots,\varepsilon_{k-1}=\pm 1} m(C^{+}_{k,\varepsilon}) \leq 2 \cdot \sum_{\varepsilon_1,\ldots,\varepsilon_{k-1}=\pm 1} m\Big(B \cap \bigcap_{j<k} \{r_j = \varepsilon_j\}\Big) \\ &= 2\cdot m(B).\end{aligned}$$

We have reached our goal: $\max_{k\leq n} m(C_k) \leq 2 \cdot m(B)$ and

$$m(C) \leq 2\cdot m(B)\cdot \max_{k\leq n} m(C_k) \leq 4\cdot m(B)^2. \qquad \text{QED}$$

Preparations are now complete:

Proof of Kahane's Inequality. If $p \geq q$, then we can obviously take $K_{p,q} = 1$; so let us consider the case $0 < p < q < \infty$.

If we assume, as we may, that $\int_0^1 \big\|\sum_{k\leq n} r_k(t) x_k\big\|^p\, dt \leq 1$, then we get

$$m\Big(\Big\|\sum_{k\leq n} r_k x_k\Big\| \geq 8^{1/p}\Big) \leq \frac{1}{8} \quad \Big(= \frac{1}{4}\cdot\frac{1}{2}\Big).$$

The preceding lemma allows us to set up an iteration:

$$m(\| \sum_{k \le n} r_k x_k \| \ge 2^\ell \cdot 8^{1/p}) \le \frac{1}{4} \cdot 2^{-2^\ell} \quad \text{for} \quad \ell = 0, 1, 2, \ldots .$$

After a little thought, integration by parts gives:

$$\int_0^1 \| \sum_{k \le n} r_k(t) x_k \|^q \, dt = \int_0^\infty q \cdot s^{q-1} \cdot m(\| \sum_{k \le n} r_k x_k \| \ge s) \, ds.$$

Consequently, if we set $\alpha_0 := 0$, $\alpha_\ell := 2^{\ell-1} \cdot 8^{1/p}$ for each $\ell = 1, 2, \ldots$ and $f(s) := q \cdot s^{q-1} \cdot m(\| \sum_{k \le n} r_k x_k \| \ge s)$ for $s \ge 0$, then we get

$$\begin{aligned} \int_0^1 \| \sum_{k \le n} r_k(t) x_k \|^q \, dt &= \sum_{\ell=0}^{\infty} \int_{\alpha_\ell}^{\alpha_{\ell+1}} f(s) \, ds \\ &\le \int_0^{\alpha_1} q \cdot s^{q-1} ds + \frac{1}{4} \cdot \sum_{\ell=0}^{\infty} 2^{-2^\ell} \cdot \int_{\alpha_{\ell+1}}^{\alpha_{\ell+2}} q \cdot s^{q-1} ds < \infty. \end{aligned}$$

QED

The constant obtained here is far from optimal. For more information, consult the Notes and Remarks.

Rademacher Sums in ℓ_r

We have already observed, using the standard unit vector basis of ℓ_1 and c_0, that in general Banach spaces the equivalent p-means $(\int_0^1 \| \sum_{k \le n} r_k(t) x_k \|^p \, dt)^{1/p}$ are not uniformly equivalent – or even uniformly comparable – to the strong ℓ_2 norms $(\sum_{k \le n} \|x_k\|^2)^{1/2}$.

However, it is worthwhile to examine just what does happen in ℓ_r. We shall find the following comparability results which will afford us an excuse to throw off our Khinchin-inspired attachment to strong ℓ_2 norms: for each $1 \le r < \infty$ there are constants $\kappa(r)$ and $\theta(r)$ such that, no matter how we choose $x_1, \ldots, x_n$ in ℓ_r,

(1) if $1 \le r \le 2$, then

$$\kappa(r) \cdot \big(\sum_{k \le n} \|x_k\|_r^2 \big)^{1/2} \le \big(\int_0^1 \| \sum_{k \le n} r_k(t) x_k \|_r^r \, dt \big)^{1/r} \le \theta(r) \cdot \big(\sum_{k \le n} \|x_k\|_r^r \big)^{1/r}$$

and

(2) if $2 \le r < \infty$, then

$$\kappa(r) \cdot \big(\sum_{k \le n} \|x_k\|_r^r \big)^{1/r} \le \big(\int_0^1 \| \sum_{k \le n} r_k(t) x_k \|_r^r \, dt \big)^{1/r} \le \theta(r) \cdot \big(\sum_{k \le n} \|x_k\|_r^2 \big)^{1/2}.$$

Both of these inequalities are sharp. Khinchin's Inequality reveals that we cannot improve comparison with the strong ℓ_2 norms. The tightness of the comparison with the strong ℓ_r norms is even easier: simply take the standard unit vectors $e_k \in \ell_r$ and observe that $(\int_0^1 \| \sum_{k\le n} r_k(t)e_k \|_r^r \, dt)^{1/r} = n^{1/r}$, whereas $(\sum_{k\le n} \|e_k\|_r^s)^{1/s} = n^{1/s}$ for any $s<\infty$.

We shall soon introduce appropriate concepts to make these statements more succinct, but first let us derive the inequalities (1) and (2). We have

$$\Big(\int_0^1 \| \sum_{k\le n} r_k(t)x_k \|_r^r \, dt\Big)^{1/r} = \Big(\int_0^1 \sum_{j=1}^{\infty} | \sum_{k\le n} r_k(t)x_k(j) |^r \, dt\Big)^{1/r}$$
$$= \Big(\sum_{j=1}^{\infty} \int_0^1 | \sum_{k\le n} r_k(t)x_k(j) |^r \, dt\Big)^{1/r}$$

and thanks to Khinchin's Inequality, this is uniformly equivalent to

$$\Big(\sum_{j=1}^{\infty}\big(\sum_{k\le n} |x_k(j)|^2\big)^{r/2}\Big)^{1/r}.$$

At this stage, the dichotomy between $r \le 2$ and $r \ge 2$ becomes apparent. In either case, Minkowski's Inequality can be applied, but the direction will change as we move r from one side of 2 to the other. When $r \le 2$, we get

$$\Big(\int_0^1 \| \sum_{k\le n} r_k(t)x_k \|_r^r \, dt\Big)^{1/r} \ge \kappa\cdot\Big(\sum_{j=1}^{\infty}\big(\sum_{k\le n} |x_k(j)|^2\big)^{r/2}\Big)^{1/r}$$
$$\ge \kappa\cdot\Big(\sum_{k\le n}\big(\sum_{j=1}^{\infty} |x_k(j)|^r\big)^{2/r}\Big)^{1/2} = \kappa\cdot\Big(\sum_{k\le n} \|x_k\|_r^2\Big)^{1/2}$$

with κ depending only on r, whereas, when $r \ge 2$, the reverse inequalities obtain.

The remaining estimates follow from the monotonicity of the ℓ_r norms. For $1 \le r \le 2$ we get, with some $\theta = \theta(r)$,

$$\Big(\int_0^1 \| \sum_{k\le n} r_k(t)x_k \|_r^r \, dt\Big)^{1/r} \le \theta\cdot\Big(\sum_{j=1}^{\infty}\big(\sum_{k\le n} |x_k(j)|^2\big)^{r/2}\Big)^{1/r}$$
$$\le \theta\cdot\Big(\sum_{j=1}^{\infty}\sum_{k\le n} |x_k(j)|^r\Big)^{1/r} = \theta\cdot\Big(\sum_{k\le n} \|x_k\|_r^r\Big)^{1/r}.$$

Again, these inequalities reverse when $r \ge 2$.

Type and Cotype

Mathematical Doctrine demands appropriate definitions.

A Banach space X has *type p* if there is a constant $\theta \ge 0$ such that, however we choose finitely many vectors $x_1, \ldots, x_n$ from X,

$$\Big(\int_0^1 \| \sum_{k=1}^{n} r_k(t)x_k \|^2 \, dt\Big)^{1/2} \le \theta\cdot\Big(\sum_{k=1}^{n} \|x_k\|^p\Big)^{1/p}.$$

If X has type p, we label the smallest of all admissible constants θ as

$$T_p(X);$$

this is the *type p constant* of X.

A Banach space X has *cotype q* if there is a constant $\kappa \geq 0$ such that no matter how we select finitely many vectors $x_1, \ldots, x_n$ from X,

$$\Big(\sum_{k=1}^{n} \|x_k\|^q\Big)^{1/q} \leq \kappa \cdot \Big(\int_0^1 \Big\|\sum_{k=1}^{n} r_k(t) x_k\Big\|^2 \, dt\Big)^{1/2}.$$

To cover the case $q = \infty$, the left hand side should be replaced by $\max_{k \leq n} \|x_k\|$. The smallest of all these constants will then be denoted by

$$C_q(X)$$

and named the *cotype q constant* of X.

Some immediate comments are in order.

11.5 Remarks:

(a) Kahane's Inequality tells us that the L_2 Rademacher average in these definitions can be replaced by the corresponding L_r average for any choice of $0 < r < \infty$, at the expense of adjusting constants, of course.

(b) Certain types and cotypes are outlawed for non-trivial Banach spaces: Khinchin's Inequality tells us that only $X = \{0\}$ can have type > 2 or cotype < 2.

(c) Every Banach space X has type p for any $0 < p \leq 1$:

$$\Big(\int_0^1 \Big\|\sum_{k \leq n} r_k(t) x_k\Big\|^2 \, dt\Big)^{1/2} \leq \sum_{k \leq n} \|x_k\| \leq \Big(\sum_{k \leq n} \|x_k\|^p\Big)^{1/p}.$$

(d) Every Banach space X has cotype ∞. Given $x_1, \ldots, x_n \in X$, choose $1 \leq j \leq n$ and $x^* \in B_{X^*}$ such that $\langle x^*, x_j \rangle = \max_{k \leq n} \|x_k\|$. Rademacher functions are orthonormal, so

$$\begin{aligned} \langle x^*, x_j \rangle &= \int_0^1 \Big\langle r_j(t) x_j^*, \sum_{k \leq n} r_k(t) x_k \Big\rangle \, dt \\ &\leq \int_0^1 \Big\|\sum_{k \leq n} r_k(t) x_k\Big\| \, dt \leq \Big(\int_0^1 \Big\|\sum_{k \leq n} r_k(t) x_k\Big\|^2 \, dt\Big)^{1/2}, \end{aligned}$$

and the statement follows.

This explains why type p is only considered for $1 \leq p \leq 2$, whereas cotype q is only considered for $2 \leq q \leq \infty$.

Note also that $T_1(X) = C_\infty(X) = 1$ for every Banach space X.

(e) If a Banach space X has type $p > 1$, it has also type $\tilde{p}$ for any $1 \leq \tilde{p} < p$, and $T_{\tilde{p}}(X) \leq T_p(X)$.

(f) If a Banach space X has cotype $q < \infty$, it also has cotype $\tilde{q}$ for any $q < \tilde{q} \leq \infty$, and $C_{\tilde{q}}(X) \leq C_q(X)$.

(g) Notice that, in view of remark (a), our motivating computations can be interpreted in our new language:

If $1 \le p \le 2$, then ℓ_p has type p and cotype 2.

If $2 \le q < \infty$, then ℓ_q has type 2 and cotype q.

Moreover, remarks (e) and (f) allow us to add that *these results are best possible.*

It is obvious that if Y is a subspace of the Banach space X and X has type p, resp. cotype q, then so does Y, and we have $T_p(Y) \le T_p(X)$, resp. $C_q(Y) \le C_q(X)$. These facts admit substantial, yet easily seen, generalizations once the finitary nature of type and cotype is recognized.

11.6 Theorem: *Suppose that the Banach space Y is λ-representable in the Banach space X for some $\lambda > 1$.*

(a) *If X has type p, then so does Y, and $T_p(Y) \le \lambda \cdot T_p(X)$.*

(b) *If X has cotype q, then so does Y, and $C_q(Y) \le \lambda \cdot C_q(X)$.*

It follows that if Y is finitely representable in X, then $T_p(Y) \le T_p(X)$ and $C_q(Y) \le C_q(X)$.

11.7 Corollary: (a) *Each $\mathcal{L}_r$-space ($1 \le r < \infty$) has type* $\min\{r, 2\}$ *and cotype* $\max\{r, 2\}$, *and this is best possible for the infinite dimensional case.*

(b) *No infinite dimensional $\mathcal{L}_\infty$-space can have type > 1 or cotype $< \infty$.*

Proof. (a) has already been observed for ℓ_r (11.5(g)), so we only need to recall that every infinite dimensional $\mathcal{L}_{r,\lambda}$-space is λ-representable in ℓ_r, and vice versa.

Similarly, for (b) it is enough to use (a) together with the observation that the universal nature of ℓ_∞ endows it with subspaces which are copies of ℓ_r for all $1 \le r < \infty$. QED

Note that the statement (a) about $\mathcal{L}_r$-spaces represents a genuine improvement of Orlicz's Theorem (see 3.12 and 10.7). Just observe that $\|(x_k)_1^n\|_1^{weak} = \max_{t\in[0,1]} \|\sum_{k\le n} r_k(t)x_k\|$ for any finite collection of vectors in any Banach space. Compare with 11.17 below.

It is also worth highlighting what happens when $r = 2$:

11.8 Corollary: *Hilbert spaces have type 2 and cotype 2, and their type 2 and cotype 2 constants are both 1.*

Although this is a corollary of 11.7, the result is simple: it follows directly from the orthonormality of the Rademacher functions. Actually, 11.8 characterizes Hilbert spaces. This, however, lies much deeper; see 12.20.

11.9 Corollary: *A Banach space has the same type or cotype as its bidual.*

This is simply because we know from the Principle of Local Reflexivity that the bidual of any Banach space is finitely representable in the space.

What about duals? The situation is only semi-satisfactory.

11.10 Proposition: *If a Banach space X has type p, then its dual X^* has cotype p^*, and $C_{p^*}(X^*) \le T_p(X)$.*

Proof. Select $x_1,\dots,x_n \in X$ and $x_1^*,\dots,x_n^* \in X^*$. Orthonormality of the Rademacher variables allows us to write

$$\begin{aligned}
\Big|\sum_{k\le n}\langle x_k^*, x_k\rangle\Big| &= \Big|\int_0^1 \Big\langle \sum_{k\le n} r_k(t)x_k^*, \sum_{k\le n} r_k(t)x_k\Big\rangle\, dt\Big| \\
&\le \Big(\int_0^1 \Big\|\sum_{k\le n} r_k(t)x_k^*\Big\|^2 dt\Big)^{1/2}\cdot\Big(\int_0^1 \Big\|\sum_{k\le n} r_k(t)x_k\Big\|^2 dt\Big)^{1/2} \\
&\le T_p(X)\cdot\Big(\sum_{k\le n}\|x_k\|^p\Big)^{1/p}\cdot\Big(\int_0^1 \Big\|\sum_{k\le n} r_k(t)x_k^*\Big\|^2 dt\Big)^{1/2}.
\end{aligned}$$

It follows that

$$\begin{aligned}
\Big(\sum_{k\le n}\|x_k^*\|^{p^*}\Big)^{1/p^*} &= \sup\Big\{\Big|\sum_{k\le n}\langle x_k^*, x_k\rangle\Big| : x_1,\dots,x_n \in X, \sum_{k\le n}\|x_k\|^p \le 1\Big\} \\
&\le T_p(X)\cdot\Big(\int_0^1 \Big\|\sum_{k\le n} r_k(t)x_k^*\Big\|^2 dt\Big)^{1/2}
\end{aligned}$$

as we desired. QED

Notice that we also could have used this duality result to obtain information about the cotype of $\mathcal{L}_{r^*}$-spaces, once the type of $\mathcal{L}_r$-spaces was known (see 11.7).

Using 11.9, we see that a Banach space X (together with X^{**}) must have cotype q once its dual is known to have type q^*. But the converse fails easily: ℓ_1 has cotype 2, but its dual ℓ_∞ does not have non-trivial type.

Recall that every separable Banach space is a quotient of ℓ_1. Since ℓ_1 has cotype 2, we see that it is impossible, for general Banach spaces X, to glean any information about the cotype of quotients X/Y from the cotype of X. An extreme example is c_0 which doesn't have any finite cotype. As is obvious from the definitions, type behaves much better in this respect.

11.11 Proposition: *If Y is a closed subspace of the Banach space X and X has type p, then X/Y also has type p, and $T_p(X/Y) \le T_p(X)$.*

So far, our investigations have turned up good permanence properties for type, but have shown cotype to be less well-behaved. Suspiciously enough, all our examples displaying cotype shortcomings have been based on ℓ_1, and we shall see later that this is no accident. Spaces with non-trivial type will turn out to be those that do not contain uniform copies of all ℓ_1^n's. If we confine ourselves to such spaces, then cotype also behaves in an exemplary manner.

Further examples are obtained by looking at vector-valued L_r-spaces. We give ourselves a *non-trivial* measure space (Ω, Σ, μ): this just means that Σ contains sets of finite positive measure. Given $1 \le r < \infty$, write

$$L_r(\mu, X)$$

for the Lebesgue-Bochner space of μ-a.e. equivalence classes of strongly measurable functions $f : \Omega \to X$ with the property that $\|f(\cdot)\|$ belongs to $L_r(\mu)$. This is a Banach space under the norm

$$\|f\|_r := \Big(\int_\Omega \|f(\omega)\|^r d\mu(\omega)\Big)^{1/r}.$$

Note that when μ is counting measure on $\mathbf{N}$, then $L_r(\mu, X)$ is nothing but $\ell_r^{strong}(X)$.

11.12 Theorem: (a) *Let $1 < p \leq 2$ and let $p \leq r < \infty$. Then X has type p if and only if $L_r(\mu, X)$ has type p.*

(b) *Let $2 \leq q < \infty$ and let $1 \leq r \leq q$. Then X has cotype q if and only if $L_r(\mu, X)$ has cotype q.*

In particular, X has type p resp. cotype q if and only if $L_2(\mu, X)$ does; and it will be apparent from the proof that in this case

$$T_p(X) = T_p\big(L_2(\mu, X)\big) \quad \text{resp.} \quad C_q(X) = C_q\big(L_2(\mu, X)\big).$$

Proof. One direction is straightforward. Since μ is a non-trivial measure, X is isomorphic to a subspace of $L_r(\mu, X)$: associate with each $x \in X$ the function $1_A(\cdot)x$ where $A \in \Sigma$ is such that $0 < \mu(A) < \infty$. Consequently, X inherits type and cotype from $L_r(\mu, X)$. – Let us investigate the other implications.

(a) Choose $f_1, \ldots, f_n \in L_r(\mu, X)$, where $p \leq r < \infty$ and X is assumed to have type p. Apply Kahane's Inequality and Fubini's Theorem:

$$\begin{aligned}\Big(\int_0^1 \Big\|\sum_{k\leq n} r_k(t) f_k\Big\|^2_{L_r(\mu,X)} dt\Big)^{1/2} &\leq K_{r,2}\cdot\Big(\int_0^1 \Big\|\sum_{k\leq n} r_k(t) f_k\Big\|^r_{L_r(\mu,X)} dt\Big)^{1/r}\\ &= K_{r,2}\cdot\Big(\int_\Omega\int_0^1 \Big\|\sum_{k\leq n} r_k(t) f_k(\omega)\Big\|^r_X dt\, d\mu(\omega)\Big)^{1/r}\\ &\leq K\cdot\Big(\int_\Omega\Big(\int_0^1 \Big\|\sum_{k\leq n} r_k(t) f_k(\omega)\Big\|^2_X dt\Big)^{r/2} d\mu(\omega)\Big)^{1/r}\end{aligned}$$

with $K = K_{r,2} \cdot K_{2,r}$. Invoke the type of X and, using $r/p \geq 1$, the Minkowski Inequality:

$$\begin{aligned}\Big(\int_0^1 \Big\|\sum_{k\leq n} r_k(t) f_k\Big\|^2_{L_r(\mu,X)} dt\Big)^{1/2} &\leq K\cdot T_p(X)\cdot\Big(\int_\Omega\Big(\sum_{k\leq n}\|f_k(\omega)\|^p_X\Big)^{r/p} d\mu(\omega)\Big)^{1/r}\\ &\leq K\cdot T_p(X)\cdot\Big(\sum_{k\leq n}\Big(\int_\Omega \|f_k(\omega)\|^r_X d\mu(\omega)\Big)^{p/r}\Big)^{1/p}\\ &= K\cdot T_p(X)\cdot\Big(\sum_{k\leq n}\|f_k\|^p_{L_r(\mu,X)}\Big)^{1/p}.\end{aligned}$$

The argument for (b) is analogous. We leave the details to the reader.

QED

SUMMING OPERATORS

It is time to relate type and cotype to our raison d'être: summing operators. For the moment, we will content ourselves with seeing how easy it is to generalize a number of results from Chapters 3 and 10.

11.13 Theorem: *Suppose that Y has cotype q where $2 \le q < \infty$. Then, for any Banach space X and all $1 \le r < \infty$, we have*

$$\Pi_r(X,Y) \subset \Pi_{q,2}(X,Y).$$

In particular, if Y has cotype 2, then for any Banach space X and all $2 < r < \infty$, $\Pi_r(X,Y) = \Pi_2(X,Y)$.

We shall actually see that $\pi_{q,2}(u) \le B_r \cdot C_q(X) \cdot \pi_r(u)$ holds for any u in $\Pi_r(X,Y)$, where B_r is taken from Khinchin's Inequality.

Notice that Theorem 3.15 is just a special case of the cotype 2 part of this theorem.

Proof. Clearly, the result is only of interest when $r > q$.

Given $u \in \Pi_r(X,Y)$, use Pietsch's Domination Theorem 2.12 to choose a probability measure $\nu \in \mathcal{C}(B_{X^*})^*$ so that, for any $x \in X$,

$$\|ux\| \le \pi_r(u) \cdot \|f_x\|_{L_r(\nu)},$$

where $f_x : B_{X^*} \to \mathbf{K}$ is defined by $f_x(x^*) := \langle x^*, x \rangle$. Then, for any finite collection $x_1, \ldots, x_n$ in X,

$$\begin{aligned}
\Big(\sum_{k\le n} \|ux_k\|^q\Big)^{1/q} &\le C_q(X) \cdot \Big(\int_0^1 \Big\|\sum_{k\le n} r_k(t) ux_k\Big\|^2 dt\Big)^{1/2} \\
&\le C_q(X) \cdot \Big(\int_0^1 \Big\|u\Big(\sum_{k\le n} r_k(t) x_k\Big)\Big\|^r dt\Big)^{1/r} \\
&\le C_q(X) \cdot \pi_r(u) \cdot \Big(\int_0^1 \Big\|\sum_{k\le n} r_k(t) f_{x_k}\Big\|_{L_r(\nu)}^r dt\Big)^{1/r} \\
&= C_q(X) \cdot \pi_r(u) \cdot \Big(\int_{B_{X^*}} \int_0^1 \Big|\sum_{k\le n} r_k(t) \langle x^*, x_k \rangle\Big|^r dt\, d\nu(x^*)\Big)^{1/r}.
\end{aligned}$$

From Khinchin's Inequality, this is

$$\begin{aligned}
&\le B_r \cdot C_q(X) \cdot \pi_r(u) \cdot \Big(\int_{B_{X^*}} \Big(\sum_{k\le n} |\langle x^*, x_k \rangle|^2\Big)^{r/2} d\nu(x^*)\Big)^{1/r} \\
&\le B_r \cdot C_q(X) \cdot \pi_r(u) \cdot \|(x_k)_1^n\|_2^{weak}.
\end{aligned}$$

QED

Our next goal is to investigate operators from $\mathcal{C}(K)$-spaces into Banach spaces of finite cotype.

11.14 Theorem: *Suppose that Y has cotype q, where $2 \leq q < \infty$, and that K is a compact Hausdorff space.*

(a) *If $q = 2$, then $\mathcal{L}(C(K), Y) = \Pi_2(C(K), Y)$.*

(b) *If $2 < q < \infty$, then $\mathcal{L}(C(K), Y) = \Pi_{q,p}(C(K), Y) = \Pi_r(C(K), Y)$ for all $p < q$ and $q < r < \infty$.*

Of course, localization lets us replace $C(K)$ in this theorem by any $\mathcal{L}_\infty$-space.

Actually, this generalization of theorems 3.5 and 10.10 turns out to be very close to a characterization of cotype; see Chapter 14.

We shall rely on a clever interpolation argument.

11.15 Lemma: *Let K be a compact Hausdorff space, Y a Banach space and, for $1 \leq p < \infty$, let $u : C(K) \to Y$ be a p-summing operator. If $p < q < \infty$, then*

$$\pi_q(u) \leq \|u\|^{1-(p/q)} \cdot \pi_p(u)^{p/q}.$$

Proof. Choose a probability measure $\mu \in C(K)^*$ to assure the usual factorization

$$u : C(K) \xrightarrow{j_q} L_q(\nu) \xrightarrow{j_{q,p}} L_p(\nu) \xrightarrow{\tilde{u}} Y$$

where j_q and $j_{q,p}$ are formal identities and $\tilde{u}$ satisfies $\|\tilde{u}\| = \pi_p(u)$. Let $v := \tilde{u}\, j_{q,p}$ and observe that $\|v\| \geq \pi_q(u)$. Recognizing that $q^* := p/(p - (p/q))$, we set $\theta := p/q$ and use Hölder's Inequality to deduce that for all $y^* \in B_{Y^*}$

$$\begin{aligned}
\|v^*y^*\|_{L_{q^*}} &= \Big(\int_K |v^*y^*|^{p(1-\theta)/(p-\theta)} \cdot |v^*y^*|^{p\theta/(p-\theta)}\, d\nu\Big)^{(p-\theta)/p} \\
&\leq \Big(\int_K |v^*y^*|\, d\nu\Big)^{1-\theta} \cdot \Big(\int_K |v^*y^*|^{p^*}\, d\nu\Big)^{\theta/p^*} \\
&= \|v^*y^*\|_{L_1}^{1-\theta} \cdot \|v^*y^*\|_{L_{p^*}}^{\theta} = \|v^*y^*\|_{L_1}^{1-\theta} \cdot \|\tilde{u}^*y^*\|_{L_{p^*}}^{\theta}.
\end{aligned}$$

Considering $L_1(\nu)$ as a subspace of $C(K)^*$, we may thus write

$$\|v^*y^*\|_{L_{q^*}} \leq \|u^*y^*\|_{C(K)^*}^{1-\theta} \cdot \|\tilde{u}^*y^*\|_{L_{p^*}}^{\theta},$$

and so

$$\pi_q(u) = \|v\| = \|v^*\| \leq \|u^*\|^{1-\theta} \cdot \|\tilde{u}^*\|^{\theta} = \|u\|^{1-\theta} \cdot \|\tilde{u}\|^{\theta} = \|u\|^{1-\theta} \cdot \pi_p(u)^{\theta},$$

as asserted. QED

We shall use this lemma with $p = 2$, $q = 4$ to get a proof of 11.14. The set-up should be familiar: think of the proof of Khinchin's Inequality.

Proof of 11.14. (a) By localization, it is enough to show that there is a uniform estimate $\pi_2(u) \leq C \cdot \|u\|$ for any $u \in \mathcal{L}(\ell_\infty^n, Y)$. But this is now easy: Lemma 11.15 gives

$$\pi_4(u) \leq \|u\|^{1/2} \cdot \pi_2(u)^{1/2},$$

whereas Theorem 11.13 tells us that

$$\pi_2(u) \leq B_4 \cdot C_2(Y) \cdot \pi_4(u).$$

Combine to get what we wanted.

(b) If we invoke Theorem 10.9, we see that we only need to prove $\mathcal{L}(C(K),Y) = \Pi_{q,1}(C(K),Y)$ to establish the first identity. Accordingly, we fix $u \in \mathcal{L}(C(K),Y)$ and select $f_1,\dots,f_n$ from $C(K)$. In the standard fashion, we define $v : \ell_\infty^n \to C(K)$ via $e_k \mapsto f_k$, so that $\|v\| = \|(f_k)_1^n\|_1^{weak}$. Thus

$$\begin{aligned}\Big(\sum_{k\le n} \|uf_k\|^q\Big)^{1/q} &\le C_q(X)\cdot\Big(\int_0^1 \Big\|\sum_{k\le n} r_k(t)uf_k\Big\|^q\,dt\Big)^{1/q}\\ &\le C_q(X)\cdot\|u\|\cdot\|v\|\cdot\Big(\int_0^1 \Big\|\sum_{k\le n} r_k(t)e_k\Big\|_{\ell_\infty^n}^2\,dt\Big)^{1/2}\\ &= C_q(X)\cdot\|u\|\cdot\|(f_k)_1^n\|_1^{weak},\end{aligned}$$

as desired.

The second identity follows as in Corollary 10.10. QED

Some coincidence results follow quickly, and these generalize nicely 3.11, 10.11, and 3.15.

11.16 Corollary: *Let X and Y be Banach spaces.*

(a) *If X has cotype 2, then $\Pi_2(X,Y) = \Pi_1(X,Y)$.*

(b) *If X has cotype $2 < q < \infty$, then $\Pi_r(X,Y) = \Pi_1(X,Y)$ for all $1 < r < q^*$.*

(c) *If X and Y both have cotype 2, then $\Pi_r(X,Y) = \Pi_1(X,Y)$ for all $1 < r < \infty$.*

Proof. We can treat (a) and (b) simultaneously. Choose a weak ℓ_1 sequence (x_n) in X and use it in the usual way to create an operator $v : c_0 \to X$. In case (a), the theorem tells us that v is 2-summing, whereas in case (b) the conclusion is that v is r^*-summing. Now appeal to the Composition Theorem 2.22.

Part (c) follows from combining (b) and 11.13(b). QED

The argument above should be familiar: we used it to prove Orlicz's Theorem 3.12. Repeating it, we get a generalization of both 3.12 and 10.7.

11.17 Corollary: *If X has cotype $2 \le q < \infty$, then every weak ℓ_1 sequence in X is a strong ℓ_q sequence. In other words, id_X is $(q,1)$-summing.*

This property is often referred to as the *Orlicz property* (in particular if $q = 2$). In Chapter 14 we shall see that it almost characterizes cotype q Banach spaces.

Theorem 11.14 also hints at deep stirrings in $\mathcal{L}_1$-spaces. We say that a Banach space has *non-trivial type* when it has type p for some $p > 1$.

11.18 Corollary: *Let μ be any finite measure and let X be a closed subspace of $L_1(\mu)$. Then X has non-trivial type if and only if it is isomorphic to a subspace of $L_r(\nu)$, for some $1 < r \le 2$ and some probability measure ν.*

The result remains true without any restriction on the measure μ. All that is needed is that $L_1(\mu)$ has an $L_\infty(\sigma)$-space as its dual. This is true but not so easy to prove.

Proof. One implication is trivial. Since $1 < r \leq 2$, we know from 11.7 that $L_r(\nu)$, along with all its subspaces, has type r.

For the converse, consider X^* as a quotient of $L_\infty(\mu) = L_1(\mu)^*$; let $q : L_\infty(\mu) \to X^*$ be the corresponding quotient map. Assume that X has type $1 < p \leq 2$. Deduce from 11.10 that X^* has cotype p^*, and appeal to 11.14 to get that q must be r^*-summing no matter how we choose $1 < r < p$. Since q's domain is an $\mathcal{L}_\infty$-space, we may invoke 3.4 to produce a probability measure ν together with operators $u : L_{r^*}(\nu) \to X^*$ and $v : L_\infty(\mu) \to L_{r^*}(\nu)$ such that $q = uv$. As q is surjective, u is, and so u is open, thanks to the Open Mapping Theorem. Consequently, X^* is reflexive and $u^* : X \to L_r(\nu)$ is an isomorphic embedding. QED

We shall encounter variations of this result in 13.19 and 15.12.

Notes and Remarks

The notions of type and cotype emerged from the work of J. Hoffmann-Jørgensen, S. Kwapień, B. Maurey and G. Pisier in the early 1970's. Chapters 11 through 14 present the basics of the theory as it relates to Banach spaces and absolutely summing operators.

This is a deep development, so it is only natural that there should be a variety of precursors. As far back as the 1930's, W. Orlicz [1933a], [1933b] gave broad hints about the cotype of L_p-spaces. G. Nordlander [1962] actually computed the type of the L_p-spaces and established that Hilbert spaces have cotype 2.

The theory of type and cotype reflects the interplay between geometry and probability in Banach spaces. A. Beck [1962] established close links between the intrinsic geometry of a Banach space and the validity of the Strong Law of Large Numbers for random variables taking values in the space. He did this through the intermediary of B-convexity, a notion which will be studied extensively in Chapter 13.

J.P. Kahane's [1968] generalization of Khinchin's Inequality (see 11.1), when taken in tandem with the Dvoretzky - Rogers Theorem, gave promise of fascinating ways of investigation relating strong p-summability and (almost) unconditional summability. Fortunately, this appeared just at the same time when the long-dormant ideas of Grothendieck were being revitalized by Lindenstrauss, Pełczyński and Pietsch. Further hints of what was on the horizon and beyond were found in E. Dubinsky, A. Pełczyński and H.P. Rosenthal [1972].

However, the spark which finally triggered the explosive developments of the 1970's was S. Kwapień's [1972a] characterization of Hilbert spaces as the only Banach spaces which simultaneously have type 2 and cotype 2; we shall prove this result in 12.20. This result which, we emphasize, predates the notions of type and cotype, relies on succinct limitations on almost summability vis-à-vis strong summability.

With these developments in hand, J.Hoffmann-Jørgensen [1972/73], B. Maurey [1972/73], [1974b], B.Maurey and G. Pisier [1973], [1976] and G. Pisier [1973] formulated explicitly the concepts of type and cotype. Let us cite some of their major achievements.

Hoffmann-Jørgensen's interests were in sharp forms of the Law of Large Numbers and other classical limit theorems of probability. His work was to contribute grandly to the study of sums of independent Banach space-valued random variables. Kwapień played a significant rôle in giving a finishing touch to this work.

Maurey had conducted profound investigations into the factoring of operators into L_0 and extended deep, but apparently unrelated, work of E.M. Nikishin [1970] and H.P. Rosenthal [1973].

G. Pisier [1975] used his penetrating insights to give a startling exposé of martingales in superreflexive Banach spaces. As a result, he was able to renorm superreflexive spaces in a uniformly convex manner with power type modulus of convexity. This completed earlier work of P. Enflo [1972] and reinforced the importance of probabilistic ideas in the study of Banach spaces.

These investigations relating the geometry of a Banach space and the behaviour of random variables taking values in that space soon led B. Maurey and G. Pisier [1976] to the discovery of their 'Great Theorem'. Much of Chapters 12 through 14 will be devoted to the pursuit of a utility grade version of their result. To be precise, we aim to show the following:

Theorem: *A Banach space has type strictly greater than* 1 *if and only if it does not contain ℓ_1^n's uniformly.*

A Banach space has finite cotype if and only if it does not contain ℓ_∞^n's uniformly.

In fact, Maurey and Pisier gave a very precise quantitative versions of the results above. Anticipating the notion of finite factorizability, which will be introduced in Chapter 14, these can be stated as follows:

Maurey - Pisier Theorem: *If X is an infinite dimensional Banach space, let*

$$p_X := \sup\{p : X \text{ has type } p\} \quad \text{and} \quad q_X := \inf\{q : X \text{ has cotype } q\}.$$

Then X finitely factors both $\ell_1 \hookrightarrow \ell_{p_X}$ and $\ell_{q_X} \hookrightarrow \ell_\infty$.

Moreover, ℓ_{p_X} and ℓ_{q_X} are both finitely representable in X; in fact, ℓ_r is finitely representable in X for every $r \in [p_X, 2]$.

We shall prove part of the first part of this theorem in Chapter 14. An important component of the second part is the highly non-trivial fact that ℓ_2 is finitely representable in every infinite dimensional Banach space: this is (a consequence of) Dvoretzky's Theorem to be proved in Chapter 19.

The remarkable inequality 11.1 is due to J.P. Kahane [1964] and we do no more than expound his proof. Lévy's Inequality 11.3, while not quite as old as probability theory, is a classical part thereof; we follow J.P. Kahane's [1968] presentation.

Kahane's Inequality is a significant generalization of Khinchin's Inequality; inevitably, it drew considerable attention. C. Borell [1979] gave a different proof which depended on the character of the Walsh system; an enlightening account of Borell's proof can be found in G. Pisier [1977/78].

B. Tomaszewski [1982], inspired by S. Szarek's [1976] determination of the best constant in the 'L_1 versus L_2' Khinchin Inequality, has given an elegant proof of Kahane's Inequality for the '$L_1(X)$ versus $L_2(X)$' case.

Providentially, as this text was going to press, a stunning new proof of Kahane's Inequality for the '$L_1(X)$ versus $L_2(X)$' case was discovered by R. Latała and K. Oleszkiewicz [1995], and shown to us by S. Kwapień. Their extraordinarily short argument even provides $\sqrt{2}$ as the best constant in both the vector-valued situation and the classical Khinchin Inequality.

The action is best viewed on the compact abelian group $D_n = \{-1,1\}^n$ in the discrete topology. Note that the Haar measure is just $\mu_n = 2^{-n}\sum_{\varepsilon\in D_n}\delta_\varepsilon$. It is important to realize that the 'Walsh functions' form an orthonormal basis for $L_2(\mu_n)$: these consist of 1, the coordinate functionals $r_1,\dots,r_n$, and all finite products of distinct coordinate functionals. We shall return to this at the beginning of Chapter 13.

For any $\varepsilon \in D_n$ write S_ε for the set of all $\eta \in D_n$ which differ from ε in just one coordinate. Define $U : L_2(\mu_n) \to L_2(\mu_n)$ by

$$Uf(\varepsilon) := \frac{n}{2}\cdot\Big(f(\varepsilon) - \frac{1}{n}\sum_{\eta\in S_\varepsilon} f(\eta)\Big).$$

One can quickly establish that if w is in W_k, the set of Walsh functions of order k, that is, the set of products of k distinct coordinate functionals, then

$$(*) \qquad Uw = k\cdot w.$$

With our goal in mind, we look at $f \in L_2(\mu_n)$ given by

$$f(\varepsilon) = \|\sum_{i\le n}\varepsilon_i x_i\|$$

where $x_1,\dots,x_n$ belong to a Banach space X. The completeness of the Walsh system tells us that

$$\int_{D_n} f^2 d\mu_n = (f\,|\,1)^2 + \sum_{i\le n}(f\,|\,r_i)^2 + \sum_{k=2}^{n}\sum_{w\in W_k}(f\,|\,w)^2.$$

Plainly, $(f\,|\,1) = \int_{D_n} f\,d\mu_n$, and oddness and evenness considerations show that $(f\,|\,r_i) = 0$ for each $1 \le i \le n$. Incorporating this information and using $(*)$, we obtain

$$\int_{D_n} f^2 d\mu_n = \Big(\int_{D_n} f\,d\mu_n\Big)^2 + \sum_{k=2}^{n}\sum_{w\in W_k}(f\,|\,w)^2 \le \Big(\int_{D_n} f\,d\mu_n\Big)^2 + \frac{1}{2}\cdot(Uf|f).$$

But, thanks to the triangle inequality, $Uf \le f$ and so

$$\int_{D_n} f^2 d\mu_n \le \Big(\int_{D_n} f\,d\mu_n\Big)^2 + \frac{1}{2}(f|f),$$

and a quick rearrangement provides us with what we wanted.

Kahane's Inequality was grafted to a Gaussian setting by X. Fernique [1970] and H.J. Landau and L.A. Shepp [1970].

Theorem: *Let (g_n) be a sequence of independent standard Gaussian random variables on a probability space (Ω, Σ, P). Then for any $0 < p, q < \infty$ there is a constant $K_{p,q} > 0$ such that*

$$\Big(\int_\Omega \Big\|\sum_{k\le n} g_k(\omega)x_k\Big\|^q\,dP(\omega)\Big)^{1/q} \le K_{p,q}\cdot\Big(\int_\Omega \Big\|\sum_{k\le n} g_k(\omega)x_k\Big\|^p\,dP(\omega)\Big)^{1/p}$$

regardless of the choice of a Banach space X and of finitely many vectors $x_1, \ldots, x_n$ from X.

Actually we can use the constant $K_{p,q}$ from 11.1.

M.B. Marcus and G. Pisier [1981] explained how to deduce this theorem from Kahane's Inequality by making careful use of the Central Limit Theorem in finite dimensions.

More recently, M. Talagrand [1988] has examined Kahane's Inequality from an isoperimetric viewpoint. Details on this and much more can be found in M. Ledoux and M. Talagrand [1991].

Naturally, the problem of computing type and cotype for concrete spaces has attracted much attention. N. Tomczak-Jaegermann [1974] carried out these computations for the Schatten - von Neumann classes $\mathcal{S}_p$. It turns out that their type and cotype is the same as for the corresponding L_p-spaces. She also showed that the predual of any von Neumann algebra has cotype 2. It can safely be said that the development of type and cotype gained much of its momentum through this work.

T. Fack [1987] extended Tomczak-Jaegermann's work by proving that certain non-commutative L_p-spaces associated with an appropriate von Neumann algebra have the same type and cotype characteristics as their commutative cousins. He prepared his way to a set of Khinchin type inequalities by establishing some rearrangement inequalities.

The type and cotype of classical Lorentz spaces $L_{p,q}$ were catalogued by J. Creekmore [1980], [1981] using ideas involving their structure as rearrangement invariant function spaces. His results will be cited in the Notes and Remarks to Chapter 16.

A. Kaminska and B. Turett [1990] showed that an Orlicz space L_Φ has cotype q if and only if its Young's function Φ satisfies $\Phi(\lambda s) \leq K \cdot \lambda^q \cdot \Phi(s)$ for some $K > 0$ and all $\lambda \geq 1$, $s \geq 0$. They also prove that L_Φ has type p if and only if Φ satisfies the Δ_2-condition and there is a constant K such that $\Phi(\lambda s) \geq K \cdot \lambda^{p^*} \cdot \Phi(s)$ for $s \geq 0$ whenever $\lambda \geq 1$. Other papers of related interest include A. Graślewicz, H. Hudzik and W. Orlicz [1986], H. Hudzik [1985], H. Hudzik and A. Kaminska [1985], H. Hudzik, A. Kaminska and W. Kurc [1987] and A. Kaminska and B. Turett [1987].

B. Beauzamy [1978] has given results relating the type of an intermediate space under real interpolation to the type of the ends of the scale. See Q. Xu [1987] for corresponding cotype results.

Our remarks and general observations from 11.5 through 11.13 belong to the standard material of the theory and can all be found in B. Maurey and G. Pisier [1976]. 11.14(a) is due to E. Dubinsky, A. Pełczyński and H.P. Rosenthal [1972] while 11.14(b) can be extracted from B. Maurey [1973/74a]. Lemma 11.15 goes back at least as far as G. Pisier [1978b], and both corollaries 11.16 and 11.17 can be gleaned from B. Maurey [1974a].

Corollary 11.18 is a variant of a deep result of H.P. Rosenthal [1973]; others will come up in 13.19 and 15.12. It originally served as an inspiration for B. Maurey's [1974a] factorization schemes and so for the theory of type and cotype.

12. Randomized Series and Almost Summing Operators

In the very first chapter, we used Khinchin's Inequality in the study of unconditionally convergent series. Type and cotype emerged in the last chapter as a means of understanding the limitations of the extension of Khinchin's Inequality to Rademacher sums in general Banach spaces. As a satisfying by-product of our study of type and cotype, we were able to upgrade many earlier results on summing operators. Now we return to basics and show how type and cotype almost (!) relate to unconditional summability.

Our path will be technical and at times tortuous, but it will ultimately lead to results of extraordinary beauty. For instance, we will see that

- up to isomorphism, Hilbert spaces are the only Banach spaces which are simultaneously of type 2 and cotype 2.

and that

- a Hahn - Banach Theorem, with uniform control of norms, is available for bounded bilinear forms on type 2 spaces.

Randomized Series

We begin with some terminology. Of course, a sequence (x_n) in a Banach space is *unconditionally summable* if and only if the Rademacher series $\sum_n r_n(t)x_n$ converges for *every* $t \in [0,1]$. We shall say that (x_n) is *almost unconditionally summable* if $\sum_n r_n(t)x_n$ converges for (Lebesgue) *almost all* $t \in [0,1]$. Although this concept is going to play a pivotal part in the theory of type and cotype, we shall need to work in a more general setting.

A *randomized series* in a Banach space X has the form

$$\sum_{k=1}^{\infty} \chi_k x_k$$

where the χ_k's are independent symmetric real-valued random variables on a probability space (Ω, Σ, P), and the x_k's are taken from X. It will be convenient to write $S_n = \sum_{k\leq n} \chi_k x_k$ for the corresponding partial sums. We say that the randomized series $\sum_k \chi_k x_k$ *converges almost surely* if $\big(S_n(\omega)\big)_n$ converges P-almost surely in X, that is, there is an event $E \in \Sigma$ with $P(E) = 1$ such that $\big(S_n(\omega)\big)_n$ converges for each $\omega \in E$.

It is handy to have some criteria for almost sure convergence.

12.1 Proposition: *Let $\sum_k \chi_k x_k$ be a randomized series in the Banach space X. The following statements are equivalent:*

(i) *$\sum_k \chi_k x_k$ converges almost surely.*

(ii) *For every $\varepsilon > 0$ and $\delta > 0$ there exists an $n_0 \in \mathbf{N}$ such that $P(\sup_{m>n} \|S_m - S_n\| \geq \varepsilon) \leq \delta$ whenever $n \geq n_0$.*

(iii) *For every $\varepsilon > 0$ and $\delta > 0$ there exists an $n_0 \in \mathbf{N}$ such that $\sup_{m>n} P(\|S_m - S_n\| \geq \varepsilon) \leq \delta$ whenever $n \geq n_0$.*

Proof. Evidently, (i)⇔(ii) and (ii)⇒(iii). To fill in the remaining implication we appeal to Lévy's Inequality (11.3):

$$\begin{aligned} P\big(\sup_{m>n} \|S_m - S_n\| \geq \varepsilon\big) &= \lim_{\ell\to\infty} P\big(\max_{n<m\leq n+\ell} \|S_m - S_n\| \geq \varepsilon\big) \\ &\leq 2 \cdot \sup_{\ell} P\big((\|S_{n+\ell} - S_n\| \geq \varepsilon\big). \end{aligned}$$

QED

Notice that the equivalence of (i) and (ii) has nothing to do with independence or symmetry; rather, (ii) is simply a useful reformulation of almost sure convergence for sums of random variables. The condition best suited for our discussions is (iii), and this is tantamount to stochastic convergence (alias convergence in probability) of (S_n); it *does* depend on the symmetry, though *not* the independence, of the χ_k's.

Rademacher Series

The randomized series that most interest us are the *Rademacher series*

$$\sum_k r_k x_k.$$

For these, the tail estimates of the last chapter (11.4) lead us to a powerful array of conditions equivalent to almost sure convergence. To gain access, we need an 'integral' version of Lévy's Inequality.

12.2 Contraction Principle: *Let $1 \leq p < \infty$ and consider the randomized sum $\sum_{k=1}^n \chi_k x_k$ in the Banach space X. Then, regardless of the choice of real numbers $a_1,\ldots, a_n$,*

$$\begin{aligned} &\Big(\int_\Omega \big\|\sum_{k\leq n} a_k\chi_k(\omega)x_k\big\|^p\, dP(\omega)\Big)^{1/p} \\ &\qquad \leq \big(\max_{k\leq n} |a_k|\big)\cdot\Big(\int_\Omega \big\|\sum_{k\leq n} \chi_k(\omega)x_k\big\|^p\, dP(\omega)\Big)^{1/p}. \end{aligned}$$

In particular, if A and B are subsets of $\{1,\ldots.,n\}$ such that $A \subset B$, then

$$\Big(\int_\Omega \big\|\sum_{k\in A} \chi_k(\omega)x_k\big\|^p\, dP(\omega)\Big)^{1/p} \leq \Big(\int_\Omega \big\|\sum_{k\in B} \chi_k(\omega)x_k\big\|^p\, dP(\omega)\Big)^{1/p}.$$

Proof. Homogeneity allows us to assume that $\max_{k\le n}|a_k| = 1$. We exploit this assumption by using the evenness and convexity of the functions

$$\mathbf{R} \longrightarrow \mathbf{R} : a \mapsto (\|x+ay\|^p + \|x-ay\|^p)^{1/p}$$

where $x, y \in X$. These functions must be increasing on $[0,1]$ and so, for any $a \in [-1,1]$,

$$(*) \qquad \|x+ay\|^p + \|x-ay\|^p \le \|x+y\|^p + \|x-y\|^p.$$

Notice that for any choice of $\varepsilon_1,\dots,\varepsilon_{n-1} = \pm 1$ the Rademacher function r_n takes values ± 1 equally often on each of the sets $\bigcap_{k<n}\{r_k = \varepsilon_k\}$. We now see that $(*)$ implies that for each $\omega \in \Omega$

$$\begin{aligned}
\int_0^1 \Big\|\sum_{k\le n} r_k(t)a_k\chi_k(\omega)x_k\Big\|^p dt
&= \sum_{\varepsilon_1,\dots,\varepsilon_{n-1}=\pm 1} 2^{-n}\cdot\Big[\Big\|\sum_{k<n}\varepsilon_k a_k\chi_k(\omega)x_k + a_n\chi_n(\omega)x_n\Big\|^p \\
&\qquad + \Big\|\sum_{k<n}\varepsilon_k a_k\chi_k(\omega)x_k - a_n\chi_n(\omega)x_n\Big\|^p\Big] \\
&\le \sum_{\varepsilon_1,\dots,\varepsilon_{n-1}=\pm 1} 2^{-n}\cdot\Big[\Big\|\sum_{k<n}\varepsilon_k a_k\chi_k(\omega)x_k + \chi_n(\omega)x_n\Big\|^p \\
&\qquad + \Big\|\sum_{k<n}\varepsilon_k a_k\chi_k(\omega)x_k - \chi_n(\omega)x_n\Big\|^p\Big] \\
&= \int_0^1 \Big\|\sum_{k<n} r_k(t)a_k\chi_k(\omega)x_k + r_n(t)\chi_n(\omega)x_n\Big\|^p dt.
\end{aligned}$$

Iterate and integrate over ω; the result is

$$\int_\Omega\int_0^1 \Big\|\sum_{k\le n} r_k(t)a_k\chi_k(\omega)x_k\Big\|^p dt\,dP(\omega) \le \int_\Omega\int_0^1 \Big\|\sum_{k\le n} r_k(t)\chi_k(\omega)x_k\Big\|^p dt\,dP(\omega).$$

To finish, we simply apply 11.2. QED

We are now set to dissect the almost sure convergence of Rademacher series.

12.3 Theorem: *The following are equivalent statements about a Rademacher series $\sum_k r_k x_k$ in the Banach space X:*

(i) *$\sum_k r_k x_k$ converges almost surely.*

(ii) *$\exp(c\cdot\|\sum_k r_k(\cdot)x_k\|)$ belongs to $L_1[0,1]$ for some $c > 0$.*

(iii) *$\sum_k r_k x_k$ converges in $L_p([0,1],X)$ for some, and then all, $0 < p < \infty$.*

Proof. (i)$\Rightarrow$(ii): The argument resembles that used to prove Kahane's Inequality.

It is convenient to write $S(t) := \sum_{k=1}^\infty r_k(t)x_k$ if the limit exists, and $S(t) := 0$ if not. Start by choosing $a > 0$ so that $m(\|S\| \ge a) \le 1/8$. Repeated application of 11.4 leads to

$$m(\|S\| \ge 2^\ell\cdot a) \le (1/4)\cdot 2^{-2^\ell} \quad (\ell = 0,1,2,\dots)$$

and so, if we partition $[0,1]$ into the events $A_0 := \{\|S\| < a\}$ and

$$A_\ell := \{2^{\ell-1}a \le \|S\| < 2^\ell a\} \quad (\ell = 1,2,3,...),$$

we can make the estimate

$$\int_0^1 \exp(c\cdot\|S(t)\|)\,dt = \sum_{\ell=0}^{\infty}\int_{A_\ell}\exp(c\cdot\|S(t)\|)\,dt \le \sum_{\ell=0}^{\infty}\exp(c\cdot a\cdot 2^\ell)\cdot m(A_\ell)$$
$$\le e^{ca} + (1/4)\cdot\sum_{\ell=0}^{\infty}\exp(c\cdot a\cdot 2^\ell)\cdot 2^{-2^{\ell-1}}.$$

The last expression, hence the first, is finite if we choose $0 < c < (\log 2)/(2a)$.
(ii)⇒(iii): It is enough to derive (iii) when $1 \le p < \infty$. If (ii) holds then, for almost all $t \in [0,1]$, $S(t) = \sum_{k=1}^{\infty} r_k(t)x_k$ exists and so $\lim_{n\to\infty}\|(S-S_n)(t)\|$ is zero. Vitali's Convergence Theorem will allow us to conclude that

$$\lim_{n\to\infty}\int_0^1 \|(S-S_n)(t)\|^p dt = 0.$$

To justify its use, first apply the hypothesis to notice that $\|S(\cdot)\|$ and all the $\|(S-S_n)(\cdot)\|$ are in $L_p[0,1]$ for each $1 \le p < \infty$, and then use the Contraction Principle to see that the $\|(S-S_n)(\cdot)\|^p$ form a bounded sequence in $L_1[0,1]$ and so are uniformly integrable.

(iii)⇒(i) is easy enough: just use Chebyshev's Inequality to derive condition (iii) of 12.1. QED

Actually, condition (ii) can be strengthened to read that $\exp(c\cdot\|S(\cdot)\|^2)$ belongs to $L_1[0,1]$ for any $c > 0$. However, this has no importance for us: we shall only be interested in condition (iii).

The almost unconditionally summable sequences (x_k) in a Banach space X plainly constitute a vector space,

$$Rad\,(X),$$

which becomes a Banach space under each of the norms

$$\|(x_n)\|_p := \Big(\int_0^1 \Big\|\sum_{n=1}^{\infty} r_n(t)x_n\Big\|^p\,dt\Big)^{1/p}$$

where $1 \le p < \infty$. (When $0 < p < 1$, $\|\cdot\|_p$ is only a p-norm.) Kahane's Inequality ensures that these quantities are all equivalent. For the sake of uniformity, we shall usually choose to equip $Rad\,(X)$ with $\|\cdot\|_2$; there are a few exceptions which will be specified explicitly.

$Rad\,(X)$ allows us to formulate type and cotype in familiar terms.

12.4 Proposition: *Let X be a Banach space and let $1 \le p \le 2 \le q \le \infty$.*

(a) *X has type p if and only if $\ell_p^{strong}(X)$ is a linear subspace of $Rad\,(X)$. In this case, the embedding is continuous and has norm $T_p(X)$.*

(b) *X has cotype q if and only if $Rad\,(X)$ is a linear subspace of $\ell_q^{strong}(X)$. In this case, the embedding is continuous and has norm $C_q(X)$.*

Notice that *Rad* (X) is isomorphic to $\ell_2^{strong}(X)$ when X is isomorphic to a Hilbert space. This 'ideal' vector-valued version of Khinchin's Inequality actually characterizes Hilbert spaces up to isomorphism, but to understand why we shall need to delve more deeply into the structure of *Rad* (X).

Almost Summing Operators

If (x_n) is almost unconditionally summable in the Banach space X, then for any $x^* \in B_{X^*}$,

$$\sum_n |\langle x^*, x_n\rangle|^2 = \int_0^1 \left|\left\langle \sum_n r_n(t)x_n, x^*\right\rangle\right|^2 dt \leq \int_0^1 \left\|\sum_n r_n(t)x_n\right\|^2 dt.$$

Thus *Rad* (X) is a linear subspace of $\ell_2^{weak}(X)$, and the norm of the embedding is at most one. The norm is actually equal to one: take a quick peek at the elements $(x, 0, 0, ...)$ of *Rad* (X).

Generally, the reverse implication holds only in finite dimensional spaces: this will emerge in 12.8 when we focus our attention on yet another variant of our central topic.

We say that a Banach space operator $u : X \to Y$ is *almost summing* if it transforms weakly 2-summable sequences in X into almost unconditionally summable sequences in Y, or, equivalently, if u gives rise to an operator

$$\hat{u} : \ell_2^{weak}(X) \longrightarrow Rad\,(X) : (x_n) \mapsto (ux_n).$$

The usual routine shows that the class

$$\Pi_{as}(X, Y)$$

of all almost summing operators $u : X \to Y$ is a linear subspace of $\mathcal{L}(X, Y)$ on which

$$\pi_{as}(u) := \|\hat{u}\|$$

defines a norm satisfying $\|u\| \leq \pi_{as}(u)$.

Again, we have encountered a concept of finitary nature: $u \in \mathcal{L}(X, Y)$ is almost summing if and only if there is a $c \geq 0$ such that

$$\left(\int_0^1 \left\|\sum_{k\leq n} r_k(t)ux_k\right\|^2 dt\right)^{1/2} \leq c \cdot \sup_{x^*\in B_{X^*}} \left(\sum_{k\leq n} |\langle x^*, x_k\rangle|^2\right)^{1/2}$$

for any finite collection $\{x_1, ..., x_n\}$ of vectors in X. The norm $\pi_{as}(u)$ is the smallest c that works in the above setup.

To some extent, our terminology is justified by the following observation:

12.5 Proposition: *Let $1 \leq p < \infty$, and let X and Y be Banach spaces. Every p-summing operator $u : X \to Y$ is almost summing with $\pi_{as}(u) \leq B_p \cdot \pi_p(u)$, where B_p is the constant from Khinchin's Inequality.*

Proof. It is enough to treat the case $p \geq 2$ (see 2.8).

If we let $u \in \Pi_p(X,Y)$, the Pietsch Domination Theorem 2.12 provides us with a probability measure $\mu \in \mathcal{C}(B_{X^*})^*$ such that

$$\|ux\| \leq \pi_p(u) \cdot \Big(\int_{B_{X^*}} |\langle x^*, x\rangle|^p d\mu(x^*)\Big)^{1/p}$$

for each $x \in X$. Choose $x_1,\dots,x_n$ in X and proceed as follows:

$$\begin{aligned}
\Big(\int_0^1 \big\|\sum_{k\leq n} r_k(t)ux_k\big\|^2 dt\Big)^{1/2} &\leq \Big(\int_0^1 \big\|u\big(\sum_{k\leq n} r_k(t)x_k\big)\big\|^p dt\Big)^{1/p} \\
&\leq \pi_p(u)\cdot\Big(\int_0^1\int_{B_{X^*}} \big|\sum_{k\leq n} r_k(t)\cdot\langle x^*, x_k\rangle\big|^p d\mu(x^*)\, dt\Big)^{1/p} \\
&= \pi_p(u)\cdot\Big(\int_{B_{X^*}}\int_0^1 \big|\sum_{k\leq n} r_k(t)\cdot\langle x^*, x_k\rangle\big|^p dt\, d\mu(x^*)\Big)^{1/p} \\
&\leq B_p\cdot\pi_p(u)\cdot\Big(\int_{B_{X^*}} \big(\sum_{k\leq n} |\langle x^*, x_k\rangle|^2\big)^{p/2} d\mu(x^*)\Big)^{1/p} \\
&\leq B_p\cdot\pi_p(u)\cdot\|(x_k)_1^n\|_2^{weak}.
\end{aligned}$$

QED

In the presence of cotype, we can say more. Here is a direct consequence of the definitions which, by 12.5 improves upon 11.13.

12.6 Proposition: *Let X and Y be Banach spaces, and suppose that Y has cotype q, with $2 \leq q < \infty$. Every almost summing operator $u : X \to Y$ is $(q,2)$-summing and $\pi_{q,2}(u) \leq C_q(Y)\cdot\pi_{as}(u)$.*

There is still more to come. In 12.29, we give a related characterization of the Banach spaces Y which satisfy $\Pi_{as}(\cdot\,,Y) \subset \Pi_{q,2}(\cdot\,,Y)$.

It is profitable to combine the last two propositions with 11.16:

12.7 Corollary: (a) *If the Banach space Y has cotype 2, then $\Pi_{as}(X,Y) = \Pi_r(X,Y)$ for every Banach space X and every $2 \leq r < \infty$.*

(b) *If the Banach spaces X and Y both have cotype 2, then $\Pi_{as}(X,Y) = \Pi_r(X,Y)$ for every $1 \leq r < \infty$.*

In all cases norms are equivalent, thanks to the Closed Graph Theorem. Notice that we have shown, in particular, that

- *the almost summing operators between Hilbert spaces are precisely the Hilbert - Schmidt operators.*

12.8 Remark: Here is another noteworthy fact:

- *The identity operator on a Banach space X is almost summing if and only if X is finite dimensional.*

We do not yet have all the tools to show why this is so, but we can give a good idea of what is going on. The missing ingredient is the Maurey - Pisier characterization of spaces having finite cotype: it will be presented in 14.1.

If X has finite cotype, there is no difficulty: Proposition 12.6 combines with the modified Dvoretzky - Rogers Theorem 10.5 to show that $id_X \in \Pi_{as}(X, X)$ forces X to be finite dimensional.

On the other hand, when X does not have finite cotype, the Maurey - Pisier Theorem 14.1, once proved, will enable us to say that, for any $\lambda > 1$ and any $N \in \mathbf{N}$, there is a λ-isomorphic copy of ℓ_∞^N inside X. But, for any $n \in \mathbf{N}$ and any $\lambda > 1$, ℓ_2^n is λ-isomorphic to a subspace of ℓ_∞^N for large enough N. Patching things together, we can arrange to find, for each $n \in \mathbf{N}$, a subspace E_n of X together with an isomorphism $u_n : \ell_2^n \to E_n$ such that $\|u_n\| \cdot \|u_n^{-1}\| \leq \lambda^2$. Thanks to 12.7, we know that $\pi_2(id_{\ell_2^n}) \leq K \cdot \pi_{as}(id_{\ell_2^n})$ for some $K > 0$ independent of n, and so

$$\begin{aligned} \sqrt{n} = \pi_2(id_{\ell_2^n}) &\leq K \cdot \pi_{as}(id_{\ell_2^n}) = K \cdot \pi_{as}(u_n^{-1} id_{E_n} u_n) \\ &\leq K \cdot \lambda^2 \cdot \pi_{as}(id_{E_n}) \leq K \cdot \lambda^2 \cdot \pi_{as}(id_X). \end{aligned}$$

This puts paid to any hope that id_X might be almost summing.

The preceding manipulations were based on some elementary, very useful, but as yet unproven facts. The following collection contains what we needed, and more.

12.9 Proposition: (a) $[\Pi_{as}, \pi_{as}]$ *is an injective, maximal Banach ideal.*

(b) $u \in \mathcal{L}(X, Y)$ *belongs to* $\Pi_{as}(X, Y)$ *if and only if* u^{**} *is in* $\Pi_{as}(X^{**}, Y^{**})$, *and then* $\pi_{as}(u) = \pi_{as}(u^{**})$.

(c) $u \in \mathcal{L}(X, Y)$ *is in* $\Pi_{as}(X, Y)$ *if and only if* $uv \in \Pi_{as}(\ell_2, Y)$ *for each* $v \in \mathcal{L}(\ell_2, X)$, *and then* $\pi_{as}(u) = \sup \pi_{as}(uv)$, *where the supremum is taken over all* $v \in \mathcal{L}(\ell_2, X)$ *with* $\|v\| \leq 1$.

(d) $u \in \mathcal{L}(X, Y)$ *is in* $\Pi_{as}(X, Y)$ *if and only if there is a constant* $c \geq 0$ *such that, for each* $n \in \mathbf{N}$ *and every* $v \in \mathcal{L}(\ell_2^n, X)$, *we have* $\pi_{as}(uv) \leq c \cdot \|v\|$. *Also,* $\pi_{as}(u)$ *is the smallest of all such constants* c.

These results are easily obtained by making appropriate modifications in the proofs of their counterparts on p-summing and (q, p)-summing operators.

In 12.8 we saw that $\Pi_{as}(X, X) = \mathcal{L}(X, X)$ characterizes X as finite dimensional. If we broaden our horizons and ask for pairs (X, Y) of Banach spaces such that $\Pi_{as}(X, Y) = \mathcal{L}(X, Y)$ the situation becomes more interesting. The case $X = \ell_1$ is particularly important. Grothendieck's Theorem 1.13 together with 12.5, for example, tells us that $\mathcal{L}(\ell_1, Y) = \Pi_{as}(\ell_1, Y)$ whenever Y is isomorphic to a Hilbert space. More generally, we may take Y to be any quotient of an $\mathcal{L}_\infty$-space which has finite cotype: the lifting property of ℓ_1 (see after 3.17) and 12.5 in combination with Theorem 11.14 leads to this conclusion.

The key issue in these affairs actually is type 2, and the next theorem is the heart of this chapter.

12.10 Theorem: *The following statements about a Banach space Y are equivalent:*

(i) *Y has type 2.*

(ii) $\mathcal{L}(\ell_1, Y) = \Pi_{as}(\ell_1, Y)$.

(iii) *For every index set I, $\mathcal{L}(\ell_1^I, Y) = \Pi_{as}(\ell_1^I, Y)$.*

(iv) *Regardless of the Banach space X, we have $\Gamma_1(X, Y) \subset \Pi_{as}(X, Y)$.*

As usual, most of these implications are easy. It is clear that (iv) implies (iii) which trivially implies (ii). To see that (ii) implies (i), first note that it is good enough to work with separable (or even finite dimensional) Y: almost summing operators and spaces of type 2 are objects whose definitions are finitary in nature. The advantage of this reduction is that it ensures the existence of a norm one surjection $q : \ell_1 \to Y$, and so, however we choose a finite collection of vectors $y_1, \ldots, y_n$ from Y, we can select $x_1, \ldots, x_n$ from ℓ_1 with $q(x_k) = y_k$ and $\|x_k\| \le 2 \cdot \|y_k\|$ for each $1 \le k \le n$. Then

$$\Big(\int_0^1 \Big\|\sum_{k\le n} r_k(t) y_k\Big\|^2 dt\Big)^{1/2} = \Big(\int_0^1 \Big\|\sum_{k\le n} r_k(t) q(x_k)\Big\|^2 dt\Big)^{1/2}$$
$$\le \pi_{as}(q) \cdot \|(x_k)\|_2^{weak} \le \pi_{as}(q) \cdot \|(x_k)\|_2^{strong} \le 2 \cdot \pi_{as}(q) \cdot \|(y_k)\|_2^{strong}.$$

We now set out to prove the remaining implication (i)$\Rightarrow$(iv). Once established, it will enable us to achieve the advertised main goals of this chapter.

At this stage it is important to abandon our strict attachment to Rademacher sums. To appreciate why, look back to (d) of Proposition 12.9. We can only hope for a good understanding of $\Pi_{as}(X, \cdot)$ if we can effectively handle the Rademacher averages $\big(\int_0^1 \|\sum_k r_k(t) u z_k\|^2 dt\big)^{1/2}$, where the z_k's are chosen from ℓ_2^n and u belongs to $\mathcal{L}(\ell_2^n, X)$. In particular, we need to have a good grip on such averages when the z_k's form an orthonormal basis of ℓ_2^n. The problem is that these averages are highly basis dependent.

It is easy to find examples, even in 2-dimensional spaces. Take u to be the formal identity $\ell_2^2 \to \ell_\infty^2$ and set $z_1 = (e_1 + e_2)/\sqrt{2}$ and $z_2 = (e_1 - e_2)/\sqrt{2}$. A straightforward calculation shows that $\int_0^1 \|\sum_{k\le 2} r_k(t) u z_k\|^2 dt = 2$, whereas $\int_0^1 \|\sum_{k\le 2} r_k(t) u e_k\|^2 dt = 1$.

We can rid ourselves of this problem – at a price. We shall need to replace Rademacher sums by Gaussian sums. Happily, the scenario is sufficiently similar for us to reap great profits.

Gaussian Variables

First, let us define our terms. A *Gaussian variable* is a (real, symmetric) random variable g on a probability space (Ω, Σ, P) whose distribution P_g has the form

$$P_g(B) = \frac{1}{\sigma\sqrt{2\pi}} \int_B e^{-t^2/(2\sigma^2)} dt$$

for all Borel sets B in $\mathbf{R}$. Here $\sigma > 0$ is a fixed number. It is traditional to call σ^2 the *variance* of g, and to say that g is *standard* when $\sigma = 1$.

It is common knowledge that the *p-th moments*

$$\Big(\int_\Omega |g(\omega)|^p dP(\omega)\Big)^{1/p} = \Big(\frac{1}{\sigma\sqrt{2\pi}} \cdot \int_{-\infty}^{\infty} |t|^p e^{-t^2/2\sigma^2} dt\Big)^{1/p}$$

of a Gaussian variable g exist for all $0 < p < \infty$, and that the square of the second moment is just g's variance.

It is even possible to compute these moments exactly. In fact, the p-th moment of a standard Gaussian variable is

$$m_p := \sqrt{2}\cdot\big(\Gamma((p+1)/2)/\sqrt{\pi}\big)^{1/p}.$$

Our main interest will be in *Gaussian sums*

$$\sum_{k=1}^{n} g_k x_k$$

where the x_k's are taken from a Banach space X and the g_k's are independent standard Gaussian variables on (Ω, Σ, P).

Let us first look at scalar-valued sums. One of the fundamental features of Gaussian variables is their stability under addition. More precisely, if $g_1,..., g_n$ are independent standard Gaussian variables on (Ω, Σ, P) and if $a_1,..., a_n$ are real numbers, then $\sum_{k\le n} a_k g_k$ is Gaussian with variance $\sum_{k\le n} a_k^2$.

This has a remarkable consequence: in quest of a Khinchin type result, we are rewarded with an identity rather than an inequality. To wit, if $\sigma = (\sum_{k\le n} a_k^2)^{1/2}$, then

$$\begin{aligned}\Big(\int_\Omega \big|\sum_{k\le n} a_k g_k(\omega)\big|^p dP(\omega)\Big)^{1/p} &= \Big(\frac{1}{\sigma\sqrt{2\pi}} \cdot \int_{-\infty}^{\infty} |t|^p e^{-t^2/(2\sigma^2)} dt\Big)^{1/p} \\ &= \sigma\cdot\Big(\frac{1}{\sqrt{2\pi}} \cdot \int_{-\infty}^{\infty} |s|^p e^{-s^2/2} ds\Big)^{1/p}\end{aligned}$$

for any $0 < p < \infty$. What this amounts to is that $(a_k) \mapsto m_p^{-1}\cdot\sum_k a_k g_k$ embeds ℓ_2 isometrically into $L_p(\Omega, \Sigma, P)$ for any $0 < p < \infty$.

The inconsiderate inconsistency of the Rademacher sums we examined above is not present with Gaussian sums. If X is a Banach space, $u : \ell_2^n \to X$ is an operator, and $e_1,..., e_n$ and $f_1,..., f_n$ are orthonormal bases for ℓ_2^n, then

$$\int_\Omega \big\|\sum_{k\le n} g_k(\omega) u e_k\big\|^2 dP(\omega) = \int_\Omega \big\|\sum_{k\le n} g_k(\omega) u f_k\big\|^2 dP(\omega)$$

whenever $g_1,..., g_n$ are independent standard Gaussian variables on (Ω, Σ, P).

To understand this, just realize that the operator $v : \ell_2^n \to \ell_2^n$ defined by $ve_k = f_k$ $(1 \le k \le n)$ has an orthogonal matrix representation $(v_{k\ell})_{k,\ell=1}^n$, and so

$$\int_\Omega \big\| \sum_{k\le n} g_k(\omega) u f_k \big\|^2 \, dP(\omega)$$
$$= (2\pi)^{-n/2} \cdot \int_{\mathbf{R}^n} \big\| \sum_{k\le n} t_k u f_k \big\|^2 \cdot \exp\Big(-\frac{1}{2}\sum_{k\le n} t_k^2\Big) \, dt_1 ... dt_n$$
$$= (2\pi)^{-n/2} \cdot \int_{\mathbf{R}^n} \big\| \sum_{\ell\le n} \big(\sum_{k\le n} v_{k\ell} t_k\big) u e_\ell \big\|^2 \cdot \exp\Big(-\frac{1}{2}\sum_{k\le n} t_k^2\Big) \, dt_1 ... dt_n.$$

Now, using the orthogonality of $(v_{k\ell})$, the change of variables $s_\ell = \sum_{k\le n} v_{k\ell} t_k$ $(1 \le \ell \le n)$ gives

$$\int_\Omega \big\| \sum_{k\le n} g_k(\omega) u f_k \big\|^2 \, dP(\omega)$$
$$= (2\pi)^{-n/2} \cdot \int_{\mathbf{R}^n} \big\| \sum_{\ell\le n} s_\ell u e_\ell \big\|^2 \cdot \exp\Big(-\frac{1}{2}\sum_{\ell\le n} s_\ell^2\Big) \, ds_1 ... ds_n$$
$$= \int_\Omega \big\| \sum_{\ell\le n} g_\ell(\omega) u e_\ell \big\|^2 \, dP(\omega).$$

Applications to Almost Summing Operators

From now on, we will make a running hypothesis in order to avoid stereotype repetitions:

- $g_1, g_2, g_3, ...$ *will henceforth denote a sequence of independent standard Gaussian variables on a fixed probability space* (Ω, Σ, P).

Our aim is to show that it makes no difference whether we use Rademacher sums or Gaussian sums to define almost summing operators. To begin the process, we establish a simple, but useful, comparison result.

12.11 Proposition: *Let $x_1, ..., x_n$ be vectors from a Banach space X. Then*

$$\Big(\int_0^1 \big\| \sum_{k=1}^n r_k(t) x_k \big\|^2 \, dt\Big)^{1/2} \le \frac{1}{m_1} \cdot \Big(\int_\Omega \big\| \sum_{k=1}^n g_k(\omega) x_k \big\|^2 \, dP(\omega)\Big)^{1/2}.$$

Proof. By 11.2, $\| \sum_{k\le n} r_k(\cdot) \cdot |g_k(\cdot)| x_k \|$ and $\| \sum_{k\le n} g_k(\cdot) x_k \|$ have the same distribution. Therefore

$$\Big(\int_0^1 \big\| \sum_{k\le n} r_k(t) x_k \big\|^2 dt\Big)^{1/2} = m_1^{-1} \cdot \Big(\int_0^1 \big\| \sum_{k\le n} r_k(t) \Big(\int_\Omega |g_k(\omega)| dP(\omega)\Big) x_k \big\|^2 dt\Big)^{1/2}$$
$$\le m_1^{-1} \cdot \Big(\int_0^1 \Big(\int_\Omega \big\| \sum_{k\le n} r_k(t) |g_k(\omega)| x_k \big\| \, dP(\omega)\Big)^2 dt\Big)^{1/2}$$
$$\le m_1^{-1} \cdot \Big(\int_0^1 \int_\Omega \big\| \sum_{k\le n} r_k(t) |g_k(\omega)| x_k \big\|^2 dP(\omega) \, dt\Big)^{1/2}$$
$$= m_1^{-1} \cdot \Big(\int_\Omega \big\| \sum_{k\le n} g_k(\omega) x_k \big\|^2 dP(\omega)\Big)^{1/2}.$$

QED

This is already enough to settle one implication in the main result of this section.

12.12 Theorem: *A Banach space operator $u : X \to Y$ is almost summing if and only if there is a constant c such that*

$$\Big(\int_\Omega \Big\| \sum_{k=1}^n g_k(\omega) u x_k \Big\|^2 \, dP(\omega)\Big)^{1/2} \le c \cdot \|(x_k)_1^n\|_2^{weak}$$

for every finite collection $\{x_1, ..., x_n\}$ of vectors from X.

For the other direction, we must establish a much more intricate relationship between Rademacher and Gaussian sums, and for this a wee bit of measure theory needs to be addressed.

Let

$$S^{n-1}$$

be the unit sphere in ℓ_2^n, and let

$$\lambda_n$$

be the normalized rotation invariant measure on S^{n-1}. Denote by

$$\mathcal{O}(n)$$

the compact topological group of all orthogonal transformations on ℓ_2^n, and let

$$o_n$$

be the normalized Haar measure on $\mathcal{O}(n)$.

12.13 Lemma: *Let $a \in S^{n-1}$, and let $\varphi_a : \mathcal{O}(n) \to S^{n-1}$ be defined by $\varphi_a(v) := v(a)$. Then φ_a is continuous and surjective, and the image measure of o_n under φ_a is λ_n.*

Proof. Only the statement about the image measure requires any explanation.

The image measure μ_a of o_n under φ_a is given by $\mu_a(B) = o_n(\varphi_a^{-1}(B))$ for any Borel set $B \subset S^{n-1}$. Evidently, $\mu_a(S^{n-1}) = 1$, so we just have to show that μ_a is rotation invariant. In other words, we need $\mu_a(B) = \mu_a(vB)$ for all $v \in \mathcal{O}(n)$ and all Borel sets $B \subset S^{n-1}$.

To this end, let $u, v \in \mathcal{O}(n)$ and let B be a Borel set in S^{n-1}. Noting that $v^{-1}\varphi_a(u) = v^{-1}u(a) = \varphi_a(v^{-1}u)$ we see that

$$u \in \varphi_a^{-1}(vB) \Leftrightarrow v^{-1}\big(\varphi_a(u)\big) \in B \Leftrightarrow \varphi_a(v^{-1}u) \in B \Leftrightarrow u \in v\big(\varphi_a^{-1}(B)\big).$$

It follows from the translation invariance of o_n that

$$o_n\big(\varphi_a^{-1}(vB)\big) = o_n\big(v\varphi_a^{-1}(B)\big) = o_n\big(\varphi_a^{-1}(B)\big). \qquad \text{QED}$$

With this in hand, we can now settle our issue:

12.14 Lemma: *Let X be a Banach space, and let $u : \ell_2^n \to X$ be an operator. Then*

$$\int_\Omega \Big\| \sum_{k=1}^n g_k(\omega) u e_k \Big\|^2 dP(\omega) = n \cdot \int_{S^{n-1}} \|u(a)\|^2 d\lambda_n(a)$$
$$= \int_{\mathcal{O}(n)} \int_0^1 \Big\| \sum_{k=1}^n r_k(t) u v e_k \Big\|^2 \, dt \, do_n(v).$$

Proof. Write $x_k = ue_k$ $(1 \le k \le n)$ and note first that

$$\int_\Omega \Big\| \sum_{k\le n} g_k(\omega) x_k \Big\|^2 dP(\omega)$$
$$= (2\pi)^{-n/2} \cdot \int_{\mathbf{R}^n} \Big\| \sum_{k\le n} t_k x_k \Big\|^2 \cdot \exp\Big(-\frac{1}{2} \sum_{k\le n} t_k^2\Big) \, dt_1 ... dt_n$$
$$= c \cdot \int_{S^{n-1}} \Big\| \sum_{k\le n} a_k x_k \Big\|^2 d\lambda_n(a) \cdot \int_0^\infty r^{n+1} e^{-r^2/2} dr,$$

where c is a constant depending only on n. The last equality comes from a change of variables from cartesian to polar coordinates. We conclude that

$$\int_\Omega \Big\| \sum_{k\le n} g_k(\omega) x_k \Big\|^2 dP(\omega) = \tilde{c} \cdot \int_{S^{n-1}} \|u(a)\|^2 d\lambda_n(a),$$

where $\tilde{c}$ depends only on n. To go further, we let $X = \ell_2^n$ and $u = id_{\ell_2^n}$. Then, thanks to our standard normalization of the g_k's,

$$n = \int_\Omega \Big\| \sum_{k\le n} g_k(\omega) e_k \Big\|_{\ell_2^n}^2 dP(\omega) = \tilde{c} \cdot \int_{S^{n-1}} \Big(\sum_{k\le n} a_k^2\Big)^{1/2} d\lambda_n(a) = \tilde{c},$$

and we have obtained the first identity.

To secure the second one, we appeal to Lemma 12.13. For almost all $t \in [0,1]$, $b(t) = n^{-1/2} \cdot \sum_{k\le n} r_k(t) e_k$ is in S^{n-1}. Thus

$$n \cdot \int_{S^{n-1}} \|ua\|^2 \, d\lambda_n(a) = n \cdot \int_{\mathcal{O}(n)} \|uvb(t)\|^2 \, do_n(v)$$
$$= \int_{\mathcal{O}(n)} \Big\| uv\Big(\sum_{k\le n} r_k(t) e_k\Big) \Big\|^2 \, do_n(v).$$

Integration with respect to t gives what we wanted. QED

Now it is easy to derive the announced characterization of almost summing operators.

Proof of Theorem 12.12. We have already observed that Proposition 12.11 takes care of one implication.

For the converse, take $u \in \Pi_{as}(X,Y)$ and select $x_1, ..., x_n \in X$. Define $w \in \mathcal{L}(\ell_2^n, X)$ by $we_k = x_k$ $(1 \le k \le n)$ and apply Lemma 12.14:

$$\Big(\int_\Omega \Big\| \sum_{k\le n} g_k(\omega) u x_k \Big\|^2 dP(\omega)\Big)^{1/2} = \Big(\int_{\mathcal{O}(n)} \int_0^1 \Big\| \sum_{k\le n} r_k(t) u w v e_k \Big\|^2 \, dt \, do_n(v)\Big)^{1/2}$$
$$\le \sup\big\{\pi_{as}(uwv) : v \in \mathcal{O}(n)\big\} \le \pi_{as}(u) \cdot \|w\| = \pi_{as}(u) \cdot \|(x_k)_1^n\|_2^{weak}.$$

QED

Denote by
$$\pi_\gamma(u)$$
the least of all constants c which work in Theorem 12.12. The proof of 12.12 shows that
$$m_1 \cdot \pi_{as}(u) \le \pi_\gamma(u) \le \pi_{as}(u).$$
Easily, π_γ is another ideal norm on Π_{as} – sometimes called the *γ-summing norm.* It has a beautiful and useful representation.

12.15 Theorem: *For every operator $u : \ell_2^n \to X$ and every orthonormal basis $\{z_1, \dots, z_n\}$ of ℓ_2^n, we have*
$$\pi_\gamma(u) = \Big(\int_\Omega \Big\| \sum_{k=1}^n g_k(\omega) u z_k \Big\|^2 dP(\omega)\Big)^{1/2}.$$

Proof. First, the very definition of π_γ gives
$$\Big(\int_\Omega \Big\| \sum_{k\le n} g_k(\omega) u z_k \Big\|^2 dP(\omega)\Big)^{1/2} \le \pi_\gamma(u).$$
Next, recall the rotational invariance property of Gaussian sums: for each $v \in \mathcal{O}(n)$,
$$\int_\Omega \Big\| \sum_{k\le n} g_k(\omega) u v z_k \Big\|^2 dP(\omega) = \int_\Omega \Big\| \sum_{k\le n} g_k(\omega) u z_k \Big\|^2 dP(\omega).$$
But every norm one operator $w : \ell_2^n \to \ell_2^n$ is a convex combination of members of $\mathcal{O}(n)$. So, for each such w,
$$\int_\Omega \Big\| \sum_{k\le n} g_k(\omega) u w z_k \Big\|^2 dP(\omega) \le \int_\Omega \Big\| \sum_{k\le n} g_k(\omega) u z_k \Big\|^2 dP(\omega).$$
It follows that if we select $x_1, \dots, x_n$ in ℓ_2^n and define the operator $w : \ell_2^n \to \ell_2^n$ via $w z_k = x_k$ $(1 \le k \le n)$, then
$$\begin{aligned}\Big(\int_\Omega \Big\| \sum_{k\le n} g_k(\omega) u x_k \Big\|^2 dP(\omega)\Big)^{1/2} &\le \|w\| \cdot \Big(\int_\Omega \Big\| \sum_{k\le n} g_k(\omega) u z_k \Big\|^2 dP(\omega)\Big)^{1/2} \\ &= \Big(\int_\Omega \Big\| \sum_{k\le n} g_k(\omega) u z_k \Big\|^2 dP(\omega)\Big)^{1/2} \cdot \|(x_k)_1^n\|_2^{weak}.\end{aligned}$$

Accordingly, $\pi_\gamma(u) \le \big(\int_\Omega \| \sum_{k\le n} g_k(\omega) u z_k \|^2 dP(\omega)\big)^{1/2}$, as we were predicting. QED

Before we proceed along our path to the proof of 12.10, we take time out to investigate a beguiling consequence of 12.15.

Recall from 7.11 that the adjoint ideal of $[\Gamma_2, \gamma_2]$ is $[\Delta_2, \Delta_2]$; a Banach space operator $u : X \to Y$ belongs to $\Delta_2(X, Y)$ if and only there exist a Banach space Z and operators $v : Z \to Y$, $w : X \to Z$ such that $u = vw$ and v^* and w are 2-summing. The norm $\Delta_2(u)$ is then the infimum of $\pi_2(v^*)\pi_2(w)$ taken over all such factorizations. A self-explanatory shorthand notation for this is $\Delta_2 = \Pi_2^d \circ \Pi_2$.

We are going to extend this as follows.

12.16 Corollary: *We have*

$$\Delta_2 = \Pi^d_{as} \circ \Pi_{as} :$$

a Banach space operator $u : X \to Y$ is 2-dominated if and only if there exist a Banach space Z and operators $v : Z \to Y$, $w : X \to Z$ with both v^* and w almost summing.

Moreover, the norm $\Delta(u)$ is equivalent to the infimum of $\pi_{as}(v^*)\pi_{as}(w)$ taken over all such factorizations.

Note that, by 12.5, we can also assert that

$$\Delta_2 = \Pi^d_r \circ \Pi_s \quad \text{for all } 2 \le r, s < \infty.$$

Proof. Since 12.5 shows that $\Pi_2 \subset \Pi_{as}$, we can infer that $\Delta_2 \subset \Pi^d_{as} \circ \Pi_{as}$.

To prove the converse, let $u \in \Pi^d_{as} \circ \Pi_{as}(X, Y)$ be given. Consider any factorization $u : X \overset{w}{\to} Z \overset{v}{\to} Y$ where v^* and w are almost summing operators. We are going to show that, no matter how we select finitely many vectors $x_1, \dots, x_n$ from X and $y^*_1, \dots, y^*_n$ from Y^*, we have

$$\sum_{k \le n} |\langle u x_k, y^*_k \rangle| \le \pi_\gamma(v^*) \cdot \pi_\gamma(w) \cdot \|(x_k)\|^{weak}_2 \cdot \|(y^*_k)\|^{weak}_2.$$

Since π_γ and π_{as} are equivalent, our claim will then follow from 9.7(ii).

It is no loss to assume that $\langle u x_k, y^*_k \rangle \ge 0$ for all $1 \le k \le n$; we can always replace each x_k by $\varepsilon_k x_k$ where $|\varepsilon_k| = 1$ and $\varepsilon_k \langle u x_k, y^*_k \rangle = |\langle u x_k, y^*_k \rangle|$. Define $w_1 : \ell^n_2 \to X$ via $w_1 e_k = x_k$ and $v_1 : Y \to \ell^n_2$ via $v^*_1 e_k = y_k$ $(1 \le k \le n)$. Then, by the orthonormality of the g_k's and 12.15,

$$\begin{aligned}
\sum_{k \le n} |\langle u x_k, y^*_k \rangle| &= \int_\Omega \Big\langle \sum_{k \le n} g_k(\omega) w w_1 e_k \,, \sum_{k \le n} g_k(\omega)(v_1 v)^* e_k \Big\rangle \, dP(\omega) \\
&\le \Big(\int_\Omega \Big\| \sum_{k \le n} g_k(\omega) w w_1 e_k \Big\|^2 dP(\omega) \Big)^{1/2} \cdot \Big(\int_\Omega \Big\| \sum_{k \le n} g_k(\omega)(v_1 v)^* e_k \Big\|^2 dP(\omega) \Big)^{1/2} \\
&= \pi_\gamma((v_1 v)^*) \cdot \pi_\gamma(w w_1) \le \pi_\gamma(v^*) \cdot \pi_\gamma(w) \cdot \|v_1\| \cdot \|w_1\| \\
&= \pi_\gamma(v^*) \cdot \pi_\gamma(w) \cdot \|(x_k)\|^{weak}_2 \cdot \|(y^*_k)\|^{weak}_2.
\end{aligned}$$

QED

Continuing along our route to the proof of 12.10, we extract a useful technical corollary from 12.15.

12.17 Corollary: *Let $u \in \mathcal{L}(\ell^m_2, \ell^n_2)$ and associate with it the matrix $(u_{ij})^m_{i=1}{}^n_{j=1}$. Then, for any vectors $x_1, \dots, x_m$ in the Banach space X, we have*

$$\Big(\int_\Omega \Big\| \sum_{j=1}^n g_j(\omega) \sum_{i=1}^m u_{ij} x_i \Big\|^2 dP(\omega) \Big)^{1/2} \le \|u\| \cdot \Big(\int_\Omega \Big\| \sum_{i=1}^m g_i(\omega) x_i \Big\|^2 dP(\omega) \Big)^{1/2}.$$

Proof. Just define $v \in \mathcal{L}(\ell_2^m, X)$ via $ve_i = x_i$ $(1 \le i \le m)$ and apply Theorem 12.15:

$$\Big(\int_\Omega \big\|\sum_{j\le n} g_j(\omega)\sum_{i\le m} u_{ij}x_i\big\|^2\,dP(\omega)\Big)^{1/2} = \Big(\int_\Omega \big\|\sum_{j\le n} g_j(\omega)vu^*e_j\big\|^2\,dP(\omega)\Big)^{1/2}$$
$$= \pi_\gamma(vu^*) \le \|u\|\cdot\pi_\gamma(v) = \|u\|\cdot\Big(\int_\Omega \big\|\sum_{i\le m} g_i(\omega)x_i\big\|^2\,dP(\omega)\Big)^{1/2}.$$

QED

One more ingredient, a refinement of Theorem 5.26, is needed for the proof of Theorem 12.10:

12.18 Lemma: *Let $1 \le p < \infty$ and let $\varepsilon > 0$. If X and Y are Banach spaces with Y finite dimensional, then every $u \in \mathcal{L}(X,Y)$ admits a finite representation*

$$u = \sum_{k=1}^N \lambda_k x_k^* \otimes y_k$$

where the x_k^*'s are taken from B_{X^*}, the y_k's in Y satisfy $\|(y_k)_{k=1}^N\|_{p^*}^{weak} \le 1$, and the scalars λ_k are such that $\big(\sum_{k\le N}|\lambda_k|^p\big)^{1/p} \le (1+\varepsilon)\cdot\iota_p(u)$.

Proof. We start by looking at a special case: $X = \ell_\infty^N$ for some $N \in \mathbf{N}$. Without loss of generality we may assume that $\iota_p(u) = 1$. Write $K = \{1,\dots,N\}$, so that $\ell_\infty^N = C(K)$, and appeal to 2.15 to find a probability measure μ on K along with a factorization

$$u : C(K) \xrightarrow{j_p} L_p(\mu) \xrightarrow{\tilde u} Y$$

such that $\|\tilde u\| = \pi_p(u) = \iota_p(u)$. Clearly, μ has the form $\mu = \sum_{k=1}^N \mu_k\delta_k$, where $\mu_k = \mu(\{k\})$, $1 \le k \le n$. We may of course assume that each μ_k is positive. Then $w : \ell_p^N \to L_p(\mu) : (\xi_k)_{k=1}^N \mapsto (\mu_k^{-1/p}\xi_k)_{k=1}^N$ is an isometric isomorphism, and $w^{-1}j_p$ is the diagonal operator $D_\lambda : \ell_\infty^N \to \ell_p^N$ induced by $\lambda = (\lambda_k)_{k=1}^N = (\mu_k^{1/p})_{k=1}^N$. Since $D_\lambda = \sum_{k=1}^N \lambda_k e_k \otimes e_k$ we get $u = \tilde u w D_\lambda = \sum_{k=1}^N \lambda_k e_k \otimes \tilde u w e_k$. As $e_k \in B_{\ell_1^N}$ for all k, and, by 2.2, $\|(\tilde u w e_k)_{k=1}^N\|_{p^*}^{weak} = \|\tilde u w\| \le 1$ and $\sum_{k\le N}|\lambda_k|^p$, we are done with the special case.

Now let X, Y be Banach spaces, with Y finite dimensional, and let $u : X \to Y$ be any operator. There is a probability measure μ and a factorization

$$u : X \xrightarrow{b} L_\infty(\mu) \xrightarrow{i_p} L_p(\mu) \xrightarrow{a} Y$$

such that $\|b\| = 1$ and $\|a\| \le \varrho\cdot\iota_p(u)$; here we let $\varrho := (1+\varepsilon)^{1/2}$.

Denote the kernel of ai_p by M. Being finite dimensional, $E := L_\infty(\mu)/M$ is the dual of some finite dimensional subspace F of $L_1(\mu)$. By 3.2, F is contained in a subspace G of $L_1(\mu)$ which has finite dimension, say N, and for which there exists an isomorphism $v : G \to \ell_1^N$ such that $\|v\| \le \varrho$ and $\|v^{-1}\| \le 1$. Moreover, by the proof of 3.2, G can be chosen norm one complemented in $L_1(\mu)$: let $P : L_1(\mu) \to G$ be the corresponding projection, and let $J : G \to L_1(\mu)$ be the canonical embedding. The operator $w := ai_pP^* : G \to Y$ satisfies $wJ^* = ai_p$.

By the first part, $wv^*: \ell_\infty^N \to Y$ has a representation $wv^* = \sum_{k=1}^N \lambda_k z_k \otimes y_k$, with $z_k \in B_{\ell_1^N}$ $(1 \le k \le N)$, $\|(y_k)_k\|_{p^*}^{weak} \le 1$ and $(\sum_k |\lambda_k|^p)^{1/p} \le \iota_p(wv^*)$. But

$$\iota_p(wv^*) = \iota_p(ai_pP^*v^*) \le \|a\|\cdot\iota_p(i_p)\cdot\|P^*\|\cdot\|v^*\| \le (1+\varepsilon)\cdot\iota_p(u)$$

and $x_k^* := b^*Jv^{-1}z_k \in B_{X^*}$ for each $1 \le k \le N$. Since

$$u = wv^*(v^{-1})^*J^*b = \sum_{k\le N} \lambda_k x_k^* \otimes y_k,$$

the proof is complete. QED

We can finally complete the proof of the fundamental Theorem 12.10.

Proof of Theorem 12.10. (i)⇒(iv): We must show that if Y has type 2 then, regardless of the Banach space X, every operator $u \in \Gamma_1(X, Y)$ is almost summing. By considering u^{**}, it suffices to consider the case where X is a space $L_1(\mu)$: after all, $T_2(Y^{**}) = T_2(Y)$.

Treading a well-worn path, we choose $x_1, \dots, x_n$ in X and define v in $\mathcal{L}(\ell_2^n, X)$ by $ve_k = x_k$ $(1 \le k \le n)$. We know from 12.18 that, given $\alpha > 1$, we can find a representation $v^* = \sum_{i\le N} \lambda_i \langle \cdot, a_i\rangle b_i$ where the a_i's are taken from $B_{X^{**}}$, the b_i's belong to ℓ_2^n and satisfy $\|(b_i)_1^N\|_2^{weak} \le 1$, and the λ_i's are scalars such that $(\sum_{i\le N} |\lambda_i|^2)^{1/2} \le \alpha\cdot\pi_2(v^*)$. Note that, by Theorem 3.7, $\pi_2(v^*) \le \kappa_G\cdot\|v^*\| = \kappa_G\cdot\|(x_k)_1^n\|_2^{weak}$, where κ_G is Grothendieck's constant. Of course, our real interest is in v, and, happily, we can write $k_X v = \sum_{i\le N} \lambda_i(\cdot\,|b_i)a_i$. Notice that the operator $w = \sum_{i\le N}(\cdot\,|b_i)e_i$ in $\mathcal{L}(\ell_2^n, \ell_2^N)$ satisfies $\|w\| = \|(b_i)_1^N\|_2^{weak} \le 1$.

Assemble all this information together:

$$\begin{aligned}
\Big(\int_\Omega \Big\|\sum_{k\le n} g_k(\omega)ux_k\Big\|_Y^2 dP(\omega)\Big)^{1/2} &= \Big(\int_\Omega \Big\|\sum_{k\le n} g_k(\omega)uve_k\Big\|_Y^2 dP(\omega)\Big)^{1/2} \\
&= \Big(\int_\Omega \Big\|\sum_{k\le n} g_k(\omega)\sum_{i\le N}(e_k|b_i)u^{**}(\lambda_i a_i)\Big\|_{Y^{**}}^2 dP(\omega)\Big)^{1/2} \\
&\le \|w^*\|\cdot\Big(\int_\Omega \Big\|\sum_{i\le N} g_i(\omega)u^{**}(\lambda_i a_i)\Big\|_{Y^{**}}^2 dP(\omega)\Big)^{1/2} \qquad \text{(by 12.17)} \\
&= \|w\|\cdot\Big(\int_\Omega\int_0^1 \Big\|\sum_{i\le N} r_i(t)g_i(\omega)u^{**}(\lambda_i a_i)\Big\|_{Y^{**}}^2 dt\,dP(\omega)\Big)^{1/2} \qquad \text{(by 11.2)} \\
&\le T_2(Y^{**})\cdot\Big(\int_\Omega \sum_{i\le N} \|g_i(\omega)u^{**}(\lambda_i a_i)\|_{Y^{**}}^2 dP(\omega)\Big)^{1/2} \\
&= T_2(Y)\cdot\Big(\sum_{i\le N} \|u^{**}(\lambda_i a_i)\|_{Y^{**}}^2\cdot\int_\Omega |g_i(\omega)|^2 dP(\omega)\Big)^{1/2} \\
&= T_2(Y)\cdot\Big(\sum_{i\le N} \|u^{**}(\lambda_i a_i)\|_{Y^{**}}^2\Big)^{1/2} \le T_2(Y)\cdot\|u\|\cdot\Big(\sum_{i\le N}|\lambda_i|^2\Big)^{1/2} \\
&\le T_2(Y)\cdot\|u\|\cdot\alpha\cdot\kappa_G\cdot\|(x_k)_1^n\|_2^{weak}.
\end{aligned}$$

The theorem is proved. QED

Some Consequences

Our work leads rapidly to some striking results.

12.19 Kwapień's Theorem: *Suppose that X is a Banach space of type 2 and that Y is a Banach space of cotype 2. Then*

$$\mathcal{L}(X,Y) = \Gamma_2(X,Y).$$

In other words, every operator from X to Y factors through a Hilbert space.

Proof. There is a quotient map $q : \ell_1^I \to X$ for some index set I. Since X has type 2, the easy part of Theorem 12.10 tells us that q is almost summing. It follows that if $u \in \mathcal{L}(X,Y)$, then $uq \in \Pi_{as}(\ell_1^I, Y)$. But Y has cotype 2, and so, using Proposition 12.6, $uq \in \Pi_2(\ell_1^I, Y)$. Consequently uq, and thus u, is 2-factorable. QED

In 17.12, we shall see that for a large class of Banach spaces it is possible to relax the condition on X by just requiring that X^* has cotype 2. Given this, we can assert, for example, that $\Gamma_\infty(\cdot\,, Y) \subset \Gamma_2(\cdot\,, Y)$ whenever Y has cotype 2.

12.20 Corollary: *The only Banach spaces which simultaneously have type 2 and cotype 2 are the isomorphic copies of Hilbert spaces.*

Another way to put this is to say that the isomorphs of Hilbert spaces are the only Banach spaces in which a vector-valued Khinchin Inequality always holds.

We are almost ready for another advertised highlight of this chapter: a powerful and exquisite improvement of Kwapień's Theorem. It is based on the general extension scheme discussed in Chapter 7 (see 7.7 - 7.9) and on the following simple consequence of Theorem 12.10.

12.21 Corollary: *If X is a Banach space of type 2, then for any Banach space Z,*

$$\Pi_2^d(Z,X) \subset \Pi_{as}(Z,X).$$

Moreover, whenever $u \in \Pi_2^d(Z,X)$, we have $\pi_{as}(u) \leq \kappa_G \cdot T_2(X) \cdot \pi_2(u^*)$.

Proof. The factorization scheme 2.16 informs us that if $u \in \Pi_2^d(Z,X)$, then $u \in \Gamma_1(Z,X)$. Theorem 12.10 now settles the issue. QED

Actually, the condition $\Pi_2^d(\cdot\,, X) \subset \Pi_{as}(\cdot\,, X)$ also characterizes type 2 spaces, but we do not need this.

12.22 Maurey's Extension Theorem: *Suppose that X has type 2 and that Y has cotype 2. Then there exists a constant $C > 0$ which depends only on X and Y such that, whenever Z is a subspace of X, every operator $u \in \mathcal{L}(Z,Y)$ admits an extension $\tilde{u} : X \to Y$ which belongs to $\Gamma_2(X,Y)$ and satisfies $\gamma_2(\tilde{u}) \leq C \cdot \|u\|$.*

Note that Kwapień's Theorem is a particular case of this result!

Proof. To get us on our way, we recall the remarks on Γ_2 in Chapter 7 (particularly 7.9) and realize that it is enough to show that whenever $(z_i)_1^m$ and $(x_j)_1^n$ are finite families in Z and X respectively, which satisfy

$$\sum_{i\le m} |\langle x^*, z_i\rangle|^2 \le \sum_{j\le n} |\langle x^*, x_j\rangle|^2 \tag{1}$$

for all $x^* \in X^*$, then

$$\sum_{i\le m} \|uz_i\|^2 \le C^2\cdot\|u\|^2\cdot\sum_{j\le n} \|x_j\|^2 \tag{2}$$

for some constant C depending only on X and Y.

Fix two such families $(z_i)_1^m$ and $(x_j)_1^n$ and use them to define operators $a : \ell_2^m \to Z$ via $ae_i = z_i$ $(1 \le i \le m)$ and $b : \ell_2^n \to X$ via $be_j = x_j$ $(1 \le j \le n)$. Writing $j : Z \to X$ for the natural embedding, we notice that (1) just says that $\|a^*j^*x^*\| \le \|b^*x^*\|$ for each $x^* \in X^*$. In other words, since b^*'s range is a Hilbert space, there is an operator $\lambda : \ell_2^n \to \ell_2^m$ such that $\|\lambda\| \le 1$ and

$$\lambda b^* = a^*j^* .$$

Our path to (2) is now clear. Writing $C := \kappa_G\cdot T_2(X)\cdot C_2(Y)$ and using 12.6, the injectivity of Π_{as} and 12.21, we have

$$\begin{aligned}\Big(\sum_{i\le m}\|uz_i\|^2\Big)^{1/2} &= \Big(\sum_{i\le m}\|uae_i\|^2\Big)^{1/2} \le \pi_2(ua) \le C_2(Y)\cdot\pi_{as}(ua)\\ &\le C_2(Y)\cdot\|u\|\cdot\pi_{as}(a) = C_2(Y)\cdot\|u\|\cdot\pi_{as}(ja) \le C\cdot\|u\|\cdot\pi_2^d(ja)\\ &\le C\cdot\|u\|\cdot\pi_2(a^*j^*) = C\cdot\|u\|\cdot\pi_2(\lambda b^*) \le C\cdot\|u\|\cdot\pi_2(b^*).\end{aligned}$$

To finish, we simply need to know that $\pi_2(b^*) \le (\sum_{j\le n}\|x_j\|^2)^{1/2}$, and this is elementary:

$$\begin{aligned}\pi_2(b^*) &= \sup\Big\{\Big(\sum_{\ell\le N}\|b^*x_\ell^*\|_{\ell_2^n}^2\Big)^{1/2} : x_1^*,\ldots,x_N^* \in X^*,\ \|(x_\ell^*)_1^N\|_2^{weak} \le 1\Big\}\\ &= \sup\Big\{\Big(\sum_{\ell\le N}\sum_{j\le n}|(b^*x_\ell^*|e_j)|^2\Big)^{1/2} : \|(x_\ell^*)_1^N\|_2^{weak} \le 1\Big\}\\ &= \sup\Big\{\Big(\sum_{\ell\le N}\sum_{j\le n}|\langle x_\ell^*, x_j\rangle|^2\Big)^{1/2} : \|(x_\ell^*)_1^N\|_2^{weak} \le 1\Big\}\\ &= \sup\Big\{\Big(\sum_{j\le n}\|x_j\|^2\cdot\sum_{\ell\le N}\Big|\Big\langle x_\ell^*, \frac{x_j}{\|x_j\|}\Big\rangle\Big|^2\Big)^{1/2} : \|(x_\ell^*)_1^N\|_2^{weak} \le 1\Big\}\\ &\le \Big(\sum_{j\le n}\|x_j\|^2\Big)^{1/2}.\end{aligned}$$

QED

Maurey's Extension Theorem has several notable consequences; we cite but three of them.

12.23 Corollary: *Let X and Y be Banach spaces of type 2. Then there is a constant $C > 0$ such that, regardless of the subspaces X_0 of X and Y_0 of Y, every bounded bilinear form $\beta_0 : X_0 \times Y_0 \to \mathbf{K}$ extends to a bounded bilinear form $\beta : X \times Y \to \mathbf{K}$ with $\|\beta\| \leq C \cdot \|\beta_0\|$.*

Proof. By 11.10, Y_0^* has cotype 2, and so by Kwapień's Theorem the operator $b_0 : X_0 \to Y_0^* : x_0 \mapsto \beta_0(x_0, \cdot)$ admits a factorization $b_0 : X_0 \xrightarrow{a_2} H \xrightarrow{a_1} Y_0^*$ for some Hilbert space H. This splitting allows us to apply Maurey's Extension Theorem twice: a_2 has an extension $b_2 \in \mathcal{L}(X, H)$ whereas, by duality, a_1 admits a lifting $b_1 \in \mathcal{L}(H, Y^*)$. The bilinear form $\beta : X \times Y \to \mathbf{K} : (x, y) \mapsto \langle b_2 x, b_1^* y \rangle$ extends β_0. All norms can be controlled nicely: β is bounded with $\|\beta\| \leq C \cdot \|\beta_0\|$ where C depends only on X and Y. QED

12.24 Corollary: *Let X be a Banach space of type 2 and let Y be a subspace of X that is isomorphic to a Hilbert space. Then Y is complemented in X.*

This is very easy: just apply Maurey's Extension Theorem to id_Y!

The last corollary is of interest even when $X = L_p[0,1]$ for $2 < p < \infty$. In this case it is a renowned result of Kadets and Pełczyński.

Here is another application of the work of Kwapień and Maurey: a pleasing complement to Corollary 10.12.

12.25 Corollary: *Suppose that $1 < p \leq 2 \leq q < \infty$ and that $p \leq r \leq q$. Let X be a subspace of an $\mathcal{L}_q$-space, and let Y be a quotient of an $\mathcal{L}_p$-space. Then every operator $u \in \mathcal{L}(X, Y)$ factors through an $L_r(\mu)$-space.*

We cannot resist one more curious consequence, for which we recall Pitt's Theorem: if $1 \leq p < q < \infty$, then every operator $u \in \mathcal{L}(\ell_q, \ell_p)$ is compact, and even approximable by finite rank operators in the uniform operator norm. Maurey's Extension Theorem entails that, when $2 < q < \infty$, every operator from a subspace X of ℓ_q to a cotype 2 space Y contains an operator $u \in \mathcal{L}(\ell_q, \ell_2)$ as a factor and so must itself be approximable by finite rank operators – even when neither X nor Y enjoys the approximation property.

Gaussian Type and Cotype

Much of our work in this chapter has been centred on the interchangeability of Rademacher sums with their more flexible Gaussian counterparts. We haven't yet looked at the effect of such replacements within the definition of type and cotype.

Let us say that a Banach space X has *Gaussian type* p $(1 \leq p \leq 2)$ if there is a constant $c \geq 0$ such that

$$\Big(\int_\Omega \Big\| \sum_{k=1}^n g_k(\omega) x_k \Big\|^2 \, dP(\omega) \Big)^{1/2} \leq c \cdot \Big(\sum_{k=1}^n \|x_k\|^p \Big)^{1/p}$$

for every finite collection $\{x_1,\ldots,x_n\}$ of vectors from X. We denote the least of all such constants c by

$$T_p^\gamma(X).$$

Likewise, we say that a Banach space X has *Gaussian cotype* q $(2\le q<\infty)$ if there is a constant $c\ge 0$ such that

$$\Big(\sum_{k=1}^n \|x_k\|^q\Big)^{1/q} \le c\cdot\Big(\int_\Omega \Big\|\sum_{k=1}^n g_k(\omega)x_k\Big\|^2\,dP(\omega)\Big)^{1/2}$$

whatever the finite collection $\{x_1,\ldots,x_n\}$ in X. The least such c is denoted by

$$C_q^\gamma(X).$$

Any attempt to develop these concepts systematically would largely be pointless: the notions of Gaussian type p and (Rademacher) type p $(1\le p\le 2)$ are equivalent up to constants, and so are the notions of Gaussian cotype q and (Rademacher) cotype q $(2\le q<\infty)$.

The type case is fairly straightforward:

12.26 Theorem: *Let $1\le p\le 2$. A Banach space X has type p if and only if it has Gaussian type p; in this case*

$$m_1\cdot T_p(X) \le T_p^\gamma(X) \le m_{2p}\cdot T_p(X).$$

Proof. Referring back to Proposition 12.11 we see at once that if X has Gaussian type p, then it has type p, with $m_1\cdot T_p(X)\le T_p^\gamma(X)$.

For the other direction, we call on Lemma 11.2. Given any finite collection $\{x_1,\ldots,x_n\}$ in a type p space X, we have

$$\begin{aligned}
\Big(\int_\Omega \Big\|\sum_{k\le n} g_k(\omega)x_k\Big\|^2 dP(\omega)\Big)^{1/2} &= \Big(\int_\Omega\int_0^1 \Big\|\sum_{k\le n} r_k(t)g_k(\omega)x_k\Big\|^2 dt\,dP(\omega)\Big)^{1/2}\\
&\le T_p(X)\cdot\Big(\int_\Omega\Big(\sum_{k\le n}\|g_k(\omega)x_k\|^p\Big)^{2/p}dP(\omega)\Big)^{1/2}\\
&\le T_p(X)\cdot\Big(\int_\Omega\Big(\sum_{k\le n}\|g_k(\omega)x_k\|^p\Big)^{2}dP(\omega)\Big)^{1/(2p)}\\
&= T_p(X)\cdot\Big(\int_\Omega\sum_{k,\ell\le n}|g_k(\omega)|^p\cdot|g_\ell(\omega)|^p\cdot\|x_k\|^p\cdot\|x_\ell\|^p dP(\omega)\Big)^{1/(2p)}\\
&\le T_p(X)\cdot\Big(\sum_{k,\ell\le n}\|g_k\|_{2p}^p\cdot\|g_\ell\|_{2p}^p\cdot\|x_k\|^p\cdot\|x_\ell\|^p\Big)^{1/(2p)}\\
&= m_{2p}\cdot T_p(X)\cdot\Big(\sum_{k\le n}\|x_k\|^p\Big)^{1/p}.
\end{aligned}$$

QED

Proposition 12.11 also tells us that if a Banach space X has cotype q, it must have Gaussian cotype q as well; moreover, $m_1\cdot C_q^\gamma(X)\le C_q(X)$. Again, the converse implication is true, but for the time being we confine ourselves to a result which is formally slightly weaker.

12.27 Theorem: *Suppose that the Banach space X has finite cotype r. Then, for any $s > r$ and any choice of finitely many vectors $x_1,\ldots,x_n$ from X,*

$$\Big(\int_\Omega \Big\|\sum_{k\le n} g_k(\omega)x_k\Big\|^2 dP(\omega)\Big)^{1/2} \le C_r(X)\cdot m_s\cdot\Big(\int_0^1 \Big\|\sum_{k\le n} r_k(t)x_k\Big\|^2 dt\Big)^{1/2}.$$

Proof. To get a handle on the Rademacher sum, fix $x_1,\ldots,x_n$ in X and define $u \in \mathcal{L}(\ell_\infty^n, L_2(X))$ by $ue_k = r_k(\cdot)x_k$ $(1 \le k \le n)$; here $L_2(X)$ is $L_2([0,1],X)$. A moment's reflection in the light of the Contraction Principle (12.2) reveals that $\|u\| = (\int_0^1 \|\sum_{k\le n} r_k(t)x_k\|^2 dt)^{1/2}$.

Conveniently, Theorem 11.12 assures us that when X has cotype r, so does $L_2(X)$; moreover for $s \ge r$ it is true that $C_s(L_2(X)) = C_s(X)$. A quick perusal of Theorem 11.14 and its proof shows that

$$\pi_s(u) \le C_s(X)\cdot\|u\| \le C_r(X)\cdot\|u\|$$

for each $s \ge r$. Since we can write $\ell_\infty^n = \mathcal{C}(K)$ where $K = \{1,\ldots,n\}$, Pietsch's Domination Theorem (2.12) provides a probability measure $\mu \in \mathcal{C}(K)^*$ such that

$$\|u(a)\|_{L_2(X)} \le C_r(X)\cdot\|u\|\cdot\Big(\int_K |a_k|^s d\mu(k)\Big)^{1/s} \quad \text{for all } a = (a_k)_{k\le n} \in \ell_\infty^n.$$

Necessarily, μ has the form $\mu = \sum_{k=1}^n \lambda_k\delta_k$ where each λ_k is non-negative and $\sum_{k\le n}\lambda_k = 1$, and so

$$\|u(a)\|_{L_2(X)} \le C_r(X)\cdot\|u\|\cdot\Big(\sum_{k\le n}\lambda_k|a_k|^s\Big)^{1/s} \quad \text{for all } a \in \ell_\infty^n.$$

The symmetry of the Gaussian variables allows us to apply Lemma 11.2 again:

$$\begin{aligned}
\Big(\int_\Omega \Big\|\sum_{k\le n} g_k(\omega)x_k\Big\|^2 dP(\omega)\Big)^{1/2} &= \Big(\int_\Omega\int_0^1 \Big\|\sum_{k\le n} r_k(t)g_k(\omega)x_k\Big\|^2 dt\, dP(\omega)\Big) \\
&= \Big(\int_\Omega \Big\| u\big((g_k(\omega))_{k=1}^n\big)\Big\|_{L_2(X)}^2 dP(\omega)\Big)^{1/2} \\
&\le C_r(X)\cdot\|u\|\cdot\Big(\int_\Omega\Big(\sum_{k\le n}\lambda_k|g_k(\omega)|^s\Big)^{2/s} dP(\omega)\Big)^{1/2} \\
&\le C_r(X)\cdot\|u\|\cdot\Big(\sum_{k\le n}\lambda_k\int_\Omega |g_k(\omega)|^s dP(\omega)\Big)^{1/s} = C_r(X)\cdot m_s\cdot\|u\| \\
&= C_r(X)\cdot m_s\cdot\Big(\int_0^1 \Big\|\sum_{k\le n} r_k(t)x_k\Big\|^2 dt\Big)^{1/2}.
\end{aligned}$$

QED

12.28 Corollary: *Suppose that the Banach space X has finite cotype and let $2 \le q < \infty$. Then X has cotype q if and only if X has Gaussian cotype q.*

The hypothesis can be relaxed: finite cotype can be replaced by finite Gaussian cotype. So, when $2 \le q < \infty$, a Banach space has cotype q if and only if it has Gaussian cotype q. At the moment, this is still beyond our grasp.

We must wait until 14.1 when we prove the Maurey - Pisier characterization of spaces having finite cotype, already alluded to in 12.8. The line of attack is first to realize that a Banach space X of finite Gaussian cotype cannot contain the ℓ_∞^n's uniformly, and then to appeal to Theorem 14.1 to deduce that X must have finite (Rademacher) cotype.

To conclude this section, we return to the train of thought begun in 12.6.

12.29 Proposition: *Let $2 \le q < \infty$. The following are equivalent statements about a Banach space X.*

(i) *Regardless of the Banach space Z, every almost summing operator from Z to X is $(q,2)$-summing.*

(ii) *Every almost summing operator from ℓ_2 to X is $(q,2)$-summing.*

(iii) *There is a constant c such that, for each $n \in \mathbf{N}$ and $u \in \mathcal{L}(\ell_2^n, X)$, we have $\pi_{q,2}(u) \le c\cdot\pi_\gamma(u)$.*

(iv) *X has (Gaussian) cotype q.*

Proof. (i)$\Rightarrow$(ii) is obvious, and the fact that (ii) implies (iii) is a simple consequence of the Closed Graph Theorem. The key to the other implications is Theorem 12.15.

Suppose that (iii) holds, and choose $x_1,\dots,x_n$ from X. In the usual way, define $u \in \mathcal{L}(\ell_2^n, X)$ via $ue_k = x_k$ $(1 \le k \le n)$. Then

$$\Big(\sum_{k\le n} \|x_k\|^q\Big)^{1/q} \le \pi_{q,2}(u) \le c\cdot\pi_\gamma(u) = c\cdot\Big(\int_\Omega \Big\|\sum_{k\le n} g_k(\omega)x_k\Big\|^2 dP(\omega)\Big)^{1/2}.$$

We have thus arrived at (iv). Continuing from here, we select Z and $u \in \Pi_{as}(Z,Y)$ and, choosing $z_1,\dots,z_n \in Z$, we define $v \in \mathcal{L}(\ell_2^n, Z)$ standardly by $ve_k = z_k$ $(1 \le k \le n)$. Again by 12.15,

$$\Big(\sum_{k\le n}\|uz_k\|^q\Big)^{1/q} \le C_q^\gamma(X)\cdot\Big(\int_\Omega \Big\|\sum_{k\le n} g_k(\omega)uz_k\Big\|^2 dP(\omega)\Big)^{1/2}$$
$$= C_q^\gamma(X)\cdot\pi_\gamma(uv) \le C_q^\gamma(X)\cdot\pi_\gamma(u)\cdot\|v\| = C_q^\gamma(X)\cdot\pi_\gamma(u)\cdot\|(z_k)_1^n\|_2^{weak}.$$

We have come full circle. QED

The Maurey - Rosenthal Theorem

Kwapień's Theorem 12.19 informs us that every operator from a type 2 space to a cotype 2 space factors through a Hilbert space. In the special case that the target space is $L_1[0,1]$, much more can be said. This is a consequence of the next result.

12.30 Maurey-Rosenthal Theorem: *Let X be a Banach space and let $v : X \to L_1[0,1]$ be an operator. Suppose that for any finite collection of vectors $x_1,\ldots,x_n \in X$,*

$$\left\| \Big(\sum_{k\le n} |vx_k|^2\Big)^{1/2} \right\|_{L_1} \le \Big(\sum_{k\le n} \|x_k\|^2\Big)^{1/2}.$$

Then v has a factorization $v : X \xrightarrow{\hat{v}} L_2[0,1] \xrightarrow{M_g} L_1[0,1]$, where the operator $\hat{v}$ has norm ≤ 1 and M_g is the multiplication operator $L_2[0,1] \to L_1[0,1] : f \mapsto f\cdot g$ induced by some $g \in L_2[0,1]$.

Before engaging in the proof, we pause to emphasize that *when X has type 2, all operators $v : X \to L_1[0,1]$ can be scaled to satisfy the hypothesis above.* Indeed, given $x_1,\ldots,x_n \in X$, we can use Khinchin's Inequality to obtain

$$\begin{aligned}\int_0^1 \Big(\sum_{k\le n} |vx_k(t)|^2\Big)^{1/2} dt &\le A_1^{-1}\cdot \int_0^1\int_0^1 \Big|\sum_{k\le n} r_k(s)vx_k(t)\Big|\, ds\, dt \\ &= A_1^{-1}\cdot \int_0^1 \Big\|\sum_{k\le n} r_k(s)vx_k\Big\|_{L_1} ds \le A_1^{-1}\cdot\|v\|\cdot \int_0^1 \Big\|\sum_{k\le n} r_k(s)x_k\Big\|_X ds \\ &\le A_1^{-1}\cdot\|v\|\cdot\Big(\int_0^1 \Big\|\sum_{k\le n} r_k(s)x_k\Big\|_X^2 ds\Big)^{1/2} \le A_1^{-1}\cdot\|v\|\cdot T_2(X)\cdot\Big(\sum_{k\le n}\|x_k\|^2\Big)^{1/2}.\end{aligned}$$

Proof. The desired form of the factorization forces our hand. We must unearth some $g \in B_{L_2[0,1]}$ such that

$$\hat{v} : X \longrightarrow L_2[0,1] : x \mapsto \frac{vx}{g}$$

is an operator with norm at most one. In other words, we must show that

$$F_x(g) := \int_0^1 \frac{|vx(t)|^2}{|g(t)|^2}\, dt \;-\; \|x\|^2 \;\le 0$$

for every $x \in X$. We shall make our choice g from the 'positive ball'

$$K := \big\{h \in B_{L_2[0,1]} : h \ge 0 \text{ a.e.}\big\},$$

a weakly compact convex subset of $L_2[0,1]$. The hard work will be done by Ky Fan's Lemma 9.10.

Inspired by the spirit of the Pietsch Domination Theorem 2.12, given any finite family $x_1,\ldots,x_n$ in X we define a measure $\mu_{x_1,\ldots,x_n} \in C[0,1]^*$ by

$$\langle \mu_{x_1,\ldots,x_n}, f\rangle := \int_0^1 f(t)\cdot\Big(\sum_{k\le n} |(vx_k)(t)|^2\Big)\, dt$$

for each $f \in C[0,1]$. Next, we specify $F_{x_1,\ldots,x_n} : K \to (-\infty,\infty]$ by

$$F_{x_1,\ldots,x_n}(h) := \int_0^1 |h(t)|^{-2} d\mu_{x_1,\ldots,x_n}(t) \;-\; \sum_{k\le n}\|x_k\|^2.$$

Thanks to Fatou's Lemma, each $F_{x_1,\ldots,x_n}$ is norm lower semicontinuous. These functions are also convex, and so they are weakly lower semicontinuous. What is more, if we set $c := \| \left(\sum_{k\le n} |vx_k|^2\right)^{1/2} \|_{L_1}^{-1/2}$ and define $h_{x_1,\dots,x_n} \in K$ by

$$h_{x_1,\dots,x_n}(t) := c\cdot\Big(\sum_{k\le n} |vx_k(t)|^2\Big)^{1/2},$$

then, as one might expect,

$$F_{x_1,\dots,x_n}(h_{x_1,\dots,x_n}) = \Big\| \Big(\sum_{k\le n} |vx_k|^2\Big)^{1/2} \Big\|_{L_1} - \sum_{k\le n} \|x_k\|^2 \le 0$$

by hypothesis and design.

Ky Fan's Lemma applies directly to supply a $g \in K$ such that for *any* $x_1,\dots, x_n$ in X we have $F_{x_1,\dots x_n}(g) \le 0$. Specializing to $n = 1$, we get what we announced. QED

The Maurey - Rosenthal Theorem will shed new light on unconditionally summable sequences in $L_1[0,1]$, but to do so it requires the collaboration of a powerful, yet deceptively simple, result from operator theory.

12.31 Dilation Theorem: *Let $u \in \mathcal{L}(H_1, H_2)$ be a Hilbert space operator with $\|u\| \le 1$. There is a Hilbert space G such that u admits a factorization*

$$u : H_1 \xrightarrow{\tilde{u}} G \oplus_2 H_2 \xrightarrow{p} H_2,$$

where $\tilde{u}$ is an isometric embedding and p is the orthogonal projection of $G \oplus_2 H_2$ onto H_2.

Proof. Since $\|u\| \le 1$ we can define a positive sesquilinear form φ on H_1 by

$$\varphi(x,y) := (x|y) - (ux|uy) \quad \text{for all} \quad x,y \in H_1.$$

The set $N := \{x \in H_1 : \varphi(x,x) = 0\}$ is a closed subspace of H_1, and we can equip H_1/N with an inner product

$$\big(x + N \,|\, y + N\big) := \varphi(x,y) \quad \text{for all} \quad x,y \in H_1.$$

Completing, we obtain our Hilbert space G.

Now define $\tilde{u} : H_1 \to G \oplus_2 H_2 : x \mapsto (x + N, ux)$. Plainly, $\tilde{u}$ is linear. It is isometric since for each $x \in H_1$

$$\|\tilde{u}x\|^2 = \|x+N\|_G^2 + \|ux\|_{H_2}^2 = \varphi(x,x) + (ux|ux) = (x|x) = \|x\|^2.$$

Evidently, $u = p\tilde{u}$, and so the proof is over. QED

Notice that the construction of G means that if $H_1 = L_2[0,1]$, then G will be separable, and so without loss of generality, can be taken to be $L_2[1,2]$. If G as constructed happens to be finite dimensional, we can always increase its size.

We shall apply this during the proof of the next result.

12.32 Bennett - Maurey - Nahoum Theorem: *Let (f_n) be an unconditionally summable sequence in $L_1[0,1]$. Then we can find $(a_n) \in \ell_2$, $g \in L_2[0,1]$ and an orthonormal sequence (g_n) in $L_2[0,2]$ such that, for each $n \in \mathbf{N}$ and almost all $t \in [0,1]$,*

$$f_n(t) = a_n\cdot g(t)\cdot g_n(t).$$

Proof. It is important to discard at the start any f_n's that happen to be zero. With this out of the way, the first step is to recall the Bounded Multiplier Test 1.6: unconditional convergence of $\sum_n f_n$ allows us to define the operator

$$u : \ell_\infty \longrightarrow L_1[0,1] : (t_n) \mapsto \sum_n t_n f_n.$$

As we have already seen in 3.7, u must be 2-summing. We shall explain how to take advantage of the simple structure of ℓ_∞ to manoeuvre the 2-summing factorization into a particularly simple form:

$$u : \ell_\infty \xrightarrow{w} \ell_2 \xrightarrow{v} L_1[0,1],$$

where $w \in B_{\mathcal{L}(\ell_\infty,\ell_2)}$ will be given by $we_k = a_k e_k$ for some $a \in B_{\ell_2}$; the commutativity of the diagram will then determine that $ve_k = f_k/a_k$.

Observe that if we set $e_0^* = 0$ and, for every $n \in \mathbf{N}$, define, $e_n^* \in B_{\ell_\infty^*}$ by $e_n^*(x) = x_n$ for each $x \in \ell_\infty$, then $K = \{e_n^* : n \geq 0\}$ is a weak$*$compact norming subset of $B_{\ell_\infty^*}$. This allows the preliminary factorization

$$u : \ell_\infty \xrightarrow{w_1} L_2(\mu) \xrightarrow{v_1} L_1[0,1],$$

where $\mu \in \mathcal{C}(K)^*$ is a probability measure and the operator w_1 is the usual evaluation map $x \mapsto \langle x, \cdot \rangle$.

As K is countable, the measure μ is completely specified by the sequence of numbers $\mu_n := \mu(\{e_n^*\})$; we have $\mu_n \geq 0$ and $\sum_{n\geq 0} \mu_n = 1$ since μ is a probability measure. Moreover, if $x \in \ell_\infty$, then

$$\|w_1 x\|_{L_2(\mu)} = \Big(\sum_{k=0}^{\infty} |\langle x, e_k^*\rangle|^2 \mu_k\Big)^{1/2} = \Big(\sum_{k=1}^{\infty} \mu_k |x_k|^2\Big)^{1/2}.$$

When $k \geq 1$, no μ_k can be zero; just note that $f_k = v_1 w_1 e_k \neq 0$ and that $\mu_k = \|w_1 e_k\|^2_{L_2(\mu)}$. We set $a_k = \mu_k^{1/2}$, $k \in \mathbf{N}$, so that $\sum_{k\geq 1} a_k^2 = 1$, and $e_k \mapsto a_k e_k$ defines an operator $w \in \mathcal{L}(\ell_\infty, \ell_2)$ with $\|w\| = 1$. Next we observe that an isometric operator $w_2 \in \mathcal{L}(\ell_2, L_2(\mu))$ is obtained by setting $w_2 e_k := a_k^{-1}\langle e_k, \cdot\rangle$ for every $k \in \mathbf{N}$; indeed, for each $x \in \ell_2$, we have

$$\|w_2 x\|_{L_2(\mu)} = \Big(\sum_{k=1}^{\infty} |a_k^{-1} x_k|^2 \mu_k\Big)^{1/2} = \Big(\sum_{k=1}^{\infty} |x_k|^2\Big)^{1/2} = \|x\|.$$

Since $w_1 = w_2 w$, we can achieve our desired factorization $u = vw$ by defining $v := v_1 w_2$. Notice that the constraints of commutativity require $ve_k = a_k^{-1} f_k$ for each $k \in \mathbf{N}$.

Now we plug $v : \ell_2 \to L_1[0,1]$ into the Maurey - Rosenthal machinery; after all, ℓ_2 certainly does have type 2. The output is a function $g \in L_2[0,1]$ and a factorization $v : \ell_2 \xrightarrow{\hat{v}} L_2[0,1] \xrightarrow{M_g} L_1[0,1]$ where M_g is the multiplication operator produced by g, and $\|\hat{v}\| \leq 1$. Invoke the Dilation Theorem: there is a Hilbert space G – which we may take to be $L_2[1,2]$ so that $G \oplus_2 L_2[0,1]$ can be identified with $L_2[0,2]$ – and there is an isometric embedding $\tilde{v} : \ell_2 \to L_2[0,2]$ such that for each $x \in \ell_2$, the restriction of $\tilde{v}x$ to $[0,1]$ coincides with $\hat{v}x$. Since

$\tilde{v}$ is isometric, it converts the orthonormal vectors e_n into the orthonormal functions $g_n = \tilde{v}e_n \in L_2[0,2]$. But on $[0,1]$, g_n is $\hat{v}e_n = a_n^{-1}\hat{v}we_n = f_n/(a_n g)$, so we just can multiply across to obtain $f_n(t) = a_n \cdot g(t) \cdot g_n(t)$ for each $n \in \mathbf{N}$ and almost all $t \in [0,1]$. QED

NOTES AND REMARKS

The Contraction Principle 12.2 is to be found in J.P. Kahane [1968]; our proof follows L. Schwartz [1981]. The Principle has been the object of close attention by J. Hoffmann-Jørgensen [1974], [1977b], and N.C. Jain and M.B. Marcus [1975].

For Theorem 12.3 we followed J. Hoffmann-Jørgensen [1974]. Out of the many further beautiful results on Banach spaces to be found in this paper we mention the following one in the impressive form which it was given by S. Kwapień [1974]:

Theorem: *Let (χ_n) be a sequence of independent Banach space-valued random variables, and write S_n for the partial sum $\sum_{k \le n} \chi_k$. The almost sure boundedness of (S_n) implies the almost sure convergence of (S_n) if and only if the Banach space contains no isomorphic copy of c_0.*

In a later paper, J. Hoffmann-Jørgensen and G. Pisier [1976] spotlighted the tightness of the relationship between type, cotype and probability, and set off a wave of activity in the study of probability in Banach spaces. There are several surveys of this topic, including J. Hoffmann-Jørgensen [1977a], M. Ledoux and M. Talagrand [1991], G. Pisier [1986a] and W. Wojczyński [1975], [1978]. For those who can track down a copy, D.J.H. Garling's 1977 Ohio State lecture notes will also be rewarding.

In their Gaussian guise, almost summing operators were first introduced by W. Linde and A. Pietsch [1974]. Under the alias ℓ, the norm π_γ was closely investigated by T. Figiel and N. Tomczak-Jaegermann in [1979]. These two papers are a good source for several further deep results related to (the Gaussian version of) type and cotype which are not covered in this book.

Some of our previous results admit generalizations in terms of almost summing operators. A good example is 2.21: the argument given there shows that an operator with values in a Hilbert space is 1-summing whenever its adjoint is almost summing.

Theorem 12.10 is a key result. It, like much of what we have presented on type and cotype, can be generalized to appropriately defined ideals of Banach space operators. H. Jarchow [1984a] has noted a version of 12.10 in this context: the 'type 2 operators' are just the 'surjective hull' of the almost summing operators. A. Pietsch's [1978] monograph is a good place to find the concepts we have just referred to.

The equivalence of the norms π_γ and π_{as} was mentioned after the proof of 12.12. This equivalence, and so the Rademacher description of almost summing operators, has been part of the folklore for many years. Lemma 12.14 follows V.D. Milman and G. Schechtman [1986]; we are indebted to A. Pietsch for pointing out the relevance of this reference in this connection.

Although we know no explicit reference for Corollary 12.16, its consequence $\Delta_2 = \Pi_r^d \circ \Pi_s$ for $2 \leq r, s < \infty$ appears as 17.4.6 in A. Pietsch's monograph [1978]. Here is a slick way to deduce this identity from Maurey's Extension Theorem 12.22.

Consider a composition $u : X \overset{v_2}{\to} Z \overset{v_1}{\to} Y$ of Banach space operators where v_1^* is r-summing and v_2 is s-summing. Next, think about the factorizations

$$v_2 : X \overset{i}{\longrightarrow} S_\infty \overset{j_2}{\hookrightarrow} S_s \overset{b}{\longrightarrow} Z$$

and

$$v_1 : Z \overset{a}{\longrightarrow} Q_{r^*} \overset{j_1}{\hookrightarrow} Q_1 \overset{q}{\longrightarrow} Y,$$

where, taking suitable probability measures μ and ν, S_∞ is a subspace of $L_\infty(\mu)$, S_s is a subspace of $L_s(\mu)$, Q_{r^*} is a quotient of $L_{r^*}(\nu)$, and Q_1 is a quotient of $L_1(\nu)$. Moreover, j_1 and j_2 are induced by the formal identities $L_{r^*}(\nu) \to L_1(\nu)$ and $L_\infty(\mu) \to L_s(\mu)$. Since S_s and $(Q_{r^*})^*$ both have type 2, Kwapień's Theorem 12.19 tells us that ab admits a factorization $ab : S_s \overset{\hat{b}}{\to} H \overset{\hat{a}}{\to} Q_{r^*}$ where H is a Hilbert space. Now Maurey's Extension Theorem enters the fray: $\hat{b}$ has an extension $\tilde{b} : L_s(\mu) \to H$, and, dually, $\hat{a}$ admits a lifting $\tilde{a} : H \to L_{r^*}(\nu)$. Finally we reach a factorization

$$u : X \longrightarrow L_\infty(\mu) \longrightarrow H \longrightarrow L_1(\nu) \longrightarrow Y,$$

and we can appeal to Grothendieck's result 3.5 to conclude the proof.

As we have mentioned before, Kwapień's Theorem 12.19 gave one of the significant impulses which started off the whole theory of type and cotype; it appeared in his paper [1972a].

It also allowed T. Figiel and G. Pisier [1974] to resolve a conjecture of J. Lindenstrauss [1963]:

Theorem: *If a uniformly convex, uniformly smooth Banach space has moduli of convexity and smoothness of the same order as Hilbert space, then it must be isomorphic to Hilbert space.*

The roots of this theorem reach back to G. Nordlander [1960]. He knew that, thanks to the parallelogram law, Hilbert spaces have modulus of convexity $1 - \sqrt{1 - (\varepsilon^2/4)}$ and he showed that no uniformly convex Banach space can have better asymptotic modulus of convexity as $\varepsilon \to 0$. Later, J. Lindenstrauss [1963] introduced another parameter associated with a Banach space X, its *modulus of smoothness* $\varrho_X(t) := \sup\{\frac{1}{2}(\|x+y\| + \|x-y\|) - 1 : \|x\| \leq 1, \|y\| \leq t\}$. By establishing a duality relation, $\varrho_{X^*}(t) = \sup_{0 \leq s \leq 2}((st/2) - \delta_X(s))$, between modulus of smoothness and modulus of convexity, he was able to demonstrate that no uniformly smooth Banach space can be any smoother than Hilbert space. He went on to show that when a uniformly convex, uniformly smooth

Banach space with unconditional basis has its moduli of convexity and smoothness of the same order as Hilbert space then it has to be isomorphic to Hilbert space. He conjectured that unconditional bases were of no relevance.

In the course of their [1974] work, Figiel and Pisier were able to show that the Lebesgue-Bochner spaces $L_p([0,1],X)$ $(1 < p < \infty)$ are as convex or smooth as $L_p[0,1]$ and X let them be. In particular, if X has modulus of convexity of order ε^2, then so does $L_2([0,1],X)$; the modulus of smoothness behaves similarly. Next came a clever trick: if $x_1,\ldots,x_n$ are in X, the vectors $r_k \otimes x_k$ $(1 \leq k \leq n)$ are unconditionally basic in $L_2([0,1],X)$, with the same norms as $x_1,\ldots,x_n$. Figiel and Pisier then used M.I. Kadets' [1956] theorem on unconditionally convergent series in uniformly convex spaces to conclude that X has cotype 2. In a similar fashion they were able to conclude that X has type 2 when its asymptotic modulus of smoothness is as in Hilbert space. Calling on Kwapień's Theorem 12.19, they laid Lindenstrauss' conjecture to rest.

The extension theorem 12.22, together with most of the corollaries presented in the text, was published by B. Maurey in [1974b].

Corollary 12.21 is the variant of 12.10 which appears most frequently in the literature; see A. Pietsch [1978] and N. Tomczak-Jaegermann [1989] for example. The equivalence is of course a consequence of 3.5.

While we are on the topic of 3.5, we note that it is fruitless to try to obtain it from Kwapień's Theorem; after all, infinite dimensional $\mathcal{L}_\infty$-spaces never have type greater than 1. On the other hand, the dual of an $\mathcal{L}_\infty$-space has cotype 2, so it is natural to ask whether the identity $\mathcal{L}(X,Y) = \Gamma_2(X,Y)$ holds when X^* and Y have cotype 2. G. Pisier [1980] showed that this is indeed the case when, in addition, X or Y has the approximation property. It follows that if X and X^* have cotype 2 and X has the approximation property, then X must be isomorphic to a Hilbert space.

The approximation property is important here. In his famous [1983] article, G. Pisier constructed Banach spaces X such that X and X^* have cotype 2 and $X\hat{\otimes}X = X\check{\otimes}X$; he actually showed that every cotype 2 space can be embedded into such a space. These spaces are not Hilbert spaces, so their identity operators cannot be 2-factorable. Necessarily, they fail the approximation property.

Our presentation of the Maurey-Rosenthal Theorem 12.30 follows B. Maurey [1974a] though its roots are in the pathfinding work of H.P. Rosenthal [1973].

The Dilation Theorem 12.31 is just one very simple example of a large family of results of similar character. The general theme of dilations is treated in some detail in C. Foiaş and B. Sz. Nagy [1970].

Theorem 12.32 was uncovered independently by G. Bennett [1976] and B. Maurey and A. Nahoum [1973]. We follow the presentation by P. Ørno [1976].

Arguments similar to that of 12.32 have been used by H. Niemi [1984] in his study of harmonizable stochastic processes.

13. K-Convexity and B-Convexity

Our objective in this chapter is to study those Banach spaces X which have *non-trivial type,* that is type > 1. The space

$$Rad\,(X)$$

of all almost unconditionally summable sequences in X will have a major part to play in the story.

One way to state Kahane's Inequality (11.1) is to say that for any Banach space X, $Rad\,(X)$ embeds into the Lebesgue - Bochner spaces $L_p([0,1],X)$ with equivalence of norms.

Now, in the scalar case, $Rad\,(\mathbf{K})$ – alias ℓ_2 – is even complemented in $L_p[0,1]$ provided that $1 < p < \infty$. In fact, we saw in 1.12 that there is a bounded linear projection $R : L_p[0,1] \to L_p[0,1]$ with range $Rad\,(\mathbf{K})$ given by

$$Rf(\cdot) = \sum_{n=1}^{\infty} r_n(\cdot)\cdot\Big(\int_0^1 r_n(t)f(t)\,dt\Big)$$

for $f \in L_p[0,1]$.

What happens in the vector-valued case? It is simple enough to come up with a *formal* analogue of the above Rademacher projection R. We shall investigate when, for a given Banach space X and $1 < p < \infty$,

$$R^X f(\cdot) := \sum_{n=1}^{\infty} r_n(\cdot)\cdot\Big(\int_0^1 r_n(t)f(t)\,dt\Big)$$

defines a bounded linear map of $L_p([0,1],X)$ into itself; of course, this time our integral is a Bochner integral. If it does, this map will certainly be a projection with range $Rad\,(X)$.

We do not have to look far for a deviant Banach space.

13.1. Example: When $X = L_1[0,1]$, R^X is *not* a bounded linear map from $L_p([0,1],X)$ into itself.

To see this, define $F_N \in L_p([0,1],X)$ by

$$F_N(t) := \prod_{k=1}^{N} \big(1 + r_k(t)r_k\big).$$

A routine computation shows that $\|F_N(t)\|_X = 1$ for each $t \in [0,1]$ and each $N \in \mathbf{N}$; hence $\|F_N\|_{L_p([0,1],X)} = 1$. By a similar computation, $(R^X F_N)(t) = \sum_{k\le N} r_k(t)r_k$ for all t and N. Since $L_1[0,1]$ has cotype 2, trouble ensues:

$$\|R^X F_N\|_{L_p([0,1],X)} = \Big(\int_0^1 \big\| \sum_{k\le N} r_k(t) r_k \big\|^p \, dt\Big)^{1/p}$$
$$\ge K\cdot\Big(\sum_{k\le N} \|r_k\|^2_{L_1[0,1]}\Big)^{1/2} = K\cdot N^{1/2}$$

where K is a constant depending on the cotype 2 constant of $X = L_1[0,1]$ and so on constants from Khinchin's Inequality.

K-Convexity

A Banach space X is called *K-convex* if R^X exists as a bounded linear projection from $L_2([0,1],X)$ to itself. As in the scalar case, the index 2 can be replaced by any $1 < p < \infty$. We shall see this later on, as a comment before 13.17; but our main interest will be in showing that

- *the K-convex spaces are the same as the spaces with non-trivial type.*

The fact that $L_1[0,1]$ is not K-convex is by no means accidental, and the argument used in 13.1 deserves careful scrutiny. In fact, a major step will be to show that, at least locally, non-K-convex Banach spaces always contain L_1.

In the sequel, it will sometimes be convenient to have another model of the Rademacher variables at our beck and call. After all, as we are hardly ever interested in anything but their distributional properties, we might as well replace the r_n's by any sequence of independent, symmetric random variables taking values in $\{-1,1\}$. What we have in mind is to take as our basic probability space the cartesian product

$$D := \{-1,1\}^{\mathbf{N}},$$

equipped with the measure

$$\mu$$

which is the product of the 'fair coin measures' $\frac{1}{2}(\delta_{-1}+\delta_1)$ on each factor. The canonical projections $D \to \{-1,1\}$ provide us with what we want: a sequence of independent, symmetric $\{-1,1\}$-valued random variables.

If we only require the first n Rademacher variables, we can limit ourselves to the probability space

$$D_n := \{-1,1\}^n$$

equipped with the measure

$$\mu_n$$

which is the n-fold product of 'fair coin' measures.

In either case, we shall use the same symbol

$$r_k$$

to denote not only the k-th Rademacher function on $[0,1]$, but also the k-th canonical projection, be it from D or D_n (for $n \ge k$, of course) to $\{-1,1\}$. Regardless of whether we are dealing with $[0,1]$, D or D_n, we have the same fundamental averaging properties: for any n vectors $x_1,\ldots,x_n$ in a Banach space,

$$\int_0^1 \Big\|\sum_{k\le n} r_k(t)x_k\Big\|^p\,dt = \int_D \Big\|\sum_{k\le n} r_k(\omega)x_k\Big\|^p\,d\mu(\omega)$$
$$= \int_{D_n} \Big\|\sum_{k\le n} r_k(\omega)x_k\Big\|^p\,d\mu_n(\omega) = 2^{-n}\cdot \sum_{(\varepsilon_k)\in D_n} \Big\|\sum_{k\le n} \varepsilon_k x_k\Big\|^p.$$

Later on, we shall find the notational conveniences offered by these measure spaces to be very useful. For now, we use the new setup to recast Example 13.1 in a way that allows us to squeeze a little more out of it. The functions F_N can be viewed as having their values in $L_1(\mu_N)$, and since these values always have $L_1(\mu_N)$ norm equal to one, the $L_p\big([0,1], L_1(\mu_N)\big)$ norm of F_N is still one. Notice now that there is an obvious isometric isomorphism between $L_1(\mu_N)$ and $\ell_1^{2^N}$. With this, the final computations of 13.1 carry over unchanged to show that if a Banach space contains isometrically isomorphic copies of all ℓ_1^n's, then it cannot be K-convex.

There is nothing special about isometric isomorphisms here; so let us agree on some terminology and then summarize our meditations.

We say that a Banach space Z *contains ℓ_p^n's λ-uniformly* or *contains λ-uniform copies of all ℓ_p^n's* (for $\lambda > 1$ and $1 \le p \le \infty$) if for each natural number n there is an n-dimensional subspace E of Z, together with an isomorphism $u_n : \ell_p^n \to E$ such that $\|u_n\|\cdot\|u_n^{-1}\| < \lambda$. It is customary to say that Z *contains ℓ_p^n's uniformly* if it contains λ-uniform copies of all ℓ_p^n's for some $\lambda > 1$.

13.2 Proposition: *If a Banach space contains ℓ_1^n's uniformly, then it cannot be K-convex.*

It is clear that Banach spaces with non-trivial type also cannot contain ℓ_1^n's uniformly. Remarkably, this is the only obstruction to K-convexity, and we shall ultimately establish the following fundamental result.

13.3 Pisier's Theorem: *The following statements about a Banach space X are equivalent:*

(i) *X is K-convex.*

(ii) *For every $\lambda > 1$, X fails to contain λ-uniform copies of all ℓ_1^n's.*

(iii) *For some $\lambda > 1$, X fails to contain λ-uniform copies of all ℓ_1^n's.*

(iv) *X has non-trivial type.*

Notice that ℓ_1 is finitely representable in X if and only if X contains ℓ_1^n's λ-uniformly for each $\lambda > 1$; see Chapter 8.

The proof of this theorem will occupy us for most of the chapter, and it will proceed in several stages, each of which will culminate in a major result.

To begin,we shall introduce a geometric notion, called B-convexity, and show that this is equivalent to (ii) and (iii). Then we will demonstrate that a Banach space is B-convex if and only if it has non-trivial type. The final link in the chain is the equivalence of B-convexity and K-convexity, and this will require sustained effort.

B-Convexity

A Banach space X is said to be *B-convex* if there exist a $\delta > 0$ and an integer $n \geq 2$ such that for any $x_1,\dots,x_n \in X$ we can choose $\varepsilon = (\varepsilon_k)_{k=1}^n \in D_n$ in such a way that

$$\Big\| \frac{1}{n} \sum_{k \leq n} \varepsilon_k x_k \Big\| \leq (1-\delta) \cdot \max_{k \leq n} \|x_k\|.$$

Attention can, of course, be restricted to x_k's in the unit ball of X. The attentive reader will soon realize that B-convexity is a real notion:

- *a complex Banach space is B-convex if and only if the underlying real Banach space is B-convex.*

In view of our program, we had better check that ℓ_1 is not B-convex. But this is very easy: when the e_k's are the usual unit basis vectors, we have $\|(1/n)\sum_{k\leq n} \varepsilon_k e_k\| = 1$ for all sign choices $\varepsilon_1,\dots,\varepsilon_n$.

B-convexity is clearly a 'super-property': every Banach space which is finitely representable in a B-convex space must itself be B-convex. This tells us that ℓ_1 can never be finitely representable in a B-convex space, which is at least consistent with with our plan of attack for Pisier's Theorem 13.3. It also allows us to assert that a Banach space is B-convex if and only if its bidual is. We shall soon see how to strengthen this substantially: a Banach space is B-convex if and only if its dual is.

To help us on our way we introduce some moduli. For a Banach space X and $n \in \mathbf{N}$, we define

$$\beta_n(X) := \sup_{x_1,\dots,x_n \in B_X} \min_{\varepsilon \in D_n} \Big\| \frac{1}{n} \sum_{k \leq n} \varepsilon_k x_k \Big\| .$$

Evidently, $0 \leq \beta_n(X) \leq 1$ for all n and $\beta_1(X) = 1$. Moreover, by considering the unit basis vectors, we see immediately that $\beta_n(\ell_1^m) = 1$ for $m \geq n$. It is important to be aware of a totally trivial truth:

13.4 Proposition: *The Banach space X fails to be B-convex if and only if $\beta_n(X) = 1$ for all $n \geq 2$.*

Despite their gruesome definition, the β_n's exhibit remarkable regularity. Key to our considerations is a submultiplicativity property.

13.5 Lemma: *For any natural numbers k, n,*

$$\beta_{kn}(X) \leq \beta_k(X) \cdot \beta_n(X).$$

Proof. Fix $x_1,\dots,x_{kn}$ in X. We gain access to $\beta_k(X)$ by breaking these elements into n blocks of k vectors. Choose $\tilde{\varepsilon} \in D_{kn}$ so that for each $0 \le j < n$,

$$\|\tilde{\varepsilon}_{kj+1}x_{kj+1} + \dots + \tilde{\varepsilon}_{k(j+1)}x_{k(j+1)}\| = \min_{\varepsilon \in D_k} \|\varepsilon_1 x_{kj+1} + \dots + \varepsilon_k x_{k(j+1)}\|.$$

Then, writing

$$y_{j+1} := \sum_{kj+1}^{k(j+1)} \tilde{\varepsilon}_i x_i \qquad (0 \le j < n),$$

we find that $\|y_{j+1}\| \le k\cdot\beta_k(X)$. It follows that

$$\min_{\eta \in D_n} \Big\|\sum_{j \le n} \eta_j y_j\Big\| \le n\cdot\beta_n(X)\cdot\max_{j \le n}\|y_j\| \le n\cdot\beta_n(X)\cdot k\cdot\beta_k(X).$$

Since each $\sum_{j \le n}\eta_j y_j$ can be expressed in the form $\sum_{i \le kn}\varepsilon_i x_i$ for an appropriate $(\varepsilon_i) \in D_{kn}$, we infer that

$$\min_{\varepsilon \in D_{kn}} \Big\|\sum_{i \le kn} \varepsilon_i x_i\Big\| \le k\cdot n\cdot\beta_k(X)\cdot\beta_n(X). \qquad \text{QED}$$

We are now ready for our first main characterization of B-convexity.

13.6 Theorem: *The following are equivalent statements about a Banach space X:*

(i) *X is B-convex.*

(ii) *X does not contain ℓ_1^n's uniformly.*

(iii) *ℓ_1 is not finitely representable in X.*

Proof. (i)⇒(ii): Assuming that X is B-convex, Proposition 13.4 (the definition!) ensures that $\beta_k(X) \le 1-\delta$ for some $k \ge 2$ and $0 < \delta < 1$. The submultiplicativity lemma 13.5 kicks in: $\beta_{k^m}(X) \le (1-\delta)^m$ for each m. We can therefore choose n so that $\beta_n(X)$ is as small as we please.

It turns out that this is impossible if (ii) is false, that is, if there is some $\lambda > 1$ such that X contains λ-uniform copies of all ℓ_1^n's. Indeed, we claim that in this case $\beta_n(X) \ge \lambda^{-1}$ for every n, contradicting the conclusion of the previous paragraph. To justify the claim note that for each $n \in \mathbf{N}$ there is an n-dimensional subsapace E_n of X together with an isomorphism $u_n : \ell_1^n \to E_n$ such that $\|u_n\| = 1$ and $\|u_n^{-1}\| < \lambda$. It follows that, for any $\varepsilon \in D_n$,

$$\begin{aligned} n = \Big\|\sum_{k \le n}\varepsilon_k e_k\Big\|_{\ell_1^n} &\le \|u_n^{-1}\|\cdot\Big\|\sum_{k \le n}\varepsilon_k u_n e_k\Big\|_X \\ &< \lambda\cdot\Big\|\sum_{k \le n}\varepsilon_k u_n e_k\Big\|_X \le \lambda\cdot n\cdot\beta_n(X). \end{aligned}$$

(ii)⇒(iii) is trivial, so we pass to (iii)⇒(i), for which we adopt the contrapositive: assuming that X is not B-convex, we show that for any $\lambda > 1$, X contains λ-uniform copies of all ℓ_1^n's.

Fix $\lambda > 1$ and $n \ge 2$. Define $0 < \delta < 1$ by $1 - \delta = \lambda^{-1}$. Since X is not B-convex, we can find $x_1,\dots,x_n$ in B_X such that $n - \delta \le \|\sum_{k \le n}\varepsilon_k x_k\|$ for all

$\varepsilon \in D_n$. We shall show that the x_k's form a basis for a λ-isomorphic copy of ℓ_1^n in X. More precisely, they have the property that however we choose scalars $a_1,\dots,a_n$ we can be sure that

$$\frac{1}{\lambda}\cdot\sum_{k\le n}|a_k| \le \Big\|\sum_{k\le n} a_k x_k\Big\| \le \sum_{k\le n}|a_k|.$$

To prove this it is good enough to normalize so that $\sum_{k\le n}|a_k| = 1$. Then

$$\begin{aligned} n-\delta &\le \Big\|\sum_{k\le n}(\operatorname{sign} a_k)x_k\Big\| = \Big\|\sum_{k\le n}\big[(\operatorname{sign} a_k)\cdot(1-|a_k|)+a_k\big]x_k\Big\| \\ &\le \sum_{k\le n}(1-|a_k|) + \Big\|\sum_{k\le n} a_k x_k\Big\| = n-1+\Big\|\sum_{k\le n} a_k x_k\Big\|. \end{aligned}$$

Rearrange to obtain

$$\frac{1}{\lambda} = 1-\delta \le \Big\|\sum_{k\le n} a_k x_k\Big\|.$$

Obviously $\|\sum_{k\le n} a_k x_k\| \le 1$, so we are done. QED

We take time out to fulfil a promise:

13.7 Corollary: *A Banach space X is B-convex if and only if X^* is.*

Proof. Suppose that X is not B-convex. The previous theorem tells us that X contains 2-uniform copies of all ℓ_1^n's: for each $n \in \mathbf{N}$ there is an n-dimensional subspace E_n of X which is 2-isomorphic to ℓ_1^n.

We shall show that X^* contains 4-uniform copies of all ℓ_1^n's, and thus is not B-convex.

Fix $n \in \mathbf{N}$ and a norm one surjection $q : \ell_1 \to \ell_\infty^n$. Pick $x_1,\dots,x_n \in \ell_1$ so that $q(x_k) = e_k$ for $k = 1,\dots,n$. Thinking of ℓ_1 as an $\mathcal{L}_1$-space, we realize that for some $N \ge n$ there is an N-dimensional subspace F_N of ℓ_1 which contains $x_1,\dots,x_n$ and is 2-isomorphic to ℓ_1^N.

By construction, E_N and F_N are 4-isomorphic. Consequently, we can compose q's restriction to F_N with a suitable isomorphism to obtain a surjective operator $q_N : E_N \to \ell_\infty^N$ of norm no greater than 4. The injectivity of ℓ_∞^N allows us to extend q_N to a surjective operator

$$\tilde{q}_N : X \longrightarrow \ell_\infty^N$$

without increasing the norm. Passing to duals, we accomplish our first objective: $\tilde{q}_N^*$ takes ℓ_1^N 4-isomorphically into X^*.

Our conclusion so far is that if X^* is B-convex, then X is too.

The converse is simple. If X is B-convex, we have already noted that X^{**} must also be B-convex. By what was just shown, X^*'s B-convexity follows suit. QED

Our next big step is to show that the B-convex Banach spaces are precisely those having non-trivial type. Two more moduli will prove useful.

Given $n \in \mathbf{N}$ and a Banach space X, we define

$$\sigma_n(X)$$

to be the smallest of all numbers $\sigma \geq 0$ such that

$$\Big(\int_0^1 \big\| \sum_{k \leq n} r_k(t) x_k \big\|^2 \, dt\Big)^{1/2} \leq \sigma \cdot n \cdot \max_{k \leq n} \|x_k\|$$

for every choice of $x_1,\dots,x_n$ in X.

We further define

$$\tau_n(X)$$

to be the smallest of all numbers $\tau \geq 0$ such that

$$\Big(\int_0^1 \big\| \sum_{k \leq n} r_k(t) x_k \big\|^2 \, dt\Big)^{1/2} \leq \tau \cdot n^{1/2} \cdot \Big(\sum_{k \leq n} \|x_k\|^2\Big)^{1/2}$$

for every choice of $x_1,\dots,x_n$ in X.

Whenever it is possible to do so without confusion, we prefer to write σ_n and τ_n instead of $\sigma_n(X)$ and $\tau_n(X)$; similarly, we write β_n for the moduli $\beta_n(X)$ introduced earlier.

The point of these painful parameters is that they can be used in the same way as the β_n's to decide X's B-convexity:

- *X is B-convex if and only if $\sigma_n < 1$ for all $n \geq 2$*

and

- *X is B-convex if and only if $\tau_n < 1$ for all $n \geq 2$.*

These facts are not obvious. To prepare the ground, we assemble some elementary properties:

13.8 Lemma: *For all positive integers m and n,*

(a) $\beta_n \leq \sigma_n \leq \tau_n \leq 1$,

(b) $n^{-1/2} \leq \tau_n$,

(c) $\tau_{mn} \leq \tau_m \tau_n$.

Proof. (a) $\sigma_n \leq \tau_n \leq 1$ follows immediately from the definitions. For the remaining inequality, select $x_1,\dots,x_n$ in B_X and note that

$$2^{-n} \cdot \sum_{\varepsilon \in D_n} \big\| \sum_{k \leq n} \varepsilon_k x_k \big\|^2 = \int_0^1 \big\| \sum_{k \leq n} r_k(t) x_k \big\|^2 \, dt \leq n^2 \cdot \sigma_n^2.$$

It follows that, for some $\tilde{\varepsilon} \in D_n$,

$$\big\| \sum_{k \leq n} \tilde{\varepsilon}_k x_k \big\|^2 \leq n^2 \cdot \sigma_n^2.$$

Hence $\beta_n \leq \sigma_n$.

(b) This is a simple consequence of the orthonormality of the Rademacher functions: examine the definition of τ_n, using vectors x_k which are all scalar multiples of some fixed non-zero vector.

(c) Let $x_1,\dots,x_{mn} \in X$ be given. For each $1 \le k \le n$ and $t \in [0,1]$ define

$$y_k(t) := \sum_{m(k-1)<j\le mk} r_j(t)x_j.$$

All depends on a good understanding of

$$I := \int_0^1\int_0^1 \Big\| \sum_{k\le n} r_k(s)y_k(t) \Big\|^2 \, ds\, dt.$$

On the one hand, the definition of the τ_j's yields

$$I \le n\cdot\tau_n^2 \cdot \sum_{k\le n}\int_0^1 \|y_k(t)\|^2 dt \le n\cdot\tau_n^2\cdot m\cdot\tau_m^2 \cdot \sum_{j\le mn}\|x_j\|^2.$$

On the other hand, thanks to 11.2 it is true that for every $s \in [0,1]$

$$\int_0^1 \Big\|\sum_{k\le n} r_k(s)y_k(t)\Big\|^2 \, dt = \int_0^1 \Big\|\sum_{k\le n} r_k(s) \sum_{m(k-1)<j\le mk} r_j(t)x_j\Big\|^2 \, dt$$
$$= \int_0^1 \Big\| \sum_{j\le mn} r_j(\theta)x_j \Big\|^2 \, d\theta.$$

We conclude that

$$\int_0^1 \Big\| \sum_{j\le mn} r_j(\theta)x_j\Big\|^2 \, d\theta = I \le m\cdot n\cdot(\tau_m\cdot\tau_n)^2\cdot \sum_{j\le mn}\|x_j\|^2.$$

It follows that $\tau_{mn} \le \tau_m\cdot\tau_n$. QED

The next lemma, which is deeper, is our main tool:

13.9 Lemma: *Let $n \in \mathbf{N}$. Either all of $\beta_n, \sigma_n, \tau_n$ equal 1, or all of them are strictly less than 1.*

Proof. We will show that $\tau_n = 1$ entails that $\sigma_n = 1$ which in turn forces $\beta_n = 1$. By virtue of 13.8(a), this is all that is required.

Suppose that $\tau_n = 1$ and fix $\delta > 0$. From the very definition of τ_n, there exist $x_1,\dots,x_n \in X$ with $\sum_{k\le n}\|x_k\|^2 = n$ and

$$\text{(1)} \qquad \int_0^1 \Big\|\sum_{k\le n} r_k(t)x_k\Big\|^2 \, dt \ge (1-\delta)^2\cdot n^2.$$

To get information about σ_n, we must be able to deduce something about $\max_{k\le n}\|x_k\|$. A surprisingly crude argument will reveal that

$$\text{(2)} \qquad \max_{k\le n}\|x_k\| \le 1 + 2\cdot(\delta n)^{1/2}.$$

When this is incorporated into (1) we find

$$\Big(\int_0^1 \Big\|\sum_{k\le n} r_k(t)x_k\Big\|^2 \, dt\Big)^{1/2} \ge \frac{1-\delta}{1+2\cdot(\delta n)^{1/2}} \cdot n \cdot \max_{k\le n}\|x_k\|.$$

In other words,

$$\sigma_n \ge (1+2\cdot(\delta n)^{1/2})^{-1}\cdot(1-\delta).$$

Letting $\delta \to 0$, we see that $\sigma_n \ge 1$, so $\sigma_n = 1$.

The missing link is the proof that (1) implies (2). We proceed as follows. We pick $1 \leq k_0 \leq n$ so that $\|x_{k_0}\| = \max_{k\leq n} \|x_k\|$ and observe that, assuming (1), our choice of the x_k's leads to

$$\begin{aligned}\sum_{k\leq n}(\|x_{k_0}\|-\|x_k\|)^2 &\leq \sum_{i,j\leq n}(\|x_i\| - \|x_j\|)^2 = 2\cdot n\cdot\sum_{k\leq n}\|x_k\|^2 - 2\cdot\big(\sum_{k\leq n}\|x_k\|\big)^2 \\ &= 2\cdot n^2 - 2\cdot\big(\sum_{k\leq n}\|x_k\|\big)^2 \leq 2\cdot n^2 - 2\cdot\int_0^1 \Big\|\sum_{k\leq n} r_k(t)x_k\Big\|^2\,dt \\ &\leq 2\cdot n^2 - 2\cdot(1-\delta)^2\cdot n^2 \leq 4\cdot\delta\cdot n^2.\end{aligned}$$

From this we find

$$\begin{aligned}n^{1/2}\cdot\|x_{k_0}\| &= \big(\sum_{k\leq n}\|x_{k_0}\|^2\big)^{1/2} \leq \big(\sum_{k\leq n}(\|x_{k_0}\| - \|x_k\|)^2\big)^{1/2} + \big(\sum_{k\leq n}\|x_k\|^2\big)^{1/2} \\ &\leq 2\cdot n\cdot\delta^{1/2} + n^{1/2}.\end{aligned}$$

This is nothing but a disguise for (2): we have completed the proof that $\tau_n = 1$ implies $\sigma_n = 1$.

Next assume that $\sigma_n = 1$:

$$n^2 = \sup \int_0^1 \Big\|\sum_{k\leq n} r_k(t)x_k\Big\|^2\,dt = \sup \frac{1}{2^n}\cdot\sum_{\varepsilon\in D_n}\Big\|\sum_{k\leq n}\varepsilon_k x_k\Big\|^2, \tag{3}$$

where the suprema are taken over all $x_1, .., x_n \in B_X$.

For the moment, fix $x_1,..., x_n \in B_X$. Pick $\tilde{\varepsilon} \in D_n$ so that

$$\Big\|\sum_{k\leq n}\tilde{\varepsilon}_k x_k\Big\| = \min_{\varepsilon\in D_n}\Big\|\sum_{k\leq n}\varepsilon_k x_k\Big\|.$$

By definition of β_n, our choice of $\tilde{\varepsilon}$ means that $\|\sum_{k\leq n}\tilde{\varepsilon}_k x_k\| \leq n\cdot\beta_n$, and so

$$\begin{aligned}\sum_{\varepsilon\in D_n}\Big\|\sum_{k\leq n}\varepsilon_k x_k\Big\|^2 &= \Big(\sum_{\varepsilon\neq\tilde{\varepsilon}}\Big\|\sum_{k\leq n}\varepsilon_k x_k\Big\|^2\Big) + \Big\|\sum_{k\leq n}\tilde{\varepsilon}_k x_k\Big\|^2 \\ &\leq (2^n - 1)\cdot n^2 + n^2\cdot\beta_n^2.\end{aligned}$$

Taking the supremum over all $x_1,..., x_n \in B_X$, and using (3), we obtain

$$n^2 \leq 2^{-n}\cdot((2^n - 1)\cdot n^2 + n^2\cdot\beta_n^2),$$

which, upon rearrangement, gives $\beta_n \geq 1$, and so $\beta_n = 1$. QED

The door is now open to the proof of the link between B-convexity and type.

13.10 Theorem: *A Banach space X is B-convex if and only if it has non-trivial type.*

Proof. One direction is done: if X has non-trivial type, it cannot contain λ-uniform copies of all ℓ_1^n's for any $\lambda > 1$ and so, thanks to 13.6, must be B-convex.

Suppose then that X is B-convex. The trivial proposition 13.4 tells us that there is an $N \geq 2$ for which $\beta_N < 1$. In view of 13.9, $\tau_N < 1$ also, and so $\tau_N = N^{-1/q}$ for some $q > 0$. Looking at 13.8(b), we realize that $q \geq 2$. We are going to show that X has type r for any $r < q^*$.

Choose $x_1, \ldots, x_n \in X$ and for brevity write $\alpha := \sum_{k \leq n} \|x_k\|^r$. Partition the set $\{1, \ldots, n\}$ using

$$M_j := \{m : \alpha \cdot N^{-j-1} < \|x_m\|^r \leq \alpha \cdot N^{-j}, \ 1 \leq m \leq n\},$$

where j can be any non-negative integer. By the definition of the σ_n's,

$$\begin{aligned} \Big(\int_0^1 \Big\|\sum_{k \leq n} r_k(t) x_k\Big\|^2 dt\Big)^{1/2} &\leq \sum_j \Big(\int_0^1 \Big\|\sum_{k \in M_j} r_k(t) x_k\Big\|^2 dt\Big)^{1/2} \\ &\leq \sum_j |M_j| \cdot \sigma_{|M_j|} \cdot \max_{m \in M_j} \|x_m\|. \end{aligned}$$

Since $\alpha \geq \sum_{m \in M_j} \|x_m\|^r \geq \alpha \cdot |M_j| \cdot N^{-j-1}$ we certainly have $|M_j| \leq N^{j+1}$ for every j. Notice now that $(n\sigma_n)_n$ increases with n. With this, we can take advantage of the features of the σ's and τ's established in 13.8:

$$|M_j| \cdot \sigma_{|M_j|} \leq N^{j+1} \cdot \sigma_{N^{j+1}} \leq N^{j+1} \cdot \tau_{N^{j+1}} \leq (N \cdot \tau_N)^{j+1} = N^{(j+1)/q^*}.$$

It follows that

$$\begin{aligned} \Big(\int_0^1 \Big\|\sum_{k \leq n} r_k(t) x_k\Big\|^2 dt\Big)^{1/2} &\leq \sum_j N^{(j+1)/q^*} \cdot \Big(\frac{\alpha}{N^j}\Big)^{1/r} \\ &= N^{1/q^*} \cdot \sum_j \big(N^{(1/q^*)-(1/r)}\big)^j \cdot \Big(\sum_{k \leq n} \|x_k\|^r\Big)^{1/r}. \end{aligned}$$

Since $r < q^*$, the infinite series $\sum_j \big(N^{(1/q^*)-(1/r)}\big)^j$ converges, and so the proof is complete.
QED

Equivalence of B- and K-Convexity

We begin to plumb the depths of this chapter: the equivalence of B- and K-convexity. One implication is known: after all, 13.2 says that a K-convex Banach space can never contain ℓ_1^n's λ-uniformly for any $\lambda > 1$, and so, thanks to 13.6, must be B-convex. To go in the other direction, we need a heavy-duty tool – the Beurling-Kato Theorem. Its description requires some definitions.

Let Y be a Banach space. A collection $(S_t)_{t \geq 0}$ of operators in $\mathcal{L}(Y)$ is a *strongly continuous semigroup of contractions* if

(1) $S_0 = id_Y$,

(2) $S_{t+t'} = S_t S_{t'}$ for any $t, t' \geq 0$,

(3) $\lim_{t \to 0} \|S_t y - y\| = 0$ for any $y \in Y$

and

(4) $\|S_t\| \leq 1$ for any $t \geq 0$.

Now consider a cone of aperture α $(0 < \alpha \leq \pi/2)$ in the complex plane:

$$V_\alpha := \{z \in \mathbf{C} : \operatorname{Re} z > 0, \, |\arg z| < \alpha\}.$$

A collection $(S_z)_{z\in V_\alpha}$ of operators in $\mathcal{L}(Y)$ is an *analytic semigroup* if

(1′) $S_0 = id_Y$,

(2′) $S_{z+z'} = S_z S_{z'}$ for every $z, z' \in V_\alpha$,

and

(3′) however we choose $y \in Y$ and $y^* \in Y^*$, the map $V_\alpha \to \mathbf{C}$: $z \mapsto \langle y^*, S_z y\rangle$ is analytic.

The Beurling - Kato Theorem asserts that in a *complex* Banach space, any strongly continuous semigroup of contractions, all of which are close enough to the identity operator, sits inside a *decent* analytic semigroup. More precisely, the theorem – which we take for granted – reads as follows:

13.11 Beurling - Kato Theorem: *For any $0 < \varrho < 2$ there are numbers $\alpha > 0$, $K > 0$ such that, regardless of the complex Banach space Y, any strongly continuous semigroup $(S_t)_{t\geq 0}$ of contractions on Y satisfying $\|S_t - id_Y\| \leq \varrho$ for every $t \geq 0$ extends to an analytic semigroup $(S_z)_{z\in V_\alpha}$ of operators on Y such that $\|S_z\| \leq K$ for all $z \in V_\alpha$.*

We shall be interested in very specific semigroups. Let $P_1, \ldots, P_N$ be commuting norm one projections in $\mathcal{L}(Y)$. For each $t \geq 0$ define $S_t \in \mathcal{L}(Y)$ by

$$S_t := \prod_{k=0}^{N} \left(P_k + e^{-t}(id_Y - P_k)\right).$$

Remembering that $P_k(id_Y - P_k) = 0$, it is easy to check that this collection satisfies the semigroup conditions (1) and (2). It is called the *semigroup generated by* $P_1, \ldots, P_N$.

To see that $(S_t)_{t\geq 0}$ is actually a strongly continuous semigroup of contractions, we introduce the notation

$$P_A := \prod_{k\in A} P_k$$

whenever A is a subset of $\{1, \ldots, N\}$, with the convention that $P_\emptyset = id_Y$. Note that

$$S_t = \prod_{k=0}^{N} \left(e^{-t} id_Y + (1 - e^{-t})P_k\right) = \sum_{k=0}^{N} e^{-(N-k)t} \cdot (1 - e^{-t})^k \sum_{|A|=k} P_A$$

and so

$$\|S_t\| \leq \sum_{k=0}^{N} e^{-(N-k)t} \cdot (1 - e^{-t})^k \cdot \binom{N}{k} = \left(e^{-t} + (1 - e^{-t})\right)^N = 1.$$

Moreover, for any $y \in Y$,

$$\begin{aligned}\|S_t y - y\| &= \Big\|\Big(\sum_{k=0}^{N} e^{-(N-k)t}\cdot(1-e^{-t})^k \sum_{|A|=k} P_A y\Big) - y\Big\| \\ &= \Big\|\Big(\sum_{k=1}^{N} e^{-(N-k)t}\cdot(1-e^{-t})^k \sum_{|A|=k} P_A y\Big) - (1-e^{-Nt})y\Big\| \\ &\le \Big(\Big[\sum_{k=1}^{N} e^{-(N-k)t}\cdot(1-e^{-t})^k \binom{N}{k}\Big] + 1 - e^{-Nt}\Big)\cdot\|y\| \\ &= 2\cdot(1-e^{-Nt})\cdot\|y\|,\end{aligned}$$

so that

$$\lim_{t\to 0}\|S_t y - y\| = 0.$$

Here is our plan of action. After a diverting tussle, we shall prove that for an appropriate finite set of commuting norm one projections on a B-convex Banach space, we can renorm the space so that, with respect to the new norm, the semigroup generated by the projections satisfies the hypothesis – and so enjoys the conclusion – of the Beurling-Kato Theorem. We then give way to expansionary urges. If Y is a B-convex Banach space, we know it must have type $p > 1$. But 11.12 tells us that for any measure ν, $L_2(\nu, Y)$ also has type p, and so is B-convex too. Choosing ν to be the measure μ_n (or μ) introduced earlier on D_n (or D), we will be able to fabricate commuting norm one projections out of conditional expectation operators and show that the semigroup they generate, when extended in the manner prescribed by 13.11, holds the key to the boundedness of the Rademacher projection R^Y.

A couple of lemmas will get the ball rolling.

13.12 Lemma: *Suppose that Y is a B-convex Banach space. Then there are a real number $0 < \varrho < 2$ and a natural number N such that any N commuting norm one projections $P_1,\ldots,P_N \in \mathcal{L}(Y)$ satisfy*

$$\Big\|\prod_{k\le N}(id_Y - P_k)\Big\| \le \varrho^N.$$

Proof. Suppose this is not so. Then, regardless of how we choose $N \in \mathbf{N}$ and $\delta > 0$, we can find commuting norm one projections $P_1,\ldots,P_N \in \mathcal{L}(Y)$ and a unit vector $y \in Y$ such that

$$\Big\|\prod_{k\in N}(id_Y - P_k)y\Big\| \ge 2^N - \delta.$$

We are going to contradict 13.6 by showing that $\{y, P_1y,\ldots,P_Ny\}$ is 2-equivalent to the unit vector basis of ℓ_1^{N+1}. To do this, we need a good understanding of the quantities $\big\| y + \sum_{k\le N}\varepsilon_k P_k y \big\|$ where $\varepsilon \in D_N$, and to achieve this we need to control the norms $\big\|\prod_{k\le N}(id_Y + \varepsilon_k P_k)y\big\|$.

Fix $\varepsilon \in D_N$ and set $A_\pm := \{1 \le k \le N : \varepsilon_k = \pm 1\}$. Since the P_k's commute, straightforward multiplication gives

$$\begin{aligned}\prod_{k\le N}(id_Y - P_k) &= \prod_{k\in A_+}(id_Y - P_k)\prod_{k\in A_-}(id_Y - P_k)\\ &= \sum_{B\subset A_+}(-1)^{|B|}P_B\prod_{k\in A_-}(id_Y - P_k).\end{aligned}$$

Rearrange to obtain

$$\begin{aligned}(-1)^{|A_+|}P_{A_+}&\prod_{k\in A_-}(id_Y - P_k)\\ &= \prod_{k\le N}(id_Y - P_k) - \sum_{B\subsetneq A_+}(-1)^{|B|}P_B\prod_{k\in A_-}(id_Y - P_k).\end{aligned}$$

This tells us that

$$\begin{aligned}\Big\|P_{A_+}\prod_{k\in A_-}(id_Y - P_k)\Big\| &\ge 2^N - \delta - \sum_{B\subsetneq A_+}\Big\|P_B\prod_{k\in A_-}(id_Y - P_k)\,y\Big\|\\ &\ge 2^N - \delta - (2^{|A_+|}-1)\cdot\Big\|\prod_{k\in A_-}(id_Y - P_k)\,y\Big\|\\ &= 2^N - \delta - (2^{|A_+|}-1)\cdot\Big\|\sum_{B\subset A_-}(-1)^{|B|}P_B\,y\Big\|\\ &\ge 2^N - \delta - (2^{|A_+|}-1)\cdot 2^{|A_-|} = 2^{|A_-|} - \delta.\end{aligned}$$

Recognizing that $P_{A_+}\prod_{k\in A_+}(id_Y + P_k) = 2^{|A_+|}P_{A_+}$, we deduce

$$\begin{aligned}\Big\|\prod_{k\le N}(id_Y + \varepsilon_k P_k)\,y\Big\| &\ge \Big\|\Big(P_{A_+}\prod_{k\in A_+}(id_Y + P_k)\prod_{k\in A_-}(id_Y - P_k)\Big)\,y\Big\|\\ &= 2^{|A_+|}\cdot\Big\|\Big(P_{A_+}\prod_{k\in A_-}(id_Y - P_k)\Big)\,y\Big\|\\ &\ge 2^{|A_+|}\cdot(2^{|A_-|}-\delta) \ge 2^N\cdot(1-\delta).\end{aligned}$$

This is the control we require. Expand $\prod_{k\le N}(id_Y + \varepsilon_k P_k)\,y$: for suitably chosen $y_{N+2},\dots,y_{2^N} \in B_Y$ we have

$$\prod_{k\le N}(id_Y + \varepsilon_k P_k)\,y = y + \sum_{k\le N}\varepsilon_k P_k y + y_{N+2} + \dots + y_{2^N}\,.$$

From this we obtain

$$\begin{aligned}\Big\|y + \sum_{k\le N}\varepsilon_k P_k y\Big\| &\ge \Big\|\prod_{k\le N}(id_Y + \varepsilon_k P_k)\,y\Big\| - \|y_{N+2}\| - \dots - \|y_{2^N}\|\\ &\ge 2^N\cdot(1-\delta) - (2^N - N - 1) = N + 1 - 2^N\cdot\delta.\end{aligned}$$

This is so regardless of $\varepsilon \in D_N$, but what we are really after is the estimate

$$(*)\qquad \sum_{k=0}^N |a_k| \ge \Big\|a_0 y + \sum_{k=1}^N a_k P_k y\Big\| \ge \frac{1}{2}\cdot\sum_{k=0}^N |a_k|,$$

valid for any choice of scalars $a_0,\dots,a_N$. To achieve it we settle on a value of δ, and $\delta = 2^{-N-1}$ will turn out to be a good choice. Naturally, there is no harm in assuming that $\sum_{k=0}^{N}|a_k| = 1$. Since B-convexity is a *real* notion, we can further assume that the a_k's are real.

The proof of $(*)$ is now very simple. The left hand inequality is obvious, and, agreeing that $P_0 = id_Y$, we can set $\varepsilon_k = \operatorname{sign} a_k$ $(0 \le k \le N)$ and get to work:

$$\begin{aligned}\Big\|\sum_{k=0}^{N} a_k P_k y\Big\| &= \Big\|\sum_{k=0}^{N} \varepsilon_k |a_k| P_k y\Big\| = \Big\|\sum_{k=0}^{N} \varepsilon_k P_k y - \sum_{k=0}^{N} \varepsilon_k \cdot (1-|a_k|) P_k y\Big\| \\ &\ge \Big\|\sum_{k=0}^{N} \varepsilon_k P_k y\Big\| - \sum_{k=0}^{N}(1-|a_k|) \ge N+1-2^N\cdot\delta - N = 1/2.\end{aligned}$$

This contradicts B-convexity. QED

This lemma easily extends to a useful result on products of arbitrary length.

13.13 Corollary: *Suppose that Y is a B-convex Banach space. Then there are real numbers $0 < \varrho < 2$ and $M > 1$ such that if $P_1,\ldots,P_n$ is any finite collection of commuting norm one projections on Y, then*

$$\Big\|\prod_{k\le n}(id_Y - P_k)\Big\| \le M\cdot\varrho^n.$$

Proof. The preceding lemma affirms that there are a $0 < \varrho < 2$ and $N \in \mathbf{N}$ such that for any commuting norm one projections $P_1,\ldots,P_N$ on Y,

$$\Big\|\prod_{k\le N}(id_Y - P_k)\Big\| \le \varrho^N.$$

Let $M := (2/\varrho)^N$ and suppose that $P_1,\ldots,P_n$ are commuting norm one projections on Y. There is an integer $m \ge 0$ such that $mN \le n < (m+1)N$. But then, as $n - mN < N$,

$$\Big\|\prod_{k\le n}(id_Y - P_k)\Big\| \le \varrho^{mN}\cdot\Big\|\prod_{k=mN+1}^{n}(id_Y - P_k)\Big\| \le \varrho^{mN}\cdot 2^{n-mN} \le M\cdot\varrho^n.$$

QED

We are nicely set up for the renorming result that will allow us to apply the Beurling - Kato Theorem.

13.14 Proposition: *Suppose that Y is a B-convex Banach space. There are real numbers $0 < \varrho < 2$ and $M > 1$ with the property that, given any finite family $P_1,\ldots,P_n$ of commuting norm one projections on Y with $P_A \ne 0$ for every $A \subset \{1,\ldots,n\}$, there is an equivalent norm $|\cdot|$ on Y with*

(a) $\|y\| \le |y| \le M\cdot\|y\|$ *for all* $y \in Y$

and such that the semigroup $(S_t)_{t\ge 0}$ generated by $P_1,\ldots,P_N$ satisfies

(b) $|S_t y| \le |y|$ *for every $y \in Y$ and $t \ge 0$*

and

(c) $|(id_Y - S_t)y| \le \varrho\cdot|y|$ *for every $y \in Y$ and $t \ge 0$.*

Proof. Naturally enough, M and ϱ come from 13.13. We start by showing that regardless of how $m \in \mathbf{N}$ and $t_1,\dots,t_m \geq 0$ are chosen,

$$\Big\| \prod_{j\leq m} (id_Y - S_{t_j}) \Big\| \leq M \cdot \varrho^m .$$

To this end, we use a 'weighted coin' model: we introduce a suitable probability space (Ω, Σ, P), and on this space we locate independent $\{0,1\}$-valued random variables $\chi_{j,k}$ $(1 \leq j \leq m,\ 1 \leq k \leq n)$ such that

$$P(\chi_{j,k} = 0) = 1 - e^{-t_j} \qquad \text{and} \qquad P(\chi_{j,k} = 1) = e^{-t_j} .$$

This allows us to express the S_{t_j}'s in terms of the vector-valued random variables

$$p_j : \Omega \longrightarrow \mathcal{L}(Y) : \omega \mapsto \prod_{k\leq n} \big(P_k + \chi_{j,k}(\omega)(id_Y - P_k)\big).$$

The relationship is simple enough since, by independence,

$$\begin{aligned} \int_\Omega p_j(\omega)\, dP(\omega) &= \prod_{k\leq n} \Big(P_k + \Big(\int_\Omega \chi_{j,k}(\omega) dP(\omega)\Big)(id_Y - P_k)\Big) \\ &= \prod_{k\leq n} \big(P_k + e^{-t_j}(id_Y - P_k)\big) = S_{t_j} . \end{aligned}$$

Since, for each $\omega \in \Omega$, $p_j(\omega) = \prod_{\{k : \chi_{j,k}(\omega) = 0\}} P_k$, the operators $p_1(\omega),\dots,p_m(\omega)$ form a commuting family of norm one projections on Y to which we can apply 13.13: there are $0 < \varrho < 2$ and $M > 1$ (independent of m and the particular collection) such that $\big\| \prod_{j\leq m} (id_Y - p_j(\omega)) \big\| \leq M \cdot \varrho^m$ for every $\omega \in \Omega$. But now, using independence again,

$$\begin{aligned} \Big\|\prod_{j\leq m} (id_Y - S_{t_j})\Big\| &= \Big\|\prod_{j\leq m} \int_\Omega (id_Y - p_j(\omega))\, dP(\omega)\Big\| \\ &= \Big\|\int_\Omega \prod_{j\leq m} (id_Y - p_j(\omega))\, dP(\omega)\Big\| \leq \int_\Omega \Big\|\prod_{j\leq m} (id_Y - p_j(\omega))\Big\|\, dP(\omega) \leq M \cdot \varrho^m . \end{aligned}$$

The climax is upon us. For $y \in Y$, we can define the desired norm by

$$|y| := \sup \Big\{ \varrho^{-m} \cdot \Big\| \prod_{j\leq m} (id_Y - S_{t_j}) \prod_{k\leq n} S_{\tau_k} y \Big\| \Big\}$$

where the supremum extends over all positive integers m and n and all real numbers $t_j, \tau_k \geq 0$ $(1 \leq j \leq m,\ 1 \leq k \leq n)$; empty products are interpreted as identity operators. Since the S_t's are contractions, $|\cdot|$ is a (semi-) norm on Y satisfying

$$\|y\| \leq |y| \leq M \cdot \|y\|$$

for every $y \in Y$. Properties (b) and (c) are totally trivial. QED

This proposition allows us to add the final link in the chain of arguments which prove Pisier's Theorem 13.3.

13.15 Theorem: *A Banach space X is K-convex if and only if it is B-convex.*

However, as we indicated in the preamble, most of the action takes place on $L_2(\mu_n, X)$ or $L_2(\mu, X)$ rather than on X itself. It behoves us to make a few remarks about these spaces.

Notice first that each $L_2(\mu_n)$ is canonically isometrically isomorphic to a subspace of $L_2(\mu)$, and to a subspace of $L_2(\mu_N)$ for each $N \geq n$. When we make this identification, $\bigcup_n L_2(\mu_n)$ is clearly dense in $L_2(\mu)$.

A useful orthonormal basis of the 2^n-dimensional space $L_2(\mu_n)$ consists of the 2^n *Walsh functions*

$$w_A := \prod_{k \in A} r_k,$$

where A runs through all subsets of $\{1,..., n\}$; we adopt the convention that $w_\emptyset$ is the constant one function. It follows that $\{w_A : A \subset \mathbf{N} \text{ is finite}\}$ is an orthonormal basis of $L_2(\mu)$.

The elements of $L_2(\mu_n, X)$ can be written in the form

$$\sum_{A \subset \{1,..,n\}} w_A(\cdot) x_A$$

where the x_A's are vectors in X. From this it is clear that, more generally, $L_2(\mu_n, X)$ is isometrically isomorphic to a subspace of $L_2(\mu, X)$ and that, under this identification, $\bigcup_n L_2(\mu_n, X)$ is still dense in $L_2(\mu, X)$. Moreover, each $L_2(\mu_n, X)$ is readily seen to be norm one complemented in $L_2(\mu, X)$.

This observation is very convenient for the study of K-convexity, where we are interested in whether or not

$$R^X : f(\cdot) \mapsto \sum_k r_k(\cdot) \int_D r_k(\omega) f(\omega) d\mu(\omega)$$

exists as a bounded linear map from $L_2(\mu, X)$ to itself.

Note that if $f \in L_2(\mu, X)$ is actually an element of $L_2(\mu_n, X)$, then there is no problem defining $R^X f$: we get the *finite* sum

$$R_n^X f(\cdot) := \sum_{k=1}^n r_k(\cdot) \int_{D_n} r_k(\omega) f(\omega) d\mu_n(\omega)$$

and so the well-defined operator

$$R_n^X : L_2(\mu_n, X) \longrightarrow L_2(\mu_n, X).$$

The fact that $\bigcup_n L_2(\mu_n, X)$ is dense in $L_2(\mu, X)$ enables us to assert that R^X exists as a bounded linear operator on $L_2(\mu, X)$ if and only if

$$\sup_n \|R_n^X\| < \infty,$$

and that in this case,

$$\|R^X\| = \sup_n \|R_n^X\|.$$

We can now settle up:

Proof of Theorem 13.15. We have seen before that K-convexity implies B-convexity.

Suppose conversely that X is a (complex) B-convex Banach space. By 13.10, X has non-trivial type $p > 1$, and as we remarked before, this allows us to assert that $L_2(\mu, X)$ and all $L_2(\mu_n, X)$ have type p and so are B-convex, too.

Now fix $n \in \mathbf{N}$. Given $\varepsilon \in D_n$ and $1 \leq j \leq n$, let

$$\varepsilon_{\pm}^{(j)} := (\varepsilon_1, ..., \varepsilon_{j-1}, \pm 1, \varepsilon_{j+1}, ..., \varepsilon_n).$$

For each $f \in L_2(\mu_n, X)$ and each $1 \leq j \leq n$, define $P_j f \in L_2(\mu_n, X)$ by

$$P_j f(\varepsilon) := \frac{1}{2} \cdot \left(f(\varepsilon_+^{(j)}) + f(\varepsilon_-^{(j)}) \right)$$

for all $\varepsilon \in D_n$. As P_j is just the conditional expectation operator relative to the σ-field generated by the events $\{\varepsilon_j = \pm 1\}$ we know that $P_1, ..., P_n$ is a commuting family of norm one projections on $L_2(\mu_n, X)$. It is convenient to view the P_j's as a commuting family of norm one projections on $L_2(\mu, X)$, and this is valid since $L_2(\mu_n, X)$ is norm one complemented in $L_2(\mu, X)$.

Let $(S_t)_{t \geq 0}$ be the semigroup on $L_2(\mu, X)$ generated by the P_j's. Since $P_A \neq 0$ for every $A \subset \{1, ..., n\}$, an appeal to Proposition 13.14 reveals that, after a suitable renorming of $L_2(\mu, X)$, we have, for some $1 \leq \varrho < 2$,

$$\|S_t\| \leq 1 \quad \text{and} \quad \|id_{L_2(\mu,X)} - S_t\| \leq \varrho$$

for all $t \geq 0$. This sets up the stage for use of the Beurling-Kato Theorem: there are numbers $K > 0$ and $0 < \alpha < \pi/2$, depending only on ϱ, such that $(S_t)_{t \geq 0}$ extends to an analytic semigroup $(S_z)_{z \in V_\alpha}$ of operators in $\mathcal{L}(L_2(\mu, X))$ with $\|S_z\| \leq K$ for all z in the cone V_α.

A quick perusal of the definition shows that each S_t is of the form

$$S_t = \sum_{k=0}^{n} e^{-kt} T_k$$

for appropriate operators T_k acting on $L_2(\mu_n, X)$, or on $L_2(\mu, X)$, whatever seems best. Our only reason for working with $L_2(\mu, X)$ was to get estimates independent of n! We shall see in a moment that $T_1 = R_n^X$. But first, let us take advantage of the analyticity we have been presented with. Clearly,

$$z \mapsto \sum_{k=0}^{n} e^{-kz} T_k$$

defines an analytic semigroup of operators on $L_2(\mu_n, X)$ whose restriction to the non-negative reals is just $(S_t)_{t \geq 0}$. The uniqueness properties of analytic functions lay bare the identity

$$S_z = \sum_{k=0}^{n} e^{-kz} T_k$$

for all $z \in V_\alpha$. The flexible domains of the T_k's allow us to view $(S_z)_{z \in V_\alpha}$ as an analytic semigroup in $\mathcal{L}(L_2(\mu_n, X))$.

We promised to prove that $T_1 = R_n^X$; here is the argument. It is good enough to show that T_1 and R_n^X agree on $w_A(\cdot)x$, however we choose $x \in X$ and $A \subset \{1, ..., n\}$.

Noticing that $w_{\{k\}} = r_k$ for each $1 \le k \le n$ and recalling the orthonormality of the r_k's, it soon becomes apparent that $R_n^X(w_A(\cdot)x) = w_A(\cdot)x$ if A is a singleton, whereas $R_n^X(w_A(\cdot)x) = 0$ otherwise. On the other hand, a perfunctory scrutiny of T_1's definition shows that

$$T_1 = \sum_{k\le n}(id_X - P_k)\prod_{j\ne k} P_j.$$

But $P_j(w_A(\cdot)x) = w_A(\cdot)x$ if $j \notin A$ and $P_j(w_A(\cdot)x) = 0$ if $j \in A$, and so we quickly realize that we always have $T_1(w_A(\cdot)x) = R_n^X(w_A(\cdot)x)$.

The final step consists estimating the norm of T_1 – alias R_n^X. Set $a := \pi/\tan\alpha$. Then $a + ib \in V_\alpha$ for any $-\pi < b < \pi$, and

$$(*) \qquad e^a \cdot \int_{-\pi}^{\pi} S_{a+ib}\, e^{ib}\, \frac{db}{2\pi} = \sum_{k=0}^{n} e^{(1-k)a}\, T_k \int_{-\pi}^{\pi} e^{i(1-k)b}\, \frac{db}{2\pi} = T_1.$$

Hence $\|T_1\| \le \sup_{-\pi<b<\pi} e^a \cdot \|S_{a+ib}\| \le K\cdot e^a$, thanks to the Beurling-Kato Theorem. We have shown that $\|R_n^X\| \le K\cdot e^{\pi/\tan\alpha}$ for every $n \in \mathbf{N}$, and so

$$\|R^X\| = \sup_n \|R_n^X\| \le K\cdot e^{\pi/\tan\alpha}.$$

Our work is done. QED

The formula $(*)$ obtained at the end of the preceding proof generalizes to give

$$e^{ka}\cdot \int_{-\pi}^{\pi} S_{a+ib}\, e^{ikb}\, \frac{db}{2\pi} = T_k$$

for each $1 \le k \le n$, so that $\|T_k\| \le K\cdot e^{ka}$.

Expand $S_t = \prod_{k=0}^{n}\big(P_k + e^{-t}(id_Y - P_k)\big)$ and use the special form of the projections P_k to find that on each $L_2(\mu_n, X)$ the T_k's are given by

$$T_k\Big(\sum_{A\subset\{1,..,n\}} w_A x_A\Big) = \sum_{|A|=k} w_A x_A \quad (1 \le k \le n).$$

Consequently, the preceding characterization of K-convexity may be extended as follows:

13.16 Theorem: *A Banach space X is K-convex if and only if (in the above notation)*

$$\sup_k \|T_k\|^{1/k} < \infty.$$

We are going to apply this in Chapter 15.

Some Consequences

In 11.10, we confronted the duality between type and cotype, but had to endure the frustration of an incomplete theory. We now relieve this frustration by showing that a perfect duality exists in spaces with non-trivial type; moreover, the equivalence of non-trivial type and K-convexity makes the proof really easy.

We start with a simple observation. If X is any Banach space, then $L_2(\mu_n, X)$ is just a weighted ℓ_2 sum of 2^n copies of X, and so the space $L_2(\mu_n, X)^*$ has a similar representation and is nothing but $L_2(\mu_n, X^*)$. The duality is in fact obtained by associating with each $g \in L_2(\mu_n, X^*)$ the linear form

$$L_2(\mu_n, X) \longrightarrow \mathbf{K} : f \mapsto \int_{D_n} \langle f(\omega), g(\omega)\rangle \, d\mu_n(\omega).$$

It follows that R_n^X's adjoint with respect to this duality is nothing but $R_n^{X^*}$, and we may conclude that X is K-convex if and only if X^* is – a restatement of 13.7 once the equivalence of B- and K-convexity is known.

If X is K-convex, then Kahane's Inequality 11.1 shows that R^X is continuous as an operator $L_p(\mu, X) \to L_p(\mu, X)$ for $2 \leq p < \infty$. The preceding observation allows us to extend this to the range $1 < p < \infty$ – as announced after the definition of K-convexity.

However, our main use for this observation is in proving:

13.17 Proposition: *Suppose that X is a K-convex Banach space and that $2 \leq q < \infty$. Then X has cotype q if and only if X^* has type q^*. Moreover,*

$$C_q(X) \leq T_{q^*}(X^*) \leq \|R^X\| \cdot C_q(X).$$

Proof. One implication is a special case of 11.10. To prove the other, fix $n \in \mathbf{N}$, choose $x_1^*, \dots, x_n^*$ in X^* and consider $F := \text{span}\{x_1^*, \dots, x_n^*\}$ as the dual of $E := X/F^\perp$. Since we may identify $L_2(\mu_n, E^*)$ with $L_2(\mu_n, E)^*$ as indicated above, we can for every $\delta > 0$ find $f \in L_2(\mu_n, E)$ with $\|f\| = 1$ such that

$$\Big(\int_{D_n} \Big\|\sum_{k\leq n} r_k(\omega) x_k^*\Big\|^2 d\mu_n(\omega)\Big)^{1/2} \leq (1+\delta) \cdot \int_{D_n} \Big\langle f(\omega), \sum_{k\leq n} r_k(\omega) x_k^*\Big\rangle \, d\mu_n(\omega).$$

Writing $x_k := \int_{D_n} f(\omega) r_k(\omega) \, d\mu_n(\omega)$ for each $1 \leq k \leq n$, we get

$$\begin{aligned}
&\Big(\int_{D_n} \Big\|\sum_{k\leq n} r_k(\omega) x_k^*\Big\|^2 d\mu_n(\omega)\Big)^{1/2} \\
&\quad \leq (1+\delta) \cdot \sum_{k\leq n} \langle x_k^*, x_k\rangle \leq (1+\delta) \cdot \Big(\sum_{k\leq n} \|x_k\|^q\Big)^{1/q} \cdot \Big(\sum_{k\leq n} \|x_k^*\|^{q^*}\Big)^{1/q^*} \\
&\quad \leq (1+\delta) \cdot C_q(X) \cdot \Big(\sum_{k\leq n} \|x_k^*\|^{q^*}\Big)^{1/q^*} \cdot \Big(\int_{D_n} \Big\|\sum_{k\leq n} r_k(\omega) x_k\Big\|^2 d\mu_n(\omega)\Big)^{1/2}.
\end{aligned}$$

But $\sum_{k\leq n} r_k(\cdot) x_k = \sum_{k\leq n} r_k(\cdot) \int_{D_n} f(\omega) r_k(\omega) \, d\mu_n(\omega) = (R_n^E f)(\cdot)$, and so

$$\begin{aligned}
\Big(\int_{D_n} \Big\|\sum_{k\leq n} r_k(\omega) x_k^*\Big\|^2 d\mu_n(\omega)\Big)^{1/2} &\leq (1+\delta) \cdot C_q(X) \cdot \|R_n^E\| \cdot \Big(\sum_{k\leq n} \|x_k^*\|^{q^*}\Big)^{1/q^*} \\
&\leq (1+\delta) \cdot C_q(X) \cdot \|R^X\| \cdot \Big(\sum_{k\leq n} \|x_k^*\|^{q^*}\Big)^{1/q^*}.
\end{aligned}$$

Since $\delta > 0$ was arbitrary and we were free to choose $x_1^*, \dots, x_n^*$ without restriction, we are done. QED

Here is an immediate consequence which should be compared with the remark preceding Proposition 11.11.

13.18 Corollary: *If X is a K-convex Banach space and has cotype q, then every quotient of X has cotype q.*

Next we return to a topic already touched upon in 11.18. Our goal is to prove the following result.

13.19 Theorem: *Let μ be any probability measure. A closed linear subspace X of $L_1(\mu)$ is reflexive if and only if it is B-convex.*

The result is true for arbitrary $L_1(\mu)$-spaces and even more general objects. We refer to Theorem 15.12 from which a second proof of 13.19 can be deduced. See also the Notes and Remarks to Chapter 15.

We know from 11.18 and 13.10 that B-convexity of a subspace of $L_1(\mu)$ entails reflexivity. The problem is to show the converse. We prepare for the proof with a lemma which sheds considerable light on the nature of almost exact replicas of ℓ_1^n's in $L_1[0,1]$. Small modifications will adapt the proof to $L_1(\mu)$ for any probability measure μ.

13.20 Lemma: *Let $0 < c < 1$ and suppose that $f_1, \ldots, f_n \in B_{L_1[0,1]}$ satisfy*

$$\Big\| \sum_{k=1}^n a_k f_k \Big\|_1 \geq c^{1/2} \cdot \sum_{k=1}^n |a_k|,$$

for all scalars $a_1, \ldots, a_n$. Then, for any $a_1, \ldots, a_n$,

$$\Big\| \max_{k \leq n} |a_k f_k| \Big\|_1 \geq c \cdot \sum_{k=1}^n |a_k|.$$

Moreover, there are pairwise disjoint Borel sets $A_1, \ldots, A_n \subset [0,1]$ such that

$$\int_{A_k} |f_k(s)|\, ds \geq c \quad \text{for all} \quad 1 \leq k \leq n.$$

Proof. The proof of the first statement consists of a chain of (in)equalities:

$$\begin{aligned}
c^{1/2} \cdot \sum_{k \leq n} |a_k| &\leq \int_0^1 \Big\| \sum_{k \leq n} a_k r_k(t) f_k \Big\|_1 dt = \int_0^1 \int_0^1 \Big| \sum_{k \leq n} a_k r_k(t) f_k(s) \Big|\, dt\, ds \\
&\leq \int_0^1 \Big(\int_0^1 \Big| \sum_{k \leq n} a_k r_k(t) f_k(s) \Big|^2 dt \Big)^{1/2} ds = \int_0^1 \Big(\sum_{k \leq n} |a_k f_k(s)|^2 \Big)^{1/2} ds \\
&\leq \int_0^1 \Big(\max_{k \leq n} |a_k f_k(s)| \Big)^{1/2} \cdot \Big(\sum_{k \leq n} |a_k f_k(s)| \Big)^{1/2} ds \\
&\leq \Big(\int_0^1 \max_{k \leq n} |a_k f_k(s)|\, ds \Big)^{1/2} \cdot \Big(\int_0^1 \sum_{k \leq n} |a_k f_k(s)|\, ds \Big)^{1/2} \\
&\leq \Big(\sum_{k \leq n} |a_k| \Big)^{1/2} \cdot \Big\| \max_{k \leq n} |a_k f_k| \Big\|_1^{1/2}.
\end{aligned}$$

We settle the second part in three steps. It is certainly enough to consider the case $f_k \geq 0$, $1 \leq k \leq n$.

As a first step we show that there exist non-negative functions $\varphi_1,\ldots,\varphi_n$ in $L_\infty[0,1]$ such that $\sum_{k\leq n} \varphi_k \leq 1$ and $\int_0^1 \varphi_k(s) f_k(s)\, ds \geq c$ for each $1 \leq k \leq n$.

To this end, consider

$$D := \{(\varphi_k)_{k=1}^n \in L_\infty[0,1]^n : \varphi_1 \geq 0,\ldots, \varphi_n \geq 0, \sum_k \varphi_k \leq 1\}$$

as a subset of $(L_\infty[0,1] \oplus \ldots \oplus L_\infty[0,1])_\infty = \big((L_1[0,1] \oplus \ldots \oplus L_1[0,1])_1\big)^*$. The set D is weak$*$compact and convex. Moreover, $(\varphi_1,\ldots,\varphi_n) \in D$ if and only if given a non-negative $f \in B_{L_1[0,1]}$ we have $\int_0^1 \varphi_k(s) f(s)\, ds \geq 0$ for each k and $\sum_{k\leq n} \int_0^1 \varphi_k(s) f(s)\, ds \leq 1$.

Define the operator

$$u : \ell_1^n \longrightarrow (L_1[0,1] \oplus \ldots \oplus L_1[0,1])_1 : (\lambda_k)_{k\leq n} \mapsto (\lambda_k f_k)_{k\leq n}.$$

$u^*(D)$ is weak$*$compact and convex in ℓ_∞^n; we claim that it has a point in common with $C := \{(c_k)_{k\leq n} \in \ell_\infty^n : c_k \geq c \text{ for all } k \leq n\}$. Once we can prove this, we need only notice that $u^*((\varphi_k)_k) = \big(\int_0^1 \varphi_k(s) f_k(s)\, ds\big)_k$ for $(\varphi_k)_k \in D$ to see that our first step has been taken.

Suppose that $u^*(D) \cap C$ is empty. Since C is plainly weak$*$closed and convex in ℓ_∞^n, the Hahn-Banach Theorem provides us with some $0 < \kappa < 1$ and $(a_k)_k \in \ell_1^n$ such that

$$\sum_{k\leq n} a_k \int_0^1 \varphi_k(s) f_k(s)\, ds \leq \kappa < 1 \leq \sum_{k\leq n} a_k c_k$$

for any $(c_k)_k \in C$ and $(\varphi_k)_k \in D$. Notice that $c \cdot \sum_{k\leq n} a_k \geq 1$. Also, for each $1 \leq j \leq n$ we have $a_j > 0$ since

$$c \cdot a_j = 1 - c \cdot \sum_{k\leq n, k\neq j} a_k \geq 1 - c > 0.$$

Let $g = \max_{k\leq n} a_k f_k$; we have shown that

$$c \cdot \sum_{k\leq n} a_k \leq \int_0^1 g(s)\, ds.$$

Break up $[0,1]$ into disjoint measurable sets $E_1,\ldots,E_n$ so that $a_k f_k(s) = g(s)$ for all $s \in E_k$ and each $1 \leq k \leq n$. Then $(1_{E_k})_{k\leq n} \in D$, and we arrive at the desired contradiction:

$$\kappa < c \cdot \sum_{k\leq n} a_k \leq \int_0^1 g(s)\, ds = \sum_{k\leq n} a_k \cdot \int_{E_k} f_k(s)\, ds \leq \kappa.$$

In the second step, we work with the functions $\varphi_k \in L_\infty[0,1]$ provided by the first step: so $\sum_{k\leq n} \varphi_k \leq 1$, $\varphi_k \geq 0$ and $\int_0^1 \varphi_k(s) f_k(s)\, ds \geq c$ for all k. We are going to find disjointly supported non-negative functions $\psi_1,\ldots,\psi_n$ in $L_\infty[0,1]$ with $\sum_{k\leq n} \psi_k \leq 1$ such that $\int_0^1 \psi_k(s) f_k(s)\, ds \geq c$ for all $1 \leq k \leq n$.

For this, look at the subset D_0 of D which consists of all $(\hat{\varphi}_k) \in D$ such that $\int_0^1 \hat{\varphi}_k(s) f_k(s)\,ds = \int_0^1 \varphi_k(s) f_k(s)\,ds$ for all $1 \le k \le n$. Certainly D_0 is a non-empty weak$*$compact convex subset of D. The Krein-Milman Theorem assures us that D_0 has an extreme point, $(\psi_k)_{k=1}^n$ say. We claim that $\psi_1,\dots,\psi_n$ are disjointly supported.

Suppose to the contrary that for some $1 \le i < j \le n$, $\psi := \psi_i \wedge \psi_j > 0$ on some set E of positive measure. Now $\{\psi \cdot 1_F : F \subset E \text{ measurable}\}$ spans an infinite dimensional subspace of $L_\infty[0,1]$ and so contains a non-negative function g with $g \le \psi$, $g = 0$ off E and, by hyperplane considerations, $\int_0^1 g(s) f_i(s)\,ds = 0 = \int_0^1 g(s) f_j(s)\,ds$. We have once more reached a contradiction since plainly $(\psi_1, \dots, \psi_{i-1}, \psi_i \pm g, \psi_{i+1}, \dots, \psi_{j-1}, \psi_j \mp g, \psi_{j+1}, \dots, \psi_n)$ belongs to D_0.

Now we can easily perform the final step. Take the ψ_k's from the second step. Then, if A_k denotes the support of ψ_k,

$$c \le \int_0^1 \psi_k(s) f_k(s)\,ds = \int_{A_k} \psi_k(s) f_k(s)\,ds \le \int_{A_k} f_k(s)\,ds,$$

as asserted. QED

In order to be able to complete our program, we recall an old standby from integration theory.

13.21 Theorem: *Let $\mu \in C(K)^*$ be a positive measure. A bounded subset W of $L_1(\mu)$ is relatively weakly compact if and only if it is uniformly integrable: for each $\varepsilon > 0$ there is a $\delta > 0$ such that whenever B is any Borel subset of K satisfying $\mu(B) \le \delta$, we have $\sup_{f \in W} \int_B |f|\,d\mu \le \varepsilon$.*

Proof of Theorem 13.19. All that remains be shown is that a reflexive subspace X of $L_1(\mu)$ is B-convex.

Our hypothesis implies that B_X is weakly compact; by 13.21 it is uniformly integrable. Accordingly, given $\varepsilon > 0$ there is a $\delta > 0$ such that for any measurable set A with $\mu(A) \le \delta$ we have $\int_A |f|\,d\mu \le \varepsilon$ for every $f \in B_X$.

Suppose X is not B-convex. Referring to 13.6, we see that X contains $(1+\varepsilon)$-uniform copies of ℓ_1^n's for any $\varepsilon > 0$. Let $0 < \varepsilon < 1$ and, for any $n \in \mathbf{N}$, locate $f_1^{(n)},\dots,f_n^{(n)} \in B_X$ satisfying $\|\sum_{k \le n} a_k f_k^{(n)}\| \ge (1+\varepsilon)^{-1} \cdot \sum_{k \le n} |a_k|$ for all scalars $a_1,\dots,a_n$. Appeal to 13.20 to find disjoint measurable sets $A_1^{(n)},\dots,A_n^{(n)}$ with the property that $\int_{A_k^{(n)}} |f_k^{(n)}|\,d\mu \ge (1+\varepsilon)^{-2}$ for each $1 \le k \le n$. If $n \cdot \delta \ge 1$ then one of the $A_k^{(n)}$'s, say $A_j^{(n)}$, must have measure $\le \delta$. By our preceding epsilonics, $(1+\varepsilon)^{-2} \le \int_{A_j^{(n)}} |f_j^{(n)}|\,d\mu < \varepsilon$. This cannot be true when ε is very small. QED

NOTES AND REMARKS

B-convexity was the discovery of A. Beck [1962]. His interests were probabilistic and he showed that the classical Strong Law of Large Numbers could be established in B-convex Banach spaces, and not in any other Banach space.

For some time it was thought that B-convexity implied reflexivity. Much was gained through the efforts to prove this. D.P. Giesy [1966], [1973a], [1973b] dedicated considerable energy to this problem, and along the way he showed that B-convex Banach lattices are reflexive. D.P. Giesy and R.C. James [1973] established the equivalence of B-convexity with non-containment of uniform copies of ℓ_1^n's. Since R.C. James [1964] had shown that uniformly non-square Banach spaces (for some $\delta > 0$, if $\|x\|, \|y\| \leq 1$, then either $(x+y)/2$ or $(x-y)/2$ has norm $\leq 1 - \delta$) are reflexive, the question of reflexivity of B-convex spaces was plainly a delicate one.

Eventually, R.C. James [1974] grew a non-reflexive B-convex space; his construction was simplified by R.C. James and J. Lindenstrauss [1974] and W.J. Davis and J. Lindenstrauss [1976] who refined it so as to produce non-reflexive spaces having type p for any $1 < p < 2$. Finally, R.C. James [1978] provided a non-reflexive space of type 2. G. Pisier and Q. Xu [1987] went in a different direction to construct non-reflexive spaces of type 2 and of cotype q for any $q > 2$; they unearthed these objects by applying Lions - Peetre interpolation to c_0 and the ℓ_1 analogue of the classical James space.

Our presentation of B-convexity follows G. Pisier's [1973/74a,b] expositions.

K-convexity was introduced by B. Maurey and G. Pisier [1976] at the very end of their paper. They raised the question of the equivalence of B-convexity and K-convexity. It was already clear from the work of Giesy and his collaborators that B-convexity follows from K-convexity; Maurey and Pisier established the converse for Banach lattices, but it was only in [1982a] that G. Pisier managed to demonstrate the general equivalence of K-convexity and B-convexity in Banach spaces. No argument has since been offered which avoids the use of the Beurling - Kato Theorem 13.11. Our presentation of this topic follows V.D. Milman and G. Schechtman [1986].

We used Rademacher variables to define the notion of K-convexity. Gaussian variables may be substituted, at the expense of changing $\|R^X\|$ by a factor $\pi/2$, as was noted by T. Figiel and N. Tomczak-Jaegermann [1979]. They also expressed K-convexity in terms of almost summing operators:

- *X is K-convex if and only if $\Pi_{as}^*(X, \ell_2)$ is contained in $\Pi_{as}^d(X, \ell_2)$.*

A good reference for such matters is N. Tomczak-Jaegermann's monograph [1989].

It should not be thought that our list of equivalences to B-convexity is exhaustive. Let us agree to say that a Banach space X has the *convex approximation property* if for every $\varepsilon > 0$ there is a natural number p such that whenever $K \subset B_X$, we have

$$\operatorname{conv}(K) \subset \Big\{\sum_{i\le p} \lambda_i x_i : x_i \in K,\ \lambda_i \ge 0,\ \textstyle\sum_{i\le p} \lambda_i = 1\Big\} + \varepsilon B_X.$$

R.E. Bruck [1981] established that a Banach space has the convex approximation property if and only if it is B-convex. To do this he called on Theorem 13.10. Bruck and, later, E. Casini and P.L. Papini [1993] applied this result to the study of the approximation of non-linear maps by maps with improved convexity properties.

In a completely different spirit, A. Pietsch [1990a] characterized the B-convexity of a Banach space in terms of the asymptotic behaviour of the eigenvalues of nuclear operators on the space.

M.I. Kadets and V.M. Kadets [1984] and V.M. Kadets [1989] showed B-convexity to be equivalent to the convexity of the sets of Riemann sums in Banach spaces.

The list goes on:

V.M. Kadets [1991] discovered that a Banach space X is B-convex if and only if there is a function $f : [0,1] \to [0,1]$ such that whenever Σ is a σ-field and $F : \Sigma \to X$ is countably additive and of total variation ≤ 1, the estimate $\|F(A)\| \le t$ for a fixed $0 \le t < 1$ and every atom $A \in \Sigma$ forces $d\big(F(\Sigma), \operatorname{conv}(F(\Sigma)\big) < f(t)$, thereby refining the Lyapunov convexity theorem.

Other orthogonal systems come in handy for the understanding of Banach spaces having non-trivial type. J. Bourgain [1982] brought harmonic analysis into the picture by showing that it is precisely in the presence of non-trivial type that a vector-valued Hausdorff-Young Inequality holds for characters on a compact abelian group.

Theorem: *Let G be a compact abelian group with dual group Γ. Denote Haar measure on G by dg. Then X has non-trivial type if and only if there exist numbers $p > 1$ and $q < \infty$ and a constant $M > 0$ such that*

$$\frac{1}{M}\cdot\Big(\sum_{\gamma\in\Gamma} \|x_\gamma\|^q\Big)^{1/q} \le \Big(\int_G \Big\|\sum_{\gamma\in\Gamma} \gamma(g) x_\gamma\Big\|^2\, dg\Big)^{1/2} \le M\cdot\Big(\sum_{\gamma\in\Gamma} \|x_\gamma\|^p\Big)^{1/p}$$

for any finitely non-zero family of vectors $(x_\gamma)_{\gamma\in\Gamma}$ in X.

Examples much like those produced early in Chapter 11 show that unless X has non-trivial type the attainment of such inequalities is hopeless.

Some years later, J.L. Rubio de Francia and J.L. Torrea [1987] continued in the spirit of Bourgain and related type and cotype to other mapping properties of the vector-valued Fourier transform.

H. König [1991] used Bourgain's work to uncover the equivalence of the K-convexity of a Banach space X with the absolute summability of the sequence $(\hat{f}(n))_{n\in\mathbf{Z}}$ of Fourier coefficients of an X-valued continuously differentiable function on the unit circle $\mathbf{T}$. He went on to use the notion of 'Fourier type p' to examine regular functions on $\mathbf{T}^m$ and $\mathbf{R}^m$. The same year, O. Blasco and A. Pełczyński [1991] showed that having non-trivial type is equivalent to having Fourier type > 1. They also established the equivalence of cotype 2 with Fourier cotype 2 and applied type and cotype ideas to the classical Hardy Inequality and Paley's Theorem for vector-valued functions.

It is interesting to note that the work of Bourgain and Rubio de Francia-Torrea traces its origins to S. Kwapień's remarkable paper [1972a] in which he characterizes Hilbert spaces by means of the behaviour of vector-valued Fourier series and transforms. Here is a typical example:

Theorem: *A complex Banach space X is isomorphic to a Hilbert space if and only if the vector-valued Fourier transform $\mathcal{F} : L_2(\mathbf{R}, X) \to L_2(\mathbf{R}, X)$ given by*

$$\mathcal{F}f(t) := \frac{1}{\sqrt{2\pi}} \cdot \int_{-\infty}^{\infty} e^{-ist} f(s)\, ds$$

exists as a bounded linear operator.

Theorem 13.19 is due to H.P. Rosenthal [1973] and we follow his proof in spirit if not in detail. In fact, the elegant Lemma 13.20 is due to L. Dor [1975b]; Rosenthal's original proof also relies on locating good copies on ℓ_1^n's. Theorem 13.21 is classical and has its roots in the work of N. Dunford [1939] though it is generally credited to N. Dunford and B.J. Pettis [1940].

14. Spaces with Finite Cotype

Having achieved a reasonable understanding of spaces with non-trivial type, we turn naturally to the study of Banach spaces with finite cotype. Suspicions that ℓ_∞^n's will play a prominent rôle in such a study are well-founded.

Consider a Banach space X which contains ℓ_∞^n's λ-uniformly for some $\lambda > 1$. It can never be graced with finite cotype: after all, for any $2 \leq q < \infty$, the unit coordinate vectors $e_1,\ldots, e_n$ of ℓ_∞^n obey $\big(\sum_{k\leq n} \|e_k\|^q\big)^{1/q} = n^{1/q}$, while $\big(\int_0^1 \|\sum_{k\leq n} r_k(t)e_k\|^q \, dt\big)^{1/q} = 1$. Further, such an X cannot have a $(q,1)$-summing identity for any finite q since the weak ℓ_1 norm of $(e_k)_{k\leq n}$ is one.

These facts are no coincidence. However, it will require some effort to put them in proper perspective. The culmination of our labours will be a most striking theorem, the central result of this chapter.

14.1 Theorem: *The following statements about the Banach space X are equivalent:*

(i) *X does not have finite cotype.*

(ii) *X contains ℓ_∞^n's λ-unformly for some (and then all) $\lambda > 1$.*

(iii) *id_X is not $(q,1)$-summing for any $2 \leq q < \infty$.*

Actually, this theorem is a special case of a more far-reaching and precise result, Theorem 14.5, which gives considerably finer information about Banach spaces having finite cotype.

Dvoretzky-Rogers Again

Recall that a Banach space X contains ℓ_∞^n's λ-uniformly if and only if for each n there are an n-dimensional subspace E_n of X and a factorization

$$id_{\ell_\infty^n} : \ell_\infty^n \xrightarrow{v_n} E_n \xrightarrow{u_n} \ell_\infty^n$$

with $\|u_n\| \leq \lambda$ and $\|v_n\| = 1$. Another way to put this is to say that for each n there are vectors $x_1,\ldots, x_n \in X$ such that

$$\lambda^{-1} \cdot \|a\|_\infty \leq \Big\| \sum_{k\leq n} a_k x_k \Big\| \leq \|a\|_\infty$$

for every $a = (a_k)_{k\leq n} \in \ell_\infty^n$.

Remarkably, we can achieve a strikingly similar, but significantly different, factorization through subspaces of *any* infinite dimensional Banach space. The next result is a profound deepening of Lemma 1.3, the key to the Dvoretzky-Rogers Theorem.

14.2 Theorem: *Let X be an infinite dimensional Banach space. For each $m \in \mathbf{N}$, we can find vectors $z_1,\ldots,z_m \in X$ such that for any $a \in \mathbf{R}^m$,*

$$\frac{1}{\sqrt{3}} \cdot \|a\|_{\ell_\infty^m} \le \Big\| \sum_{j \le m} a_j z_j \Big\| \le \|a\|_{\ell_2^m}.$$

In other words, we claim that for each n there are an n-dimensional subspace E_n of X together with two operators $v_n : \ell_2^n \to E_n$ and $u_n : E_n \to \ell_\infty^n$ such that $\|u_n\| \le \sqrt{3}$, $\|v_n\| = 1$, and $u_n v_n$ is the formal identity $\ell_2^n \to \ell_\infty^n$.

Do not think that the book is closed on this story. In our final chapter, we shall derive an astounding strengthening of the theorem above and prove Dvoretzky's famous theorem that ℓ_2 is finitely representable in any infinite dimensional Banach space.

The scheme of the proof of 14.2 bears some resemblence to the procedure we adopted for Lemma 1.3. However, we need to take on board our knowledge of the John ellipsoid (Chapter 6), and weather the complications that ensue.

Proof of 14.2. We shall actually prove a stronger result:

- *Let E be an n-dimensional normed space, and let $m \in \mathbf{N}$ be such that $m \le \sqrt{n}/4$. There are vectors $z_1,\ldots,z_m \in E$ such that, for every $a \in \mathbf{R}^m$,*

$$\frac{1}{\sqrt{3}} \cdot \|a\|_{\ell_\infty^m} \le \Big\| \sum_{j \le m} a_j z_j \Big\| \le \|a\|_{\ell_2^m}.$$

We present the proof in several steps.

First, Theorem 6.30 furnishes us with an isomorphism $u : \ell_2^n \to E$ having $\|u\| = 1$ and $\iota_1(u^{-1}) = n$. Thanks to 6.29, we know more: u^{-1} has a representation

$$u^{-1} = \sum_{j \le N} \lambda_j x_j^* \otimes y_j \tag{1}$$

where $N \le n^2 + 1$, the x_j^*'s and y_j's are unit vectors in E^* and ℓ_2^n, respectively, and the non-negative λ_j's sum to n.

Even more useful information can be extracted. Define

$$x_j := u y_j \quad (1 \le j \le N).$$

We shall establish that, for each $1 \le j \le N$,

$$\|x_j\|_E = \|x_j^*\|_{E^*} = \langle x_j^*, x_j \rangle = 1 \tag{2}$$

and

$$u^* x_j^* = y_j. \tag{3}$$

To confirm (2) we examine id_E.

$$\begin{aligned} n = \operatorname{tr}(id_E) &= \operatorname{tr}(uu^{-1}) = \sum_{j \le N} \lambda_j \langle x_j^*, u y_j \rangle \\ &\le \sum_{j \le N} \lambda_j \cdot \|x_j^*\|_{E^*} \cdot \|u y_j\|_E \le \sum_{j \le N} \lambda_j \cdot \|x_j^*\|_{E^*} \cdot \|y_j\|_2 = n. \end{aligned}$$

Equality must prevail everywhere in this chain. Consequently,

$$\|uy_j\|_E = \|y_j\|_2 = 1 \quad \text{and} \quad \langle x_j^*, uy_j\rangle = 1$$

for each $1 \le j \le N$, which is (2).

An absolutely analogous analysis of $\mathrm{tr}\,(id_{\ell_2^n}) = \mathrm{tr}\,(u^{-1}u)$ gives

$$\|u^* x_j^*\| = \left(u^* x_j^* \,\middle|\, y_j\right) = 1$$

for each j, and this implies (3) because we are in a Hilbert space.

In the next step, we describe an inductive process to select vectors $\hat{y}_1, \ldots, \hat{y}_n$ from $y_1, \ldots, y_N$ and orthogonal vectors $w_1, \ldots, w_n$ in ℓ_2^n so that for each $1 \le j \le n$

$$\hat{y}_j - w_j \in \mathrm{span}\left\{w_1, \ldots, w_{j-1}\right\}, \tag{4}$$

$$\|w_j\|_2^2 \ge 1 - \frac{j-1}{n}, \tag{5}$$

$$\|\hat{y}_j - w_j\|_2^2 \le \frac{j-1}{n}. \tag{6}$$

When $j = 1$ we interpret (4) to mean $\hat{y}_1 - w_1 = 0$. To start the induction, set $w_1 = y_1$. Since $\|y_1\|_2 = 1$, relations (4) and (5) hold trivially when $j = 1$.

Next assume that $1 \le k < n$ and that we have successfully discovered $\hat{y}_1, \ldots, \hat{y}_k$ and $w_1, \ldots, w_k$ satisfying (4), (5) and (6). Let $P_{k+1} \in \mathcal{L}(\ell_2^n)$ be the orthogonal projection onto $\mathrm{span}\,\{w_1, \ldots, w_k\}^\perp$. Locate an index $1 \le j_{k+1} \le N$ such that

$$\|P_{k+1} y_{j_{k+1}}\|_2 = \max_{j \le N} \|P_{k+1} y_j\|_2 \tag{7}$$

and set $\hat{y}_{k+1} := y_{j_{k+1}}$ and $w_{k+1} := P_{k+1}\hat{y}_{k+1}$. By design, w_{k+1} is orthogonal to $w_1, \ldots, w_k$. To show that $\hat{y}_{k+1}$ and w_{k+1} obey (4), simply observe that $P_{k+1}(\hat{y}_{k+1} - w_{k+1}) = P_{k+1}\hat{y}_{k+1} - P_{k+1}^2 \hat{y}_{k+1} = 0$, and so $\hat{y}_{k+1} - w_{k+1}$ belongs to $\ker P_{k_1} = \mathrm{span}\,\{w_1, \ldots, w_k\}$.

Inequality (5) is based on relations (1) and (3). Since $u^{-1}u = id_{\ell_2^n}$, we can use them to write each $y \in \ell_2^n$ in the form $y = \sum_{j \le N} \lambda_j (y_j | y) y_j$, and so

$$P_{k+1} = \sum_{j \le N} \lambda_j y_j \otimes P_{k+1} y_j.$$

Using (7) and the properties of orthogonal projections,

$$\begin{aligned} n - k = \mathrm{tr}\,(P_{k+1}) &= \sum_{j \le N} \lambda_j \left(y_j \,\middle|\, P_{k+1} y_j\right) = \sum_{j \le N} \lambda_j \|P_{k+1} y_j\|_2^2 \\ &\le \sum_{j \le N} \lambda_j \|P_{k+1}\hat{y}_{k+1}\|_2^2 = n \cdot \|w_{k+1}\|_2^2, \end{aligned}$$

and this reduces to (5). The remaining inequality (6) follows easily. Since (4) ensures that $\hat{y}_{k+1} - w_{k+1}$ and w_{k+1} are orthogonal,

$$1 = \|\hat{y}_{k+1}\|_2^2 = \|\hat{y}_{k+1} - w_{k+1}\|_2^2 + \|w_{k+1}\|_2^2 \ge \|\hat{y}_{k+1} - w_{k+1}\|_2^2 + 1 - (k/n);$$

this rearranges to give (6).

For the final step, we fix $m \le \sqrt{n}/4$, set $v_j := w_j/\|w_j\|_2$ and define

$$z_j := uv_j \quad (1 \le j \le m).$$

Obviously, when $a \in \mathbf{R}^m$,

$$\Big\|\sum_{j\le m} a_j z_j\Big\|_E \le \|u\|\cdot\Big\|\sum_{j\le m} a_j v_j\Big\|_2 = \|a\|_{\ell_2^m}.$$

We claim that

$$\|(u^{-1})^* v_j\|_{E^*} \le \sqrt{3} \quad (1 \le j \le m). \tag{8}$$

Granting this, we obtain for any $a \in \mathbf{R}^m$ and any $1 \le k \le m$,

$$\sqrt{3}\cdot\Big\|\sum_{j\le m} a_j z_j\Big\|_E \ge \Big|\Big\langle\sum_{j\le m} a_j z_j\,,\,(u^{-1})^* v_k\Big\rangle\Big| = \Big|\Big\langle\sum_{j\le m} a_j u^{-1} z_j\,,\,v_k\Big\rangle\Big| = |a_k|,$$

and hence

$$\Big\|\sum_{j\le m} a_j z_j\Big\|_E \ge \frac{1}{\sqrt{3}}\cdot\|a\|_{\ell_\infty^m},$$

which was what we wanted.

The end is nigh. We again use induction to supply the missing link (8). It is useful to rewrite (3) in the form

$$x_j^* = (u^*)^{-1} y_j = (u^{-1})^* y_j \quad (1 \le j \le N). \tag{9}$$

To begin the induction, note that $v_1 = w_1 = y_1$, and so

$$\big\|(u^{-1})^* v_1\big\|_{E^*} = \|x_1^*\|_{E^*} \le \sqrt{3}\,.$$

Now assume that for some $1 \le k < m$ we know that $\|(u^{-1})^* v_j\|_{E^*} \le \sqrt{3}$ for all $1 \le j \le k$. The norm of the restriction of $(u^{-1})^*$ to span $\{v_1,\dots,v_k\}$ is then at most $\sqrt{3k}$. In particular, because of (4), we get

$$\big\|(u^{-1})^*(\hat{y}_{k+1} - w_{k+1})\big\|_{E^*} \le \sqrt{3k}\cdot\|\hat{y}_{k+1} - w_{k+1}\|_2$$

and so, applying (6),

$$\big\|(u^{-1})^*(\hat{y}_{k+1} - w_{k+1})\big\|_{E^*} \le \sqrt{3/n}\cdot k.$$

Hence

$$\begin{aligned}\big\|(u^{-1})^* w_{k+1}\big\|_{E^*} &\le \big\|(u^*)^{-1}\hat{y}_{k+1}\big\|_{E^*} + \big\|(u^*)^{-1}(\hat{y}_{k+1} - w_{k+1})\big\|_{E^*}\\ &\le \|x_{j_{k+1}}\|_{E^*} + \sqrt{3/n}\cdot k \;=\; 1 + \sqrt{3/n}\cdot k.\end{aligned}$$

When we incorporate (5), our restriction $k \le m \le \sqrt{n}/4$ leads to

$$\big\|(u^{-1})^* v_{k+1}\big\|_{E^*} \le \frac{1 + \sqrt{3/n}\cdot k}{\|w_{k+1}\|_2} \le \frac{1 + \sqrt{3/n}\cdot k}{\sqrt{1 - (k/n)}} \le \sqrt{3}\,. \qquad \text{QED}$$

Factoring Formal Identities

The preceding theorem calls for a definition. If $2 \le q < \infty$ then we say that the Banach space X *finitely factors (the formal identity)* $\ell_q \hookrightarrow \ell_\infty$ *for* $0 < \delta < 1$ if for every $n \in \mathbf{N}$ we can find $x_1,\dots,x_n \in X$ such that

$$(1-\delta)\cdot\|a\|_\infty \le \Big\|\sum_{k\le n} a_k x_k\Big\| \le \|a\|_q$$

for every $a \in \ell_q^n$.

No distortion of $\ell_q \hookrightarrow \ell_\infty$ produced by X is ever really bad; in fact, if X finitely factors $\ell_q \hookrightarrow \ell_\infty$ for some $0 < \delta < 1$, then it does so for all of them. But this requires some work, and to carry it out we continue with our program of defining well-behaved moduli.

Let X be a Banach space and let $2 \le q < \infty$. For each $n \in \mathbf{N}$, define

$$\alpha_n(q, X) = \alpha_n(q)$$

to be the smallest non-negative α such that, however we choose $x_1,\dots,x_n \in X$,

$$\min_{k\le n} \|x_k\| \le \alpha \cdot \sup\Big\{\Big\|\sum_{k\le n} \lambda_k x_k\Big\| : \lambda \in B_{\ell_q^n}\Big\}.$$

14.3 Lemma: *Let X be a Banach space and let $2 \le q \le \infty$. For any natural numbers m, n we have*

(a) $0 \le \alpha_{m+n}(q) \le \alpha_m(q) \le \dots \le \alpha_1(q) = 1$

and

(b) $\alpha_{mn}(q) \le \alpha_m(q)\cdot\alpha_n(q)$.

Proof. (a) falls after a few moments spent contemplating the definition of the moduli $\alpha_n(q)$.

For (b) we fix vectors $x_1,\dots,x_{mn} \in X$ and assign scalars $t_1,\dots,t_{mn}$ so that, for each $1 \le k \le n$,

$$\Big\|\sum_{(k-1)m<j\le km} t_j x_j\Big\| = \sup\Big\{\Big\|\sum_{(k-1)m<j\le km} \lambda_j x_j\Big\| : \lambda \in B_{\ell_q^m}\Big\}.$$

Applying the definition of $\alpha_n(q)$ to the vectors

$$y_k := \sum_{(k-1)m<j\le km} t_j x_j,$$

we get

$$\min_{(k-1)m<j\le km} \|x_j\| \le \alpha_m(q)\cdot\|y_k\| \quad (1 \le k \le n).$$

It follows that

$$\begin{aligned}\min_{k\le mn} \|x_k\| &\le \alpha_m(q)\cdot\min_{k\le n}\|y_k\| \le \alpha_m(q)\cdot\alpha_n(q)\cdot\sup\Big\{\Big\|\sum_{k\le n}\lambda_k y_k\Big\| : \lambda \in B_{\ell_q^n}\Big\}\\ &= \alpha_m(q)\cdot\alpha_n(q)\cdot\sup\Big\{\Big\|\sum_{k\le n}\sum_{(k-1)m<j\le km}\lambda_k t_j x_j\Big\| : \lambda \in B_{\ell_q^n}\Big\}\\ &\le \alpha_m(q)\cdot\alpha_n(q)\cdot\sup\Big\{\Big\|\sum_{k\le mn} s_k x_k\Big\| : s \in B_{\ell_q^{mn}}\Big\}.\end{aligned}$$

QED

Indifference to δ is now a delicate diversion.

14.4 Proposition: *Let $2 \le q \le \infty$. If the Banach space X finitely factors $\ell_q \hookrightarrow \ell_\infty$ for some $0 < \delta < 1$, then it does so for all of them.*

Proof. For accurate book-keeping, let us assume that X finitely factors $\ell_q \hookrightarrow \ell_\infty$ for $0 < \delta_0 < 1$. This allows us to select, for each $n \in \mathbf{N}$, vectors $x_1,\dots,x_n \in X$ satisfying

$$(1-\delta_0)\cdot\|a\|_\infty \le \Big\|\sum_{k\le n} a_k x_k\Big\| \le \|a\|_q$$

for every $a = (a_1,\dots,a_n) \in \ell_q^n$.

The first stage of the argument is to exploit this information through use of the lemma to show that $\alpha_n(q) = 1$ for every $n \in \mathbf{N}$.

First notice that if $\|x_{k_0}\| = \min_{k\le n}\|x_k\|$, then

$$\begin{aligned}1-\delta_0 &\le \Big\|\sum_{k\le n}\delta_{k,k_0}x_k\Big\| = \|x_{k_0}\| \\ &\le \alpha_n(q)\cdot\sup\Big\{\Big\|\sum_{k\le n}a_k x_k\Big\| : a \in B_{\ell_q^n}\Big\} \le \alpha_n(q).\end{aligned}$$

Taking this with (a) of the lemma, we deduce that $\lim_{n\to\infty}\alpha_n(q)$ exists and is at least $1-\delta_0 > 0$.

Could it be that some $\alpha_N(q) < 1$? If so, since part (b) of the lemma entails that $\alpha_{n^k}(q) \le \alpha_n(q)^k$ for each n, we would get the contradiction $\lim_{n\to\infty}\alpha_n(q) = \lim_{k\to\infty}\alpha_{N^k}(q) = 0$. We conclude, as we asserted above, that each $\alpha_n(q) = 1$.

This is just what is needed to prove that X finitely factors $\ell_q \hookrightarrow \ell_\infty$ for any $0 < \delta < 1$.

Fix $0 < \delta < 1$ and $n \in \mathbf{N}$ and, with great foresight, set $\Delta = n^{-1}(\delta/2)^{q^*}$. Certainly $0 < \Delta < 1$, and so, since $\alpha_n(q) = 1$, we can find $y_1,\dots,y_n \in X$ such that

$$\sup\Big\{\Big\|\sum_{k\le n}a_k y_k\Big\| : a \in B_{\ell_q^n}\Big\} = 1 \tag{1}$$

and

$$\min_{k\le n}\|y_k\| \ge (1-\Delta)^{1/q^*}. \tag{2}$$

Thanks to (1) we simply have to show that for any $a \in \ell_q^n$ we have

$$(1-\delta)\cdot\|a\|_\infty \le \Big\|\sum_{k\le n}a_k y_k\Big\|. \tag{3}$$

By homogeneity, it is enough to prove this for a's with $\|a\|_q = 1$. Fix one such a, and suppose that $|a_{k_1}| = \|a\|_\infty$. Select $y_1^* \in X^*$ with $\|y_1^*\| = 1$ and $\langle y_{k_1}, y_1^*\rangle = \|y_{k_1}\|$. Since

$$\sup\Big\{\Big\|\sum_{k\le n}t_k y_k\Big\| : t \in B_{\ell_q^n}\Big\} = \sup\Big\{\Big(\sum_{k\le n}|\langle y_k, x^*\rangle|^{q^*}\Big)^{1/q^*} : x^* \in B_{X^*}\Big\}$$

we can apply (2) to find

$$1 \ge \|y_{k_1}\|^{q^*} + \sum_{k\ne k_1}|\langle y_k, y_1^*\rangle|^{q^*} \ge (1-\Delta) + \sum_{k\ne k_1}|\langle y_k, y_1^*\rangle|^{q^*}.$$

Rearranging, we get $\sum_{k\ne k_1}|\langle y_k, y_1^*\rangle|^{q^*} \le \Delta$. This explains the following chain of events.

$$\begin{aligned}
\Big\|\sum_{k\le n} a_k y_k\Big\| &\ge \Big|\sum_{k\le n} a_k\langle y_k, y_1^*\rangle\Big| \ge |a_{k_1}|\cdot\|y_{k_1}\| - \sum_{k\ne k_1}|a_k|\cdot|\langle y_k, y_1^*\rangle| \\
&\ge \|a\|_\infty\cdot\Big((1-\Delta)^{1/q^*} - \sum_{k\ne k_1}|\langle y_k, y_1^*\rangle|\Big) \\
&\ge \|a\|_\infty\cdot\Big((1-\Delta)^{1/q^*} - n^{1/q}\cdot\Big(\sum_{k\ne k_1}|\langle y_k, y_1^*\rangle|^{q^*}\Big)^{1/q^*}\Big) \\
&\ge \|a\|_\infty\cdot\Big((1-\Delta)^{1/q^*} - n^{1/q}.\Delta^{1/q^*}\Big) \ge \|a\|_\infty\cdot\Big(1-\Delta^{1/q^*} - n^{1/q}.\Delta^{1/q^*}\Big) \\
&\ge \|a\|_\infty\cdot\Big(1-2\cdot n^{1/q^*}.\Delta^{1/q^*}\Big) = \|a\|_\infty\cdot(1-\delta).
\end{aligned}$$

The proof is complete. QED

THE MAIN THEOREM

To discuss the main theorem of this chapter, we require three parameters related to the various concepts we have introduced so far. Given a Banach space X, we define q_X, r_X, $s_X \in [2,\infty]$ by

$$\begin{aligned}
q_X &:= \inf\{2\le q\le\infty : X \text{ has cotype } q\}, \\
r_X &:= \sup\{2\le q\le\infty : X \text{ finitely factors } \ell_q \hookrightarrow \ell_\infty\}, \\
s_X &:= \inf\{2\le q\le\infty : id_X \in \Pi_{q,1}(X)\ \}.
\end{aligned}$$

Here we agree that $\Pi_{\infty,1} = \mathcal{L}$, which is certainly consistent with our earlier convention $\Pi_\infty = \mathcal{L}$.

These notions are uninteresting when X is finite dimensional, so we assume for the rest of the chapter that

- *X is an infinite dimensional Banach space.*

In fact, such a proliferation of parameters is quite unnecessary:

14.5 Theorem: *For every infinite dimensional Banach space X,*

$$q_X = r_X = s_X.$$

Theorem 14.1 is just the case $q_X = \infty$ of this theorem.

Our strategy for proving Theorem 14.5 is very straightforward: we shall show that $r_X \le s_X$, $s_X \le q_X$ and $q_X \le r_X$. Two of these are reasonably easy to handle, so we dispose of them separately.

14.6 Proposition: *$r_X \le s_X$ and $s_X \le q_X$.*

Proof. To show that $r_X \le s_X$ we work by contradiction. Suppose that $s_X < r < q < r_X$. The definition of r_X tells us that X finitely factors $\ell_q \hookrightarrow \ell_\infty$ and so, for every $n \in \mathbf{N}$, 14.4 provides us with vectors $x_1,\ldots,x_n \in X$ such that

$$\frac{1}{2}\cdot\|a\|_\infty \le \Big\|\sum_{k\le n} a_k x_k\Big\| \le \|a\|_q$$

for any $a \in \ell_q^n$. In particular, each x_k has norm $\geq 1/2$, and so

$$\Big(\sum_{k\leq n} \|x_k\|^r\Big)^{1/r} \geq \frac{1}{2}\cdot n^{1/r}.$$

However,

$$\|(x_k)\|_1^{weak} = \sup_{\varepsilon\in D_n} \Big\|\sum_{k\leq n} \varepsilon_k x_k\Big\| \leq n^{1/q}.$$

Since $r < q$, the map id_X cannot be $(r,1)$-summing and the choice of $s_X < r$ is contradicted.

The other inequality is even easier. Thanks to Kahane's Inequality 11.1, when X has cotype q, there is a $c > 0$ such that

$$\Big(\sum_{k\leq n} \|x_k\|^q\Big)^{1/q} \leq c\cdot \int_0^1 \Big\|\sum_{k\leq n} r_k(t)x_k\Big\|\, dt$$

for any choice of $x_1,\ldots,x_n \in X$. But $\sup_{0\leq t\leq 1} \big\|\sum_{k\leq n} r_k(t)x_k\big\| = \|(x_k)\|_1^{weak}$, and so id_X must be $(q,1)$-summing. Thus $s_X \leq q$ whenever X has cotype q, and $s_X \leq q_X$ follows. QED

Since $r_X \geq 2$ for every Banach space X, nothing more needs to be proved when $q_X = 2$. We take advantage of this situation and give ourselves room for manoeuvre.

- *From now on we assume that* $q_X > 2$.

To settle the third inequality $q_X \leq r_X$ we need much stronger tools; we shall spend several pages honing these. We start by observing that, thanks to our running hypothesis, we can get away with proving that if $2 \leq q < q_X$, then X finitely factors $\ell_q \hookrightarrow \ell_\infty$. The next result still falls short of what we require, but it is a significant stepping stone on our way.

14.7 Proposition: *Let $2 \leq q < q_X$. There is a constant $C > 0$ such that for each β satisfying $0 < \beta < q^{-1} - q_X^{-1}$ and each $n \in \mathbf{N}$ there are norm one vectors $x_1,\ldots,x_n \in X$ with the property*

$$\Big(\int_0^1 \Big\|\sum_{k\leq n} a_k r_k(t)x_k\Big\|^q\, dt\Big)^{1/q} \leq C\cdot\beta^{-1/q}\cdot\|a\|_{\ell_{q,1}^n}$$

for all $a \in \ell_{q,1}^n$.

To get a good grip on things it is useful to introduce variants of the constants appearing in the definition of cotype q. We take advantage of Kahane's Inequality and replace L_2-Rademacher averages by L_q-averages which turn out to be more suited to the manipulations we intend to perform. More precisely, given a Banach space X and $2 \leq q \leq \infty$, the *n-th cotype q number* of X,

$$c_q(n, X) \quad \text{or} \quad c_q(n),$$

designates the smallest number c such that, no matter how we choose n vectors $x_1,\ldots,x_n$ from X,

$$\Big(\sum_{k\le n}\|x_k\|^q\Big)^{1/q} \le c\cdot\Big(\int_0^1 \Big\|\sum_{k\le n} r_k(t)x_k\Big\|^q\,dt\Big)^{1/q}.$$

The basic properties of our new parameters shouldn't come as a surprise.

14.8 Lemma: *Let $2 \le q \le \infty$. Then $c_q(1) = 1$ and for all $m, n \in \mathbf{N}$,*

(a) $c_q(n) \le c_q(n+1)$,

(b) $c_q(n) \le n^{1/q}$

and

(c) $c_q(mn) \le c_q(m)\cdot c_q(n)$.

Proof. We might as well assume that $q < \infty$. Then (a) yields to a cursory examination of the definition, whereas (b) follows almost as quickly from the Contraction Principle 12.2: however $x_1,\dots,x_n$ are selected from X,

$$\Big(\sum_{k\le n}\|x_k\|^q\Big)^{1/q} \le n^{1/q}\cdot\max_{k\le n}\|x_k\| \le n^{1/q}\cdot\Big(\int_0^1 \Big\|\sum_{k\le n} r_k(t)x_k\Big\|^q dt\Big)^{1/q}.$$

Only (c) requires a modicum of effort. Choose $x_1,\dots,x_{mn}$ from X. Partitioning in the natural way, we can use the definition together with the equidistribution of the Rademacher functions to derive

$$\begin{aligned}
\sum_{k\le mn}\|x_k\|^q &= \sum_{j\le n}\Big(\sum_{(j-1)m<k\le jm}\|x_k\|^q\Big)\\
&\le c_q(m)^q\cdot\int_0^1\sum_{j\le n}\Big\|\sum_{(j-1)m<k\le jm} r_{k-(j-1)m}(t)x_k\Big\|^q\,dt\\
&= c_q(m)^q\cdot\int_0^1\sum_{j\le n}\Big\|\sum_{(j-1)m<k\le jm} r_k(t)x_k\Big\|^q\,dt\\
&\le c_q(m)^q\cdot c_q(n)^q\cdot\int_0^1\int_0^1\Big\|\sum_{j\le n} r_j(s)\sum_{(j-1)m<k\le jm} r_k(t)x_k\Big\|^q\,dt\,ds\\
&= c_q(m)^q\cdot c_q(n)^q\cdot\int_0^1\Big\|\sum_{k\le mn} r_k(t)x_k\Big\|^q\,dt;
\end{aligned}$$

the last equation is due to the symmetry of the Rademacher sequence. QED

Of course, our interest in the $c_q(n)$'s lies deeper. We see next that (b) in the previous lemma can be nicely complemented.

14.9 Lemma: *Let $2 \le q \le q_X$. Then*

(a) $c_q(n) \ge n^{(1/q)-(1/q_X)}$ *for every* $n \in \mathbf{N}$

and

(b) $\limsup_n n\cdot\Big(\dfrac{c_q(n)}{c_q(n-1)} - 1\Big) \ge \dfrac{1}{q} - \dfrac{1}{q_X}$.

Proof. (a) Since $c_\infty(n)$ is always 1, we can assume that $q < \infty$.

Fix $n \geq 2$. From (b) of the previous lemma we know that $c_q(n) \leq n^{1/q}$ and so we can assert that $c_q(n) = n^{(1/q)-(1/r)}$ for some $q \leq r \leq \infty$. We will show that X has cotype s for any $s > r$; this will alert us to the fact that $q_X \leq r$, and so $c_q(n) \geq n^{(1/q)-(1/q_X)}$, as required.

The statement about cotype is vacuously true if $r = \infty$; assume therefore that $r < \infty$. Choose $m \in \mathbf{N}$ and select $x_1,\ldots,x_m$ from X, ordering them, without prejudice to our cause, so that $\|x_1\| \geq \|x_2\| \geq \ldots \geq \|x_m\|$. Apply the Contraction Principle 12.2 to see that for each $1 \leq k \leq m$,

$$\|x_k\| \leq k^{-1/q}\cdot\big(\sum_{j\leq k}\|x_j\|^q\big)^{1/q} \leq k^{-1/q}\cdot c_q(k)\cdot\big(\int_0^1 \|\sum_{j\leq m} r_j(t)x_j\|^q\,dt\big)^{1/q}.$$

To proceed further, notice that every k determines an $\ell \in \mathbf{N}$ such that $n^{\ell-1} \leq k < n^\ell$. The various parts of the preceding lemma join to give

$$c_q(k) \leq c_q(n^\ell) \leq c_q(n)^\ell = c_q(n)\cdot\big(n^{(1/q)-(1/r)}\big)^{\ell-1} \leq c_q(n)\cdot k^{(1/q)-(1/r)}.$$

Plug this into our previous estimate to obtain, for each $1 \leq k \leq m$,

$$\|x_k\| \leq c_q(n)\cdot k^{-1/r}\cdot\big(\int_0^1 \|\sum_{j\leq m} r_j(t)x_j\|^q\,dt\big)^{1/q}.$$

From this it follows that

$$\big(\sum_{k\leq m}\|x_k\|^s\big)^{1/s} \leq c_q(n)\cdot\big(\sum_{k=1}^{\infty}k^{-s/r}\big)^{1/s}\cdot\big(\int_0^1 \|\sum_{j\leq m} r_j(t)x_j\|^q\,dt\big)^{1/q}.$$

The infinite sum converges when $s > r$ and so, referring to Kahane's Inequality, we realize that we have proved what we wanted.

(b) The monotonicity of the $c_q(n)$'s allows us to suppose that $L := q^{-1} - q_X^{-1}$ is positive. Choose any $0 < \varepsilon < L$. Using (a) we can write

$$c_q(n) = a_n\cdot n^{L-\varepsilon}$$

where $\lim_{n\to\infty} a_n = \infty$. Accordingly, $a_n/a_{n-1} \geq 1$ must be true for infinitely many n's, and for these we have

$$n\cdot\Big(\frac{c_q(n)}{c_q(n-1)} - 1\Big) = n\cdot\Big(\frac{a_n}{a_{n-1}}\big(\frac{n^{L-\varepsilon}}{(n-1)^{L-\varepsilon}} - 1\big) + \big(\frac{a_n}{a_{n-1}} - 1\big)\Big)$$
$$\geq n\cdot\Big(\big(1-\frac{1}{n}\big)^{-L+\varepsilon} - 1\Big) \geq L - \varepsilon.$$

Our conclusion is immediate. QED

Actually, it is possible to make this lemma much more precise. It is even true that $\lim_n c_q(n)\cdot n^{(1/q_X)-(1/q)} = 1$ when $q \leq q_X$. We do not need this fact, so we do not stop to prove it. Perhaps, however, it is worth noting that if $q_X = \infty$, then $c_q(n) = n^{1/q}$ for all q and all n.

ΘBecause we are working with L_q-Rademacher averages, we can help ourselves to a useful stability result (compare also with Theorem 11.12).

14.10 Lemma: *If (Ω, Σ, μ) is a non-trivial measure space, then the Banach spaces X and $L_q(X) = L_q(\Omega, \Sigma, \mu, X)$ have the same cotype q numbers:*

$$c_q(n, X) = c_q(n, L_q(X)).$$

Proof. The inequality $c_q(n, X) \leq c_q(n, L_q(X))$ is easy: just identify X with a subspace of $L_q(X)$ consisting of functions constant on a fixed set of finite non-zero measure and zero outside.

For the reverse, take any $f_1, ..., f_n$ in $L_q(X)$. Then

$$\begin{aligned}\Big(\sum_{k\leq n} \|f_k\|_{L_q(X)}\Big)^{1/q} &= \Big(\int_\Omega \sum_{k\leq n} \|f_k(\omega)\|_X^q \, d\mu(\omega)\Big)^{1/q} \\ &\leq c_q(n, X)\cdot\Big(\int_\Omega \int_0^1 \Big\|\sum_{k\leq n} r_k(t) f_k(\omega)\Big\|_X^q \, dt\, d\mu(\omega)\Big)^{1/q} \\ &= c_q(n, X)\cdot\Big(\int_0^1 \Big\|\sum_{k\leq n} r_k(t) f_k\Big\|_{L_q(X)}^q \, dt\Big)^{1/q}.\end{aligned}$$

We are done. QED

There is one more ingredient to the proof of Proposition 14.7. Although we have suppressed all reference to $(q, 1)$-summing operators, 14.6 makes it clear that they are always lurking somewhere in the background. This means that operators in $\mathcal{L}(\ell_\infty^n, X)$ are never far from sight. The missing link is a clever result which shows how knowledge of the $(q, 1)$-summing norm of an operator $u : \ell_\infty^n \to X$ gives good control of $\|u(a)\|$ provided $a \in \ell_\infty^n$ has suitably restricted support.

14.11 Lemma: *Let $2 \leq q \leq \infty$ and suppose that $u \in \mathcal{L}(\ell_\infty^n, X)$ satisfies*

$$\Big(\sum_{k\leq n} \|u(a^{(k)}\|^q\Big)^{1/q} \leq \|(a^{(k)})\|_1^{weak}$$

for every $a^{(1)}, ..., a^{(n)} \in \ell_\infty^n$. There is a subset A of $\{1, ..., n\}$ and a constant K, independent of n, such that

(a) *A has at least $n/2$ elements*

and

(b) $\|u(a1_A)\| \leq K \cdot n^{-1/q} \cdot \|a\|_{\ell_{q,1}^n}$ *for all $a \in \ell_{q,1}^n$.*

Proof. Ignoring the trivial case $q = \infty$, we begin by aiming for a somewhat different result in which (b) is replaced by

(c) $\|u(\varepsilon 1_B)\| \leq \Big(\dfrac{2\cdot|B|}{n}\Big)^{1/q}$ for all $\varepsilon \in D_n$ and $B \subset A$.

Here $|B|$ is the cardinality of the set B.

Consider the collection $\mathcal{C}$ of all non-empty subsets C of $\{1, ..., n\}$ satisfying

$$\|u(\varepsilon 1_C)\| > \Big(\frac{2\cdot|C|}{n}\Big)^{1/q}$$

for some $\varepsilon \in D_n$. It is of course possible that $\mathcal{C}$ is empty. In this case we set $A = \{1, ..., n\}$ to be done with (a) and (c).

The interesting case is when C is non-empty. It must then contain a maximal system $\{C_1,\ldots,C_m\}$ of pairwise disjoint members. Clearly, $m \leq n$. We are going to show that

$$(*) \qquad \left|\bigcup_{m\leq n} C_k\right| < \frac{n}{2}\,.$$

First note that for each $1 \leq k \leq m$ there is an $\varepsilon^{(k)} \in D_n$ such that

$$\|u(\varepsilon^{(k)}1_{C_k})\| > \left(\frac{2}{n}\right)^{1/q} \cdot |C_k|^{1/q}.$$

Disjointness of the C_k's tells us that

$$\Big(\sum_{k\leq m} \|u(\varepsilon^{(k)}1_{C_k})\|^q\Big)^{1/q} > \left(\frac{2}{n}\right)^{1/q} \cdot \left|\bigcup_{k\leq m} C_k\right|^{1/q},$$

while our hypothesis on u in tandem with our knowledge that $m \leq n$ informs us that

$$\Big(\sum_{k\leq m} \|u(\varepsilon^{(k)}1_{C_k})\|^q\Big)^{1/q} \leq \|(\varepsilon^{(k)}1_{C_k})\|_1^{weak} = 1.$$

Piecing these two inequalities together we get $(*)$.

We set $A := \{1,\ldots,n\} \setminus \bigcup_{k\leq m} C_k$. Certainly (a) holds, thanks to $(*)$. The maximality of the system $\{C_1,\ldots,C_m\}$ ensures that (c) is also satisfied.

We can now make short work of the lemma. Since there is an absolute constant α such that

$$|B|^{1/q} \leq \alpha \cdot \sum_{k\leq |B|} k^{(1/q)-1},$$

(c) implies

$$\|u(\varepsilon 1_B)\| \leq \alpha \cdot \left(\frac{2}{n}\right)^{1/q} \cdot \|\varepsilon 1_B\|_{\ell_{q,1}^n}$$

for all $B \subset A$ and all $\varepsilon \in D_n$. This is almost what we want. To obtain condition (b) just recall that we can replace $\|\cdot\|_{\ell_{q,1}^n}$ with an equivalent norm and that all vectors in $B_{\ell_\infty^n}$ are convex combinations of elements of D_n. QED

Now our preparations are complete.

Proof of 14.7. Recall that we are working with numbers $2 \leq q < q_X$ and $0 < \beta < q^{-1} - q_X^{-1}$. We know from 14.9(b) that

$$\frac{1}{q} - \frac{1}{q_X} \leq \limsup_n n \cdot \left(\frac{c_q(n)}{c_q(n-1)} - 1\right),$$

so we can find an infinite set $\mathbf{M} \subset \mathbf{N}$ such that $\beta < m \cdot \left(\frac{c_q(m)}{c_q(m-1)} - 1\right)$ and hence

$$\left(1 + \frac{\beta}{m}\right)^{1/q} \leq 1 + \frac{\beta}{m} < \frac{c_q(m)}{c_q(m-1)}$$

for all $m \in \mathbf{M}$. Fix $n \in \mathbf{N}$ and select $m \in \mathbf{M}$ with $m > 2n$ and $m \geq \beta$. There is a number $0 < \varrho_m < 1$ with

$$\left(1 + \frac{\beta}{m}\right)^{1/q} = \frac{1}{1+\varrho_m} \cdot \frac{c_q(m)}{c_q(m-1)},$$

and there are vectors $y_1,\ldots,y_m \in X$ such that

$$c_q(m)\cdot\Big(\int_0^1 \big\|\sum_{k\le m} r_k(t)y_k\big\|^q\,dt\Big)^{1/q} \le (1+\varrho_m)\cdot\Big(\sum_{k\le m}\|y_k\|^q\Big)^{1/q}.$$

It is harmless, but helpful, to assume that $\sum_{k\le m}\|y_k\|^q = m$, and this gives

$$\Big(\int_0^1 \big\|\sum_{k\le m} r_k(t)y_k\big\|^q\,dt\Big)^{1/q} \le (1+\varrho_m)\cdot m^{1/q}\cdot c_q(m)^{-1}.$$

The vectors we want will come from amongst the normalized y_k's. The set A of the previous lemma will play a key part in the selection process.

Let us first estimate $\min_{k\le m}\|y_k\|$. Supposing that the minimum is attained at $k=k_0$, we can use the Contraction Principle 12.2 to obtain

$$\begin{aligned}(m-\|y_{k_0}\|^q)^{1/q} &= \Big(\sum_{k\ne k_0}\|y_k\|^q\Big)^{1/q} \le c_q(m-1)\cdot\Big(\int_0^1 \big\|\sum_{k\le m} r_k(t)y_k\big\|^q\,dt\Big)^{1/q}\\ &\le \frac{c_q(m-1)}{c_q(m)}\cdot(1+\varrho_m)\cdot m^{1/q} = \Big(1+\frac{\beta}{m}\Big)^{-1/q}\cdot m^{1/q} = \Big(\frac{m^2}{\beta+m}\Big)^{1/q}.\end{aligned}$$

Rearranging,

$$\min_{k\le m}\|y_k\| = \|y_{k_0}\| \ge \Big(m-\frac{m^2}{\beta+m}\Big)^{1/q} = \Big(\frac{\beta m}{\beta+m}\Big)^{1/q} \ge \Big(\frac{\beta}{2}\Big)^{1/q}.$$

To set things up for application of Lemma 14.11, pass to the unit vectors $z_k = y_k/\|y_k\|$ $(1\le k\le m)$ and consider the operator

$$u_m : \ell_\infty^m \longrightarrow L_q([0,1],X) : (a_k)_{k\le m} \mapsto \sum_{k\le m} a_k r_k(\cdot)z_k.$$

Use the Contraction Principle 12.2 to estimate its norm:

$$\begin{aligned}\|u_m(a)\| &= \Big(\int_0^1 \big\|\sum_{k\le m} a_k r_k(t)z_k\big\|^q\,dt\Big)^{1/q}\\ &\le \Big(\max_{k\le m}\frac{|a_k|}{\|y_k\|}\Big)\cdot\Big(\int_0^1 \big\|\sum_{k\le m} r_k(t)y_k\big\|^q\,dt\Big)^{1/q}\\ &\le \Big(\frac{2}{\beta}\Big)^{1/q}\cdot\frac{(1+\varrho_m)\cdot m^{1/q}}{c_q(m)}\cdot\|a\|_\infty.\end{aligned}$$

Thus

$$\|u_m\| \le \Big(\frac{2}{\beta}\Big)^{1/q}\cdot\frac{(1+\varrho_m)m^{1/q}}{c_q(m)} \le \frac{\pi}{c_q(m)}\cdot\Big(\frac{m}{\beta}\Big)^{1/q}.$$

What about the hypothesis of 14.11? Using our knowledge from 14.10 we get for any choice of $a^{(1)},\ldots,a^{(m)}$ from ℓ_∞^m,

$$\begin{aligned}\Big(\sum_{k\le m}\|u_m(a^{(k)}\|^q\Big)^{1/q} &\le c_q(m)\cdot\Big(\int_0^1 \big\|\sum_{k\le m} r_k(s)u_m(a^{(k)})\big\|^q\,ds\Big)^{1/q}\\ &\le c_q(m)\cdot\|u_m\|\cdot\Big(\int_0^1 \big\|\sum_{k\le m} r_k(s)a^{(k)}\big\|^q\,ds\Big)^{1/q} \le c_q(m)\cdot\|u_m\|\cdot\|(a^{(k)})\|_1^{weak}.\end{aligned}$$

Lemma 14.11 now kicks in to give us a subset A_m of $\{1,...,m\}$ with at least $m/2$ elements such that whenever $a \in \mathbf{K}^m$ has its support in A_m,

$$\Big(\int_0^1 \big\|\sum_{k\leq m} a_k r_k(t) z_k\big\|^q\, dt\Big)^{1/q} = \|u_m(a)\|$$
$$\leq K\cdot c_q(m)\cdot\|u_m\|\cdot m^{-1/q}\cdot\|a\|_{\ell^m_{q,1}} \leq \pi\cdot K\cdot\beta^{-1/q}\cdot\|a\|_{\ell^m_{q,1}}.$$

Now write the elements of A_m in increasing order: $k_1 < k_2 < ....$ Since $|A_m| \geq m/2 \geq n$, we can validly select $x_j = z_{k_j}$, $1 \leq j \leq n$. The quantity $\|\cdot\|_{\ell^n_{q,1}}$ is 'invariant under spreading', that is,

$$\big\|\sum_{j\leq n} b_j e_{k_j}\big\|_{\ell^n_{q,1}} = \big\|\sum_{j\leq n} b_j e_j\big\|_{\ell^n_{q,1}}$$

for all (b_j) and all n. Consequently, with $C = \pi\cdot K$, we have shown that

$$\Big(\int_0^1 \big\|\sum_{j\leq n} a_j r_j(t) x_j\big\|^q\, dt\Big)^{1/q} \leq C\cdot\beta^{-1/q}\cdot\|a\|_{\ell^n_{q,1}}$$

for all $a \in \mathbf{K}^n$. This was our goal. QED

To be able to prove the final inequality needed to complete our main Theorem 14.5, we need a way to manufacture unconditional bases from bounded sequences. The packaging of this process may appear unpalatable but its proof is surprisingly tasty; it can stand by itself and is delayed until after we have seen its effects.

14.12 Lemma: *No matter how we choose integers $k > 1$ and m and real numbers $\varepsilon, \delta > 0$, there always exists an integer $n > \max\{k, 2m\}$ such that the following is true in any Banach space X: if $x_1,..., x_n \in B_X$ satisfy*

$P(k,\delta)$: *Whenever $A \subset \{1,...,n\}$ contains k elements, there exist $i, j \in A$ such that $\|x_i - x_j\| \geq \delta$,*

then they also satisfy

$Q(\varepsilon,\delta,m)$: *There is a subsequence $(x_{i_j})_{j=1}^{2m}$ of $(x_1,..., x_n)$ such that $\|x_{i_{2j}} - x_{i_{2j-1}}\| \geq \delta/2$ for all $1 \leq j \leq m$ and the sequence $(x_{i_{2j}} - x_{i_{2j-1}})_{j\leq m}$ is $(1+\varepsilon)$-unconditional.*

Let us see where this is headed.

14.13 Proposition: *In any Banach space X, we have $q_X \leq r_X$.*

Proof. We know that we can assume $q_X > 2$. Choose $2 < r < q_X$. Our task is to show that X finitely factors $\ell_r \hookrightarrow \ell_\infty$ and so, for any $m \in \mathbf{N}$, we must locate vectors $y_1,..., y_m \in X$ and numbers c_1, c_2 independent of m such that

$$c_1\cdot\|a\|_\infty \leq \big\|\sum_{j\leq m} a_j y_j\big\| \leq c_2\cdot\|a\|_r$$

for all $a \in \mathbf{R}^m$. The y_j's will appear as differences of vectors whose existence is granted by 14.7 and which satisfy the hypotheses of 14.12.

Pick $r < q < q_X$ and settle on some $0 < \beta < q^{-1} - q_X^{-1}$. As $q > 2$, we can find an integer $k > 1$ with $k^{1/2} > C\cdot(k/\beta)^{1/q}$ where C is the constant from 14.7. For any integer $n \geq k$, consider the vectors $x_1,...,x_n$ supplied by the aforesaid 14.7. We shall show that these vectors satisfy $P(k,\delta)$ for $\delta := k^{-1}\cdot(k^{1/2} - C\cdot(k/\beta)^{1/q})$.

Choose a k-element subset A from $\{1,...,n\}$. Then thanks to 14.7,

$$\Big(\int_0^1 \big\| \sum_{i\in A} r_i(t)x_i \big\|^q \, dt\Big)^{1/q} \leq C\cdot\beta^{-1/q}\cdot\|1_A\|_{\ell_{q,1}^n} \leq C\cdot\beta^{-1/q}.k^{1/q}.$$

To dispel any qualms about the last inequality, recall that the identity $\ell_p^n \to \ell_{q,1}^n$ has norm 1 whenever $p < q$.

On the other hand, we can derive an inequality going the other way. For each $i_0 \in A$,

$$\begin{aligned}
&\Big(\int_0^1 \big\| \sum_{i\in A} r_i(t)x_i \big\|^q \, dt\Big)^{1/q} \\
&\geq \Big(\int_0^1 \big\| \sum_{i\in A} r_i(t)x_{i_0} \big\|^q \, dt\Big)^{1/q} - \Big(\int_0^1 \big\| \sum_{i\in A} r_i(t)(x_i - x_{i_0}) \big\|^q \, dt\Big)^{1/q} \\
&\geq \Big(\int_0^1 \big| \sum_{i\in A} r_i(t) \big|^2 \, dt\Big)^{1/2} - k\cdot \max_{i,j\in A} \|x_i - x_j\| = k^{1/2} - k\cdot \max_{i,j\in A} \|x_i - x_j\|.
\end{aligned}$$

Combining the two estimates, we obtain

$$\max_{i,j\in A} \|x_i - x_j\| \geq \delta.$$

Now let $\varepsilon = 1$. With k and δ as described above and an arbitrary m, Lemma 14.12 supplies us with an integer $n > \max\{k, 2m\}$, and this is what we work with. Since we have just shown that the vectors $x_1,...,x_n$ from 14.7 satisfy $P(k,\delta)$, it follows that they also satisfy $Q(1,\delta,m)$. So there is a subsequence $(x_{i_j})_{j=1}^{2m}$ such that the $y_j := x_{i_{2j}} - x_{i_{2j-1}}$, $1 \leq j \leq m$, form a 2-unconditional basic sequence in X with each $\|y_j\| \geq \delta/2$.

The end is swift. First, using 2-unconditionality and then 14.7, we get for each $a \in \mathbf{R}^m$,

$$\begin{aligned}
\big\| \sum_{j\leq m} a_j y_j \big\| &\leq 2\cdot \int_0^1 \big\| \sum_{j\leq m} a_j r_j(t) y_j \big\| \, dt \leq 2\cdot\Big(\int_0^1 \big\| \sum_{j\leq m} a_j r_j(t) y_j \big\|^q dt\Big)^{1/q} \\
&\leq 2\cdot\Big(\int_0^1 \big\| \sum_{j\leq m} a_j r_j(t) x_{i_{2j}} \big\|^q dt\Big)^{1/q} + 2\cdot\Big(\int_0^1 \big\| \sum_{j\leq m} a_j r_j(t) x_{i_{2j-1}} \big\|^q dt\Big)^{1/q} \\
&\leq 4\cdot C\cdot\beta^{-1/q}\cdot\|a\|_{\ell_{q,1}^m} \leq 4\cdot C\cdot\beta^{-1/q}\cdot\|a\|_{\ell_r^m}.
\end{aligned}$$

This is one half of what we require. The other half is a simple consequence of 2-unconditionality and the lower bound for the $\|y_j\|$'s. In fact, if $a \in \mathbf{R}^m$,

then
$$\frac{\delta}{4}\cdot\|a\|_{\ell_\infty^m} = \frac{\delta}{4}\cdot\max_{j\le m}|a_j| \le \frac{1}{2}\cdot\max_{j\le m}\|a_j y_j\|$$
$$= \frac{1}{2}\cdot\max_{j\le m}\Big\|\sum_{i\le m} a_i\delta_{i,j}y_i\Big\| \le \Big\|\sum_{i\le m} a_i y_i\Big\|,$$
and all is well. QED

Brunel - Sucheston Affairs

Our proof of 14.12 will rely on a powerful combinatorial result. A little notation will come in handy. Let k be a positive integer or ∞. When $\mathbf{M}\subset\mathbf{N}$, we write
$$\mathcal{P}_k(\mathbf{M})$$
for the collection of all subsets of $\mathbf{M}$ with k elements.

14.14 Ramsey's Theorem: *Let k be a positive integer. However we choose $\mathcal{A}\subset\mathcal{P}_k(\mathbf{N})$, there is always some $\mathbf{M}\in\mathcal{P}_\infty(\mathbf{N})$ such that either $\mathcal{P}_k(\mathbf{M})\subset\mathcal{A}$ or $\mathcal{P}_k(\mathbf{M})\cap\mathcal{A}=\emptyset$.*

Proof. We shall just give the details for the case $k=2$, leaving the general case to the dutiful reader.

A little terminology will help to simplify the formulation. For any $n\in\mathbf{N}$ and $\mathbf{M}\in\mathcal{P}_\infty(\mathbf{N})$ we say that the pair $(n,\mathbf{M})$ is *good* if $\{n,m\}$ is an element of $\mathcal{A}$ for every $m\in\mathbf{M}$, otherwise we say that it is *bad*.

Begin by fixing any $m_1\in\mathbf{N}$. Clearly there is an $\mathbf{M}_1\in\mathcal{P}_\infty(\mathbf{N})$ such that $(m_1,\mathbf{M}_1)$ is either good or bad. In either case, proceed by choosing $m_2\in\mathbf{M}_1$ with $m_2>m_1$. Again there is an $\mathbf{M}_2\in\mathcal{P}_\infty(\mathbf{M}_1)$ such that $(m_2,\mathbf{M}_2)$ is either good or bad.

Continue ad nauseam. At least one of the sets
$$\{i\in\mathbf{N}:\ (m_i,\mathbf{M}_i)\text{ is good}\}\qquad\text{and}\qquad\{i\in\mathbf{N}:\ (m_i,\mathbf{M}_i)\text{ is bad}\}$$
must be infinite. Choose one that is infinite and label it I. The set
$$\mathbf{M}:=\{m_i:\ i\in I\}$$
is just what we want, since $\mathcal{P}_2(\mathbf{M})\subset\mathcal{A}$ if $(m_i,\mathbf{M}_i)$ is bad for every $i\in I$, whereas $\mathcal{P}_2(\mathbf{M})\cap\mathcal{A}=\emptyset$ if all the $(m_i,\mathbf{M}_i)$ are bad. QED

The next result, which relies on Ramsey's Theorem, is a cornerstone for our project to generate unconditional basic sequences from bounded ones.

14.15 Brunel - Sucheston Theorem: *Let (x_n) be a bounded sequence in the separable Banach space X. There is a subsequence (y_n) of (x_n) with the property that*
$$\lim_{n_k>\ldots>n_1\to\infty}\|x+a_1y_{n_1}+\ldots+a_ky_{n_k}\|$$
exists for all $x\in X$, all $k\in\mathbf{N}$, and all k-tuples $(a_1,\ldots,a_k)$ of scalars.

The proof will be a little more digestible if we have decent notation at our disposal. We shall have occasion to consider maps $\chi : \mathcal{P}_k(\mathbf{N}) \to [0,\infty)$. For efficient handling let us agree that if $\mathbf{M} \in \mathcal{P}_k(\mathbf{N})$ is the set $\{n_1,\ldots,n_k\}$ and if $n_1 < \ldots < n_k$, then we shall often write

$$(n_1 < \ldots < n_k) \text{ for } \mathbf{M} \qquad \text{and} \qquad \chi(n_1 < \ldots < n_k) \text{ for } \chi(\mathbf{M}).$$

Proof. As a first step, we limit ourselves to proving the existence of a sequence (y_n) that has the desired property for just one $x \in X$, one $k \in \mathbf{N}$, and one k-tuple $(a_1,\ldots,a_n) \in \mathbf{R}^n$. Define

$$\chi : \mathcal{P}_k(\mathbf{N}) \longrightarrow [0,\infty) \quad \text{by} \quad \chi(n_1 < \ldots < n_k) := \Big\| x + \sum_{i \le k} a_i x_{n_i} \Big\|$$

and set

$$C := \|x\| + \big(\sup_n \|x_n\|\big) \cdot \sum_{i \le k} |a_i|.$$

Then evidently $\chi(n_1 < \ldots < n_k) \le C$ for every $(n_1 < \ldots < n_k) \in \mathcal{P}_k(\mathbf{N})$. Closer examination of the behaviour of χ, connected with a diagonalization procedure, will lead to what we want.

First partition $\mathcal{P}_k(\mathbf{N})$ into the complementary sets

$$\mathcal{A}_1 := \big\{(n_1 < \ldots < n_k) : \chi(n_1 < \ldots < n_k) \le C/2\big\}$$

and

$$\mathcal{B}_1 := \big\{(n_1 < \ldots < n_k) : C/2 < \chi(n_1 < \ldots < n_k) \le C\big\}.$$

Ramsey's Theorem provides an $\mathbf{M}_1 \in \mathcal{P}_\infty(\mathbf{N})$ such that either $\mathcal{P}_k(\mathbf{M}_1) \subset \mathcal{A}_1$ or $\mathcal{P}_k(\mathbf{M}_1) \subset \mathcal{B}_1$.

We suppose that $\mathcal{P}_k(\mathbf{M}_1) \subset \mathcal{A}_1$; the alternative can be treated using a very similar argument. Consequently, we know that $\chi(m_1 < \ldots < m_k) \le C/2$ for any $(m_1 < \ldots < m_k)$ in $\mathcal{P}_k(\mathbf{M}_1)$. To continue, we partition $\mathcal{P}_k(\mathbf{M}_1)$ into the sets

$$\mathcal{A}_2 := \big\{(m_1 < \ldots < m_k) \in \mathcal{P}_k(\mathbf{M}_1) : \chi(m_1 < \ldots < m_k) \le C/4\big\}$$

and

$$\mathcal{B}_2 := \big\{(m_1 < \ldots < m_k) \in \mathcal{P}_k(\mathbf{M}_1) : C/4 < \chi(m_1 < \ldots < m_k) \le C/2\big\}.$$

This time, Ramsey's Theorem churns out an $\mathbf{M}_2 \in \mathcal{P}_\infty(M_1)$ such that either $\mathcal{P}_k(\mathbf{M}_2) \subset \mathcal{A}_2$ or $\mathcal{P}_k(\mathbf{M}_2) \subset \mathcal{B}_2$. Elect one to consider: the other will surely yield to a parallel treatment.

Continue ad infinitum. We end up with a nested sequence $(\mathbf{M}_n)$ of members of $\mathcal{P}_\infty(\mathbf{N})$ from which we extract the diagonal sequence $\mathbf{M}$. Cantor's nested interval theorem ensures that

$$\lim_{\substack{m_k > \ldots > m_1 \to \infty \\ (m_1 < \ldots < m_k) \in \mathcal{P}_k(\mathbf{M})}} \chi(m_1 < \ldots < m_k) = \lim_{\substack{m_k > \ldots > m_1 \to \infty \\ (m_1 < \ldots < m_k) \in \mathcal{P}_k(\mathbf{M})}} \Big\| x + \sum_{i \le k} a_i x_{m_i} \Big\|$$

exists.

We now build on this success, reiterating the same technique. The result will be a subsequence (y_n) of (x_n) with the desired property for all x in a countable dense subset D of X, for all $k \in \mathbf{N}$, and for all $(a_1,\ldots,a_k) \in \mathbf{Q}^k$.

For things to proceed smoothly, we want an enumeration (z_n, R_n) of the countable set $D \times \mathcal{F}(\mathbf{Q})$, where $\mathcal{F}(\mathbf{Q})$ designates the collection of all finite-tuples from $\mathbf{Q}$.

Refer back to the first step. Put z_1 in place of x and R_1 in place of $(a_1, \ldots, a_k)$. We may extract a subsequence $(x_n^{(1)})$ from (x_n) such that

$$\lim_{n_k > \ldots > n_1 \to \infty} \left\| z_1 + \sum_i a_i x_{n_i}^{(1)} \right\|$$

exists. Now that we have identified this new bounded sequence, we turn the handle once more and extract a subsequence $(x_n^{(2)})$ from $(x_n^{(1)})$ which has the desired property for z_2 and R_2. Continue, and then diagonalize.

To finish, approximate with gusto. QED

We stop to moralize. A *good sequence* in a Banach space X is any bounded sequence (y_n) for which

$$L(x, a) := \lim_{n_k > \ldots > n_1 \to \infty} \|x + a_1 y_{n_1} + \ldots + a_k y_{n_k}\|$$

exists for all $x \in X$, all $k \in \mathbf{N}$ and all $a = (a_1, \ldots, a_k) \in \mathbf{R}^k$. Obviously, $L(\cdot, \cdot)$ is a seminorm on $X \times \mathbf{R}^{(\mathbf{N})}$ whenever (y_n) is a good sequence. When (y_n) fails to converge, we shall see that $L(\cdot, \cdot)$ is even a norm. We can use it to define a norm $L(0, \cdot)$ on $\mathbf{R}^n$. It will eventually become clear that $(e_{2n} - e_{2n+1})_n$ is a 1-unconditional basic sequence in $\mathbf{R}^{(\mathbf{N})}$ under this norm.

That is our basic strategy. Let us put it into effect.

14.16 Lemma: *A good sequence (y_n) in the Banach space X fails to converge if and only if $L(\cdot, \cdot)$ is a norm on $X \times \mathbf{R}^{(\mathbf{N})}$.*

Proof. Let us take the trivial direction first. If (y_n) converges to y, then $L(y, -1) = \lim_{n \to \infty} \|y - y_n\| = 0$, yet $(y, -1)$ is certainly non-zero in $X \times \mathbf{R}^{(\mathbf{N})}$.

Suppose next that (y_n) is *not* convergent in X. Let $x \in X$ and $a = (a_1, \ldots, a_k) \in \mathbf{R}^k$ be such that $L(x, a) = 0$. This means that given any $\varepsilon > 0$, there is some $n_\varepsilon \in \mathbf{N}$ with the property that if $n_\varepsilon < n_1 < \ldots < n_k$, then

$$\left\| x + \sum_{i \le k} a_i y_{n_i} \right\| < \varepsilon.$$

So, for $n_\varepsilon < m < n < n_2 < \ldots < n_k$ we have

$$|a_1| \cdot \|y_m - y_n\| \\ = \|(x + a_1 y_m + a_2 y_{n_2} + \ldots + a_k y_{n_k}) - (x + a_1 y_n + a_2 y_{n_2} + \ldots + a_k y_{n_k})\| < 2\varepsilon.$$

Since we are assuming that (y_n) is not convergent, we must conclude that $a_1 = 0$. Continue this process to find that all the a_i's are zero. This leaves x little choice: it too must be zero. We are done. QED

So, given a good non-convergent sequence (y_n) in a Banach space, we can define a norm

$$|||\cdot||| := L(0, \cdot)$$

on $\mathbf{R}^{(\mathbf{N})}$. This norm, true to its origins, is well-behaved.

14.17 Corollary: *The norm* $|\!|\!|\cdot|\!|\!|$ *on* $\mathbf{R}^{(\mathbf{N})}$ *associated with a good non-convergent sequence* (y_n) *is 'invariant under spreading':*

$$|\!|\!|\sum_{i\le k} a_i e_{m_i}|\!|\!| = |\!|\!|\sum_{i\le k} a_i e_i|\!|\!|$$

for all $k \in \mathbf{N}$, *all natural numbers* $m_1 < \ldots < m_k$ *and all scalars* $a_1,\ldots,a_k$.

Proof. Just think about the definition:

$$|\!|\!|\sum_{i\le k} a_i e_i|\!|\!| = \lim_{n_k>\ldots>n_1\to\infty} \Big\|\sum_{i\le k} a_i y_{n_i}\Big\| .$$

QED

Here is the most important consequence of all this.

14.18 Theorem: *Let* $|\!|\!|\cdot|\!|\!|$ *be the norm on* $\mathbf{R}^{(\mathbf{N})}$ *associated with a good non-convergent sequence in a Banach space. Then* $(e_{2n} - e_{2n+1})_n$ *is a 1-unconditional basic sequence in* $[\mathbf{R}^{(\mathbf{N})}, |\!|\!|\cdot|\!|\!|]$.

Proof. Choose scalars $a_1,\ldots,a_n$ and fix $1 \le k \le n$ and $m \in \mathbf{N}$. Using 14.17 to spread our muck enough, we can compare $\big\|\sum_{i\le n} a_i(e_{2i} - e_{2i+1})\big\|$ with $\big\|\sum_{i\ne k} a_i(e_{2i} - e_{2i+1})\big\|$ in a way that suits us well:

$$\begin{aligned}
m\cdot|\!|\!|\sum_{i\le n} a_i(e_{2i} - e_{2i+1})|\!|\!| \\
&= \sum_{j\le m}|\!|\!|\sum_{i<k} a_i(e_{2i} - e_{2i+1}) + a_k(e_{2k} - e_{2k+1}) + \sum_{i>k} a_i(e_{2i} - e_{2i+1})|\!|\!| \\
&= \sum_{j\le m}|\!|\!|\sum_{i<k} a_i(e_{2i} - e_{2i+1}) + a_k(e_{2k+j} - e_{2k+j+1}) \\
&\qquad + \sum_{i>k} a_i(e_{2i+m} - e_{2i+m+1})|\!|\!| \\
&\ge |\!|\!|m\cdot\sum_{i<k} a_i(e_{2i} - e_{2i+1}) + a_k\cdot\sum_{j\le m}(e_{2k+j} - e_{2k+j+1}) \\
&\qquad + m\cdot\sum_{i>k} a_i(e_{2i+m} - e_{2i+m+1})|\!|\!| \\
&\ge m\cdot|\!|\!|\sum_{i<k} a_i(e_{2i} - e_{2i+1}) + \sum_{i>k} a_i(e_{2i+m} - e_{2i+m+1})|\!|\!| \\
&\qquad - |a_k|\cdot|\!|\!|\sum_{j\le m}(e_{2k+j} - e_{2k+j+1})|\!|\!| \\
&= m\cdot|\!|\!|\sum_{i\ne k} a_i(e_{2i} - e_{2i+1})|\!|\!| - |a_k|\cdot|\!|\!|e_{2k+1} - e_{2k+m+1}|\!|\!|.
\end{aligned}$$

We conclude that, for every $m \in \mathbf{N}$,

$$|\!|\!|\sum_{i\ne k} a_i(e_{2i} - e_{2i+1})|\!|\!| \le |\!|\!|\sum_{i\le n} a_i(e_{2i} - e_{2i+1})|\!|\!| + \frac{|a_k|}{m}\cdot|\!|\!|e_{2k+1} - e_{2k+m+1}|\!|\!|$$

and so

$$|\!|\!|\sum_{i\ne k} a_i(e_{2i} - e_{2i+1})|\!|\!| \le |\!|\!|\sum_{i\le n} a_i(e_{2i} - e_{2i+1})|\!|\!|.$$

As this holds for every $k \le n$, the 1-unconditionality follows. QED

We can now round off the chapter by supplying the long-awaited proof of 14.12.

Proof of 14.12. We argue by contradiction. Assume we can find integers m and $k > 1$ and positive numbers ε, δ such that for each $n > \max\{k, 2m\}$ there is a Banach space X_n having vectors $x_1^{(n)},\dots,x_n^{(n)}$ in its unit ball which satisfy $P(k,\delta)$ but not $Q(\varepsilon,\delta,m)$. Notation is simpler if we agree to set $x_i^{(n)} = 0$ for each $i > n$.

Our first priority is to establish a link with the Brunel - Sucheston Theorem so that we can start creating unconditional basic sequences to throw doubt on the failure of $Q(\varepsilon,\delta,m)$.

Fix a non-trivial ultrafilter $\mathcal{U}$ on $\mathbf{N}$. For each $c = (c_i) \in \mathbf{R}^{(\mathbf{N})}$ the bound $\|\sum_i c_i x_i^{(n)}\|_{X_n} \leq \sum_i |c_i|$ is independent of n. Accordingly,

$$p(c) := \lim_{\mathcal{U}} \|\sum_i c_i x_i^{(n)}\|_{X_n}$$

exists for all $c \in \mathbf{R}^{(\mathbf{N})}$. Since p is clearly a seminorm on $\mathbf{R}^{(\mathbf{N})}$, we can use it to define a norm on $\mathbf{R}^{(\mathbf{N})}/p^{-1}(0)$; completion gives a Banach space $[X, \|\cdot\|_X]$. Let us agree to use $\bar{c}$ to label the image of $c \in \mathbf{R}^{(\mathbf{N})}$ in X.

Evidently, $\|\bar{e}_i\|_X \leq 1$ for each i; so the Brunel - Sucheston Theorem gives us a way to extract a good subsequence $(f_i) = (\bar{e}_{\ell_i})$ from $(\bar{e}_i)$. Then we know that

$$|\!|\!| c |\!|\!| := \lim_{\substack{k_1 \to \infty \\ k_1 < k_2 < \dots}} \|\sum_i c_i f_{k_i}\|_X$$

defines a seminorm on $\mathbf{R}^{(\mathbf{N})}$. To show that $|\!|\!|\cdot|\!|\!|$ is actually a norm, we follow up 14.16. It is enough to prove that (f_i) is not a Cauchy sequence in X, and here is where $P(k,\delta)$ helps out. If $A \subset \{1,\dots,n\}$ has k elements, then we have $\max_{i,j\in A} \|x_i^{(n)} - x_j^{(n)}\|_{X_n} \geq \delta$. Recall the definition of $|\!|\!|\cdot|\!|\!|$ to see that this implies $\max_{i,j\in A} |\!|\!|\bar{e}_i - \bar{e}_j|\!|\!| \geq \delta$. Consequently, no matter how far we go along the sequence, we will always find f_i's whose distance from others is at least δ.

Invoking 14.18 we conclude that the sequence $(e_{2i} - e_{2i+1})$ is 1-unconditional in $[\mathbf{R}^{(\mathbf{N})}, |\!|\!|\cdot|\!|\!|]$. This allows us finally to approach the contradiction we are seeking.

Let Y_m be the span of $e_1,\dots,e_{2m}$ in $[\mathbf{R}^{(\mathbf{N})}, |\!|\!|\cdot|\!|\!|]$, and for each $i,n \in \mathbf{N}$ define the operator $u_{i,n} : Y_m \to X_n$ via $u_{i,n}(e_j) = x_{\ell_{i+j}}^{(n)}$ for all $1 \leq j \leq 2m$. Spreading allows us to nail things down: for each $y \in Y_m$,

$$|\!|\!| y |\!|\!| = \lim_{k_{2m} > \dots > k_1 \to \infty} \|\sum_{j\leq 2m} y_j f_{k_j}\|_X = \lim_{i\to\infty} \|\sum_{j\leq 2m} y_j f_{i+j}\|_X \,.$$

This means that

$$|\!|\!| y |\!|\!| = \lim_{i\to\infty} \lim_{\mathcal{U}} \|u_{i,n}(y)\|_{X_n} \,.$$

Since Y_m is finite dimensional, B_{Y_m} is compact, and so the $\|u_{i,n}(\cdot)\|_{X_n}$ form an equicontinuous family in $\mathcal{C}(B_{Y_m})$. Accordingly, for any $0 < \varrho < 1$, we can arrange for

$$(1+\varrho)^{-1}\cdot|\!|\!| y |\!|\!| \leq \|u_{i,n}(y)\|_{X_n} \leq (1+\varrho)\cdot|\!|\!| y |\!|\!|$$

for all $y \in B_{Y_m}$, simply by taking i and n large enough. Consider this arranged.

Now let us examine $z_j := u_{i,n}e_{2j} - u_{i,n}e_{2j+1}$. We find that for all scalars a_j and all signs $\varepsilon_j = \pm 1$,

$$\begin{aligned} \Big\| \sum_{j\leq m} \varepsilon_j a_j z_j \Big\|_{X_n} &= \Big\| u_{i,n}\Big(\sum_{j\leq m} \varepsilon_j a_j (e_{2j} - e_{2j+1})\Big) \Big\|_{X_n} \\ &\leq (1+\varrho)\cdot\Big|\!\Big|\!\Big| \sum_{j\leq m} \varepsilon_j a_j (e_{2j} - e_{2j+1})\Big|\!\Big|\!\Big|. \end{aligned}$$

Unconditionality enters to yield

$$\Big\| \sum_{j\leq m} \varepsilon_j a_j z_j \Big\|_{X_n} \leq (1+\varrho)\cdot\Big|\!\Big|\!\Big| \sum_{j\leq m} a_j (e_{2j} - e_{2j+1})\Big|\!\Big|\!\Big| \leq (1+\varrho)^2\cdot \Big\| \sum_{j\leq m} a_j z_j \Big\|_{X_n} .$$

By selecting ϱ so that $(1+\varrho)^2 \leq 1+\varepsilon$, we can make (z_j) $(1+\varepsilon)$-unconditional.

To contradict the failure of $Q(\varepsilon, \delta, m)$, we need just one more thing: $\|z_j\|_{X_n} \geq \delta/2$ for all $j \leq m$. But

$$\|z_j\|_{X_n} = \|u_{i,n}(e_{2j} - e_{2j+1})\| \geq \frac{1}{1+\varrho}\cdot|\!|\!| e_{2j} - e_{2j+1} |\!|\!| \geq \frac{\delta}{1+\varrho} \geq \frac{\delta}{2}\,. \qquad \text{QED}$$

Notes and Remarks

In the main we follow the lead of B. Maurey and G. Pisier [1976] in the derivation of Theorem 14.1 and the tools needed to prove it. Our treatment of the refined Dvoretzky - Rogers Theorem 14.2, however, was strongly influenced by A. Pełczyński's [1980] lectures.

The other central ingredients in the proof of 14.1 are Lemma 14.11 and the Brunel - Sucheston technique of constructing 'spreading models' of a given Banach space. Lemma 14.11 was inspired by E.M. Nikishin's [1970] work. When combined with the powerful Lemma 14.12, it provides the crucial combinatorial link needed to establish Theorem 14.5, the most difficult step in the proof of 14.1.

Ramsey's Theorem 14.14 goes back to F.P. Ramsey's seminal paper [1929]. Its use by A. Brunel and L. Sucheston [1973] in proving 14.15 and 14.18 alerted abstract analysts to a powerful new combinatorial tool. E. Odell's [1980] survey gives an idea of how this has changed the face of Banach space theory.

Like many remarkable results, 14.1 has a variety of complements. One of the most attractive is a dichotomy relating finite cotype and harmonic analysis. Let G be a compact abelian group and consider a subset Λ of the dual group Γ. The closed subspace $\mathcal{C}_\Lambda(G)$ of $\mathcal{C}(G)$ consists of all those continuous functions $G \to \mathbf{C}$ whose Fourier transforms are supported by Λ. The set Λ is a *Sidon set* if there is a constant $C \geq 0$ such that for any finitely non-zero scalar family

$(a_\gamma)_{\gamma\in\Gamma}$ one has $\sum_{\gamma\in\Gamma}|a_\gamma| \le C\cdot\|\sum_{\gamma\in\Gamma} a_\gamma\gamma\|_{\mathcal{C}(G)}$. Thus the Sidon condition establishes a natural isomorphism between $\mathcal{C}_\Lambda(G)$ and ℓ_1^Λ.

N.T. Varopoulos [1975/76] showed that if $\mathcal{C}_\Lambda(G)$ is an $\mathcal{L}_1$-space, then Λ is a Sidon set. Soon after, G. Pisier [1977/78b] extended this by showing that if $\mathcal{C}_\Lambda(G)$ has cotype 2 then Λ is a Sidon set. Finally, J. Bourgain and V.D. Milman [1985], using a marvellous mix of ideas from harmonic analysis and Banach space theory, came up with a remarkable necessary and sufficient condition:

- *Λ is a Sidon set if and only if $\mathcal{C}_\Lambda(G)$ has finite cotype.*

In a completely different area, the study of irreducible domains in complex Banach spaces, S. Dineen and R.M. Timoney [1987] discovered that it is precisely when the Banach space X fails to contain a copy of c_0 that all of its bounded domains are biholomorphically equivalent to a finite product of irreducible complex manifolds. Making heavy use of 14.1, they went on to give a quantitative version of this result in terms of the cotype of X.

Very recently, P. Abraham [1992] and S. Saeki [1992] gave yet another criterion for a Banach space X to have finite cotype, namely that any sequence (μ_n) of countably additive X-valued measures with σ-field domain Σ must be uniformly countably additive whenever $\lim_n \|\mu_n(E)\|$ exists for each $E \in \Sigma$. Specialization to the real line uncovers a striking variation on the classical Vitali - Hahn - Saks - Nikodým convergence theorem. S.J. Dilworth and M. Girardi [1995] related the degree of non-differentiability of indefinite Pettis integrals in a Banach space X to the cotype of X.

S. Dineen [1995] has shown that the space of n-homogeneous polynomials on a Banach space of infinite dimension never has finite cotype when $n \ge 2$.

A few remarks on the quantities q_X, r_X and s_X are in order. Although r_X was defined to be an supremum, it is not hard to show that any Banach space X will finitely factor $\ell_{r_X} \hookrightarrow \ell_\infty$; this can be found in B. Maurey - G. Pisier [1976].

The quantities q_X and s_X are, in general, true infima. The reasons lead us into some interesting developments. M. Talagrand [1992b] proved that 11.17 has a partial converse.

- *When $2 < q < \infty$, a Banach space X has cotype q if and only if id_X is $(q,1)$-summing.*

Weaker versions of this can be found in 16.8 and 17.4 where we establish its validity for a wide class of Banach spaces which, for example, contains all Banach lattices. Consequently, when $2 < q_X < \infty$, the Banach space X has cotype q_X if and only if id_X is $(q_X, 1)$-summing. This is no longer true when $q_X = 2$. M. Talagrand [1992a] produced a Banach lattice whose identity is $(2,1)$-summing, but which fails to have cotype 2.

There are methods, by now standard, for obtaining Banach spaces X which fail to have cotype q_X. Best known are those which are based on an example of

B.S. Tsirelson [1974] who produced the first infinite dimensional Banach space which contains no copy of c_0 or ℓ_p for any $1 \le p < \infty$. T. Figiel and W.B. Johnson [1974] came up with a description of the dual of Tsirelson's original space, and they called it T. It is the completion of $\mathbf{R}^{(\mathbf{N})}$ under a norm $\|\cdot\|_T$ which is obtained as follows: given $x = (x_k) \in \mathbf{R}^{(\mathbf{N})}$, set $\|x\|_0 := \|x\|_{c_0}$ and, for each $n = 0, 1, 2, \ldots$, define

$$(*) \qquad \|x\|_{n+1} := \max\Big\{\|x\|_n, \frac{1}{2}\cdot \sup \sum_{j=1}^{N} \|E_j x\|_n\Big\},$$

where the supremum extends over all choices of $N \in \mathbf{N}$ and non-empty finite subsets $E_1, \ldots, E_N$ of $\mathbf{N}$ satisfying $N \le \min E_1$ and $\max E_j < \min E_{j+1}$ for $1 \le j < N$; by $E_j x$ we mean the vector with coordinates x_k for $k \in E_j$ and 0 otherwise. This gives a sequence $(\|\cdot\|_n)$ of norms on $\mathbf{R}^{(\mathbf{N})}$ which is increasing and bounded above by the ℓ_1 norm. Thus

$$\|x\|_T := \lim_{n\to\infty} \|x\|_n$$

exists for each $x \in \mathbf{R}^{(\mathbf{N})}$: this is the norm to look for.

The gruesome computations needed to calculate $\|x\|_T$ can be delegated to a computer, thanks to the work of J.W. Baker, O.A. Slotterbeck and R. Aron; their work is an appendix to the Lecture Notes [1989] by P.G. Casazza and T.J. Shura on Tsirelson's space and its many ramifications.

The standard unit vectors e_n form a normalized 1-unconditional basis in T. Moreover, T is reflexive and so cannot contain a copy of c_0 or ℓ_1; it doesn't contain any copy of ℓ_p for any $1 < p < \infty$ either; in fact, no infinite dimensional subspace of T even admits an equivalent uniformly convex norm.

Replacing the factor 1/2 in $(*)$ by any $0 < \theta < 1$ does not produce qualitative changes, but does lead to a scale of mutually non-isomorphic 'Tsirelson-like' spaces; we refer to P.G. Casazza and T.J. Shura [1989] for details.

Let $1 < q < \infty$. The *q-convexification* of T is the space $T^{(q)}$ which consists of all scalar sequences $x = (x_n)_n$ for which $(|x_n|^q)_n$ belongs to T. This is a Banach space with respect to the norm

$$\|x\|_{(q)} := \Big\|\sum_n |x_n|^q e_n\Big\|_T^{1/q}.$$

Although $T^{(q)}$ is uniformly convex it, like T, fails to contain copies of c_0 and ℓ_p $(1 \le p < \infty)$. When $2 \le q < \infty$, $T^{(q)}$ has type 2 and, most significantly for the present discussion, has cotype $q + \varepsilon$ for each $\varepsilon > 0$ but fails to have cotype q. For $q = 2$, the reason for the last statement is easy: if $T^{(2)}$ had cotype 2, Kwapień's Theorem (12.20) would force it to be a Hilbert space – but it doesn't even contain a copy of ℓ_2.

This shows that not all Banach spaces X have cotype q_X. Similarly for type: if we set $p_X := \sup\{p : X \text{ has type } p\}$, then the supremum is not necessarily attained either. Just note that $(T^{(q)})^*$ has type $q^* - \varepsilon$ for every $\varepsilon > 0$ but fails to have type q^*.

For various reasons, the most important of these spaces is the 2-convexified Tsirelson space. It was the first non-Hilbertian example of an infinite dimensional Banach space with the property that every subspace of each of its quotients has a basis; see W.B. Johnson [1980]. Nowadays it is the most prominent example of what is called a 'weak Hilbert space.'

To introduce this notion, let us return to our discussion in Chapter 6 of the John ellipsoid $\mathcal{E}_E$ of an n-dimensional normed space E; recall that $\mathcal{E}_E$ is the ellipsoid of maximal volume inside E's unit ball B_E. We know that $\mathcal{E}_E \subset B_E \subset \sqrt{n}\cdot\mathcal{E}_E$ and that no general improvement is possible. The quantity

$$vr(E) := \left(\frac{\text{vol }(B_E)}{\text{vol }(\mathcal{E}_E)}\right)^{1/n},$$

can therefore always be bounded by $n^{1/2}$; $vr(E)$ is called the *volume ratio* of the space E.

Important information about an infinite dimensional Banach space X can be obtained by looking at the volume ratios of its finite dimensional subspaces. A key result is due to J. Bourgain and V.D. Milman [1987]:

- *If the Banach space X has cotype 2 then there is a constant C such that $vr(E) \leq C$ for each finite dimensional subspace E of X.*

A characterization of Banach spaces X such that $\sup_{E\in\mathcal{F}_X} vr(E) < \infty$ was given shortly after by V.D. Milman and G. Pisier [1986]. To state it, we need some background.

Recall from 12.29 that X has cotype $q \in [2,\infty)$ if and only if there is a constant c such that for each $n\in\mathbf{N}$ and $u\in\mathcal{L}(\ell_2^n, X)$ we have $\pi_{q,2}(u)\leq c\cdot\pi_\gamma(u)$. Using trace duality this can be rephrased to read that

(a) *X has cotype q if and only if there is a constant c such that $\sigma_1(uv) \leq c\cdot\pi_{q,2}^*(u)\cdot\pi_\gamma(v)$ for all $n \in \mathbf{N}$ and all $\ell_2^n \xrightarrow{v} X \xrightarrow{u} \ell_2^n$.*

Analogously,

(b) *X has type p if and only if there is a constant c such that $\sigma_1(uv) \leq c\cdot\pi_\gamma^*(u)\cdot\pi_{p^*,2}^*(v^*)$ for all $n \in \mathbf{N}$ and all $\ell_2^n \xrightarrow{v} X \xrightarrow{u} \ell_2^n$.*

Recall that $w \in \mathcal{L}(\ell_2)$ belongs to $\mathcal{S}_1(\ell_2)$ if and only if the approximation numbers $a_k(w)$ form an ℓ_1 sequence and that in this case we have $\sigma_1(w) = \sum_k a_k(w)$. We shall say that w belongs to $\mathcal{S}_{1,\infty}(\ell_2)$ whenever $(a_k(w))_k$ belongs to the Lorentz sequence space $\ell_{1,\infty}$ (also called 'weak ℓ_1'), or, in other words, when $\sigma_{1,\infty}(w) := \sup_k k\cdot a_k(w) < \infty$. With respect to $\sigma_{1,\infty}$, $\mathcal{S}_{1,\infty}(\ell_2)$ is a Banach space, containing $\mathcal{S}_1(\ell_2)$ and contained in $\mathcal{S}_p(\ell_2)$ for all $p > 1$.

Here is one of the key results of Milman and Pisier.

Theorem: *Let X be a Banach space. The set $\{vr(E) : E \in \mathcal{F}_X\}$ is bounded if and only if there is a constant c such that, regardless of how we choose $n \in \mathbf{N}$, $v \in \mathcal{L}(\ell_2^n, X)$ and $u \in \mathcal{L}(X, \ell_2^n)$, we have*

$$\sigma_{1,\infty}(uv) \leq c\cdot\pi_2(u)\cdot\pi_\gamma(v).$$

Spaces satisfying the conditions of this theorem are said to have *weak cotype* 2. The concepts of *weak cotype* q ($2 \leq q < \infty$) and *weak type* p ($1 < p \leq 2$) are defined through similar modifications of the characterizations (a) and (b). If a Banach space X has weak type p (weak cotype q) then it has type $p - \varepsilon$ (cotype $q + \varepsilon$) for all $\varepsilon > 0$.

A. Pajor [1987] has given a characterization of weak type 2 spaces in terms of volume ratios which is dual to the Milman - Pisier theorem:

- *a Banach space X has weak type 2 if and only if there is a constant c such that $vr(Q) \leq c$ for all finite dimensional quotient spaces Q of X.*

This is also equivalent to the existence of a constant C such that

$$\left(\frac{\text{vol}\,(\mathcal{E}_E^0)}{\text{vol}\,(B_E)}\right)^{1/\dim E} \leq C$$

for every finite dimensional subspace E of X; here $\mathcal{E}_E^0$ is the polar body of $\mathcal{E}_E$, that is the ellipsoid of minimal volume which contains B_E. – Generalizations to weak type p spaces were obtained by V. Mascioni [1988a,b].

When a Banach space X simultaneously has weak type 2 and weak cotype 2 it is called a *weak Hilbert space.* The results above show that X is a weak Hilbert space if and only if

$$\sup_{E \in \mathcal{F}_X} \left(\frac{\text{vol}\,(\mathcal{E}_E^0)}{\text{vol}\,(\mathcal{E}_E)}\right)^{1/\dim E} < \infty.$$

Weak Hilbert spaces are reflexive and have no subspaces of quotients which fail the approximation property. The 2-convexified Tsirelson space $T^{(2)}$ is a non-Hilbertian weak Hilbert space. It is just one of a whole family obtained by varying the basic Tsirelson construction.

While the elementary theory of weak type and weak cotype closely resembles that of general type and cotype, beware: there are pitfalls. For example: the Banach space X has type 2 (cotype 2) if and only if $L_2([0,1], X)$ (or $\ell_2^{strong}(X)$, or *Rad*(X) for that matter) has weak type 2 (weak cotype 2). Moreover, P.G. Casazza has shown that the theory of weak Hilbert spaces is essentially a theory of separable spaces: if X is a non-separable weak Hilbert space then $X = Y \oplus Z$ where Y is a separable weak Hilbert space and Z is a Hilbert space. Once again, P.G. Casazza and T.J. Shura [1989] is the appropriate source for details.

If we only allow vectors of norm one (or of equal length) in the definition of type p and cotype q for a Banach space X, then we arrive at concepts L. Tzafriri [1979] called 'equal norm type p' and 'equal norm cotype q'. It was shown by G. Pisier (see R.C. James [1978]) that equal norm type 2 is the same as type 2 and that equal norm cotype 2 is the same as cotype 2. This is in sharp contrast to a result of V. Mascioni and U. Matter [1988]: if $1 < p < 2$ then equal norm type p coincides with weak type p; similarly, equal norm cotype q is the same as weak cotype q when $2 < q < \infty$.

Further results on weak type, weak cotype and generalizations can be found in M. Defant and M. Junge [1990], S. Geiss [1987],[1990a,b,c], V. Mascioni [1988a,b], A. Pietsch [1990b], and G. Pisier [1988], [1989].

Tsirelson-like constructions, due to T. Schlumprecht [1991], have recently led to some spectacular results in the structure theory of Banach spaces. For example, W.T. Gowers and B. Maurey [1993] found a Banach space without an unconditional basic sequence; W.T. Gowers [1995a] uncovered an infinite dimensional Banach space that neither contains a copy of c_0 or ℓ_1 nor has an infinite dimensional reflexive subspace. There are Banach spaces which are not isomorphic to any of their proper closed subspaces (in particular not to any of their closed hyperplanes), and it is now known that ℓ_2 is, up to isomorphism, the only Banach space which is isomorphic to all of its closed infinite dimensional subspaces. We refer to the work of W.T. Gowers and B. Maurey [1993], [1994], R. Komorowski and N. Tomczak-Jaegermann [1994] and W.T. Gowers [1994a,b] for details on this and much more, and for further references.

15. WEAKLY COMPACT OPERATORS ON $\mathcal{C}(K)$-SPACES

Our goal in this chapter is to show that weakly compact operators with domain a $\mathcal{C}(K)$-space enjoy a number of remarkable properties, properties which are intimately linked with summing operators, with type and cotype, and with the validity of Grothendieck's Theorem for operators into Hilbert space when the domain is not an $\mathcal{L}_1$-space.

Throughout, K will be a compact Hausdorff space and Y a Banach space. Since an operator $u : \mathcal{C}(K) \to Y$ is weakly compact if and only if $W := u^*(B_{Y^*})$ is weakly compact in $\mathcal{C}(K)^*$, the space of all regular Borel measures on K, it shouldn't come as a surprise that our discussion is going to be based on two classical weak compactness results. Here is the first.

15.1 Theorem: *A subset W of $\mathcal{C}(K)^*$ is (relatively) weakly compact if and only if there is a positive measure $\mu \in \mathcal{C}(K)^*$ such that each $\nu \in W$ is μ-absolutely continuous and the corresponding densities form a (relatively) weakly compact subset of $L_1(\mu)$.*

Any measure μ with the announced property is called a *control measure* for W. Of course, we can always normalize so that μ is a probability measure.

The second classical result that we shall exploit in the next few pages is the characterization 13.21 of weakly compact subsets of $L_1(\mu)$-spaces as those sets which are bounded and uniformly integrable.

CHARACTERIZATION OF WEAKLY COMPACT OPERATORS

We begin by using these tools to deduce the following characterization of weakly compact operators on $\mathcal{C}(K)$.

15.2 Theorem: *The operator $u : \mathcal{C}(K) \to Y$ is weakly compact if and only if for some (and then every) $1 \le p < \infty$ there exists a probability measure $\mu \in \mathcal{C}(K)^*$ with the property that for each $\varepsilon > 0$ we can find an $N(\varepsilon) > 0$ such that, for all $f \in \mathcal{C}(K)$,*

$$(*) \qquad \|uf\| \le N(\varepsilon)\cdot\Big(\int_K |f|^p d\mu\Big)^{1/p} + \varepsilon\cdot\|f\|.$$

Proof. Suppose first that u is weakly compact. Then $W := u^*(B_{Y^*})$ is weakly compact in $\mathcal{C}(K)^*$ and can, by 15.1, be considered to be a weakly compact subset of $L_1(\mu)$ for some probability measure $\mu \in \mathcal{C}(K)^*$. Accordingly, u^* can be regarded to be a weakly compact operator $Y^* \to L_1(\mu)$.

We shall verify $(*)$ for $p = 1$; the general case follows suit.

Suppose what we claim is false: we can find $\varepsilon > 0$ and norm one functions $f_n \in C(K)$ such that

(1) $$\|uf_n\| > n \cdot \int_K |f_n|\, d\mu + \varepsilon$$

for all $n \in \mathbf{N}$. Notice that

(2) $$\lim_{n\to\infty} \int_K |f_n|\, d\mu = 0.$$

Since $\|uf_n\| > \varepsilon$, we can find $y_n^* \in B_{Y^*}$ such that $\langle y_n^*, uf_n\rangle > \varepsilon$. Consequently the functions $g_n := u^* y_n^* \in W \subset L_1(\mu)$ satisfy

(3) $$\int_K f_n g_n\, d\mu > \varepsilon$$

for all $n \in \mathbf{N}$.

Thanks to the weak compactness of W, 13.21 allows us to conclude that

(4) $$\lim_{\mu(B)\to 0} \int_B |g_n|\, d\mu = 0,$$

uniformly in $n \in \mathbf{N}$.

Let $\delta > 0$ be arbitrary, and consider the sets $B_{n,\delta} := \{|f_n| \geq \delta\}$, $n \in \mathbf{N}$. As $\mu(B_{n,\delta}) \leq \delta^{-1} \cdot \int_K |f_n|\, d\mu$ for each n, (2) yields

(5) $$\lim_{n\to\infty} \mu(B_{n,\delta}) = 0.$$

According to (4), we may choose $0 < \delta \leq \varepsilon/(2\cdot\|u\|+1)$ so that $\int_B |g_n|\, d\mu \leq \varepsilon/2$ for any n and any Borel set $B \subset K$ with $\mu(B) \leq \delta$. Fix such a δ, and use (5) to find an $n_\delta \in \mathbf{N}$ such that $\mu(B_{n,\delta}) \leq \delta$ for every $n \geq n_\delta$. Now look at (3) to realize that

$$\begin{aligned}\varepsilon < \int_K f_n g_n\, d\mu &\leq \int_{B_{n,\delta}} |g_n|\, d\mu + \delta \cdot \int_{K\setminus B_{n,\delta}} |g_n|\, d\mu \\ &\leq \frac{\varepsilon}{2} + \delta \cdot \|u^* y_n^*\| \leq \frac{\varepsilon}{2} + \delta \cdot \|u\| \leq \varepsilon\end{aligned}$$

whenever $n \geq n_\delta$. Contradiction.

It remains to show that u is weakly compact whenever $(*)$ is satisfied for some $1 \leq p < \infty$. To this end, set $Z := L_p(\mu) \oplus_1 C(K)$ and, for $\varepsilon > 0$, define $a_\varepsilon \in \mathcal{L}(C(K), Z)$ by $a_\varepsilon(f) := (N(\varepsilon) j_p f, \varepsilon f)$, where j_p is the usual natural map $C(K) \to L_p(\mu)$. As a_ε is injective, we can define $b_\varepsilon : a_\varepsilon(C(K)) \to Y$ by $b_\varepsilon(a_\varepsilon f) = uf$. The map b_ε is clearly linear and continuous on $a_\varepsilon(C(K))$ in the Z-norm topology and has norm ≤ 1.

Choose a set J large enough ($J = B_{Y^*}$ will do) that there is an isometric embedding $i_Y : Y \to \ell_\infty^J$. Since ℓ_∞^J is an injective Banach space, $i_Y b_\varepsilon$ is the restriction of some $\widetilde{b_\varepsilon} \in \mathcal{L}(Z, \ell_\infty^J)$ with $\|\widetilde{b_\varepsilon}\| = \|b_\varepsilon\| \leq 1$. Clearly, $i_Y u = \widetilde{b_\varepsilon} a_\varepsilon$.

Now consider the operators $v_\varepsilon, w_\varepsilon : C(K) \to \ell_\infty^J$ given by

$$v_\varepsilon(f) := \widetilde{b_\varepsilon}(N(\varepsilon) j_p f, 0) \quad \text{and} \quad w_\varepsilon(f) := \widetilde{b_\varepsilon}(0, \varepsilon f).$$

As v_ε has j_p as a factor, it is p-summing, hence weakly compact. By design, $i_Y u = v_\varepsilon + w_\varepsilon$ and so $\|i_Y u - v_\varepsilon\| = \|w_\varepsilon\| \leq \varepsilon$. Letting $\varepsilon \to 0$, we obtain that $i_Y u$ is weakly compact. The conclusion follows. QED

An Approximation Scheme

It will be profitable to take a closer look at the argument above.

Given Banach spaces X and Y, let

$$\mathcal{H}(X,Y)$$

be the collection of all operators $u \in \mathcal{L}(X,Y)$ for which one can find a $p \in [1,\infty)$, a Banach space Z, an operator $v \in \Pi_p(X,Z)$ and, for each $\varepsilon > 0$, an $N(\varepsilon) > 0$ such that

$$(*) \qquad \|ux\| \leq N(\varepsilon)\cdot\|vx\| + \varepsilon\cdot\|x\|$$

for all $x \in X$.

Theorem 15.2 just states that $\mathcal{W}(\mathcal{C}(K),Y) = \mathcal{H}(\mathcal{C}(K),Y)$.

Note that in the definition of $\mathcal{H}(X,Y)$ the particular choice of p doesn't really matter, it will only result in a change of the parameters in $(*)$. To see this, first observe that, by Pietsch's Factorization Theorem 2.13, we may choose v in $(*)$ to be the composition of canonical maps $X \overset{i_X}{\to} \mathcal{C}(B_{X^*}) \overset{j_p}{\to} L_p(\mu)$ for some probability measure μ; then apply Theorem 15.2 to j_p.

Our next result shows that the operators in $\mathcal{H}$ can be approximated nicely.

15.3 Theorem: *An operator $u : X \to Y$ belongs to $\mathcal{H}(X,Y)$ if and only if for some (and then all) finite $p \geq 1$ there exist a set J and an isometric embedding $i_Y : Y \to \ell_\infty^J$ such that for some sequence (v_n) in $\Pi_p(X,\ell_\infty^J)$ we have*

$$\lim_{n\to\infty} \|i_Y u - v_n\| = 0.$$

Proof. If u belongs to $\mathcal{H}(X,Y)$ then, no matter how we choose $1 \leq p < \infty$, the construction in 15.2 produces operators $v_n \in \Pi_p(X,\ell_\infty^J)$ with the desired properties.

Passing to the converse, suppose we have found J, $i_Y : Y \to \ell_\infty^J$ and (v_n) in $\Pi_p(X,\ell_\infty^J)$ as announced. We want to construct a Banach space Z along with an operator $v \in \Pi_p(X,Z)$ and a function $N : [0,\infty) \to [0,\infty)$ such that, for all $x \in X$ and $\varepsilon > 0$,

$$(1) \qquad \|ux\| \leq N(\varepsilon)\cdot\|vx\| + \varepsilon\cdot\|x\|.$$

Replacing (v_n) by a subsequence if necessary, we may assume that

$$(2) \qquad \|ux\| \leq \|v_n x\| + 2^{-n}\cdot\|x\|$$

holds for all $x \in X$ and $n \in \mathbf{N}$. We may also assume that each $v_n \neq 0$.

We take Z to be the ℓ_1 direct sum $\left(\oplus_n \ell_\infty^J\right)_1$ of countably many copies of ℓ_∞^J. For each $k \in \mathbf{N}$, let i_k denote the k-th canonical injection of ℓ_∞^J into Z, and let $w_k : X \to Z$ be the operator $w_k = 2^{-k}\cdot\pi_p(v_k)^{-1} i_k v_k$. Since $\pi_p(w_k) \leq 2^{-k}$ for each k, $v = \sum_k w_k$ exists in the Banach space $\Pi_p(X,Z)$. By (2), we get for any $n \in \mathbf{N}$ and $x \in X$,

$$\begin{aligned}\|ux\| &\le \|v_n x\| + 2^{-n}\cdot\|x\| = 2^n\cdot\pi_p(v_n)\cdot\|w_n x\| + 2^{-n}\cdot\|x\| \\ &\le 2^n\cdot\pi_p(v_n)\cdot\sum_k \|w_k x\| + 2^{-n}\cdot\|x\| = 2^n\cdot\pi_p(v_n)\cdot\|vx\|_Z + 2^{-n}\cdot\|x\|.\end{aligned}$$

If $0 < \varepsilon < 1$ is given, choose n so that $2^{-n} < \varepsilon$ and let $N(\varepsilon) := 2^n\cdot\pi_p(v_n)$; (1) follows. QED

15.4 Corollary: *Under the uniform norm, $\mathcal{H}$ is a closed, injective operator ideal. If $u \in \mathcal{H}(X,Y)$, then u is weakly compact and completely continuous.*

This follows immediately from 15.3; after all, 15.3 says that $i_Y u$ is the uniform limit of a sequence of p-summing operators, and these are weakly compact and completely continuous, by 2.17. Now just recall that weakly compact operators as well as completely continuous operators form closed and injective ideals.

15.5 Corollary: *An operator $u : X \to Y$ belongs to $\mathcal{H}$ if and only if $u^{**} : X^{**} \to Y^{**}$ does.*

Proof. The injectivity of $\mathcal{H}$ ensures that u is in $\mathcal{H}$ if u^{**} is. As for the converse, let $u \in \mathcal{H}(X,Y)$ and choose J, i_Y and (v_n) in $\Pi_p(X,\ell_\infty^J)$ as in 15.3 so that $\lim_{n\to\infty}\|i_Y u - v_n\| = 0$. Weak compactness entitles us to consider each v_n^{**} as an element of $\Pi_p(X^{**},\ell_\infty^J)$ and u^{**} as an element of $\mathcal{L}(X^{**},Y)$. Accordingly, identify $(i_Y u)^{**}$ and $i_Y u^{**}$, and note that

$$\lim_{n\to\infty}\|i_Y u^{**} - v_n^{**}\| = \lim_{n\to\infty}\|(i_Y u - v_n)^{**}\| = \lim_{n\to\infty}\|i_Y u - v_n\| = 0. \quad \text{QED}$$

Our next result is a preliminary one; it will make life a little bit easier in the discussions to come.

15.6 Proposition: *An operator $u : X \to Y$ belongs to $\mathcal{H}(X,Y)$ if and only if its restriction to any separable subspace E of X belongs to $\mathcal{H}(E,Y)$.*

Proof. If $u \in \mathcal{H}(X,Y)$ and if E is any (separable) subspace of X, then certainly $u\big|_E \in \mathcal{H}(E,Y)$.

Suppose now that $u : X \to Y$ is such that $u\big|_E \in \mathcal{H}(E,Y)$ for every separable subspace E of X. So, given E, there is a probability measure $\mu \in \mathcal{C}(B_{E^*})^*$ and a function N of $(0,\infty)$ into itself such that

$$(*) \qquad \|ux\| \le N(\varepsilon)\cdot\int_{B_E^*} |\langle x^*, x\rangle|\, d\mu(x^*) + \varepsilon\cdot\|x\| \quad \forall x \in E,\ \forall \varepsilon > 0.$$

In the canonical manner, $\mathcal{C}(B_{E^*})$ is a subspace of $\mathcal{C}(B_{X^*})$ and μ extends to a probability measure in $\mathcal{C}(B_{X^*})^*$; we denote the extension by μ as well and have

$$(**) \qquad \|ux\| \le N(\varepsilon)\cdot\int_{B_X^*} |\langle x^*, x\rangle|\, d\mu(x^*) + \varepsilon\cdot\|x\| \quad \forall x \in E,\ \forall \varepsilon > 0.$$

Now let W_E denote the set of all functions $N : (0,\infty) \to (0,\infty)$ for which there exists a probability measure $\mu \in \mathcal{C}(B_{X^*})$ satisfying $(**)$. Define

$$N_E(\varepsilon) := \inf_{N\in W_E} N(\varepsilon), \quad \varepsilon > 0.$$

We claim that $\sup\{N_E(\varepsilon) : E \subset X \text{ separable}\}$ is finite for each $\varepsilon > 0$.

Assume to the contrary that there are separable subspaces $E_n \subset X$, $n \in \mathbf{N}$, and an $\varepsilon_0 > 0$ such that $\lim_{n\to\infty} N_{E_n}(\varepsilon_0) = \infty$. Let $E \subset X$ be a separable subspace which contains all the E_n's, and let $N \in W_E$. Naturally, $E_n \subset E$ implies $N \in W_{E_n}$, and so $N_{E_n}(\varepsilon_0) < N(\varepsilon_0) < \infty$ for each n: a contradiction.

Now set $M(\varepsilon) := \sup_E N_E(\varepsilon)$ for each $\varepsilon > 0$. For every separable subspace E of X and $\varepsilon > 0$, there is an $N \in W_E$ so that $N(\varepsilon) \le N_E(\varepsilon) + \varepsilon \le M(\varepsilon) + \varepsilon$ and, with N in hand, we find a probability measure $\mu_{E,\varepsilon} \in \mathcal{C}(B_{X^*})^*$ such that, for all $x \in B_E$,

$$(\bullet) \qquad \begin{aligned} \|ux\| &\le N(\varepsilon)\cdot \int_{B_{X^*}} |\langle x^*, x\rangle| \, d\mu_{E,\varepsilon}(x^*) + \varepsilon \\ &\le \big(M(\varepsilon)+\varepsilon\big)\cdot \int_{B_{X^*}} |\langle x^*, x\rangle| \, d\mu_{E,\varepsilon}(x^*) + \varepsilon \\ &\le M(\varepsilon)\cdot \int_{B_{X^*}} |\langle x^*, x\rangle| \, d\mu_{E,\varepsilon}(x^*) + 2\cdot\varepsilon. \end{aligned}$$

For each $\varepsilon > 0$, the $\mu_{E,\varepsilon}$'s form a net in $B_{\mathcal{C}(B_{X^*})^*}$; let $\mu_\varepsilon \in \mathcal{C}(B_{X^*})^*$ be a corresponding weak$*$limit point. Then μ_ε is clearly a probability measure, and from $(\bullet)$ we obtain, for all $x \in X$,

$$\|ux\| \le M(\varepsilon)\cdot \int_{B_{X^*}} |\langle x^*, x\rangle| \, d\mu_\varepsilon(x^*) + 2\cdot\varepsilon\cdot\|x\|.$$

We now pass to the probability measure $\mu := \sum_{n=1}^\infty 2^{-n}\mu_{1/n}$, and we define $N(\varepsilon) := 2\cdot M(1)$ if $\varepsilon \ge 1$, and $N(\varepsilon) := 2^{n+1}\cdot M(1/(n+1))$ if $1/(n+1) \le \varepsilon < 1/n$. For each $x \in X$ and each $\varepsilon > 0$,

$$\|ux\| \le N(\varepsilon)\cdot \int_{B_{X^*}} |\langle x^*, x\rangle| \, d\mu(x^*) + 2\cdot\varepsilon\cdot\|x\|. \qquad \text{QED}$$

Ultrapower Stability

It follows from 15.4 that the identity operator on an infinite dimensional Banach space can never be in $\mathcal{H}$. But just as in the case of almost summing operators (see 12.10), it can happen that every operator from ℓ_1 to a Banach space Y is in $\mathcal{H}(\ell_1, Y)$; Banach spaces with this property are said to enjoy *property (H)*.

Hilbert spaces have this property, thanks to Grothendieck's Theorem. More generally, every reflexive quotient Y of a $\mathcal{C}(K)$-space enjoys property (H). Indeed, if $q : \mathcal{C}(K) \to Y$ is a quotient map and $u : \ell_1 \to Y$ is any operator, then the lifting property of ℓ_1 (see after 3.17) supplies us with an operator $v \in \mathcal{L}(\ell_1, \mathcal{C}(K))$ such that $u = qv$. By 15.2, q belongs to $\mathcal{H}$; it follows that u does, too, and so Y must have property (H).

Using 15.6, we can see that property (H) is equivalent to the property

$$\mathcal{L}(\ell_1^I, Y) = \mathcal{H}(\ell_1^I, Y) \quad \text{for every index set } I.$$

In fact, (H) is certainly weaker than this property, and the converse follows from 15.6 and the observation that if I is infinite and X_0 is a separable subspace of ℓ_1^I, then X_0 embeds into ℓ_1: just pick a dense sequence (x_n) in X_0, write $x_n = \sum_{i\in I} \xi_{i,n} e_i$ for each n, and note that $J := \bigcup_n \{i \in I : \xi_{i,n} \neq 0\}$ is countable and that X_0 is a closed subspace of $\ell_1^J \subset \ell_1$.

The following theorem is the main result of this section.

15.7 Theorem: *If the Banach space Y has property (H), then every Banach space which is finitely representable in Y has property (H) as well.*

In other words, (H) is a super property.

We shall derive 15.7 from the following general result.

15.8 Theorem: *The ideal $\mathcal{H}$ is stable under the formation of ultrapowers.*

Proof. Let $u \in \mathcal{H}(X,Y)$ be given. Our task is to show that, whatever index set I and ultrafilter $\mathcal{U}$ on I we take, the operator

$$u^{\mathcal{U}} : X^{\mathcal{U}} \longrightarrow Y^{\mathcal{U}} : (x_i)_{\mathcal{U}} \mapsto (ux_i)_{\mathcal{U}}$$

belongs to $\mathcal{H}(X^{\mathcal{U}}, Y^{\mathcal{U}})$; see Chapter 8 for notation and background.

To this end, fix a Banach space Z, an operator $v \in \Pi_2(X,Z)$ and a function $\varepsilon \mapsto N(\varepsilon)$ such that

$$\|ux\| \leq N(\varepsilon)\cdot\|vx\| + \varepsilon\cdot\|x\|$$

for all $x \in X$ and $\varepsilon > 0$. In view of Corollary 2.16 we may assume that Z is $L_2(\mu)$ for some probability measure $\mu \in C(B_{X^*})^*$ and that v is the composition $X \xrightarrow{i_X} C(B_{X^*}) \xrightarrow{j_2} L_2(\mu)$ of canonical maps. Then $v^{\mathcal{U}}$ is the composition of $j_2^{\mathcal{U}} : C(B_{X^*})^{\mathcal{U}} \to L_2(\mu)^{\mathcal{U}}$ and $i_X^{\mathcal{U}} : X^{\mathcal{U}} \to C(B_{X^*})^{\mathcal{U}}$. But, by 8.7, $C(B_{X^*})^{\mathcal{U}}$ is a $C(K)$-space and $L_2(\mu)^{\mathcal{U}}$ is a Hilbert space. Consequently $j_2^{\mathcal{U}}$ is 2-summing, and so is $v^{\mathcal{U}}$.

Now let $(x_i)_{i\in I}$ be a bounded family in X, and set $x = (x_i)_{\mathcal{U}} \in X^{\mathcal{U}}$. For each $\varepsilon > 0$ we have

$$\begin{aligned}\|u^{\mathcal{U}}x\| &= \lim_{\mathcal{U}} \|ux_i\| \leq \lim_{\mathcal{U}} \big(N(\varepsilon)\cdot\|vx_i\| + \varepsilon\cdot\|x_i\|\big)\\ &\leq N(\varepsilon)\cdot\|v^{\mathcal{U}}x\| + \varepsilon\cdot\|x\|.\end{aligned}$$

We have reached our destination. QED

Do not be tempted to think that $\mathcal{H}$ is stable with respect to the formation of arbitrary ultraproducts! If it were, it would be maximal by 8.10 and therefore equal to $\mathcal{L}$, which, as we noted in 6.11, is the only ideal which is maximal *and* uniformly closed.

Proof of 15.7. By 8.12, every Banach space which is finitely representable in X appears as a subspace of some ultrapower of X. Since $\mathcal{H}$ is injective, it suffices to show that any ultrapower $X^{\mathcal{U}}$ has property (H) once X does.

So, let X have property (H), choose an index set I so that X is a quotient of ℓ_1^I, and let $q : \ell_1^I \to X$ be a quotient map. We know that q is in $\mathcal{H}(\ell_1^I, X)$, and so 15.8 tells us that $q^{\mathcal{U}}$ is in $\mathcal{H}((\ell_1^I)^{\mathcal{U}}, X^{\mathcal{U}})$. Now $q^{\mathcal{U}}$ is clearly surjective and ℓ_1 has the lifting property, so that each $u \in \mathcal{L}(\ell_1, X^{\mathcal{U}})$ has the form $u = q^{\mathcal{U}} \hat{u}$ for some $\hat{u} \in \mathcal{L}(\ell_1, (\ell_1^I)^{\mathcal{U}})$. It follows that u is in $\mathcal{H}(\ell_1, X^{\mathcal{U}})$, and all is well.

QED

Combining 15.3 with 15.7 and using Grothendieck's Theorem, we can now state:

15.9 Corollary: *The following are equivalent statements about the Banach space Y:*

(i) *Y has property (H).*

(ii) *If $q : \ell_1^I \to Y$ is a quotient map and $j : Y \to \ell_\infty^J$ is an isometric embedding, then there is a sequence (u_n) in $\Gamma_2(\ell_1^I, \ell_\infty^J)$ such that $\lim_{n\to\infty} \|jq - u_n\| = 0$.*

(iii) *For each $v \in \mathcal{L}(Y, \ell_\infty)$ and $w \in \mathcal{L}(\ell_1, Y)$ there is a sequence (u_n) in $\Gamma_2(\ell_1, \ell_\infty)$ such that $\lim_{n\to\infty} \|vw - u_n\| = 0$.*

(iv) *Y^* has property (H).*

It is clear that every Banach space which satisfies (H) must be reflexive. In particular, it cannot contain a copy of ℓ_1 or of c_0. Theorem 15.7 allows a fundamental strengthening when we combine it with 13.7, 13.15 and 11.10.

15.10 Corollary: *A Banach space with property (H) does not contain ℓ_1^n's uniformly and so has non-trivial type and cotype.*

We are now in position to prove a dual version of a celebrated theorem due to H.P.Rosenthal. Compare also with 11.18 and 13.19.

15.11 Theorem: *Suppose that Y is a reflexive quotient of a $C(K)$-space. Then Y has type 2; indeed, there exist a number $2 \leq r < \infty$ and a probability measure $\mu \in C(K)^*$ such that Y appears in a natural fashion as a quotient of $L_r(\mu)$.*

Proof. It follows from 15.2 and the lifting property of ℓ_1 that Y has property (H). We have just observed that Y must therefore have finite cotype. We may proceed as in the proof of 11.18 to achieve our goal. QED

By duality, we obtain the original version of Rosenthal's Theorem:

15.12 Theorem: *If X is a reflexive subspace of some $\mathcal{L}_1$-space, then there exist a probability measure μ and a number $1 < p \leq 2$ such that X is isomorphic to a subspace of $L_p(\mu)$.*

In particular, we obtain once more the hard part of Theorem 13.19: reflexive subspaces of $L_1(\mu)$-spaces must be B-convex.

Spaces Verifying Grothendieck's Theorem

So far we have concentrated on reflexive Banach spaces of the form $C(K)/X$. The special properties of such spaces make it natural to expect that we can say something interesting about X as well. The central result is the following theorem which signifies that X must indeed be a fairly 'big' subspace of $C(K)$.

15.13 Theorem: *Suppose that X is a subspace of $C(K)$ and that $C(K)/X$ is reflexive. Then every operator with domain X and range a cotype 2 space is 2-summing.*

Consequently, referring to the Π_2-Extension Theorem 4.15, we see that given any cotype 2 space Y and any Banach space $\tilde{X}$ which contains X as a subspace, each $u \in \mathcal{L}(X,Y)$ admits an extension $\tilde{u} : \tilde{X} \to Y$ which even belongs to $\Pi_2(\tilde{X},Y)$.

Our proof depends on two results which are of independent interest.

15.14 Kisliakov's Lemma: *Let X, X_1, Y be Banach spaces and suppose that X is a subspace of X_1. Let $u : X \to Y$ be any operator. Then there are a Banach space Y_1 and an operator $u_1 : X_1 \to Y_1$ such that the following hold:*

(a) *Y is a subspace of Y_1,*

(b) *$u_1 x = ux$ for all $x \in X$,*

(c) *$\|u_1\| = \|u\|$,*

(d) *Y_1/Y is isometrically isomorphic to X_1/X.*

Proof. We may of course assume that $\|u\| = 1$.

Let Z be the ℓ_1 direct sum of X_1 and Y: so Z is $X_1 \times Y$ with the norm $\|(x_1, y)\| = \|x_1\| + \|y\|$ for each $x_1 \in X_1$, $y \in Y$. Consider $W := \{(x, -ux) : x \in X\}$ and note that this is a closed subspace of Z. We claim that $Y_1 := Z/W$ is the space that will do what we want it to do.

Let $q : Z \to Y_1$ be the quotient map. Then $j : Y \to Y_1 : y \mapsto q(0, y)$ is an isometric embedding. In fact, for each $y \in Y$, we plainly have $\|jy\| \leq \|y\|$ and, since $\|u\| = 1$, we also have

$$\begin{aligned}\|jy\| &= \|q(0,y)\| = \inf\{\|(0,y) + w\| : w \in W\} \\ &= \inf\{\|(x, y - ux)\| : x \in X\} = \inf\{\|x\| + \|y - ux\| : x \in X\} \\ &\geq \|y\| + \inf\{\|x\| - \|ux\| : x \in X\} \geq \|y\|.\end{aligned}$$

Notice that if $x \in X$, then

$$jux = q(0, ux) = q\big((0, ux) + (x, -ux)\big) = q(x, 0),$$

so that if we define the operator $u_1 : X_1 \to Y_1$ by $u_1 x_1 := q(x_1, 0)$ for each $x_1 \in X_1$ then $ju = u_1|_X$. By design, $\|u_1\| \leq 1$; so $\|u_1\| = 1$ since u_1 extends u.

We are left with the proof of (d).

Let $p_X : X_1 \to X_1/X$ and $p_Y : Y_1 \to Y_1/Y$ be the canonical quotient maps. For each $x \in X$, we have $u_1 x \in Y$ $(= jY)$; therefore there is a well-defined linear map $v : X_1/X \to Y_1/Y$ such that $vp_X = p_Y u_1$. Moreover, v is isometric: for each $x_1 \in X_1$ we have

$$\begin{aligned}
\|vp_X x_1\|_{Y_1/Y} &= \inf_{y \in Y} \|u_1 x_1 + jy\|_{Y_1} = \inf_{y \in Y} \|q(x_1, y)\|_{Y_1} \\
&= \inf \big\{ \|x_1 + x, y - ux)\|_Z : y \in Y,\, x \in X \big\} \\
&= \inf_{x \in X} \big(\|x_1 + x\|_{X_1} + \inf_{y \in Y} \|y - ux\|_Y \big) = \|p_X x_1\|_{X_1/X}.
\end{aligned}$$

Finally, to show that v is onto, take any $y_1 \in Y_1$ and let $z \in Z$ be such that $y_1 = qz$. Write $z = (x_1, y)$ where $x_1 \in X_1$ and $y \in Y$. Since $vp_X x_1 = p_Y u_1 x_1 = p_Y q(x_1, 0) = p_Y qz = p_Y y_1$, we are done. QED

Next we show that having finite cotype is a 'three space property':

15.15 Lemma: *Suppose that Z is a Banach space and that W is a closed subspace of Z. If W and Z/W both have finite cotype, then Z has finite cotype as well.*

Proof. We only need to consider the case when Z is infinite dimensional. As a first step we show that if Z has a subspace Y which is isomorphic to c_0, then W or Z/W must contain an isomorphic copy of c_0.

If $Y \cap W$ is infinite dimensional, then, like all closed infinite dimensional subspaces of c_0, $Y \cap W$ contains a copy of c_0. Consequently, if we assume that W does not contain a copy of c_0, then $Y \cap W$ must be finite dimensional and so complemented in Y. Let Y_0 be a complement of $Y \cap W$ in Y. Like any infinite dimensional complemented subspace (of an isomorphic copy) of c_0, Y_0 must itself be isomorphic to c_0. But $Y_0 \cap W = \{0\}$, so the quotient map $Z \to Z/W$ takes Y_0 isomorphically into Z/W, and Z/W contains a copy of c_0.

Now suppose that W and Z/W both have finite cotype. We want to show that Z has finite cotype. Let us assume this is not the case. Then, by 14.1, Z contains the ℓ_∞^n's uniformly, whereas neither W nor Z/W has this property. By 8.12, some ultrapower $Z^{\mathcal{U}}$ contains a copy of c_0, whereas this is not true for $W^{\mathcal{U}}$ and $(Z/W)^{\mathcal{U}}$. But $W^{\mathcal{U}}$ is canonically isomorphic to a subspace of $Z^{\mathcal{U}}$, and $Z^{\mathcal{U}}/W^{\mathcal{U}}$ is canonically isomorphic to $(Z/W)^{\mathcal{U}}$; we run into a contradiction with what we just showed. QED

We are now ready for our main goal.

Proof of 15.13. Let Y be a Banach space of cotype 2, and let $u \in \mathcal{L}(X, Y)$ be any operator. By 15.14, there exists a Banach space Y_1 together with an operator $u_1 \in \mathcal{L}(\mathcal{C}(K), Y_1)$ such that Y is a subspace of Y_1, u_1 extends u, and Y_1/Y is isometrically isomorphic to $\mathcal{C}(K)/X$. By 15.11 Y_1/Y has type 2, and therefore it must have finite cotype. Apply 15.15 to conclude that Y_1 must have finite cotype, and 11.14 to obtain that u_1 must be r-summing for some $2 \leq r < \infty$. It follows that u is also r-summing. But since Y has cotype 2, 11.13 informs us that u is even 2-summing. QED

The preceding results reconnect us with one of our earliest topics: Grothendieck's Theorem.

15.16 Corollary: *Let μ be any measure and let X be a reflexive subspace of $L_1(\mu)$. Then every operator $u : L_1(\mu)/X \to \ell_2$ is 1-summing.*

As an example, the Rademacher functions span a copy R of ℓ_2 inside $L_1[0,1]$, so that $L_1[0,1]/R$ has the announced property.

Proof. Let X be reflexive, and let $u \in \mathcal{L}(L_1(\mu)/X, \ell_2)$ be given. According to 2.7 we must show that, regardless of how we choose $v \in \mathcal{L}(c_0, L_1(\mu)/X)$, the composition $uv : c_0 \to \ell_2$ is 1-summing.

We use the fact that $L_1(\mu)^*$, while not necessarily identifiable with $L_\infty(\mu)$, is $L_\infty(\nu)$ for some measure ν, and is thus a known $\mathcal{C}(K)$-space. Since X is reflexive, X^* is too. But $X^* = L_\infty(\nu)/X^\perp$, so we are in position to apply 15.13. It tells us that v^* takes $X^\perp \subset L_\infty(\nu)$ to ℓ_1 in a 2-summing manner. By 4.15, v^* extends to some $w : L_\infty(\nu) \to \ell_1$. It follows that $(uv)^* = v^*u^*$ belongs to $\Gamma_\infty(\ell_2, \ell_1)$, and so, invoking 7.2, $uv \in \Gamma_1(c_0, \ell_2)$. By Grothendieck's Theorem (3.4), uv is 1-summing. QED

Frequently, Banach spaces Z such that $\mathcal{L}(Z, \ell_2) = \Pi_1(Z, \ell_2)$ are said *to verify Grothendieck's Theorem.* We know that $\mathcal{L}_1$-spaces have this property, and 15.16 informs us that there are potentially others. To realize the potential, we shall show that when X is a reflexive subspace of $L_1(\mu)$ the quotient space $L_1(\mu)/X$ cannot be an $\mathcal{L}_1$-space unless X is finite dimensional.

Suppose $X \subset L_1(\mu)$ is reflexive and $L_1(\mu)/X$ is an $\mathcal{L}_1$-space. By 8.14(b), $(L_1(\mu)/X)^{**}$ is then a complemented subspace of some $L_1(\mu)$, and so $X^\perp = (L_1(\mu)/X)^*$, which is complemented in its bidual, is a complemented subspace of $L_1(\mu)^*$. Now $Z := L_1(\mu)^*$ is known to be a $\mathcal{C}(K)$-space: this can be seen referring forward to Kakutani's Theorem 16.1. (One may even show that Z is an $L_\infty(\nu)$-space for some measure ν.) Noting that the complement of $X^\perp$ in Z is isomorphic to $Z/X^\perp$, a space which is isomorphic with X^*, we can use 15.2 to deduce that $id_{X^*} \in \mathcal{H}$. Using 15.4 we see that X^*, and with it X, must be finite dimensional.

Notes and Remarks

The subject of weakly compact operators on $\mathcal{C}(K)$-spaces is an astoundingly broad one and we touch on only a very small part of it in this chapter.

In tandem with 13.21, Theorem 15.1 is a cornerstone of the classical theory; 13.21 is due to N. Dunford [1939], whereas 15.1 is due to R.G. Bartle,

N. Dunford and J.T. Schwartz [1955]. Both 13.21 and 15.1 as well as much of the classical theory of operators on $\mathcal{C}(K)$-spaces are treated in the monographs of J. Diestel and J.J. Uhl, Jr. [1977] and N. Dunford and J.T. Schwartz [1958].

Theorem 15.2 is due to C.P. Niculescu [1975], [1979]; the proof we present is taken from H. Jarchow [1981] and was found in collaboration with A. Pełczyński. The general scheme was developed further by H. Jarchow and U. Matter [1985], [1988], U. Matter [1987], [1989], and F. Räbiger [1991]. Being able to find an $r > 0$ and a $C > 0$ so that the crucial function N satisfies $N(\varepsilon) \leq C \cdot \varepsilon^{-r}$ for all $\varepsilon > 0$ brings interpolation theory into play.

The approach is flexible enough to be extended beyond $\mathcal{C}(K)$-spaces. The following generalization of Theorem 15.2 was proved by H. Jarchow [1986]:

Theorem: *Let A be a C^*-algebra and Y a Banach space. For any operator $u : A \to Y$ the following statements are equivalent.*

(i) *u is weakly compact.*

(ii) *One can find a state $\varphi \in A^*$ and a function $N : (0, \infty) \to (0, \infty)$ such that for each $x \in A$ and $\varepsilon > 0$*
$$\|ux\| \leq N(\varepsilon) \cdot \varphi\big(|x|^2\big)^{1/2} + \varepsilon \cdot \|x\|.$$

(iii) *There exist a Hilbert space H, an operator $v : A \to H$ and a function $N : (0, \infty) \to (0, \infty)$ such that for any $x \in A$ and $\varepsilon > 0$*
$$\|ux\| \leq N(\varepsilon) \cdot \|vx\| + \varepsilon \cdot \|x\|.$$

(iv) *There is a sequence (v_n) in $\Gamma_2(A, \ell_\infty^{B_{Y^*}})$ such that, viewing u as an operator into $\ell_\infty^{B_{Y^*}}$,*
$$\lim_{n\to\infty} \|u - v_n\|_{\mathcal{L}(A, \ell_\infty^{B_{Y^*}})} = 0.$$

A few words filling in some background are in order before we discuss the proof of this result. If A is a C^*-algebra, then $|x| := \big((xx^* + x^*x)/2\big)^{1/2}$ for any $x \in A$, where $x \mapsto x^*$ is the involution of A. States are norm one functionals φ on A such that $\varphi(x^*x) \geq 0$ for any $x \in A$.

It is a stirring fact, uncovered by C.A. Akemann [1967], that

- *a bounded subset K of the dual of a C^*-algebra A is relatively weakly compact if and only if there is a state $\varphi \in A^*$ such that given $\varepsilon > 0$ there is a $\delta > 0$ such that for any $x \in B_A$ with $\varphi(|x|^2)^{1/2} \leq \delta$ and any $\kappa \in K$ we have $|\langle \kappa, x \rangle| \leq \varepsilon$.*

Of course, in the present setup this theorem provides a convenient substitute for Theorem 15.1, with states taking over the rôle of the 'control measures'.

Accordingly, to prove (i)$\Rightarrow$(ii) of the above theorem, just proceed as in the proof of 15.2. Use the Akemann result to produce a state $\varphi \in A^*$ that 'controls' the weakly compact subset $K = u^*(B_{Y^*})$ and check by arguing contrapositively that φ fulfills the demands of (ii).

To pass from (ii) to (iii), define a sesquilinear form $(\cdot|\cdot)_\varphi$ on $A \times A$ by $(x|y)_\varphi := \varphi(xy^* + y^*x)/2$. Factoring out those x's such that $(x|x)_\varphi = 0$ we get a pre-Hilbert space. Its completion, H, is the space to look for: the canonical map $v : A \to H$ is primed to satisfy (iii).

That (iii) implies (iv) is also obtained by copying the corresponding argument in the proof of 15.2, and since (i) patently follows from (iv), we are done.

It was shown by C.A. Akemann, P.G. Dodds and J.L.B. Gamlen [1972] that if the Banach space X does not contain a copy of c_0 then every operator from a C^*-algebra A into X is weakly compact; for $A = \mathcal{C}(K)$ the result is a classic due to A. Pełczyński [1960]. This applies in particular when X does not contain the ℓ_∞^n's uniformly, that is, when X has finite cotype. Now duals of C^*-algebras have cotype 2; see N. Tomczak-Jaegermann [1974]. As in 15.11 it follows that quotients of C^*-algebras have type 2 whenever they are reflexive (or merely do not contain a copy of c_0). Using Kwapień's Theorem 12.19, we thereby arrive at the following generalization of Theorem 3.9:

- *If a Banach space X is simultaneously isomorphic to a subspace of the dual of a C^*-algebra and to a quotient of a C^*-algebra, then it is isomorphic to a Hilbert space.*

Another interesting consequence of the preceding theorem was observed by G. Pisier [1992b]:

- *If X is a complemented reflexive subspace of a C^*-algebra A, then X is isomorphic to a Hilbert space.*

In fact, applying (iii) to the projection from A onto X, we find an operator $v : A \to H$ where H is a Hilbert space and a number $N > 0$ so that $\|x\| \le N \cdot \|vx\| + \|x\|/2$ for each $x \in X$. The assertion follows from $\|x\|/2 \le N \cdot \|vx\| \le N \cdot \|v\| \cdot \|x\|$ for all $x \in X$.

If $A = \mathcal{C}(K)$ then complemented reflexive subspaces are even finite dimensional, thanks to the fact that $\mathcal{C}(K)$-spaces have the Dunford-Pettis property.

While mentioning the Pełczyński-Akemann-Dodds-Gamlen results we would be amiss if we did not bring up the corresponding results for individual operators. A. Pełczyński [1962] showed that *if the operator $u : \mathcal{C}(K) \to X$ takes weak ℓ_1 sequences to unconditionally summable sequences, then u is weakly compact.* It was only recently that H. Pfitzner [1992] showed that Pełczyński's result holds for C^*-algebras as well.

C.P. Niculescu [1979] used 15.2 to derive a theorem of J. Diestel and C.J. Seifert [1979] concerning the Banach-Saks nature of weakly compact operators on $\mathcal{C}(K)$-spaces. Using the theorem cited above, Jarchow extended the Diestel-Seifert theorem to general C^*-algebras:

- *If $u : A \to Y$ is a weakly compact operator on the C^*-algebra A, then every bounded sequence $(a_n)_n$ in A has a subsequence $(a_{n_k})_k$ such that the arithmetic means of $(ua_{n_k})_k$ are norm convergent.*

By the way, similar results can be derived for other $\mathcal{C}(K)$-like spaces, most particularly, the disk algebra. Also, the above theorem was quickly generalized to Jordan triples by C.H. Chu and B. Iochum [1988]; see also C.H. Chu, B. Iochum and G. Loupias [1989].

The first systematic investigation of the ideal $\mathcal{H}$ and its relatives is due to U. Matter [1987], [1989]. Some basic results, however, can already be found in C.P. Niculescu's work [1979]: Proposition 15.6, for example, is due to him. The essential results 15.7 and 15.8 are due to U. Matter [1987], [1989] and can be proved in much greater generality. It should be pointed out that in 15.9 we may replace Γ_2 by Γ_p for any $1 < p < \infty$.

Theorems 15.11 and 15.12 are due to H.P. Rosenthal [1973]. It was his penetrating analysis of subspaces of L_p $(1 \leq p < 2)$ that provided much of the functional analytic inspiration which led B. Maurey [1974a] to his profound analysis of factorization of operators into L_0 and so, in the end, to the Maurey-Pisier Theorem 14.1. It is worth observing that G. Pisier, incorporating the results just discussed in his study of factorization properties of operators on C^*-algebras, generalized 15.11 to reflexive quotients of C^*-algebras.

Theorem 15.13 is due to S.V. Kisliakov [1978] and G. Pisier [1978b]. Only very recently, S.V. Kisliakov [1991] found further examples of Banach spaces verifying Grothendieck's Theorem among duals of subspaces of $\mathcal{C}(\mathbf{T})$ which satisfy a Hadamard lacunarity condition for the Fourier coefficients of their members indexed by negative integers.

The deceptively innocent Lemma 15.14 is due to S.V. Kisliakov [1978]. It was designed for the same purposes for which it is used here, but has found use in several other spectacular constructions. We mention the paper of J. Bourgain and G. Pisier [1983] where the lemma was a valuable tool in constructing $\mathcal{L}_\infty$-spaces that defy naive intuition, but answered a number of long-standing problems related to the stability of weak sequential completeness and absence of copies of c_0 in projective tensor products. We also mention the remarkable [1983] paper by G. Pisier where he succeeded in proving the existence of infinite dimensional Banach spaces P such that $P\hat{\otimes}P = P\check{\otimes}P$ and used this to give a partial answer to another old problem raised by A. Grothendieck [1953a]: are there Banach spaces X and Y such that all compact operators $X \to Y$ are nuclear? Pisier proved that all approximable operators (that is, uniform limits of finite rank operators) from P to P are nuclear. In [1990], K. John observed that every compact operator that factors through Hilbert space is approximable by finite rank operators; using this he showed that every compact operator from P to P^* is nuclear. A modification of Grothendieck's question remains open: does there exist an infinite dimensional Banach space X such that every compact operator $X \to X$ is nuclear?

Actually, Pisier proved more than we have mentioned. Using an inductive construction based on Kisliakov's Lemma, he showed that every space having cotype 2 can be embedded in a space P as above which, together with its dual, has cotype 2. If one starts with a separable space, then the construction ends with a separable space. The spaces P so constructed do not have the approximation property nor are they K-convex – otherwise 13.17 in combination with Kwapień's Theorem (12.20) would grant such a space the status of a Hilbert space!

Consequences of Pisier's construction continue to be uncovered. R.J. Kaiser and J.R. Retherford [1983] showed (more or less) that any ℓ_2 sequence is the sequence of eigenvalues of a nuclear endomorphism of Pisier's space. Using this as a catalyst they went on to show the same is so for several classical Banach spaces. H. Jarchow and K. John [1995] used such spaces to obtain examples of non-nuclear Fréchet - Schwartz spaces on which all bounded bilinear forms are nuclear. H.G. Dales and H. Jarchow [1995] have shown that the Banach algebra of approximable operators on P has discontinuous point derivations. P.G. Casazza and H. Jarchow [1995] used Pisier's space to study Banach spaces in which all compact subsets are generates by compact endomorphisms.

Lemma 15.15 is due to S.A. Rakov [1976], but our proof does not follow his.

It is not presently known whether a Banach space verifying Grothendieck's Theorem must have cotype 2. In this direction it is worth establishing that quotients of $\mathcal{L}_1$-spaces modulo reflexive subspaces do have cotye 2.

The basic result is due to G. Pisier [1982b]:

Theorem: *Let X be a closed subspace of the Banach space Z, and let $q : Z \to Z/X$ be the corresponding quotient map. If X is K-convex, then $(z_k) \mapsto (qz_k)$ induces a surjective mapping $Rad\,(Z) \to Rad\,(Z/X)$.*

It is clear that $(z_k) \mapsto (qz_k)$ *always* induces a norm one bounded linear map $Rad\,(Z) \to Rad\,(Z/X)$. The theorem asserts that when X is K-convex, there is a constant $C > 0$ such that, no matter how we select finitely many vectors $y_1,\dots, y_n$ from Z/X, we can find vectors $z_1,\dots, z_n$ in Z which satisfy $q(z_k) = y_k$ for each $1 \le k \le n$ and

$$(*)\quad \Big(\int_{D_n} \Big\| \sum_{k\le n} r_k(\varepsilon) z_k \Big\|^2 d\mu_n(\varepsilon)\Big)^{1/2} \le C\cdot\Big(\int_{D_n} \Big\| \sum_{k\le n} r_k(\varepsilon) y_k \Big\|^2 d\mu_n(\varepsilon)\Big)^{1/2}.$$

Before entering the details of the proof, let us check that this theorem provides the information we are looking for.

Corollary: *If Z is a Banach space of cotype r and if X is a K-convex subspace of Z, then Z/X also has cotype r.*

Proof. Given $y_1,\dots, y_n \in Z/X$ such that $\big(\int_{D_n} \| \sum_{k\le n} r_k(\varepsilon) y_k\|^2 d\mu_n(\varepsilon)\big)^{1/2} \le 1$, use the theorem – alias $(*)$ – to produce $z_1,\dots, z_n \in Z$ such that $qz_k = y_k$ for each $1 \le k \le n$ and $\big(\int_{D_n} \| \sum_{k\le n} r_k(\varepsilon) z_k\|^2 d\mu_n(\varepsilon)\big)^{1/2} \le C$. Since Z has cotype r, we get that $(\sum_{k\le n} \|z_k\|^r)^{1/r} \le C\cdot C_r(Z)$, and so

$$\Big(\sum_{k\leq n}\|y_k\|^r\Big)^{1/r} = \Big(\sum_{k\leq n}\|qz_k\|^r\Big)^{1/r} \leq \Big(\sum_{k\leq n}\|z_k\|^r\Big)^{1/r} \leq C\cdot C_r(Z).$$

It follows that Z/X has cotype r, with $C_r(Z/X)\leq C\cdot C_r(Z)$. QED

Our main interest is in the case $r=2$:

Corollary: *If X is a K-convex subspace of the $\mathcal{L}_1$-space Z, then Z/X has cotype 2.*

A reflexive subspace of an $\mathcal{L}_1$-space is K-convex: in fact, by Rosenthal's Theorem 15.12, it occurs as a subspace of an $L_p(\mu)$-space for some $1<p\leq 2$.

The converse is also true. Suppose that X is a K-convex subspace of the $\mathcal{L}_1$-space Z. Then X^* has finite cotype and is a quotient of the $\mathcal{L}_\infty$-space Z^*. By 11.14, the quotient map is r-summing for some finite r, and so X^* appears as a quotient of some $\mathcal{L}_r$-space and must thus be reflexive.

Now we proceed towards the proof of the theorem.

Recall that for each $n\in\mathbf{N}$ the Walsh functions

$$w_A = \prod_{k\in A} r_k\,, \quad A\subset\{1,2,\ldots,n\}$$

form an orthonormal basis of the Hilbert space $L_2(\mu_n)$ $[=L_2(D_n,\mu_n)]$ and that, when Z is a Banach space, each $f\in L_2(\mu_n,Z)$ can be represented in the form

$$f = \sum_{A\subset\{1,\ldots,n\}} w_A z_A$$

where $z_A=\int_{D_n} w_A(\omega)f(\omega)d\mu_n(\omega)$. Formally, this is a Bochner integral, but in our setup we are really just manipulating finite sums.

Fix $n\in\mathbf{N}$, and for each $1\leq k\leq n$ consider the projection

$$T_k : L_2(\mu_n,Z)\longrightarrow L_2(\mu_n,Z) : \sum_{A\subset\{1,\ldots,n\}} w_A z_A \mapsto \sum_{|A|=k} w_A z_A.$$

We know from 13.16 that if Z happens to be K-convex, then there is a constant $C=C(Z)$ such that $\|T_k\|\leq C^k$ for all $n\in\mathbf{N}$ and $1\leq k\leq n$. We will be interested in operators on $L_2(\mu_n,Z)$ which are of the form

$$Q_s := \sum_{k\leq n} s^k T_k,$$

where s is a scalar. More precisely, we shall need that

$$(\bullet)\qquad \|Q_s\|\leq 1 \quad \text{whenever } \ 0\leq s\leq 1.$$

This can be seen as follows. Consider the function $g\in L_1(\mu_n)$ given by

$$g := \prod_{k=1}^{n}(1+sr_k).$$

Since $g=\sum_{A\subset\{1,\ldots,n\}} s^{|A|}w_A$ is non-negative, we have

$$\|g\|_{L_1(\mu_n)} = \sum_{A\subset\{1,\ldots,n\}} s^{|A|}\int_{D_n} w_A(\omega)\,d\mu_n(\omega) = 1.$$

Take any $B \subset \{1,\dots,n\}$, $z \in Z$ and $\omega \in D_n$. Then

$$Q_s(w_B z)(\omega) = \Big(\sum_{k\le n} s^k T_k(w_B z)\Big)(\omega) = s^{|B|} w_B(\omega) z = (w_B * g)(\omega) z.$$

Consequently, if we define convolution of $f = \sum_A w_A z_A \in L_2(\mu_n, Z)$ and g by

$$f * g = \sum_{A\subset\{1,\dots,n\}} (w_A * g) z_A,$$

then, thanks to linearity, $Q_s f = f * g$, and straightforward generalizations of classical scalar estimates lead to

$$\|Q_s f\|_{L_2(\mu_n,Z)} = \|f * g\|_{L_2(\mu_n,Z)} \le \|f\|_{L_2(\mu_n,Z)} \cdot \|g\|_{L_1(\mu_n)} = \|f\|_{L_2(\mu_n,Z)}.$$

This proves (•).

We are now set to complete our program.

Proof of the theorem. Let $y_1,\dots,y_n \in Z/X$ be given. By homogeneity, it is no loss of generality to assume that

$$\int_{D_n} \Big\|\sum_{k\le n} r_k(\omega) y_k\Big\|^2 d\mu_n(\omega) < 1.$$

We want to show that there exist a constant c (independent of n) and elements $z_1,\dots,z_n \in Z$ such that $y_k = q z_k$ for each $1 \le k \le n$ and

$$\int_{D_n} \|\sum_{k\le n} r_k(\omega) z_k\|^2 d\mu_n(\omega) \le c.$$

Recall that our integrals are finite sums. So pointwise lifting can be applied to the function

$$\varphi : D_n \longrightarrow Z/X : \omega \mapsto \sum_{k\le n} r_k(\omega) y_k$$

to produce another function $\phi : D_n \to Z$ such that

$$q(\phi) = \varphi \quad \text{and} \quad \int_{D_n} \|\phi(\omega)\|_Z^2 d\mu_n(\omega) \le 1 \quad \forall\, \omega \in D_n.$$

The function φ is certainly odd (that is, $\varphi(-\omega) = -\varphi(\omega)$ for all $\omega \in D_n$). By replacing ϕ, if necessary, by $\omega \mapsto (\phi(\omega) - \phi(-\omega))/2$ we may assume that ϕ is odd as well.

Consider ϕ as a member of $L_2(\mu_n, Z)$ and write

$$\phi = \sum_{A\subset\{1,\dots,n\}} w_A z_A, \tag{1}$$

where $z_A = \int_{D_n} w_A(\omega)\phi(\omega) d\mu_n(\omega)$. Oddness of ϕ implies

$$z_A = 0 \quad \text{whenever } |A| \text{ is even,} \tag{2}$$

and from $q\,\phi = \varphi$ we infer that

$$q(z_A) = 0 \quad (\text{or } z_A \in X) \quad \text{if} \quad |A| \ne 1, \tag{3}$$

whereas

$$q(z_{\{k\}}) = y_k \quad \text{for all } 1 \le k \le n. \tag{4}$$

We claim that the vectors $z_k := z_{\{k\}}$ $(1 \le k \le n)$ are the ones we are looking for.

To this end, define $\phi_k \in L_2(\mu_n, Z)$ by $\phi_k := T_k(\phi)$, $1 \le k \le n$. Note that $\phi = \sum_{k=1}^n \phi_k$ and that (2) implies that $\phi_k = 0$ if k is even. We need to estimate $\|\phi_1\|^2_{L_2(\mu_n,Z)} = \int_{D_n} \|\sum_{k\le n} r_k(\omega) z_k\|^2 d\mu_n(\omega)$.

Let $0 < s < 1$ be given. Since

$$s\phi_1 = Q_s(\phi) - \sum_{k\ge 3} s^k \phi_k,$$

we see that

$$\|s\phi_1\|_{L_2(\mu_n,Z)} \le 1 + \Big\| \sum_{k\ge 3} s^k \phi_k \Big\|_{L_2(\mu_n,Z)}. \tag{5}$$

Now we may view $\sum_{k\ge 3} s^k \phi_k = Q_s(\phi) - s\phi_1$ as a member of $L_2(\mu_n, X)$. But X is K-convex, and so Theorem 13.16 furnishes us with a constant $C = C(X)$ such that each T_k has norm $\le C^k$ as an operator on $L_2(\mu_n, X)$. If $0 < s < 1/C$, then we may write

$$\Big\|\sum_{k\ge 3} s^k\phi_k\Big\|_{L_2(\mu_n,Z)} = \Big\|\sum_{k\ge 1} s^k T_k(\phi-\phi_1)\Big\|_{L_2(\mu_n,Z)} = \Big\|\sum_{k\ge 3} s^k T_k(\phi-\phi_1)\Big\|_{L_2(\mu_n,X)}$$

$$\le \sum_{k\ge 3} s^k C^k \|\phi-\phi_1\|_{L_2(\mu_n,X)} \le \frac{(sC)^3}{1-sC}\cdot\|\phi-\phi_1\|_{L_2(\mu_n,Z)}.$$

Hence

$$\Big\|\sum_{k\ge 3} s^k\phi_k\Big\|_{L_2(\mu_n,Z)} \le \frac{(sC)^3}{1-sC}\cdot\big(1+\|\phi_1\|_{L_2(\mu_n,Z)}\big). \tag{6}$$

Now combine (5) and (6); the result is

$$s\cdot\|\phi_1\|_{L_2} \le 1 + \frac{(sC)^3}{1-sC}\cdot\big(1+\|\phi_1\|_{L_2}\big).$$

In particular, this is true if we choose $s = C^{-3/2}/2$. Then $1 - s\cdot C \ge 1/2$, and so

$$(C^{-3/2}/2)\cdot\|\phi_1\|_{L_2} \le 1 + (C^{-3/2}/4)\cdot(1+\|\phi_1\|_{L_2}).$$

Rearrange to find that

$$\Big(\int_{D_n} \Big\|\sum_{k\le n} r_k(\omega) z_k\Big\|^2 \, d\mu_n(\omega)\Big)^{1/2} = \|\phi_1\|_{L_2} \le 4\cdot C^{3/2} + 1.$$

The theorem is proved. QED

While the condition that X be a K-convex subspace of L_1 is sufficient for L_1/X to have cotype 2 and verify Grothendieck's Theorem, it is hardly necessary. In fact, J. Bourgain [1984c] showed that $L_1(\mathbf{T})/H_0^1$ has cotype 2 and verifies Grothendieck's Theorem. This result solved a problem of A. Pełczyński [1976b] affirmatively and offered a substantial improvement of several results of S. Kwapień and A. Pełczyński [1978]. Along his way, Bourgain broke a few backs with his hard analytical techniques. Other approaches to these exceedingly deep, difficult results emerged. J. Bourgain and W.J. Davis [1986] offered martingale arguments. S.V. Kisliakov [1991] had a go at the cotype 2 result and obtained, in addition, delicate details about (q,p)-summing operators on the disk algebra. G. Pisier [1992a] applied ideas from interpolation theory; as a consequence, he showed that $L_p(\mathbf{T})/H^p$ has cotype 2 for any $0 < p \le 1$.

16. Type and Cotype in Banach Lattices

Type and cotype take especially pleasing form in Banach lattices; nagging ambiguities disappear, to be replaced by sharp alternatives. Old friends, like the Khinchin and Grothendieck Inequalities, are given new meaning and contribute to the elegance of the theory. Of course, for so many good things to happen, a modicum of hard work is required.

If $x_1,\dots,x_n$ are in some $\mathcal{C}(K)$ or $L_q(\mu)$ $(1 \le q \le \infty)$, expressions like

$$(*) \qquad \Big(\sum_{k\le n} |x_k|^p\Big)^{1/p} \qquad (1 \le p < \infty)$$

have an obvious meaning: they are defined pointwise, almost everywhere if necessary. The principal chore will be making sense of expressions even more complicated than $(*)$ in mathematical objects – general Banach lattices – that ostensibly lack multiplicative structure of any sort. This behind us, many of the beautiful classical relations adapt well to a new environment. There are particularly satisfying consequences for summing operators.

Functional Calculus

We shall have to develop a process known as the functional calculus. The general idea runs as follows. If $(t_1,\dots,t_n) \mapsto h(t_1,\dots,t_n)$ is a real-valued function on $\mathbf{R}^n$ given by a 'reasonable' algebraic expression, then, when $x_1,\dots,x_n$ are elements of a Banach lattice, it should be possible to define $h(x_1,\dots,x_n)$ by simply substituting each x_k for the corresponding t_k. The functions

$$h_p(t_1,\dots,t_n) := \Big(\sum_{k\le n} |t_k|^p\Big)^{1/p} \quad (1 \le p < \infty)$$

are the ones we will be most interested in, but they are not 'reasonable' enough to be handled straightforwardly.

The strategy is to come up with a sizeable class of 'reasonable' functions and then to develop a limiting process which will allow us to create the desired objects. In the end, we will be able to deal with any continuous function on the unit sphere of ℓ_∞^n.

It is by no means obvious that the substitution process we mentioned above is (well-) defined in sufficiently general circumstances. To address the problem, we need to broach the representation theory of Banach lattices.

Recall that a Banach lattice M is an *abstract M-space* if

$$\|x \vee y\| = \|x\| \vee \|y\|$$

for any non-negative members $x, y \in M$. We shall only be interested in abstract M-spaces M *with unit*, that is, we require the unit ball B_M to contain a largest element. Much of our work will be based on a fundamental classical result.

16.1 Kakutani's Representation Theorem: *Every abstract M-space with unit is isometrically isomorphic, as a Banach lattice, to $\mathcal{C}(K)$ for some compact Hausdorff space K.*

It is clear that this isomorphism takes M's unit to the constant one function in $\mathcal{C}(K)$.

The crucial observation is that positive elements in Banach lattices can always be used to generate abstract M-spaces. To see how, we start by recalling that a linear subspace I of a Banach lattice L is an *ideal* if every $y \in L$ which satisfies $|y| \leq |x|$ for some $x \in I$ necessarily belongs to I. For example, given any $x \in L$,

$$I(x) := \{y \in L : |y| \leq \lambda \cdot |x| \text{ for some } 0 < \lambda < \infty\},$$

is an ideal; this is the *ideal generated by* x. Of course, $I(x) = \{0\}$ if and only if $x = 0$.

When $x \neq 0$, we can set up a useful norm on $I(x)$: for each $y \in I(x)$, define

$$\|y\|_\infty := \inf\{\lambda > 0 : |y| \leq \frac{\lambda}{\|x\|} \cdot |x|\}.$$

Note that $\|x\| = \|x\|_\infty$ and $\|y\| \leq \|y\|_\infty$ for each $y \in I(x)$. The unit ball of $[I(x), \|\cdot\|_\infty]$ is closed in L; therefore $[I(x), \|\cdot\|_\infty]$ is a Banach space. Most important for us, however, is that $[I(x), \|\cdot\|_\infty]$ is a Banach lattice with respect to the order inherited from L; it is even an abstract M-space with unit $|x|/\|x\|$. Consequently, $[I(x), \|\cdot\|_\infty]$ can be identified, as a Banach lattice, with a space $\mathcal{C}(K)$ for a suitably chosen compact Hausdorff space K.

Now consider, for each $n \in \mathbf{N}$, the family

$$\mathcal{C}_n$$

of all functions $\mathbf{R}^n \to \mathbf{R}$ which are obtained from the coordinate functions $(t_1,..., t_n) \mapsto t_k$ $(1 \leq k \leq n)$ by finitely many operations of addition, scalar multiplication, finite infimum and finite supremum. This is about as straightforward a non-trivial class as one can get, and we shall first show that if $h, \tilde{h} \in \mathcal{C}_n$ satisfy $h(t_1,..., t_n) = \tilde{h}(t_1,..., t_n)$ for all scalars $t_1,..., t_n$, then the substitution process makes $h(x_1,..., x_n) = \tilde{h}(x_1,..., x_n)$ for any elements $x_1,..., x_n$ in the Banach lattice L.

Look at $x := |x_1| \vee ... \vee |x_n| \in L$. As $h, \tilde{h} \in \mathcal{C}_n$, both $h(x_1,..., x_n)$ and $\tilde{h}(x_1,..., x_n)$ evidently belong to $I(x)$, alias $\mathcal{C}(K)$ for some compact Hausdorff space K. Since the algebraic and order relations in $\mathcal{C}(K)$ are defined pointwise, having $h(t_1,..., t_n) = \tilde{h}(t_1,..., t_n)$ for all scalars $t_1,..., t_n$ ensures that $h(x_1,..., x_n) = \tilde{h}(x_1,..., x_n)$ in $I(x)$, and so in L. To summarize, we have a well-defined map

$$\jmath : \mathcal{C}_n \longrightarrow L : h \mapsto h(x_1,..., x_n).$$

Notice that $\mathcal{C}_n$ is plainly a vector lattice, and that $\jmath$ is linear and order preserving.

Further progress depends on being able to approximate functions like $h_p(t_1,\ldots,t_n) = (\sum_{k\le n}|t_k|^p)^{1/p}$ by members of $\mathcal{C}_n$. But all of the h_p's and all elements of $\mathcal{C}_n$ belong to the space

$$\mathcal{H}_n$$

of all continuous functions $f : \mathbf{R}^n \to \mathbf{R}$ which are positively homogeneous of degree one, that is, which satisfy $f(\lambda t_1,\ldots,\lambda t_n) = \lambda\cdot f(t_1,\ldots,t_n)$ for any $\lambda \ge 0$ and $(t_1,\ldots,t_n) \in \mathbf{R}^n$. Clearly, $\mathcal{H}_n$ is also a vector lattice – another model of $\mathcal{C}(S_n)$ where S_n is the unit sphere of ℓ_∞^n. Indeed, $\mathcal{H}_n \to \mathcal{C}(S_n) : f \mapsto f\,|_{S_n}$ is an isomorphism of vector lattices.

Evidently, $\mathcal{C}_n$ is a sublattice of $\mathcal{H}_n = \mathcal{C}(S_n)$. The real issue is that it is a very special sublattice: it separates the points of S_n, and it contains the functions which are constant on S_n. So $\mathcal{C}_n$ is dense in $\mathcal{C}(S_n)$, by Kakutani's lattice version of the Stone-Weierstrass Theorem.

Now everything falls into place beautifully. The map $\jmath$ that we introduced above is *continuous* when we give $\mathcal{C}_n$ the norm induced by $\mathcal{C}(S_n)$. To see why, notice first that the positive homogeneity of $h \in \mathcal{C}_n$ gives

$$|h(t_1,\ldots,t_n)| \le \|h\|_{\mathcal{C}(S_n)}\cdot\max_{k\le n}|t_k|$$

for all $t_1,\ldots,t_n \in \mathbf{R}$. Consequently, with $x = |x_1|\vee\ldots\vee|x_n|$ as before,

$$|h(x_1,\ldots,x_n)| \le \|h\|_{\mathcal{C}(S_n)}\cdot x$$

and hence

$$\|h(x_1,\ldots,x_n)\|_L \le \|x\|_L\cdot\|h\|_{\mathcal{C}(S_n)}.$$

Since $\mathcal{C}_n$ is dense in $\mathcal{H}_n = \mathcal{C}(S_n)$, the map $\jmath$ admits a unique continuous extension

$$\tilde{\jmath} : \mathcal{H}_n = \mathcal{C}(S_n) \longrightarrow L$$

which is linear and preserves order. It is natural and convenient to write

$$h(x_1,\ldots,x_n) := \tilde{\jmath}(h).$$

The whole process of taking $h \in \mathcal{H}_n$ and $x_1,\ldots,x_n \in L$ and assigning the element $h(x_1,\ldots,x_n) \in L$ is traditionally called the *functional calculus* in L.

To provide evidence of the power of the functional calculus, we demonstrate how to obtain smooth generalizations of some of our favourite scalar relations involving the expressions $\left(\sum_{k\le n}|x_k|^p\right)^{1/p}$.

16.2 Proposition: *Let $x_1,\ldots,x_n$ be elements of the Banach lattice L.*

(a) *For $1\le p<\infty$, $\left(\sum_{k\le n}|x_k|^p\right)^{1/p} = \sup\left\{\sum_{k\le n}a_kx_k : a \in B_{\ell_{p^*}^n}\right\}$.*

(b) *$\sup_{k\le n}|x_k| = \sup\left\{\sum_{k\le n}a_kx_k : a\in B_{\ell_1^n}\right\}$.*

Proof. (a) Fix $1\le p<\infty$ and $x_1,\ldots,x_n \in L$. We already know that $x := \left(\sum_{k=1}^n|x_k|^p\right)^{1/p}$ exists in L. Our claim is trivially true in $I(x) = \mathcal{C}(K)$, and so x is certainly an upper bound for $S := \{\sum_{k=1}^n a_kx_k : a \in B_{\ell_{p^*}^n}\}$ in L.

Let $y \in L$ be any upper bound for S in L. Clearly, y is an upper bound for S in $I(x \vee y)$, and $z := \sup S$ exists in $I(x \vee y)$. Since $I(x) \subset I(x \vee y)$ we have $z \leq x$. Since $I(x)$ is an ideal, we may conclude that $z \in I(x)$. It follows that $z = x$ and therefore $x \leq y$, both in $I(x \vee y)$ and in L. We have proved that x is the supremum of the set S in L.

(b) just requires some notational modifications. QED

It is equally simple to extend Khinchin's Inequality:

16.3 Proposition: *Let $0 < p < \infty$. There are constants A_p, $B_p > 0$ such that, for any choice of finitely many vectors $x_1,\dots, x_n$ in a Banach lattice,*

$$A_p \cdot \Big(\sum_{k\leq n} |x_k|^2\Big)^{1/2} \leq \Big(\int_0^1 \Big|\sum_{k\leq n} r_k(t) x_k\Big|^p \, dt\Big)^{1/p} \leq B_p \cdot \Big(\sum_{k\leq n} |x_k|^2\Big)^{1/2}.$$

Just observe that the functions $(s_1,\dots, s_n) \mapsto \big(\int_0^1 \big|\sum_{k\leq n} r_k(t) s_k\big|^p \, dt\big)^{1/p}$ belong to $\mathcal{H}_n$ and apply the functional calculus to Khinchin's Inequality 1.10.

Inserting norms into these inequalities is a more delicate matter. However, one direction is straightforward:

16.4 Corollary: *No matter how we choose finitely many vectors $x_1,\dots, x_n$ from a Banach lattice and $0 < p < \infty$,*

$$A_1 \cdot \Big\|\Big(\sum_{k\leq n} |x_k|^2\Big)^{1/2}\Big\| \leq \Big(\int_0^1 \Big\|\sum_{k\leq n} r_k(t) x_k\Big\|^p \, dt\Big)^{1/p}.$$

Proof. We have just seen that $A_1 \cdot (\sum_{k\leq n} |x_k|^2)^{1/2} \leq \int_0^1 |\sum_{k\leq n} r_k(t) x_k| \, dt$. Hence

$$\begin{aligned} A_1 \cdot \Big\| \Big(\sum_{k\leq n} |x_k|^2\Big)^{1/2} \Big\| &\leq \Big\| \int_0^1 \Big|\sum_{k\leq n} r_k(t) x_k\Big| \, dt \Big\| \\ &\leq \int_0^1 \Big\| \Big|\sum_{k\leq n} r_k(t) x_k\Big| \Big\| \, dt = \int_0^1 \Big\|\sum_{k\leq n} r_k(t) x_k\Big\| \, dt. \end{aligned}$$

QED

This inequality is necessarily one-sided. In the course of this chapter, we shall see that the two-sided version is valid if, and only if, the Banach lattice has finite cotype.

Before we begin our main development, we should emphasize that we have been working with *real* Banach lattices. However, given a Banach lattice L, there is a natural way to complexify. The product space

$$L_{\mathbf{C}} := L \times L$$

is a complex linear space with respect to the usual coordinatewise addition and the scalar multiplication

$$(a + ib)(x, y) := (ax - by, ay + bx)$$

for $x, y \in L$ and $a, b \in \mathbf{R}$. Our recent work allows us to define $|\cdot|$ on $L_{\mathbf{C}}$ by

$$|(x, y)| := \big(|x|^2 + |y|^2\big)^{1/2}.$$

Since $|(x,y)| \in L$, we can set

$$\|(x,y)\|_{L_{\mathbb{C}}} := \|\,|(x,y)|\,\|_L.$$

This is easily seen to be a norm. Indeed, $[L_{\mathbb{C}}, \|\cdot\|_{L_{\mathbb{C}}}]$ is a Banach space, the *complexification* of L. Any time the term *complex Banach lattice* is used, it refers to the complexification of a real Banach lattice in the sense above.

In what follows, all our Banach lattices will be assumed to be real. The complexification procedure usually allows a smooth and easy transfer of our results to the complex case, but we shall not take the time to investigate.

(q,p)-Concave Operators

Now that the preliminaries are behind us, we can start to think again about summing operators and their relatives. Notice that if $1 \le q < \infty$ and if $x_1,\dots,x_n$ lie in some Banach lattice L, the following inequality stems from 16.2:

$$\begin{aligned}\|(x_k)\|_p^{weak} &= \sup\Big\{\Big\|\sum_{k\le n} a_k x_k\Big\| : a \in B_{\ell_{p^*}^n}\Big\}\\ &\le \Big\|\sup\Big\{\sum_{k\le n} a_k x_k : a \in B_{\ell_{p^*}^n}\Big\}\Big\| = \Big\|\Big(\sum_{k\le n}|x_k|^p\Big)^{1/p}\Big\|.\end{aligned}$$

This suggests that we should add new classes of operators to our repertoire. Take a Banach lattice L, a Banach space Y, and numbers $1 \le p,q < \infty$. We say that an operator $u : L \to Y$ is *(q,p)-concave* if there is a constant C such that, for any choice of finitely many vectors $x_1,\dots,x_n \in L$, we have

$$\Big(\sum_{k\le n}\|ux_k\|^q\Big)^{1/q} \le C\cdot\Big\|\Big(\sum_{k\le n}|x_k|^p\Big)^{1/p}\Big\|.$$

We write

$$K_{q,p}(u)$$

for the least constant C that works. Normally, the (q,q)-concave operators are called *q-concave,* and $K_q(u)$ is used instead of $K_{q,q}(u)$. The usual one-dimensional rigmarole confirms that when $p > q$, the only (q,p)-concave operator is the zero operator. We therefore always assume that

$$p \le q.$$

Our observations about the weak ℓ_p norm entail that

- *every (q,p)-summing operator defined on a Banach lattice is (q,p)-concave.*

Of course, the converse is true when the Banach lattice is a $\mathcal{C}(K)$-space.

The combined weight of the functional calculus and Kakutani's Theorem are needed to reach a crucial characterization.

16.5 Theorem: *Let L be a Banach lattice and Y a Banach space. Suppose that $1 \le p \le q < \infty$ and $C > 0$. An operator $u : L \to Y$ is (q,p)-concave with $K_{q,p}(u) \le C$ if and only if, for each compact Hausdorff space and every positive operator $v : \mathcal{C}(K) \to L$, the composition $uv : \mathcal{C}(K) \to Y$ is (q,p)-summing with $\pi_{q,p}(uv) \le C\cdot\|v\|$.*

Proof. Let $v : \mathcal{C}(K) \to L$ be any positive operator, and select $f_1,\dots,f_n$ from $\mathcal{C}(K)$. Applying 16.2, we find

$$\begin{aligned}\Big(\sum_{k\le n} |vf_k|^p\Big)^{1/p} &= \sup\big\{v\big(\sum_{k\le n} a_k f_k\big) : a \in B_{\ell_{p^*}^n}\big\} \le v\big(\sup\big\{\sum_{k\le n} a_k f_k : a \in B_{\ell_{p^*}^n}\big\}\big)\\ &= v\Big(\sum_{k\le n} |f_k|^p\Big)^{1/p}.\end{aligned}$$

Hence, if u is (q,p)-concave and $K_{q,p}(u) \le C$, then

$$\begin{aligned}\Big(\sum_{k\le n} \|uvf_k\|^q\Big)^{1/q} &\le C\cdot \Big\|\Big(\sum_{k\le n} |vf_k|^p\Big)^{1/p}\Big\|_L \le C\cdot\|v\|\cdot\Big\|\Big(\sum_{k\le n}|f_k|^p\Big)^{1/p}\Big\|_{\mathcal{C}(K)}\\ &= C\cdot\|v\|\cdot\|(f_k)\|_p^{weak}.\end{aligned}$$

This shows that uv is (q,p)-summing with $\pi_{q,p}(uv) \le C\cdot\|v\|$.

For the converse, take $x_1,\dots,x_n \in L$. We are interested in homogeneous inequalities, so we can assume that $x := (\sum_{k\le n} |x_k|^p)^{1/p}$ has norm one. We saw before that x generates an ideal $[I(x), \|\cdot\|_\infty]$ and that $\|y\| \le \|y\|_\infty$ for each $y \in I(x)$. Also, $I(x)$ is an abstract M-space and so, by Kakutani's Theorem, may be identified with a $\mathcal{C}(K)$-space. Under this identification, x corresponds to the constant one function on K.

Let $j : I(x) = \mathcal{C}(K) \to L$ be the canonical embedding. Then

$$\begin{aligned}\Big(\sum_{k\le n} \|ux_k\|^q\Big)^{1/q} &= \Big(\sum_{k\le n}\|ujx_k\|^q\Big)^{1/q} \le \pi_{q,p}(uj)\cdot\Big\|\Big(\sum_{k\le n}|x_k|^p\Big)^{1/p}\Big\|_{\mathcal{C}(K)}\\ &= \pi_{q,p}(uj)\cdot\|x\|_\infty \le C,\end{aligned}$$

and so we can rest our case. QED

We are now in a position to reveal that for each q there are only two distinct classes of non-trivial (q,p)-concave operators.

16.6 Corollary: *Let $1 \le p < q < \infty$. An operator from a Banach lattice to a Banach space is (q,p)-concave if and only if it is $(q,1)$-concave.*

The other class of course consists of the q-concave operators.

Proof. Theorem 10.9 made us aware that, when $1 \le p < q < \infty$, Banach space operators from a $\mathcal{C}(K)$-space are (q,p)-summing if and only if they are $(q,1)$-summing. Apply the previous theorem to get what we want. QED

More can be said when $q > 2$.

16.7 Corollary: *Let $2 < q < \infty$. The following are equivalent statements about an operator u from a Banach lattice L to a Banach space Y.*

(i) *u is $(q,1)$-summing.*

(ii) *u is (q,p)-concave for some (and then all) $1 \le p < q$.*

(iii) *There is a positive constant C such that for any finite collection $x_1,\dots,x_n$ from L we have*

$$\Big(\sum_{k\le n} \|ux_k\|^q\Big)^{1/q} \le C\cdot\Big(\int_0^1 \Big\|\sum_{k\le n} r_k(t)x_k\Big\|^2\,dt\Big)^{1/2}.$$

Proof. Since $(q,1)$-summing operators are always $(q,1)$-concave, the previous corollary gives (i)⇒(ii).

To show that (ii) implies (iii), we take $p = 2$ and apply 16.4:

$$\Big(\sum_{k\le n}\|ux_k\|^q\Big)^{1/q} \le K_{q,2}(u)\cdot\Big\|\Big(\sum_{k\le n}|x_k|^2\Big)^{1/2}\Big\|$$
$$\le A_1^{-1}\cdot K_{q,2}(u)\cdot\Big(\int_0^1\Big\|\sum_{k\le n} r_k(t)x_k\Big\|^2\,dt\Big)^{1/2}.$$

For the final implication (iii)⇒(i), it suffices to observe that

$$\Big(\int_0^1\Big\|\sum_{k\le n} r_k(t)x_k\Big\|^2\,dt\Big)^{1/2} \le \|(x_k)\|_1^{weak}:$$

think about it! QED

When we specialize to identity operators, we find that we have stumbled on a very pleasing characterization of cotype q Banach lattices.

16.8 Corollary: *Let $2 < q < \infty$. A Banach lattice L has cotype q if and only if id_L is $(q,1)$-summing.*

16.9 Corollary: *The following statements about a Banach lattice L are equivalent:*

(i) *L has cotype 2.*

(ii) *id_L is 2-concave.*

(iii) *Every operator from a $\mathcal{C}(K)$-space to L is 2-summing.*

Proof. In Theorem 11.14 we showed that (i)⇒(iii) even for general Banach spaces. The implication (iii)⇒(ii) is covered by 16.5. To establish (ii)⇒(i) it is enough to copy the proof of (ii)⇒(iii) in 16.7, working of course with the identity operator. QED

It is perhaps worth recalling that general Banach space theory (11.16 and 6.20) allows us to add yet another equivalence:

- *A Banach lattice L has cotype 2 if and only if every 2-summing operator from L into any Banach space is 1-summing.*

We say that a Banach lattice L is *q-concave* if id_L is a q-concave operator. What we have just seen is that for Banach lattices 2-concavity is the same as cotype 2. The equivalence breaks down when $q > 2$, but it is still true that

- *q-concave Banach lattices have cotype q.*

A nice feature of q-concavity ($2 \le q < \infty$) is that it leads to a Banach lattice version of Khinchin's Inequality. This follows from a general result on q-concave operators.

16.10 Proposition: *Let L be a Banach lattice and Y a Banach space. If $u : L \to Y$ is a q-concave operator ($1 \le q < \infty$), then there is a $C > 0$ such that, however we select finitely many vectors $x_1,\dots,x_n$ from L,*

$$\int_0^1\Big\|\sum_{k\le n} r_k(t)ux_k\Big\|\,dt \le C\cdot\Big\|\Big(\sum_{k\le n}|x_k|^2\Big)^{1/2}\Big\|.$$

Proof. Remember that Rademacher integrals are finite sums and apply 16.3:

$$\int_0^1 \Big\| \sum_{k\le n} r_k(t) u x_k \Big\| \, dt \le \Big(\int_0^1 \Big\| \sum_{k\le n} r_k(t) u x_k \Big\|^q \, dt \Big)^{1/q}$$
$$= \Big(2^{-n} \cdot \sum_{\varepsilon\in D_n} \Big\| u\Big(\sum_{k\le n} \varepsilon_k x_k\Big) \Big\|^q \Big)^{1/q} \le K_q(u) \cdot \Big\| \Big(2^{-n} \cdot \sum_{\varepsilon\in D_n} \Big| \sum_{k\le n} \varepsilon_k x_k \Big|^q \Big)^{1/q} \Big\|$$
$$= K_q(u) \cdot \Big\| \Big(\int_0^1 \Big| \sum_{k\le n} r_k(t) x_k \Big|^q \, dt \Big)^{1/q} \Big\| \le K_q(u) \cdot B_q \cdot \Big\| \Big(\sum_{k\le n} |x_k|^2 \Big)^{1/2} \Big\| .$$

QED

When we specialize to the identity operator, a call on Kahane's Inequality allows us to incorporate 16.4 as well.

16.11 Maurey - Khinchin Inequality: *Let $1 \le q < \infty$. If L is a q-concave Banach lattice, then, for each $0 < p < \infty$, there is a constant $C_{p,q}$ such that for every choice of $n \in \mathbf{N}$ and $x_1,..., x_n \in L$, we have*

$$A_1^{-1} \cdot \Big\| \Big(\sum_{k\le n} |x_k|^2 \Big)^{1/2} \Big\| \le \Big(\int_0^1 \Big\| \sum_{k\le n} r_k(t) x_k \Big\|^p \, dt \Big)^{1/p} \le C_{p,q} \cdot \Big\| \Big(\sum_{k\le n} |x_k|^2 \Big)^{1/2} \Big\| .$$

The Rôle of Disjointness

The conclusion of the last result is actually characteristic of Banach lattices with finite concavity. Finite concavity is even equivalent to finite cotype, but to arrive at these conclusions, we must exploit the concept of disjointness for all it's worth. Everything hinges on the next result, which tells us that we can test for $(q, 1)$-summing operators $\mathcal{C}(K) \to Y$ by concentrating our attention on disjoint vectors in $\mathcal{C}(K)$.

To begin, notice that when $f_1,..., f_n$ are disjointly supported in $\mathcal{C}(K)$,

$$\|(f_k)\|_1^{weak} = \Big\| \sum_{k\le n} f_k \Big\| = \max_{k\le n} \|f_k\|.$$

16.12 Proposition: *Let $1 \le q < \infty$, and let K be a compact Hausdorff space. An operator $u : \mathcal{C}(K) \to Y$ is $(q, 1)$-summing with $\pi_{q,1}(u) \le 1$ if and only if regardless of how we select finitely many disjointly supported functions $f_1,..., f_n \in B_{\mathcal{C}(K)}$, we have*

$$\Big(\sum_{k\le n} \|u f_k\|^q \Big)^{1/q} \le 1.$$

The heart of the matter is a technical lemma; the proposition then follows from a routine approximation argument in tandem with localization.

16.13 Lemma: *Let K be a compact Hausdorff space and let $n \in \mathbf{N}$. Denote by A the set of all $u \in B_{\mathcal{L}(\ell_\infty^n, \mathcal{C}(K))}$ such that the $u(e_k)$'s $(1 \le k \le n)$ are disjointly supported. The convex hull of A is $B_{\mathcal{L}(\ell_\infty^n, \mathcal{C}(K))}$.*

Proof. Let $v \in \mathcal{L}(\ell_1^n, \mathcal{C}(K)^*)$. Since $\mathrm{tr}\,(u^* v) = \sum_{k\le n} \langle u e_k, v e_k \rangle$ for any u in $\mathcal{L}(\ell_\infty^n, \mathcal{C}(K))$, we can summon Theorem 6.16 and assert that the 1-integral norm

$$\iota_1(v) = \sup\Big\{\sum_{k\le n}\langle ue_k, ve_k\rangle : u \in B_{\mathcal{L}(\ell_\infty^n, \mathcal{C}(K))}\Big\}.$$

Consequently, we can prove what we want by showing that

$$\iota_1(v) = \sup_{u\in A}\sum_{k\le n}\langle ue_k, ve_k\rangle$$

for any $v \in \mathcal{L}(\ell_1^n, \mathcal{C}(K)^*)$. But we need a manageable way to compute $\iota_1(v)$.

Since v has finite rank, it takes its values in $L_1(\mu)$ for some non-negative measure $\mu \in \mathcal{C}(K)^*$: indeed, $\mu := \sup_{k\le n}|ve_k|$ does the job. Naturally, $L_1(\mu)$ is norm one complemented in $\mathcal{C}(K)^*$, so we can conclude that $\iota_1(v)$ coincides with the 1-integral norm of v as an operator $\ell_1^n \to L_1(\mu)$. Order boundedness comes into play: using Theorem 5.19, we may write

$$\iota_1(v) = \big\|\sup\{|va| : a \in B_{\ell_1^n}\}\big\|_{L_1(\mu)}.$$

However, by Proposition 16.2,

$$\sup\big\{|va| : a\in B_{\ell_1^n}\big\} = \sup\big\{\big|\sum_{k\le n} a_k ve_k\big| : a \in B_{\ell_1^n}\big\} = \sup_{k\le n}|ve_k|.$$

Consequently,

$$\iota_1(v) = \big\|\sup_{k\le n}|ve_k|\,\big\|_{L_1(\mu)}.$$

To avoid heavy-handed notation, let us set $f_k := ve_k$ $(1\le k\le n)$ and $f := |f_1|\vee\ldots\vee|f_n|$. This allows us to write $\iota_1(v) = \int_K f\,d\mu$.

Now define $2n$ pairwise disjoint sets as follows. Put

$$A_1^+ := \{f_1 = f\}\cup\{f=0\} \quad\text{and}\quad A_1^- := \{-f_1 = f\}\setminus\{f=0\},$$

and then continue inductively so that for each $1\le k<n$

$$A_{k+1}^+ := \{f_{k+1} = f\}\setminus \textstyle\bigcup_{j\le k}(A_j^+\cup A_j^-)$$

and

$$A_{k+1}^- := \{-f_{k+1} = f\}\setminus \textstyle\bigcup_{j\le k}(A_j^+\cup A_j^-).$$

Notice that for all k,

$$A_k^+ \subset \{f_k\ge 0\} \quad\text{and}\quad A_k^-\subset\{f_k<0\}.$$

Now, set $f_k^+ := f_k\cdot 1_{\{f_k\ge 0\}}$ and $f_k^- := -f_k\cdot 1_{\{f_k<0\}}$ for each $1\le k\le n$. Of course, $f_k = f_k^+ - f_k^-$. Define functions $g_k^\pm := f_k^\pm\cdot 1_{A_k^\pm}$ $(1\le k\le n)$ and remark that by design

$$f = \sum_{k\le n} g_k^+ + g_k^-.$$

Next, use the regularity of μ to construct compact sets $B_k^\pm$, $C_k^\pm$ inside $A_k^\pm$ according to the following recipes:

If $\int_{A_k^\pm} f_k^\pm d\mu = 0$, set $B_k^\pm = \emptyset$; if not, select a proper compact subset $B_k^\pm$ of $A_k^\pm$ so that

$$\Big|\int_{A_k^\pm\setminus B_k^\pm} f_k^\pm\,d\mu\Big| \le \varepsilon/(4n).$$

If $\int_{B_k^\pm} f_k^\pm d\mu = 0$, set $C_k^\pm = \emptyset$; if not, select a proper compact subset $C_k^\pm$ of $B_k^\pm$ so that

$$\Big|\int_{B_k^\pm\setminus C_k^\pm} f_k^\pm d\mu\Big| \le \varepsilon/(4n).$$

The compact sets $B_k^\pm$ are mutually disjoint, and the $C_k^\pm$'s are compact subsets of the $B_k^\pm$'s, properly when $B_k^\pm \neq \emptyset$. So Urysohn's Lemma provides us with continuous functions $\varphi_k^\pm : K \to [0,1]$ $(1 \le k \le n)$ such that

- $\varphi_k^\pm(x) = 1$ if $x \in C_k^\pm$,
- $\varphi_k^\pm(x) = 0$ if $x \in B_k^\mp$,
- $\varphi_k^\pm$ has support in $B_k^\pm$.

Finally, for each $1 \le k \le n$, define $\varphi_k \in B_{C(K)}$ by $\varphi_k := \varphi_k^+ - \varphi_k^-$. We have arranged the construction so that the φ_k's obviously have mutually disjoint supports. Also, we can define an operator $u_0 \in A$ by setting $u_0 e_k := \varphi_k$ for every $1 \le k \le n$. Then

$$\begin{aligned}\sum_{k\le n}\langle u_0 e_k, v e_k\rangle &= \sum_{k\le n}\langle \varphi_k, f_k\rangle \\ &= \sum_{k\le n}\big(\langle \varphi_k^+, f_k^+\rangle + \langle \varphi_k^-, f_k^-\rangle\big) - \sum_{k\le n}\big(\langle \varphi_k^+, f_k^-\rangle + \langle \varphi_k^-, f_k^+\rangle\big).\end{aligned}$$

But $\langle \varphi_k^+, f_k^-\rangle = 0$: we have

$$0 \le \langle \varphi_k^+, f_k^-\rangle = \int_K \varphi_k^+ f_k^- \, d\mu = \int_{B_k^+} \varphi_k^+ f_k^- \, d\mu \le \int_{B_k^+} f_k^- \, d\mu = 0$$

by the properties of the φ_k^+'s and since $\{f_k < 0\} \cap B_k^+$ is empty. Similarly, $\langle \varphi_k^-, f_k^+\rangle = 0$, and so $\sum_{k\le n}\langle u_0 e_k, v e_k\rangle = \sum_{k\le n}\big(\langle \varphi_k^+, f_k^+\rangle + \langle \varphi_k^-, f_k^-\rangle\big)$. Another application of the properties of the φ_k^+'s gives

$$\begin{aligned}\langle \varphi_k^+, f_k^+\rangle &= \int_{A_k^+} \varphi_k^+ f_k^+ \, d\mu = \int_{A_k^+} f_k^+ \, d\mu - \int_{A_k^+} (1-\varphi_k^+) f_k^+ \, d\mu \\ &= \int_K g_k^+ \, d\mu - \int_{A_k^+} (1-\varphi_k^+) f_k^+ \, d\mu = \int_K g_k^+ \, d\mu - \int_{A_k^+\setminus C_k^+} (1-\varphi_k^+) f_k^+ \, d\mu \\ &\ge \int_K g_k^+ \, d\mu - \int_{A_k^+\setminus C_k^+} f_k^+ \, d\mu = \int_K g_k^+ \, d\mu - \int_{A_k^+\setminus B_k^+} f_k^+ \, d\mu - \int_{B_k^+\setminus C_k^+} f_k^+ \, d\mu \\ &\ge \int_K g_k^+ \, d\mu \; - \; \varepsilon/(2n).\end{aligned}$$

In just the same way, $\langle \varphi_k^-, f_k^-\rangle \ge \int_K g_k^- \, d\mu - \varepsilon/(2n)$. Sum everything up to get

$$\sum_{k\le n}\langle u_0 e_k, v e_k\rangle \ge \int_K \Big(\sum_{k\le n} g_k^+ + g_k^-\Big)\, d\mu - \varepsilon = \int_K f \, d\mu - \varepsilon = \iota_1(v) - \varepsilon.$$

As $\varepsilon > 0$ was arbitrary, we conclude that $\sup_{u\in A}\sum_{k\le n}\langle u e_k, v e_k\rangle \ge \iota_1(v)$, and this is good enough. QED

The fruits of our labours are scrumptious.

16.14 Theorem: *Let $2 < q < \infty$. An operator u from a Banach lattice L to a Banach space Y is $(q,1)$-summing if and only if there is a constant C such that for all choices of finitely many disjoint vectors $x_1, \ldots, x_n$ from L,*

$$\Big(\sum_{k\le n} \|u x_k\|^q\Big)^{1/q} \le C \cdot \Big\| \sum_{k\le n} x_k \Big\| .$$

Proof. Assume first that our condition holds. Thanks to 16.7, we can show that u is $(q,1)$-summing by proving that it is $(q,1)$-concave. According to 16.5, we can accomplish this by demonstrating that whenever $v : \mathcal{C}(K) \to L$ is a positive operator the composition $uv : \mathcal{C}(K) \to Y$ is $(q,1)$-summing. But we have just shown in 16.12 that it is sufficient to do this by testing uv against disjointly supported functions $f_1,\ldots, f_n$ from $\mathcal{C}(K)$. The very nature of v ensures that $vf_1,\ldots, vf_n$ are disjoint members of L, so our hypothesis signals that

$$\Big(\sum_{k\le n} \|uvf_k\|^q\Big)^{1/q} \le C\cdot\Big\|\sum_{k\le n} vf_k\Big\| \le C\cdot\|v\|\cdot\Big\|\sum_{k\le n} f_k\Big\| = C\cdot\|v\|\cdot\|(f_k)\|_1^{weak}.$$

To pass in the other direction, we resort to 16.7 again. It tells us that if u is $(q,1)$-summing, then there is a constant C such that, however we select a finite family of vectors $x_1,\ldots, x_n \in L$,

$$\Big(\sum_{k\le n} \|ux_k\|^q\Big)^{1/q} \le C\cdot\Big(\int_0^1 \Big\|\sum_{k\le n} r_k(t)x_k\Big\|^2 dt\Big)^{1/2}.$$

But, when the x_k's are disjoint and $\varepsilon \in D_n$,

$$\begin{aligned}\Big\|\sum_{k\le n} \varepsilon_k x_k\Big\| &= \Big\|\,\Big|\sum_{k\le n} \varepsilon_k x_k\Big|\,\Big\| = \Big\|\sup_{k\le n} |\varepsilon_k x_k|\,\Big\| \\ &= \Big\|\sup_{k\le n} |x_k|\,\Big\| = \Big\|\,\Big|\sum_{k\le n} x_k\Big|\,\Big\| = \Big\|\sum_{k\le n} x_k\Big\|.\end{aligned}$$

Consequently,

$$\Big(\int_0^1 \Big\|\sum_{k\le n} r_k(t)x_k\Big\|^2 dt\Big)^{1/2} = \Big\|\sum_{k\le n} x_k\Big\|.$$

The conclusion is immediate. QED

A quick look at 16.8 is all that is needed for the next result.

16.15 Corollary: *A Banach lattice L has cotype $2 < q < \infty$ if and only if there is a constant C such that, for all finite sets of disjoint vectors $x_1,\ldots, x_n$ in L, we have*

$$\Big(\sum_{k\le n} \|x_k\|^q\Big)^{1/q} \le C\cdot\Big\|\sum_{k\le n} x_k\Big\|.$$

Our next goal will be to establish the equivalence of finite cotype and finite concavity for Banach lattices. This will depend on a careful scrutiny of the inequality in the last corollary. Notice that when it holds $\|\sum_{k\le n} x_k\|$ must be 'large'. Indeed, if $x_1,\ldots, x_n$ are disjoint unit vectors in a cotype q Banach lattice, we must have $\|\sum_{k\le n} x_k\| \ge n^{1/q}/C$ where C is a constant. This crude observation is the forerunner of a remarkably sharp dichotomy.

16.16 Proposition: *If L is a Banach lattice, then precisely one of the following is true:*

(I) *There is some $2 \le q < \infty$ such that id_L is $(q,1)$-summing.*

(II) *For each $n \in \mathbf{N}$ and $\varepsilon > 0$ there are disjoint unit vectors $x_1,\ldots, x_n \in L$ such that $\|\sum_{k\le n} x_k\| \le 1+\varepsilon$.*

It is convenient to introduce yet another sequence of moduli. When L is a Banach lattice and $n \in \mathbf{N}$, we write

$$\delta_n(L)$$

for the smallest of all numbers $\delta \geq 0$ with the property that, regardless of the choice of n pairwise disjoint (non-zero) vectors $x_1,\dots,x_n$, we have

$$\min_{k\leq n} \|x_k\| \leq \delta\cdot \Big\| \sum_{k\leq n} x_k \Big\| .$$

Each $\delta_n(L)$ belongs to $[0,1]$ since plainly

$$|x_j| \leq \sum_{k\leq n} |x_k| = \Big|\sum_{k\leq n} x_k\Big| \quad (1 \leq j \leq n)$$

whenever $x_1,\dots,x_n$ are disjoint in L. Moreover, it is clear that

$$\delta_{n+1}(L) \leq \delta_n(L) \leq \dots \leq \delta_1(L) = 1$$

for every $n \in \mathbf{N}$. Finally, it is straightforward that our new numbers have the submultiplicativity property, that is,

$$\delta_{mn}(L) \leq \delta_m(L)\cdot\delta_n(L)$$

for every $m, n \in \mathbf{N}$.

Proof of 16.16. It may happen that $\delta_n(L) = 1$ for all $n \in \mathbf{N}$. In this case, for any $\varepsilon > 0$ and any n, there are disjoint non-zero vectors $x_1,\dots,x_n \in L$ such that $\| \sum_{k\leq n} x_k\| \leq (1+\varepsilon)\cdot\min_{k\leq n} \|x_k\|$. It follows that

$$\begin{aligned} \Big\|\sum_{k\leq n} \frac{x_k}{\|x_k\|}\Big\| &= \Big\|\sup_{k\leq n} \frac{|x_k|}{\|x_k\|}\Big\| \leq \big(\min_{k\leq n} \|x_k\|\big)^{-1}\cdot \Big\|\sup_{k\leq n} |x_k|\Big\| \\ &= \big(\min_{k\leq n} \|x_k\|\big)^{-1}\cdot \Big\|\sum_{k\leq n} x_k\Big\| . \end{aligned}$$

This is (II).

The alternative is that $\delta_N(L) < 1$ for some $N \geq 2$, and in this case we may find $\delta > 0$ with $\delta_N(L) \leq N^{-\delta}$. It will follow that

$$\delta_n(L) \leq (n/N)^{-\delta}$$

for every $n \in \mathbf{N}$. To understand why, settle on an $n \in \mathbf{N}$ and pick $\ell \in \mathbf{N}$ so that $N^{\ell-1} \leq n < N^\ell$. Then the monotonicity and submultiplicativity properties of our moduli show that

$$\delta_n(L) \leq \delta_{N^{\ell-1}}(L) \leq \delta_N(L)^{\ell-1} \leq N^{-\delta(\ell-1)} \leq (n/N)^{-\delta}.$$

This is enough to show that id_L is $(q,1)$-summing for every $q > 2$ satisfying $q > 1/\delta$. Indeed, choose disjoint elements $x_1,\dots,x_n \in L$ and arrange them so that $\|x_1\| \geq \|x_2\| \geq \dots \geq \|x_n\|$. For each $1 \leq m \leq n$,

$$\begin{aligned} \|x_m\| &= \min_{k\leq m} \|x_k\| \leq \delta_m(L)\cdot\Big\|\sum_{k\leq m} x_k\Big\| \\ &\leq (m/N)^{-\delta}\cdot\Big\|\sum_{k\leq m} x_k\Big\| \leq (m/N)^{-\delta}\cdot\Big\|\sum_{k\leq n} x_k\Big\| . \end{aligned}$$

Hence
$$\Big(\sum_{m\leq n}\|x_m\|^q\Big)^{1/q} \leq \Big(\sum_{m=1}^{\infty}(m/N)^{-\delta q}\Big)^{1/q}\cdot\Big\|\sum_{k\leq n}x_k\Big\|\,.$$
Thanks to Corollary 16.15, we have arrived at (I). QED

Pay careful attention to what the last proposition is saying. Since (I) holds exactly when the Banach lattice L has finite cotype, (II) must hold exactly when L *fails* to have finite cotype. In this case we have actually shown that ℓ_∞ is finitely representable in L – and in a very special way. This gives a much easier proof of some of the material we struggled with in the context of Banach spaces.

Let us summarize what we know so far about Banach *lattices* with finite cotype.

16.17 Scholium: *The following are equivalent statements about a Banach lattice L:*

(i) *L has finite cotype.*

(ii) *id_L is $(q,1)$-summing for some $2 \leq q < \infty$.*

(iii) *There exist an $\varepsilon > 0$ and an $n \in \mathbf{N}$ such that all n-tuples of disjoint vectors $x_1,\ldots,x_n$ in L satisfy $\|\sum_{k\leq n}x_k\| > (1+\varepsilon)\cdot\min_{k\leq n}\|x_k\|$.*

(iv) *For some $2 \leq r < \infty$, $\mathcal{L}(\mathcal{C}(K),L) = \Pi_r(\mathcal{C}(K),L)$ for every compact Hausdorff space K.*

(v) *L has finite concavity.*

Proof. The equivalence of (i), (ii) and (iii) is covered by 16.8 and 16.16. Next, (iv) follows from (i) even for general Banach spaces (11.14), whereas 16.5 allows us to pass from (iv) to (v). Finally, the trivial implication (v)$\Rightarrow$(i) was mentioned casually just before 16.10. QED

We fulfill a promise and show that the Maurey - Khinchin-type inequality 16.11 gives a beautiful characterization of Banach lattices with finite cotype.

16.18 Theorem: *A Banach lattice L has finite cotype if and only if there is a constant $K \geq 1$ such that, for any finite collection $x_1,\ldots,x_n$ from L,*
$$\frac{1}{K}\cdot\Big\|\Big(\sum_{k\leq n}|x_k|^2\Big)^{1/2}\Big\| \leq \int_0^1\Big\|\sum_{k\leq n}r_k(t)x_k\Big\|\,dt \leq K\cdot\Big\|\Big(\sum_{k\leq n}|x_k|^2\Big)^{1/2}\Big\|\,.$$

Proof. We have just seen that, in Banach lattices, finite cotype is equivalent to finite concavity, and we have been aware since 16.11 that the inequalities hold when L has finite concavity.

We must investigate what happens when L fails to have finite cotype. The left hand inequality will hold: this is nothing other than 16.3. So we had better show that the right hand one fails. The dichotomy of 16.16 informs us that we can, for each $n \in \mathbf{N}$, find disjoint unit vectors $y_1,\ldots,y_{2^n} \in L$ such that $\|\sum_{k\leq 2^n}y_k\| \leq 2$. Now introduce sign groupings to define

$$\begin{aligned}
x_1 &= y_1 - y_2 + y_3 - y_4 + - \dots + y_{2^n-1} - y_{2^n} \\
x_2 &= (y_1 + y_2) - (y_3 + y_4) + (y_5 + y_6) - (y_7 + y_8) + - \dots - (y_{2^n-1} + y_{2^n}) \\
&\dots\dots\dots\dots\dots\dots\dots\dots\dots\dots \\
x_n &= (y_1 + \dots + y_{2^{n-1}}) - (y_{2^{n-1}+1} + \dots + y_{2^n}) \,.
\end{aligned}$$

Disjointness makes it clear that $|x_j| = \big| \sum_{k\le 2^n} y_k \big|$ for each $1 \le j \le n$. It follows easily from this that

$$(*) \qquad \Big\| \big(\sum_{j\le n} |x_j|^2\big)^{1/2} \Big\| = \sqrt{n} \cdot \Big\| \sum_{k\le 2^n} y_k \Big\| \,.$$

But the x_j's were designed in such a way that the Rademacher averages are simple to estimate. For every $\varepsilon \in D_n$ there is some $1 \le i \le n$ such that

$$\Big| \sum_{j\le n} \varepsilon_j x_j \Big| \ge n \cdot |x_i| = n \cdot \Big| \sum_{k \le 2^n} y_k \Big| \,.$$

Incorporating $(*)$,

$$\int_0^1 \Big\| \sum_{j\le n} r_j(t) x_j \Big\| \, dt = \int_0^1 \Big\| \Big| \sum_{j\le n} r_j(t) x_j \Big| \Big\| \, dt \ge \int_0^1 \Big\| n \cdot \Big| \sum_{k\le 2^n} y_k \Big| \Big\| \, dt$$

$$= n \cdot \Big\| \sum_{k\le 2^n} y_k \Big\| = \sqrt{n} \cdot \Big\| \big(\sum_{j\le n} |x_j|^2 \big)^{1/2} \Big\| \,. \qquad \text{QED}$$

We can even characterize finite cotype lattices in terms of properties of order bounded operators (see Chapter 5):

16.19 Theorem: *A Banach lattice L has finite cotype if and only if every order bounded operator $\ell_2 \to L$ is almost summing.*

Proof. Suppose first that L has finite cotype. If $u \in \mathcal{L}(\ell_2, L)$ is order bounded, then it maps ℓ_2 into the ideal $I(a)$ generated by some positive element $a \in L$. By Kakutani's Theorem 16.1, $I(a)$ is a $\mathcal{C}(K)$-space. Since L has finite cotype, the canonical map $I(a) \hookrightarrow L$ is q-summing for some $2 \le q < \infty$ (11.14), so u is q-summing and therefore almost summing (12.5).

Conversely, assume that every order bounded operator $u : \ell_2 \to L$ is almost summing. We show that L must have finite cotype by verifying the right hand inequality in 16.18.

Assume that this inequality fails. We can produce a partition of $\mathbf{N}$ consisting of sets $I_n = \{k \in \mathbf{N} : k_{n-1} \le k < k_n\}$ where (k_n) is an increasing sequence of integers with $k_1 = 1$, and we can find a sequence (x_k) in L such that $\big\| \big(\sum_{k\in I_n} |x_k|^2\big)^{1/2} \big\| < 2^{-n}$ but $\int_0^1 \big\| \sum_{k\in I_n} r_k(t) x_k \big\| \, dt > 2^n$. Notice that $\sum_{n=1}^\infty \big(\sum_{k\in I_n} |x_k|^2\big)^{1/2}$ exists as an element of L. Use the x_k's to define an operator $u : \ell_2 \to L$ via $ue_k = x_k$ for every $k \in \mathbf{N}$. When $a = (a_k) \in B_{\ell_2}$,

$$|ua| = \Big| \sum_{k=1}^\infty a_k x_k \Big| \le \sum_{n=1}^\infty \big(\sum_{k\in I_n} |a_k| \cdot |x_k| \big) \le \sum_{n=1}^\infty \big(\sum_{k\in I_n} |x_k|^2 \big)^{1/2}.$$

This informs us that u is order bounded. The definition of the x_k's rules out any possibility that u is almost summing, giving a contradiction to end the story. QED

TYPE AND CONVEXITY

Predictably, the theory of type in Banach lattices is clean and crisp.

16.20 Theorem: *Let $1 < p \leq 2$. A Banach lattice L has type p if and only if it has finite cotype and there is a constant C such that, regardless of how we choose finitely many vectors $x_1, \ldots, x_n \in L$,*

$$\Big\| \Big(\sum_{k \leq n} |x_k|^2\Big)^{1/2} \Big\| \leq C \cdot \Big(\sum_{k \leq n} \|x_k\|^p\Big)^{1/p}.$$

Proof. If L has type p, Banach space theory grants it finite cotype. With this, we can immediately derive both implications from the Maurey-Khinchin Inequality 16.11. QED

As before, there is an operator theoretic concept behind this characterization. It will lead us to a duality result for type and cotype in Banach lattices which improves upon 13.17.

Let $1 \leq p \leq q < \infty$. An operator u from a Banach space X to a Banach lattice L is called (p,q)-*convex* if there is a constant C such that irrespective of the finite collection of vectors $x_1, \ldots, x_n \in X$,

$$\Big\| \Big(\sum_{k \leq n} |ux_k|^q\Big)^{1/q} \Big\| \leq C \cdot \Big(\sum_{k \leq n} \|x_k\|^p\Big)^{1/p}.$$

We say that L is (p,q)-*convex* if id_L has this property. In this language, the last theorem states that a Banach lattice has type $1 < p \leq 2$ if and only if it has finite concavity and is $(p,2)$-convex.

We shall shortly show that taking adjoints converts (p,q)-convexity into (p^*, q^*)-concavity, but some preparations are necessary. Let $1 \leq r \leq \infty$, and let L be a Banach lattice. Denote by

$$\tilde{\ell}_r(|L|)$$

the space of all sequences (x_n) in L such that

$$\|\!|(x_n)|\!\|_r := \sup_n \Big\| \Big(\sum_{k \leq n} |x_k|^r\Big)^{1/r} \Big\|$$

is finite. It is routine to show that $[\tilde{\ell}_r(|L|), \|\!| \cdot |\!\|_r]$ is a Banach space. A (generally) smaller space also comes in handy. We write

$$\ell_r(|L|)$$

for the closure of $L^{(\mathbf{N})}$ in $\tilde{\ell}_r(|L|)$. Boring hackwork shows that $\tilde{\ell}_r(|L|)$ is even a Banach lattice and that $\ell_r(|L|)$ is an ideal therein.

Our concepts of (p,q)-convexity and (q,p)-concavity slip nicely into the general setup that was so profitable in the study of (q,p)-summing affairs.

Saying that $u \in \mathcal{L}(X, L)$ is (p, q)-convex is the same as saying that $(x_n) \mapsto (ux_n)$ induces a bounded linear map

$$\breve{u} : \ell_p^{strong}(X) \longrightarrow \ell_q(|L|).$$

Analogously, the statement that $v \in \mathcal{L}(L, Y)$ is (q, p)-concave can be reformulated to read that $(x_n) \mapsto (vx_n)$ induces an operator

$$\hat{v} : \ell_p(|L|) \longrightarrow \ell_q^{strong}(Y).$$

For $1 \leq r < \infty$, ℓ_r^* and ℓ_{r^*} are isometrically isomorphic not only as Banach spaces but also as Banach lattices. One can (but we refuse to) mimic the proof to arrive at the isometric isomorphism as Banach lattices of $\ell_r(|L|)^*$ and $\tilde{\ell}_{r^*}(|L^*|)$. Here is the punchline.

16.21 Theorem: *Let $1 < p \leq q < \infty$, and suppose that L is a Banach lattice and Z is a Banach space.*

(a) *$u \in \mathcal{L}(Z, L)$ is (p, q)-convex iff $u^* \in \mathcal{L}(L^*, Z^*)$ is (p^*, q^*)-concave.*

(b) *$v \in \mathcal{L}(L, Z)$ is (q, p)-concave iff $v^* \in \mathcal{L}(Z^*, L^*)$ is (q^*, p^*)-convex.*

Proof. If $u : Z \to L$ is (p, q)-convex, then the adjoint of $\breve{u}$ is an operator from $\ell_q(|L|)^*$ to $\ell_p^{strong}(Z)^*$, that is, from $\tilde{\ell}_{q^*}(|L^*|)$ to $\ell_{p^*}^{strong}(Z^*)$. Since it takes each (x_n^*) to $(u^* x_n^*)$ it must restrict to $\widehat{u^*}$. Hence u^* is (p^*, q^*)-concave. The other implications are just as simple. QED

Specializing, we arrive at a good duality theory for type and cotype of Banach lattices which complements 13.17. It is, however, just a forerunner of an even stronger result to be proved in 17.13.

16.22 Corollary: *Let $1 < p \leq 2$ and let L be a Banach lattice with finite cotype. Then L has type p if and only if L^* has cotype p^*.*

Proof. No lattice structure is needed to show that L^* has cotype p^* whenever L has type p: we have already proved this in 11.10.

For the converse, suppose that L has finite cotype and that L^* has cotype p^*. We know from 16.8 (for $p < 2$) and 16.9 (for $p = 2$) that id_L must be $(p^*, 2)$-concave. By the theorem, id_{L^*} has to be $(p, 2)$-convex. Apply 16.20 to deduce that L has type p. QED

Notes and Remarks

Most of what we say in this chapter is a plain derivative of the lectures of J.L. Krivine and B. Maurey found in the 1973/74 Séminaire Maurey - Schwartz.

We have more or less followed Krivine's presentation of the operational calculus. We ought to mention that his goal was to give a Banach lattice version of Grothendieck's Inequality. He succeeded by proving the following:

Theorem: *Let L, $\tilde{L}$ be Banach lattices and $u : L \to \tilde{L}$ an operator. Then, for any $x_1,..., x_n \in L$,*

$$\Big\| \big(\sum_{k\leq n} |ux_k|^2\big)^{1/2} \Big\| \leq \kappa_G \cdot \|u\| \cdot \Big\| \big(\sum_{k\leq n} |x_k|^2\big)^{1/2} \Big\| .$$

An alternative approach to the operational calculus and Grothendieck's Inequality was proposed by T.K. Carne [1980]. Both approaches place considerable reliance on S. Kakutani's famous representation for L-spaces [1941a] and for M-spaces [1941b].

Propositions 16.2 and 16.3 are from J.L. Krivine [1973/74], as is Corollary 16.4. Likewise, Krivine introduced (q,p)-concave operators, albeit under another name. Theorem 16.5 and Corollaries 16.6, 16.7, 16.8 and 16.9 are due to B.Maurey [1973/74b]. For 16.9, see also E. Dubinski, A. Pełczyński and H.P. Rosenthal [1972]. In fact, 16.8 and 16.9 are as good as it gets. M. Talagrand [1992a], in a probabilistic tour de force, showed that

- *there is a Banach lattice which does not have cotype* 2 *but whose identity is* $(2,1)$*-summing.*

Talagrand's paper is full of information about delicate classification of cotype q operators on $\mathcal{C}(K)$-spaces. It builds on previous work of S.J. Montgomery-Smith [1988], [1989] who proved that

- *for a probability measure $\mu \in \mathcal{C}(K)^*$ the natural map of $\mathcal{C}(K)$ into the Lorentz - Orlicz space $L_{t^2 \log t, 2}(\mu)$ is of cotype* 2.

Very recently M. Talagrand [1994] was even able to construct a Banach space with a symmetric basis which fails cotype 2 but whose identity is (2, 1)-summing.

Theorems 16.10 and 16.11 are straight from Maurey's exposés, and so are 16.12 through 16.18, except, maybe, 16.17. These results, in particular 16.15, have proved very useful in computations of cotype for classical Banach lattices.

For instance, J. Creekmore [1980], [1981] calculated the precise type and cotype of the classical Lorentz spaces $L_{p,q}$ using 16.15 and 16.20 to good effect. His results:

- *If $1 \leq q < p < \infty$, then $L_{p,q}$ has type* $\min\{2,q\}$, *and cotype* $\max\{2,p\}$ *except for $p = 2$ when $L_{2,q}$ is of cotype $2+\varepsilon$ for each $\varepsilon > 0$.*
- *If $1 < p < q < \infty$, then $L_{p,q}$ has cotype* $\max\{2,q\}$, *and type* $\min\{2,p\}$ *except for $p = 2$ when $L_{2,q}$ is of type $2-\varepsilon$ for each $\varepsilon > 0$.*

As in the case of cotype and concavity, the results presented on type and convexity are due to B. Maurey, and we have again followed his [1973/74b] exposition.

Variations in notions related to p-summing operators are natural and fruitful for the finer study of Banach lattices. The earliest such generalizations, 'cone absolutely summing' and 'majorizing' operators, were introduced by U. Schlotterbeck [1969]. Cone absolutely summing operators are given a

Pietsch-type domination theorem and majorizing operators are related to lattice bounded operators. A careful exposition of Schlotterbeck's work is found in H.H. Schaefer's [1974] treatise on positive operators in which the theory of tensor products of Banach lattices is interwoven with these notions as well.

A variation of the p-summing notion arises for operators with values in a Banach lattice: an operator $u : X \to L$ from a Banach space X to a Banach lattice L is said to be *p-lattice summing* $(1 \leq p \leq \infty)$ if there is a constant $C \geq 0$ such that, no matter how we select finitely many vectors $x_1,..., x_n \in X$,

$$\| \big(\sum_{k\leq n} |ux_k|^p\big)^{1/p} \| \leq C \cdot \sup_{x^* \in B_{X^*}} \big(\sum_{k\leq n} |\langle x^*, x_k\rangle|^p\big)^{1/p}.$$

This applies when $1 \leq p < \infty$; in case $p = \infty$ we require

$$\| \bigvee_{k\leq n} |ux_k| \| \leq C \cdot \sup_{k\leq n} \|x_k\|.$$

The least C that works is the p-lattice summing norm of u and is denoted by $\lambda_p(u)$. Predictably, the collection $\Lambda_p(X, L)$ of all p-lattice summing operators from X to L is a Banach space. It was first studied by L.P. Yanovskii [1979] in case $p = 1$ and by N.J. Nielsen and J. Szulga [1984] when $p > 1$.

Though the very definition hints at an analogous development to that of Π_p, such just is not in the cards. Nielsen and Szulga show that *if* $1 \leq p \leq q \leq 2$, *then* $\Lambda_p(X, L) \subset \Lambda_q(X, L)$; *if* $p > 2$, *then* $\Lambda_p(X, L) \subset \Lambda_2(X, L)$, *too.* They rely on their understanding of order bounded operators for much of their sharpest analysis. For instance, $u \in \Lambda_\infty(X, L)$ *if and only if* $k_L u : X \to L^{**}$ *is order bounded. Consequently,* $u \in \Lambda_p(X, L)$ $(1 \leq p < \infty)$ *if and only if, for every* $v \in \mathcal{L}(\ell_{p^*}, X)$, $uv \in \Lambda_\infty(\ell_{p^*}, L)$.

The relationship between Λ_1 and Π_1 is studied by J. Szulga [1983a,b]. He proved that $\Lambda_1(X, L) = \Pi_1(X, L)$ if and only if there is $1 \leq r \leq 2$ such that L is r-concave and $\Pi_{r^*}(\mathcal{L}_\infty, X) = \mathcal{L}(\mathcal{L}_\infty, X)$. Szulga ([1984], [1985]) goes on to study the case $p > 1$ and again finds the concavity of L to play a crucial role in including Π_p within Λ_p.

Another variation of p-summing operators has been extensively studied and applied by O. Blasco ([1986], [1987], [1988a,b], [1989]). An operator $u : L \to X$ from a Banach lattice L to a Banach space is called *positive p-summing* if there is a $C \geq 0$ so that, given finitely many vectors $f_1,..., f_n \geq 0$ in L,

$$\big(\sum_{k\leq n} \|uf_k\|^p\big)^{1/p} \leq C \cdot \sup_{\ell^* \in B_{L^*}} \big(\sum_{k\leq n} |\langle \ell^*, f_k\rangle|^p\big)^{1/p}.$$

The least C that works makes the collection $\mathcal{P}_p(L, X)$ of all positive p-summing operators from L to X a Banach space. Among other things, Blasco shows that $\mathcal{P}_p(L, X) \subset \mathcal{P}_q(L, X)$ if $1 \leq p \leq q < \infty$ and that the norm of this embedding is ≤ 1, that every operator from L_p to X is in $\mathcal{P}_p$, that $\mathcal{P}_p(\mathcal{C}(K), X)$ and $\Pi_p(\mathcal{C}(K), X)$ coincide, and that $\mathcal{P}_1(L_{p^*}, X) = \mathcal{P}_p(L_{p^*}, X)$. He applies all this to study convolution operators and boundary value problems for classes of vector-valued harmonic functions.

17. LOCAL UNCONDITIONALITY

The precise character of the theory of type and cotype in Banach lattices suggests that they possess some very special local properties. In this chapter we identify two such properties, the second more general than the first, which make good sense in the absence of lattice structure.

Let X be a Banach space with a basis (x_n), and let (x_n^*) be the corresponding biorthogonal sequence in X^*. Recall that (x_n) is an *unconditional basis* if for each $x \in X$ the series $\sum_n \langle x_n^*, x\rangle x_n$ converges unconditionally to x. We know from 1.9 that this is equivalent to requiring the convergence of $\sum_n t_n \langle x_n^*, x\rangle x_n$ for every $t = (t_n) \in \ell_\infty$, and that there is a constant $\lambda \geq 1$ such that

$$\Big\| \sum_n t_n \langle x_n^*, x\rangle x_n \Big\| \leq \lambda \cdot \Big\| \sum_n \langle x_n^*, x\rangle x_n \Big\|$$

for every $t \in B_{\ell_\infty}$. More precisely, we say that the basis (x_n) is λ-*unconditional* in such a case and write $\lambda_{(x_n)}$ for the smallest such λ. The *unconditional basis constant* of X is

$$ub(X) := \inf\big\{ \lambda_{(x_n)} : (x_n) \text{ is an unconditional basis of } X \big\}.$$

If X is a real Banach space and if (x_n) is an unconditional basis of X, normalized so that $\|x_n\| = 1$ for each n, we can build a natural lattice structure on X by defining

$$x \leq y \;:\Leftrightarrow\; \langle x_n^*, x\rangle \leq \langle x_n^*, y\rangle \quad \text{for all } n.$$

By setting

$$|\!|\!| x |\!|\!| := \sup\Big\{ \Big\| \sum_n t_n \langle x_n^*, x\rangle x_n \Big\| : (t_n) \in B_{\ell_\infty} \Big\}$$

we get an equivalent norm on X with respect to which it is a Banach lattice.

Passing to the complex case is easy. If (x_n) is a normalized unconditional basis for a complex Banach space X, the collection of all series $\sum_n \mathrm{Re}\langle x_n^*, x\rangle \cdot x_n$, where $x \in X$, is a real Banach space $X_{\mathbf{R}}$ in which (x_n) is also an unconditional basis. If we renorm $X_{\mathbf{R}}$ as above to make it a Banach lattice, the complexification of $X_{\mathbf{R}}$, as described after 16.4, is actually isomorphic to X.

LOCAL UNCONDITIONAL STRUCTURE

Although we have seen in 3.14 that one of the most familiar Banach lattices fails to have an unconditional basis, we shall soon discover that it does enjoy a local property which goes a long way towards making up for this deficiency.

As usual, let $\mathcal{F}_X$ denote the collection of all finite dimensional subspaces of the Banach space X. We say that X has *local unconditional structure – l.u.st.* for short – if there is a constant $\Lambda \geq 1$ such that for all $E \in \mathcal{F}_X$ the canonical embedding $E \hookrightarrow X$ has a factorization $E \xrightarrow{v} Y \xrightarrow{u} X$, where Y is a Banach space with unconditional basis, and u and v are operators satisfying $\|u\| \cdot \|v\| \cdot ub(Y) \leq \Lambda$. The smallest of all such Λ's is called the *l.u.st. constant* of X; we denote it by

$$\Lambda(X).$$

Plainly, every Banach space X which has an unconditional basis also has l.u.st. and $\Lambda(X) \leq ub(X)$. Moreover,

- *any $\mathcal{L}_p$-space $(1 \leq p \leq \infty)$ has local unconditional structure.*

Indeed, if X is an $\mathcal{L}_{p,\lambda}$-space, each $E \in \mathcal{F}_X$ is contained in some $F \in \mathcal{F}_X$ for which there is an isomorphism $u : \ell_p^{\dim F} \to F$ satisfying $\|u\| \cdot \|u^{-1}\| \leq \lambda$. This allows the trivial factorization $E \hookrightarrow F \xrightarrow{u^{-1}} \ell_p^{\dim F} \xrightarrow{u} F \hookrightarrow X$ of the natural embedding of E into X. Glance at the unit vector basis of $\ell_p^{\dim F}$ to appreciate that $ub(\ell_p^{\dim F}) = 1$. It is now apparent that X has l.u.st. with $\Lambda(X) \leq \lambda$.

In particular, any $L_p(\mu)$-space $(1 \leq p \leq \infty)$ and any $C(K)$-space have l.u.st., and their l.u.st. constants are always one. Those with time on their hands might prefer to draw these conclusions from our first significant result.

17.1 Theorem: *Every Banach lattice has local unconditional structure.*

A simple observation begins the proof. If $x_1, ..., x_n$ are disjoint non-zero vectors in a Banach lattice, then they are a 1-unconditional basis for their span. For, if $a_1, ..., a_n$ are arbitrary scalars, then no matter how we select $(t_k)_1^n$ in $B_{\ell_\infty^n}$, disjointness leads to

$$\Big\| \sum_{k \leq n} t_k a_k x_k \Big\| = \Big\| \Big| \sum_{k \leq n} t_k a_k x_k \Big| \Big\| = \Big\| \sum_{k \leq n} |t_k a_k x_k| \Big\|$$
$$\leq \Big\| \sum_{k \leq n} |a_k x_k| \Big\| = \Big\| \Big| \sum_{k \leq n} a_k x_k \Big| \Big\| = \Big\| \sum_{k \leq n} a_k x_k \Big\| .$$

Now it would be too much to hope that every finite dimensional subspace of a Banach lattice could be embedded into the span of finitely many disjoint vectors – the proof of the theorem isn't that easy! Nevertheless, such hopes are not too far from the truth: the Principle of Local Reflexivity combined with the next perturbation lemma will enable us to bridge the gap.

17.2 Lemma: *Let $x_1, ..., x_n$ be elements of a Banach lattice L, and let $\delta > 0$. There are vectors $\hat{x}_1, ..., \hat{x}_n$ in L^{**} such that*

(a) $\|\hat{x}_k - x_k\| \leq \delta \quad (1 \leq k \leq n)$

and

(b) $\text{span}\{\hat{x}_1, ..., \hat{x}_n\}$ *is generated by disjoint vectors from L^{**}.*

Proof. To simplify our writing, we suppose that $x := \sum_{k \leq n} |x_k|$ has norm one. Recall from Chapter 16 that the ideal $I(x) := \bigcup_{\lambda > 0} \{y \in L : |y| \leq \lambda x\}$ generated by x is an abstract M-space with norm $\|y\|_\infty := \inf\{\lambda > 0 : |y| \leq \lambda x\}$ and so, by Kakutani's Theorem 16.1, is isometrically isomorphic as a Banach lattice to $\mathcal{C}(K)$ for a suitable compact Hausdorff space K. As the order in $I(x)$ is just that of L, we may consider the inclusion of $I(x)$ into L as an injective norm one lattice homomorphism $v : \mathcal{C}(K) \to L$. Notice that $v^{**} : \mathcal{C}(K)^{**} \to L^{**}$ is a norm one lattice homomorphism as well. By design, each x_k belongs to $I(x) = \mathcal{C}(K)$; we shall identify x_k with $v(x_k) \in L$ and $v^{**}(x_k) \in L^{**}$.

The key to obtaining the approximations we seek is that $\mathcal{C}(K)^{**}$ contains all bounded Borel functions on K. Consequently, we can find pairwise disjoint Borel subsets $B_1, \ldots, B_m$ of K and simple functions $h_1, \ldots, h_n$ on K taking constant values on these sets such that $\|h_k - x_k\|_{\mathcal{C}(K)^{**}} \leq \delta$ for each $1 \leq k \leq n$. Let $b_i^{**} := v^{**}1_{B_i}$ $(1 \leq i \leq m)$ and $\hat{x}_k := v^{**}h_k$ $(1 \leq k \leq n)$. Naturally, each $\hat{x}_k$ is in the span of the b_i^{**}'s, and since v^{**} is a norm one lattice homomorphism, the b_i^{**}'s are disjoint in L^{**} with

$$\|\hat{x}_k - x_k\| = \|v^{**}(h_k - x_k)\| \leq \|h_k - x_k\|_{\mathcal{C}(K)^{**}} \leq \delta. \qquad \text{QED}$$

Our next lemma indicates that, at least locally, unconditional basic sequences behave stably under small perturbations.

17.3 Lemma: *Let $\{x_1, \ldots, x_n\}$ be a basis for the finite dimensional normed space E. Given $0 < \varepsilon < 1$ there is a $\delta > 0$ such that if X is any Banach space containing E and if $\hat{x}_1, \ldots, \hat{x}_n \in X$ satisfy $\|\hat{x}_k - x_k\| \leq \delta$ $(1 \leq k \leq n)$, then there exists an operator $u \in \mathcal{L}(X)$ such that*

(a) $u\hat{x}_k = x_k \quad (1 \leq k \leq n)$

and

(b) $(1 - \varepsilon)\cdot\|x\| \leq \|ux\| \leq (1 + \varepsilon)\cdot\|x\|$ *for each $x \in X$.*

Notice that (b) contains the information that u is invertible and that $\|u\| \leq 1 + \varepsilon$ and $\|u^{-1}\| \leq (1 - \varepsilon)^{-1}$.

Proof. We can choose $K \geq 1$ such that, for any scalars $a_1, \ldots, a_n$,

$$K^{-1}\cdot\max_{k\leq n} |a_k| \leq \big\|\textstyle\sum_{k\leq n} a_k x_k\big\| \leq K\cdot\max_{k\leq n} |a_k|.$$

Fix $0 < \varepsilon < 1$ and select a Banach space X containing E. Pick $\hat{x}_1, \ldots, \hat{x}_n$ in X so as to satisfy $\|x_k - \hat{x}_k\| \leq \varepsilon/(2nK)$ $(1 \leq k \leq n)$. This ensures that

$$\frac{1}{K}\cdot\max_{k\leq n}|a_k| \leq \Big\|\sum_{k\leq n} a_k\hat{x}_k\Big\| + \sum_{k\leq n}|a_k|\cdot\|x_k - \hat{x}_k\| \leq \Big\|\sum_{k\leq n} a_k\hat{x}_k\Big\| + \frac{\varepsilon}{2K}\cdot\max_{k\leq n}|a_k|.$$

A slight adjustment yields

$$\max_{k\leq n}|a_k| \leq 2K\cdot\Big\|\sum_{k\leq n} a_k\hat{x}_k\Big\|.$$

This inequality has a dual rôle. First, it ensures the independence of $\hat{x}_1, \ldots, \hat{x}_n$. Secondly, it also forces the biorthogonal functionals $\hat{x}_j^*$ acting on $\operatorname{span}\{\hat{x}_1, \ldots, \hat{x}_n\}$ via $\sum_{k\leq n} a_k\hat{x}_k \mapsto a_j$ to have norm at most $2K$ for each $1 \leq j \leq n$. Select norm preserving extensions of these functionals in X^* and continue to call them $\hat{x}_j^*$ $(1 \le j \le n)$. The sought-after operator $u \in \mathcal{L}(X)$ can now be defined by setting $ux := x + \sum_{k \le n} \langle \hat{x}_k^*, x \rangle (x_k - \hat{x}_k)$ for each $x \in X$. Property (a) is obvious, and property (b) follows at once from the observation that $\|ux - x\| \le \varepsilon \|x\|$ for each $x \in X$. QED

Armed with these two lemmas and the Principle of Local Reflexivity, we can now prove the theorem.

Proof of 17.1. Choose a finite dimensional subspace E of the Banach lattice L, and settle on a basis $x_1, \ldots, x_n$ of E. Fix $0 < \varepsilon < 1$. Our lemmas entail that we can find vectors $\hat{x}_1, \ldots, \hat{x}_n$ in L^{**} and an operator $u : L^{**} \to L^{**}$ such that

$$(1) \qquad u\hat{x}_k = x_k \quad (1 \le k \le n)$$

and

$$(2) \qquad (1 - \varepsilon) \cdot \|x^{**}\| \le \|ux^{**}\| \le (1 + \varepsilon) \cdot \|x^{**}\| \quad \text{for all } x^{**} \in L^{**}.$$

Moreover, we can select the $\hat{x}_k$'s to lie in the span of disjoint non-zero vectors b_i^{**} of L^{**} $(1 \le i \le m)$. We have already recorded that, for all scalars $a_1, \ldots, a_m$ and all $(t_i)_{i=1}^m \in B_{\ell_\infty^m}$,

$$\Big\| \sum_{i \le m} t_i a_i b_i^{**} \Big\| \le \Big\| \sum_{i \le m} a_i b_i^{**} \Big\| .$$

It follows from (2) that

$$\Big\| \sum_{i \le m} t_i a_i u b_i^{**} \Big\| \le \frac{1 + \varepsilon}{1 - \varepsilon} \cdot \Big\| \sum_{i \le m} a_i u b_i^{**} \Big\| .$$

In other words, the ub_i^{**}'s $(1 \le i \le m)$ form an unconditional basis for their span F, and $ub(F) \le (1 + \varepsilon) \cdot (1 - \varepsilon)^{-1}$.

We are tantalizingly close to our objective. Thanks to (2) we can factor the natural embedding $E \hookrightarrow L^{**}$ as $E \xrightarrow{j} F \xrightarrow{i} L^{**}$ where i and j are themselves natural embeddings. Moreover, $\|i\| \cdot \|j\| \cdot ub(F) \le (1 + \varepsilon) \cdot (1 - \varepsilon)^{-1}$. It is just a minor irritation that we end up in L^{**} rather than in L. But this can be taken care of by the Principle of Local Reflexivity 8.16. It provides us with an operator $w : F \to L$ such that $wy = y$ for each $y \in L \cap F$ and $\|w\| \le 1 + \varepsilon$.

Now consider the composition $E \xrightarrow{i} F \xrightarrow{w} L$. It too is the natural embedding, and $\|i\| \cdot \|w\| \cdot ub(F) \le (1 + \varepsilon)^2 \cdot (1 - \varepsilon)^{-1}$. Since $E \in \mathcal{F}_L$ was arbitrary, we have shown that $\Lambda(L) \le (1 + \varepsilon)^2 \cdot (1 - \varepsilon)^{-1}$. QED

The proof shows that $\Lambda(L) = 1$ whenever L is a Banach lattice.

Close inspection shows that in the above argument we have proved more than we set out to do: the space F that we constructed is finite dimensional. In fact, a close perusal of the above proof reveals:

17.4 Corollary: *A Banach space X has l.u.st. if and only if there is a constant $\Lambda \ge 1$ such that for each $E \in \mathcal{F}_X$ we can find a finite dimensional normed space F along with operators $u \in \mathcal{L}(F, X)$ and $v \in \mathcal{L}(E, F)$ such that $uv : E \to X$ is the natural inclusion and $\|u\| \cdot \|v\| \cdot ub(F) \le \Lambda$.*

Again, $\Lambda(X)$ is the least of all the Λ's which work in this statement.

The link between Banach lattices and Banach spaces with l.u.st. is even closer than what we have already seen. Recall that any Banach space is a subspace of a Banach lattice – the Banach lattice of continuous functions on the unit ball of its dual (equipped with the weak$*$topology). Complementation is another matter.

17.5 Theorem: *A Banach space X has local unconditional structure if and only if X^{**} is isomorphic to a complemented subspace of a Banach lattice.*

Proof. Assume first that X^{**} is complemented in a Banach lattice. Since l.u.st. is evidently inherited by complemented subspaces, we infer from 17.1 that X^{**} has l.u.st. We shall use the last corollary to transfer this property to X.

We know that for any $\lambda > \Lambda(X^{**})$ and $E \in \mathcal{F}_X$ we can find a finite dimensional normed space F and operators $u \in \mathcal{L}(F, X^{**})$ and $v \in \mathcal{L}(E, F)$ such that $uv : E \to X^{**}$ is the canonical embedding and $\|u\| \cdot \|v\| \cdot ub(F) < \lambda$. If we fix $\varepsilon > 0$, we can invoke Local Reflexivity to obtain an operator $w : u(F) \to X$ of norm $\leq 1 + \varepsilon$ such that $wx = x$ whenever $x \in X \cap u(F)$. Putting everything together we see that the natural embedding $E \hookrightarrow X$ has a factorization $E \overset{v}{\to} F \overset{u}{\to} u(F) \overset{w}{\to} X$. Since $\|wu\| \cdot \|v\| \cdot ub(F) \leq (1+\varepsilon) \cdot \lambda$, we conclude that X has l.u.st. and $\lambda(X) \leq \lambda(X^{**})$.

We settle the converse by an ultraproduct argument. To establish notation, let X be a Banach space with l.u.st. and pick $\lambda > \Lambda(X)$. For each E in $\mathcal{F}_X$ we can find a Banach space L_E with unconditional basis and operators u_E in $\mathcal{L}(L_E, X)$ and v_E in $\mathcal{L}(E, L_E)$ such that $u_E v_E : E \to X$ is the natural embedding and $\|u_E\| \cdot \|v_E\| \cdot ub(L_E) < \lambda$. In addition, we may arrange to have $\|u_E\| < \sqrt{\lambda}$ and $\|v_E\| < \sqrt{\lambda}$. In the introductory remarks to this chapter we saw that L_E can be viewed as a Banach lattice.

Once more we consider $\mathcal{F}_X$ as a directed set with respect to inclusion, and we fix an ultrafilter $\mathcal{U}$ on $\mathcal{F}_X$ which refines the corresponding order filter. As in Proposition 8.6, the ultraproduct $L := (\prod L_E)_{\mathcal{U}}$ has a natural Banach lattice structure.

We aim to exhibit a factorization $k_X : X \overset{v}{\to} L \overset{u}{\to} X^{**}$. Once this is achieved, the identity $id_{X^{**}} = (k_{X^*})^* \, (k_X)^{**}$ reveals X^{**} to be a complemented subspace of the Banach lattice L^{**}.

A familiar strategy can be used to fabricate the desired operator $v : X \to L$ from the v_E's. First extend each v_E to a (non-linear!) map $\widetilde{v_E} : X \to L_E$ by setting $\widetilde{v_E}(x) := v_E(x)$ if $x \in E$ and $\widetilde{v_E}(x) := 0$ otherwise. Then define $v : X \to L$ by $vx := (\widetilde{v_E} x)_{\mathcal{U}}$ for each $x \in X$. This map is linear and satisfies $\|v\| \leq \sup_{E \in \mathcal{F}_X} \|v_E\| \leq \sqrt{\lambda}$.

To complete the factorization, notice that for each $y = (y_E)_{\mathcal{U}} \in L$ the family $(u_E y_E)_{\mathcal{F}_X}$ is bounded in X. Consequently, the weak$*$limit $uy := \lim_{\mathcal{U}} u_E y_E$ exists in X^{**}. A bounded linear map $u : L \to X^{**}$ is thus obtained, with $\|u\| \leq \sup_{E \in \mathcal{F}_X} \|u_E\| \leq \sqrt{\lambda}$.

Our construction was set up to ensure that uv is the canonical embedding $k_X : X \hookrightarrow X^{**}$, and so we are done. QED

As a by-product of this development, we can now assert that a Banach space X has l.u.st. if and only if X^{**} does. Even more is true:

17.6 Corollary: *A Banach space X has local unconditional structure if and only if X^* has it.*

Proof. If X has l.u.st. we know that X^{**} is complemented in a Banach lattice L. Passing to duals, X^{***} is complemented in the Banach lattice L^*. Since X^* is naturally complemented in X^{***}, it must have l.u.st. as well.

If X^* has l.u.st., we know that X^{**} has l.u.st., and thanks to our preliminary remark, X follows suit. QED

THE GORDON - LEWIS INEQUALITY

According to the Pietsch Factorization Theorem 2.13, every 1-summing operator factors through a subspace of a suitable $L_1(\mu)$-space. There are notable situations where we can make the factorization through the whole of an $L_1(\mu)$-space.

- *If X is a subspace of $L_1(\mu)$ for some measure μ, and if $u : X \to Y$ is 1-summing, then u factors through $L_1(\mu)$.*

Indeed, as u is 2-summing, 4.15 tells us that it has a (2-summing) extension $\tilde{u} : L_1(\mu) \to Y$.

There is a dual result, but in this case the work cannot be hidden so successfully.

- *If X is a quotient of $\mathcal{C}(K)$, where K is a compact Hausdorff space, and if $u : X \to Y$ is 1-summing, then u factors through some $L_1(\mu)$-space.*

Let $q : \mathcal{C}(K) \to X$ be the quotient map. Then, thanks to 2.15, the 1-summing operator uq enjoys a factorization $uq : \mathcal{C}(K) \overset{w}{\to} L_2(\mu) \overset{j}{\to} L_1(\mu) \overset{v}{\to} Y$, where μ is a suitable probability measure, j is the formal identity, and v, w are appropriate operators. It is advantageous to make an adjustment by factoring vj through the Hilbert space $H = \ker(vj)^\perp$. Then $uq : \mathcal{C}(K) \overset{\hat{w}}{\to} H \overset{\hat{v}}{\to} Y$, where $\hat{v} = vj\,\big|_H$ is injective. This injectivity in combination with the Closed Graph Theorem permits us to well-define an operator $\tilde{w} : X \to H$ by setting $\tilde{w}(qf) = \hat{w}f$ for all $f \in \mathcal{C}(K)$. As $u = \hat{v}\tilde{w}$, we can unravel to reveal what we have been seeking:

$$u : X \overset{\tilde{w}}{\longrightarrow} H \hookrightarrow L_2(\mu) \overset{j}{\longrightarrow} L_1(\mu) \overset{v}{\longrightarrow} Y.$$

These examples are important but special. Later in this chapter we show that there are 1-summing operators (on $\mathcal{L}(\ell_2)$, for example) that are not 1-factorable. However, when spaces as different as quotients of $\mathcal{C}(K)$'s and subspaces of $L_1(\mu)$'s enjoy a common property, that property is worthy of our attention. Further interest is piqued by the next fundamental result.

17.7 Gordon-Lewis Inequality: *Let X and Y be Banach spaces and suppose that X has local unconditional structure. Then every 1-summing operator $u : X \to Y$ is 1-factorable, with*

$$\gamma_1(u) \leq \Lambda(X) \cdot \pi_1(u).$$

Proof. As $[\Gamma_1, \gamma_1]$ is a maximal Banach ideal, we can estimate $\gamma_1(u)$ by finding a common upper bound for $\gamma_1(u|_E)$ when $E \in \mathcal{F}_X$. So fix $\lambda > \Lambda(X)$, take $E \in \mathcal{F}_X$, and pick a Banach space Z with unconditional basis, along with operators $v : Z \to X$ and $w : E \to Z$ such that $vw : E \to X$ is the canonical embedding and $\|v\| \cdot \|w\| \cdot ub(Z) < \lambda$. We shall show how to factor $uv : Z \to Y$ in a controlled way through ℓ_1.

Select any μ with $ub(Z) < \mu < \lambda \cdot \|v\|^{-1} \cdot \|w\|^{-1}$ and find an unconditional basis (z_n) for Z such that, for any $(c_n) \in B_{\ell_\infty}$ and any $\sum_n a_n z_n \in Z$,

$$\Big\| \sum_n c_n a_n z_n \Big\| \leq \mu \cdot \Big\| \sum_n a_n z_n \Big\| .$$

This entails that $\|(a_n z_n)\|_1^{weak} \leq \mu \cdot \| \sum_n a_n z_n \|$, and so

$$\sum_n \|uv(a_n z_n)\| \leq \pi_1(uv) \cdot \|(a_n z_n)\|_1^{weak} \leq \mu \cdot \pi_1(u) \cdot \|v\| \cdot \Big\| \sum_n a_n z_n \Big\| .$$

Consequently,

$$t : Z \longrightarrow \ell_1 : \sum_n a_n z_n \mapsto \big(a_n \cdot \|uv(z_n)\| \big)_n$$

is a well-defined operator and $\|t\| \leq \mu \cdot \pi_1(u) \cdot \|v\|$.

To finish things off, notice that by setting

$$s((b_n)) := \sum_{uvz_n \neq 0} b_n \cdot \frac{uvz_n}{\|uvz_n\|}$$

we obtain an operator $s : \ell_1 \to Y$ with $\|s\| \leq 1$, satisfying $stz_n = uvz_n$ for every n. Thus $st = uv$, and so $u|_E = uvw = stw$. Moreover, the construction ensures that

$$\gamma_1(u|_E) \leq \|s\| \cdot \|t\| \cdot \|w\| \leq \mu \cdot \pi_1(u) \cdot \|v\| \cdot \|w\| \leq \lambda \cdot \pi_1(u).$$

Since $\lambda > \Lambda(X)$ was arbitrary, we are done. QED

Note that the argument actually shows that if Z is a Banach space with unconditional basis and $u : Z \to Y$ is 1-summing, then for every $\lambda > ub(Z)$ there is a factorization $u : Z \xrightarrow{w} \ell_1 \xrightarrow{v} Y$ with $\|v\| \cdot \|w\| \leq \lambda \cdot \pi_1(u)$.

GL-SPACES

We say that a Banach space X has the *Gordon-Lewis property, (GL)* for short, or that X is a *GL-space,* if every 1-summing operator from X to ℓ_2 is 1-factorable. In other words, we require the existence of a constant c such that

$$\gamma_1(u) \leq c \cdot \pi_1(u)$$

for every $u \in \Pi_1(X, \ell_2)$. We label the smallest of these constants

$$gl(X).$$

It will turn out that property (GL) is well-suited to the study of the finer unconditional structure of some classical Banach spaces that are not $\mathcal{L}_p$-spaces.

We begin with something simple.

17.8 Proposition: *Let X be a Banach space. Then X is a GL-space if and only if X^{**} is, and in this case $gl(X) = gl(X^{**})$.*

Proof. Everything hangs on 2.19 and 7.2.

First assume that X is a GL-space and select $u \in \Pi_1(X^{**}, \ell_2)$. Noting that $(uk_X)^{**} = u$, we obtain

$$\pi_1(u) = \pi_1(uk_X) \le gl(X)\cdot\gamma_1(uk_X) = gl(X)\cdot\gamma_1((uk_X)^{**}) = gl(X)\cdot\gamma_1(u)$$

and see that X^{**} is a GL-space with $gl(X^{**}) \le gl(X)$.

Conversely, if X^{**} is a GL-space and $u \in \Pi_1(X, \ell_2)$, we have

$$\pi_1(u) = \pi_1(u^{**}) \le gl(X^{**})\cdot\gamma_1(u^{**}) = gl(X^{**})\cdot\gamma_1(u)$$

and so X is a GL-space with $gl(X) \le gl(X^{**})$. QED

Much more is true.

17.9 Proposition: *A Banach space X is a GL-space if and only if X^* is, and then $gl(X) = gl(X^*)$.*

Proof. What remains to be shown is that whenever X is a GL-space so is X^*, and then $gl(X^*) \le gl(X)$. To this end, fix $u \in \Pi_1(X^*, \ell_2)$; we want to conclude that $u \in \Gamma_1(X^*, \ell_2)$. By 9.5, this amounts to proving that $u \in \left(\Pi_1^d\right)^*(X^*, \ell_2)$. So trace duality enters, and we shall be done if we can demonstrate that for all compositions of operators $E \xrightarrow{v} X^* \xrightarrow{u} \ell_2 \xrightarrow{t} F \xrightarrow{s} E$ with E and F finite dimensional we have $|\mathrm{tr}\,(stuv)| \le C\cdot\|t\|\cdot\|v\|\cdot\pi_1(s^*)$ where $C = gl(X)$. Since v has finite rank, we can actually extend this chain of compositions – without changing the overall effect – to $E \xrightarrow{v} X^* \xrightarrow{u} \ell_2 \xrightarrow{b} \ell_2^n \xrightarrow{a} \ell_2 \xrightarrow{t} F \xrightarrow{s} E$ where $n = \dim E$ and a and b are the canonical inclusion and projection. This allows us to use 6.14 and 5.16:

$$|\mathrm{tr}(stuv)| = |\mathrm{tr}(stabuv)| = |\mathrm{tr}(buvsta)| \le \pi_1(bu)\cdot\gamma_\infty(vsta) \le \pi_1(u)\cdot\gamma_\infty(vst).$$

But now we can use the previous proposition. Since X^{**} is a GL-space with $gl(X^{**}) = gl(X)$, we have

$$\begin{aligned}\pi_1(u)\cdot\gamma_\infty(vst) &= \pi_1(u)\cdot\gamma_1(t^*s^*v^*) \le gl(X^{**})\cdot\pi_1(u)\cdot\pi_1(t^*s^*v^*)\\ &\le gl(X)\cdot\pi_1(u)\cdot\|v\|\cdot\|t\|\cdot\pi_1^d(s).\end{aligned}$$

QED

Note that 17.9 can be rephrased by saying that X is a GL-space if and only if $\Pi_1^d(\ell_2, X)$ is contained in $\Gamma_\infty(\ell_2, X)$.

GL-Spaces and Cotype

It is especially rewarding to study the Gordon - Lewis property in the presence of finite cotype. In particular we shall find that GL-spaces of cotype 2 enjoy most of the properties of cotype 2 Banach lattices that we uncovered in Chapter 16. The preceding results ease our access to satisfying extensions of our introductory examples. Here are immediate consequences of our next proposition:

- If $1 \le p \le 2$, any subspace of an $\mathcal{L}_p$-space is a GL-space.
- If $2 \le q \le \infty$, any quotient of an $\mathcal{L}_q$-space is a GL-space.

17.10 Proposition: *Let X be a GL-space.*

(a) *If X has cotype 2 then every subspace of X is a GL-space.*

(b) *If X^* has cotype 2, then every quotient of X is a GL-space.*

Proof. (b) follows from (a) by duality, with the help of 17.9.

To prove (a), let Z be a subspace of X and let $u : Z \to \ell_2$ be 1-summing. By 4.15, u admits a 2-summing extension $\tilde{u} : X \to \ell_2$. But X has cotype 2, so 11.16 informs us that $\tilde{u}$ is actually 1-summing. We can finish by applying our hypothesis: $\tilde{u}$, and so u, is 1-factorable. QED

We have just taken advantage of the fact that 2-summing operators acting on a cotype 2 space are automatically 1-summing. In the addendum to 16.9 we saw that for a Banach lattice L the coincidence of $\Pi_1(L, \cdot)$ and $\Pi_2(L, \cdot)$ obliged L to have cotype 2. We are ready for an extension.

17.11 Theorem: *A GL-space X has cotype 2 if and only if every 2-summing operator with domain X is 1-summing.*

Proof. We only need to show that X has cotype 2 whenever 2-summing operators with domain X are 1-summing; the other direction is covered by 11.16. Referring to 12.29, we realize that we can meet our needs by proving that $\Pi_{as}(\ell_2, X) \subset \Pi_2(\ell_2, X)$.

As a first step, we show that for all $u \in \Pi_2(X, \ell_2)$ and $v \in \Pi_{as}(\ell_2, X)$, the composition uv is 1-integral. Wielding our twin assumptions that X is a GL-space and that $\Pi_2(X, \ell_2) = \Pi_1(X, \ell_2)$, we obtain a factorization $u : X \overset{u_2}{\to} L_1(\mu) \overset{u_1}{\to} \ell_2$ for some measure μ. Now, recalling that $L_1(\mu)$ has cotype 2, we can invoke 12.7 to infer that $u_2 v$ is 2-summing. Grothendieck's Theorem 3.1 tells us that u_1 is also 2-summing. Consequently, by 5.9 and 5.16, $uv = u_1(u_2 v)$ is 1-integral. Naturally, there is also a constant C, depending only on X, such that $\iota_1(uv) \le C \cdot \pi_2(u) \cdot \pi_{as}(v)$.

To finish things off, we bring in trace duality. By 6.17, Π_2 is its own trace dual, so if we take u to be a composition of operators $X \to F \to E \to \ell_2$ with E and F finite dimensional, a *careful* use of 6.14 enables us to conclude that v is 2-summing. QED

The Gordon - Lewis property also sheds light on Kwapień's Theorem 12.19.

17.12 Theorem: *Let X and Y be Banach spaces, at least one of which is a GL-space. If both X^* and Y have cotype 2, then*

$$\mathcal{L}(X,Y) = \Gamma_2(X,Y).$$

Proof. Contemplate 9.11 for a moment. To accomplish our goal it is good enough to show that for any $v \in \Pi_2^d(\ell_2, X)$ and $u \in L(X,Y)$, the operator $uv : \ell_2 \to Y$ is 2-summing.

Since X^* has cotype 2, 11.16 reveals that v^*, and so v^*u^*, is 1-summing. But then, as one or other of X^* and Y^* is a GL-space (use 17.9), we find that v^*u^* is in $\Gamma_1(Y^*, \ell_2)$, or equivalently, that uv is in $\Gamma_\infty(\ell_2, Y)$. Bring the fact that Y has cotype 2 into play: 11.14 informs us that $\Gamma_\infty(\ell_2, Y) = \Pi_2(\ell_2, Y)$. We can rest our case. QED

In particular, a GL-space X with the property that both it and its dual have cotype 2 must be isomorphic to a Hilbert space. It is tempting to think that we have come across a partial generalization of Kwapień's characterization 12.20 of Hilbert spaces. Not so: in the presence of (GL), there is almost complete duality between type and cotype. Compare the next result with 16.22.

17.13 Theorem: *Let X be a GL-space with finite cotype and let $p > 1$. Then X has type p if and only if X^* has cotype p^*.*

Proof. Recall that 11.10 disposes of one implication without the need for any assumptions about X. To handle the other direction, it is convenient to exploit 12.26 and to show that if X^* has cotype p^*, then X has Gaussian type p. In other words, we wish to prove that there is a constant C such that, however we choose finitely many vectors $x_1, \ldots, x_n$ in X, we always have

$$\Big(\int_\Omega \big\|\sum_{k\le n} g_k(\omega) x_k\big\|^2 \, dP(\omega)\Big)^{1/2} \le C\cdot\big(\sum_{k\le n} \|x_k\|^p\big)^{1/p}.$$

By 12.15, another way to express this is to say that we must show that

$$\pi_\gamma(u) \le C\cdot\big(\sum_{k\le n} \|ue_k\|^p\big)^{1/p}$$

for every $n \in \mathbf{N}$ and every operator $u : \ell_2^n \to X$. There is no harm in normalizing so that $\sum_{k\le n} \|ue_k\|^p = 1$.

To reach this conclusion, we first show that $\pi_1(u^*)$ is bounded by a constant depending only on X and on p. To see why this helps, observe that, by 12.12 and 12.5, $\pi_\gamma(u) \le \pi_{as}(u) \le B_r \cdot \pi_r(u)$ for every $1 \le r < \infty$; B_r is the constant from Khinchin's Inequality. Next, we call on 11.14: since X has finite cotype, there are numbers $2 \leq r < \infty$ and $c_1 > 0$, depending only on X, such that $\pi_r(u) \leq c_1 \cdot \gamma_\infty(u)$. Finally, we can use the fact that X^* is a GL-space to conclude that, for some c_2 depending only on X,

$$\pi_\gamma(u) \leq B_r \cdot c_1 \cdot \gamma_\infty(u) = B_r \cdot c_1 \cdot \gamma_1(u^*) \leq B_r \cdot c_1 \cdot c_2 \cdot \pi_1(u^*).$$

Our task then is to show that $\pi_1(u^*)$ can be bounded above in an appropriate way. It is helpful to recall first from 2.7 that for any $\lambda > 1$ we can find $v \in \mathcal{L}(\ell_\infty^n, X^*)$ with $\|v\| \leq 1$ and $\pi_1(u^*) \leq \lambda \cdot \pi_1(u^* v)$. Since Π_1 and Γ_∞ are in trace duality (6.16), there is a $w \in \mathcal{L}(\ell_2^n, \ell_\infty^n)$ with $\|w\| \leq 1$ such that $\pi_1(u^* v) \leq \lambda \cdot |\mathrm{tr}\,(w u^* v)| = \lambda \cdot |\mathrm{tr}\,(v w u^*)|$. Now if $u e_k = x_k$ $(1 \leq k \leq n)$, we can write $u^* = \sum_{k \leq n} x_k \otimes e_k$ to get

$$\begin{aligned} \pi_1(u^*) &\leq \lambda^2 \cdot \big|\, \mathrm{tr}\,\big(\sum_{k \leq n} x_k \otimes v w e_k\big) \big| = \lambda^2 \cdot \big| \sum_{k \leq n} \langle x_k, v w e_k \rangle \big| \\ &\leq \lambda^2 \cdot \sum_{k \leq n} \|x_k\| \cdot \|v w e_k\|. \end{aligned}$$

Our normalization surfaces:

$$\pi_1(u^*) \leq \lambda^2 \cdot \big(\sum_{k \leq n} \|v w e_k\|^{p^*}\big)^{1/p^*} \leq \lambda^2 \cdot \pi_{p^*,2}(v w) \leq \lambda^2 \cdot \pi_{p^*,2}(v).$$

Theorem 11.14 reveals that $\pi_{p^*,2}(v) \leq K \cdot \|v\| \leq K$ where K is a constant depending only on X and p. QED

Actually, the theorem we just used (11.14) has an attractive complement in the context of GL-spaces.

17.14 Theorem: *Let X be a GL-space and let $2 \leq q < \infty$. If every operator from c_0 to X is q-summing, then X has cotype q.*

Of course, c_0 can be replaced by any infinite dimensional $\mathcal{L}_\infty$-space: after all, we know (by using 3.3) that our hypothesis is equivalent to the existence of a constant C such that, regardless of how we choose $N \in \mathbf{N}$ and an operator $v : \ell_\infty^N \to X$, we have $\pi_q(v) \leq C \cdot \|v\|$.

Proof. Pick finitely many vectors $x_1, \ldots, x_n \in X$ and consider the operator $u : \ell_2^n \to X$ which is given by $u e_k = x_k$ $(1 \leq k \leq n)$. The remark following 17.9 provides us with a constant K depending only on X such that $\gamma_\infty(u) \leq K \cdot \pi_1^d(u)$. Passing to any $K' > K$, we can, for an appropriate measure μ, settle on a factorization $k_X u : \ell_2^n \xrightarrow{w} L_\infty(\mu) \xrightarrow{v} X^{**}$ such that $\|v\| \cdot \|w\| \leq K' \cdot \pi_1^d(u)$. Looking at 3.3, we realize that we may take $L_\infty(\mu) = \ell_\infty^N$ for some $N \in \mathbf{N}$, so that, by using our hypothesis, we arrive at

$$\begin{aligned} \big(\sum_{k \leq n} \|x_k\|^q\big)^{1/q} &= \big(\sum_{k \leq n} \|u e_k\|^q\big)^{1/q} \leq \pi_q(u) = \pi_q(k_X u) \\ &= \pi_q(v w) \leq \pi_q(v) \cdot \|w\| \leq C \cdot \|v\| \cdot \|w\| \leq C \cdot K' \cdot \pi_1^d(u). \end{aligned}$$

But now, since $u^* = \sum_{k\le n} x_k \otimes e_k$, Khinchin's Inequality gives

$$\|u^*x^*\| = \Big(\sum_{k\le n} |\langle x^*, x_k\rangle|^2\Big)^{1/2} \le A_1^{-1}\cdot \int_0^1 \Big|\Big\langle \sum_{k\le n} r_k(t)x_k, x^*\Big\rangle\Big|\, dt$$

for every $x^* \in X^*$. Consequently

$$\pi_1^d(u) = \pi_1(u^*) \le A_1^{-1}\cdot \int_0^1 \Big\|\sum_{k\le n} r_k(t)x_k\Big\|\, dt.$$

Piecing things together, we find that X has cotype 2. QED

Using trace duality, Theorem 17.14 can be rephrased to say that a GL-space X satisfying $\Pi_{q^*}(X,\,\cdot\,) = \Pi_1(X,\,\cdot\,)$ for some $2 \le q < \infty$ must be of cotype q (compare also with 11.16). In case $q = 2$, the theorem actually characterizes GL-spaces of cotype 2; see 17.11. In case $2 < q < \infty$, however, we are not even close to a characterization: as we observed in the Notes and Remarks concerning 10.10, there are operators $c_0 \to \ell_q$ which fail to be q-summing.

Λ(2)-Sets

The (GL)-property has important applications, for example in harmonic analysis. We shall work on the unit circle $\mathbf{T} = \{z \in \mathbf{C} : |z| = 1\}$ and consider the classical spaces $L_p(\mathbf{T})$ $(1 \le p \le \infty)$ defined with respect to the normalized Lebesgue (Haar) measure on $\mathbf{T}$. Identifying functions on $\mathbf{T}$ with their 2π-periodic counterparts on $\mathbf{R}$, we can take $L_p(\mathbf{T})$ to consist of all (classes of) functions f on $[-\pi, \pi]$ for which

$$\|f\|_p := \Big(\int_{-\pi}^{\pi} |f(t)|^p \,\frac{dt}{2\pi}\Big)^{1/p}$$

is finite, with the usual modification when $p = \infty$. Each $f \in L_p(\mathbf{T})$ gives rise to a sequence of Fourier coefficients

$$\hat{f}(n) = \int_{-\pi}^{\pi} f(t)e^{-int}\,\frac{dt}{2\pi}\,, \quad n \in \mathbf{Z}.$$

Now let Λ be a subset of $\mathbf{Z}$. As in 5.13, we write

$$L_{p,\Lambda}(\mathbf{T})$$

for the collection of all $f \in L_p(\mathbf{T})$ satisfying $\hat{f}(n) = 0$ for all $n \notin \Lambda$. This is clearly a closed subspace of $L_p(\mathbf{T})$.

We say that Λ is a Λ(2)-*set* if $L_{1,\Lambda}(\mathbf{T}) = L_{2,\Lambda}(\mathbf{T})$, and we shall write

$$\Lambda_2$$

for the smallest of all constants λ which satisfy $\|f\|_2 \le \lambda \cdot \|f\|_1$ for every $f \in L_{1,\Lambda}(\mathbf{T})$.

It is a remarkable fact that $\Lambda = \{n_k : k \in \mathbf{N}\}$ is a Λ(2)-set whenever $\liminf_k n_{k+1}/n_k > 1$; the mathematical public shows a strong tendency to prefer $n_k = 2^k$.

For each $a \in \mathbf{R}$, the *translation operator* $f(\cdot) \mapsto f(\cdot + a)$ defines a bounded linear map

$$T_a : L_p(\mathbf{T}) \longrightarrow L_p(\mathbf{T}).$$

Without proof, we summarize the essential properties of these operators.

17.15 Lemma: *Let $1 \le p \le \infty$.*

(a) *For each $a \in \mathbf{R}$, T_a is an isometric isomorphism of $L_p(\mathbf{T})$ onto itself; it also induces an isometric isomorphism of $L_{p,\Lambda}(\mathbf{T})$ onto itself.*

(b) *If $f \in L_p(\mathbf{T})$ and $g \in L_{p^*}(\mathbf{T})$, then $h : t \mapsto \langle g, T_t f\rangle$ defines an element of $L_\infty(\mathbf{T})$, and $\hat{h}(n) = \hat{f}(n) \cdot \int_{-\pi}^{\pi} g(t)e^{int}dt/(2\pi) = \hat{f}(n) \cdot \hat{g}(-n)$ for all $n \in \mathbf{Z}$.*

17.16 Theorem: *Let $2 \le p < \infty$ and suppose that $\Lambda \subset \mathbf{Z}$ is a $\Lambda(2)$-set. If $L_{p,\Lambda}(\mathbf{T})$ is a GL-space, then it has an unconditional basis.*

Proof. For convenience, set $X = L_{p,\Lambda}(\mathbf{T})$ and write $\chi_n(t) = e^{int}$, $n \in \mathbf{Z}$. Note that X is reflexive (since $2 \le p < \infty$).

If $f \in X$, the series $\sum_{n\in\Lambda} \hat{f}(n)\chi_n$ certainly converges to f in the norm of X; the χ_n's, $n \in \Lambda$, form a basis of X. We will be done if we can show that convergence is unconditional. We shall prove that there is a constant C such that, for each $f^* \in X^*$,

$$\sum_{n\in M} |\hat{f}(n)|\cdot|\langle f^*, \chi_n\rangle| \le C \cdot \|f\| \cdot \|f^*\|$$

for every finite set $M \subset \Lambda$; since X is reflexive, the Omnibus Theorem 1.9 will settle the case.

Fix then $f \in X$, $f^* \in X^*$ and a finite subset M of Λ. Thanks to X's translation invariance, we can define for each $g \in X$ a function $G \in L_{\infty,\Lambda}(\mathbf{T})$ by

$$G(t) := \langle f^*, T_t g\rangle \quad \text{for each } \ t \in \mathbf{T}.$$

The preceding lemma shows that

$$\hat{G}(n) = \langle f^*, \chi_n\rangle \hat{g}(n) \quad \text{for each } \ n \in \mathbf{Z};$$

here we have taken the liberty of using f^* to denote some fixed representative in $L_{p^*}(\mathbf{T})$ of $f^* \in X^*$.

Also, if $g^* \in X^*$, we define $G^* \in L_{\infty,\Lambda}(\mathbf{T})$ by

$$G^*(t) := \langle g^*, T_t f\rangle \quad \text{for each } \ t \in \mathbf{T}.$$

Using the same convention as before, we have

$$\widehat{G^*}(n) = \langle g^*, \chi_n\rangle \hat{f}(n) \quad \text{for each } \ n \in \mathbf{Z}.$$

For the purpose of removing absolute values from the scene, it is useful to introduce the sequence $(\varepsilon_n)_{n\in\mathbf{Z}}$ by setting $\varepsilon_n := \operatorname{sign} \hat{f}(n)\langle f^*, \chi_n\rangle$ if $n \in M$ and $\varepsilon_n := 0$ otherwise. As G^* belongs to $L_2(\mathbf{T})$, the sequence $(\varepsilon_n \widehat{G^*}(n))$ is in ℓ_2 and so, by Plancherel's Theorem, is the sequence of Fourier coefficients of some function in $L_2(\mathbf{T})$ – and even, after a cursory inspection, in $L_{2,\Lambda}(\mathbf{T})$. We can therefore define a linear map $u : X^* \longrightarrow L_{2,\Lambda}(\mathbf{T})$ by specifying $\widehat{ug^*}(n) = \varepsilon_n \widehat{G^*}(n)$ for every $n \in \mathbf{Z}$ and $g^* \in X^*$. The $\Lambda(2)$ property of Λ allows us to estimate its norm: if $g \in X^*$ then

$$\begin{aligned}\|ug^*\|_{L_{2,\Lambda}(\mathbf{T})} &= \|\widehat{ug^*}\|_{\ell_2} \leq \|\widehat{G^*}\|_{\ell_2} = \|G^*\|_{L_{2,\Lambda}(\mathbf{T})} \\ &\leq \Lambda_2 \cdot \|G^*\|_{L_{1,\Lambda}(\mathbf{T})} = \Lambda_2 \cdot \int_{-\pi}^{\pi} |\langle g^*, T_t f\rangle| \, \frac{dt}{2\pi}.\end{aligned}$$

This allows us to estimate u's 1-summing norm: if $g_1^*, \ldots, g_k^*$ are in X^*, then thanks to the isometric nature of the translations T_t,

$$\begin{aligned}\sum_{i \leq k} \|ug_i^*\| &\leq \Lambda_2 \cdot \int_{-\pi}^{\pi} \sum_{i \leq k} |\langle g_i^*, T_t f\rangle| \, \frac{dt}{2\pi} \\ &\leq \Lambda_2 \cdot \|(g_i^*)\|_1^{weak} \cdot \int_{-\pi}^{\pi} \|T_t f\| \, \frac{dt}{2\pi} = \Lambda_2 \cdot \|f\| \cdot \|(g_i^*)\|_1^{weak}.\end{aligned}$$

We infer that $\pi_1(u) \leq \Lambda_2 \cdot \|f\|$.

Next, we use a similar procedure to set up an operator

$$v : X \longrightarrow L_{2,\Lambda}(\mathbf{T});$$

this time we specify $\widehat{vg}(n) = \hat{G}(n)$ for each $n \in \mathbf{Z}$ and $g \in X$. Again, v is well-defined and linear, and the above arguments require hardly any changes to yield that v is also 1-summing with $\pi_1(v) \leq \Lambda_2 \cdot \|f^*\|$. In fact, we find as before that if $g \in X$, then $\|vg\| \leq \Lambda_2 \cdot \int_{-\pi}^{\pi} |\langle f^*, T_t g\rangle| \, dt/(2\pi)$, and so, given $g_1, \ldots, g_k \in X$,

$$\begin{aligned}\sum_{i \leq k} \|vg_i\| &\leq \Lambda_2 \cdot \int_{-\pi}^{\pi} \sum_{i \leq k} |\langle T_t^* f^*, g_i\rangle| \, \frac{dt}{2\pi} \\ &\leq \Lambda_2 \cdot \|(g_i)\|_1^{weak} \cdot \int_{-\pi}^{\pi} \|T_t^* f^*\| \, \frac{dt}{2\pi} \leq \Lambda_2 \cdot \|f^*\| \cdot \|(g_i)\|_1^{weak}.\end{aligned}$$

To prove what we are aiming for, we now examine the composition vu^*. Observe first that for each $g^* \in X^*$ and each $n \in \Lambda$

$$\langle u^* \chi_{-n}, g^* \rangle = \langle ug^*, \chi_{-n} \rangle = \varepsilon_n \widehat{G^*}(n) = \langle \varepsilon_n \hat{f}(n) \chi_n, g^* \rangle$$

and so $u^* \chi_{-n} = \varepsilon_n \hat{f}(n) \chi_n$. Similarly $v \chi_n = \langle f^*, \chi_n \rangle \chi_n$, and consequently

$$vu^* \chi_{-n} = \varepsilon_n \hat{f}(n) \langle f^*, \chi_n \rangle \chi_n$$

for each $n \in \Lambda$. We can therefore view $\sum_{n \in M} |\hat{f}(n) \langle f^*, \chi_n \rangle|$ as the trace of the composition

$$L_{2,M}(\mathbf{T}) \xrightarrow{i} L_{2,-\Lambda}(\mathbf{T}) \xrightarrow{u^*} X \xrightarrow{v} L_{2,\Lambda}(\mathbf{T}) \xrightarrow{q} L_{2,M}(\mathbf{T})$$

where i is the embedding given by $\chi_m \mapsto \chi_{-m}$ $(m \in M)$ and q is the natural quotient map. Apply trace duality and the Gordon-Lewis property of X to complete the proof:

$$\begin{aligned}\sum_{n \in M} |\hat{f}(n) \langle f^*, \chi_n \rangle| &= \operatorname{tr}(qvu^*i) \leq \gamma_\infty(u^*) \cdot \pi_1(v) \\ &\leq gl(X) \cdot \pi_1(u) \cdot \pi_1(v) \leq gl(X) \cdot \Lambda_2^2 \cdot \|f\| \cdot \|f^*\|.\end{aligned}$$

QED

Banach Spaces Failing (GL)

We have shown that the Gordon - Lewis property bestows much the same benefits on a Banach space as a lattice structure. So far, however, we have neglected the task of exhibiting a Banach space which is not a GL-space. We give an important, yet moderately elementary, example. A more general and more sophisticated approach can be found in the Notes and Remarks.

For simplicity, we are going to deal with real Banach spaces.

17.17 Theorem: *$\mathcal{L}(\ell_2)$ does not have the Gordon - Lewis property.*

This will be an immediate consequence of a quantitative finite dimensional result. Let $n \in \mathbf{N}$ be fixed. We consider the usual orthonormal basis of ℓ_2^n and represent the elements a of $\mathcal{L}(\ell_2^n)$ as matrices $(a_{ij})_{i,j=1}^n$ with respect to this basis. Being a space of dimension n^2, $\mathcal{L}(\ell_2^n)$ is linearly isomorphic to $\ell_2^{n^2}$; if $\{e_{ij} : 1 \leq i,j \leq n\}$ is an orthonormal basis of $\ell_2^{n^2}$, then a specific isomorphism is given by the 'natural identity'

$$I_n : \mathcal{L}(\ell_2^n) \longrightarrow \ell_2^{n^2} : (a_{ij})_{i,j=1}^n \mapsto \sum_{i,j=1}^n a_{ij} e_{ij}.$$

It will be instrumental to analyze various guises of I_n in the course of the proof of the theorem. The result itself follows immediately as soon as the following estimates are available.

17.18 Theorem: *There are constants $A, B > 0$ such that for each $n \in \mathbf{N}$*

(a) $n \leq \pi_1(I_n) \leq A \cdot n$

and

(b) $B \cdot n^{3/2} \leq \gamma_1(I_n) \leq n^{3/2}$.

Naturally, the upper bound for $\pi_1(I_n)$ and the lower bound for $\gamma_1(I_n)$ are the significant ones. The other bounds are simple to establish, so we include them for aesthetic reasons. We start by proving the upper estimate in (a).

17.19 Lemma: *Let $n \in \mathbf{N}$. Then $\pi_1(I_n) \leq A_1^{-2} \cdot n$, where A_1 is the constant from Khinchin's Inequality.*

Proof. This gives us our first opportunity to work with a modification of the map I_n: we consider $\widehat{I}_n : \mathcal{L}(\ell_\infty^n, \ell_1^n) \to \ell_2^{n^2} : (a_{ij}) \mapsto \sum_{i,j \leq n} a_{ij} e_{ij}$ and prove that

$$(*) \qquad \pi_1(\widehat{I}_n) \leq A_1^{-2}.$$

Our claim then follows from factoring I_n as $\mathcal{L}(\ell_2^n) \xrightarrow{id} \mathcal{L}(\ell_\infty^n, \ell_1^n) \xrightarrow{\widehat{I}_n} \ell_2^{n^2}$ and noting that the norm of the identity $id : \mathcal{L}(\ell_2^n) \to \mathcal{L}(\ell_\infty^n, \ell_1^n)$ is no more than n: in fact, each $a \in \mathcal{L}(\ell_\infty^n, \ell_1^n)$ has a formal decomposition $\ell_\infty^n \xrightarrow{id} \ell_2^n \xrightarrow{a} \ell_2^n \xrightarrow{id} \ell_1^n$, and the norm of both identities in the chain is $\sqrt{n}$.

To prove $(*)$ we need to show that if we take finitely many elements $a^{(1)},\ldots,a^{(N)}$ from $\mathcal{L}(\ell_\infty^n,\ell_1^n)$, then

$$\sum_{\ell\le N}\|\widehat{I}_n a^{(\ell)}\| \le A_1^{-2}\cdot\|(a^{(\ell)})_1^N\|_1^{weak}.$$

The left hand side can be written explicitly in terms of the components of $a^{(\ell)}=(a_{ij}^{(\ell)})_{i,j=1}^n$ $(1\le\ell\le N)$:

$$\sum_{\ell\le N}\|\widehat{I}_n(a^{(\ell)})\| = \sum_{\ell\le N}\Big(\sum_{i,j=1}^n|a_{ij}^{(\ell)}|^2\Big)^{1/2}.$$

As for the right hand side, we can use the fact that $D_m=\{-1,1\}^m$ is the set of extreme points of $B_{\ell_\infty^m}$ $(m\in\mathbf{N})$ to glean

$$\begin{aligned}\|(a^{(\ell)})_1^N\|_1^{weak} &= \sup\Big\{\Big\|\sum_{\ell\le N}\varepsilon_\ell a^{(\ell)}\Big\|:\ \varepsilon\in D_N\Big\}\\ &= \sup\Big\{\Big|\sum_{i,j,\ell}\varepsilon_\ell\lambda_i\mu_j a_{ij}^{(\ell)}\Big|:\ \varepsilon\in D_N,\ \lambda,\mu\in D_n\Big\}\\ &= \sup\Big\{\sum_\ell\Big|\sum_{i,j}\lambda_i\mu_j a_{ij}^{(\ell)}\Big|:\ \lambda,\mu\in D_n\Big\}.\end{aligned}$$

In particular, for all $s,t\in[0,1]$,

$$\|(a^{(\ell)})_1^N\|_1^{weak}\ge\sum_\ell\Big|\sum_{i,j}a_{ij}^{(\ell)}r_j(s)r_i(t)\Big|$$

and so, using Khinchin's Inequality, then Minkowski's Inequality and finally Khinchin's Inequality again,

$$\begin{aligned}\|(a^{(\ell)})_1^N\|_1^{weak} &\ge \sum_\ell\int_0^1\!\!\int_0^1\Big|\sum_i\Big(\sum_j a_{ij}^{(\ell)}r_j(s)\Big)r_i(t)\Big|\,dt\,ds\\ &\ge A_1\cdot\sum_\ell\int_0^1\Big(\sum_i\Big|\sum_j a_{ij}^{(\ell)}r_j(s)\Big|^2\Big)^{1/2}ds\\ &\ge A_1\cdot\sum_\ell\Big(\sum_i\Big(\int_0^1\Big|\sum_j a_{ij}^{(\ell)}r_j(s)\Big|\,ds\Big)^2\Big)^{1/2}\\ &\ge A_1^2\cdot\sum_\ell\Big(\sum_{i,j}|a_{ij}^{(\ell)}|^2\Big)^{1/2} = A_1^2\cdot\sum_\ell\|\widehat{I}_n a^{(\ell)}\|.\end{aligned}$$

QED

The proof of the remaining significant inequality in 17.18 takes us on an interesting excursion. We wish to estimate $\int_0^1\|\big(r_{ij}(t)\big)_{i,j}\|_{\mathcal{L}(\ell_2^n)}dt$ where the r_{ij}'s are distinct Rademacher functions. We begin with a touch of geometry.

17.20 Lemma: *Let $\varepsilon>0$. The closed unit ball of ℓ_2^n can be covered by no more than $\big(1+(2/\varepsilon)\big)^n$ closed balls, each with radius ε and centre in $B_{\ell_2^n}$.*

Proof. Let V be the volume of $B_{\ell_2^n}$. Then any ball in ℓ_2^n whose radius is r has volume $r^n\cdot V$.

Consider the family of all collections of disjoint closed balls in ℓ_2^n having radius $\varepsilon/2$ and centre in $B_{\ell_2^n}$. Let $\{B_1,\ldots,B_N\}$ be a maximal collection of this sort; for future reference, write $\widetilde{B_k}$ $(1 \le k \le N)$ for the closed ball with the same centre as B_k, but with double the radius. Clearly $\bigcup_{k=1}^N B_k \subset (1 + (\varepsilon/2))B_{\ell_2^n}$, so we have

$$N \cdot (\varepsilon/2)^n \cdot V = \sum_{k=1}^N \text{vol}\,(B_k) \le \text{vol}\,\big((1 + (\varepsilon/2))B_{\ell_2^n}\big) = (1 + (\varepsilon/2))^n \cdot V.$$

Hence $N \le (1 + (2/\varepsilon))^n$.

We can accomplish our task by showing that $B_{\ell_2^n}$ is contained in $\bigcup_{k=1}^N \widetilde{B_k}$. Write $x^{(k)}$ for the centre of B_k $(1 \le k \le N)$. Were our claim not true we would be able to find some $x \in B_{\ell_2^n}$ with $\|x - x^{(k)}\| > \varepsilon$ for each $1 \le k \le N$. But then the closed ball with centre x and radius $\varepsilon/2$ would be disjoint from each B_k – a contradiction of our maximality assumption. QED

17.21 Lemma: *Let $n \in \mathbf{N}$ and let $r = (r_{ij})_{i,j=1}^n$ be an $n \times n$ matrix of distinct Rademacher functions. For any $x, y \in B_{\ell_2^n}$ and any $\alpha > 0$,*

$$m\big(\{t \in [0,1] : |(r(t)x \,|\, y)| > \alpha\}\big) \le 2 \cdot e^{-\alpha^2/2}.$$

Here m denotes Lebesgue measure.

Proof. The independence of the Rademacher functions guides our way. Let λ be a real number to be specified later and let $x, y \in B_{\ell_2^n}$. Then

$$\begin{aligned}
\int_0^1 \exp\,(\lambda(r(t)x \,|\, y))\, dt &= \int_0^1 \exp\,\Big(\lambda \cdot \sum_{i,j=1}^n r_{ij}(t)x_iy_j\Big)\, dt \\
&= \int_0^1 \prod_{i,j=1}^n \exp\,(\lambda r_{ij}(t)x_iy_j)\, dt = \prod_{i,j=1}^n \int_0^1 \exp\,(\lambda r_{ij}(t)x_iy_j)\, dt \\
&= \prod_{i,j=1}^n \cosh\,(\lambda x_iy_j).
\end{aligned}$$

Since $2^p \cdot p! \le (2p)!$ for each $p \in \mathbf{N}$, a comparison of Taylor series shows that $\cosh s \le e^{s^2/2}$ for any real s. We can continue:

$$\int_0^1 \exp(\lambda(r(t)x \,|\, y))dt \le \prod_{i,j=1}^n \exp\,(\lambda^2x_i^2y_j^2/2) = \exp\,\Big((\lambda^2/2) \cdot \sum_{i,j} x_i^2y_j^2\Big) \le e^{\lambda^2/2}.$$

Calling on Chebyshev's Inequality we discover that, for any $\alpha > 0$,

$$m\big(\{t \in [0,1] : \lambda \cdot (r(t)x \,|\, y) > \alpha \cdot |\lambda|\}\big) \le \exp\,\big((\lambda^2/2) - \alpha \cdot |\lambda|\big).$$

By considering positive and negative λ's separately we find

$$m\big(\{t \in [0,1] : |(r(t)x \,|\, y)| > \alpha\}\big) \le 2 \cdot \exp\,\big((\lambda^2/2) - \alpha \cdot |\lambda|\big).$$

Choose $|\lambda| = \alpha$ to minimize the right hand side and reach our conclusion.

QED

These results put us within reach of our declared goal.

17.22 Lemma: *Let $r = (r_{ij})_{i,j=1}^n$ be an $n \times n$ matrix of distinct Rademacher functions. Then*

$$\int_0^1 \|r(t)\|_{\mathcal{L}(\ell_2^n)} \le 25 \cdot \sqrt{n}\,.$$

Proof. Using 17.20, cover $B_{\ell_2^n}$ with $N = 9^n$ closed balls of radius $1/4$ whose centres $x^{(1)},\dots,x^{(N)}$ lie in $B_{\ell_2^n}$. For each $t \in [0,1]$, define

$$|||r(t)||| := \max_{1\le k,\ell\le N} |(r(t)x^{(k)} \,|\, x^{(\ell)})|.$$

We begin our campaign by showing that

$$(*) \qquad \|r(t)\|_{\mathcal{L}(\ell_2^n)} \le 2 \cdot |||r(t)|||.$$

Fix $t \in [0,1]$ and find $x, y \in B_{\ell_2^n}$ so that $\|r(t)\|_{\mathcal{L}(\ell_2^n)} = |(r(t)x \,|\, y)|$. Pick $1 \le k, \ell \le N$ so that $\|x - x^{(k)}\| \le 1/4$ and $\|y - x^{(\ell)}\| \le 1/4$. Then

$$\begin{aligned}\|r(t)\|_{\mathcal{L}(\ell_2^n)} &\le |(r(t)(x - x^{(k)}) \,|\, y)| + |(r(t)x^{(k)} \,|\, y - x^{(\ell)})| + |(r(t)x^{(k)} \,|\, x^{(\ell)})| \\ &\le 2 \cdot \frac{1}{4} \cdot \|r(t)\|_{\mathcal{L}(\ell_2^n)} + |||r(t)|||.\end{aligned}$$

Rearrange to reach $(*)$.

It is relatively simple to obtain distributional inequalities for $|||r(t)|||$. Select $\alpha > 0$. Basic measure theory together with 17.21 yields

$$\begin{aligned}m\big(\{t \in [0,1] : |||r(t)||| > \alpha\}\big) &= m\big(\{t \in [0,1] : \max_{1\le k,\ell\le N} |(r(t)x^{(k)} \,|\, x^{(\ell)})| > \alpha\}\big) \\ &\le \sum_{k,\ell=1}^{N} m\big(\{t \in [0,1] : |(r(t)x^{(k)} \,|\, x^{(\ell)})| > \alpha\}\big) \le 2 \cdot N^2 \cdot e^{-\alpha^2/2}.\end{aligned}$$

From this and $(*)$ we can conclude that

$$m\big(\{t \in [0,1] : \|r(t)\|_{\mathcal{L}(\ell_2^n)} > \alpha\}\big) \le 2 \cdot N^2 \cdot e^{-\alpha^2/8}.$$

Consequently

$$\begin{aligned}\int_0^1 \|r(t)\|_{\mathcal{L}(\ell_2^n)}\, dt &= \int_0^\infty m\big(\{t \in [0,1] : \|r(t)\|_{\mathcal{L}(\ell_2^n)} > \alpha\}\big)\, d\alpha \\ &\le \int_0^{16\sqrt{n}} d\alpha + \int_{16\sqrt{n}}^\infty m\big(\{t \in [0,1] : \|r(t)\|_{\mathcal{L}(\ell_2^n)} > \alpha\}\big)\, d\alpha \\ &\le 16 \cdot \sqrt{n} + 2 \cdot 9^{2n} \cdot \int_{16\sqrt{n}}^\infty e^{-\alpha^2/8}\, d\alpha.\end{aligned}$$

Notice that when $\alpha \ge 16 \cdot \sqrt{n}$ we have $9^{2n} \cdot e^{-\alpha^2/8} \le e^{-\alpha^2/16}$. It follows that

$$\begin{aligned}\int_0^1 \|r(t)\|_{\mathcal{L}(\ell_2^n)}\, dt &\le 16 \cdot \sqrt{n} + 2 \cdot \int_{16\sqrt{n}}^\infty e^{-\alpha^2/16}\, d\alpha \\ &\le 16 \cdot \sqrt{n} + 2 \cdot 2 \cdot \sqrt{\pi} < 25 \cdot \sqrt{n}.\end{aligned}$$

QED

To continue we need to look at yet another version of our mapping I_n. We can identify $\mathcal{L}(\ell_2^n)$ with its dual (trace duality!), and so each $b \in \mathcal{L}(\ell_2^n)^*$ has a matrix representation $(b_{ij})_{i,j=1}^n$. The duality between $\mathcal{L}(\ell_2^n)^*$ and $\mathcal{L}(\ell_2^n)$ is then given by

$$\langle b, a \rangle = \sum_{i,j=1}^n a_{ij} b_{ij},$$

and we have a natural identity

$$J_n : \mathcal{L}(\ell_2^n)^* \longrightarrow \ell_2^{n^2} : b \mapsto \sum_{i,j} b_{ij} e_{ij}$$

which is of course nothing but $(I_n^*)^{-1}$. Lemma 17.22 enables us to estimate $\pi_1(J_n)$.

17.23 Lemma: *Let $n \in \mathbf{N}$. Then*

$$\pi_1(J_n) \leq 25 \cdot A_1^{-1} \cdot n^{1/2}.$$

Proof. Choose a finite family $\{b^{(1)}, \ldots, b^{(N)}\}$ from $\mathcal{L}(\ell_2^n)^*$. Our task is to demonstrate that

$$\sum_{k \leq N} \|J_n b^{(k)}\|_{\ell_2^{n^2}} \leq 25 \cdot A_1^{-1} \cdot n \cdot \|(b^{(k)})_1^N\|_1^{weak}.$$

Now

$$\sum_{k \leq N} \|J_n b^{(k)}\|_{\ell_2^{n^2}} = \sum_{k \leq N} \Big(\Big| \sum_{i,j \leq n} b_{ij}^{(k)} \Big|^2 \Big)^{1/2} = \sum_{k \leq N} \Big(\int_0^1 \Big| \sum_{i,j \leq n} b_{ij}^{(k)} r_{ij}(t) \Big|^2 \, dt \Big)^{1/2}$$

$$\leq A_1^{-1} \cdot \sum_{k \leq N} \int_0^1 \Big| \sum_{i,j \leq n} b_{ij}^{(k)} r_{ij}(t) \Big| \, dt = A_1^{-1} \cdot \int_0^1 \sum_{k \leq N} |\langle b^{(k)}, r(t) \rangle| \, dt$$

$$\leq A_1^{-1} \cdot \|(b^{(k)})_1^N\|_1^{weak} \cdot \int_0^1 \|r(t)\|_{\mathcal{L}(\ell_2^n)} \, dt.$$

The preceding lemma implies the conclusion we want. QED

Our preparations are complete.

Proof of 17.18. (a) We know from 17.19 that $\pi_1(I_n) \leq A_1^{-2} \cdot n$. The other inequality can be broached as follows:

$$n = \pi_2(id_{\ell_2^{n^2}}) \leq \pi_1(id_{\ell_2^{n^2}}) = \pi_1(I_n I_n^{-1}) \leq \pi_1(I_n) \cdot \|I_n^{-1}\|.$$

The factorization $I_n^{-1} : \ell_2^{n^2} \to \Pi_2(\ell_2^n) \to L(\ell_2^n)$, where both factors are natural identities, shows that $\|I_n^{-1}\| \leq 1$. It follows that $\pi_1(I_n) \geq n$.

(b) A similar line gives the lower estimate for $\gamma_1(I_n)$. Using 6.16, we get

$$n^2 = \operatorname{tr}(id_{\ell_2^{n^2}}) = \operatorname{tr}(J_n J_n^{-1}) \leq \gamma_\infty(J_n^{-1}) \cdot \pi_1(J_n)$$

$$= \gamma_\infty(I_n^*) \cdot \pi_1(J_n) = \gamma_1(I_n) \cdot \pi_1(J_n).$$

Harvesting the conclusion of 17.23,

$$\gamma_1(I_n) \geq (A_1/25) \cdot n^{3/2}.$$

The upper estimate is much more elementary. Introduce an artificial factorization

$$I_n : \mathcal{L}(\ell_2^n) \xrightarrow{I_n} \ell_2^{n^2} \xrightarrow{i_2} \ell_1^{n^2} \xrightarrow{i_1} \ell_2^{n^2}$$

where i_1 and i_2 are canonical identities; i_1 has norm 1, and i_2 has norm n. We only have to show that $\|I_n\| \leq \sqrt{n}$ to be able to deduce that $\gamma_1(I_n) \leq n^{3/2}$. But this is easy. If $a = (a_{ij}) \in \mathcal{L}(\ell_2^n)$,

$$\sum_{j=1}^n |a_{ij}|^2 = \|a e_i\|^2 \leq \|a\|_{\mathcal{L}(\ell_2^n)}^2$$

for each $1 \leq i \leq n$, and so

$$\|I_n a\|^2 = \sum_{i,j=1}^{n} |a_{ij}|^2 \leq n \cdot \|a\|^2_{\mathcal{L}(\ell_2^n)},$$

as asserted. QED

Theorem 17.17 can easily be generalized.

17.24 Theorem: *For $p \neq 2$, $\mathcal{S}_p(\ell_2)$ is not a GL-space.*

Proof. By duality, it is enough to consider the case $2 < p \leq \infty$.

Denote by $\mathcal{S}_p^n$ the space $\mathcal{L}(\ell_2^n)$ equipped with the $\mathcal{S}_p$ norm. Note that the equivalence $n^{-1/p} \cdot \|\cdot\|_{\ell_p^n} \leq \|\cdot\|_{\ell_\infty^n} \leq \|\cdot\|_{\ell_p^n}$ can, by the definition of the $\mathcal{S}_p$ norm, be transferred to obtain

$$n^{-1/p} \cdot \|u\|_{\mathcal{S}_p^n} \leq \|u\|_{\mathcal{L}(\ell_2^n)} \leq \|u\|_{\mathcal{S}_p^n}$$

for each $u \in \mathcal{L}(\ell_2^n)$. Consequently, we can follow the setup we used to prove that $\mathcal{L}(\ell_2)$ is not a GL-space. Write

$$I_n^{(p)} : \mathcal{S}_p^n \longrightarrow \mathcal{L}(\ell_2^n) \longrightarrow \ell_2^{n^2}$$

for the 'natural identity' and note that $\pi_1(I_p^{(n)}) \leq \pi_1(I_n) \leq A \cdot n$. In addition write

$$J_n^{(p)} : (\mathcal{S}_p^n)^* \longrightarrow (\mathcal{L}(\ell_2^n))^* \longrightarrow \ell_2^{n^2}$$

for the 'natural identity' which satisfies $J_n^{(p)} = \big((I_n^{(p)})^*\big)^{-1}$. Then

$$\pi_1(J_n^{(p)}) \leq n^{1/p} \cdot \pi_1(J_n) \leq 25 \cdot A_1^{-1} \cdot n^{(1/2)+(1/p)}.$$

Now, just as in the proof of 17.18, $\gamma_1(I_n^{(p)}) \geq (A_1/25) \cdot n^{(3/2)-(1/p)}$. Since $p > 2$, we have $(3/2) - (1/p) > 1$, and our result is proved. QED

Notes and Remarks

A concept of local unconditional structure was first isolated by E. Dubinsky, A. Pełczyński and H.P. Rosenthal [1972]. Their definition: X has l.u.st. if there is a $\lambda > 1$ such that for each $E \in \mathcal{F}_X$ there is an $F \in \mathcal{F}_X$ so that $E \subset F$ and $ub(F) < \lambda$. When we refer in the sequel to this definition of l.u.st. we will call it DPR l.u.st. We have followed Y. Gordon and D.R. Lewis [1974] in our treatment and take note that spaces with DPR l.u.st have l.u.st. T. Figiel, W.B. Johnson and L. Tzafriri [1975] have shown that l.u.st. and DPR l.u.st. are equivalent when a space fails to have finite cotype. But the general problem of whether l.u.st. and DPR l.u.st are equivalent remains open. See Y. Gordon [1980] and C. Schütt [1978] for more on the presence of ℓ_∞^n's and local unconditional structures.

Our presentation of Theorem 17.1 as well as our treatment of the subsequent results 17.2 through 17.6 follows B. Maurey [1973/74b]. The basic inequality 17.7 is a central result found in the fundamental paper [1974] by Y. Gordon and D.R. Lewis. It was a key to their solution of one of Grothendieck's Résumé questions: does every absolutely summing operator factor through an L_1-space?

The isolation of GL-spaces was essentially effected by S. Reisner [1979], but his GL-property for X is defined by requiring that $\Pi_1(X,Y) \subset \Gamma_1(X,Y)$ should hold for every Banach space Y. It seems to be unknown if this is really more restrictive than our definition. Propositions 17.9 and 17.10 and Theorem 17.11 are due to Reisner. Theorems 17.12 and 17.13 can be found in G. Pisier [1985], while Theorem 17.14 is due to B. Maurey [1973/74b].

Having property (GL) is *not* the same as having l.u.st. This was first demonstrated by W.B. Johnson, J. Lindenstrauss and G. Schechtman [1980]; later T. Ketonen [1981] found subspaces of $L_p[0,1]$ $(1 \leq p < 2)$ lacking l.u.st.

Theorem 17.16 is once more a result of G. Pisier [1978c]; he gave the theorem in much more general context but, in principle, the same idea of proof works.

The notion of $\Lambda(p)$-set has its roots in the properties of lacunary sets. R.E.A.C. Paley [1933] proved that these are always $\Lambda(2)$-sets, but it was only in [1960] that W. Rudin formally introduced the concept of a general $\Lambda(p)$-set. Their study has been a fascinating chapter in harmonic analysis, and there was in particular a long struggle to determine whether the $\Lambda(p)$-sets form different classes for each p. The question was settled positively by J. Bourgain [1989].

Theorem 17.17 is due to Y. Gordon and D.R. Lewis [1974], as is the plan of attack we pursue, if not some of the particular steps taken. The space $\mathcal{L}(\ell_2)$ is far from alone among the classical operator spaces lacking l.u.st. As we have seen in 17.24, all the Schatten classes $\mathcal{S}_p(\ell_2)$, $1 \leq p \leq \infty$, $p \neq 2$ share this property. Of course, the Hilbert-Schmidt class, being a Hilbert space itself, has l.u.st.; but, as D.R. Lewis [1975] has shown, the Hilbert-Schmidt class is essentially unique in this regard among the unitarily invariant spaces of compact operators on ℓ_2 (or any other infinite dimensional Hilbert space).

The proof of Theorem 17.17, and so of Theorem 17.24, was based on a number of elementary, but technical, lemmas. We now present an alternative approach due to C. Schütt [1978] and G. Pisier [1978c]. The strategy is based on the observation that if a Banach space doesn't have an unconditional basis, it ought to be easier to show this directly rather than show that it fails the (GL)-condition. This approach has the merit of extracting a positive result from a counterexample. Our presentation follows G. Pisier [1985].

To begin our developments, we recall from 4.5 that the finite rank operators $\ell_2 \to \ell_2$ form a dense subspace in each $\mathcal{S}_p(\ell_2)$. In the usual way, we identify this subspace with $\ell_2 \otimes \ell_2$, the algebraic tensor product of ℓ_2 with itself. The central result is about certain norms on the algebraic tensor product of Banach spaces with unconditional basis.

Let X and Y be Banach spaces with 1-unconditional bases $(x_n)_{n\in\mathbf{N}}$ and $(y_n)_{n\in\mathbf{N}}$, respectively. Let α be a uniform cross norm on the algebraic tensor product $X \otimes Y$ of X and Y, that is, α is a norm on $X \otimes Y$ for which $\alpha(x \otimes y) = \|x\| \cdot \|y\|$ for all $x \in X$ and $y \in Y$ and, given any two operators $u \in \mathcal{L}(X)$ and

$v \in \mathcal{L}(Y)$, their tensor product $u \otimes v$ is a bounded linear operator on $[X \otimes Y, \alpha]$ with

$$(1) \qquad \|u \otimes v\| \leq \|u\| \cdot \|v\|.$$

We denote by $X \widetilde{\otimes}_\alpha Y$ the Banach space completion of $[X \otimes Y, \alpha]$.

We shall use a classical result due to B. Gelbaum and J. Gil de Lamadrid [1961] according to which the $x_m \otimes y_n$'s taken in the rectangular ordering form a basis in $X \widetilde{\otimes}_\alpha Y$. We note further that both X and Y are isometrically isomorphic to complemented subspaces of $X \widetilde{\otimes}_\alpha Y$.

Theorem: *$X \widetilde{\otimes}_\alpha Y$ is a GL-space if and only if $(x_m \otimes y_n)_{(m,n) \in \mathbf{N} \times \mathbf{N}}$ is an unconditional basis for $X \widetilde{\otimes}_\alpha Y$.*

Proof. We only have to worry about the 'only if' part; we are going to show that if $Z := X \widetilde{\otimes}_\alpha Y$ has (GL), then the vectors

$$z_{mn} := x_m \otimes y_n \quad (m, n \in \mathbf{N})$$

form an unconditional basis in Z.

We shall prove the existence of a constant $\lambda \geq 1$ such that regardless of the choice of signs $\varepsilon_{mn} = \pm 1$ and of finitely supported sequences (a_{mn}) of scalars we have

$$(2) \qquad \alpha\Big(\sum_{m,n} \varepsilon_{mn} a_{mn} z_{mn}\Big) \leq \lambda \cdot \alpha\Big(\sum_{m,n} a_{mn} z_{mn}\Big).$$

This signifies that (z_{mn}) is an unconditional basic sequence in $X \widetilde{\otimes}_\alpha Y$; it is obviously total, and so we shall be done.

Fix a finitely supported sequence (a_{mn}) of scalars, and let $N \in \mathbf{N}$ be such that $a_{mn} = 0$ if $m > N$ or $n > N$. Each $\eta \in D_N = \{-1, 1\}^N$ gives rise to norm one operators

$$u_\eta : X \longrightarrow X : \Big(\sum_n a_n x_n\Big) \mapsto \sum_{n \leq N} \eta_n a_n x_n$$

and

$$v_\eta : Y \longrightarrow Y : \Big(\sum_n b_n y_n\Big) \mapsto \sum_{n \leq N} \eta_n b_n y_n.$$

Also, u_η's range, X_0, is the span of $x_1, \ldots, x_N$, and v_η's range, Y_0, is the span of $y_1, \ldots, y_N$. We are going to consider $Z_0 := X_0 \otimes Y_0$ as a subspace of $[X \otimes Y, \alpha]$.

Let $\eta, \eta' \in D_N$ be given. We infer from (1) that

$$\alpha\big((u_\eta \otimes v_{\eta'})(z)\big) \leq \alpha(z) \quad \text{for all } z \in X \otimes Y,$$

so that $u_\eta \otimes v_{\eta'}$ extends to a norm one operator $X \widetilde{\otimes}_\alpha Y \to X \widetilde{\otimes}_\alpha Y$; we continue to denote it by $u_\eta \otimes v_{\eta'}$.

Let (x_n^*) and (y_n^*) be the sequences biorthogonal to (x_n) and (y_n), respectively. The functionals $z_{mn}^* := x_m^* \otimes y_n^* \in Z^*$ are biorthogonal to the z_{mn}'s. For $\eta, \eta' \in D_N$, $z \in Z_0$ and $z^* \in Z_0^*$ we have

$$\begin{aligned} \Big| \sum_{m,n \leq N} \eta_m \cdot \eta'_n \cdot \langle z_{mn}^*, z \rangle \cdot \langle z^*, z_{mn} \rangle \Big| &= \big| \langle z^*, (u_\eta \otimes v_{\eta'})(z) \rangle \big| \\ &\leq \|(u_\eta \otimes v_{\eta'})(z)\|_{Z_0} \cdot \|z^*\|_{Z_0^*}, \end{aligned}$$

and hence

$$(3) \qquad \Big| \sum_{m,n\leq N} \eta_m \cdot \eta'_n \cdot \langle z^*_{mn}, z\rangle \cdot \langle z^*, z_{mn}\rangle \Big| \leq \|z\|_{Z_0} \cdot \|z^*\|_{Z_0^*}.$$

Fix $z_0 \in Z_0$ and $z_0^* \in Z_0^*$. For each $m, n \in \{1,..., N\}$ put

$$\zeta_{mn} := \langle z^*_{mn}, z_0\rangle \quad \text{and} \quad \zeta^*_{mn} := \langle z_0^*, z_{mn}\rangle$$

and note that $z_0 = \sum_{m,n\leq N} \zeta_{mn} z_{mn}$, whereas $z_0^* = \sum_{m,n\leq N} \zeta^*_{mn} z^*_{mn}$. Write $Q := \{1,..., N\} \times \{1,..., N\}$, and define linear operators

$$A : Z \to \ell_2^Q : z \mapsto \big(\zeta^*_{mn}\langle z^*_{mn}, z\rangle\big)_{m,n\leq N}$$

and

$$B : Z^* \to \ell_2^Q : z^* \mapsto \big(\zeta_{mn}\langle z^*, z_{mn}\rangle\big)_{m,n\leq N}.$$

We claim that there is a constant $K > 0$ such that

$$(4) \qquad \pi_1(A) \leq K \cdot \|z_0^*\| \qquad \text{and} \qquad \pi_1(B) \leq K \cdot \|z_0\|.$$

It will turn out that $K = A_1^{-2}$ does the job, where A_1 is once more the constant from Khinchin's Inequality 1.10.

Let c_{mn} $(1 \leq m, n \leq N)$ be arbitrary scalars. Use Khinchin's Inequality and Minkowski's Inequality to write

$$\begin{aligned}
\Big(\sum_{m,n=1}^N |c_{mn}|^2\Big)^{1/2} &= \Big(\int_0^1\int_0^1 \Big|\sum_{m,n} c_{mn} r_m(s) r_n(t)\Big|^2 \, ds\, dt\Big)^{1/2} \\
&\leq A_1^{-1} \cdot \Big(\int_0^1 \Big(\int_0^1 \Big|\sum_{m,n} c_{mn} r_m(s) r_n(t)\Big| \, ds\Big)^2 dt\Big)^{1/2} \\
&\leq A_1^{-1} \cdot \int_0^1 \Big(\int_0^1 \Big|\sum_{m,n} c_{mn} r_m(s) r_n(t)\Big|^2 \, dt\Big)^{1/2} ds.
\end{aligned}$$

A further application of Khinchin's Inequality leads to

$$(5) \qquad \Big(\sum_{m,n=1}^N |c_{mn}|^2\Big)^{1/2} \leq A_1^{-2} \cdot \int_0^1\int_0^1 \Big|\sum_{m,n=1}^N c_{mn} r_m(s) r_n(t)\Big| \, ds\, dt.$$

For $s = 1, \infty$ let Σ_s designate the span of $r_m(\cdot) r_n(\cdot)$ $(1 \leq m, n \leq N)$ in $L_s[0,1]^2$. Plainly, $\pi_1(j) = 1$, where $j : \Sigma_\infty \to \Sigma_1$ is the formal identity.

Consider the operator

$$w : Z_0 \longrightarrow \Sigma_\infty : z \mapsto \sum_{m,n\leq N} \zeta^*_{mn} \langle z^*_{mn}, z\rangle r_m(\cdot) r_n(\cdot)$$

and note that, thanks to (3), $\|wz\|_{\Sigma_\infty} \leq \|z\|_{Z_0} \cdot \|z_0^*\|_{Z_0^*}$ and hence

$$\|w\| \leq \|z_0^*\|_{Z_0^*}.$$

Now we use (5) to see that, for any $z \in Z_0$,

$$\|Az\|_{\ell_2^Q} = \Big(\sum_{m,n=1}^{N} |\zeta_{mn}^* \langle z_{mn}^*, z\rangle|^2 \Big)^{1/2}$$
$$\le K \cdot \int_0^1 \int_0^1 \Big| \sum_{m,n} \zeta_{mn}^* \langle z_{mn}^*, z\rangle r_m(s) r_n(t) \Big| \, ds\, dt$$
$$= K \cdot \int_0^1 \int_0^1 \big| (wz)(s,t) \big| \, ds\, dt = K \cdot \|jwz\|_{\Sigma_1}.$$

It follows that $\pi_1(A) \le K \cdot \pi_1(jw) \le K \cdot \|z_0^*\|$. This is the first inequality in (4). The second is obtained in an analogous fashion.

Next, given $\varepsilon_{mn} = \pm 1$ $(1 \le m, n \le N)$, define the norm one operator

$$d_\varepsilon : \ell_2^Q \longrightarrow \ell_2^Q : (a_{mn})_{m,n} \mapsto (\varepsilon_{mn} a_{mn})_{m,n}.$$

Consider the composition

$$\ell_2^Q \xrightarrow{A^*} Z^* \xrightarrow{B} \ell_2^Q \xrightarrow{d_\varepsilon} \ell_2^Q$$

and use trace duality to obtain

$$\iota_1(d_\varepsilon B A^*) \le \gamma_\infty(A^*) \cdot \pi_1(d_\varepsilon B) \le \gamma_1(A) \cdot \pi_1(B).$$

By hypothesis, Z is a GL-space; so there is a constant $C = C(Z)$ such that $\gamma_1(A) \le C \cdot \pi_1(A)$. Combining this with our earlier estimates, we arrive at

$$\iota_1(d_\varepsilon B A^*) \le C \cdot K^2 \cdot \|z_0\| \cdot \|z_0^*\|.$$

So, with $\lambda := C \cdot K^2$, we find

$$\Big| \Big\langle z_0^*, \sum_{m,n} \varepsilon_{mn} \zeta_{mn} z_{mn} \Big\rangle \Big| = \big| \mathrm{tr}\,(d_\varepsilon B A^*) \big| \le \iota_1(d_\varepsilon B A^*) \le \lambda \cdot \|z_0\| \cdot \|z_0^*\|.$$

Recall that $z_0 \in Z_0$ and $z_0^* \in Z_0^*$ have been chosen arbitrarily. We may conclude that

$$\Big\| \sum_{m,n} \varepsilon_{mn} \zeta_{mn} z_{mn} \Big\| \le \lambda \cdot \|z_0\|$$

and, since $z_0 = \sum_{m,n} \zeta_{mn} z_{mn}$, we are done. QED

Let us show how to get 17.24 from this theorem.

Second Proof of 17.24. It suffices to deal with $2 < p \le \infty$; Proposition 17.9 will then take care of the remaining cases. We shall consider the case of complex scalars; the real case follows suit.

We identify $\ell_2 \otimes \ell_2$ with the dense subspace of finite rank operators in $[\mathcal{S}_p(\ell_2), \sigma_p]$. It is trivial that $[\ell_2 \otimes \ell_2, \sigma_p]$ satisfies condition (1), and clearly the standard unit vector basis (e_n) is 1-unconditional in ℓ_2. In view of the preceding theorem we will be done if we can show that $(e_m \otimes e_n)_{\mathbf{N}\times\mathbf{N}}$ cannot be an unconditional basis in $\mathcal{S}_p(\ell_2)$.

Fix $n \in \mathbf{N}$. By setting

$$u_{k\ell} := n^{-1/2} \cdot e^{2\pi i (k-\ell)/n}$$

for all $1 \le k, \ell \le n$ we define a unitary matrix $u_n = (u_{k\ell})_{k,\ell \le n}$. We consider

u_n as an element of $\ell_2^n \otimes \ell_2^n \subset \ell_2 \otimes \ell_2 \subset \mathcal{S}_p(\ell_2)$ and write accordingly

$$u_n = \sum_{k,\ell \le n} u_{k\ell} e_k \otimes e_\ell.$$

Next we define

$$v_n := \sum_{k,\ell \le n} |u_{k\ell}| e_k \otimes e_\ell.$$

Writing $v_n = n^{-1/2} \cdot \left(\sum_{k \le n} e_k\right) \otimes \left(\sum_{\ell \le n} e_\ell\right)$ we see that

$$\sigma_p(v_n) = n^{-1/2} \cdot \| \sum_{k \le n} e_k \|_2^2 = n^{1/2}.$$

Using 10.3, 2.2, and the fact that u_n is unitary we find that, if $2 < p < \infty$,

$$\begin{aligned} \sigma_p(u_n) = \pi_{p,2}(u_n) &= \sup\left\{ \left(\sum_{k \le n} \|u_n w e_k\|_2^p\right)^{1/p} : w \in \mathcal{L}(\ell_2^n),\ \|w\| \le 1 \right\} \\ &= \sup\left\{ \left(\sum_{k \le n} \|w e_k\|_2^p\right)^{1/p} : w \in \mathcal{L}(\ell_2^n),\ \|w\| \le 1 \right\} = n^{1/p}, \end{aligned}$$

whereas $\sigma_\infty(u_n) = \|u_n\| = 1$ is a triviality.

Suppose now that there exists a $C > 0$ such that

$$\sigma_p\left(\sum_{k,\ell} \eta_{k\ell} a_{k\ell} e_k \otimes e_\ell\right) \le C \cdot \sigma_p\left(\sum_{k,\ell} \eta_{k\ell} a_{k\ell} e_k \otimes e_\ell\right)$$

for all $\sum_{k,\ell} a_{k\ell} e_k \otimes e_\ell \in \mathcal{S}_p(\ell_2)$ and all $|\eta_{k\ell}| = 1$. Then we would be forced to conclude that $n^{1/2} \le C \cdot n^{1/p}$ (or $n^{1/2} \le C$ if $p = \infty$) for all n which cannot be reconciled with $p > 2$. QED

Theorem 17.24 can even be extended as follows:

- *If $p \ne 2$, then $\mathcal{S}_p(\ell_2)$ cannot be embedded into any GL-space Z of finite cotype.*

For the full proof, within the framework of tensor products $X \tilde{\otimes}_\alpha Y$ of Banach spaces with unconditional basis, see G. Pisier [1978c]. We provide a short argument to prove the result when $1 \le p < 2$. One must know, however, that $\mathcal{S}_p(\ell_2)$ has cotype 2, a highly non-trivial fact proved by N. Tomczak-Jaegermann [1974].

For notational convenience, let us write $\mathcal{S}_p$ for $\mathcal{S}_p(\ell_2)$.

Suppose that $\mathcal{S}_p$ is contained in a GL-space Z of finite cotype. By 17.9, Z^*, too, is a GL-space, so $\Pi_1(Z^*, \ell_2) \subset \Gamma_1(Z^*, \ell_2)$ and $\Pi_1^d(\ell_2, Z) \subset \Gamma_\infty(\ell_2, Z)$. Since Z has finite cotype, 11.14 informs us that there is a $2 \le q < \infty$ such that $\Gamma_\infty(\ell_2, Z) \subset \Pi_q(\ell_2, Z)$. It follows that $\Pi_1^d(\ell_2, Z) \subset \Pi_q(\ell_2, Z)$, and this implies $\Pi_1^d(\ell_2, \mathcal{S}_p) \subset \Pi_q(\ell_2, \mathcal{S}_p)$, by injectivity of Π_q. Since $\mathcal{S}_p$ has cotype 2 we have $\Pi_1^d(\ell_2, \mathcal{S}_p) \subset \Pi_2(\ell_2, \mathcal{S}_p)$ by 11.13. Moreover, the Closed Graph Theorem provides us with a constant $K > 0$ such that $\pi_2(w) \le K \cdot \pi_1^d(w)$ for each $w \in \Pi_1^d(\ell_2, \mathcal{S}_p)$.

Now let $u \in \Pi_1(\mathcal{S}_p, \ell_2)$ be given. Combining the preceding observation with the Composition Theorem 5.16, we may state that uw is 1-integral for all $w \in \Pi_1^d(\ell_2, \mathcal{S}_p)$. We claim that this leads to the desired conclusion.

To see why, let E and F be finite dimensional normed spaces and let $s \in \mathcal{L}(E,F)$, $t \in \mathcal{L}(\ell_2, E)$ and $v \in \mathcal{L}(F, \mathcal{S}_p)$ be any operators. Then, a careful use of 5.16 gives

$$\begin{aligned} |\mathrm{tr}\,(stuv)| &= |\mathrm{tr}\,(uvst)| \leq \pi_2(u)\cdot\pi_2(vst) \\ &\leq K\cdot\pi_2(u)\cdot\pi_1^d(vst) \leq K\cdot\pi_1(u)\cdot\|v\|\cdot\|t\|\cdot\pi_1^d(s). \end{aligned}$$

In view of 9.5, we conclude that u is in $\Gamma_1(\mathcal{S}_p, \ell_2) = (\Pi_1^d)^*(\mathcal{S}_p, \ell_2)$ (with $\gamma_1(u) \leq K\cdot\pi_1(u)$). But by 17.24, this is only possible when $p = 2$. QED

More generally, the proof works for any cotype 2 tensor product $X\tilde{\otimes}_\alpha Y$ of Banach spaces with unconditional basis.

In [1974] A. Pełczyński proved that

- *the disk algebra $A(D)$ is not a GL-space.*

He did this by showing that the natural inclusion of $A(D)$ into the Hardy space H^1, which is absolutely summing, cannot factor through any L_1-space. P. Wojtaszczyk [1977] generalized Pełczyński's result to uniform algebras with separable annihilators. In each case, a classical representation theorem of Dunford and Pettis played a crucial rôle.

Curiously, Pełczyński and Wojtaszczyk had already noticed that, by a theorem of R.E.A.C. Paley [1933], there is a 1-summing surjection $q \in \mathcal{L}(A(D), \ell_2)$. Such a map cannot be factorized through any $L_1(\mu)$-space. Indeed, suppose there is a measure μ together with operators $v : L_1(\mu) \to \ell_2$ and $w : A(D) \to L_1(\mu)$ such that $u = vw$. We know that v is 1-summing, hence completely continuous. The crucial step is to show that w is weakly compact; this will force q to be compact and so lead us to the desired contradiction. Thanks to the F. & M. Riesz Theorem, $A(D)^*$ can be identified with the ℓ_1 direct sum of L_1/H_0^1 and the space of measures on the boundary $\mathbf{T}$ of D which are singular with respect to Lebesgue measure. By the Mooney - Havin Theorem, L_1/H_0^1 is weakly sequentially complete and so $A(D)^*$ is weakly sequentially complete, too. But A. Grothendieck [1953a] has shown that every operator from a $\mathcal{C}(K)$-space – like $L_1(\mu)^*$ – to a weakly sequentially complete space is weakly compact. In particular, $w^* : L_1(\mu)^* \to A(D)^*$ is weakly compact, and we are done.

The ingredients of the above argument are just the kind used in S.V. Kisliakov's remarkable solution [1989] of 'Glicksberg's problem':

- *if A is a proper uniform algebra modelled on the compact space K, then A is uncomplemented in $\mathcal{C}(K)$.*

Recall that a uniform algebra is a closed subalgebra of the Banach algebra $\mathcal{C}(K)$ of continuous complex-valued functions on the compact Hausdorff space K that separates the points of K and contains the constant functions. Kisliakov's solution to Glicksberg's problem relies on the following result:

Theorem: *Let A be a proper uniform algebra in $C(K)$. Then there exist operators $v : A(D) \to A^{**}$ and $u : A \to \ell_2$ such that u is 1-summing but $u^{**}v : A(D) \to \ell_2$ is not compact.*

Proof. Look at $\overline{A} := \{\overline{a} : a \in A\}$ where $^-$ indicates complex conjugation. This is a closed linear subspace of $C(K)$. Because A is a proper closed subalgebra of $C(K)$, the Stone-Weierstrass Theorem tells us that $\overline{A}$ cannot be contained in A. Hence there is a $\mu \in C(K)^*$ such that $\langle \mu, a \rangle = 0$ for all $a \in A$ but $\varphi := \mu|_{\overline{A}} \in \overline{A}^*$ is non-zero. By scaling μ appropriately, we may assume that $\|\varphi\| = 1$, and then use the Hahn-Banach Theorem to extend φ to a norm one functional $\lambda \in C(K)^*$. Set $\nu := \lambda - \mu$ and notice that ν vanishes on $\overline{A}$. Pick $a_n \in B_A$ so that $\lim_{n\to\infty} \langle \lambda, \overline{a}_n \rangle = 1$; we can assume that (a_n) converges weakly to some F in $L_2(|\mu| + |\nu|)$ and we can even assume that (a_n) converges to F $(|\mu|+|\nu|)$-a.e. It follows that $|F| \leq 1$ $(|\mu|+|\nu|)$-a.e. But $\overline{a}_n \to \overline{F}$ $(|\mu|+|\nu|)$-a.e., hence λ-a.e., so that $\int \overline{F}\, d\lambda = \lim_n \int \overline{a}_n\, d\lambda = 1$ by the Bounded Convergence Theorem. Consequently, $|F| = 1$ λ-a.e.

Since ν vanishes on $\overline{A}$, $\overline{\nu}$ vanishes on A. Hence each $a_n^2 \overline{\nu}$ vanishes on A, and so $F^2 \overline{\nu}$ vanishes on A, too.

We shall work with

$$\rho := \mu + F^2 \overline{\nu}.$$

Let

$$A_\rho^\perp := A^\perp \cap L_1(\rho)$$

be the subspace of $C(K)^*$ which consists of all measures σ that vanish on A and are absolutely continuous with respect to ρ. Here is a key point of the proof:

Lemma: $\sigma \mapsto \left(\int \overline{F}^{2^k}\, d\sigma\right)_{k\in\mathbf{N}}$ *defines a bounded linear map*

$$w : A_\rho^\perp \longrightarrow \ell_2.$$

We take time out to give a proof of this lemma before applying it to produce the operators u and v of the theorem. Some analytic know-how now becomes indispensible.

Proof of the Lemma. Given $f \in C(\mathbf{T})$, let $H_f : \overline{D} \to \mathbf{C}$ be the function obtained by juxtaposing f with its harmonic extension to D. For each $n \in \mathbf{N}$,

$$v_n : C(\mathbf{T}) \longrightarrow C(K) : f \mapsto H_f \circ a_n$$

is obviously well-defined, linear and bounded with $\|v_n\| \leq 1$. Thanks to the holomorphic calculus, it maps $A(D)$ into A. Its adjoint $v_n^* : C(K)^* \to C(\mathbf{T})^*$ takes each measure $\gamma \in C(K)^*$ to the measure whose value at $f \in C(\mathbf{T})$ is given by $\int_K H_f \circ a_n(\omega)\, d\gamma(\omega)$. Naturally, v_n^* takes $A^\perp$ into $A(D)^\perp$ which, by the F. & M. Riesz Theorem, is just $H_0^1 \cdot m$; here m is normalized Lebesgue measure on the circle $\mathbf{T}$. If $\gamma \in A^\perp$, then

$$\left(\widehat{v_n^*\gamma}(2^k)\right)_k = \left(\int_{\mathbf{T}} \overline{z}^{2^k}\, d(v_n^*\gamma)\right)_k = \left(\int_K \overline{a_n}^{2^k}(\omega)\, d\gamma(\omega)\right)_k \in \ell_2$$

by Paley's Theorem [1933] and moreover, with some constant C,

$$\Big\|\Big(\int_K \overline{a_n}^{2^k}(\omega)\,d\gamma(\omega)\Big)_k\Big\|_{\ell_2} \le C\cdot\|v_n^*\gamma\|_{C(\mathbf{T})^*} \le C\cdot\|\gamma\|_{C(K)^*}.$$

In other words, $\sum_k \big|\int_K \overline{a}_n^{2^k}(\omega)d\gamma(\omega)\big|^2 \le C^2\cdot\|\gamma\|^2$. By choice of F, it follows that $\sum_k \big|\int_K \overline{F}(\omega)^{2^k}d\gamma(\omega)\big|^2 \le C^2\cdot\|\gamma\|^2$, too. QED

Now we are ready to define the operators u and v of the theorem. The first one is easy. Consider $M_\rho : A \to A_\rho^\perp : a \mapsto a\cdot\rho$; this is just the operator induced from the canonical map $\mathcal{C}(K) \to L_1(\rho)$; it is therefore 1-summing. We let $u : A \to \ell_2$ be $w \circ M_\rho$.

For v we return to the operators $v_n : \mathcal{C}(\mathbf{T}) \to \mathcal{C}(K)$ which arose in the lemma; as noted there, each v_n takes $A(D)$ into A. Pass to the operators $k_A v_n : A(D) \to A^{**}$, refine $\mathbf{N}$'s order filter by an ultrafilter $\mathcal{U}$, and for each $f \in A(D)$, define $vf := \lim_{\mathcal{U}} k_A v_n f$ in A^{**}'s weak$*$topology. Thus is born a bounded linear operator $v : A(D) \to A^{**}$.

First notice that for each $j \in \mathbf{N}$, $(u^{**}v)(z^j) \in \ell_2$ is a weak limit point of $(w(a_n^j\cdot\rho))_n$. This comes right out of the definitions of u, v und w. But F is the $(|\mu| + |\nu|)$-a.e. limit of $(a_n)_n$, and all the a_n's and F are $(|\mu| + |\nu|)$-a.e. bounded by one in modulus. So the Bounded Convergence Theorem ensures that $F^j\cdot\rho$ is the norm limit in $A_\rho^\perp$ of $(a_n^j\cdot\rho)_n$. It follows that

$$w(F^j\cdot\rho) = (u^{**}v)(z^j)$$

for each j. To see that this is just what is needed, observe that by definition

$$\big(w(F^{2^n-1}\rho)\,\big|\,e_n\big) = \int F^{2^n-1}\overline{F}^{2^n}\,d\rho = \int (F\overline{F})^{2^n-1}\overline{F}\,d\rho$$

for each n, and so

$$\lim_{n\to\infty}\big(w(F^{2^n-1}\rho)\mid e_n\big) = \int_{\{|F|=1\}} \overline{F}\,d\rho.$$

Now recall that $\rho = \mu + F^2\overline{\nu}$, that $\mu + \nu = \lambda$, and that λ is supported by $\{|F| = 1\}$. Therefore

$$\begin{aligned}\int_{\{|F|=1\}} \overline{F}\,d\rho &= \int_{\{|F|=1\}} \overline{F}\,d\mu + \int_{\{|F|=1\}} \overline{F}F^2\,d\overline{\nu}\\ &= \int_{\{|F|=1\}} \overline{F}\,d\mu + \int_{\{|F|=1\}} F\,d\overline{\nu}\\ &= \int_{\{|F|=1\}} \overline{F}\,d\lambda + 2i\cdot\mathrm{Im}\Big(\int_{\{|F|=1\}} F\,d\overline{\nu}\Big)\\ &= \int \overline{F}\,d\lambda + 2i\cdot\mathrm{Im}\Big(\int_{\{|F|=1\}} F\,d\overline{\nu}\Big) \ne 0\end{aligned}$$

since $\int\overline{F}\,d\lambda = 1$. Were $u^{**}v$ a compact operator, then $\{u^{**}v(z^j) : j \in \mathbf{N}\}$ would be relatively compact in ℓ_2 and so the weakly null sequence (e_n) in ℓ_2 should go to zero uniformly along $\{u^{**}v(z^j) : j \in \mathbf{N}\}$ by the Arzelà - Ascoli Theorem. This does not happen and so $u^{**}v$ is not compact. QED

Corollary: *Proper uniform algebras cannot be GL-spaces.*

Proof. Let A be the uniform algebra in question. Appeal to the theorem to produce $v \in \mathcal{L}(A(D), A^{**})$ and $u \in \Pi_1(A, \ell_2)$ such that $u^{**}v : A(D) \to \ell_2$ is not compact. Suppose that A is a GL-space. Then $u^{**}v$ is in $\Gamma_1(A(D), \ell_2)$, and so there is a factorization $u^{**}v : A(D) \xrightarrow{b} L_1(\mu) \xrightarrow{a} \ell_2$ for some measure μ. By Pełczyński's argument, $u^{**}v$ is compact: contradiction. QED

In particular, since (GL) is inherited by complemented subspaces, no such algebra can ever be complemented in any $C(K)$-space.

Even worse, D.J.H. Garling [1989] showed that a proper uniform algebra is not the continuous linear image of any C^*-algebra. His work relies heavily on Kisliakov's ideas discussed above and on G. Pisier's proof that L^1/H_0^1 is not uniformly PL-convexifiable (see Theorem 6.3 in W.J. Davis, D.J.H. Garling and N. Tomczak-Jaegermann [1984]).

18. Summing Algebras

By a *Banach algebra* we mean a *complex* Banach space X equipped with a continuous multiplication $X \times X \to X : (x,y) \mapsto xy$ which is associative (with respect to both vectors and scalars) and distributive. The continuity of multiplication is equivalent to the existence of a constant $C > 0$ for which

(1) $$\|xy\| \leq C \cdot \|x\| \cdot \|y\| \quad \text{for all } x, y \in X.$$

It is well-known that X can be renormed so as to make $C = 1$ and, should the algebra have a multiplicative identity e, to have $\|e\| = 1$, too. However, this will be of no importance here.

As Banach algebras have their own traditional notation, in which for example * denotes involution rather than duality, we shall modify some of our standard practices in this chapter. Generic Banach spaces will continue to be X, Y,..., their elements will be x, y,..., and their dual spaces X^*, Y^*,..., but the elements of these duals will be φ, ψ,... .

Our fundamental examples are the Banach algebras

- $\mathcal{C}(K)$ of all continuous $\mathbf{C}$-valued functions on a compact Hausdorff space K

and

- $\mathcal{L}(H)$ of all bounded operators on a Hilbert space H

with their usual norms. Multiplication in $\mathcal{C}(K)$ is just pointwise multiplication of functions, whereas multiplication in $\mathcal{L}(H)$ is given by composition of operators. $\mathcal{C}(K)$ is commutative but $\mathcal{L}(H)$ fails to be commutative as soon as H has dimension > 1.

p-Summing Algebras

Let X be a Banach algebra. An immediate consequence of (1) and Hölder's Inequality is that, no matter how we choose $1 \leq p \leq \infty$ and finitely many vectors $x_1,..., x_n$ and $y_1,..., y_n$ from X, we have

$$\Big\| \sum_{k \leq n} x_k y_k \Big\| \leq C \cdot \|(x_k)\|_{p^*}^{strong} \cdot \|(y_k)\|_p^{strong}.$$

Our primary goal is to study Banach algebras satisfying a stronger condition. We say that X is a *p-summing algebra* $(1 \leq p < \infty)$ if there is a constant $c > 0$ such that

(2) $$\Big\| \sum_{k \leq n} x_k y_k \Big\| \leq c \cdot \|(x_k)\|_{p^*}^{strong} \cdot \|(y_k)\|_p^{weak}$$

regardless of the choice of finite collections of vectors $x_1,\dots,x_n$ and $y_1,\dots,y_n$ from X.

The terminology becomes compelling when we shift our perspective. Notice that if X is a Banach algebra and if $\varphi \in X^*$, we can define an operator

$$\widetilde{\varphi} : X \longrightarrow X^* \quad \text{by} \quad \langle \widetilde{\varphi}(x), y \rangle = \langle \varphi, yx \rangle$$

for each $x, y \in X$. Evidently, $\widetilde{\varphi}$ is bounded with $\|\widetilde{\varphi}\| \leq C \cdot \|\varphi\|$ where C is taken from (1).

18.1 Proposition: *Let $1 \leq p < \infty$. A Banach algebra X is p-summing if and only if the operator $\widetilde{\varphi} : X \to X^*$ is p-summing for each $\varphi \in X^*$. In this case, there is a constant $c > 0$ such that*

$$\pi_p(\widetilde{\varphi}) \leq c \cdot \|\varphi\| \tag{3}$$

for each $\varphi \in X^*$. Moreover, the constants c in (2) and (3) can be taken to be the same.

Proof. Assuming that all the operators in question are p-summing and satisfy (3), we select $x_1,\dots,x_n$ and $y_1,\dots,y_n$ from X and choose $\varphi \in B_{X^*}$ so that

$$\Big\|\sum_{k\leq n} x_k y_k\Big\| = \langle \varphi, \sum_{k\leq n} x_k y_k \rangle = \sum_{k\leq n} \langle \widetilde{\varphi} y_k, x_k \rangle.$$

Then

$$\begin{aligned}\Big\|\sum_{k\leq n} x_k y_k\Big\| &\leq \Big(\sum_{k\leq n} \|x_k\|^{p^*}\Big)^{1/p^*} \cdot \Big(\sum_{k\leq n} \|\widetilde{\varphi} y_k\|^p\Big)^{1/p} \\ &\leq \pi_p(\widetilde{\varphi}) \cdot \|(x_k)\|_{p^*}^{strong} \cdot \|(y_k)\|_p^{weak}.\end{aligned}$$

Inequality (2) follows.

Conversely, assume (2) to be true and choose $\varphi \in X^*$ and $y_1,\dots,y_n \in X$. Fixing $\varepsilon > 0$ we can find, for each $1 \leq k \leq n$, a norm one element x_k of X such that

$$\|\widetilde{\varphi} y_k\|^p \leq |\langle \widetilde{\varphi} y_k, x_k \rangle|^p + \frac{\varepsilon^p}{n} = |\langle \varphi, x_k y_k \rangle|^p + \frac{\varepsilon^p}{n}.$$

Next, there is a $\lambda \in B_{\ell_{p^*}^n}$ with

$$\begin{aligned}\Big(\sum_{k\leq n} |\langle \varphi, x_k y_k \rangle|^p\Big)^{1/p} &= \Big|\sum_{k\leq n} \lambda_k \langle \varphi, x_k y_k \rangle\Big| = \Big|\langle \varphi, \sum_{k\leq n} \lambda_k x_k y_k \rangle\Big| \\ &\leq \|\varphi\| \cdot \Big\|\sum_{k\leq n} (\lambda_k x_k) y_k\Big\| \leq c \cdot \|\varphi\| \cdot \|(\lambda_k x_k)\|_{p^*}^{strong} \cdot \|(y_k)\|_p^{weak} \\ &\leq c \cdot \|\varphi\| \cdot \|(y_k)\|_p^{weak}.\end{aligned}$$

Thus $(\sum_{k\leq n} \|\widetilde{\varphi} y_k\|^p)^{1/p} \leq c \cdot \|\varphi\| \cdot \|(y_k)\|_p^{weak} + \varepsilon$. Letting $\varepsilon \to 0$ we get what we wanted. QED

It follows at once from the definition that

- *closed subalgebras of p-summing algebras are p-summing algebras.*

Something which is not immediately apparent from the original definition but which is obvious from Proposition 18.1 is that

- *if $p_1 < p_2$ every p_1-summing algebra is also a p_2-summing algebra.*

Some basic examples are in order.

18.2 Example: (a) *$\mathcal{C}(K)$ is a p-summing algebra for every $1 \le p < \infty$.*
(b) *If we make $\mathcal{C}(K)$ into a Banach algebra by using a non-standard multiplication, it will always be a 2-summing algebra.*

Proof. (a) It is good enough to work with $p = 1$. If $f_1,\dots, f_n$ and $g_1,\dots, g_n$ are in $\mathcal{C}(K)$, then

$$\begin{aligned}\Big\|\sum_{k\le n} f_k g_k\Big\| &= \sup_{\omega\in K}\Big|\sum_{k\le n} f_k(\omega)g_k(\omega)\Big| \le \sup_{\omega\in K}\Big(\sum_{k\le n}|f_k(\omega)|\Big)\cdot\max_{k\le n}\|g_k\| \\ &= \sup_{\omega\in K}\Big(\sum_{k\le n}|\langle\delta_\omega, f_k\rangle|\Big)\cdot\|(g_k)\|_\infty^{strong} \le \|(f_k)\|_1^{weak}\cdot\|(g_k)\|_\infty^{strong}.\end{aligned}$$

(b) This follows from 18.1 since every bounded linear operator from $\mathcal{C}(K)$ to $\mathcal{C}(K)^*$ is 2-summing. QED

The sequence space ℓ_p $(1 \le p < \infty)$ can be made into a Banach algebra by using coordinatewise multiplication: $xy = (x_n y_n)$ when $x = (x_n)$ and $y = (y_n)$.

18.3 Example: *ℓ_p is a 1-summing algebra for $p \le 2$, and a p^*-summing algebra for $2 < p < \infty$.*

In fact, if $\varphi = (\varphi_n)$ is in ℓ_{p^*}, then the map $\widetilde{\varphi}$ is just the diagonal operator $\ell_p \to \ell_{p^*} : (x_n) \mapsto (\varphi_n x_n)$. Notice that $\widetilde{\varphi}$ actually takes values in ℓ_1. Since the natural inclusion $\ell_1 \hookrightarrow \ell_2$ is 1-summing, $\widetilde{\varphi}$ will be 1-summing whenever $p \le 2$. If $2 < p < \infty$, then we may consider $\widetilde{\varphi}$ to be the restriction of a diagonal operator $\ell_\infty \to \ell_{p^*}$ and refer to 2.9(e).

To save the ambitious optimist some time, we mention that if H is an infinite dimensional Hilbert space, then

- *$\mathcal{L}(H)$ cannot be a p-summing algebra for any $1 \le p < \infty$.*

However, this is a delicate result which will be proved much later in the chapter (see 18.22).

Polynomial Inequalities

The definition of a p-summing algebra uses products of pairs of elements. We shall need to keep track of lengthier products, and this leads us to investigate some polynomial inequalities.

Any polynomial q of degree d in n complex variables can be expressed as a sum $q = q_0 + q_1 + \dots + q_d$ where each q_k is a k-homogeneous polynomial of the form

$$q_k(z_1,\dots, z_n) = \sum_{j_1,\dots,j_k=1}^{n} t_{j_1\dots j_k} z_{j_1}\cdots z_{j_k}.$$

There is no loss in assuming that the coefficients $t_{j_1\dots j_k}$ are *symmetric*, that is, that they are unaffected by any reordering of the subscripts. It is often convenient to write $q(z)$ for $q(z_1,\dots, z_n)$ and $q_k(z)$ for $q_k(z_1,\dots, z_n)$ where $z = (z_1,\dots, z_n) \in \mathbf{C}^n$.

We shall study various norms of these polynomials. For $1 \leq p \leq \infty$, set

$$\|q\|_p := \sup\{|q(z)| : z = (z_1,..., z_n) \in B_{\ell_p^n}\}.$$

18.4 Lemma: *If $q = q_0 + ... + q_d$ is a polynomial of degree d in n variables with q_k k-homogeneous for every $0 \leq k \leq d$, then*

$$\|q_k\|_p \leq \|q\|_p$$

for every $0 \leq k \leq d$.

Proof. If $f(\lambda) = \sum_{k=0}^d a_k\lambda^k$, then for every $0 \leq k \leq d$,

$$|a_k| = \left| \int_0^{2\pi} e^{-ik\theta} f(e^{i\theta}) \frac{d\theta}{2\pi} \right| \leq \sup\{|f(e^{i\theta})| : 0 \leq \theta \leq 2\pi\},$$

and the Maximum Principle lets us affirm that $|a_k| \leq \sup\{|f(\lambda)| : |\lambda| \leq 1\}$. Now fix $(z_1,..., z_n) \in B_{\ell_p^n}$ and apply this observation to

$$f(\lambda) = q(\lambda z_1,..., \lambda z_n) = \sum_{k=0}^d \lambda^k q_k(z_1,..., z_n).$$

Then for each $0 \leq k \leq d$,

$$|q_k(z_1,..., z_n)| \leq \sup_{|\lambda|\leq 1} |q(\lambda z_1,..., \lambda z_n)| \leq \|q\|_p.$$

Taking the supremum over all $(z_1,..., z_n)$ in $B_{\ell_p^n}$ gives $\|q_k\|_p \leq \|q\|_p$. QED

Our next step is to associate with each k-homogeneous polynomial in n variables a symmetric k-linear form on $\mathbf{C}^n$. If

$$\hat{t}(z) = \hat{t}(z_1,..., z_n) = \sum_{j_1,...,j_k=1}^n t_{j_1...j_k} z_{j_1} \cdots z_{j_k}$$

with the coefficients $t_{j_1...j_k}$ symmetric, define

$$t : \mathbf{C}^n \times ... \times \mathbf{C}^n \longrightarrow \mathbf{C} \quad (k \text{ factors})$$

by

$$t(z^{(1)},..., z^{(k)}) := \sum_{j_1,...,j_k=1}^n t_{j_1...j_k} z_{j_1}^{(1)} \cdots z_{j_k}^{(k)}.$$

The symmetry of the coefficients means that t is symmetric in the sense that $t(z^{(1)},..., z^{(k)})$ is unaffected by a reordering of the arguments.

The norm of t as a k-linear form on ℓ_p^n is

$$\|t\|_p := \sup\{|t(z^{(1)},..., z^{(k)})| : z^{(1)},..., z^{(k)} \in B_{\ell_p^n}\}.$$

Remarkably, it is easy to compare the norm of a symmetric k-linear form and the norm of the associated k-homogeneous polynomial.

18.5 Lemma: *If t is a symmetric k-linear form on ℓ_p^n associated to the k-homogeneous polynomial $\hat{t}$, then*

$$\|\hat{t}\|_p \leq \|t\|_p \leq e^k \cdot \|\hat{t}\|_p.$$

Proof. The first inequality holds trivially. To obtain the second one, we call yet again on the Rademacher functions. Choose $z^{(1)},..., z^{(k)}$ in $B_{\ell_p^n}$ and observe that the multilinearity of t leads to a multinomial type of expansion for $\hat{t}$: for any $0 \leq s \leq 1$,

$$\hat{t}\Big(\sum_{j\le k} r_j(s)z^{(j)}\Big) = t\Big(\sum_{j\le k} r_j(s)z^{(j)},\dots,\sum_{j\le k} r_j(s)z^{(j)}\Big)$$
$$= \sum \frac{k!}{n_1!\cdots n_k!}\cdot r_1(s)^{n_1}\cdots r_k(s)^{n_k}\cdot t\big(\overbrace{z^{(1)},\dots,z^{(1)}}^{n_1 \text{ copies}},\dots,\overbrace{z^{(k)},\dots,z^{(k)}}^{n_k \text{ copies}}\big)$$

where the summation extends over all integers $n_1,\dots,n_k \ge 0$ such that $\sum_{j\le k} n_j = k$. Multiply through by $r_1(s)\cdots r_k(s)$ and integrate; the orthonormality of the Rademacher functions gives

$$k!\cdot t(z^{(1)},\dots,z^{(k)}) = \int_0^1 r_1(s)\cdots r_k(s)\cdot \hat{t}\big(r_1(s)z^{(1)}+\dots+r_k(s)z^{(k)}\big)\,ds.$$

Now $\|\sum_{j\le k} r_j(s)z^{(j)}\| \le \sum_{j\le k}\|z^{(j)}\| \le k$ for every $0\le s\le 1$, and so

$$|t(z^{(1)},\dots,z^{(k)})| \le \frac{1}{k!}\cdot\int_0^1 \|\hat{t}\|_p\cdot\|\sum_{j\le k} r_j(s)z^{(j)}\|^k ds \le \frac{k^k}{k!}\cdot\|\hat{t}\|_p \le e^k\cdot\|\hat{t}\|_p.$$

QED

The point of this development is that although polynomials are attractive objects, their norms are inherently difficult to estimate. This difficulty lessens when homogeneity is imposed. Decoupling of variables, made possible by considering multilinear forms, starts to make norm estimations tractable. The cost of this passage is loss of precision, but we shall find it worth paying.

If q is a polynomial in n complex variables and X is a *commutative* Banach algebra, then we can make perfect sense of $q(x_1,\dots,x_n)\in X$ by simply replacing each complex variable z_k by the vector x_k in the formula for q. (Interpretational problems arise if we do not have commutativity.)

18.6 Proposition: *A quotient algebra X of a commutative p-summing algebra satisfies each of the following equivalent conditions:*

(i) *There is a constant C_1 such that whenever q is a polynomial in n variables with no constant term and $x_1,\dots,x_n \in X$ satisfy $\|(x_j)\|_{p^*}^{strong} \le C_1$, we have*
$$\|q(x_1,\dots,x_n)\| \le \|q\|_{p^*}.$$

(ii) *There is a constant C_2 such that whenever q_k is a k-homogeneous polynomial in n variables and $x_1,\dots,x_n\in X$ satisfy $\|(x_j)\|_{p^*}^{strong}\le 1$, we have*
$$\|q_k(x_1,\dots,x_n)\| \le C_2^k\cdot\|q_k\|_{p^*}.$$

(iii) *There is a constant C_3 such that whenever t is a symmetric k-linear form on $\mathbf{C}^n$ and $x_1,\dots,x_n\in X$ satisfy $\|(x_j)\|_{p^*}^{strong}\le 1$, we have*
$$\Big\|\sum_{j_1,\dots,j_k=1}^{n} t_{j_1\dots j_k}x_{j_1}\cdots x_{j_k}\Big\| \le C_3^k\cdot\|t\|_{p^*}.$$

Proof. It is enough to establish these conditions when X is a commutative p-summing algebra, since each is plainly passed from the algebra to any quotient algebra.

We begin by showing that (iii) holds, and then prove that (i), (ii), (iii) are in fact equivalent conditions.

Take any symmetric k-linear form t on $\mathbf{C}^n$ and any $x_1,\dots,x_n \in X$ with $\|(x_j)\|_{p^*}^{strong} \leq 1$. The definition of a p-summing algebra tells us that, for some $C > 0$,

$$\Big\| \sum_{j_1,\dots,j_k=1}^{n} t_{j_1\dots j_k} x_{j_1} \cdots x_{j_k} \Big\| = \Big\| \sum_{j_k=1}^{n} \Big(\sum_{j_1,\dots,j_{k-1}=1}^{n} t_{j_1\dots j_k} x_{j_1} \cdots x_{j_{k-1}} \Big) x_{j_k} \Big\|$$
$$\leq C \cdot \sup\Big\{ \Big\| \sum_{j_1,\dots,j_{k-1}=1}^{n} \Big(\sum_{j_k=1}^{n} \lambda_{j_k} t_{j_1\dots j_k} \Big) x_{j_1} \cdots x_{j_{k-1}} \Big\| : \lambda \in B_{\ell_{p^*}^n} \Big\}.$$

Iterating the procedure we get down to

$$\Big\| \sum_{j_1,\dots,j_k=1}^{n} t_{j_1\dots j_k} x_{j_1} \cdots x_{j_k} \Big\|$$
$$\leq C^{k-1} \cdot \sup\Big\{ \Big\| \sum_{j_1=1}^{n} \Big(\sum_{j_2,\dots,j_k=1}^{n} t_{j_1\dots j_k} \lambda_{j_2}^{(2)} \cdots \lambda_{j_k}^{(k)} \Big) x_{j_1} \Big\| : \lambda^{(2)},\dots,\lambda^{(k)} \in B_{\ell_{p^*}^n} \Big\}$$
$$= C^{k-1} \cdot \sup\Big\{ \Big| \sum_{j_1,\dots,j_k=1}^{n} t_{j_1\dots j_k} \langle x_{j_1}, \varphi \rangle \lambda_{j_2}^{(2)} \cdots \lambda_{j_k}^{(k)} \Big| \Big\}$$

where the final supremum is taken over all $\lambda^{(2)},\dots,\lambda^{(k)} \in B_{\ell_{p^*}^n}$ and all $\varphi \in B_{X^*}$.

Now for the equivalences. By specialization, (i)$\Rightarrow$(ii) with $C_2 = C_1^{-1}$. Trivially, (ii)$\Rightarrow$(iii) with $C_3 = C_2$, whereas Lemma 18.5 gives (iii)$\Rightarrow$(ii) with $C_2 = e \cdot C_3$. Finally, that (ii)$\Rightarrow$(i) with $C_1 = (2 \cdot C_2)^{-1}$ follows from Lemma 18.4, since $\|(x_j)\|_{p^*}^{strong} \leq (2 \cdot C_2)^{-1}$ ensures that $\|q_k(x_1,\dots,x_n)\| \leq 2^{-k} \cdot \|q_k\|_{p^*}$, and consequently,

$$\|q(x_1,\dots,x_n)\| \leq \sum_{k \leq d} \|q_k(x_1,\dots,x_n)\| \leq \sum_{k \leq d} 2^{-k} \cdot \|q_k\|_{p^*} \leq \|q\|_{p^*}. \qquad \text{QED}$$

Q-Algebras and Operator Algebras

A Banach algebra which is isometrically isomorphic, as a Banach algebra, to a closed subalgebra of $\mathcal{C}(K)$ for some compact Hausdorff space K is called a *uniform algebra.* By a *Q-algebra* we mean a commutative Banach algebra isomorph of a quotient algebra of a uniform algebra.

- Notice that in this chapter we do not require a uniform algebra to contain constants or separate points.

18.7 Theorem: *The following are equivalent statements about a commutative Banach algebra X:*

(i) *X is a Q-algebra.*

(ii) *X is isomorphic, as a Banach algebra, to a quotient algebra of a commutative 1-summing algebra.*

(iii) *There is a constant $c > 0$ such that for any $n \in \mathbf{N}$, any polynomial q in n variables without constant term and any $x_1, \ldots, x_n \in X$ of norm at most c, we have*
$$\|q(x_1, \ldots, x_n)\| \leq \|q\|_\infty.$$

Proof. We see from 18.2(a) that uniform algebras are 1-summing algebras, and so that (i) implies (ii). That (iii) is a consequence of (ii) follows from Proposition 18.6, so we only need to show how (i) is derived from (iii).

We base our Q-algebra X on a compact space K found as follows. Let $\Gamma = \{x \in X : \|x\| \leq c\}$, consider ℓ_∞^Γ as the dual of ℓ_1^Γ, and let K be $B_{\ell_\infty^\Gamma}$ in its weak$*$topology. Note that each $x \in \Gamma$ defines $e_x \in \mathcal{C}(K)$ where $e_x(\omega) = \langle \delta_\omega, x \rangle$ for all $\omega \in K$. The set
$$Y_0 := \bigcup_{n \in \mathbf{N}} \{q(e_{x_1}, \ldots, e_{x_n}) : q \text{ a polynomial in } n \text{ variables}, x_1, \ldots, x_n \in \Gamma\}$$
is a subalgebra of $\mathcal{C}(K)$. We denote its closure in $\mathcal{C}(K)$ by Y: this is a uniform algebra. Now we notice that $q(e_{x_1}, \ldots, e_{x_n}) \mapsto q(x_1, \ldots, x_n)$ establishes a well-defined map $u : Y_0 \to X$ which is clearly linear, multiplicative and onto. Thanks to (iii), u is continuous and so is the restriction of a continuous, linear, multiplicative surjection $Y \to X$. By the Open Mapping Theorem, X is isomorphic to a quotient algebra of the uniform algebra Y. QED

If K is a compact Hausdorff space and $\mu \in \mathcal{C}(K)^*$ is any probability measure, then $\mathcal{C}(K)$ is isometrically isomorphic as a Banach algebra to a closed subalgebra of $\mathcal{L}(L_2(\mu))$: simply associate to each $f \in \mathcal{C}(K)$ the operator $L_2(\mu) \to L_2(\mu) : g \mapsto fg$.

A Banach algebra is called an *operator algebra* if it is isomorphic, as a Banach algebra, to a closed subalgebra of $\mathcal{L}(H)$ for some Hilbert space H. In this jargon, then, uniform algebras are operator algebras.

Actually, even Q-algebras are operator algebras and, in fact, a more precise result is available:

18.8 Theorem: *A quotient algebra X of a uniform algebra is isometrically isomorphic as a Banach algebra to a closed subalgebra of $\mathcal{L}(H)$ for an appropriate Hilbert space H.*

Proof. Let $X = Y/N$ where Y is a closed subalgebra of $\mathcal{C}(K)$ for some compact Hausdorff space K and N is a closed ideal in Y.

It is essential for our argument below to assume that Y contains 1_K, the identically one function. This causes no harm since if Y_1 denotes the closed subalgebra of $\mathcal{C}(K)$ generated by Y and 1_K, then N is also a closed ideal of Y_1 and X is a closed subalgebra of Y_1/N. Consequently, if we can prove our theorem for Y_1/N, it also follows for X.

We begin by outlining our general strategy. If $\mu \in \mathcal{C}(K)^*$ is a probability measure, we write Y_μ for the closure of Y and N_μ for the closure of N in $L_2(\mu)$. Let H_μ be N_μ's orthogonal complement inside Y_μ and designate $p_\mu : Y_\mu \to H_\mu$

to be the corresponding orthogonal projection. The Hilbert space we require is the ℓ_2 direct sum

$$H := \left(\bigoplus\nolimits_{\mu\in W} H_\mu\right)_2$$

indexed by the set W of all probability measures $\mu \in C(K)^*$.

Our task is to construct a norm preserving algebra homomorphism

$$u : X \longrightarrow \mathcal{L}(H).$$

We shall build this up from multiplicative operators $u_\mu : X \to \mathcal{L}(H_\mu)$, each with norm ≤ 1, by setting

$$u(x)\big((h_\mu)_{\mu\in W}\big) := \big(u_\mu(x)h_\mu\big)_{\mu\in W}$$

for every $x \in X$ and $(h_\mu)_{\mu\in W} \in H$. It is a mindless exercise to verify that u really is a multiplicative linear map and that $\|ux\| \leq \|x\|$ for each $x \in X$. The final step will be to show that for each $x \in X$ there is a $\mu_x \in W$ such that $\|u_{\mu_x}(x)\| = \|x\|$, and so that u is indeed norm preserving.

To embark on the details of this program, we define a product

$$yz \in Y_\mu$$

for each $y \in Y$ and $z \in Y_\mu$. To achieve this, notice that z is the limit in $L_2(\mu)$ of a sequence (z_n) in Y. The sequence (yz_n) in $Y \subset Y_\mu$ is a Cauchy sequence in Y_μ since

$$\begin{aligned}\|yz_m - yz_n\|_{Y_\mu}^2 &= \int_K |y(\omega)|^2 \cdot |z_m(\omega) - z_n(\omega)|^2 d\mu(\omega)\\ &\leq \|y\|_Y^2 \cdot \|z_m - z_n\|_{Y_\mu}^2.\end{aligned}$$

Accordingly – after appeasing our conscience on the matter of well-definedness – we can set

$$yz := \lim_{n\to\infty} yz_n,$$

the limit being taken in Y_μ. The above computations naturally extend to give

$$\|yz\|_{Y_\mu} \leq \|y\|_Y \cdot \|z\|_{Y_\mu}.$$

Now if $y \in Y$ we can define $v_\mu(y) \in \mathcal{L}(H_\mu)$ by

$$v_\mu(y)(h_\mu) := p_\mu(yh_\mu) \quad \text{for all} \quad h_\mu \in H_\mu.$$

Since it is clear that $\|v_\mu(y)\| \leq \|y\|$, we can assert that the map

$$v_\mu : Y \longrightarrow \mathcal{L}(H_\mu) : y \mapsto v_\mu(y)$$

is an operator with norm at most one.

This operator is also multiplicative. To see why, first observe that if $y_1, y_2 \in Y$, then

$$v_\mu(y_1)v_\mu(y_2)(h_\mu) = v_\mu(y_1)(p_\mu(y_2h_\mu)) = p_\mu(y_1p_\mu(y_2h_\mu))$$

for all $h_\mu \in H_\mu$. So, by writing $y_2h_\mu = p_\mu(y_2h_\mu) + z \in H_\mu \oplus N_\mu$, we obtain

$$v_\mu(y_1)v_\mu(y_2)(h_\mu) = p_\mu(y_1y_2h_\mu - y_1z) = p_\mu(y_1y_2h_\mu) = v_\mu(y_1y_2)(h_\mu).$$

The ideal property of N entails that v_μ vanishes on N, and so that

$$u_\mu : X \longrightarrow \mathcal{L}(H_\mu) : y + N \mapsto v_\mu(y)$$

is well-defined. It is straightforward to establish that u_μ is a multiplicative operator with $\|u_\mu\| \leq 1$.

The operators u_μ are what we have been seeking. All that remains is to produce the measures μ_x, and for this it is convenient to normalize and to take $\|x\| = 1$. Then the Hahn-Banach Theorem provides us with a $\nu \in X^*$ satisfying $\|\nu\| = \langle \nu, x\rangle = 1$. Remarking that $X^* = (Y/N)^*$ is the annihilator of N in Y^*, we can view ν as a measure in $\mathcal{C}(K)^*$ with the property that $\int_K z\, d\nu = 0$ for every $z \in N$. The measure μ_x is just $|\nu|$.

Let $x = y + N$. By definition of the quotient norm, we can extract a sequence (z_n) from N such that $\lim_{n\to\infty} \|y + z_n\|_Y = 1$. By construction, the sequence (z_n) is bounded in $L_\infty(|\nu|)$ and hence it has a weak$*$limit point $\hat{z} \in L_\infty(|\nu|)$ and $\|y + \hat{z}\|_{L_\infty(|\nu|)} \leq 1$. The function $y + \hat{z}$ has useful properties:

(a) $(y + \hat{z})\cdot\nu = |\nu|$,

(b) $|y + \hat{z}| = 1$ ν-almost everywhere,

(c) $p_{|\nu|}(y) = y + \hat{z}$.

Properties (a) and (b) both follow from the observation that $1 = \int_K y\, d\nu = \int_K (y + \hat{z})\, d\nu \leq 1$ is in fact a chain of equalities. Property (c) is immediate from (a) and (b) via the identity $(y + \hat{z} \mid z)_{L_2(|\nu|)} = \int_K \overline{(y + \hat{z})} z\, d|\nu| = \int_K z\, d\nu = 0$ for every $z \in N$.

Exploiting the presence of constant functions in Y, we can now assert that $u_{|\nu|}(x)(1_K) = v_{|\nu|}(y)(1_K) = p_{|\nu|}(y) = y + \hat{z}$ and so

$$\|u_{|\nu|}(x)\|^2 \geq \|y + \hat{z}\|^2_{L_2(|\nu|)} = \int_K (y + \hat{z})\overline{(y + \hat{z})}\, d|\nu| = \int_K (y + \hat{z})\, d\nu = 1.$$

We are done. QED

A Commutative Non-Operator Algebra

After this effort, it is sensible to ask if we have really proved anything. Could it be that all commutative Banach algebras are operator algebras? Luckily, the answer is no, but surprisingly the examples needed are quite delicate. It is, however, very easy to see that commutative Banach algebras need not be uniform algebras. Just take a Banach space X, equip it with the trivial multiplication where all products are zero, and observe that $\|x\|^2 = \|x^2\|$ if and only if $x \in X$ is the zero vector. To get a handle on operator algebras, we call once again on Grothendieck's Inequality:

18.9 Proposition: *If X is an operator algebra, then there is a constant $C > 0$ such that*

$$\Big\| \sum_{i,j=1}^{n} t_{ij} x_i y_j \Big\| \leq C \cdot \|t\|_\infty,$$

no matter how we choose the integer n, the $n \times n$ matrix (t_{ij}), and the elements $x_1,\dots, x_n$ and $y_1,\dots, y_n$ of B_X.

Proof. Let X be a subalgebra of $\mathcal{L}(H)$, where H is a Hilbert space. Then

$$\Big\|\sum_{i,j=1}^{n} t_{ij}x_iy_j\Big\| = \sup_{h,k\in B_H}\Big|\sum_{i,j=1}^{n} t_{ij}(x_iy_jh|k)\Big| = \sup_{h,k\in B_H}\Big|\sum_{i,j=1}^{n} t_{ij}(y_jh|x_i^*k)\Big|$$

$$\leq \sup\Big\{\Big|\sum_{i,j=1}^{n} t_{ij}(h_j|k_i)\Big| : h_1,\dots,h_n,k_1,\dots,k_n \in B_H\Big\} \leq \kappa_G\cdot\|t\|_\infty,$$

where κ_G is Grothendieck's constant. QED

Notice the similarity of this condition with condition (iii) of 18.6.

Let $\mathbf{T} := \{z \in \mathbf{C} : |z| = 1\}$ be the unit circle in the complex plane. The *Wiener algebra*

$$W(\mathbf{T})$$

consists of all 2π-periodic functions $f : \mathbf{R} \to \mathbf{C}$ of the form

$$f(\theta) = \sum_{n=-\infty}^{\infty} a_n e^{in\theta}$$

with

$$\|f\| := \sum_{n=-\infty}^{\infty} |a_n| < \infty;$$

it is thus isometrically isomorphic, as a Banach space, to $\ell_1^{\mathbf{Z}}$. $W(\mathbf{T})$ is a commutative Banach algebra with respect to pointwise multiplication $(fg)(\theta) = f(\theta)g(\theta)$; this multiplication corresponds to the formation of Cauchy products in $\ell_1^{\mathbf{Z}}$.

To show that $W(\mathbf{T})$ is not an operator algebra, we introduce some notable matrices. These have been used in a different way in 5.13, and they have similar properties to the majority of 'Rademacher matrices' considered in 17.22.

18.10 Lemma: *Let $t = (t_{jk})_{j,k=1}^n$ be given by $t_{jk} = \exp(2\pi ijk/n)$. Then*

$$\|t\|_2 \leq n^{1/2} \quad \textit{and} \quad \|t\|_\infty \leq n^{3/2}.$$

Proof. It is enough to show that $\|t\|_2 \leq n^{1/2}$, since if $x, y \in B_{\ell_\infty^n}$, then $n^{-1/2}x,\ n^{-1/2}y \in B_{\ell_2^n}$ and so

$$|t(x,y)| = n\cdot|t(n^{-1/2}x, n^{-1/2}y)| \leq n\cdot\|t\|_2.$$

Consequently, $\|t\|_\infty \leq n\cdot\|t\|_2$.

Note that $\|t\|_2 = \sup\{(\sum_{j\leq n}|\sum_{k\leq n} t_{jk}x_k|^2)^{1/2} : x \in B_{\ell_2^n}\}$. If $x \in B_{\ell_2^n}$, then, using the properties of n-th roots of unity, we get

$$\sum_j\Big|\sum_k t_{jk}x_k\Big|^2 = \sum_{j,k,m} t_{jk}\overline{t_{jm}}x_k\overline{x_m} = \sum_{k,m}\Big(\sum_j e^{2\pi ij(k-m)/n}\Big)x_k\overline{x_m}$$

$$= \sum_{k,m} n\cdot\delta_{km}x_k\overline{x_m} = n\cdot\|x\|_2^2 \leq 1.$$

Our conclusion is now immediate. QED

18.11 Proposition: *The Wiener algebra is not an operator algebra.*

Proof. Define $x_j \in W(\mathbf{T})$ by $x_j(\theta) = \exp(ij\theta)\cdot\exp(\pi ij^2/n)$ $(1 \leq j \leq n)$. Then, if (t_{jk}) is the matrix introduced above,

$$\Big\| \sum_{j,k=1}^{n} t_{jk} x_j x_k \Big\| = \Big\| \sum_{j,k=1}^{n} e^{2\pi ijk/n} e^{\pi i(j^2+k^2)/n} e^{i(j+k)\theta} \Big\|$$
$$= \Big\| \sum_{j,k=1}^{n} e^{\pi i(j+k)^2/n} e^{i(j+k)\theta} \Big\| = \Big\| \sum_{m=2}^{2n} a_m e^{\pi i m^2/n} e^{im\theta} \Big\|$$

where $a_m = \sum_{j+k=m} 1$. Evidently, $a_m = m-1$ for $2 \le m \le n+1$, and so

$$\Big\| \sum_{j,k=1}^{n} t_{jk} x_j x_k \Big\| = \sum_{m=2}^{2n} \big| a_m e^{\pi i m^2/n} \big| \ge \sum_{m=2}^{n+1} (m-1) = \frac{n(n+1)}{2} .$$

However, Lemma 18.10 shows that $\|t\|_\infty \le n^{3/2}$, and so, letting $n \to \infty$, we find that the condition of Proposition 18.9 cannot be satisfied. QED

Notice that the last result shows directly that commutative Banach algebras need not be Q-algebras since it gives an example where condition (iii) of Proposition 18.7 is violated.

FAILURE OF THE MANY VARIABLE VON NEUMANN INEQUALITY

The next natural question is whether all commutative operator algebras must be Q-algebras. At first sight this is believable: a famous inequality of von Neumann asserts that

$$\|q(u)\| \le \|q\|_\infty$$

whenever q is a polynomial and u belongs to $B_{\mathcal{L}(H)}$, and this is just the ticket for condition (iii) of 18.7 when $n = 1$.

When $n = 2$ things still look good: another famous result, this time due to Andô, asserts that

$$\|q(u_1, u_2)\| \le \|q\|_\infty$$

whenever q is a polynomial in two variables and u_1, u_2 are *commuting* elements of $B_{\mathcal{L}(H)}$. However, a strikingly simple argument using our knowledge that Grothendieck's constant κ_G is strictly greater than one shows that there is a limit to such good behaviour.

18.12 Proposition: *For large n, there are commuting norm one operators $u_1, \dots, u_n$ on a Hilbert space H and a 2-homogeneous polynomial q in n variables such that $\|q(u_1, \dots, u_n)\| > \|q\|_\infty$.*

Proof. Let H be the ℓ_2 direct sum $H = [e] \oplus_2 \ell_2 \oplus_2 [f]$ where e and f are any norm one vectors. If $y = (y_n) \in \ell_2$, define $\overline{y} = (\overline{y}_n) \in \ell_2$; trivially $\|y\|_2 = \|\overline{y}\|_2$. Now for each $x \in \ell_2$, define $u_x \in \mathcal{L}(H)$ via

$$u_x e = x \,, \quad u_x f = 0 \ \text{ and } \ u_x y = (x|\overline{y}) f \text{ for } y \in \ell_2 .$$

Then $\|u_x\| \le \|x\|$ for every $x \in \ell_2$ since

$$\|u_x(\lambda e + y + \mu f)\|^2 = \|\lambda x + (x|\overline{y}) f\|^2 = |\lambda|^2 \cdot \|x\|^2 + |(x|\overline{y})|^2$$
$$\le (|\lambda|^2 + \|y\|^2) \cdot \|x\|^2 \le \|\lambda e + y + \mu f\|^2 \cdot \|x\|^2$$

for every $y \in \ell_2$ and $\lambda, \mu \in \mathbf{C}$. Moreover, u_{x_1} and u_{x_2} commute for each $x_1, x_2 \in \ell_2$; indeed,

$$u_{x_1} u_{x_2} e = u_{x_1}(x_2) = (x_1|\overline{x}_2) f = (x_2|\overline{x}_1) f = u_{x_2}(x_1) = u_{x_2} u_{x_1} e$$

and, for any $z \in \ell_2 \oplus [f]$, $u_{x_1} u_{x_2} z = 0 = u_{x_2} u_{x_1} z$.

Now use Grothendieck's Inequality to produce an $n \times n$ matrix (t_{jk}) and norm one vectors $x_1,..., x_n, y_1,..., y_n$ in ℓ_2 so that

$$\Big| \sum_{j,k=1}^{n} t_{jk}(x_j|\overline{y}_k) \Big| > \|t\|_\infty$$
$$= \sup \Big\{ \Big| \sum_{j,k=1}^{n} t_{jk} \lambda_j \mu_k \Big| : |\lambda_j| \le 1, \ |\mu_k| \le 1, \ 1 \le j, k \le n \Big\}.$$

Set $w_j = u_{x_j}$, $w_{n+j} = u_{y_j}$, $(1 \le j \le n)$, and define

$$q(z_1,..., z_n) = \sum_{j,k=1}^{n} t_{jk} z_j z_{k+n}.$$

Then

$$\|q(w_1,..., w_{2n})\| \ge \|q(w_1,..., w_{2n})e\| = \Big\| \sum_{j,k} t_{jk} w_j w_{k+n} e \Big\|$$
$$= \Big\| \big(\sum_{j,k} t_{jk} u_{x_j} u_{y_k}\big) e \Big\| = \Big\| \sum_{j,k} t_{jk}(x_j|\overline{y}_k) f \Big\| > \|t\|_\infty = \|q\|_\infty. \quad \text{QED}$$

Although this does not in itself violate condition (iii) of 18.7, it indicates that there might be commutative operator algebras which are not Q-algebras. Before showing that this is in fact so, we give a positive result to illustrate the delicacy of the situation.

18.13 Proposition: *Let X be a commutative Banach algebra. There is a constant $C > 0$ such that, no matter how we choose finitely many vectors $x_1,..., x_n$ in X satisfying $(\sum_{j \le n} \|x_j\|^2)^{1/2} \le C$ and a polynomial q in n variables with no constant term, we have*

$$\|q(x_1,..., x_n)\| \le \|q\|_\infty.$$

The proof follows the familiar homogeneity ploy, and rests on a much more concrete piece of information.

18.14 Lemma: *If t is any k-linear form on ℓ_∞^n, then its coefficients satisfy*

$$\Big(\sum_{j_1,...,j_k=1}^{n} |t_{j_1...j_k}|^2 \Big)^{1/2} \le A_1^{-1} \cdot \|t\|_\infty$$

where A_1 is the constant from Khinchin's Inequality.

Proof. It is helpful to note that

$$\|t\|_\infty = \sup |t(a_1,..., a_k)|,$$

where the supremum is taken over all choices of extreme points $a_1,..., a_k$ of $B_{\ell_\infty^n}$. These extreme points have the property that each coordinate has absolute value one.

So fix extreme points $a_1,\dots,a_k$ in $B_{\ell_\infty^n}$ and for each $s_1 \in [0,1]$ define

$$a_1(s_1) := \left(r_j(s_1)a_{1j}\right)_{j=1}^n \in B_{\ell_\infty^n},$$

where $r_1,\dots,r_n$ are the first n Rademacher functions. Since

$$\|t\|_\infty \geq \left|t(a_1(s_1),a_2,\dots,a_k)\right|$$

for each $s_1 \in [0,1]$, we certainly have the coarser inequality

$$\begin{aligned}\|t\|_\infty^2 &\geq \left(\int_0^1 \left|t(a_1(s_1),a_2,\dots,a_k)\right|\, ds_1\right)^2 \\ &= \left(\int_0^1 \Big|\sum_{j_1=1}^n\Big(\sum_{j_2,\dots,j_k=1}^n t_{j_1\dots j_k}a_{2j_2}\cdots a_{kj_k}\Big)a_{1j_1}r_{j_1}(s_1)\Big|\, ds_1\right)^2.\end{aligned}$$

We seize the opportunity to use Khinchin's Inequality. In light of $|a_{1j_1}| = 1$ $(1 \leq j_1 \leq n)$, this tells us that

$$\|t\|_\infty^2 \geq A_1^2\cdot\sum_{j_1=1}^n\Big|\sum_{j_2,\dots,j_k=1}^n t_{j_1\dots j_k}a_{2j_2}\cdots a_{kj_k}\Big|^2.$$

Now if we define

$$a_2(s_2) := \left(r_j(s_2)a_{2j}\right)_{j=1}^n \in B_{\ell_\infty^n},$$

then the above procedure shows that

$$\|t\|_\infty^2 \geq A_1^2\cdot\sum_{j_1=1}^n\Big|\sum_{j_2,\dots,j_k=1}^n t_{j_1\dots j_k}a_{2j_2}(s_2)a_{3j_3}\cdots a_{kj_k}\Big|^2$$

for every $s_2 \in [0,1]$. Integrating again we obtain

$$\begin{aligned}\|t\|_\infty^2 &\geq A_1^2\cdot\sum_{j_1=1}^n\int_0^1\Big|\sum_{j_2=1}^n\Big(\sum_{j_3,\dots,j_k=1}^n t_{j_1\dots j_k}a_{3j_3}\cdots a_{kj_k}\Big)a_{2j_2}r_{j_2}(s_2)\Big|^2\, ds_2 \\ &= A_1^2\cdot\sum_{j_1=1}^n\sum_{j_2=1}^n\Big|\sum_{j_3,\dots,j_k=1}^n t_{j_1\dots j_k}a_{3j_3}\cdots a_{kj_k}\Big|^2.\end{aligned}$$

Iterating, we eventually arrive where we wanted to be. QED

Proof of 18.13. Assume for convenience that $\|xy\| \leq \|x\|\cdot\|y\|$ for every $x,y \in X$. Then choose a k-homogeneous polynomial q_k in n variables and select $x_1,\dots,x_n$ in X satisfying $(\sum_{j\leq n}\|x_j\|^2)^{1/2} \leq 1/(2e)$.

If $q_k(z_1,\dots,z_n) = \sum_{j_1,\dots,j_k\leq n} t_{j_1\dots j_k}z_{j_1}\cdots z_{j_k}$ with $(t_{j_1\dots j_k})$ symmetric, then

$$\|(q_k(x_1,\dots,x_n)\| \leq \Big(\sum_{j_1,\dots,j_k=1}^n |t_{j_1\dots j_k}|^2\Big)^{1/2}\cdot\Big(\sum_{j_1,\dots,j_k=1}^n \|x_{j_1}\|^2\cdots\|x_{j_k}\|^2\Big)^{1/2}$$

by the Cauchy-Schwarz Inequality. Invoke the lemma and then 18.5 to get

$$\begin{aligned}\|q_k(x_1,\dots,x_n)\| &\leq A_1^{-1}\cdot\|t\|_\infty\cdot\Big(\sum_{j\leq n}\|x_j\|^2\Big)^{k/2} \\ &\leq A_1^{-1}\cdot e^k\cdot\|q_k\|_\infty\cdot(2e)^{-k} \leq 2^{-k}\cdot\|q_k\|_\infty.\end{aligned}$$

Finally, if q is any polynomial in n variables without constant term, write it in the form $q = q_1+\dots+q_d$ where q_k $(1 \leq k \leq d)$ is homogeneous of degree k. Apply Lemma 18.4 to obtain

$$\|q(x_1,\dots,x_n)\| \le \sum_{k\le d} \|q_k(x_1,\dots,x_n)\| \le \sum_{k\le d} 2^{-k}\cdot \|q_k\|_\infty$$
$$\le \Big(\sum_{k\le d} 2^{-k}\Big)\cdot \|q\|_\infty \le \|q\|_\infty . \qquad \text{QED}$$

A Commutative Non-Q Operator Algebra

To construct a commutative operator algebra which is not a Q-algebra, it is convenient to concentrate on condition (iii) of 18.7 when $k = 3$. (In view of the fact that Q-algebras are operator algebras (18.8) it follows from 18.9 that it is fruitless to consider the case $k = 2$.)

Let t be a trilinear form on ℓ_2^n and assume that $t_{j_1 j_2 j_3} = \pm 1$ for each $1 \le j_1, j_2, j_3 \le n$. Then

$$\|t\|_2 \ge \sup\{|t(\xi, e_1, e_1)| : \|\xi\|_2 \le 1\} = \sup\{|\sum_{j\le n} t_{j11}\xi_j| : \|\xi\|_2 \le 1\} = \sqrt{n}\ .$$

In fact, simple natural modifications of the proof of 17.22, which we leave to the reader, show that a great many symmetric trilinear forms on ℓ_2^n of the form just described have norm close to this lower bound.

18.15 Proposition: *Let $n \in \mathbf{N}$ and let $r = (r_{jkm})_{j,k,m=1}^n$ be an $n \times n \times n$ array whose entries are Rademacher functions satisfying $r_{jkm} = r_{j'k'm'}$ if and only if $\{j,k,m\} = \{j',k',m'\}$. There is a constant C, independent of n, such that*

$$\int_0^1 \|r(s)\|_2\, ds \le C\cdot\sqrt{n}\ .$$

We can now construct what we have been promising for a while.

18.16 Theorem: *There is a commutative operator algebra which is not a Q-algebra.*

Proof. We begin by constructing commuting contractions $u_1,\dots,u_n$ in $\mathcal{L}(\ell_2^{2n+2})$ and a symmetric trilinear form t on ℓ_∞^n so that

$$\Big\|\sum_{j,k,m=1}^{n} t_{jkm} u_j u_k u_m\Big\| \ge \frac{1}{C^2}\cdot\sqrt{n}\cdot \|t\|_\infty ,$$

where C is the constant from 18.15. Let $\{e, e_1,\dots,e_n, f_1,\dots,f_n, f\}$ be an orthonormal basis of ℓ_2^{2n+2}. Use 18.15 to locate a symmetric trilinear form t on ℓ_2^n with $t_{jkm} = \pm 1$ for each $1 \le j,k,m \le n$ and $\|t\|_2 \le C\cdot\sqrt{n}$. Consider t as a trilinear form on ℓ_∞^n; since $\|t\|_\infty \le n^{3/2}\cdot\|t\|_2$, we have $\|t\|_\infty \le C\cdot n^2$.

To define the u_j's, $1 \le j \le n$, set

$$u_j e := e_j\ ,\ u_j f := 0 \text{ and } u_j e_k := \frac{1}{C\cdot n^{1/2}} \sum_{m\le n} t_{jkm} f_m\ ,\ u_j f_k := \delta_{jk} f$$

for each $1 \le k \le n$. Thanks to the symmetry of t, the u_j's form a commuting family in $\mathcal{L}(\ell_2^{2n+2})$. Checking that $\|u_j\| \le 1$ is tantamount to

$$\begin{aligned}\|u_j(\sum_{k\le n}\lambda_k e_k)\| &= \frac{1}{C\cdot n^{1/2}}\cdot \|\sum_{k,m\le n} t_{jkm}\lambda_k f_m\| \\ &= \frac{1}{C\cdot n^{1/2}}\cdot \sup\{|\sum_{k,m\le n} t_{jkm}\lambda_k\mu_m| : \|\mu\|_2\le 1\} \\ &\le \frac{1}{C\cdot n^{1/2}}\cdot \|t\|_2\cdot (\sum_{k\le n}|\lambda_k|^2)^{1/2} \le \|\sum_{k\le n}\lambda_k e_k\|\,.\end{aligned}$$

The triple product $u_j u_k u_m$ kills off all the basis vectors except e, and $u_j u_k u_m e = (1/C\cdot n^{1/2})t_{jkm}f$. Consequently,

$$\|\sum_{j,k,m} t_{jkm}u_j u_k u_m\| \le \frac{1}{C\cdot n^{1/2}}\cdot\Big(\sum_{j,k,m=1}^{n} t_{jkm}^2\Big) = \frac{n^{5/2}}{C} \ge \frac{n^{1/2}}{C^2}\cdot\|t\|_\infty.$$

The operators $u_1,\ldots,u_n$ generate a commutative Banach algebra X_n of operators on ℓ_2^{2n+2}. Our counterexample is the Banach algebra X of all operators v on $\left(\oplus_{n=1}^\infty \ell_2^{2n+2}\right)_2$ which are of the form $v((h_n)_{n=1}^\infty) = (v_n h_n)_{n=1}^\infty$ where the $v_n\in X_n$ satisfy $\sup_n \|v_n\| < \infty$. It is plain that X is a commutative operator algebra and violates condition (iii) of Theorem 18.7. QED

2-Summing Algebras and Operator Algebras

Our construction gives considerable flexibility and is adaptable effortlessly to other situations which we now investigate. Our next project is to show that 2-summing algebras must be operator algebras. We rely on a fixed point argument.

18.17 Ky Fan's Fixed Point Theorem: *Let E be a Hausdorff topological vector space, let $C\neq\emptyset$ be a compact convex subset of E, and let $K(C)$ denote the collection of all non-empty compact convex subsets of C. If $\Phi: C\to K(C)$ has its graph $\{(c,D): c\in C,\ D\in\Phi(c)\}$ closed in $C\times C$, then there exists a $c_0\in C$ with $c_0\in\Phi(c_0)$.*

It is technically convenient to work with algebras with identity. If X is a Banach algebra, it can always be viewed as a subalgebra of a Banach algebra X_1 with identity: set $X_1 := X\times \mathbf{C}$ and define the multiplication on X_1 by $(x,\lambda)(y,\mu) := (xy+\lambda y+\mu x, \lambda\mu)$. The identity is $(0,1)$, and the map $X\to X_1 : x\mapsto (x,0)$ is an isometric embedding of Banach algebras. This construction also works for p-summing algebras.

18.18 Lemma: *Let $1\le p<\infty$. If X is p-summing algebra, then so is X_1.*

Proof. Choose (x_j,λ_j) and (y_j,μ_j) in X_1, $1\le j\le n$, and notice that $\|(\lambda_j)\|_{p^*}$ and $\|(x_j)\|_{p^*}^{strong}$ are both dominated by $\|((x_j,\lambda_j))\|_{p^*}^{strong}$, whereas $\|(\mu_j)\|_p$ and $\|(y_j)\|_p^{weak}$ are both less than or equal to $\|((y_j,\mu_j))\|_p^{weak}$. This allows us to affirm that for some C determined by the p-summing algebra X,

$$\begin{aligned}\|\sum_{j\le n}(x_j,\lambda_j)(y_j,\mu_j)\| &\le \|\sum_{j\le n} x_j y_j\| + \|\sum_{j\le n}\lambda_j\cdot(y_j,\mu_j)\| + \|\sum_{j\le n}\mu_j x_j\| \\ &\le C\cdot\|(x_j)\|_{p^*}^{strong}\cdot\|(y_j)\|_p^{weak} + \|(\lambda_j)\|_{p^*}\cdot\|((y_j,\mu_j))\|_p^{weak} \\ &\qquad + \|(\mu_j)\|_p\cdot\|(x_j)\|_{p^*}^{strong} \\ &\le (C+2)\cdot\|((x_j,\lambda_j))\|_{p^*}^{strong}\cdot\|((y_j,\mu_j))\|_p^{weak}.\end{aligned}$$

This is precisely what was needed. QED

18.19 Theorem: *Every 2-summing algebra is an operator algebra.*

Proof. Let X be a 2-summing algebra. By the preceding lemma we may assume that X has an identity. Our task is to find a Hilbert space $\hat{H}$ and a Banach algebra isomorphism from X onto a closed subalgebra of $\mathcal{L}(\hat{H})$. To begin with, however, we just construct a Hilbert space H together with a particular multiplicative operator $u : X \to \mathcal{L}(H)$.

By 18.1, there is a $c > 0$ such that $\pi_2(\widetilde{\varphi}) \le c$ for each $\varphi \in B_{X^*}$; recall that $\widetilde{\varphi} : X \to X^*$ is related to $\varphi \in X^*$ via $\widetilde{\varphi}(x)(y) = \langle \varphi, yx\rangle$, for all $x, y \in X$. It follows that, regardless of the probability measure $\mu \in \mathcal{C}(B_{X^*})^*$, we have

$$(1) \qquad \sum_{k\le n}\int_{B_{X^*}} \|\widetilde{\varphi}(x_k)\|^2 d\mu(\varphi) \le c^2\cdot \sup_{\psi\in B_{X^*}} \sum_{k\le n} |\langle\psi, x_k\rangle|^2$$

for every finite collection $\{x_1,\dots,x_n\}$ in X. The manoeuvre brings to light a 2-summing map $v : X \to L_2(\mu, X^*)$ defined by

$$v(x) : B_{X^*} \longrightarrow X^* : \varphi \mapsto \widetilde{\varphi}(x).$$

In terms of v, (1) just tells us that $\pi_2(v) \le c$. Hence, the Pietsch Domination Theorem 2.12 guarantees the existence of a probability measure $\nu \in \mathcal{C}(B_{X^*})^*$ such that

$$\int_{B_{X^*}} \|\widetilde{\varphi}(x)\|^2 d\mu(\varphi) \le c^2\cdot\int_{B_{X^*}} |\langle\varphi, x\rangle|^2 d\nu(\varphi);$$

let us denote by $\Phi(\mu)$ the collection of all such measures ν.

The set $W(B_{X^*})$ of all probability measures in $\mathcal{C}(B_{X^*})^*$ is convex and weak$*$compact, and it is straightforward to check that $\mu \mapsto \Phi(\mu)$ satisfies the conditions required to apply Ky Fan's Fixed Point Theorem 18.17. So we deduce the existence of some $\omega \in W(B_{X^*})$ with the property that

$$\int_{B_{X^*}} \|\widetilde{\varphi}(x)\|^2 d\omega(\varphi) \le c^2\cdot\int_{B_{X^*}} |\langle\varphi, x\rangle|^2 d\omega(\varphi)$$

holds for each x in X. It follows that, for any $x, y \in X$,

$$(2) \qquad \int_{B_{X^*}} |\langle\varphi, yx\rangle|^2 d\omega(\varphi) \le c^2\cdot\|y\|^2\cdot\int_{B_{X^*}} |\langle\varphi, x\rangle|^2 d\omega(\varphi).$$

This leads us to define the Hilbert space H as the closure of $i_X(X)$ in $L_2(\omega)$, where i_X is the usual embedding $X \to \mathcal{C}(B_{X^*})$. The sought-after multiplicative operator $u : X \to \mathcal{L}(H)$ is obtained from $u(y)(i_X x) := i_X(yx)$

(for $x, y \in X$) by natural extension; the algebraic properties of u are trivial, and (2) expresses the fact that $\|u\| \leq c$.

Of course, nothing guarantees that u will be a Banach algebra isomorphism onto its range. Therefore we modify the construction of u to ensure 'local invertibility' and then resort to a direct sum argument to achieve our final goal.

We start by showing that for each $y \in X$ we can find a Hilbert space H_y and a multiplicative operator $u_y : X \to \mathcal{L}(H_y)$ satisfying

$$\|u_y\| \leq \sqrt{2} \cdot c \quad \text{and} \quad \|u_y(y)\| \geq \frac{1}{\sqrt{2}} \cdot \|y\|.$$

Fixing $y \in X$, we take $\eta \in B_{X^*}$ with $\langle \eta, y \rangle = \|y\|$ and use it to define the weak$*$compact convex subset

$$W(y) := \{(\delta_\eta + \mu)/2 : \mu \in W(B_{X^*})\}$$

of $W(B_{X^*})$. Recall the definition of Φ and observe that for each $\lambda = (\delta_\eta + \mu)/2$ in $W(y)$ and each $\nu \in \Phi(\lambda)$ we have, for all $x \in X$,

$$\begin{aligned} \int_{B_{X^*}} \|\widetilde{\varphi}(x)\|^2 d\lambda(\varphi) &\leq c^2 \cdot \int_{B_{X^*}} |\langle \varphi, x \rangle|^2 d\nu(\varphi) \\ &\leq 2 \cdot c^2 \cdot \int_{B_{X^*}} |\langle \varphi, x \rangle|^2 \, d((\delta_\eta + \nu)/2)(\varphi). \end{aligned}$$

Prompted by this cue, we define

$$\Psi(\lambda) := (\delta_\eta + \Phi(\lambda))/2;$$

it is apparent that Ψ is a map on $W(y)$ which satisfies the requirements needed by 18.17 to conclude that there is a fixed point $\omega_y \in W(y)$ such that

$$\int_{B_{X^*}} \|\widetilde{\varphi}(x)\|^2 d\omega_y(\varphi) \leq 2 \cdot c^2 \cdot \int_{B_{X^*}} |\langle \varphi, x \rangle|^2 d\omega_y(\varphi)$$

for each $x \in X$. It follows that for any $x \in X$ and $z \in B_X$ we have

$$\int_{B_{X^*}} |\langle \varphi, zx \rangle|^2 d\omega_y(\varphi) \leq 2 \cdot c^2 \cdot \int_{B_{X^*}} |\langle \varphi, x \rangle|^2 d\omega_y(\varphi).$$

Moreover, since $\delta_\eta/2 \leq \omega_y$, the very definition of η alerts us to the fact that

$$\int_{B_{X^*}} |\langle \varphi, y \rangle|^2 d\omega_y(\varphi) \geq \|y\|^2/2. \tag{3}$$

Naturally, we are led to define the Hilbert space H_y as the closure in $L_2(\omega_y)$ of $i_X(X)$ and, just as before, we find that the map

$$u_y : X \longrightarrow \mathcal{L}(H_y) \text{ given by } u_y(z)(i_X x) = i_X(zx)$$

is linear, multiplicative, and bounded with $\|u_y\| \leq c \cdot \sqrt{2}$. Our assumption that X has an identity, say e, allows us to deduce, after referring to (3), that

$$\|u_y(y)\| \geq \|u_y(y)(i_X(e))\| = \|i_X(y)\| \geq \frac{1}{\sqrt{2}} \cdot \|y\|.$$

Everything can now be wrapped up tidily. We introduce the ℓ_2 direct sum of the H_y's, $\hat{H} := (\bigoplus_{y\in X} H_y)_2$, and we define $\hat{u} : X \to \mathcal{L}(\hat{H})$ by setting

$$\hat{u}(x)\big((h_y)_{y\in X}\big) := (u_y(x)h_y)_{y\in X}$$

for all $(h_y)_{y\in X} \in H$ and $x \in X$. By design, $\hat{u}$ is linear and multiplicative, and it satisfies

$$\frac{1}{\sqrt{2}} \cdot \|x\| \leq \|\hat{u}(x)\| \leq c \cdot \sqrt{2} \cdot \|x\|$$

for each $x \in X$.

All is well. QED

A modification of the proof of Theorem 18.8 shows that every quotient algebra of an operator algebra is an operator algebra. So we may state:

18.20 Corollary: *Every quotient algebra of a 2-summing algebra is an operator algebra.*

The converse fails:

18.21 Theorem: *There is a commutative operator algebra which is not isomorphic to a quotient of 2-summing algebra.*

Proof. The construction is almost identical to that in the proof of Theorem 18.16. We simply replace the operaors $u_1,\ldots,u_n$ constructed there by $v_k := n^{-1/2}u_k$ $(1 \leq k \leq n)$. Then $\sum_{k\leq n} \|v_k\|^2 \leq 1$ and

$$\Big\| \sum_{j,k,m=1}^{n} t_{jkm} v_j v_k v_m \Big\| = \frac{1}{C\cdot\sqrt{n}} \cdot n^{-3/2} \cdot \Big(\sum_{j,k,m=1}^{n} t_{jkm}^2 \Big) = \frac{n}{C} \geq \frac{\sqrt{n}}{C^2} \cdot \|t\|_2 .$$

This is impossible to reconcile with (iii) of Proposition 18.6. QED

The same argument exhibits a commutative operator algebra which is not a quotient of any p-summing algebra $(1 \leq p < \infty)$. As a consequence, we may confirm what was announced after 18.3:

18.22 Corollary: *If H is an infinite dimensional Hilbert space, then $\mathcal{L}(H)$ can never be a p-summing algebra for any $1 \leq p < \infty$.*

Strictly p-Summing Algebras

We close this chapter by investigating p-summing algebras X which have a norm one identity and satisfy $\pi_p(\widetilde{\varphi}) \leq \|\varphi\|$ for every $\varphi \in X^*$. Such algebras will be called *strictly p-summing algebras,* and we intend to prove:

18.23 Theorem: *Let $1 \leq p < \infty$. Every strictly p-summing algebra is a uniform algebra and so, in particular, is commutative.*

Notice that the existence of an identity is critical here. For example, ℓ_1 with pointwise multiplication is strictly 1-summing but is not a uniform algebra since $\|x\|^2 \neq \|x^2\|$ in general.

To get a handle on this theorem, we need another basic result from Banach algebras:

18.24 Lemma: *Let X be a Banach algebra with identity. Then X is a uniform algebra if and only if $\|x^2\| = \|x\|^2$ for each $x \in X$.*

Proof. We only need to prove that if $\|x^2\| = \|x\|^2$ for each $x \in X$, then X is a uniform algebra.

To begin, we establish commutativity, and for this we need a little machinery from Banach algebra theory. For $x \in X$, the spectral radius $\rho(x)$ is given by

$$\rho(x) = \lim_{n\to\infty} \|x^n\|^{1/n}.$$

Our hypothesis implies that $\|x^{2^n}\| = \|x\|^{2^n}$ for each n, and so $\rho(x) = \|x\|$ follows. Now recall that $\rho(xy) = \rho(yx)$ for any $x, y \in X$, hence $\|xy\| = \|yx\|$. To check that this in fact implies that $xy = yx$ we use a nice complex variable argument. Notice that if $\varphi \in X^*$, then

$$f(z) := \langle \varphi, e^{-zx} y e^{zx} \rangle$$

defines an entire function. This function is bounded:

$$|f(z)| \le \|\varphi\| \cdot \|e^{-zx}(ye^{zx})\| = \|\varphi\| \cdot \|(ye^{zx})e^{-zx}\| = \|\varphi\| \cdot \|y\|.$$

By Liouville's Theorem, f must therefore be constant. In particular, the coefficient $\langle \varphi, yx - xy \rangle$ of z in f's Taylor expansion must be zero. Since $\varphi \in X^*$ was arbitrary, $xy = yx$ follows.

Commutativity opens up Gelfand theory for us. Let $\mathcal{M}_X$ be the weak$*$ compact subset of S_{X^*} which consists of all (non-trivial) multiplicative linear forms on X. Since we know that $\rho(x) = \|x\|$ for each $x \in X$, the Gelfand transform $X \to \mathcal{C}(\mathcal{M}_X) : x \mapsto \langle \cdot, x \rangle$ is an isometric embedding, and so we have what we wanted. QED

Proof of Theorem 18.23. Let $\varphi \in B_{X^*}$. The Pietsch Factorization Theorem 2.13 provides us with a probability measure $\mu \in \mathcal{C}(B_{X^*})^*$ for which $\widetilde{\varphi}$ admits a factorization of the form $\widetilde{\varphi} : X \xrightarrow{j} X_p \xrightarrow{\hat{\varphi}} X^*$ where X_p is the closure of $i_X(X)$ in $L_p(\mu)$, $j : X \to X_p$ is the canonical map, and $\hat{\varphi}$ satisfies $\|\hat{\varphi}\| = \pi_p(\widetilde{\varphi}) \le \|\varphi\|$.

We choose to exploit this factorization by noting that, for any $x, y \in X$,

$$\langle \varphi, xy \rangle = \langle \widetilde{\varphi} y, x \rangle = \langle \hat{\varphi} j y, x \rangle = \langle jy, \hat{\varphi}^* x \rangle.$$

The last duality is between X_p and $X_p{}^*$. Since X_p is a subspace of $L_p(\mu)$, $\hat{\varphi}^* x$ is the restriction of some $f_x \in L_p(\mu)^* = L_{p^*}(\mu)$ such that $\|f_x\|_{L_{p^*}} = \|\hat{\varphi}^* x\| \le \|\varphi\| \cdot \|x\|$. As a consequence,

$$\langle \varphi, xy \rangle = \int_{B_{X^*}} \langle \psi, y \rangle f_x(\psi) \, d\mu(\psi).$$

So, if e denotes the identity of X, we have

$$\langle \varphi, y \rangle = \int_{B_{X^*}} \langle \psi, y \rangle f_e(\psi) \, d\mu(\psi)$$

for any $y \in X$.

This apparently complicated representation will now be exploited to show that each extreme point φ of B_{X^*} is actually a multiple of a multiplicative linear functional.

Let $\mu \in \mathcal{C}(B_{X^*})^*$ correspond to an extreme point φ of B_{X^*} as in the above procedure. If S is a μ-measurable subset of B_{X^*}, then

$$\varphi_S(y) := \int_S \langle\psi, y\rangle f_e(\psi)\, d\mu(\psi)$$

defines a new element $\varphi_S \in X^*$. Of course, $\varphi = \varphi_S + \varphi_{S^c}$ and so

$$1 = \|\varphi\| \le \|\varphi_S\| + \|\varphi_{S^c}\| \le \|f_e\|_{L_1(\mu)} \le \|f_e\|_{L_{p^*}(\mu)} \le \|e\| = 1.$$

It follows that $\|f_e\|_{L_1(\mu)} = \|f_e\|_{L_{p^*}(\mu)} = 1$ and so $|f_e(\psi)| = 1$ μ-almost everywhere. Notice that $\|\varphi_S\| \le \int_S |f_e(\psi)| d\mu(\psi) = \mu(S)$ and similarly $\|\varphi_{S^c}\| \le \mu(S^c)$. So the inequalities

$$1 = \|\varphi\| \le \|\varphi_S\| + \|\varphi_{S^c}\| \le \mu(S) + \mu(S^c) = 1$$

are really equalities, and, when $0 < \mu(S) < 1$,

$$\varphi = \mu(S) \cdot \frac{\varphi_S}{\mu(S)} + \mu(S^c) \cdot \frac{\varphi_{S^c}}{\mu(S^c)}$$

is a convex combination of norm one elements of X^*. Since φ is an extreme point of B_{X^*}, we must have $\varphi_S = \mu(S) \cdot \varphi$ – even when $\mu(S)$ is 0 or 1. So, for any $y \in X$,

$$\int_S \langle\psi, y\rangle f_e(\psi)\, d\mu(\psi) = \int_S \langle\varphi, y\rangle\, d\mu(\psi).$$

As this obtains for every measurable set $S \subset B_{X^*}$ and every $y \in X$, we can deduce that μ-almost everywhere

$$\varphi = f_e(\psi)\psi.$$

From this identity we derive that, for any $x, y \in X$,

$$\begin{aligned}
\langle\varphi, e\rangle \cdot \langle\varphi, xy\rangle &= \langle\varphi, e\rangle \cdot \int_{B_{X^*}} \langle\psi, y\rangle f_x(\psi)\, d\mu(\psi) \\
&= \int_{B_{X^*}} \langle\psi, e\rangle f_e(\psi)\langle\psi, y\rangle f_x(\psi)\, d\mu(\psi) = \int_{B_{X^*}} \langle\psi, e\rangle\langle\varphi, y\rangle f_x(\psi) d\mu(\psi) \\
&= \langle\varphi, y\rangle \cdot \int_{B_{X^*}} \langle\psi, e\rangle f_x(\psi)\, d\mu(\psi) = \langle\varphi, x\rangle \cdot \langle\varphi, y\rangle.
\end{aligned}$$

In fact, $|\langle\varphi, e\rangle| = 1$. It is clear that $|\langle\varphi, e\rangle| \le 1$, and the opposite inequality results from the observation that, for each $x \in B_X$,

$$|\langle\varphi, x\rangle|^2 = |\langle\varphi, e\rangle| \cdot |\langle\varphi, x^2\rangle| \le |\langle\varphi, e\rangle|.$$

So what we actually have is $|\langle\varphi, xy\rangle| = |\langle\varphi, x\rangle| \cdot |\langle\varphi, y\rangle|$ for any $x, y \in X$. It follows that $|\langle\varphi, x\rangle|^2 = |\langle\varphi, x^2\rangle|$ holds for any $x \in X$ and any extreme point $\varphi \in B_{X^*}$. Using the Krein-Milman Theorem, we deduce that $\|x^2\| = \|x\|^2$ for every $x \in X$. An appeal to Lemma 18.24 completes the proof. QED

NOTES AND REMARKS

In the late 1960s, N.T. Varopoulos asked whether or not every commutative Banach algebra is an operator algebra. The question was answered in the [1969] lecture notes of J. Wermer, where he proved that the Wiener algebra is not a Q-algebra, a slightly weaker result than 18.11. B. Cole played an important part: he supplied 18.8, which Wermer used in conjunction with a result related to the von Neumann Inequality [1951].

Theorem: *Let X be a Q-algebra and let $x \in X$. If there is a constant $C > 0$ such that $\|x^n\| \leq C$ for every $n \in \mathbf{Z}$, then for every polynomial p we have $\|p(x)\| \leq C^2 \cdot \|p\|_\infty$.*

Actually, it follows from a result of B. Sz. Nagy [1947] that this theorem is valid even when X is a general operator algebra. What Sz. Nagy proved was that if H is a Hilbert space and $C > 0$, then every $u \in \mathcal{L}(H)$ satisfying $\|u^n\| \leq C$ for all $n \in \mathbf{Z}$ is *similar to a contraction,* that is, there exist $v, w \in \mathcal{L}(H)$ with v invertible and $\|w\| \leq 1$ for which $u = v^{-1}wv$.

An operator u on a Hilbert space H is said to be *polynomially bounded* if there is a constant M such that $\|p(u)\| \leq M \cdot \|p\|_\infty$ for every polynomial p. Thanks to von Neumann's Inequality, all Hilbert space operators which are similar to a contraction must be polynomially bounded. It is unknown whether or not the converse holds, although J. Bourgain [1986] has shown that this cannot be far from the truth.

A. Bernard [1971] extended the above theorem to polynomials in several variables. Also, in [1973], he showed that there is a non-commutative version of 18.8:

- *every quotient algebra of an operator algebra is an operator algebra.*

P.G. Dixon [1976a], [1976/77], [1977] took up this theme and continued in more abstract directions.

I.G. Craw is responsible for the equivalences of (i) and (iii) in 18.7. His result was presented by A.M. Davie [1973]. In this paper, Davie also took the important step of reducing to homogeneous polynomials; the $p = \infty$ case of 18.4, 18.5 and 18.6 can be found there. N.T. Varopoulos [1972a] should also be consulted in such instances.

It is worth noting that the constants which appear in 18.5 are by no means best possible. For example, S. Banach [1938] proved that

- *the norm of a k-homogeneous polynomial on a Hilbert space equals the norm of the associated symmetric k-linear form.*

Very recently, C. Benítez and Y. Sarantopoulos [1995] showed that this characterizes real Hilbert spaces. An illuminating discussion of improvements to 18.5

can be found in L.Harris' commentary to problems 73 and 74 in the Scottish Book (see R.D. Mauldin [1981]). For more recent work consult Y. Sarantopoulos [1986], [1987], A.M. Tonge [1984] and Y. Sarantopoulos and A.M. Tonge [1995].

A.M. Davie [1973] noted Banach algebra properties of Q-algebras, such as Arens regularity, and gave several examples of Q-algebras and non-Q-algebras, some of which were completed by N.T. Varopoulos [1972a,b]. In [1972a], Varopoulos introduced the notion of an injective algebra: a Banach algebra X is an *injective algebra* if the multiplication induces a continuous linear map of the injective tensor product $X \check{\otimes} X$ into X. He showed that commutative injective algebras are Q-algebras, and in [1972b] succeeded in characterizing commutative injective algebras as the quotient algebras of uniform algebras by complemented ideals. Further work on injective algebras was undertaken by S. Kaijser [1976]. In particular, he established a precursor to 18.23.

Injective algebras are closely related to p-summing algebras: one can view a p-summing algebra X as a Banach algebra for which the multiplication induces a continuous linear map $[X \otimes X, d_p] \to X$. Here d_p is the tensor norm induced by the dual norm of the p-integral norm via the inclusions $X \otimes X \hookrightarrow X^{**} \otimes X \hookrightarrow [\mathcal{I}_p^d(X^*, X), \iota_p^d]$. This norm was first defined by P. Saphar [1970]; see A. Defant and K. Floret [1993] for more details.

It was not long before P. Charpentier [1974] isolated the class of 1-summing algebras and established the equivalence of (i) and (ii) in 18.7. He also took note of 18.9. The failure of the converse was a delicate question which was resolved by T.K. Carne [1979] using methods from harmonic analysis.

Theorem 18.19 is an extension of a result of N.T. Varopoulos [1975], who showed that an $\mathcal{L}_\infty$-space which is a Banach algebra must be an operator algebra. The same is true for the disk algebra or H^∞, thanks to the deep work of J. Bourgain [1984c], but it is at present unknown whether or not $\mathcal{L}(H)$ can be given a Banach algebra structure without making it an operator algebra. The original proof of 18.19 can be found in A.M. Tonge [1976], but the proof we present is due to B. Maurey, and was originally published in A.M. Tonge [1975/76].

Ky Fan's Fixed Point Theorem 18.17 is closely related to 9.10. An account can be found in C. Berge [1959].

Theorem 18.19 was the starting point for a systematic study of p-summing algebras which was carried out by A.M. Tonge [1975/76], [1976] and by A.M. Mantero and A.M. Tonge [1979], [1980]. We have taken 18.1, 18.4, 18.6 and most of the material from 18.13 onwards from these papers. Further work was done by G. Racher [1980] and, in the context of completely bounded operators, by D.P. Blecher, Z.J. Ruan and A.M. Sinclair [1990] and D.P. Blecher [1990], [1995a,b].

Theorem 18.16 gives a simplified account of a result first proved by N.T.

Varopoulos [1974]. Another treatment was given by P.G. Dixon [1976b]. These results were the culmination of a long investigation into the validity of J. von Neumann's [1951] inequality when translated into the situation of many variables. Although S. Brehmer [1961], with a theorem similar in spirit to 18.13, and T. Andô [1963] had achieved partial success, N.T. Varopoulos [1974] dashed all hopes of total success. Simple counterexamples followed swiftly; see the appendix to Varopoulos' paper, written jointly with S. Kaijser, and M.J. Crabb and A.M. Davie [1975]. Perhaps the most attractive counterexample, which we give in 18.12, appeared in N.T. Varopoulos [1976]; it is the only instance we know of where the fact that Grothendieck's constant is greater than one has been of any use.

The many variable von Neumann inequality has continued to attract attention. We cite as examples S.W. Drury [1978], [1983], P.G. Dixon and S.W. Drury [1986], B. Cole, K. Lewis and J. Wermer [1995] and A.M. Tonge [1978]. There remains a stubborn open problem: if $N \geq 3$ is an integer, does there exist a constant K_N such that for any polynomial p in N complex variables and any N-tuple $u_1,\dots,u_N$ of commuting contractions on a Hilbert space, we have $\|p(u_1,\dots,u_N)\| \leq K_N \cdot \|p\|_\infty$?

Even the single variable von Neumann inequality has excited interest in the context of L_p-spaces. We mention the work of V.V. Peller [1976a,b], [1978a,b], [1982], R.R. Coifman, R. Rochberg and G. Weiss [1978], M.A. Akcoğlu and P.E. Kopp [1977] and M.A. Akcoğlu and L. Sucheston [1975], [1976], [1977].

19. Dvoretzky's Theorem

We return to a recurring theme which first appeared right at the beginning of this text: the Dvoretzky - Rogers Theorem 1.2. In 14.2, we extended Lemma 1.3, a major component of the proof of 1.2:

- *all infinite dimensional Banach spaces contain subspaces F of all finite dimensions with the property that the formal identity $\ell_2^{\dim F} \to \ell_\infty^{\dim F}$ has a factorization $\ell_2^{\dim F} \xrightarrow{v} F \xrightarrow{u} \ell_\infty^{\dim F}$ with $\|u\| \cdot \|v\| \leq \sqrt{3}$.*

  Had we taken a little more care we could have replaced $\sqrt{3}$ by $1+\varepsilon$ for any $\varepsilon > 0$. In fact, there is much more waiting to be extracted from this fruitful line of thought. We now aim to show that

- *all infinite dimensional Banach spaces contain subspaces F, of dimension as large as you please, with the property that for any $\varepsilon > 0$ the identity of $\ell_2^{\dim F}$ has a factorization $\ell_2^{\dim F} \xrightarrow{v} F \xrightarrow{u} \ell_2^{\dim F}$ with $\|u\| \cdot \|v\| \leq 1+\varepsilon$.*

  In other words, every infinite dimensional Banach space X contains $(1+\varepsilon)$-isomorphic copies of ℓ_2^n for each $n \in \mathbf{N}$; alternatively, ℓ_2 is finitely representable in X. An even more precise result can be established:

19.1 Dvoretzky's Theorem: *Let k be a natural number and let $\varepsilon > 0$. There is a positive integer $n_\varepsilon(k)$ with the property that each Banach space of dimension $n_\varepsilon(k)$ contains a k-dimensional subspace which is $(1+\varepsilon)$-isomorphic to ℓ_2^k.*

Once this theorem has been proved we shall focus on two remarkable results which give profound new insights into topics discussed earlier.

The first of these involves Hilbert - Schmidt operators. In 4.12 we proved that a Hilbert space operator is Hilbert - Schmidt if and only if it factors through an $\mathcal{L}_\infty$-space (or an $\mathcal{L}_1$-space), thereby giving a non-Hilbertian characterization of a purely Hilbertian concept! Now we shall go much further.

19.2 Theorem: *A Hilbert space operator $u: H_1 \to H_2$ is Hilbert - Schmidt if and only if, for any infinite dimensional Banach space Z, there are operators $v \in \mathcal{L}(Z, H_2)$ and $w \in \mathcal{L}(H_1, Z)$ with $u = vw$.*

Moreover, we can choose w to be compact and v compact and 2-summing.

Notice that if Z contains ℓ_2, the Π_2-Extension Theorem 4.15 allows us to factorize Hilbert - Schmidt operators through Z in the way we describe. However, not all Banach spaces enjoy this property, and we shall need to develop additional machinery to provide a proof of 19.2.

We finish the chapter with a result that harks back to Maurey's Extension Theorem 12.22. If the infinite dimensional Banach space X has type 2 then there is a constant C such that each of the $(1+\varepsilon)$-isomorphic copies of ℓ_2^n that Dvoretzky's Theorem provides inside X is complemented in X by a projection $P \in \mathcal{L}(X)$ with $\|P\| \leq C$; for brevity we shall say that the ℓ_2^n's are *C-complemented* in X. While what we have just pointed out can hardly be described as superficial, it is certainly just a beginning.

19.3 Theorem: *The following are equivalent statements about an infinite dimensional Banach space X:*

(i) *X is K-convex.*

(ii) *There exists a constant $C \geq 1$ and, for each $n \in \mathbf{N}$, an $N \in \mathbf{N}$ such that every N-dimensional subspace of X has an n-dimensional subspace which is C-complemented in X and C-isomorphic to ℓ_2^n.*

(iii) *There exists a constant $C \geq 1$ and, for each $n \in \mathbf{N}$ and $\varepsilon > 0$, an $N \in \mathbf{N}$ such that every N-dimensional subspace of X has an n-dimensional subspace which is C-complemented in X and $(1+\varepsilon)$-isomorphic to ℓ_2^n.*

Fréchet Derivatives of Convex Functions

Our approach to Dvoretzky's Theorem relies on constructions involving Fréchet derivatives of norms on finite dimensional Banach spaces. Norms are convex functions, and it is convenient to develop what we need in this wider context.

We begin by fixing our notation. Let a be an element of an open subset U of a real Banach space X, and let $f : U \to \mathbf{R}$ be a function. If there is an element $\nabla f(a)$ of X^* such that

$$\lim_{x \to 0} \frac{f(x+a) - f(a) - \langle \nabla f(a), x \rangle}{\|x\|} = 0$$

we say that f is *Fréchet differentiable* at a, and we refer to

$$\nabla f(a)$$

as the *Fréchet derivative* of f at a.

If $X = \mathbf{R}^n$, f's Fréchet derivative at a, when it exists, can always be identified, via the scalar product, with the gradient vector

$$\nabla f(a) = \Big(\frac{\partial f}{\partial x_1}(a), \dots, \frac{\partial f}{\partial x_n}(a)\Big).$$

In general, mere existence of all partial derivatives of f at some point in $\mathbf{R}^n$ is not enough to ensure that f is Fréchet differentiable at that point. However, there is no problem when f is convex (and so continuous).

19.4 Lemma: *Let $f : \mathbf{R}^n \to \mathbf{R}$ be convex. If all the partial derivatives of f exist at $a \in \mathbf{R}^n$, then f is Fréchet differentiable at a.*

Proof. We know what we are looking for, so it is convenient to label

$$g(a) := \Big(\frac{\partial f}{\partial x_1}(a),\dots,\frac{\partial f}{\partial x_n}(a)\Big).$$

Considering the auxiliary function

$$F : \mathbf{R}^n \longrightarrow \mathbf{R} \quad \text{given by} \quad F(x) := f(x+a) - f(a) - \langle g(a), x\rangle,$$

we can lighten the notational burden: F is convex too, and

$$\frac{\partial F}{\partial x_i}(0) = 0 \quad \text{for all} \quad 1 \le i \le n.$$

As we are working with $\mathbf{R}^n$, the particular choice of norm is irrelevant, but it will be convenient to work with $\|\cdot\|_1$, the norm of ℓ_1^n. We need to show that

$$\lim_{x\to 0} \frac{F(x)}{\|x\|_1} = 0.$$

To this end, define

$$F_k : \mathbf{R} \longrightarrow \mathbf{R} \quad \text{by} \quad F_k(t) := \max\Big\{\frac{F(te_k)}{t}, \frac{F(-te_k)}{t}\Big\},\ 1 \le k \le n,$$

with $\dfrac{0}{0} = 0$ being sacrosanct. Noticing that $F(t) \ge 0$ for $t \ge 0$ and

$$\lim_{t\to 0} F_k(t) = \frac{\partial F}{\partial x_k}(0) = 0 \quad (1 \le k \le n),$$

we appreciate that it will be enough to show that

$$\frac{|F(x)|}{\|x\|_1} \le \max_{k\le n} F_k(\|x\|_1).$$

For this, first observe that if $x = (x_1,\dots,x_n)$, then convexity gives

$$\begin{aligned} F(x) &= F\Big(\sum_{k\le n} x_k e_k\Big) \le \sum_{k\le n} \frac{|x_k|}{\|x\|_1} \cdot F\big((\operatorname{sign} x_k)\|x\|_1 e_k\big) \\ &\le \max_{k\le n} \|x\|_1 \cdot F_k(\|x\|_1). \end{aligned}$$

But convexity also implies that $0 = F(0) \le \frac{1}{2}\big(F(x) + F(-x)\big)$, and so

$$F(x) \ge -F(-x) \ge -\max\nolimits_{k\le n} \|-x\|_1 \cdot F_k(\|-x\|_1).$$

Assembling, we find what we sought, namely

$$|F(x)| \le \|x\|_1 \cdot \max_{k\le n} F_k(\|x\|_1).$$

QED

Later, we shall want to perform integrations involving Fréchet derivatives of norms. Our next result makes this feasible.

19.5 Theorem: *If $f : \mathbf{R}^n \to \mathbf{R}$ is convex, then f is Fréchet differentiable almost everywhere.*

Proof. When $n = 1$, this is elementary. After all, proceeding from left to right, a convex function is monotonic non-increasing up to a point where it becomes monotonic non-decreasing. Lebesgue's Theorem on almost everywhere differentiability of monotonic functions is all we need to reach our conclusion.

With a little care, the more general situation can be reduced to the one dimensional case. For each $1 \le k \le n$ and for each $a \in \mathbf{R}^n$, the function

$$\mathbf{R} \longrightarrow \mathbf{R} : t \mapsto f(a + te_k)$$

is convex and so, as we have just decided, almost everywhere differentiable. Also, for each $1 \le k \le n$,

$$E_k := \left\{ b \in \mathbf{R}^n : \frac{\partial f}{\partial x_k}(b) \text{ is not defined} \right\}$$

is a Borel subset of $\mathbf{R}^n$. Its intersection with any line $\{a + te_k : t \in \mathbf{R}\}$ has measure zero. So, by Fubini's Theorem, E_k itself must have measure zero. Consequently, outside a set of measure zero, all partial derivatives of f will exist, and we can apply the previous lemma to get the conclusion we want.

QED

It is good not to have to worry too much about the existence of Fréchet derivatives of norms on $\mathbf{R}^n$, but we also need tools to bend them to our will. Limits exhibit gratifying malleability, even away from the finite dimensional setting.

19.6 Proposition: *Let U be an open convex subset of a Banach space X, and let $(f_n)_n$ be a sequence of convex functions $U \to \mathbf{R}$ converging uniformly on U to $f : U \to \mathbf{R}$. If f and all the f_n's are Fréchet differentiable at $a \in U$ then*

$$\lim_{n\to\infty} \|\nabla f_n(a) - \nabla f(a)\| = 0.$$

Proof. We might as well apply a constant linear shift to all our functions to arrange that $\nabla f(a) = 0$. Then, for a fixed $\varepsilon > 0$, we may choose $r > 0$ so that for each $\|x\| \le r$ we have $a + x \in U$ and

$$(1) \qquad |f(a+x) - f(a)| < \frac{\varepsilon\|x\|}{3} \le \frac{\varepsilon r}{3}.$$

The uniform convergence of $(f_n)_n$ allows us to find $N \in \mathbf{N}$ such that, for each $\|x\| \le r$ and each $n \ge N$,

$$(2) \qquad |f_n(a+x) - f(a+x)| < \frac{\varepsilon r}{3}.$$

Convexity of f_n tells us that if $0 < \lambda \le 1$ and $\|y\| = r$ then

$$f_n(a + \lambda y) \le \lambda \cdot f_n(a+y) + (1-\lambda) \cdot f_n(a).$$

Rearranging, this becomes

$$\frac{f_n(a+\lambda y) - f_n(a)}{\lambda} \le f_n(a+y) - f_n(a).$$

So, provided $n \ge N$, we can take advantage of (2) and then (1) to obtain

$$\frac{f_n(a+\lambda y) - f_n(a)}{\lambda} < f(a+y) - f(a) + \frac{2\cdot\varepsilon\cdot r}{3} < \varepsilon\cdot r = \varepsilon\cdot\|y\|.$$

Passing with λ to zero, we get

$$|\langle y, \nabla f_n(a)\rangle| \le \varepsilon\cdot\|y\|.$$

It follows that $\|\nabla f_n(a)\| \le \varepsilon$ whenever $n \ge N$. QED

To wrap up this discussion, we add that for the convex functions we have in mind – norms – we have some serious control over the size of their Fréchet derivatives, when they exist. After all, if p is a norm which is Fréchet differentiable at a,

$$|\langle x, \nabla p(a)\rangle| = \lim_{\lambda \to 0+} \left| \frac{p(a+\lambda x) - p(a)}{\lambda} \right| \leq p(x).$$

Group Actions and Invariant Measures

The compact group of orthogonal transformations on a finite dimensional real inner product space will figure prominently in our proof of Dvoretzky's Theorem. We need some background material, and it is simplest to develop this in abstract form.

Let G be a compact topological group, and let Ω be a compact Hausdorff space. We say that G *acts transitively on* Ω if there is a continuous map $G \times \Omega \to \Omega : (g, \omega) \mapsto g(\omega)$ satisfying

(i) when e is the identity of G, $e(\omega) = \omega$ for all $\omega \in \Omega$,

(ii) $(g_1 g_2)(\omega) = g_1(g_2(\omega))$ for all $g_1, g_2 \in G$ and all $\omega \in \Omega$,

(iii) for every $\omega_1, \omega_2 \in \Omega$ there exists $g \in G$ such that $g(\omega_1) = \omega_2$.

It is interesting to note that each $g \in G$ can be viewed as a homeomorphism of Ω onto itself: the map $\Omega \to \Omega : \omega \mapsto g(\omega)$ is clearly continuous, and $\Omega \to \Omega : \omega \mapsto (g^{-1})(\omega)$ is its continuous inverse.

Recall that on any compact group there is a unique translation invariant probability measure – the Haar measure. Our interest in transitive group actions stems from the fact that they allow us to transfer G's Haar measure to a probability measure μ on the compact space Ω which is uniquely determined by its *G-invariance:* if B is any Borel subset of Ω then $\mu(g(B)) = \mu(B)$ for each $g \in G$.

In proving these results, one particular type of transitive group action comes in very handy.

19.7 Example: Suppose that H is a closed subgroup of the compact group G. Consider also the set G/H of all left cosets of H in G. This is a compact Hausdorff space when viewed as a quotient space of G via the natural map $q : G \to G/H : g \mapsto gH$, and the map

$$G \times (G/H) \longrightarrow G/H : (g_1, g_2 H) \mapsto g_1 g_2 H$$

gives us a transitive action of G on G/H.

Notice that in this example G/H need not have any group structure – we are *not* assuming that H is an invariant subgroup of G.

Our example turns out to be the prototype for all transitive actions of compact groups on compact spaces, but to appreciate this we need one more concept. Suppose that the compact group G acts transitively on the compact Hausdorff spaces Ω_1 and Ω_2. We say that Ω_1 and Ω_2 are *isomorphic under the action of* G if there is a homeomorphism $\varphi : \Omega_1 \to \Omega_2$ with the property that

$$g(\varphi(\omega)) = \varphi(g(\omega))$$

for every $g \in G$ and $\omega \in \Omega_1$.

19.8 Theorem: *Let the compact group G act transitively on the compact Hausdorff space Ω. There is a closed subgroup H of G such that Ω and G/H are isomorphic under the action of G.*

Proof. We plunge in and identify H. Fixing $\omega_0 \in \Omega$, we set

$$H := \{g \in G : g(\omega_0) = \omega_0\}.$$

This is evidently a closed subgroup of G.

A natural candidate for a suitable homeomorphism between G/H and Ω thrusts itself upon us. Tentatively, we try to define $\varphi : G/H \to \Omega$ by

$$\varphi(gH) := g(\omega_0).$$

The tests for well-definition and injectivity can be done simultaneously. For $g_1, g_2 \in G$,

$$\begin{aligned} g_1(\omega_0) = g_2(\omega_0) &\Leftrightarrow g_1^{-1}(g_2(\omega_0)) = g_1^{-1}(g_1(\omega_0)) = e(\omega_0) = \omega_0 \\ &\Leftrightarrow g_1^{-1}g_2 \in H \Leftrightarrow g_1 H = g_2 H. \end{aligned}$$

The transitivity of G's action guarantees that φ is surjective. Also, φ satisfies the isomorphism requirement since, if $g_1, g_2 \in G$, then

$$g_1(\varphi(g_2 H)) = g_1(g_2(\omega_0)) = (g_1 g_2)(\omega_0) = \varphi(g_1(g_2 H)).$$

It just remains to convince ourselves that φ is continuous – after all, a continuous bijection between compact Hausdorff spaces is automatically a homeomorphism. Fix $g \in G$ and pick an open set $V \subset \Omega$ containing $g(\omega_0)$. Thanks to the continuity of $(h, \omega) \mapsto h(\omega)$, there must be an open set $U \subset G$ which contains g and has the property that $g'(\omega_0) \in V$ for all $g' \in U$. But $q(U)$ is open in G/H's quotient topology because $q^{-1}(q(U)) = \bigcup_{h \in H} hU$ is open in G. Since $q(U) \subset \varphi^{-1}(V)$, the proof is complete. QED

This theorem gives us access to one of our main tools.

19.9 Theorem: *Suppose that the compact group G acts transitively on the compact Hausdorff space Ω. There is a unique G-invariant regular Borel probability measure on Ω.*

Proof. In view of the last result, we need look no further than the action of G on G/H where H is a closed subgroup of G.

The existence argument is short and simple. As usual, write $q : G \to G/H$ for the quotient map. If μ is normalized Haar measure on G we can define the image measure $\mu_{G/H}$ on G/H by setting

$$\mu_{G/H}(B) := \mu(q^{-1}(B))$$

for all Borel subsets B of G/H. Notice that for any $g \in G$,

$$g(q^{-1}(B)) = \{gx : xH \in B\} = \{gx : gxH \in gB\} = q^{-1}(gB).$$

It follows that $\mu_{G/H}$ is G-invariant:

$$\mu_{G/H}(gB) = \mu(q^{-1}(gB)) = \mu(g(q^{-1}(B))) = \mu(q^{-1}(B)) = \mu_{G/H}(B).$$

Naturally, $\mu_{G/H}$ is a regular Borel probability measure, so existence is settled.

To decide the issue of uniqueness, we need to look more closely at the action of regular Borel measures on continuous function spaces. Start with $\varphi \in \mathcal{C}(G)$, and for each $g \in G$ define the translate

$$\varphi_g \in \mathcal{C}(G) \quad \text{by} \quad \varphi_g(x) = \varphi(gx).$$

Denote normalized Haar measure on H by μ_H. Take note of the fact that the map $G \to \mathcal{C}(G) : g \mapsto \varphi_g$ is uniformly continuous so that

$$\hat{\varphi}(g) := \int_H \varphi(gh)\, d\mu_H(h) \qquad (g \in G)$$

defines a function $\hat{\varphi} \in \mathcal{C}(G)$.

If g_1, g_2 are in the same left coset of H, then $g_1^{-1}g_2 \in H$ and so the translation invariance of μ_H gives

$$\begin{aligned}\hat{\varphi}(g_1) &= \int_H \varphi_{g_1}(h)\, d\mu_H(h) = \int_H \varphi_{g_1}(g_1^{-1}g_2h)\, d\mu_H(h)\\ &= \int_H \varphi_{g_2}(h)\, d\mu_H(h) = \hat{\varphi}(g_2).\end{aligned}$$

The function $\hat{\varphi}$ is therefore constant on left cosets of H, and we may use it to define

$$\check{\varphi} \in \mathcal{C}(G/H) \quad \text{by} \quad \check{\varphi}(gH) := \hat{\varphi}(g)$$

for each $g \in G$.

Remarkably, all elements of $\mathcal{C}(G/H)$ arise in this way. Indeed, if $\psi \in \mathcal{C}(G/H)$, then $\psi\circ q \in \mathcal{C}(G)$ and for each $g \in G$

$$\begin{aligned}(\psi\circ q)\check{}\,(gH) &= (\psi\circ q)\hat{}\,(g) = \int_H (\psi\circ q)(gh)\, d\mu_H(h)\\ &= \int_H \psi(ghH)\, d\mu_H(h) = \int_H \psi(gH)\, d\mu_H(h) = \psi(gH).\end{aligned}$$

In other words, $\psi = (\psi\circ q)\check{}$.

Now for the action. Take any G-invariant regular Borel probability measure ν on G/H. For $\varphi \in \mathcal{C}(G)$ define

$$\lambda(\varphi) := \int_{G/H} \check{\varphi}(gH)\, d\nu(gH).$$

Evidently, λ is a probability measure in $\mathcal{C}(G)^*$. Moreover, it is translation invariant since for each $x \in G$

$$\begin{aligned}\langle \lambda, \varphi_x\rangle &= \int_{G/H} \widehat{\varphi_x}(g)\, d\nu(gH) = \int_{G/H} \hat{\varphi}(xg)\, d\nu(gH)\\ &= \int_{G/H} \check{\varphi}(xgH)\, d\nu(gH) = \int_{G/H} \check{\varphi}(gH)\, d\nu(gH) = \langle \lambda, \varphi\rangle.\end{aligned}$$

We must conclude that λ is nothing other than the normalized Haar measure on G. Hence, if ν_1, ν_2 are G-invariant regular Borel probability measures on G/H and if $\psi = (\psi \circ q)^{\vee} \in \mathcal{C}(G/H)$, then $\langle \nu_1, \psi \rangle = \langle \lambda, \psi \circ q \rangle = \langle \nu_2, \psi \rangle$. In other words, ν_1 and ν_2 are the same. QED

ACTIONS OF THE ORTHOGONAL GROUP

Our interest in group actions lies in applications of the abstract material we have just developed to very specific situations.

Let E be a finite dimensional real inner product space. Recall that an operator $u \in \mathcal{L}(E)$ is said to be an *orthogonal transformation* if it preserves inner products, that is if $(ux|uy) = (x|y)$ holds for all $x, y \in E$. We denote the collection of all orthogonal transformations in $\mathcal{L}(E)$ by

$$\mathcal{O}(E)$$

and note that it is a group under composition of operators. We give $\mathcal{O}(E)$ the topology inherited from $\mathcal{L}(E)$. Since E is finite dimensional, $\mathcal{O}(E)$ is then a compact group.

Our first group action is now very natural.

19.10 Example: *$\mathcal{O}(E)$ acts transitively on the unit sphere S_E of E.*

The action $\mathcal{O}(E) \times S_E \to S_E$ is in fact given by $(u, x) \mapsto ux$. It is mostly routine to check that this really does fulfill all the requirements of a transitive group action. We pause only to remark that if x_1 and x_1' are in S_E, then we may consider them as members of orthonormal bases $\{x_1, \ldots, x_n\}$ and $\{x_1', \ldots, x_n'\}$; it is simple to verify that the linear map $u : E \to E$ defined by $x_k \mapsto x_k'$ $(1 \leq k \leq n)$ is an orthogonal transformation.

With geometry in mind, we refer to the $\mathcal{O}(E)$-invariant measures as *rotation invariant measures*. We denote the rotation invariant probability measure on S_E by

$$\lambda_E,$$

but, as in the discussion before 12.13, when $E = \ell_2^n$ it will be convenient to replace $\mathcal{O}(E)$, S_E and λ_E by

$$\mathcal{O}_n, \quad S^{n-1} \quad \text{and} \quad \lambda_n.$$

Our first result can actually be viewed as a reworking of 12.14 for the case of scalar-valued maps.

19.11 Proposition: *If $f \in \ell_2^n$, then* $\displaystyle\int_{S^{n-1}} |(x|f)|^2 d\lambda_n(x) = \frac{1}{n} \cdot \|f\|_2.$

Proof. Of course, we may assume that $f \neq 0$. For any $1 \leq k \leq n$, we can find an orthogonal transformation u which rotates $f/\|f\|_2$ into e_k. Then

$$\int_{S^{n-1}} |(x|f)|^2 d\lambda_n(x) = \|f\|_2^2 \cdot \int_{S^{n-1}} \left|\left(x \middle| \frac{f}{\|f\|_2}\right)\right|^2 d\lambda_n(x)$$
$$= \|f\|_2^2 \cdot \int_{S^{n-1}} \left|\left(u^*x \middle| \frac{f}{\|f\|_2}\right)\right|^2 d\lambda_n(x) = \|f\|_2^2 \cdot \int_{S^{n-1}} |(x|e_k)|^2 d\lambda_n(x).$$

Summing over k, we obtain

$$n \cdot \int_{S^{n-1}} |(x|f)|^2 d\lambda_n(x) = \|f\|_2^2 \cdot \int_{S^{n-1}} \|x\|_2^2 \, d\lambda_n(x) = \|f\|_2^2. \qquad \text{QED}$$

Our next example of a group action will lead to a very significant result. Consider the set

$$\Sigma(E) := \{(x,y) \in S_E \times S_E : (x|y) = 0\}.$$

As $(x,y) \in \Sigma(E)$ precisely when $\|\lambda x + \mu y\|^2 = \lambda^2 + \mu^2$ for all real scalars λ, μ, it is clear that $\Sigma(E)$ is closed in $S_E \times S_E$, and thus is a compact Hausdorff space.

19.12 Example: *$\mathcal{O}(E)$ acts transitively on $\Sigma(E)$. The action is given by*

$$\mathcal{O}(E) \times \Sigma(E) \longrightarrow \Sigma(E) : \big(u, (x,y)\big) \mapsto (ux, uy).$$

Checking the requirements of a transitive group action requires no new ideas and very little effort.

We denote the rotation invariant probability measure on $\Sigma(E)$ by

$$\sigma_E,$$

but when $E = \ell_2^n$, we prefer to use the notation

$$\Sigma_n \quad \text{and} \quad \sigma_n.$$

The uniqueness of rotation invariant probability measures leads immediately to a useful identity. If f is a bounded Borel measurable function on $\Sigma(E)$ then

$$\text{(A)} \qquad \int_{\Sigma_E} f(x,y)\, d\sigma_E(x,y) = \int_{S_E} \Big(\int_{S_{\{x\}^\perp}} f(x,y)\, d\lambda_{\{x\}^\perp}(y) \Big)\, d\lambda_E(x).$$

We now can state one of our principal objectives, a cornerstone of this proof of Dvoretzky's theorem.

19.13 Theorem: *Let $f \geq 0$ be a bounded measurable function on Σ_n. Then for each natural number $1 \leq k \leq n$ there is a k-dimensional subspace E of ℓ_2^n such that*

$$\int_{\Sigma(E)} f(x,y)\, d\sigma_E(x,y) \leq \int_{\Sigma_n} f(x,y)\, d\sigma_n(x,y).$$

To prove this, we require the services of yet another transitive group action which finds its origins in a modified version of our last example. Let E be an n-dimensional real inner product space and take an integer $1 \leq m \leq n$. Denote by

$$\Sigma^{(m)}(E)$$

the subset of $(S_E)^m$ consisting of all m-tuples $(x_1, ..., x_m)$ of orthonormal vectors. (Notice that $\Sigma^{(1)}(E) = S_E$ and $\Sigma^{(2)}(E) = \Sigma(E)$.) It is mindless to repeat the arguments we omitted for Example 19.12 to find that $\Sigma^{(m)}(E)$ is a compact Hausdorff space and that $(u, (x_1,..., x_m)) \mapsto (ux_1,..., ux_m)$ defines a transitive action of $\mathcal{O}(E)$ on $\Sigma^{(m)}(E)$.

Next consider the set

$$\mathcal{G}_m(E)$$

of all m-dimensional subspaces of E. We abbreviate this to

$$\mathcal{G}_m$$

when $E = \ell_2^n$ for some $n \geq m$. There is a natural surjective map

$$\Sigma^{(m)}(E) \longrightarrow \mathcal{G}_m(E) \quad \text{given by} \quad (x_1,..., x_m) \mapsto \operatorname{span}\{x_1,..., x_m\}.$$

Equip $\mathcal{G}_m(E)$ with the quotient topology and check that a compact Hausdorff space is obtained in this way.

19.14 Example: *$\mathcal{O}(E)$ acts transitively on $\mathcal{G}_m(E)$.*

The action $\mathcal{O}(E) \times \mathcal{G}_m(E) \to \mathcal{G}_m(E)$ is given by

$$(u, \operatorname{span}\{x_1,..., x_m\}) \mapsto \operatorname{span}\{ux_1,..., ux_m\}$$

whenever $(x_1,..., x_m) \in \Sigma^{(m)}(E)$. Checking that the various requirements for a transitive group action are met is routine hackwork, which we omit. However, it may be helpful to observe that, loosely speaking, vector spaces F, F' in $\mathcal{G}_m(E)$ are close when F has an orthonormal basis $\{x_1,..., x_m\}$ and F' has an orthonormal basis $\{x'_1,..., x'_m\}$ with the property that for each $1 \leq k \leq m$, x_k is close to x'_k in E.

We shall denote the rotation invariant probability measure on $\mathcal{G}_m$ by

$$\gamma_m.$$

The infrastructure is now prepared.

Proof of 19.13. The idea is to show that for any bounded Borel function $f \geq 0$ on Σ_n and any $1 \leq k \leq n$

$$\text{(B)} \qquad \int_{\Sigma_n} f(x,y)\, d\sigma_n(x,y) = \int_{\mathcal{G}_k} \Big(\int_{\Sigma(E)} f(x,y)\, d\sigma_E(x,y) \Big)\, d\gamma_k(E).$$

As γ_k is a probability measure, it will be impossible to have

$$\int_{\Sigma(E)} f(x,y)\, d\sigma_E(x,y) > \int_{\Sigma_n} f(x,y)\, d\sigma_n(x,y)$$

for every $E \in \mathcal{G}_k$, and the conclusion will follow.

To prove (B) we exploit the uniqueness of the rotation invariant probability measure on Σ_n. Let $h \in C(\Sigma_n)$ and define $\varphi \in C(\mathcal{G}_k)$ by

$$\varphi(E) := \int_{\Sigma(E)} h(x,y)\, d\sigma_E(x,y)$$

for each $E \in \mathcal{G}_k$. To see why φ is continuous note that if E, $E' \in \mathcal{G}_k$, there is certainly some $u = u_{E,E'} \in \mathcal{O}_n$ such that $E' = u(E)$. Then

$$\varphi(E') = \int_{\Sigma(u(E))} h(\xi,\eta)\, d\sigma_{u(E)}(\xi,\eta) = \int_{\Sigma(E)} h(ux,uy)\, d\sigma_E(x,y).$$

The topology on $\mathcal{G}_k$ allows us to conclude that as $E' \to E$, we have $u_{E,E'}x \to x$ uniformly on S_E. Continuity is now clear.

Since γ_k is rotation invariant,

$$\int_{\mathcal{G}_k} \varphi(E)\, d\gamma_k(E) = \int_{\mathcal{G}_k} \varphi(vE)\, d\gamma_k(E)$$

for every $v \in \mathcal{O}_n$. It follows that the map $\mathcal{C}(\Sigma_n) \to \mathbf{R}$ given by

$$h \mapsto \int_{\mathcal{G}_k} \int_{\Sigma(E)} h(x,y)\, d\sigma_E(x,y)\, d\gamma_k(E)$$

defines a rotation invariant probability measure on Σ_n, and so must be σ_n in disguise. This is the content of (B). QED

Proof of Dvoretzky's Theorem

Although the preliminaries are over, we shall still need substantial effort to reach our goal.

We require an effective way to measure when a k-dimensional Banach space is close to ℓ_2^k, and to do this we work with a restricted class of norms. Let us say that a norm p on $\mathbf{R}^k$ is a *Dvoretzky - Rogers norm,* and write

$$p \in DR(k),$$

when

(a) the identity map $\ell_2^k \to [\mathbf{R}^k, p]$ has norm 1,
(b) the identity map $[\mathbf{R}^k, p] \to \ell_\infty^k$ has norm at most $\sqrt{3}$.

Notice that the construction of 14.2 shows that for every $k \in \mathbf{N}$, each infinite dimensional Banach space induces a Dvoretzky - Rogers norm on some k-dimensional subspace.

For $p \in DR(k)$, we set

$$v_k(p) := \int_{\Sigma_k} \frac{\left|(y \,|\, \nabla p(x))\right|^2}{p(x)^2}\, d\sigma_k(x,y).$$

This makes perfectly good sense. Indeed, the integrand is defined almost everywhere, thanks to 19.5, and it is measurable since it is almost everywhere the pointwise limit of continuous functions. Moreover, it is bounded by one, because of the final remarks we made in the section on convex functions. The significance of $v_k(p)$ is apparent from the next result.

19.15 Theorem: *Let k be a positive integer. For each $\varepsilon > 0$, there is a $\delta = \delta(k,\varepsilon) > 0$ such that $d\big([\mathbf{R}^k, p], \ell_2^k\big) < 1 + \varepsilon$ whenever p is a Dvoretzky - Rogers norm satisfying $v_k(p) < \delta$.*

Here $d(X,Y)$ is the *Banach - Mazur distance* between two isomorphic Banach spaces X and Y:

$$d(X,Y) := \inf\{\|v\| \cdot \|v^{-1}\| : v : X \to Y \text{ is an isomorphism}\}.$$

Proof: We argue by contradiction. If the result fails, there are an exceptional $\varepsilon_0 > 0$ and a sequence (p_n) in $DR(k)$ such that

$$\lim_{n\to\infty} v_k(p_n) = 0 \quad \text{but} \quad d([\mathbf{R}^k, p_n], \ell_2^k) \geq 1 + \varepsilon_0.$$

Since p_n is a Dvoretzky-Rogers norm, the identity map $\ell_2^k \to [\mathbf{R}^k, p_n]$ has norm one. Accordingly, its inverse must have norm at least $1 + \varepsilon_0$. In other words, for each n,

$$\max_{\|x\|_2=1} p_n(x) = 1 \quad \text{and} \quad \max_{p_n(x)=1} \|x\|_2 \geq 1 + \varepsilon_0.$$

This can be reworked into the form

$$(*) \qquad \max_{\|x\|_2=1} p_n(x) = 1 \quad \text{and} \quad \min_{\|x\|_2=1} p_n(x) \leq \frac{1}{1+\varepsilon_0}.$$

After all, there must be an $x \in \mathbf{R}^k$ with $p_n(x) = 1$ but $\|x\|_2 \geq 1 + \varepsilon_0$. Looking at $y = x/\|x\|_2$, we see that $\|y\|_2 = 1$ and $p_n(y) = p_n(x)/\|x\|_2 \leq 1/(1+\varepsilon_0)$.

We push $(*)$ one stage further by invoking the Arzelà-Ascoli Theorem. Since the p_n's are Dvoretzky-Rogers norms, they are equicontinuous and uniformly bounded, say on $2 \cdot B_{\ell_2^k}$. Thus we can, passing to a subsequence if necessary, assume that the p_n's converge uniformly to a function p on $2 \cdot B_{\ell_2^k}$. This limit p can evidently be extended to an element of $DR(k)$, and plainly

$$\max_{\|x\|_2=1} p(x) = 1 \quad \text{and} \quad \min_{\|x\|_2=1} p(x) \leq \frac{1}{1+\varepsilon_0}.$$

This will prove to be our sticking point. We shall soon see that p is constant on S^{k-1}, and from this we are forced to an absurdity:

$$1 = \max_{\|x\|_2=1} p(x) = \min_{\|x\|_2=1} p(x) \leq \frac{1}{1+\varepsilon_0}.$$

Naturally, we must exploit the consequences of $\lim_{n\to\infty} v_k(p_n) = 0$. Let us first show that

$$(y \,|\, \nabla p(x)) = 0$$

for σ_k-almost all $(x,y) \in \Sigma_k$. For this we consider the set

$$A := \{x \in S^{k-1} : p \text{ and each } p_n \text{ are Fréchet differentiable at } x\}.$$

Since p and each p_n are positively homogeneous, their Fréchet differentiability at some $x \in S^{k-1}$ is tantamount to their Fréchet differentiability at tx for all $t \neq 0$. In view of 19.5, it follows that $\lambda_k(A) = 1$, and, thanks to the remarks preceding 19.13, that $\sigma_k((A \times S^{k-1}) \cap \Sigma_k) = 1$. Now Fatou's Theorem tells us that

$$\int_{\Sigma_k} \liminf_{n\to\infty} \frac{|(y \,|\, \nabla p_n(x))|^2}{p_n(x)^2} \, d\sigma_k(x,y) \leq \lim_{n\to\infty} v_k(p_n) = 0.$$

Consequently, for σ_k-almost all (x,y) in $(A \times S^{k-1}) \cap \Sigma_k$, we obtain $\liminf_n (y \,|\, \nabla p_n(x)) \cdot p_n(x)^{-1} = 0$, and so $\liminf_n (y \,|\, \nabla p_n(x)) = 0$. As we already know from 19.6 that for each $x \in A$, $\lim_{n\to\infty} \|\nabla p_n(x) - \nabla p(x)\| = 0$, we can certainly affirm that

$$(y \,|\, \nabla p(x)) = 0 \quad \text{for } \sigma_k\text{-almost all } (x,y) \in (A \times S^{k-1}) \cap \Sigma_k.$$

We can squeeze a little more information by noticing that, for each $x \in A$, $y \mapsto \big(y \,|\, \nabla p(x)\big)$ is continuous in y. So if

$$B := \{x \in A : \big(y \,|\, \nabla p(x)\big) = 0 \text{ for all } y \perp x\},$$

then $\lambda_k(B) = \lambda_k(A) = 1$.

To finish the proof that p must be constant on S^{k-1}, select $a, b \in S^{k-1}$ and a smooth curve $g : [0,1] \to S^{k-1}$ joining a and b. Since S^{k-1} can be covered by disjoint curves of this form, it is possible to select g so that almost all of its values lie in B. Do so. As $\|g(t)\|$ is constant for all $0 \leq t \leq 1$, we have $\big(g'(t) \,|\, g(t)\big) = 0$ for every $0 < t < 1$. Should $g(t)$ find itself inside B – as we have supposed it almost always does – then $\big(g'(t) \,|\, \nabla p(g(t))\big) = 0$ also.

Now $p \circ g$ is absolutely continuous, so the Lebesgue-Vitali Theorem, in tandem with the chain rule, warns us that

$$p(b) - p(a) = p(g(1)) - p(g(0)) = \int_0^1 (p\, g)'(t)\, dt = \int_0^1 \big(g'(t) \,|\, \nabla p(g(t))\big)\, dt = 0.$$

QED

To make effective use of the last theorem, we must be able to select Dvoretzky-Rogers norms p with $v_k(p) < \delta$. The next result shows that this is never a problem when k is large enough.

19.16 Theorem: $\lim_{k\to\infty} \big(\sup_{p \in DR(k)} v_k(p)\big) = 0.$

Proof. For $p \in DR(k)$,

$$\begin{aligned} v_k(p) &= \int_{\Sigma_k} \frac{\big|\big(y \,|\, \nabla p(x)\big)\big|^2}{p(x)^2}\, d\sigma_k(x,y) \\ &= \int_{S^{k-1}} \frac{1}{p(x)^2} \cdot \Big(\int_{S_{\{x\}^\perp}} \big|\big(y \,|\, \nabla p(x)\big)\big|^2\, d\lambda_{\{x\}^\perp}(y) \Big)\, d\lambda_k(x). \end{aligned}$$

Thanks to 19.11,

$$v_k(p) = \int_{S^{k-1}} \frac{1}{p(x)^2} \cdot \frac{1}{k-1} \cdot \big\| \nabla p(x) \,|_{\{x\}^\perp} \big\|_2^2\, d\lambda_k(x).$$

However, since we have already observed that $|(z, \nabla p(x))| \leq p(z)$, the Dvoretzky-Rogers nature of p gives

$$v_k(p) \leq \frac{1}{k-1} \cdot \int_{S^{k-1}} \frac{1}{p(x)^2}\, d\lambda_k(x) \leq \frac{3}{k-1} \cdot \int_{S^{k-1}} \frac{1}{\|x\|_\infty} d\lambda_k(x).$$

To proceed further, we call on a sequence $(g_i)_{i \in \mathbf{N}}$ of independent standard Gaussian random variables on a probability space (Ω, Σ, P). Fix $k \in \mathbf{N}$, set

$$h_j := \frac{g_j}{\big(\sum_{i\leq k} |g_i|^2\big)^{1/2}} \quad (1 \leq j \leq k),$$

and observe that $\omega \mapsto (h_i(\omega))_{i=1}^k$ defines a map $\Omega \to S^{k-1}$. The corresponding image measure on S^{k-1} of P is rotation invariant, thanks to the fact that rows of orthogonal matrices have ℓ_2 norm one and that $\sum_{i\leq k} a_i g_i$ is again a standard Gaussian variable when $\sum_{i \leq k} |a_i|^2 = 1$. It follows that the image measure is nothing other than λ_k, and so

$$\int_{S^{k-1}} \frac{1}{\|x\|_\infty}\, d\lambda_k(x) = \int_\Omega \frac{1}{\max_{i\le k}|h_i(\omega)|^2}\, dP(\omega)$$
$$= \sum_{j\le k}\int_\Omega \frac{|g_j(\omega)|^2}{\max_{i\le k}|g_i(\omega)|^2}\, dP(\omega) = k\cdot\int_\Omega \frac{|g_1(\omega)|^2}{\max_{i\le k}|g_i(\omega)|^2}\, dP(\omega).$$

We can therefore assert that

$$\sup_{p\in DR(k)} v_k(p) \le \frac{3k}{k-1}\cdot\int_\Omega \frac{|g_1(\omega)|^2}{\max_{i\le k}|g_i(\omega)|^2}\, dP(\omega),$$

and the conclusion will follow if we can show that the integral converges to zero as $k\to\infty$. This we shall do by proving that the integrand converges to zero in probability as $k\to\infty$ and applying the Bounded Convergence Theorem.

Notice that for any $c>0$ the independence of the g_i's gives

$$P\Big(\frac{1}{\max_{i\le k}|g_i(\omega)|^2} > c\Big) = P\big(\max_{i\le k}|g_i(\omega)| < 1/\sqrt{c}\big)$$
$$= \prod_{i\le k} P\big(|g_i(\omega)| < 1/\sqrt{c}\big) = \Big(P\big(|g_1(\omega)| < 1/\sqrt{c}\big)\Big)^k,$$

and this converges to zero as $k\to\infty$.

Now fix ε, $\delta>0$. Then, for any $c>0$,

$$P\Big(\frac{|g_1(\omega)|^2}{\max_{i\le k}|g_i(\omega)|^2} > \delta\Big) \le P\Big(\frac{1}{\max_{i\le k}|g_i(\omega)|^2} > c\Big) + P\big(|g_1(\omega)|^2 > \delta/c\big).$$

Choose c sufficiently small to arrange for $P\big(|g_1(\omega)|^2 > \delta/c\big) < \varepsilon/2$, and then record that $P\big((1/\max_{i\le k}|g_i(\omega)|^2) > c\big) < \varepsilon/2$ for all sufficiently large k. Convergence to zero in probability has been established. QED

All that remains is a cut and paste job.

Proof of 19.1. Fix $k\in\mathbf{N}$ and $\varepsilon>0$. Refer to 19.15 and find $\delta=\delta(k,\varepsilon)>0$ such that $d\big([\mathbf{R}^k,p],\ell_2^k\big) < 1+\varepsilon$ for each $p\in DR(k)$ satisfying $v_k(p)<\delta$. Next, turn to 19.16 to find an $m\ge k$ such that $\sup_{p\in DR(m)} v_m(p) < \delta$.

Thanks to Theorem 14.2, we can now choose $n=n_\varepsilon(k)\ge(4m)^2$ such that any Banach space of dimension n has the property that its norm when restricted to an appropriate m-dimensional subspace is equivalent to a Dvoretzky-Rogers norm on $\mathbf{R}^m$.

Select an n-dimensional Banach space $[X,p]$, take an m-dimensional subspace F such that p is a Dvoretzky-Rogers norm on F, and appeal to 19.13 to find a k-dimensional subspace E of F such that

$$\int_{\Sigma(E)} \frac{\big|\big(y\,\big|\,\nabla p(x)\big)\big|^2}{p(x)^2}\, d\sigma_E(x,y) \le v_m(p).$$

In other words, $v_k(p)\le v_m(p)<\delta$; so we are done. QED

To show how Dvoretzky's Theorem may be used we include a relatively straightforward application.

19.17 Proposition: *Let X and Y be infinite dimensional Banach spaces.*

(a) *The space $\mathcal{K}(X, Y)$ of all compact operators from X to Y does not have finite cotype.*

(b) *The space $\mathcal{N}_1(X, Y)$ of all nuclear operators from X into Y does not have non-trivial type.*

Proof. (a) By the easy part of Theorem 14.1, all we need to show is that $\mathcal{K}(X, Y)$ contains uniform copies of all ℓ_∞^n's.

For this, use Theorem 19.1 to obtain, for each $n \in \mathbf{N}$, isomorphic embeddings $i_n : \ell_2^n \to X^*$ and $j_n : \ell_2^n \to Y$ such that $\|i_n\|, \|j_n\| \leq 2$ and $\|i_n^{-1}\| = \|j_n^{-1}\| = 1$. Notice that

$$v_n : \ell_\infty^n \longrightarrow \mathcal{L}(\ell_2^n) : (\xi_m) \mapsto \sum_{j \leq n} \xi_m e_m \otimes e_m$$

is an isometric embedding and that

$$u_n : \mathcal{L}(\ell_2^n) \longrightarrow \mathcal{K}(X, Y) : \sum_{k \leq n} g_k \otimes h_k \mapsto \sum_{k \leq n} i_n(g_k) \otimes j_n(h_k)$$

is one-to-one and satisfies $\|u_n\| \cdot \|u_n^{-1}\| \leq 4$. The composition $u_n \circ v_n$ produces a 4-isomorphic copy of ℓ_∞^n inside $\mathcal{K}(X, Y)$.

(b) For this case we appeal to the easy part of Pisier's Theorem 13.3: all we need to explain is why $\mathcal{N}_1(X, Y)$ contains uniform copies of all ℓ_1^n's.

Once again call on 19.1 to obtain 2-isomorphic embeddings, but this time $i_n : \ell_2^n \to X$ and $j_n : \ell_2^n \to Y^*$. The maps $i_n^* : X^* \to \ell_2^n$ and $j_n^* k_Y : Y \to \ell_2^n$ are both (!) surjective, and they produce quotients of X^* and Y which are 2-isomorphic to ℓ_2^n.

For each $n \in \mathbf{N}$, the operator

$$\mathcal{N}_1(X, Y) \longrightarrow \mathcal{N}_1(\ell_2^n) : \sum_k x_k^* \otimes y_k \mapsto \sum_k i_n^*(x_k^*) \otimes j_n^*(y_k)$$

displays $\mathcal{N}_1(\ell_2^n)$ as a space which is 4-isomorphic to a quotient of $\mathcal{N}_1(X, Y)$. By 6.4, $\mathcal{N}_1(\ell_2^n)$ is the trace dual of $\mathcal{L}(\ell_2^n)$ and so, by part (a), ℓ_1^n is 4-isomorphic to a quotient of $\mathcal{N}_1(X, Y)$. The lifting property of ℓ_1-spaces (see after 3.17) reveals that each ℓ_1^n is actually 5-isomorphic to a complemented subspace of $\mathcal{N}_1(X, Y)$. This is more than enough. QED

Basic Sequences

We turn our attention to the proof of the characterization 19.2 of Hilbert-Schmidt operators. In order to set up the factorization, we shall need to construct basic sequences in infinite dimensional Banach spaces in such a way that they can be cut into blocks whose elements resemble the unit basis vectors of arbitrarily large ℓ_2^n's. The precise description is unavoidably involved.

We shall make use of the fact that every infinite dimensional Banach space contains a basic sequence – a sequence which is a basis of its closed linear span. Recall that a sequence (x_k) of non-zero vectors in a Banach space is a *basic sequence* if there is a constant $C \geq 1$ such that, for any choice of scalars α_k $(k \in \mathbf{N})$ and of $m, n \in \mathbf{N}$ we have

$$\Big\| \sum_{k \leq n} \alpha_k x_k \Big\| \leq C \cdot \Big\| \sum_{k \leq n+m} \alpha_k x_k \Big\| .$$

The least C that works is the *basis constant* of the basic sequence.

A general process for using one basic sequence to construct another will pave our way.

19.18 Proposition: *Let (x_n) be a basic sequence in the Banach space X. Let $(k_n)_{n=1}^{\infty}$ and $(m_n)_{n=0}^{\infty}$ be sequences in $\mathbf{N}$ such that $m_{n-1} < m_n$ and $k_n \leq m_n - m_{n-1}$ for all $n \in \mathbf{N}$. For each $n \in \mathbf{N}$, let $(z_1^{(n)}, \ldots, z_{k_n}^{(n)})$ be a basic sequence in* span $\{x_i \colon m_{n-1} < i \leq m_n\}$ *with basis constant c_n. If* $\sup_n c_n < \infty$, *then*

$$(z_1^{(1)}, \ldots, z_{k_1}^{(1)}, z_1^{(2)}, \ldots, z_{k_2}^{(2)}, z_1^{(3)}, \ldots)$$

is a basic sequence in X.

Proof. We introduce a little more helpful notation. For each $n \in \mathbf{N}$, let $M_n := \{i \in \mathbf{N} \colon m_{n-1} < i \leq m_n\}$ and $N_n := \{i \in M_n \colon i \leq m_{n-1} + k_n\}$. Also set $\widetilde{\mathbf{N}} := \bigcup_{n=1}^{\infty} N_n$. It is convenient to relabel the z's: for each $n \in \mathbf{N}$ write

$$z_{m_{n-1}+j} := z_j^{(n)} \quad \text{for } 1 \leq j \leq k_n .$$

Now let k, ℓ be positive integers with $\ell \in N_r$ and $\ell + k \in N_q$ for some $r, q \in \mathbf{N}$, and let (α_i) be a sequence in $\mathbf{K}^{\widetilde{\mathbf{N}}}$. By the definition of the z_j's, we can find a sequence $(\beta_i) \in \mathbf{K}^{\mathbf{N}}$ with

$$\sum_{j \in N_p} \alpha_j z_j = \sum_{i \in M_p} \beta_i x_i \text{ for each } 1 \leq p < q, \text{ and } \sum_{j=m_{q-1}+1}^{\ell+k} \alpha_j z_j = \sum_{i \in M_q} \beta_i x_i .$$

If we take C to be the maximum of $\sup_n c_n$ and the basis constant of (x_n), we obtain

$$\begin{aligned}
\Big\| \sum_{j \leq \ell, j \in \widetilde{\mathbf{N}}} \alpha_j z_j \Big\| &\leq \Big\| \sum_{p<r} \sum_{j \in N_p} \alpha_j z_j \Big\| + \Big\| \sum_{j=m_{r-1}+1}^{\ell} \alpha_j z_j \Big\| \\
&\leq \Big\| \sum_{i \leq m_{r-1}} \beta_i x_i \Big\| + C \cdot \Big\| \sum_{j \in N_r} \alpha_j z_j \Big\| = \Big\| \sum_{i \leq m_{r-1}} \beta_i x_i \Big\| + C \cdot \Big\| \sum_{i \in M_r} \beta_i x_i \Big\| \\
&\leq C \cdot \Big\| \sum_{i \leq m_q} \beta_i x_i \Big\| + C \cdot \Big\| \sum_{i \leq m_r} \beta_i x_i - \sum_{i \leq m_{r-1}} \beta_i x_i \Big\| \\
&\leq C \cdot \Big\| \sum_{i \leq m_q} \beta_i x_i \Big\| + 2 \cdot C^2 \cdot \Big\| \sum_{i \leq m_q} \beta_i x_i \Big\| \\
&\leq 3 \cdot C^2 \cdot \Big\| \sum_{i \leq m_q} \beta_i x_i \Big\| = 3 \cdot C^2 \cdot \Big\| \sum_{j \leq \ell+k, j \in \widetilde{\mathbf{N}}} \alpha_j z_j \Big\| .
\end{aligned}$$

The z_j's thus form a basic sequence with basis constant at most $3 \cdot C^2$. QED

The next result is designed to be the tool for carving out what we really need.

19.19 Theorem: *Let X be an infinite dimensional Banach space, let $\kappa > 1$ and let (k_n) be a sequence in $\mathbf{N}$. There are a sequence (N_n) of finite subsets of $\mathbf{N}$ and a basic sequence (z_j) in X with the following properties:*

(a) $|N_n| = k_n$ *and* $\max N_n < \min N_{n+1}$ *for each* $n \in \mathbf{N}$,

(b) *for any* $(\alpha_j) \in \mathbf{K}^{\mathbf{N}}$ *and* $n \in \mathbf{N}$,

$$\frac{1}{\kappa}\cdot\Big(\sum_{j\in N_n} |\alpha_j|^2\Big)^{1/2} \leq \|\sum_{j\in N_n} \alpha_j z_j\| \leq \kappa\cdot\Big(\sum_{j\in N_n} |\alpha_j|^2\Big)^{1/2}.$$

Proof. Dvoretzky's Theorem 19.1 provides us with a function $\nu : \mathbf{N} \to \mathbf{N}$ such that, for each $n \in \mathbf{N}$, every $\nu(n)$-dimensional subspace of X contains an n-dimensional subspace which is κ^2-isomorphic to ℓ_2^n.

We start by selecting an increasing sequence $(m_n)_{n=0}^{\infty}$ of integers with $m_0 = 0$ and $\nu(k_n) \leq m_n - m_{n-1}$ for every $n \in \mathbf{N}$. Next we fix a basic sequence (x_n) in X. By design, each $X_n := \operatorname{span}\{x_i : m_{n-1} < i \leq m_n\}$ has dimension $\geq \nu(k_n)$, and so there is an isomorphic embedding $u_n : \ell_2^{k_n} \to X_n$ which satisfies $\|u_n\| = \|u_n^{-1}\| \leq \kappa$.

The vectors $z_{m_{n-1}+i} := u_j e_j$ $(1 \leq j \leq k_n)$ form a basic sequence in X with basis constant at most κ^2 since, if $\alpha_1,\ldots,\alpha_{k_n}$ are scalars and if $1 \leq p < q \leq k_n$, then

$$\begin{aligned}
\Big\|\sum_{j\leq p}\alpha_j z_{m_{n-1}+j}\Big\| &= \Big\|u_n\Big(\sum_{j\leq p}\alpha_j e_j\Big)\Big\| \leq \|u_n\|\cdot\Big\|\sum_{j\leq p}\alpha_j e_j\Big\| \\
&\leq \|u_n\|\cdot\Big\|\sum_{j\leq q}\alpha_j e_j\Big\| = \|u_n\|\cdot\Big\|\sum_{j\leq q}\alpha_j u_n^{-1} z_{m_{n-1}+j}\Big\| \\
&\leq \|u_n\|\cdot\|u_n^{-1}\|\cdot\Big\|\sum_{j\leq q}\alpha_j z_{m_{n-1}+j}\Big\| \leq \kappa^2\cdot\Big\|\sum_{j\leq q}\alpha_j z_{m_{n-1}+j}\Big\|\,.
\end{aligned}$$

This opens the door to 19.18. If we put $N_n := \{j \in \mathbf{N} : m_{n-1} < j \leq m_{n-1}+k_n\}$ for each $n \in \mathbf{N}$ and set $\widetilde{\mathbf{N}} := \bigcup_{n=1}^{\infty} N_n$, then $(z_j)_{j\in\widetilde{\mathbf{N}}}$ is a basic sequence in X. Modulo reindexing of the z_j's, the inequalities we want are just a reflection of the fact that $\|u_n\| = \|u_n^{-1}\| \leq \kappa$. QED

Factorization

We are now close to a proof of Theorem 19.2 but we need one more result to help us out. To facilitate its statement, let us say that a (compact) Banach space operator $u : X \to Y$ *factors compactly* through the Banach space Z if there are compact operators $v : Z \to Y$ and $w : X \to Z$ satisfying $u = vw$.

It is also worth recalling that any sequence $\tau = (\tau_n)$ in ℓ_∞ induces a diagonal operator $D_\tau : \ell_2 \to \ell_2$. The operator is compact if and only if $\tau \in c_0$. Naturally, D_τ is said to be *positive* if each τ_n is non-negative.

19.20 Theorem: *Let Z be an infinite dimensional Banach space. Any compact operator $u : H_1 \to H_2$ between Hilbert spaces factors compactly through some subspace of Z which has a basis.*

Proof. It follows in the standard fashion from Theorem 4.1 that u has a factorization $u : H_1 \xrightarrow{b} \ell_2 \xrightarrow{D_\tau} \ell_2 \xrightarrow{a} H_2$ where a and b are bounded and D_τ is a compact diagonal operator. It clearly suffices to prove our statement for D_τ, and there is certainly no loss in assuming that $\|\tau\|_\infty = 1$ and $\tau_n \geq 0$ for all $n \in \mathbf{N}$.

We first use 19.19 to produce an appropriate basic sequence in Z. Set $\lambda_n := \tau_n^{1/4}$ for each $n \in \mathbf{N}$ and choose a strictly increasing sequence (k_n) of integers so that $k_0 = 0$ and $\lambda_j \leq 2^{-n}$ whenever $j \geq k_n$. Then 19.19 ensures the existence of a basic sequence (z_n) in Z with the property that, for any scalar sequence (α_n) and for any $m \in \mathbf{N}$,

$$\frac{1}{2} \cdot \Big(\sum_{k_{m-1}<j\leq k_m} |\alpha_j|^2 \Big)^{1/2} \leq \Big\| \sum_{k_{m-1}<j\leq k_m} \alpha_j z_j \Big\| \leq 2 \cdot \Big(\sum_{k_{m-1}<j\leq k_m} |\alpha_j|^2 \Big)^{1/2}.$$

Let Z_0 be the Banach space generated by (z_n). We shall show that it is possible to define operators

$$a : \ell_2 \longrightarrow Z_0 : (\xi_n) \mapsto \sum_n \lambda_n \xi_n z_n \quad \text{and} \quad b : Z_0 \to \ell_2 : \sum_n \eta_n z_n \mapsto (\lambda_n \eta_n)_n.$$

If we take this on trust we can easily construct the desired factorization: we have $D_\tau = vw$ where $v := D_\lambda\, b : Z_0 \to \ell_2$ and $w := a\, D_\lambda : \ell_2 \to Z_0$ are compact operators.

Let us show that a is what we claim. Fix $m \in \mathbf{N}$ and let $n_0 \in \mathbf{N}$ be such that $k_{n_0-1} < m \leq k_{n_0}$. Then, if $\xi = (\xi_n) \in \ell_2$,

$$\begin{aligned}
\Big\| \sum_{j\leq m} \lambda_j \xi_j z_j \Big\| &\leq \sum_{n\leq n_0-1} \Big\| \sum_{k_{n-1}<j\leq k_n} \lambda_j \xi_j z_j \Big\| + \Big\| \sum_{k_{n_0-1}<j\leq m} \lambda_j \xi_j z_j \Big\| \\
&\leq 2 \cdot \sum_{n\leq n_0} \Big(\sum_{k_{n-1}<j\leq k_n} |\lambda_j \xi_j|^2 \Big)^{1/2} \leq 2 \cdot \sum_{n\leq n_0} 2^{-n+1} \cdot \Big(\sum_{k_{n-1}<j\leq k_n} |\xi_j|^2 \Big)^{1/2} \\
&\leq 2 \cdot \sum_{n\leq n_0} 2^{-n+1} \cdot \|\xi\|_2 \leq 8 \cdot \|\xi\|_2.
\end{aligned}$$

It follows that a is properly defined, and $\|a\| \leq 8$.

Now for b. Take $y = \sum_n \eta_n z_n$ in Z_0, fix $m \in \mathbf{N}$ and, as before, locate $n_0 \in \mathbf{N}$ such that $k_{n_0-1} < m \leq k_{n_0}$. Then if C is the basis constant for (z_n),

$$\begin{aligned}
\|(\lambda_n \eta_n)_1^m\|_{\ell_2} &\leq \sum_{n\leq n_0-1} \Big\| \sum_{k_{n-1}<j\leq k_n} \lambda_j \eta_j e_j \Big\|_{\ell_2} + \Big\| \sum_{k_{n_0-1}<j\leq m} \lambda_j \eta_j e_j \Big\|_{\ell_2} \\
&\leq \sum_{n\leq n_0-1} 2^{-n+1} \cdot \Big(\sum_{k_{n-1}<j\leq k_n} |\eta_j|^2 \Big)^{1/2} + 2^{-n_0+1} \cdot \Big(\sum_{k_{n_0-1}<j\leq m} |\eta_j|^2 \Big)^{1/2}
\end{aligned}$$

$$\leq \sum_{n\leq n_0-1} 2^{-n+2}\cdot\Big\| \sum_{k_{n-1}<j\leq k_n} \eta_j z_j \Big\| + 2^{-n_0+2}\cdot\Big\| \sum_{k_{n_0-1}<j\leq m} \eta_j z_j \Big\|$$
$$\leq \sum_{n\leq n_0-1} 2^{-n+3}\cdot C\cdot\Big\| \sum_{j\leq m} \eta_j z_j \Big\| + 2^{-n_0+3}\cdot C\cdot\Big\| \sum_{j\leq m} \eta_j z_j \Big\|$$
$$\leq 12\cdot C\cdot\Big\| \sum_{j\leq m} \eta_j z_j \Big\| .$$

Again, b is well-defined, and $\|b\| \leq 12\cdot C$. QED

The proof of our second major result is now within easy reach

Proof of Theorem 19.2. Let $u = \sum_n \tau_n g_n \otimes h_n$ be any ON-representation of our Hilbert-Schmidt operator $u : H_1 \to H_2$. Take sequences $(\alpha_n), (\beta_n) \in c_0$ and $(\sigma_n) \in \ell_2$ such that $\alpha_n \sigma_n \beta_n = \tau_n$ for each $n \in \mathbf{N}$. Accordingly, we can write $u = aD_\sigma b$ where $a := \sum_n \alpha_n e_n \otimes h_n : \ell_2 \to H_2$ and $b := \sum_n \beta_n g_n \otimes e_n : H_1 \to \ell_2$ are compact and the diagonal operator $D_\sigma : \ell_2 \to \ell_2$ is still Hilbert-Schmidt. By 19.20, b has a compact factorization $b : H_1 \xrightarrow{b_2} Z_0 \xrightarrow{b_1} \ell_2$ through a subspace Z_0 of Z. Consider D_σ as the restriction of a diagonal operator $D : \ell_\infty \to \ell_2$ and let $i : \ell_2 \to \ell_\infty$ be the formal identitity. By injectivity of ℓ_∞ (see 2.13 and 4.14), ib_1 has an extension $\widetilde{b}_1 : Z \to \ell_\infty$. Then $v := aD\widetilde{b}_1$ is compact and 2-summing, b_2 considered as an operator $w : H_1 \to Z$ is compact, and clearly $u = vw$. QED

With only a little extra effort, Theorem 19.20 can be extended significantly to show that any compact operator into or out of a Hilbert space factors compactly through a subspace of any Banach space. Here is the key.

19.21 Proposition: *Let X be a Banach space.*

(a) *Every compact operator $u : X \to \ell_2$ has a factorization $u : X \xrightarrow{w} \ell_2 \xrightarrow{v} \ell_2$ where v and w are compact and, in addition, v is a positive diagonal operator.*

(b) *Every compact operator $u : \ell_2 \to X$ has a factorization $u : \ell_2 \xrightarrow{w} \ell_2 \xrightarrow{v} X$ where v and w are compact and, in addition, w is a positive diagonal operator.*

Proof. (a) For each $n \in \mathbf{N}$, let $P_n : \ell_2 \to \ell_2^n$ be the canonical projection. We begin by showing that $\lim_{n\to\infty} \|u - P_n u\| = 0$.

Certainly, if $x \in B_X$, then $\lim_{n\to\infty} \|ux - P_n ux\| = 0$. The relative compactness of $u(B_X)$ allows us to pass from pointwise convergence to uniform convergence. Here is the straightforward argument.

Fix $\varepsilon > 0$ and select an $(\varepsilon/3)$-net $ux_1,\dots, ux_k$ in $u(B_X)$. Then there is an $N \in \mathbf{N}$ such that $\|ux_j - P_n ux_j\| \leq \varepsilon/3$ for all $n \geq N$ and all $1 \leq j \leq k$. Now select $x \in B_X$ and locate $1 \leq j \leq k$ with $\|ux - ux_j\| \leq \varepsilon/3$. Then, if $n \geq N$, $\|ux - P_n ux\| \leq \|ux - ux_j\| + \|ux_j - P_n ux_j\| + \|P_n(ux_j - ux)\| \leq \varepsilon$. Thus $\lim_{n\to\infty} \|u - P_n u\| = 0$.

Next, set $n_0 = 0$ and for each $k \in \mathbf{N}$ construct an integer n_k so that $n_k > n_{k-1}$ and $\|u - P_n u\| \leq 2^{-k}$ for every $n \geq n_k$. Define a scalar null sequence $\lambda = (\lambda_n)$ by setting $\lambda_n := k^{-1}$ for $n_{k-1} < n \leq n_k$, $k \in \mathbf{N}$. We use this to obtain the operators v and w we are looking for.

The operator v is just D_λ; it is compact since $\lambda \in c_0$. There is an obvious candidate for w, but since we need to prove compactness, we take an indirect route.

Notice that if $x \in X$, then

$$\sum_n |\lambda_n^{-2}(ux|e_n)|^2 = \sum_{k=1}^{\infty} \sum_{n_{k-1}<i\leq n_k} k^4 \cdot |(ux|e_i)|^2 = \sum_{k=1}^{\infty} k^4 \cdot \|P_{n_k}ux - P_{n_{k-1}}ux\|^2$$

$$\leq \sum_{k=1}^{\infty} k^4 \cdot \left(\|P_{n_k}ux - ux\| + \|P_{n_{k-1}}ux - ux\|\right)^2 \leq 9 \cdot \sum_{k=1}^{\infty} k^4 \cdot 2^{-2k} \cdot \|x\|^2.$$

Since the series converges, we can interpret this as saying that $w_0 : X \to \ell_2 : x \mapsto \sum_n \lambda_n^{-2}(ux|e_n)e_n$ is a bounded linear map. But now notice that $u = v(vw_0)$ is the desired composition of two compact operators.

(b) By duality we conclude from (a) that $k_X u : \ell_2 \xrightarrow{D_\lambda} \ell_2 \xrightarrow{v} X^{**}$ where v is compact and D_λ is the diagonal operator induced by a sequence $\lambda = (\lambda_n) \in c_0$ with $\lambda_n > 0$ for each $n \in \mathbf{N}$. But then $ve_n = \lambda_n^{-1}v(\lambda_n e_n) = \lambda_n^{-1}vD_\lambda e_n = \lambda_n^{-1}k_X ue_n \in k_X(X)$ for each n, and so v takes its values in X. This proves the statement. QED

We can now deduce the following result directly from 19.20:

19.22 Corollary: *Let $u : X \to Y$ be a compact Banach space operator. If X or Y is a Hilbert space then, no matter how we choose the infinite dimensional Banach space Z, u must factor compactly through a subspace of Z (with a basis).*

When Z has special properties, we can even factor through the whole of Z and Z^*.

19.23 Corollary: *Let X and Z be infinite dimensional Banach spaces, with Z having type 2. All operators in $\mathcal{K}(X, \ell_2)$ and $\mathcal{K}(\ell_2, X)$ factor compactly through Z and through Z^*.*

Proof. By 19.21 it is enough to consider the case $X = \ell_2$; so let $u \in \mathcal{L}(\ell_2)$ be compact. We have just proved that u has a compact factorization $\ell_2 \xrightarrow{w} Z_0 \xrightarrow{v} \ell_2$ where Z_0 is some subspace of Z. But Maurey's Extension Theorem 12.22 assures us that v is the restriction of some $\tilde{v} \in \mathcal{L}(Z, \ell_2)$, and, after a little adjustment for compactness, the assertion about factoring through Z follows.

Apply duality to obtain factorization through Z^*. QED

COMPLEMENTATION

We now begin our preparations for the proof of Theorem 19.3. We shall make extensive use of a device that was introduced in Chapter 14. Recall that, for each $n \in \mathbf{N}$, the n-th cotype 2 number

$$c_2(n, X)$$

of a Banach space X is the smallest of all constants $c \geq 0$ such that, regardless of how we select n vectors $x_1, \ldots, x_n$ from X,

$$\Big(\sum_{k\leq n} \|x_k\|^2\Big)^{1/2} \leq c \cdot \Big(\int_0^1 \Big\|\sum_{k\leq n} r_k(t) x_k\Big\|^2 dt\Big)^{1/2}.$$

If $2 \leq q < \infty$, we have $\big(\sum_{k\leq n} \|x_k\|^2\big)^{1/2} \leq n^{(1/2)-(1/q)} \cdot \big(\sum_{k\leq n} \|x_k\|^q\big)^{1/q}$. Hence

$$c_2(n, X) \leq n^{(1/2)-(1/q)} \cdot C_q(X)$$

whenever X has cotype $2 \leq q < \infty$. So if X has finite cotype,

$$\text{(C)} \qquad \lim_{n\to\infty} \Big(\frac{\log n}{n}\Big)^{1/2} \cdot c_2(n, X) = 0.$$

In 14.10 we saw that for any Banach space X, any $n \in \mathbf{N}$ and any non-trivial measure μ,

$$c_2(n, X) = c_2(n, L_2(\mu, X)).$$

We now need a companion result.

19.24 Proposition: *Let X be a Banach space. For all $k, n \in \mathbf{N}$,*

$$c_2(k, X) = c_2\big(k, [\mathcal{L}(\ell_2^n, X), \pi_\gamma]\big).$$

Proof. To ease our notational burden, write $L_n := \mathcal{L}(\ell_2^n, X)$. Let P denote standard Gaussian measure on $\mathbf{R}^n$ and observe that the coordinate functions $g_i : \mathbf{R}^n \to \mathbf{R}$ $(1 \leq i \leq n)$ form a sequence of independent standard Gaussian variables. Thanks to 14.10 we only need to show that

$$c_2(k, X) \leq c_2(k, [L_n, \pi_\gamma]) \leq c_2(k, L_2(P, X)).$$

For the left inequality, choose scalars $\alpha_1, \ldots, \alpha_n$ with $\sum_{k\leq n} |\alpha_k|^2 = 1$ and consider the linear map $X \to L_n : x \mapsto u_x$ defined by $u_x(\xi) := \big(\sum_{k\leq n} \alpha_k \xi_k\big) x$ for all $\xi = (\xi_k)_1^n \in \ell_2^n$. By 12.15,

$$\begin{aligned}
\pi_\gamma(u_x)^2 &= \int_{\mathbf{R}^n} \Big\|\sum_{k\leq n} g_k(\omega) u_x(e_k)\Big\|^2 dP(\omega) \\
&= \|x\|^2 \cdot \int_{\mathbf{R}^n} \Big|\sum_{k\leq n} \alpha_k g_k(\omega)\Big|^2 dP(\omega) = \|x\|^2.
\end{aligned}$$

The map is therefore an isometric embedding of X in $[L_n, \pi_\gamma]$; the inequality is immediate.

The other inequality succumbs to a similar strategy. Define a linear map $L_n \to L_2(P,X) : u \mapsto \tilde{u}$ by $\tilde{u} := \sum_{k\le n} g_k(\,\cdot\,)ue_k$. Again by 12.15, this map takes $[L_n, \pi_\gamma]$ isometrically into $L_2(P,X)$, which was what we wanted. QED

The Hilbert space projections to be considered in the next two statements are all assumed to be orthogonal.

19.25 Corollary: *Let k and n be positive integers and let $u \in \mathcal{L}(\ell_2^n, X)$. If $P_1,\dots,P_k$ are projections on ℓ_2^n which are pairwise orthogonal, then*

$$\min_{j\le k} \pi_\gamma(uP_j) \le k^{-1/2}\cdot c_2(k,X)\cdot \pi_\gamma(u).$$

Proof. Note that $\|\sum_{j\le k} \varepsilon_j P_j\| = 1$ for $(\varepsilon_j) \in \{-1,1\}^k$ and use the proposition to estimate:

$$\begin{aligned}
k\cdot \min_{j\le k} \pi_\gamma(uP_j)^2 &\le \sum_{j\le k} \pi_\gamma(uP_j)^2 \\
&\le c_2(k, [\mathcal{L}(\ell_2^n, X), \pi_\gamma])^2\cdot \int_0^1 \pi_\gamma\Big(\sum_{j\le k} r_j(t)uP_j\Big)^2 dt \\
&= c_2(k,X)^2\cdot \int_0^1 \pi_\gamma\Big(u\circ\Big(\sum_{j\le k} r_j(t)P_j\Big)\Big)^2 dt \\
&\le c_2(k,X)^2\cdot\pi_\gamma(u)^2\cdot \int_0^1 \Big\|\sum_{j\le k} r_j(t)P_j\Big\|^2 dt = c_2(k,X)^2\cdot\pi_\gamma(u)^2.
\end{aligned}$$

QED

It goes without saying that if $u \in \mathcal{L}(\ell_2^n, X)$, then $\|u\| \le \pi_\gamma(u)$. We are now in a position to prove that when X has finite cotype, ℓ_2^n has subspaces E of relatively large dimension for which $\|u\,|_E\|$ is substantially smaller than $\pi_\gamma(u)$.

19.26 Proposition: *Let X be a Banach space, and let $k,n \in \mathbf{N}$ satisfy $k \le n$. If $u \in \mathcal{L}(\ell_2^n, X)$, then there is a subspace E of ℓ_2^n such that*

$$\dim E > n-k \quad \text{and} \quad \|u\,|_E\| \le k^{-1/2}\cdot c_2(k,X)\cdot\pi_\gamma(u).$$

Proof. The key is to realize that $\|u\,|_E\| = \sup \|uP\|$ where the supremum runs over all rank one projections $P \in \mathcal{L}(\ell_2^n)$ with $P(\ell_2^n) \subset E$.

Let $P_1,\dots,P_m \in \mathcal{L}(\ell_2^n)$ be pairwise orthogonal rank one projections having the property that for each $1 \le j \le m$

$$\pi_\gamma(uP_j) > k^{-1/2}\cdot c_2(k,X)\cdot\pi_\gamma(u).$$

The previous corollary shows that $m < k$.

Let $P_1,\dots,P_{m_0}$ be a maximal collection of the type just described. The E we want can be taken to be the orthogonal complement of $\bigoplus_{i\le m_0} P_i(\ell_2^n)$.

To see this, first observe that $\dim E = n - m_0 \ge n-k$. Next, if P is a rank one projection in ℓ_2^n with $P(\ell_2^n) \subset E$, it must be orthogonal to $P_1,\dots,P_{m_0}$. Thanks to maximality, $\pi_\gamma(uP) \le k^{-1/2}\cdot c_2(k,X)\cdot\pi_\gamma(u)$. But π_γ is an ideal norm and uP has rank ≤ 1, and so $\|uP\| = \pi_\gamma(uP)$. Taking the appropriate supremum, we get what we want. QED

The estimates provided by this proposition open the door to just what we need – the construction of finite rank projections in a Banach space.

19.27 Proposition: *Let X_n be an n-dimensional subspace of the Banach space X, let $u : \ell_2^n \to X_n$ be an isomorphism and let $v : X \to \ell_2^n$ be an operator extending u^{-1}. For any subspace E of ℓ_2^n there is a projection $P \in \mathcal{L}(X)$ with range $u(E)$ such that*

$$\|P\| \leq \|u|_E\| \cdot \|v^*|_E\|.$$

Proof. Let $Q : \ell_2^n \to E$ be the orthogonal projection and define $P \in \mathcal{L}(X)$ as $P := uQv$. Clearly, P is a projection with range $u(E)$. As for the norm, begin by noting that for each $x \in X$,

$$\begin{aligned} \|Qvx\|^2 &= (Qvx|Qvx) = (vx|Qvx) = \langle x, v^*Qvx\rangle \\ &\leq \|x\| \cdot \|v^*Qvx\| \leq \|x\| \cdot \|v^*|_E\| \cdot \|Qvx\|. \end{aligned}$$

It follows that $\|Qvx\| \leq \|v^*|_E\| \cdot \|x\|$ and so

$$\|Px\| = \|uQvx\| \leq \|u|_E\| \cdot \|Qvx\| \leq \|u|_E\| \cdot \|v^*|_E\| \cdot \|x\|.$$ QED

Fortunately, the assumption of the existence of an extension of u^{-1} is not a problem for the norms we are working with.

19.28 Proposition: *Let E be a finite dimensional subspace of the Banach space X. Then every $w \in \mathcal{L}(E, \ell_2^n)$ has an extension $\tilde{w} \in \mathcal{L}(X, \ell_2^n)$ satisfying $\pi_\gamma^*(\tilde{w}) = \pi_\gamma^*(w)$.*

Proof. It is important to note that π_γ is an injective (ideal) norm. This allows us to view $[\mathcal{L}(\ell_2^n, E), \pi_\gamma]$ as a subspace of $[\mathcal{L}(\ell_2^n, X), \pi_\gamma]$. Trace duality exhibits w as an element of $[\mathcal{L}(\ell_2^n, E), \pi_\gamma]^*$, and so there is a norm preserving Hahn - Banach extension $W \in [\mathcal{L}(\ell_2^n, X), \pi_\gamma]^*$.

If X is finite dimensional, trace duality allows us to think of W as an element of $\mathcal{L}(X, \ell_2^n)$ with $\pi_\gamma^*(W) = \pi_\gamma^*(w)$, and we are done.

The general case requires a further irksome argument. Let F be a finite dimensional subspace of X containing E. Trace duality enables us to identify $W|_{\mathcal{L}(\ell_2^n, F)}$ with an operator $w_F \in \mathcal{L}(F, \ell_2^n)$. Clearly, w_F extends w and satisfies $\pi_\gamma^*(w_F) = \pi_\gamma^*(w)$.

But each $x \in X$ belongs to a finite dimensional subspace F of X which contains E. Since $\tilde{w}(x) := w_F(x)$ does not depend on the particular choice of F, we have set up a linear map $\tilde{w} : X \to \ell_2^n$ which extends w. Moreover, the maximality of π_γ^* (see 6.10) combines with the Biadjoint Criterion 6.8 to ensure that $\pi_\gamma^*(\tilde{w}) = \pi_\gamma^*(w)$. QED

Notice that π_γ could have been replaced by any injective (ideal) norm.

We need to add just one more ingredient: a link between the norms π_γ and $(\pi_\gamma^*)^d$. We begin by describing a useful procedure for obtaining a lower bound for π_γ^*.

It will again be convenient to think of the first n Rademacher functions as being defined on $D_n = \{-1,1\}^n$ rather than on $[0,1]$, just as discussed before 13.2. Let $u \in \mathcal{L}(\ell_2^n, E)$ where E is finite dimensional, and let $w \in \mathcal{O}_n$. Motivated by 12.14 we define

$$\varrho_w(u) := \Big\| \sum_{k\le n} r_k(\cdot) u w e_k \Big\|_{L_2(D_n,E)} .$$

It is quick and easy to check that ϱ_w is a norm on $\mathcal{L}(\ell_2^n, E)$, and 12.14 alerts us to the fact that

$$\pi_\gamma(u) \le \sup_{w\in\mathcal{O}_n} \varrho_w(u).$$

So, by trace duality,

$$\inf_{w\in\mathcal{O}_n} \varrho_w^*(v) \le \pi_\gamma^*(v)$$

for all $v \in \mathcal{L}(E, \ell_2^n)$.

19.29 Proposition: *Let $2 \le q < s < \infty$ and let $n \in \mathbf{N}$. If E is a finite dimensional normed space and $u \in \mathcal{L}(\ell_2^n, E)$, then*

$$\pi_\gamma(u) \le C_q(E)\cdot\|R^E\|\cdot m_s\cdot\pi_\gamma^*(u^*).$$

Here R^E is the Rademacher projection discussed at the beginning of Chapter 13, and m_s is the s-th moment of a standard Gaussian variable.

Proof. Let $w \in \mathcal{O}_n$. First of all we shall show that

$$(*) \qquad \varrho_w(u) \le \|R^E\|\cdot\varrho_w^*(u^*).$$

Choose a norm one element f of $L_2(D_n,E)^* = L_2(D_n,E^*)$ to satisfy

$$\varrho_w(u) = \Big\langle \sum_{k\le n} r_k(\cdot) u w e_k, f \Big\rangle.$$

Now f has the form

$$f(\cdot) = \sum_{A\subset\{1,\dots,n\}} w_A(\cdot) x_A^*$$

where the w_A's are the Walsh functions and the x_A^*'s are in E^*. For $k \le n$ write $x_k^* = x_{\{k\}}^*$ and notice that $\varrho_w(u) = \sum_{k\le n} \langle u w e_k, x_k^*\rangle$. Define $v \in \mathcal{L}(\ell_2^n, E^*)$ via $v(we_k) = x_k^*$ $(k \le n)$, so that

$$\varrho_w(u) = \sum_{k\le n} \langle u w e_k, v w e_k\rangle = \sum_{k\le n} \langle w e_k, u^* v w e_k\rangle = \mathrm{tr}\,(u^*v) \le \varrho_w^*(u^*)\cdot\varrho_w(v).$$

To obtain $(*)$ we need to explain why $\varrho_w(v) \le \|R^E\|$. For this observe that

$$x_k^* = \int_{D_n} r_k(\omega) f(\omega)\, d\mu_n(\omega)$$

for each $k \le n$. It follows that

$$\varrho_w(v) = \Big\| \sum_{k\le n} r_k(\cdot) x_k^* \Big\| = \|R_n^{E^*} f\| \le \|R_n^{E^*}\| = \|R_n^E\| \le \|R^E\|.$$

With $(*)$ secured, we now take $2 \le q < s < \infty$ and use 12.15 and 12.27 to get

$$\pi_\gamma(u) = \Big(\int_\Omega \Big\| \sum_{k\le n} g_k(\omega) u w e_k \Big\|^2 dP(\omega) \Big)^{1/2} \le C_q(E)\cdot m_s\cdot\varrho_w(u).$$

Combine this with $(*)$ and the remark above to reach the desired conclusion.

QED

K-CONVEXITY

We now head directly for the characterization 19.3 of K-convexity. This is still by no means easy and requires considerable machinery in addition to what we have provided so far. In particular, we shall call on a deep result, the Isoperimetric Inequality, in the course of the proof.

As before, let S^{n+1} be the Euclidean unit sphere in $\mathbf{R}^{n+2}$ and let λ_{n+2} be its normalized rotation invariant measure. Equip S^{n+1} with the *geodesic metric* d, so that if $x, y \in S^{n+1}$, then $d(x, y)$ is the angle between the lines joining x, y to the origin. Given any set $A \subset S^{n+1}$ and $\varepsilon > 0$, we set

$$A_\varepsilon := \{s \in S^{n+1} : d(s, A) < \varepsilon\}.$$

19.30 Isoperimetric Inequality: *Let $A \subset S^{n+1}$ satisfy $\lambda_{n+2}(A) \geq 1/2$. Then for all $\varepsilon > 0$,*

$$\lambda_{n+2}(A_\varepsilon) \geq 1 - 3 \cdot \exp(-\frac{n\varepsilon^2}{2}) .$$

We delay the proof of 19.30 until the end of this chapter. However, we use it straight away to obtain a useful estimate for the *median* of a continuous function $f : S^{n+1} \to \mathbf{R}$; this is the number M_f with the property that

$$\lambda_{n+2}(\{f \leq M_f\}) \geq \frac{1}{2} \quad \text{and} \quad \lambda_{n+2}(\{f \geq M_f\}) \geq \frac{1}{2} .$$

The continuity of f ensures that M_f is unique.

We are mainly interested in the following corollary.

19.31 Lévy's Lemma: *Let $f : S^{n+1} \to \mathbf{R}$ be a continuous function and set $A := \{f = M_f\}$. Then, for all $\varepsilon > 0$,*

$$\lambda_{n+2}(A_\varepsilon) \geq 1 - 6 \cdot \exp(-\frac{n\varepsilon^2}{2}).$$

Proof. The key is to observe that $A_\varepsilon = \{f \leq M_f\}_\varepsilon \cap \{f \geq M_f\}_\varepsilon$. One inclusion is evident. For the other, select $y \in \{f \leq M_f\}_\varepsilon \cap \{f \geq M_f\}_\varepsilon$. Then there exist $x_1, x_2 \in S^{n+1}$ with $d(x_1, y) < \varepsilon$, $d(x_2, y) < \varepsilon$ and $f(x_1) \leq M_f \leq f(x_2)$. By the Intermediate Value Theorem, any arc in $\{x \in S^{n+1} : d(x, y) < \varepsilon\}$ joining x_1 to x_2 contains some x with $f(x) = M_f$. Hence $y \in A_\varepsilon$.

The conclusion now falls easily from the Isoperimetric Inequality:

$$\lambda_{n+2}(A_\varepsilon^c) \leq \lambda_{n+2}(\{f \leq M_f\}_\varepsilon^c) + \lambda_{n+2}(\{f \geq M_f\}_\varepsilon^c) \leq 6 \cdot \exp(-\frac{n\varepsilon^2}{2}).$$

QED

Next, we shall deduce from Lévy's Lemma that when $u : \ell_2^n \to X$ is any operator, $\|us\|/\pi_\gamma(u)$ is small for most $s \in S^{n+1}$. Actually a much more precise result is available.

19.32 Proposition: *Let $F = \ell_2^m$ be a subspace of ℓ_2^n, $m, n \in \mathbf{N}$. Write M for the median of $s \mapsto \|us\|$ on S^{m-1}. Then, provided that $m \geq 3$ and $C \geq 2^{-1/2}$,*

$$\lambda_m\big(\{s \in S^{m-1} \colon \|us\| > \frac{3C}{\sqrt{m}}\pi_\gamma(u)\}\big) \leq 6 \cdot \exp\Big(-\frac{C^2\pi_\gamma(u)^2}{6 \cdot \|u|_F\|^2}\Big).$$

Proof. First of all, 12.14 shows us that

$$\pi_\gamma(u|_F) = \sqrt{m} \cdot \Big(\int_{S^{m-1}} \|us\|^2 d\lambda_m(s)\Big)^{1/2} \geq \sqrt{m} \cdot \Big(\int_{\|us\| \geq M} \|us\|^2 d\lambda_m(s)\Big)^{1/2}$$
$$\geq M \cdot \sqrt{m} \cdot \lambda_m\big(\{s \in S^{m-1} : \|us\| \geq M\}\big)^{1/2} \geq M \cdot \sqrt{\frac{m}{2}}\,.$$

So, provided that $C \geq 2^{-1/2}$,

$$M \leq \frac{2C}{\sqrt{m}} \cdot \pi_\gamma(u).$$

Now, writing $A := \{s \in S^{m-1} : \|us\| = M\}$, and, provided that $m \geq 3$, we can use 19.31 to derive

$$\begin{aligned}
&\lambda_m\big(\{s \in S^{m-1} \colon \|us\| > \frac{3C}{\sqrt{m}}\pi_\gamma(u)\}\big) \\
&\quad \leq \lambda_m\big(\{s \in S^{m-1} \colon \|us\| > M + \frac{C}{\sqrt{m}}\pi_\gamma(u)\}\big) \\
&\quad = 1 - \lambda_m\big(\{s \in S^{m-1} \colon \|us\| - M \leq \frac{C}{\sqrt{m}}\pi_\gamma(u)\}\big) \\
&\quad \leq 1 - \lambda_m\big(\{s \in S^{m-1} \colon d(s, A) \leq \frac{C \cdot \pi_\gamma(u)}{\sqrt{m} \cdot \|u|_F\|}\}\big) \\
&\quad \leq 6 \cdot \exp\Big(-\frac{m-2}{2} \cdot \frac{C^2\pi_\gamma(u)^2}{m \cdot \|u|_F\|^2}\Big) \leq 6 \cdot \exp\Big(-\frac{C^2 \cdot \pi_\gamma(u)^2}{6 \cdot \|u|_F\|^2}\Big).
\end{aligned}$$

QED

We now have all the prerequisites for the proof of 19.3.

Proof of Theorem 19.3. (i)⇒(ii): Given $n \in \mathbf{N}$, select any n-dimensional subspace X_n of X and use Lewis' Theorem 6.25 to obtain an isomorphism $u : \ell_2^n \to X_n$ for which $\pi_\gamma(u) \cdot \pi_\gamma^*(u^{-1}) = n$. Proposition 19.28 allows us to extend u^{-1} to an operator $v : X \to \ell_2^n$ with $\pi_\gamma^*(v) = \pi_\gamma^*(u^{-1})$.

Our aim is to show how to produce a constant C, independent of n, and a subspace E_n of ℓ_2^n in such a way that

$$\text{(1)} \qquad \lim_{n \to \infty} \dim E_n = \infty \quad \text{and} \quad \|u|_{E_n}\| \cdot \|v^*|_{E_n}\| \leq C.$$

Once this is done, we can refer to 19.27 to infer that $u(E_n)$ is C-complemented in X, and keep careful track of u^{-1} and v^* to notice that

$$\|u|_{E_n}\| \cdot \|u^{-1}|_{u(E_n)}\| \leq \|u|_{E_n}\| \cdot \|v^*|_{E_n}\| \leq C$$

and so that $u(E_n)$ is C-isomorphic to $\ell_2^{\dim E_n}$. With this, (ii) follows at once.

Note, by the way, that it is good enough to arrange for (1) to hold for all sufficiently large n.

We begin the search for E_n by applying 19.26 with $k = [\frac{n}{4}]$ to the operators $u : \ell_2^n \to X_n$ and $v^* : \ell_2^n \to X^*$. This furnishes us with subspaces F_1 and F_2 of ℓ_2^n, each of dimension $\geq \frac{3n}{4}$, for which

$$\|u|_{F_1}\| \leq \frac{4}{\sqrt{n}} \cdot c_2(n, X_n) \cdot \pi_\gamma(u) \quad \text{and} \quad \|v^*|_{F_2}\| \leq \frac{4}{\sqrt{n}} \cdot c_2(n, X^*) \cdot \pi_\gamma(v^*),$$

at least for $n \geq 4$. Set $F := F_1 \cap F_2$ and observe that $m := \dim F \geq n/2$ and

$$\|u|_F\| \leq \frac{4}{\sqrt{n}} \cdot c_2(n, X_n) \cdot \pi_\gamma(u) \quad \text{and} \quad \|v^*|_F\| \leq \frac{4}{\sqrt{n}} \cdot c_2(n, X^*) \cdot \pi_\gamma(v^*).$$

Apply our hypothesis that X is K-convex: 13.7 and 13.17 inform us that both X and X^* have finite cotype, and so, if we set

$$c_n := 4 \cdot \max\{c_2(n, X), c_2(n, X^*)\}$$

and glance at (C) just before 19.24, we find

$$\lim_{n\to\infty} n^{-1/2} c_n = 0 \tag{2}$$

and

$$\|u|_F\| \leq n^{-1/2} \cdot c_n \cdot \pi_\gamma(u) \quad \text{and} \quad \|v^*|_F\| \leq n^{-1/2} \cdot c_n \cdot \pi_\gamma(v^*). \tag{3}$$

We shall shortly show how to obtain E_n as a subspace of F satisfying

$$\dim E_n \geq \frac{n}{5 \cdot c_n^2 \cdot \log n} \tag{4}$$

and

$$\|u|_{E_n}\| \leq \frac{5}{\sqrt{n}} \cdot \pi_\gamma(u) \quad \text{and} \quad \|v^*|_{E_n}\| \leq \frac{5}{\sqrt{n}} \cdot \pi_\gamma(v^*). \tag{5}$$

Given these, we can use 19.29 to derive (1). Here's how.

Let $G \subset X$ be a finite dimensional subspace with $X_n \subset G$, and let $i_G : G \to X$ be the canonical embedding. Choose $q > 2$ so that X^* has cotype q and set $K := C_q(X^*) \cdot \|R^X\| \cdot m_{2q}$. Then, by 19.29, $\pi_\gamma(i_G^* v^*) \leq K \cdot \pi_\gamma^*(v i_G)$. Using maximality of π_γ, we get

$$\pi_\gamma(v^*) = \sup_G \pi_\gamma(i_G^* v^*) \leq K \cdot \sup_G \pi_\gamma^*(v i_G) \leq K \cdot \pi_\gamma^*(v).$$

Combining this with (5) we obtain

$$\begin{aligned} \|u|_{E_n}\| \cdot \|v^*|_{E_n}\| &\leq \frac{25}{n} \cdot \pi_\gamma(u) \cdot \pi_\gamma(v^*) \leq \frac{25 \cdot K}{n} \cdot \pi_\gamma(u) \cdot \pi_\gamma^*(v) \\ &= \frac{25 \cdot K}{n} \cdot \pi_\gamma(u) \cdot \pi_\gamma^*(u^{-1}) = 25 \cdot K. \end{aligned}$$

To broach (4) and (5), we consider the set

$$A := \left\{ s \in S_F : \|us\| \leq \frac{3}{\sqrt{m}} \cdot \pi_\gamma(u) \text{ and } \|v^* s\| \leq \frac{3}{\sqrt{m}} \cdot \pi_\gamma(v^*) \right\}.$$

The existence of E_n hangs on the fact that when n is large, so is $\lambda_m(A)$, and even $\lambda_k(A \cap E^{(k)})$ for suitably chosen k-dimensional subspaces $E^{(k)}$ of F. Indeed, if we first use 19.32 and then (3), we find

$$\begin{aligned}\lambda_m(A) &\geq 1 - \lambda_m\big(\{s \in S_F : \|us\| > \frac{3}{\sqrt{m}} \cdot \pi_\gamma(u)\}\big)\\ &\quad - \lambda_m\big(\{s \in S_F : \|v^*s\| > \frac{3}{\sqrt{m}} \cdot \pi_\gamma(v^*)\}\big)\\ &\geq 1 - 6 \cdot \exp\big(-\frac{\pi_\gamma(u)^2}{6 \cdot \|u\,|_F\|^2}\big) - 6 \cdot \exp\big(-\frac{\pi_\gamma(v^*)^2}{6 \cdot \|v^*\,|_F\|^2}\big)\\ &\geq 1 - 12 \cdot \exp\big(-\frac{n}{6 \cdot c_n^2}\big)\,.\end{aligned}$$

Naturally, (2) guarantees the smallness of the exponential when n is large.

To produce the desired subspaces $E^{(k)}$ of F, we work with the identities (A) after 19.12 and (B) in the proof of 19.13 applied to the function $f(x,y) = 1_A(x)$. Then, for any $1 \leq k \leq m = \dim F$,

$$\lambda_m(A) = \int_{\mathcal{G}_k} \lambda_E(A \cap E)\, d\gamma_k(E),$$

where γ_k is the normalized rotational invariant measure on the compact Hausdorff space $\mathcal{G}_k$ of k-dimensional subspaces of F. For each $1 \leq k \leq m$, there must therefore be a k-dimensional subspace $E^{(k)}$ of F for which

$$\lambda_{E^{(k)}}(A \cap E^{(k)}) \geq 1 - 13 \cdot \exp\big(-\frac{n}{6 \cdot n^2}\big)\,. \tag{6}$$

The remainder of the proof is devoted to showing that for suitable values of k, $S_{E^{(k)}}$ has a $\frac{1}{2\sqrt{n}}$-net, all of whose elements lie in $E^{(k)} \cap A$. If this is known (5) follows with little effort, since for any $x \in S_{E^{(k)}}$ we can find a y in the net with $\|x - y\| \leq \frac{1}{2\sqrt{n}}$ and so

$$\|ux\| \leq \|ux - uy\| + \|uy\| \leq \|u\| \cdot \frac{1}{2 \cdot \sqrt{n}} + \frac{3}{\sqrt{m}} \cdot \pi_\gamma(u) \leq \frac{5}{\sqrt{n}} \cdot \pi_\gamma(u).$$

The analogous inequality $\|v^*x\| \leq (5/\sqrt{n}) \cdot \pi_\gamma(v^*)$ follows in just the same way.

So, provided that we can find such a net in $E^{(k_0)} \cap A$ for some $k_0 \geq n/(5 \cdot c_n^2 \cdot \log n)$, we can set $E_n = E^{(k_0)}$ and our argument will be complete. Actually, for large enough n, such nets will exist for *any* $k \leq n/(4 \cdot c_n^2 \cdot \log n)$.

Finding *some* $(2 \cdot \sqrt{n})^{-1}$-net in $S_{E^{(k)}}$ is not a problem. The argument of Lemma 17.20 even tells us that there will be such a net N_k with no more than $(3 \cdot \sqrt{n})^k$ members. The proper location of N_k is what matters; it relies on the rotation invariance of $\lambda_{E^{(k)}}$ and the fact that if o_k is $\mathcal{O}_k$'s Haar measure and B is any Borel subset of $S_{E^{(k)}}$, then $o_k(\{u \colon ux \in B\})$ is independent of $x \in S_{E^{(k)}}$. Consequently, for any $x \in S_{E^{(k)}}$ we can use (6) to derive

$$\begin{aligned}o_k(\{u \in \mathcal{O}_k \colon ux \in A \cap E^{(k)}\}) &= \int_{S_{E^{(k)}}} \int_{\mathcal{O}_k} 1_{A \cap E^{(k)}}(uy)\, do_k(u)\, d\lambda_{E^{(k)}}(y)\\ &= \int_{\mathcal{O}_k} \int_{S_{E^{(k)}}} 1_{A \cap E^{(k)}}(uy)\, d\lambda_{E^{(k)}}(y)\, do_k(u) = \int_{\mathcal{O}_k} \lambda_{E^{(k)}}\big(u(A \cap E^{(k)})\big)\, do_k(u)\\ &= \lambda_{E^{(k)}}(A \cap E^{(k)}) \geq 1 - 13 \cdot \exp\big(-\frac{n}{6 \cdot c_n^2}\big)\,.\end{aligned}$$

As a result,

$$o_k\big(\{u \in \mathcal{O}_k : u(N_k) \subset A \cap E^{(k)}\}\big) \geq 1 - 13\cdot|N_k|\cdot\exp\big(-\frac{n}{6\cdot n^2}\big)$$
$$\geq 1 - 13\cdot(3\cdot\sqrt{n})^k\cdot\exp\big(-\frac{n}{6\cdot n^2}\big),$$

and this is positive if $k < \dfrac{\frac{n}{6c_n^2} - \log 13}{\frac{1}{2}\log n + \log 3}$. This means that, for large n, we can rotate N_k into a subset of $A \cap E^{(k)}$, and everything is in order.

(ii)⇒(iii): We maintain the notation of the previous implication. Choose a natural number k and some $0 < \varepsilon < 1$. Refer to Dvoretzky's Theorem 19.1 and select a space E_n, as in (i)⇒(ii), with dimension at least $n_\varepsilon(k)$. Then E_n has a k-dimensional subspace E, say, which is $(1+\varepsilon)$-isomorphic to ℓ_2^k. Moreover, if P is the orthogonal projection of $\ell_2^{\dim E_n}$ onto $u^{-1}(E)$, then $uPu^{-1} : E_n \to E$ is a projection of norm $\leq C$, and so E is C^2-complemented in X.

(iii)⇒(i): We argue by contradiction: we suppose there is a Banach space X satisfying (iii) and containing the ℓ_1^N's uniformly. Then for each $N \in \mathbf{N}$ there is an N-dimensional subspace F_N of X together with an isomorphism $u_N : F_N \to \ell_1^N$ such that $\|u_N\|\cdot\|u_N^{-1}\| \leq 2$.

Fix $k \in \mathbf{N}$ and use Dvoretzky's Theorem 19.1 to select N sufficiently large that F_N has a k-dimensional subspace H_k which is 2-isomorphic to ℓ_2^k and C-complemented in X, and so in F_N. Write $P : F_N \to H_k$ for the corresponding projection and $J : H_k \to F_N$ for the canonical injection. We obtain our contradiction from 1.13 and 4.17:

$$\sqrt{k} = \pi_2(id_{H_k}) \leq 2\cdot\pi_2(Pu_N^{-1}u_NJ) \leq 2\cdot\pi_2(Pu_N^{-1})\cdot\|u_NJ\|$$
$$\leq 2\cdot\pi_1(Pu_N^{-1})\cdot\|u_N\| \leq 4\cdot\kappa_G\cdot C.$$

This completes the proof. QED

The Isoperimetric Inequality

Recall that we are equipping S^{n-1} with the geodesic distance. The intersection of a half-space in $\mathbf{R}^n$ with S^{n-1} is called a *cap*. Caps are precisely the closed balls with respect to the geodesic metric.

Our immediate objective is to prove a *qualitative* version of the Isoperimetric Inequality.

19.33 Theorem: *Let $n \in \mathbf{N}$, let A be a closed subset of S^{n-1} and let C be a cap of S^{n-1} such that $\lambda_n(C) = \lambda_n(A)$. Then for every $\varepsilon > 0$,*

$$\lambda_n(C_\varepsilon) \leq \lambda_n(A_\varepsilon).$$

Our proof will revolve around three propositions and will proceed by induction on n, an induction which comes into play in the second proposition only. First, however, we need to introduce the Hausdorff metric and some of its basic properties.

Let (Ω, d) be a metric space and denote by $\mathcal{K}(\Omega)$ the collection of all non-empty closed subsets of Ω. For $C, D \in \mathcal{K}(\Omega)$, define

$$\delta(C, D) := \inf\{\varepsilon > 0 : C \subset D_\varepsilon \text{ and } D \subset C_\varepsilon\}.$$

It is routine to check that δ defines a metric – the *Hausdorff metric* – on $\mathcal{K}(\Omega)$.

The following classical result is fundamental to the study of the Hausdorff metric and its applications.

19.34 Blaschke's Selection Theorem: *If (Ω, d) is a compact metric space, then so is $\mathcal{K}(\Omega)$ when it is equipped with the Hausdorff metric δ.*

Proof. We begin by establishing the completeness of $\mathcal{K}(\Omega)$. Let $(A_n)_n$ be a Cauchy sequence in $\mathcal{K}(\Omega)$ and set

$$A_0 := \{x \in \Omega : x \text{ is a limit point of some } (x_n) \text{ where each } x_n \in A_n\}.$$

We shall see that (A_n) converges to A_0 in $[\mathcal{K}(\Omega), \delta]$.

The first thing to check is that $A_0 \in \mathcal{K}(\Omega)$. The non-emptiness of A_0 is a quick consequence of the compactness of Ω and the non-emptiness of each A_n. To see that A_0 is closed, suppose that $(x^{(k)})$ is a sequence in A_0 converging to some $x^{(0)} \in \Omega$. Now each $x^{(k)}$ is a limit point of a sequence $(x_n^{(k)})_n$ where $x_n^{(k)} \in A_n$ for each n. So, for each $k \in \mathbf{N}$ we can choose an integer $i(k)$ so that $d(x_{i(k)}^{(k)}, x^{(k)}) < k^{-1}$. Without a doubt, we can assume that $i(1) < i(2) < i(3) < ....$ Next, write

$$x_n^{(0)} = \begin{cases} x_{i(k)}^{(k)} & \text{if } n = i(k) \\ x_n^{(1)} & \text{if } n \notin \{i(1), i(2), ...\} . \end{cases}$$

It is plain that $x_n^{(0)} \in A_n$ for each n and that $x^{(0)}$ is a limit point of $(x_n^{(0)})$. We have shown that $x^{(0)} \in A_0$, and so that A_0 is closed.

We turn to our real objective, the proof that (A_n) converges to A_0 in the Hausdorff metric. Fix $\varepsilon > 0$ and choose $p \in \mathbf{N}$ so large that $\delta(A_m, A_n) < \varepsilon/2$ whenever $m, n \geq p$. We shall show that $\delta(A_n, A_0) < \varepsilon$ for all $n \geq p$.

Pick a point $x_0 \in A_0$. There exists $m \geq p$ such that $x_0 \in (A_m)_{\varepsilon/2}$. But for any $n \geq p$ we have $A_m \subset (A_n)_{\varepsilon/2}$ and so $x_0 \in (A_n)_\varepsilon$. It follows that $A_0 \subset (A_n)_\varepsilon$ for all $n \geq p$.

On the other hand, if $n \geq p$ and $x_n \in A_n$, we can find a sequence (y_k) with $y_k \in A_k$ for each k and $d(y_k, x_n) < \varepsilon/2$ for $k \geq p$. As Ω is compact, (y_k) has a limit point y, which is *a fortiori* a member of A_0 and satisfies $d(y, x_n) \leq \varepsilon/2$. From this we see that $A_n \subset (A_0)_\eta$ for any $\eta > \varepsilon/2$; in particular $A_n \subset (A_0)_\varepsilon$.

Now that we know that $\mathcal{K}(\Omega)$ is complete, we can show it is compact by establishing total boundedness. To this end, fix $\varepsilon > 0$ and, using Ω's compactness, cover Ω with open balls $W_1, ..., W_n$, each of radius $\varepsilon/2$. Let $\mathcal{P}$ denote the collection of all pairs (S, T) of disjoint subsets of $\{1, ..., n\}$ with $S \neq \emptyset$ and $S \cup T = \{1, ..., n\}$. For $P = (S, T) \in \mathcal{P}$ write

$$\kappa(P) = \{K \in \mathcal{K}(\Omega) : K \cap W_s \neq \emptyset \text{ if } s \in S,\ K \cap W_t = \emptyset \text{ if } t \in T\}.$$

The $\kappa(P)$'s cover $\mathcal{K}(\Omega)$: indeed, if $K \in \mathcal{K}(\Omega)$, then $K \in \kappa(S,T)$ where $S = \{i : 1 \le i \le n,\ K \cap W_i \neq \emptyset\}$ and $T = \{i : 1 \le i \le n,\ K \cap W_i = \emptyset\}$. To finish, we show that each $\kappa(P)$ has δ-diameter no more than ε.

Accordingly, fix $P = (S,T) \in \mathcal{P}$ and select $U, V \in \kappa(P)$. If $x \in U$, then $x \in W_s$ for some $s \in S$. But $V \cap W_s \neq \emptyset$ and so since W_s has diameter ε we find that $x \in V_\varepsilon$. We deduce that $U \subset V_\varepsilon$. A symmetric argument gives $V \subset U_\varepsilon$, and we are done. QED

We hasten to add that a by-product of the proof above is the identification of the limit A_0 of a sequence (A_n) in $\mathcal{K}(\Omega)$: A_0 must be the set $\bigcap_n \overline{\bigcup_{m \ge n} A_m}$.

Two elementary properties of the Hausdorff metric on $\mathcal{K}(S^{n-1})$ pave the way to our first proposition. For $A \in \mathcal{K}(S^{n-1})$, define its radius to be

$$r(A) := \inf\{r > 0 : K \subset \{x\}_r \text{ for some } x \in S^{n-1}\}.$$

19.35 Lemma: (a) *The map $\mathcal{K}(S^{n-1}) \to [0,\infty) : A \mapsto r(A)$ is continuous.*
(b) *The map $\mathcal{K}(S^{n-1}) \to [0,\infty) : A \mapsto \lambda_n(A)$ is upper semicontinuous.*

Proof. (a) Fix $\varepsilon > 0$ and select $A, A' \in \mathcal{K}(S^{n-1})$ with $\delta(A, A') < \varepsilon/2$. We shall show that $|r(A) - r(A')| \le \varepsilon$.

We can always find some $x \in S^{n-1}$ for which $A \subset \{x\}_{r(A)+\varepsilon/2}$. It follows that $A' \subset A_{\varepsilon/2} \subset \{x\}_{r(A)+\varepsilon}$ and so that $r(A') \le r(A) + \varepsilon$.

A symmetric argument gives $r(A) \le r(A') + \varepsilon$ and so the conclusion.

(b) Let (A_j) be a sequence in $\mathcal{K}(S^{n-1})$ with limit $A_0 \in \mathcal{K}(S^{n-1})$. Since $A_0 = \bigcap_j \overline{\bigcup_{m \ge j} A_m}$, we have

$$\lambda_n(A_0) = \lim_j \lambda_n(\overline{\bigcup_{m \ge j} A_m}) \ge \lim_j \sup_{m \ge j} \lambda_n(A_m) = \limsup_j \lambda_n(A_j).$$

This is just what we needed to show. QED

A little more notation makes life easier. When A is a closed subset of S^{n-1}, we write

$$M(A) := \{C \in \mathcal{K}(S^{n-1}) : \lambda_n(C) = \lambda_n(A) \text{ and } \lambda_n(C_\varepsilon) \le \lambda_n(A_\varepsilon)\ \forall\, \varepsilon > 0\}.$$

19.36 Lemma: *Let A be a closed subset of S^{n-1}. There exists $B \in M(A)$ such that $r(B) = \min\{r(C) : C \in M(A)\}$.*

Proof. We have just shown that the radius is continuous, so it is enough to prove that $M(A)$ is closed in $\mathcal{K}(S^{n-1})$. To this end let (C_j) be a sequence in $M(A)$ converging to $C_0 \in \mathcal{K}(S^{n-1})$.

Let $\varepsilon \ge 0$. If $\eta > 0$, then for large enough k we have $C_0 \subset (C_k)_\eta$ and so $(C_0)_\varepsilon \subset (C_k)_{\eta+\varepsilon}$. Consequently, $\lambda_n((C_0)_\varepsilon) \le \lambda_n((C_k)_{\eta+\varepsilon}) \le \lambda_n(A_{\eta+\varepsilon})$ and so

$$\lambda_n((C_0)_\varepsilon) \le \inf_{\eta>0} \lambda_n(A_{\eta+\varepsilon}) = \lambda_n(\bigcap_{\eta>0} A_{\eta+\varepsilon}) = \lambda_n(A_\varepsilon).$$

Also, if $\varepsilon = 0$ we have $\lambda_n(C_0) \leq \lambda_n(A)$, whereas upper semicontinuity of $K \to \lambda_n(K)$ gives

$$\lambda_n(C_0) \geq \limsup_j \lambda_n(C_j) = \limsup_j \lambda_n(A) = \lambda_n(A).$$

Hence C_0 fulfills all requirements for membership in $M(A)$. QED

To proceed further, we need to describe a procedure of spherical symmetrization. Let γ be a half-circle on S^{n-1} joining a 'north pole' x_0 to a 'south pole' $-x_0$. For $y \in \gamma^0 := \gamma \setminus \{x_0, -x_0\}$ we set

$$S^{n-2,y}$$

to be the intersection of S^{n-1} with the hyperplane orthogonal to $\mathrm{span}\{x_0\}$ and passing through y. As $S^{n-2,y}$ is a sphere in $\mathbf{R}^{n-1}$, it has a normalized rotation invariant measure

$$\lambda_{n-1,y}.$$

When $n = 3$, $S^{1,y}$ is a circle of latitude: draw a picture!

Next, whenever $A \in \mathcal{K}(S^{n-1})$ and $y \in \gamma^0$, we write

$$A^y := A \cap S^{n-2,y}$$

and denote by C_A^y the cap in $S^{n-2,y}$ with centre y satisfying

$$\lambda_{n-1,y}(C_A^y) = \lambda_{n-1,y}(A^y).$$

To accommodate the end-points of γ, we set

$$A^{\pm x_0} = C_A^{\pm x_0} := \begin{cases} \{\pm x_0\} & \text{if } \pm x_0 \in A \\ \emptyset & \text{if not .} \end{cases}$$

The set

$$\sigma_\gamma(A) := \bigcup_{y \in \gamma} C_A^y$$

is called the *spherical symmetrization of A relative to the half-circle γ*. Draw a picture!

It will also be useful to work with 'projections along meridians'. If γ has midpoint m, then for each $y \in \gamma^0$ we define a homeomorphism $\tau_y : S^{n-2,y} \to S^{n-2,m}$: if $x \in S^{n-2,y}$, then $\tau_y(x)$ is the point of $S^{n-2,m}$ which belongs to the half-circle joining x_0 and $-x_0$ and passing through x. Draw a picture!

A little thought about the angles involved makes it clear that if $y_1, y_2 \in \gamma^0$, $x_1, \widetilde{x}_1 \in S^{n-2,y_1}$ and $x_2, \widetilde{x}_2 \in S^{n-2,y_2}$, then $d(x_1, x_2) = d(\widetilde{x}_1, \widetilde{x}_2)$ if and only if $d\big(\tau_{y_1}(x_1), \tau_{y_2}(x_2)\big) = d\big(\tau_{y_1}(\widetilde{x}_1), \tau_{y_2}(\widetilde{x}_2)\big)$. This lies at the heart of a simple but useful observation. If $\varepsilon > 0$ and if $a, y \in \gamma^0$ satisfy $d(a, y) < \varepsilon$, then there is some $\eta = \eta(a, y, \varepsilon) > 0$ such that for every closed subset B of $S^{n-2,a}$

$$(*) \qquad \tau_y\big((B_\varepsilon)^y\big) = \big(\tau_a(B)\big)_\eta.$$

Moreover, η goes to zero with ε. Draw a picture!

We are now in a position to prove the inductive ingredient of the proof of 19.33.

19.37 Proposition: *Let $n > 1$. Assume that for every closed subset B of S^{n-2}, every cap C in S^{n-2} with $\lambda_{n-1}(C) = \lambda_{n-1}(B)$, and every $\varepsilon > 0$ we have $\lambda_{n-1}(C_\varepsilon) \leq \lambda_{n-1}(B_\varepsilon)$. Then, if $A \in \mathcal{K}(S^{n-1})$ we have $\sigma_\gamma(A) \in M(A)$ for every half-circle γ in S^{n-1}.*

Proof. Select a half-circle γ with north pole x_0, south pole $-x_0$ and 'equatorial point' m. Choose $A \in \mathcal{K}(S^{n-1})$.

We first show that $Z := \sigma_\gamma(A) \in \mathcal{K}(S^{n-1})$. Certainly $Z \neq \emptyset$, since $A \neq \emptyset$. To show that Z is closed in S^{n-1}, we give ourselves a sequence (z_k) in Z which converges in S^{n-1} to z, say. Of course, we must show that z is in $\sigma_\gamma(A)$.

This is no problem if there is some $y \in \gamma$ with z_k in the closed set C_A^y for infinitely many k. To settle the other case, we may assume that there is a sequence (y_k) of distinct points of γ such that $z_k \in C_A^{y_k}$ for each k. The compactness of γ tells us that (a subsequence of) (y_k) converges to some $y \in \gamma$. We shall show that $z \in C_A^y \subset Z$ by establishing that

$$\lambda_{n-1,y}(C_A^y) \geq \limsup_k \lambda_{n-1,y_k}(C_A^{y_k}).$$

Draw a picture!

For this, observe that compactness of $\mathcal{K}(S^{n-1})$ entails that (a subsequence of) $(A^{y_k})_k$ converges to some $D \in \mathcal{K}(S^{n-1})$. Clearly, D is a subset of $S^{n-2,y}$, and as A is closed, we even have $D \subset A^y$. The properties of τ ensure that $\left(\tau_{y_k}(A^{y_k})\right)_k$ converges to $\tau_y(D)$ in $\mathcal{K}(S^{n-2,m})$. The upper semicontinuity property of $\lambda_{n-1,m}$ derived in 19.35(b) gets us what we want:

$$\begin{aligned} \lambda_{n-1,y}(C_A^y) &= \lambda_{n-1,y}(A^y) = \lambda_{n-1,m}(\tau_y A^y) \geq \lambda_{n-1,m}(\tau_y D) \\ &\geq \limsup_k \lambda_{n-1,m}(\tau_{y_k} A^{y_k}) = \limsup_k \lambda_{n-1,y_k}(A^{y_k}) \\ &= \limsup_k \lambda_{n-1,y_k}(C_A^{y_k}). \end{aligned}$$

Now that we have proved $Z \in \mathcal{K}(S^{n-1})$, we fix $\varepsilon > 0$ and aim to show that $\lambda_{n-1}(Z_\varepsilon) \leq \lambda_{n-1}(A_\varepsilon)$. For this it is helpful to note that for any closed $F \in S^{n-1}$ and any $y \in \gamma^0$ we have

$$(F_\varepsilon)^y = \bigcup\{((F^a)_\varepsilon)^y : a \in \gamma^0,\ d(a,y) < \varepsilon\}.$$

It follows from $(*)$ that

$$\tau_y((F_\varepsilon)^y) = \bigcup\{(\tau_a F^a)_{\eta(a,y,\varepsilon)} : a \in \gamma^0,\ d(a,y) < \varepsilon\}.$$

Now, for $a \in \gamma^0$, we note that $\tau_a C_A^a$ is a cap in $S^{n-2,m}$ for which $\lambda_{n-1,m}(\tau_a C_A^a) = \lambda_{n-1,m}(\tau_a A^a)$ and apply the (inductive) hypothesis to get

$$\lambda_{n-1,m}((\tau_a C_A^a)_\eta) \leq \lambda_{n-1,m}((\tau_a A^a)_\eta)$$

for any η. Take special note of the fact that all the caps $(\tau_a C_A^a)_\eta$ have the same centre, m. Then, for $y \in \gamma^0$,

$$\begin{aligned}\lambda_{n-1,m}(\tau_y(Z_\varepsilon)^y) &= \lambda_{n-1,m}\Big(\bigcup\{(\tau_a C_A^a)_{\eta(a,y,\varepsilon)} : a \in \gamma^0,\, d(a,y) < \varepsilon\}\Big) \\ &= \sup\{\lambda_{n-1,m}((\tau_a C_A^a)_{\eta(a,y,\varepsilon)}) : a \in \gamma^0,\, d(a,y) < \varepsilon\} \\ &\leq \sup\{\lambda_{n-1,m}((\tau_a A^a)_{\eta(a,y,\varepsilon)}) : a \in \gamma^0,\, d(a,y) < \varepsilon\} \\ &\leq \lambda_{n-1,m}(\tau_y(A_\varepsilon)^y).\end{aligned}$$

This translates into

$$\lambda_{n-1,y}((Z_\varepsilon)^y) \leq \lambda_{n-1,y}((A_\varepsilon)^y),$$

and as this is valid for all $y \in \gamma^0$, Cavalieri's Principle can be invoked to give

$$\lambda_{n-1}(Z^\varepsilon) \leq \lambda_{n-1}(A^\varepsilon),$$

just as we wanted. QED

We require one more ingredient for the proof of 19.33.

19.38 Proposition: *Let A be a non-empty closed subset of S^{n-1} which is not a cap. There is a finite collection $\gamma_0,\ldots,\gamma_k$ of half-circles on S^{n-1} such that*

$$r\big(\sigma_{\gamma_k}\sigma_{\gamma_{k-1}}\cdots\sigma_{\gamma_0}(A)\big) < r(A).$$

Proof. To simplify notation, we write r for $r(A)$. The definition of r provides us with an $m \in S^{n-1}$ such that A is a subset of $B(m,r) := \{x \in S^{n-1} : d(x,m) \leq r\}$.

We shall symmetrize several times. Each symmetrization will be with respect to a half-circle with m as its equatorial point and so will leave $B(m,r)$ invariant.

To begin, we perform a spherical symmetrization with respect to an arbitrary half-circle γ_0 with equatorial point m. Since A is not a cap, $\sigma_{\gamma_0}(A)$ cannot be all of $B(m,r)$ and so certainly does not contain all of $\partial B(m,r)$. We shall continue symmetrizing with the goal of finding a closed set of the form $\sigma_{\gamma_k}\sigma_{\gamma_{k-1}}\cdots\sigma_{\gamma_0}(A)$ which is disjoint from $\partial B(m,r)$. This will ensure that $r\big(\sigma_{\gamma_k}\sigma_{\gamma_{k-1}}\cdots\sigma_{\gamma_0}(A)\big) < r$. Predictably, a compactness argument will be at the core of our attack.

The approach will be two-pronged and will rely on repeated application of two sharp observations which are best understood geometrically.

First, if Z is a closed subset of $B(m,r)$ and if $x \in \partial B(m,r) \setminus Z$, then $x \notin \sigma_\gamma(Z)$ for every half-circle γ with equatorial point m. Keep drawing pictures!

Second, if G is a relatively open subset of $\partial B(m,r)$, then for each x in $\partial B(m,r)$ there exist a relatively open set G_x in $\partial B(m,r)$ with $x \in G_x$ and a half circle γ_x with equatorial point m such that whenever $Z \subset B(m,r)$ is closed with $Z \cap G = \emptyset$ we have $\sigma_{\gamma_x}(Z) \cap G_x = \emptyset$.

When $x \in G$ the first observation allows us to take $G_x = G$; so let's concentrate on the case $x \notin G$.

A pair of diagrams will help in understanding the following words of explanation. We start by selecting any $p \in G$ and connecting it to x with a geodesic Γ. Next, we select γ_x to be the half-circle which is normal to Γ and which has m on its equator. Let T be an open 'temperate zone' with Γ as its median. There is clearly a relatively open set $G_x \subset T \cap \partial B(m,r)$ which has all the properties we mentioned.

We can now set to work. Using the second of our observations and applying it to $G = \partial B(m,r) \backslash \sigma_{\gamma_0}(A)$, we exploit the compactness of $\partial B(m,r)$ to produce $x_1,..., x_k \in \partial B(m,r)$ such that G and $G_{x_1},..., G_{x_k}$ cover $\partial B(m,r)$. Naturally, the half-circles $\gamma_i := \gamma_{x_i}$ all have m on their equators and all have the property that $\sigma_{\gamma_i}(Z) \cap G_{x_i} = \emptyset$ whenever Z is a closed subset of $B(m,r) \setminus G$.

Now take any $x \in \partial B(m,r)$. If $x \notin \sigma_{\gamma_0}(A)$, then our first observation applied to the closed subset $\sigma_{\gamma_0}(A)$ of $B(m,r)$ ensures that $x \notin \sigma_{\gamma_1}\sigma_{\gamma_0}(A)$. The same observation, this time applied to $\sigma_{\gamma_1}\sigma_{\gamma_0}(A)$, tells us that $x \notin \sigma_{\gamma_2}\sigma_{\gamma_1}\sigma_{\gamma_0}(A)$, and so on. We eventually find that $x \notin \sigma_{\gamma_k} \ldots \sigma_{\gamma_0}(A)$.

If $x \in \sigma_{\gamma_0}(A)$, then $x \notin G$ and so there is a smallest integer $1 \le i \le k$ for which $x \in G_{x_i}$. Since $\sigma_{\gamma_0}(A) \cap G = \emptyset$, the second observation gives $\sigma_{\gamma_i}\sigma_{\gamma_0}(A) \cap G_{x_i} = \emptyset$, and so $x \notin \sigma_{\gamma_i}\sigma_{\gamma_0}(A)$. We are now in a position to apply the procedure of the previous case.

The proof is finished. QED

The *Proof of Theorem 19.33* is now a simple combination of the last three propositions, provided we observe that the theorem is easily true when $n = 1$.

To wrap things up we need to pass from the qualitative result 19.33 to the quantitative 19.30. This is elementary.

19.39 Proposition: *Let $C \subset S^{n+1}$ be a cap with $\lambda_{n+2}(C) \ge \frac{\pi}{2}$. Then, for all $\varepsilon > 0$,*

$$\lambda_{n+2}(C_\varepsilon) \ge 1 - 3\cdot\exp\left(-\frac{n\cdot\varepsilon^2}{2}\right).$$

Proof. There is no loss in assuming that C is a half-sphere and we can think of it as $\{x \in S^{n+1} \colon d(x,x_0) \le \pi/2\}$ for some x_0.

For our purposes it is best to fix m in S^{n+1} with $d(m,x_0) = \pi/2$ and to consider C_ε as the set of points in S^{n+1} making angles between $-\varepsilon$ and $\pi/2$ with $S^{n,m}$. Notice that if θ is the angle for $y \in C_\varepsilon$, then $\cos\theta$ is the radius of $S^{n,y}$. A diagram will make things clear! Consequently

$$\lambda_{n+2}(C_\varepsilon) = \frac{\int_{-\varepsilon}^{\pi/2} \cos^n t\, dt}{\int_{-\pi/2}^{\pi/2} \cos^n t\, dt} = 1 - \frac{\int_{\varepsilon}^{\pi/2} \cos^n t\, dt}{2\cdot\int_0^{\pi/2} \cos^n t\, dt}.$$

Now we just need elementary calculus. It is routine to check that $\cos t \ge e^{-t^2}$ for $0 \le t \le \pi/4$ and that $\cos t \le 2\cdot e^{-t^2/2}$ for $0 \le t \le \pi/2$. So

$$\int_0^{\pi/2} \cos^n t\, dt \ge \int_0^{\pi/4} e^{-nt^2} dt \ge n^{-1/2}\cdot\int_0^{\pi/4} e^{-s^2} ds \ge \frac{\pi}{4\cdot\sqrt{n}} e^{-\pi^2/16}$$

and

$$\int_{\varepsilon}^{\pi/2} \cos^n t\, dt \leq 2 \cdot \int_{\varepsilon}^{\pi/2} e^{-nt^2/2} dt$$
$$\leq 2 \cdot e^{-n\varepsilon^2/2} \cdot \int_0^{\infty} e^{-ns^2/2} ds = \sqrt{\frac{2\pi}{n}} \cdot e^{-n\varepsilon^2/2}.$$

This gives

$$\lambda_{n+2}(C_\varepsilon) \geq 1 - \sqrt{\frac{8}{\pi}} \cdot e^{\pi^2/16} \cdot e^{-n\varepsilon^2/2} \geq 1 - 3 \cdot e^{-n\varepsilon^2/2}. \qquad \text{QED}$$

NOTES AND REMARKS

The Dvoretzky Theorem was a natural outgrowth of the work of A. Dvoretzky and C.A. Rogers [1950] which we have highlighted at various points. As we mentioned earlier, A. Grothendieck was fascinated by the Dvoretzky-Rogers Lemma, so much so that he devoted most of his [1953b] sequel to the Résumé to studying its consequences. Grothendieck was actually led to conjecture the Dvoretzky Theorem, and it was soon after Dvoretzky reviewed Grothendieck's paper that he proved the result.

In his original proof of 19.1, A. Dvoretzky [1961] employed techniques from integral geometry; these are sometimes difficult to digest without the supplements provided by T. Figiel [1972].

Despite its difficulty, the depth and beauty of Dvoretzky's result attracted close scrutiny. Alternative approaches evolved and more palatable proofs emerged. V.D. Milman [1971] gave a geometric proof based on Lévy's Isoperimetric Inequality (19.30). Along the way he improved Dvoretzky's estimates of how big a nearly Euclidean space will fit inside a given finite dimensional normed space. See V.D. Milman and G. Schechtman [1986] for an up to date approach along these lines.

Measure theoretic ideas were next to come to the fore. A. Szankowski [1974] and T. Figiel [1976b] gave full rein to invariant integrals. In the text we have followed Figiel's lead, but were helped substantially by insights drawn from the magnificent book of V.M. Kadets and M.I. Kadets [1991].

Functional analysis also furnishes a proof of Dvoretzky's Theorem. As early as [1974], L. Tzafriri noted that his techniques, taken in tandem with the Brunel-Sucheston circle of ideas discussed in Chapter 14, proved that every infinite dimensional Banach space contains the ℓ_2^n's uniformly. Soon after, J.L. Krivine [1976] showed that infinite dimensional Banach spaces containing ℓ_p^n's uniformly must also contain them almost isometrically, so completing a proof of Dvoretzky's Theorem in an infinite dimensional setting.

T. Figiel, J. Lindenstrauss and V.D. Milman [1977] carefully scrutinized Milman's [1971] proof of 19.1. Not only did they come up with a fresh proof of Dvoretzky's Theorem, but they also related the estimates directly to the K-convexity and cotype constants of the spaces involved. Incidentally, they also presented a succinct and complete proof of Lévy's classical Isoperimetric Inequality, a proof we follow herein.

Most recently, Y. Gordon [1985] gave a probabilistic proof of Dvoretzky's Theorem, using a refinement of a classical lemma of Slepian on Gaussian processes. For another 'Gaussian' proof of Theorem 19.1 and related information we refer to G. Pisier's lectures [1986a] and to his book [1989].

D.G. Larman and P. Mani [1975] extended Dvoretzky's Theorem to non-symmetric convex bodies.

N.J. Kalton initiated the study of Theorem 19.1 in infinite dimensional quasinormed spaces, though his work has not been published. S.J. Dilworth [1985] used ideas from V.D. Milman [1971] and T. Figiel, J. Lindenstrauss and V.D. Milman [1977] to find k-dimensional Euclidean subspaces of n-dimensional continuously quasinormed spaces. Finally, Y. Gordon and D.R. Lewis [1991] showed that Dvoretzky's Theorem holds in any continuously quasinormed space and that the estimates are as sharp as those in Banach spaces. Part of their arguments extend those of T. Figiel, J. Lindenstrauss and V.D. Milman [1977] and Y. Gordon [1985].

J. Lindenstrauss' [1992] survey is laced with commentary giving insight about the development of Dvoretzky's Theorem and its historical importance, with applications and new research developments. An ideal complement to the Lindenstrauss survey is that of V.D. Milman [1992].

Our discussion of convex functions is standard and can be found in texts such as A.W. Roberts and D.E. Varberg [1973].

Theorem 19.5 follows easily from H. Rademacher's [1918] proof of the almost everywhere Fréchet differentiability of Lipschitz functions; the issue is that convex functions are locally Lipschitz. The proof we follow seems to be due to R.D. Anderson and V.L. Klee [1952] who obtain even more.

Invariant measures naturally figure prominently in measure theory, and they are a source of many of measure theory's most important applications. L. Nachbin's charming [1965] book provides an excellent introduction to the theory of invariant integrals. The results on invariant integrals that we discuss in the text are due to A. Weil [1940].

Theorems 19.19 and 19.20 are due to S. Bellenot [1975]. This paper was our source for 19.2. Bellenot took note that results similar to 19.19 appeared in C. Stegall and J.R. Retherford [1972] and W.J. Davis and W.B. Johnson [1974].

The notion formulated in (ii) of Theorem 19.3 is due to C. Stegall and J.R. Retherford [1972]; they used 'sufficiently Euclidean' as a label for spaces having this property. Later, A. Pełczyński and H.P. Rosenthal [1976] isolated the spaces enjoying (iii) of 19.3 and called them 'locally π-Euclidean'. They showed that if $1 < p < \infty$, then $L_p(\mu)$ (any μ) is locally π-Euclidean.

T. Figiel and N. Tomczak-Jaegermann [1979], building on the work of T. Figiel, J. Lindenstrauss and V.D. Milman [1977], proved that K-convex spaces are locally π-Euclidean, while G. Pisier [1982a] established the converse.

Our presentation of 19.3 is an expansion of V.D. Milman's and G. Schechtman's [1986] account. In their book, Dvoretzky's Theorem 19.1 is also derived from 19.30.

Here is an interesting complement to 19.3. Theorem 4.20 can be seen as a characterization of all Banach spaces Z such that Hilbert space operators which factor through Z must be Hilbert - Schmidt operators. Among other things we have seen there that the class of such 'Hilbert - Schmidt spaces' is self-dual. Finite dimensional spaces, all $\mathcal{L}_\infty$- and all $\mathcal{L}_1$-spaces are in this class, but it is much larger: all Banach spaces 'verifying Grothendieck's Theorem' (see Chapter 15) have this property. In particular, H^∞, the disk algebra and related spaces from harmonic analysis are Hilbert - Schmidt spaces. From 19.3 we may conclude that

- *no infinite dimensional K-convex Banach space can be a Hilbert - Schmidt space.*

The proof is essentially a repetition of the argument used to get (iii)$\Rightarrow$(i) in 19.3. Suppose that X is an infinite dimensional Hilbert - Schmidt space which is K-convex. Then there exist, for each $n \in \mathbf{N}$, operators $i_n : \ell_2^n \to X$ and $p_n : X \to \ell_2^n$ such that $p_n i_n = id_{\ell_2^n}$ and $C := \sup_n \|p_n\| \cdot \|i_n\| < \infty$. On the other hand, there is a $c > 0$ such that $\pi_2(p_n) \le c \cdot \|p_n\|$ for all $n \in \mathbf{N}$. It follows that $n^{1/2} = \pi_2(id_{\ell_2^n}) \le \pi_2(p_n) \cdot \|i_n\| \le c \cdot C$ for all n in $\mathbf{N}$: a contradiction.

It is easy to see that a Banach space X is a Hilbert - Schmidt space if and only if its identity is in the ideal $\Pi_{2,2,2}$ mentioned in Notes and Remarks of Chapter 9: regardless of how we choose weak ℓ_2 sequences $(x_n)_n$ in X and $(x_n^*)_n$ in X^*, the scalar sequence $(\langle x_n^*, x_n\rangle)_n$ belonges to ℓ_2. Recall that if we were to replace ℓ_2 by c_0 here, then we would arrive at A. Grothendieck's classical characterization of the Dunford - Pettis property.

REFERENCES

P. Abraham

[1992] *Saeki's improvement of the Vitali-Hahn-Saks-Nikodym theorem holds precisely for Banach spaces having cotype.* Proc. Amer. Math. Soc. **116** (1992) 171-173.

M.A. Akcoğlu & P.E. Kopp

[1977] *Construction of positive dilations of positive contractions on L_p-spaces.* Math. Zeitschr. **155** (1977) 119-127.

M.A. Akcoğlu & L. Sucheston

[1975] *Remarks on dilations in L_p-spaces.* Proc. Amer. Math. Soc. **53** (1975) 80-82.

[1976] *On positive dilations to isometries in L_p spaces.* Lecture Notes in Math. **541** (1976) 389-401.

[1977] *Dilations of positive contractions on L_p spaces.* Canad. Math. Bull. **20** (1977) 285-292.

C.A. Akemann

[1967] *The dual space of an operator algebra.* Trans. Amer. Math. Soc. **126** (1967) 286-302.

C.A. Akemann, P.G. Dodds & J.L.B. Gamlen

[1972] *Weak compactness in the dual space of a C^*-algebra.* Journ. Funct. Anal. **10** (1972) 446-450.

D.E. Allahverdiev

[1957] *On the completeness of a system of eigenvalues and adjoined elements of non-selfadjoint operators close to normal ones.* Dokl. Akad. Nauk SSSR **115** (1957) 207-210.

D. Alspach, P. Enflo & E. Odell

[1977] *On the structure of separable $\mathcal{L}^p$-spaces* $(1 < p < \infty)$. Studia Math. **60** (1977) 79-90.

Z. Altshuler

[1980] *The moduli of convexity of Lorentz and Orlicz sequence spaces.* Notes in Banach Spaces, University of Texas Press (Austin) (1980) 359-378.

R.D. Anderson & V.L. Klee, Jr.

[1952] *Convex functions and upper semi-continuous collections.* Duke Math. Journ. **19** (1952) 349-357.

T. Andô

[1963] *On a pair of commuting contractions.* Acta Sci. Math. (Szeged) **24** (1963) 88-90.

F.G. Arutjunjan

[1962] *Bases of Banach spaces that contain a subspace isomorphic to $L_1[0,1]$.* Izv. Akad. Nauk Armyanskoi SSR, Ser. Math. **7** (1962)

S. Banach

[1922] *Sur les opérations dans les ensembles abstraits et leurs applications aux équations intégrales.* Fundamenta Math. **3** (1922) 131-181.

[1932] *Théorie des opérations linéaires.* PWN, 1932.

[1938] *Über homogene Polynome in (L^2).* Studia Math. **7** (1938) 36-44.

W. Banaszczyk

[1987] *The Steinitz constant of the plane.* Journ. Reine Angew.Math. **373** (1987) 218-220.

[1990] *The Steinitz theorem on rearrangement of series in nuclear spaces.* Journ. Reine Angew. Math. **403** (1990) 187-200.

[1993] *Rearrangement of series in nonnuclear spaces.* Studia Math. **107** (1993) 213-222.

R.G. Bartle

[1956] *A general bilinear vector integral.* Studia Math. **15** (1956) 337-352.

R.G. Bartle, N. Dunford & J.T. Schwartz

[1955] *Weak compactness and vector measures.* Canad. Journ. Math. **7** (1955) 289-305.

T. Barton & Y. Friedman

[1987] *Grothendieck's inequality for JB^*-triples and applications.* Journ. London Math. Soc. **36** (1987) 513-523.

T. Barton & X.T. Yu

[1995] *A generalized principle of local reflexivity.* Quaestiones Math.

B. Beauzamy

[1976] *Opérateurs de convolution r-sommants sur un groupe compact abélien.* Thèse Univ. Paris VII, Chap. II (1976).

[1978] *Espaces d'Interpolation Réels: Topologie et Géométrie.* Lecture Notes in Math. **666**, 1978.

B. Beauzamy & J.T. Lapresté

[1984] *Modèles étalés des espaces de Banach.* Hermann, 1984

B. Beauzamy & B. Maurey

[1973] *Opérateurs de convolution r-sommants sur un groupe compact abélien.* C. R. Acad. Sci. Paris **A277** (1973) 521-523.

A. Beck

[1962] *A convexity condition in Banach spaces and the strong law of large numbers.* Proc. Amer. Math. Soc. **13** (1962) 329-334.

E. Behrends

[1986] *A generalization of the principle of local reflexivity.* Rev. Roum. Math. Pures Appl. **31** (1986) 293-296.

[1991] *On the principle of local reflexivity.* Studia Math. **100** (1991) 109-128.

A. Bélanger & J. Diestel

[1986] *A remark on weak convergence in the dual of a C^*-algebra.* Proc. Amer. Math. Soc. **98** (1986) 185-186.

A. Bélanger & P.N. Dowling

[1988] *Two remarks on absolutely summing operators.* Math. Nachr. **136** (1988) 229-232.

S.F. Bellenot

[1975] *The Schwartz-Hilbert variety.* Mich. Math. Journ. **22** (1975) 373-377.

[1984] *Local reflexivity of normed spaces.* Journ. Funct. Anal. **59** (1984) 1-11.

A. Benedek & R. Panzone

[1961] *The spaces L^p with mixed norm.* Duke Math. Journ. **28** (1961) 301-324.

C. Benítez & Y. Sarantopoulos

[1995] *Characterization of real inner product spaces by means of symmetric bilinear forms.* Journ. Math. Anal. Appl.

G. Bennett

[1973] *Inclusion mappings between ℓ^p spaces.* Journ. Funct. Anal. **12** (1973) 420-427.

[1975/76] *Some ideals of operators on Hilbert space.* Studia Math. **55** (1975/76) 27-40

[1976] *Unconditional convergence and almost everywhere convergence.* Zeitschr. f. Wahrsch. theorie **34** (1976) 135-155.

[1977] *Schur multipliers.* Duke Math. Journ. **44** (1977) 609-639.

[1980] *Lectures on matrix transformations of ℓ^p spaces.* Notes in Banach Spaces, University of Texas Press (Austin) (1980) 39-80.

G. Bennett, L.E. Dor, V. Goodman, W.B. Johnson & C.M. Newman

[1977] *On uncomplemented subspaces of L_p, $1 < p < 2$.* Israel Journ. Math. **26** (1977) 178 - 187.

G. Bennett, V. Goodman & C.M. Newman

[1975] *Norms of random matrices.* Pacific Journ. Math. **59** (1975) 359-365.

C. Berge

[1959] *Espaces Topologiques, Fonctions Multivoques.* Dunod 1959.
Engl. transl.: *Topological Spaces.* Macmillan 1963.

A. Bernard

[1971] *Algèbres quotients d'algèbres uniformes.* C. R. Acad. Sci.Paris **A272** (1971) 1101-1104.

[1973] *Quotients of operator algebras.* Seminar on Uniform Algberas, University of Aberdeen (1973).

S.J. Bernau

[1980] *A unified approach to the principle of local reflexivity.* Notes in Banach Spaces, University of Texas Press (Austin) (1980) 427-439.

C. Bessaga & A. Pełczyński

[1958] *On bases and unconditional convergence of series in Banach spaces.* Studia Math. **17** (1958) 151 - 164.

O. Blasco

[1986] *A class of operators from a Banach lattice into a Banach space.* Collect. Math. **37** (1986) 13 - 22.

[1987] *Positive p-summing operators on L_p-spaces.* Proc. Amer. Math. Soc. **100** (1987) 275 - 280.

[1988a] *Convolutions of operators and applications.* Math. Zeitschr. **199** (1988) 109 - 114.

[1988b] *Positive p-summing operators, vector measures and tensor products.* Proc. Edinburgh Math. Soc. **31** (1988) 179 - 184.

[1989] *Vector valued harmonic functions and cone absolutely summing operators.* Illinois Journ. Math. **33** (1989) 292 - 300.

O. Blasco & A. Pełczyński

[1991] *Theorems of Hardy and Paley for vector-valued analytic functions and related classes of Banach spaces.* Trans. Amer. Math. Soc. **323** (1991) 335 - 367.

D.P. Blecher

[1988] *Geometry of the tensor product.* Math. Proc. Camb. Phil. Soc. **104** (1988) 119 - 127.

[1990] *Commutativity in operator algebras.* Proc. Amer. Math. Soc. **109** (1990) 709 - 715.

[1995a] *A completely bounded characterization of operator algebras.* Math. Ann.

[1995b] *Factorization in universal operator spaces and algebras.* Rocky Mountain Journ. Math.

D.P. Blecher, Z.J. Ruan & A.M. Sinclair

[1990] *A characterization of operator algebras.* Journ. Funct. Anal. **89** (1990) 188 - 201.

R.C. Blei

[1977] *A uniformity property for $\Lambda(2)$ sets and Grothendieck's inequality.* Sympos. Math. **22** (1977) 321 - 336.

[1979] *Multi-dimensional extensions of Grothendieck's inequality and applications.* Arkiv för Mat. **17** (1979) 51 - 68.

[1980] *Interpolation sets and extensions of the Grothendieck inequality.* Illinois Journ. Math. **24** (1980) 180 - 187.

[1983] *Functions of bounded variation and fractional dimension.* Lecture Notes in Math. **992** (1983) 309 - 322.

[1987] *An elementary proof of the Grothendieck inequality.* Proc. Amer. Math. Soc. **100** (1987) 58 - 60.

[1988] *Multilinear measure theory and the Grothendieck factorization theorem.* Proc. London Math. Soc. **56** (1988) 529 - 546.

[1989] *Multi-linear measure theory and multiple stochastic integration.* Probability Theory and Related Fields **81** (1989) 569 - 584.

S.V. Bočkariev

[1975] *Fourier series divergent on a set of positive measure for arbitrary bounded orthonormal systems.* Mat. Sbornik **98** (1975) 436 - 449.

C. Borell

[1979] *On the integrability of Banach space valued Walsh polynomials.* Lecture Notes in Math. **721** (1979) 1 - 3.

J. Bourgain

[1980] *Espaces $\mathcal{L}^1$ ne vérifiant pas la propriété de Radon-Nikodym.* C. R. Acad. Sci. Paris **A291** (1980) 343 - 345.

[1981a] *A counter example to a complementation problem.* Compositio Math. **43** (1981) 133 - 144.

[1981b] *New Classes of $\mathcal{L}^p$-Spaces.* Lecture Notes in Math. **889** 1981.

[1982] *A Hausdorff-Young inequality for B-convex Banach spaces.* Pacific Journ. Math. **101** (1982) 255 - 262.

[1984a] *Bilinear forms on H^∞ and bounded bianalytic functions.* Trans. Amer. Math. Soc. **286** (1984) 313 - 337.

[1984b] *The dimension conjecture for polydisc algebras.* Israel Journ. Math. **48** (1984) 289 - 304.

[1984c] *New Banach space properties of the disc algebra and H^∞.* Acta Math. **152** (1984) 1 - 48.

[1984d] *The Dunford - Pettis property for the ball algebras, the polydisk algebras and the Sobolev spaces.* Studia Math. **77** (1984) 245 - 254.

[1985] *Applications of the spaces of homogeneous polynomials to some problems on the ball algebra.* Proc. Amer. Math. Soc. **93** (1985) 277 - 283.

[1986] *On the similarity problem for polynomially bounded operators on Hilbert space.* Israel Journ. Math. **54** (1986) 227 - 241.

[1989] *Bounded orthogonal systems and the $\Lambda(p)$-set problem.* Acta Math. **162** (1989) 227 - 245.

J. Bourgain & W.J. Davis

[1986] *Martingale transforms and complex uniform convexity.* Trans. Amer. Math. Soc. **294** (1986) 501 - 510.

J. Bourgain & F. Delbaen

[1980] *A class of special $\mathcal{L}^\infty$-spaces.* Acta Math. **145** (1980) 155 - 176.

J. Bourgain & V.D. Milman

[1985] *Dichotomie du cotype pour les espaces invariants.* C. R. Acad. Sci. Paris **A300** (1985) 263 - 265.

[1987] *New volume ratio properties for convex symmetric bodies in $\mathbf{R}^n$.* Inventiones Math. **88** (1987) 319 - 340.

J. Bourgain, V.D. Milman & H. Wolfson

[1986] *On type of metric spaces.* Trans. Amer. Math. Soc. **294** (1986) 295 - 317.

J. Bourgain & G. Pisier

[1983] *A construction of $\mathcal{L}_\infty$-spaces and related Banach spaces.* Bol. Soc. Brasil Mat. **14** (1983) 109 - 123.

J. Bourgain, H.P. Rosenthal & G. Schechtman

[1981] *An ordinal L^p-index for Banach spaces, with applications to complemented subspaces of L^p.* Ann. of Math. **114** (1981) 193 - 228.

M. Bożejko

[1987] *Littlewood functions, Hankel multipliers and power bounded operators on a Hilbert space.* Colloq. Math. **51** (1987) 35 - 42.

S. Brehmer

[1961] *Über vertauschbare Kontraktionen des Hilbertschen Raumes.* Acta Sci. Math. (Szeged) **22** (1961) 106 - 109.

R.E. Bruck

[1981] *On the convex approximation property and the asymptotic behavior of non linear contractions in Banach spaces.* Israel Journ. Math. **38** (1981) 304 - 314.

A. Brunel & L. Sucheston

[1973] *On B-convex Banach spaces.* Math. System Theory **7** (1973) 294 - 299.

J.W. Bunce

[1981] *The similarity problem for representations of C^*-algebras.* Proc. Amer. Math. Soc. **81** (1981) 409 - 414.

C. Carathéodory

[1907] *Über den Variabilitätsbereich der Koeffizienten von Potenzreihen, die gegebene Werte nicht annehmen.* Math. Ann. **64** (1907) 95 - 115.

[1911] *Über den Variabilitätsbereich der Fourier'schen Konstanten von positiven harmonischen Funktionen.* Rend. Circ. Mat. Palermo **32** (1911) 193 - 217.

C.S. Cardassi

[1989] *Strictly p-integral and p-nuclear operators.* Analyse harmonique: Groupe de Travail sur les Espaces de Banach Invariants par Translation. Exp. II, Publ. Math. Orsay 1989.

B. Carl

[1974] *Absolut $(p,1)$-summierende identische Operatoren von ℓ_u nach ℓ_v.* Math. Nachr. **63** (1974) 353 - 360.

[1976] *A remark on p-integral and p-absolutely summing operators from ℓ_u into ℓ_v.* Studia Math. **57** (1976) 257 - 262.

[1980/81] *Inequalities between (p,q)-summing norms.* Studia Math. **69** (1980/81) 143 - 148.

B. Carl & A. Defant

[1992] *Tensor products and Grothendieck type inequalities of operators in L_p-spaces.* Trans. Amer. Math. Soc. **331** (1992) 55-76.

B. Carl, B. Maurey & J. Puhl

[1978] *Grenzordnungen von absolut (r,p)-summierenden Operatoren.* Math. Nachr. **82** (1978) 205-218.

T.K. Carne

[1979] *Not all H'-algebras are operator algebras.* Math. Proc. Camb. Phil. Soc. **86** (1979) 243-249.

[1980] *Banach lattices and extensions of Grothendieck's inequality.* Journ. London Math. Soc. **21** (1980) 496-516.

N. Carothers

[1981] *Rearrangement invariant subspaces of Lorentz function spaces.* Israel Journ. Math. **40** (1981) 217-228.

[1982] *Symmetric structures in Lorentz spaces.* Ph.D. dissertation, Ohio State University 1982.

[1987] *Rearrangement invariant subspaces of Lorentz function spaces II.* Rocky Mountain Journ. Math. **17** (1987) 607-616.

P.G. Casazza & H. Jarchow

[1995] *Self-induced compactness in Banach spaces.* Proc. Roy. Soc. Edinburgh.

P.G. Casazza & T.J. Shura

[1989] *Tsirelson's Space.* Lecture Notes in Math. **1363** 1989.

E. Casini & P.L. Papini

[1993] *A counterexample to the infinite version of the Hyers and Ulam stability theorem.* Proc. Amer. Math. Soc. **118** (1993) 885-890.

P. Charpentier

[1974] *Sur les quotients d'algèbres uniformes.* C. R. Acad. Sci. Paris **A278** (1974) 929-932.

S.A. Chobanyan

[1984] *Structure of the set of sums of a conditionally convergent series in a Banach space.* Dokl. Akad. Nauk SSSR **278** (1984) 556-559; Engl. transl.: Soviet Math. Dokl. **30** (1984) 438-441.

E. Christensen

[1981] *On non self-adjoint representations of C^*-algebras.* Amer. Journ. Math. **103** (1981) 817-833.

[1982] *Extensions of derivations, II.* Math. Scand. **50** (1982) 111-122.

E. Christensen & A.M. Sinclair

[1989] *A survey of completely bounded operators.* Bull. London Math. Soc. **21** (1989) 417-448.

C.H. Chu & B. Iochum

[1988] *Weakly compact operators on Jordan triples.* Math. Ann. **281** (1988) 451-458.

C.H. Chu, B. Iochum & G. Loupias

[1989] *Grothendieck's theorem and factorization of operators in Jordan triples.* Math. Ann. **284** (1989) 41-53.

I.I. Chuchaev (see V.A. Geiler)

J.S. Cohen

[1970] *A characterization of inner-product spaces using absolutely 2-summing operators.* Studia Math. **38** (1970) 271-276.

[1973] *Absolutely p-summing, p-nuclear operators and their conjugates.* Math. Ann. **201** (1973) 177-200.

R.R. Coifman, R. Rochberg & G. Weiss

[1978] *Applications of transference: the L_p version of von Neumann's inequality and the Littlewood-Paley-Stein theory.* Birkhäuser, Linear Spaces and Approximation (Oberwolfach 1977) **40** (1978) 53-67.

B. Cole, K. Lewis & J. Wermer

[1995] *Pick conditions on a uniform algebra and von Neumann inequalities.*

J. Conway

[1969] *The inadequacy of sequences.* Amer. Math. Monthly **76** (1969) 68-69.

M. Cotlar & S. Sadosky

[1991] *Toeplitz liftings of Hankel forms bounded by non-Toeplitz norms.* Integral Equations and Operator Theory **14** (1991) 501-532.

M. Cowling

[1978] *Some applications of Grothendieck's theory of topological tensor products in harmonic analysis.* Math. Ann. **232** (1978) 273-285.

M.J. Crabb & A.M. Davie

[1975] *Von Neumann's inequality for Hilbert space operators.* Bull. London Math. Soc. **7** (1975) 49-50.

J. Creekmore

[1980] *Geometry and linear topological properties of Lorentz function spaces.* Ph.D. dissertation, Kent State University 1980.

[1981] *Type and cotype in Lorentz L_{pq} spaces.* Indag. Math. **43** (1981) 145-152.

B. Cuartero & M.A. Triana

[1986] *(p,q)-convexity in quasi-Banach lattices and applications.* Studia Math. **84** (1986) 113-124.

D. Dacunha-Castelle & J.L. Krivine

[1972] *Applications des ultraproduits à l'étude des espaces et algèbres de Banach.* Studia Math. **41** (1972) 315-334.

H.G. Dales & H. Jarchow

[1995] *Continuity of homomorphisms and derivations from algebras of approximable and nuclear operators.* Math. Proc. Camb. Phil. Soc.

L. Danzer, B. Grünbaum & V.L. Klee, Jr.

[1963] *Helly's theorem and its relatives.* Proc. Symp. Pure Math. (Convexity) **7** (1963) 101-180.

A.M. Davie (see also M.J. Crabb)

[1973] *Quotient algebras of uniform algebras.* Journ. London Math. Soc. **7** (1973) 31-40.

W.J. Davis (see J. Bourgain)

W.J. Davis, D.J.H. Garling & N. Tomczak-Jaegermann

[1984] *The complex convexity of quasi-normed linear spaces.* Journ. Funct. Anal. **55** (1984) 110-150.

W.J. Davis & W.B. Johnson

[1974] *Compact, non-nuclear operators.* Studia Math. **51** (1974) 81-85.

W.J. Davis & J. Lindenstrauss

[1976] *The ℓ_1^n-problem and degrees of non-reflexivity, II.* Studia Math. **58** (1976) 179-196.

M.M. Day

[1947] *Polygons circumscribed about closed convex curves.* Trans. Amer. Math. Soc. **62** (1947) 315-319.

D.W. Dean

[1973] *The equation $L(E,X^{**}) = L(E,X)^{**}$ and the principle of local reflexivity.* Proc. Amer. Math. Soc. **40** (1973) 146-148.

A. Defant (see B. Carl)

A. Defant & K. Floret

[1993] *Tensor Norms and Operator Ideals.* North-Holland, 1993.

M. Defant & M. Junge

[1990] *Grothendieck type inequalities and weak Hilbert spaces.* London Math. Soc. Lect. Notes **158** (1990) 71-87.

D. Dehay

[1987] *Strong law of large numbers for weakly harmonizable processes.* Stoch. Proc. Appl. **24** (1987) 259-267.

F. Delbaen (see J. Bourgain)

J. Diestel (see also A. Bélanger, C. Cardassi)

[1972] *An elementary characterization of absolutely summing operators.* Math. Ann. **196** (1972) 101-105.

[1984] *Sequences and Series in Banach Spaces.* Springer-Verlag, 1984.

J. Diestel & C.J. Seifert

[1979] *The Banach-Saks ideal, I. Operators acting on $\mathcal{C}(\Omega)$.* Comment. Math. tom. spec. (1979) 109-118 & 343-344.

J. Diestel & J.J. Uhl, Jr.

[1977] *Vector Measures.* Amer. Math. Soc., 1977.

S.J. Dilworth

[1985] *The dimension of Euclidean subspaces of quasi-normed spaces.* Math. Proc. Camb. Phil. Soc. **97** (1985) 311-320.

[1986] *Complex convexity and the geometry of Banach spaces.* Math. Proc. Camb. Phil. Soc. **99** (1986) 495-506.

S.J. Dilworth & M. Girardi

[1995] *Nowhere weak differentiability of the Pettis integral.* Preprint

S. Dineen

[1986a] *Complete holomorphic vector fields on the second dual of a Banach space.* Math. Scand. **59** (1986) 131-142.

[1986b] *The second dual of a JB^* triple system.* North-Holland Math. Studies **125** (1986) 67-69.

[1995] *A Dvoretsky theorem for polynomials.* To appear.

S. Dineen & R.M. Timoney

[1987] *Irreducible domains in Banach spaces.* Israel Journ. Math. **57** (1987) 327-346.

U. Dini

[1868] *Sui prodotti infiniti.* Annali di Matem. **2** (1868) 28-38.

J.P.G.L. Dirichlet

[1837] *Mathematische Werke, Band I.* reprinted Chelsea Publ. Comp. 1969.

P.G. Dixon

[1976a] *Varieties of Banach algebras.* Quarterly J. Math. Oxford **27** (1976) 481-487.

[1976b] *The von Neumann inequality for polynomials of degree greater than two.* Journ. London Math. Soc. **14** (1976) 369-375.

[1976/77] *A characterization of closed subalgebras of $B(H)$.* Proc. Edinburgh Math. Soc. **20** (1976/77) 215-217.

[1977] *Classes of algebraic systems defined by universal Horn sentences.* Algebra Universalis **7** (1977) 313-339.

P.G. Dixon & S.W. Drury

[1986] *Unitary dilations, polynomial identities and the von Neumann inequality.* Math. Proc. Camb. Phil. Soc. **99** (1986) 115-122.

P.G. Dodds (see C.A. Akemann)

P.Domański

[1990] *Principle of local reflexivity for operators and quojections.* Archiv Math. **54** (1990) 567-575.

L.E. Dor (see also G. Bennett)

[1975a] *On sequences spanning a complex ℓ_1-space.* Proc. Amer. Math. Soc. **47** (1975) 515-516.

[1975b] *On projections in L_1.* Ann. of Math. **102** (1975) 363-374.

L.E. Dor & T. Starbird

[1979] *Projections of L_p onto subspaces spanned by independent random variables.* Compositio Math. **39** (1979) 141-175.

P.N. Dowling (see A. Bélanger)

S.W. Drury (see also P.G. Dixon)

[1978] *A generalization of von Neumann's inequality to the complex ball.* Proc. Amer. Math. Soc. **68** (1978) 300-304.

[1983] *Remarks on von Neumann's inequality.* Lecture Notes in Math. **995** (1983) 14-32.

E. Dubinsky, A. Pełczyński & H.P. Rosenthal

[1972] *On Banach spaces X for which $\Pi_2(\mathcal{L}_\infty, X) = B(\mathcal{L}_\infty, X)$.* Studia Math. **44** (1972) 617-648.

N. Dunford (see also R.G. Bartle)

[1938] *Uniformity in linear spaces.* Trans. Amer. Math. Soc. **44** (1938) 305-356.

[1939] *A mean ergodic theorem.* Duke Math. Journ. **5** (1939) 635-646.

N. Dunford & B.J. Pettis

[1940] *Linear operators on spaces of summable functions.* Trans. Amer. Math. Soc. **47** (1940) 323-392.

N. Dunford & J.T. Schwartz

[1958] *Linear Operators. Part I.* Interscience Publ., 1958.

A. Dvoretzky

[1961] *Some results on convex bodies and Banach spaces.* Proc. Intern. Symp. on Linear Spaces, Jerusalem (1961) 123-160.

[1963] *Some near-sphericity results.* Proc. Symp. Pure Math. (Convexity) Amer. Math. Soc. (1963) 203-210.

A. Dvoretzky & C.A. Rogers

[1950] *Absolute and unconditional convergence in normed linear spaces.* Proc. Nat. Acad. Sci. USA **36** (1950) 192-197.

E.G. Effros

[1988] *Amenability and virtual diagonals for von Neumann algebras.* Journ. Funct. Anal. **78** (1988) 137-153.

H.G. Eggleston

[1958] *Convexity.* Cambridge University Press, 1958.

P. Enflo (see also D. Alspach)

[1969] *On the non existence of uniform homeomorphisms between L_p-spaces.* Arkiv för Mat. **8** (1969) 103-105.

[1973] *A counterexample to the approximation property in Banach spaces.* Acta Math. **130** (1973) 309-317

T. Fack

[1987] *Type and cotype inequalities for non-commutative L^p-spaces.* Journ. Operator Theory **17** (1987) 255-279.

X. Fernique

[1970] *Intégrabilité des vecteurs gaussiens.* C. R. Acad. Sci. Paris **A270** (1970) 1698-1699.

T. Figiel

[1972] *Some remarks on Dvoretzky's theorem on almost spherical sections of convex bodies.* Colloq. Math. **24** (1972) 241-252.

[1976a] *A short proof of Dvoretzky's theorem about spherical sections.* Compositio Math. **33** (1976) 297-301.

[1976b] *On the moduli of convexity and smoothness.* Studia Math. **56** (1976) 121-155.

T. Figiel & W.B. Johnson

[1973] *The approximation property does not imply the bounded approximation property.* Proc. Amer. Math. Soc. **41** (1973) 197-200.

[1974] *A uniformly convex space which contains no ℓ_p.* Compositio Math. **29** (1974) 179-190.

T. Figiel, W.B. Johnson & L. Tzafriri

[1975] *On Banach lattices and spaces having local unconditional structures, with applications to Lorentz function spaces.* Journ. Approx. Theory **13** (1975) 395-412.

T. Figiel, S. Kwapień & A. Pełczyński

[1977] *Sharp estimates for the constant of local unconditional structure of Minkowski spaces.* Bull. Polon. Acad. Sci. Math. **25** (1977) 1221-1226.

T. Figiel, J. Lindenstrauss & V.D. Milman

[1977] *The dimension of almost spherical sections of convex bodies.* Acta Math. **139** (1977) 53-94.

T. Figiel & G. Pisier

[1974] *Séries aléatoires dans les espaces uniformément convexes ou uniformément lisses.* C. R. Acad. Sci. Paris **A279** (1974) 611 - 614.

T. Figiel & N. Tomczak-Jaegermann

[1979] *Projections onto Hilbertian subspaces of Banach spaces.* Israel Journ. Math. **33** (1979) 155 - 171.

W. Filter & I. Labuda

[1990] *Essays on the Orlicz-Pettis Theorem, I.* Real Analysis Exchange **16** (1990-91) 393 - 403.

K. Floret (see A. Defant)

C. Foiaş & B. Sz. Nagy

[1970] *Harmonic Analysis of Operators on Hilbert Space.* North-Holland 1970.

J.J. Fournier

[1979] *On a theorem of Paley and the Littlewood conjecture.* Arkiv för Mat. **17** (1979) 199 - 216.

[1981] *Multilinear Grothendieck inequalities via the Schur algorithm.* Canad. Math. Soc. Conf. Proc. **1** (1981) 189 - 211.

Y. Friedman (see T. Barton)

J.L.B. Gamlen (see C.A. Akemann)

D.J.H. Garling (see also W.J. Davis)

[1970] *Absolutely p-summing operators in Hilbert space.* Studia Math. **38** (1970) 319 - 331.

[1974] *Diagonal mappings between sequence spaces.* Studia Math. **51** (1974) 129 - 138.

[1975] *Lattice bounding, Radonifying and summing mappings.* Math. Proc. Camb. Phil. Soc. **77** (1975) 327 - 333.

[1989] *On the dual of a proper uniform algebra.* Bull. London Math. Soc. **21** (1989) 279 - 284.

D.J.H. Garling & Y. Gordon

[1971] *Relations between some constants associated with finite dimensional Banach spaces.* Israel Journ. Math. **9** (1971) 346 - 361.

D.J.H. Garling & N. Tomczak-Jaegermann

[1983] *The cotype and uniform convexity of unitary ideals.* Israel Journ. Math. **45** (1983) 175 - 197.

V.A. Geiler & I.I. Chuchaev

[1982] *General principle of local reflexivity and its applications to the theory of duality of cones.* Sibirsk. Math. Zh. **23** (1982) 32 - 43; Engl. transl.: Siberian Math. Journ. **23** (1982) 24 - 32

S. Geiss

[1987] *Grothendieck-Zahlen linearer und stetiger Operatoren auf Banach-Räumen.* Dissertation, Universität Jena (1987).

[1990a] *Grothendieck numbers of linear and continuous operators on Banach spaces.* Math. Nachr. **148** (1990) 65 - 79.

[1990b] *Grothendieck numbers and volume ratios of operators on Banach spaces.* Forum Math. **2** (1990) 323 - 340.

[1990c] *Some relations between s-numbers of operators on Banach spaces.* Monatsh. Math. **110** (1990) 217 - 230.

B. Gelbaum & J. Gil de Lamadrid

[1961] *Bases of tensor products of Banach spaces.* Pacific Journ. Math. **11** (1961) 1281 - 1286.

I.M. Gelfand

[1938] *Abstrakte Funktionen und lineare Operatoren.* Mat. Sbornik **46** (1938) 235 - 286.

N. Ghoussoub, G. Godefroy, B. Maurey & W. Schachermayer

[1987] *Some topological and geometric structures in Banach spaces.* Memoirs Amer. Math. Soc. **70**, 1987.

D.P. Giesy

[1966] *On a convexity condition in normed linear spaces.* Trans. Amer. Math. Soc. **125** (1966) 114 - 146.

[1973a] *B-convexity and reflexivity.* Israel Journ. Math. **15** (1973) 430 - 436.

[1973b] *The completion of a B-convex normed Riesz space is reflexive.* Journ. Funct. Anal. **12** (1973) 188 - 198.

D.P. Giesy & R.C. James

[1973] *Uniformly non-$\ell_n^{(1)}$ and B-convex Banach spaces.* Studia Math. **48** (1973) 61-69.

J.E. Gilbert

[1977] *Harmonic analysis and the Grothendieck fundamental theorem.* Symposia Math. **22** (1977) 393-420.

[1979] *Nikišin-Stein theory and factorization with applications.* Proc. Symp. Pure Math. Amer. Math. Soc. (1979) 233-267.

J.E. Gilbert, T. Ito & B.M. Schreiber

[1985] *Bimeasure algebras on locally compact groups.* Journ. Funct. Anal. **64** (1985) 134-162.

J. Gil de Lamadrid (see B. Gelbaum)

I. Glicksberg

[1968] *Some uncomplemented function algebras.* Trans. Amer. Math. Soc. **111** (1968) 121-137.

J. Globevnik

[1975] *On complex strict and uniform convexity.* Proc. Amer. Math. Soc. **47** (1975) 175-178.

E.D. Gluskin

[1978] *Norm estimates of certain p-summing operators.* Funktsional Anal. i Prilozhen **12** (1978) 24-31; Engl. transl.: Funct. Anal. and Appl. **12** (1978) 94-101.

E.D. Gluskin, S.V. Kisliakov & O.I. Reinov

[1979] *Tensor products of p-summing operators and right (I_p, N_p)-multipliers.* Sem. Leningrad Otdel Mat. Inst. Steklov (LOMI) **92** (1979) 85-102 & 319-320.

G. Godefroy (see N. Ghoussoub)

I.C. Gohberg & M.G. Krein

[1969] *Introduction to the Theory of Linear Nonselfadjoint Operators in Hilbert Space.* Translations Amer. Math. Soc. **18**, 1969.

V. Goodman (see G. Bennett)

D.B. Goodner

[1950] *Projections in normed linear spaces.* Trans. Amer. Math. Soc. **69** (1950) 89-108.

Y. Gordon (see also D.J.H. Garling)

[1968] *On the projection and Macphail constants of ℓ_n^p-spaces.* Israel Journ. Math. **6** (1968) 295-302.

[1969] *On p-absolutely summing constants of Banach spaces.* Israel Math.Journ. **7** (1969) 151-163.

[1973] *Asymmetry and projection constants of Banach spaces.* Israel Journ. Math. **14** (1973) 50-62.

[1980] *p-local unconditional structure of Banach spaces.* Compositio Math. **41** (1980) 189-205.

[1985] *Some inequalities for Gaussian processes and applications.* Israel Journ. Math. **50** (1985) 265-289.

Y. Gordon & D.R. Lewis

[1974] *Absolutely summing operators and local unconditional structure.* Acta Math. **133** (1974) 27-48.

[1991] *Dvoretzky's theorem for quasi-normed spaces.* Illinois Journ. Math. **35** (1991) 250-259.

Y. Gordon, D.R. Lewis & J.R. Retherford

[1972] *Banach ideals of operators with applications to the finite dimensional structure of Banach spaces.* Israel Journ. Math. **13** (1972) 348-360.

W.T. Gowers

[1995a] *A space not containing c_0, ℓ_1 or a reflexive subspace.* Preprint.

[1995b] *A new dichotomy for Banach spaces.* Preprint.

W.T. Gowers & B. Maurey

[1993] *The unconditional basic sequence problem.* Journ. Amer. Math. Soc. **6** (1993) 851-874.

[1995] *Banach spaces with small spaces of operators.* Preprint.

C.C. Graham & B.M. Schreiber

[1984] *Bimeasure algebras and LCA groups.* Pacific Journ. Math. **115** (1984) 91-127.

R. Graślewicz, H. Hudzik & W. Orlicz

[1986] *Uniformly non-$\ell_n^{(1)}$ property in some normed spaces.* Bull. Polon. Acad. Sci. Math. **34** (1986) 161-171.

R. Grigorieff

[1991] *A note on von Neumann's trace inequality.* Math. Nachr. **151** (1991) 327-328.

W. Gross

[1917] *Bedingt konvergente Reihen.* Monat. Math. Physik **28** (1917) 221-237.

A. Grothendieck

[1953a] *Résumé de la théorie métrique des produits tensoriels topologiques.* Bol. Soc. Mat. São Paulo **8** (1953/1956) 1-79.

[1953b] *Sur certaines classes des suites dans les espaces de Banach, et le théorème de Dvoretzky-Rogers.* Bol. Soc. Mat. São Paulo **8** (1953/1956) 81-110.

[1953c] *Sur les applications linéaires faiblement compactes d'espaces du type $C(K)$.* Canad. Journ. Math. **5** (1953) 129-173.

[1955] *Produits tensoriels topologiques et espaces nucléaires.* Memoirs Amer. Math. Soc. **16**, 1955.

B. Grünbaum (see L. Danzer)

U. Haagerup

[1978] *Les meilleurs constantes de l'inégalité de Khintchine.* C. R. Acad. Sci. Paris **A286** (1978) 259-262.

[1982] *The best constants in the Khintchine inequality.* Studia Math. **70** (1982) 231-283.

[1983a] *All nuclear C^*-algebras are amenable.* Inventiones Math. **74** (1983) 305-319.

[1983b] *Solution of the similarity problem for cyclic representations of C^*-algebras.* Ann. of Math. **118** (1983) 215-240.

[1985] *The Grothendieck inequality for bilinear forms on C^*-algebras.* Advances in Math. **56** (1985) 93-116.

[1987] *A new upper bound for the complex Grothendieck constant.* Israel Journ. Math. **60** (1987) 199-224.

U. Haagerup & G. Pisier

[1989] *Factorization of analytic functions with values in non-commutative L_1-spaces and applications.* Canad. Journ. Math. **41** (1989) 882-906.

H. Hadwiger

[1941] *Über die Konvergenzarten unendlicher Reihen im Hilbertschen Raum.* Math. Zeitschr. **47** (1941) 325-329.

J. Hagler

[1973] *Some more Banach spaces which contain ℓ^1.* Studia Math. **46** (1973) 35-42.

P.R. Halmos & V.S. Sunder

[1978] *Bounded integral operators on L^2 spaces.* Springer-Verlag 1978.

O. Hanner

[1956] *On the uniform convexity of L^p and ℓ^p.* Arkiv för Mat. **3** (1956) 239-244.

M. Hasumi

[1958] *The extension property of complex Banach spaces.* Tôhoku Math. Journ. **10** (1958) 135-142.

S. Heinrich

[1980] *Ultraproducts in Banach space theory.* Journ. Reine Angew. Math. **313** (1980) 72-104.

S. Heinrich & P. Mankiewicz

[1982] *Applications of ultrapowers to the uniform and Lipschitz classification of Banach spaces.* Studia Math. **73** (1982) 225-251.

S. Heinrich, N.J. Nielsen & G.H. Olsen

[1981] *Order bounded operators and tensor products of Banach lattices.* Math. Scand. **49** (1981) 99-127.

E. Helly

[1921] *Über Systeme linearer Gleichungen mit unendlich vielen Unbekannten.* Monatsh. f. Math. u. Phys. **31** (1921) 60-91.

G.M. Henkin

[1967] *Non-isomorphism of some spaces of functions of different numbers of variables.* Funktsional Anal. i Prilozhen **1** (1967) 57-68; Engl. transl.: Funct. Anal. Applic. **1** (1967) 306-315.

[1968] *Banach spaces of analytic functions on the ball and on the bicylinder are not isomorphic.* Funktsional Anal. i Prilozhen **2** (1968) 82-91; Engl. transl.: Funct. Anal. Applic. **2** (1968) 334-341.

C.W. Henson

[1974] *The isomorphism property in nonstandard analysis and its use in the theory of Banach spaces.* Journ. Symb. Logic **39** (1974) 717-731.

[1975] *Ultraproducts of Banach spaces.* The Altgeld Book 1975-76, University of Illinois, Functional Analysis Seminar.

[1976] *Nonstandard hulls of Banach spaces.* Israel Journ. Math. **25** (1976) 108-144.

C.W. Henson & L.C. Moore, Jr.

[1972] *The nonstandard theory of topological vector spaces.* Trans. Amer. Math. Soc. **172** (1972) 405-435.

[1974a] *Nonstandard hulls of the classical Banach spaces.* Duke Math. Journ. **41** (1974) 277-284.

[1974b] *Subspaces of the nonstandard hull of a normed space.* Trans. Amer. Math. Soc. **197** (1974) 131-143.

D. Hilbert

[1904] *Grundzüge einer allgemeinen Theorie der linearen Integralgleichungen I.* Nachr. Wiss. Ges. Göttingen, Math.-phys. Klasse (1904) 49-91.

[1906a] *Grundzüge einer allgemeinen Theorie der linearen Integralgleichungen II.* Nachr. Wiss. Ges. Göttingen, Math.-phys. Klasse (1906) 157-227.

[1906b] *Grundzüge einer allgemeinen Theorie der linearen Integralgleichungen III.* Nachr. Wiss. Ges. Göttingen, Math.-phys. Klasse (1906) 439-462.

[1909] *Wesen und Ziele einer Analysis der unendlich vielen unabhängigen Variablen.* Rend. Circ. Mat. Palermo **27** (1909) 59-74.

[1910] *Grundzüge einer allgemeinen Theorie der linearen Integralgleichungen IV.* Nachr. Wiss. Ges. Göttingen, Math.-phys. Klasse (1910) 595-618.

T.H. Hildebrandt

[1940] *On unconditional convergence in normed vector spaces.* Bull. Amer. Math. Soc. **46** (1940) 959-962.

[1953] *Integration in abstract spaces.* Bull. Amer. Math. Soc. **59** (1953) 111-139.

J. Hoffmann-Jørgensen

[1972/73] *Sums of independent Banach space valued random variables.* Aarhus Univ. Preprint Series **15** (1972/73).

[1974] *Sums of independent Banach space valued random variables.* Studia Math. **52** (1974) 159-186.

[1977a] *Probability in Banach spaces.* Lecture Notes in Math. **598** (1977) 1-186.

[1977b] *Integrability of seminorms, the 0-1 law and the affine kernel for product measures.* Studia Math. **56** (1977) 137-159.

J. Hoffmann-Jørgensen & G. Pisier

[1976] *The law of large numbers and the central limit theorem in Banach spaces.* Ann. of Prob. **4** (1976) 587-599.

C. Houdré

[1990a] *Harmonizability, V-boundedness, $(2,p)$-boundedness of stochastic processes.* Probability Theory and Related Fields **84** (1990) 39-54.

[1990b] *Linear Fourier and stochastic analysis.* Probability Theory and Related Fields **87** (1990) 167-188.

H. Hudzik (see also R. Graślewicz)

[1985] *Uniformly non-$\ell_n^{(1)}$ Orlicz spaces with Luxemburg norm.* Studia Math. **81** (1985) 271-284.

[1987] *An estimation of the modulus of convexity in a class of Orlicz spaces.* Math. Japon. **32** (1987) 227-237.

[1991] *Lower and upper estimates of the modulus of convexity in some Orlicz spaces.* Archiv Math. **57** (1991) 80-87.

H. Hudzik & A. Kamińska

[1985] *On uniformly convexifiable and B-convex Musielak-Orlicz spaces.* Comment. Math. Prace Mat. **25** (1985) 59-75.

H. Hudzik, A. Kamińska & W. Kurc

[1987] *Uniformly non-$\ell_n^{(1)}$ Musielak-Orlicz spaces.* Bull. Polon. Acad. Sci. Math. **35** (1987) 441-448.

H. Hunziker

[1989] *Kompositionsoperatoren auf klassischen Hardyräumen.* Dissertation, Universität Zürich (1989).

H. Hunziker & H. Jarchow

[1991] *Composition operators which improve integrability.* Math. Nachr. **152** (1991) 83-99.

H. Hunziker, H. Jarchow & V. Mascioni

[1990] *Some topologies on the space of analytic self-maps of the unit disk.* London Math. Soc. Lecture Note Series **158** (1990) 133-148.

B. Iochum (see C.H. Chu)

T. Ito (see J.E. Gilbert)

N.C. Jain & M.B. Marcus

[1975] *Integrability of infinite sums of independent vector-valued random variables.* Trans. Amer. Math. Soc. **212** (1975) 1-36.

R.C. James (see also D.P. Giesy)

[1950] *Bases and reflexivity of Banach spaces.* Ann. of Math. **52** (1950) 518-527.

[1964] *Uniformly non square Banach spaces.* Ann. of Math. **80** (1964) 542-550.

[1972] *Some self-dual properties of normed linear spaces.* Annals of Math. Studies **69** (1972) 159-175.

[1974] *A non reflexive Banach space that is uniformly non octahedral.* Israel Journ. Math. **18** (1974) 145-155.

[1978] *Nonreflexive spaces of type* 2. Israel Journ. Math. **30** (1978) 1-13.

R.C. James & J. Lindenstrauss

[1974] *The octahedral problem for Banach spaces.* Proceedings of the Seminar on Random Series, Convex Sets and Geometry of Banach Spaces, Aarhus 1974.

G.J.O. Jameson

[1985] *The interpolation proof of Grothendieck's inequality.* Proc. Edinburgh Math. Soc. **28** (1985) 217-223.

[1987] *Summing and Nuclear Norms in Banach Space Theory.* Cambridge University Press, 1987.

H. Jarchow (see also P.G. Casazza, H.G. Dales, H. Hunziker)

[1981] *Locally Convex Spaces.* Teubner-Verlag, Stuttgart 1981.

[1982] *On Hilbert-Schmidt spaces.* Rend. Circ. Mat. Palermo (Suppl.) **II** (2) (1982) 153-160.

[1984a] *On operator ideals related to type 2 and cotype 2 operators.* Proc. Conf. Leipzig 1983, Teubner-Texte zur Math. **67** (1984) 164-172.

[1984b] *Remarks on a characterization of nuclearity.* Archiv Math. **43** (1984) 469-472.

[1986] *On weakly compact operators on C^*-algebras.* Math. Ann. **273** (1986) 341-343.

[1992] *Some factorization properties of composition operators.* North-Holland Mathematical Studies **170** (1992) 405-413.

[1993] *Absolutely summing composition operators.* Functional Analysis, Marcel Dekker Inc. (1993) 193-202.

[1995] *Some functional analytic properties of composition operators.* Quaestiones Math.

H. Jarchow & K. John

[1995] *Bilinear forms and nuclearity.* Czech. Math. Journ. **44** (1994) 367-373.

H. Jarchow & U. Matter

[1985] *On weakly compact operators on $C(K)$-spaces.* Lecture Notes in Math. **1166** (1985) 80 - 88.

[1988] *Interpolative constructions for operator ideals.* Note di Mat. **8** (1988) 45 - 56.

H. Jarchow & R. Ott

[1982] *On trace ideals.* Math. Nachr. **108** (1982) 23 - 37.

H. Jarchow & R .Riedl

[1995] *Factorization properties of composition operators through Bloch type spaces.* Illinois Journ. Math.

F. John

[1948] *Extremum problems with inequalities as subsidiary conditions.* Courant Anniversary Volume. Interscience (1948) 187 - 204.

K. John (see also H. Jarchow)

[1983] *Counterexample to a conjecture of Grothendieck.* Math. Ann. **265** (1983) 169 - 179.

[1990] *On the compact non-nuclear problem.* Math. Ann. **287** (1990) 509 - 514.

W.B. Johnson (see also G. Bennett, W.J. Davis, T. Figiel)

[1974] *On finite dimensional subspaces of Banach spaces with local unconditional structure.* Studia Math. **51** (1974) 223 - 238.

[1980] *Banach spaces all of whose subspaces have the approximation property.* Special Topics of Applied Mathematics, North-Holland (1980) 15 - 26.

W.B. Johnson, H. König, B. Maurey & J.R. Retherford

[1979] *Eigenvalues of p-summing and ℓ_p-type operators in Banach spaces.* Journ. Funct. Anal. **32** (1979) 353 - 380.

W.B. Johnson, J. Lindenstrauss & G. Schechtman

[1980] *On the relation between several notions of local unconditional structure.* Israel Journ. Math. **37** (1980) 120 - 129.

W.B. Johnson, B. Maurey, G. Schechtman & L. Tzafriri

[1979] *Symmetric structures in Banach spaces.* Memoirs Amer. Math. Soc. **217**, 1979.

W.B. Johnson, H.P. Rosenthal & M. Zippin

[1971] *On bases, finite dimensional decompositions and weaker structures in Banach spaces.* Israel Journ. Math. **9** (1971) 488 - 506.

W.B. Johnson & G. Schechtman

[1994] *Computing p-summing norms with few vectors.* Israel Journ. Math. **87** (1994) 19 - 31.

W.B. Johnson & L. Tzafriri

[1977] *Some more Banach spaces which do not have l.u.st.* Houston Journ. Math. **3** (1977) 55 - 60.

M. Junge (see M. Defant)

M.I. Kadets (see also V.M. Kadets)

[1956] *Unconditional convergence of series in uniformly convex spaces.* Uspehi. Mat. Nauk. **11** (1956) 185 - 190.

M.I. Kadets & A .Pełczyński

[1962] *Bases, lacunary sequences and complemented subspaces in the spaces L_p.* Studia Math. **21** (1962) 161 - 176.

M.I. Kadets & M.G. Snobar

[1971] *Certain functionals on the Minkowski compactum.* Mat. Zametki **10** (1971) 453 - 458.

M.I. Kadets & K. Woźniakowski

[1989] *On series whose permutations have only two sums.* Boll. Polon. Acad. Sci. **37** (1989) 15 - 21.

V.M. Kadets

[1986] *On a problem of Banach, 'Problem 106 in the Scottish Book'.* Funktsional Anal. i Prilozhen **20** (1986) 74 - 75; Engl. transl.: Funct. Anal. Appl. **20** (1986) 317 - 319.

[1989] *Characterization of B-convex spaces in terms of the behavior of integral sums.* Current Analysis and its Applications, Naukova Dumka, Kiev 1989, 56 - 62.

[1991] *A remark on Lyapunov's theorem on a vector measure.* Funksional Anal. i Prilozhen **25** (1991) 78 - 80; Engl.transl: Funct. Anal. Appl. **25** (1991) 295 - 297.

V.M. Kadets & M.I. Kadets

[1984] *Conditions for convexity of the set of limits of Riemann sums of a vector-valued function.* Mat. Zametki **35** (1984) 161 - 167; Engl. transl.: Math. Notes **35** (1984) 85 - 88.

[1991] *Rearrangement of Series in Banach Spaces.* Amer. Math. Soc., 1991.

J.P. Kahane

[1964] *Sur les sommes vectorielles $\sum \pm u_n$.* C. R. Acad. Sci. Paris **259** (1964) 2577 - 2580.

[1968] *Some random series of functions.* Heath Math. Monographs 1968; 2nd ed. Cambridge University Press 1985.

S. Kaijser

[1973] *A note on the Grothendieck constant with an application to harmonic analysis.* Univ. Uppsala DM Report No. 1973:10.

[1975/76] *An application of Grothendieck's inequality to a problem in harmonic analysis.* École Polyt. Palaiseau, Sém. Maurey - Schwartz 1975/76, Exp. V.

[1976] *Some remarks on injective Banach algebras.* Lecture Notes in Math. **512** (1976) 84 - 95.

[1978] *Some results in the metric theory of tensor products.* Studia Math. **63** (1978) 157 - 170.

[1983] *A simple-minded proof of the Pisier - Grothendieck inequality.* Lecture Notes in Math. **995** (1983) 33 - 55.

S. Kaijser & A.M. Sinclair

[1984] *Projective tensor products of C^*-algebras.* Math. Scand. **55** (1984) 161 - 187.

R.J. Kaiser & J.R. Retherford

[1983] *Eigenvalue distribution of nuclear operators: a survey.* Proc. Conf. Funct. Analysis, Universität Essen (1982) 245 - 287.

S. Kakutani

[1941a] *Concrete representation of abstract (L)-spaces and the mean ergodic theorem.* Ann. of Math. **42** (1941) 523 - 537.

[1941b] *Concrete representation of abstract (M)-spaces (A characterization of the space of continuous functions).* Ann. of Math. **42** (1941) 994 - 1024.

N.J. Kalton

[1980] *The Orlicz - Pettis theorem.* Contemp. Math. **2** (1980) 91 - 100.

A. Kamińska (see H. Hudzik)

A. Kamińska & B. Turett

[1987] *Uniformly non-$\ell_n^{(1)}$ Orlicz - Bochner Spaces.* Bull. Polon. Acad. Sci. Math. **35** (1987) 263 - 270.

[1990] *Type and cotype in Musielak - Orlicz spaces.* London Math. Soc. Lect. Notes **158** (1990) 165 - 180.

S. Karni & E. Merzbach

[1990] *On the extension of bimeasures.* Journ. d'Analyse Math. **55** (1990) 1 - 17.

Y. Katznelson & O.C. McGehee

[1974] *Conditionally convergent series in $\mathbf{R}^\infty$.* Mich. Math. Journ. **21** (1974) 97 - 106.

J.L. Kelley

[1952] *Banach spaces with the extension property.* Trans. Amer. Math. Soc. **72** (1952) 323 - 326.

T. Ketonen

[1981] *On unconditionality in L_p-spaces.* Ann. Acad. Sci. Fenn. **35** (1981) 1 - 42.

R. Khalil

[1980] *Trace-class norm multipliers.* Proc. Amer. Math. Soc. **79** (1980) 379 - 387.

A. Khinchin

[1923] *Über dyadische Brüche.* Math. Zeitschr. **18** (1923) 109 - 116.

A. Khinchin & A.N. Kolmogorov

[1925] *Über Konvergenz von Reihen, deren Glieder durch den Zufall bestimmt werden.* Mat. Sbornik **32** (1925) 668 - 677.

S.V. Kisliakov (see also E.D. Gluskin)

[1975] *Sobolev imbedding operators and the non isomorphism of certain Banach spaces.* Funktsional Anal. i Prilozhen **9** (1975) 22-27. Engl. transl. Funct. Anal. Appl. **9** (1975) 290 - 294.

[1977] *There is no local unconditional structure in the space of continuously differentiable functions.* Sem. Leningrad Otdel. Mat. Inst. Steklov (LOMI) (1977).

[1978] *On spaces with "small" annihilators.* Sem. Leningrad Otdel. Mat. Inst. Steklov (LOMI) **73** (1978) 91-101.

[1981a] *What is needed for 0-absolutely summing operators to be nuclear?* Lecture Notes in Math. **864** (1981) 336-364.

[1981b] *Two remarks on the equality* $\Pi_p(X, \cdot) = I_p(X, \cdot)$. Sem. Leningrad Otdel. Mat. Inst. Steklov (LOMI) **113** (1981) 135-148.

[1989] *Proper uniform algebras are uncomplemented.* Dokl. Akad. Nauk SSSR **309** (1989) 795-798; Engl. transl.: Soviet Math. Dokl. **40** (1989) 584-586.

[1991] *Absolutely summing operators on the disk algebra.* Algebra and Analysis **3** (1991) 1-79; Engl. transl.: St. Petersburg Math. J. **3** 705-774.

S.V. Kisliakov & N.G. Sidorenko

[1988] *The non-existence of local unconditional structure in anisotropic spaces of smooth functions.* Sibir. Math. Journ. **29** (1988) 64-77; Engl. transl.: Siberian Math. Journ. **29** (1988) 384-394.

V.L. Klee, Jr. (see R.D. Anderson, L. Danzer)

A.N. Kolmogorov (see A. Khinchin)

R.A. Komorowski & N. Tomczak-Jaegermann

[1995] *Banach spaces without local unconditional structure.* Israel Journ. Math.

H. König (see also W.B. Johnson)

[1975] *Diagonal and convolution operators as elements of operator ideals.* Math. Ann. **218** (1975) 97-106.

[1980] *A Fredholm determinant theory for p-summing maps in Banach spaces.* Math. Ann. **247** (1980) 255-274.

[1985] *Spaces with large projection constant.* Israel Journ. Math. **50** (1985) 181-188.

[1986] *Eigenvalue Distribution of Compact Operators.* Birkhäuser-Verlag, 1986.

[1990] *On the complex Grothendieck constant in the n-dimensional case.* London Math. Soc. Lect. Notes **158** (1990) 181-198.

[1991] *On the Fourier-coefficients of vector-valued functions.* Math. Nachr. **152** (1991) 215-227.

[1995] *Projections onto symmetric spaces.* Quaestiones Math.

H. König & D.R .Lewis

[1987] *A strict inequality for projection constants.* Journ. Funct. Anal. **73** (1987) 328-332.

H.König, D.R. Lewis & P.K. Lin

[1983] *Finite dimensional projection constants.* Studia Math. **75** (1983) 341-358.

H. König, J.R. Retherford & N. Tomczak-Jaegermann

[1980] *On the eigenvalues of (p, 2)-summing operators and constants associated to normed spaces.* Journ. Funct. Anal. **37** (1980) 149-168.

H. König & N. Tomczak-Jaegermann

[1990] *Bounds for projection constants and 1-summing norms.* Trans. Amer. Math. Soc. **320** (1990) 799-823.

[1995] *Norms of minimal projections.* Journ. Funct. Anal.

P.E. Kopp (see M.A. Akcoğlu)

P.A. Kornilov

[1980] *On rearrangements of conditionally convergent function series.* Mat. Sbornik **113** (1980) 598-616; Engl. transl.: Math. USSR Sb. **41** (1982).

[1988a] *Structure of the sums set of a functional series.* Vestnik Mosk. Univ. Ser. I (1988), no.4, 9-13; Engl. transl.: Mosc. Univ. Math. Bull. **43**, no.4, (1988) 6-11.

[1988b] *On the set of sums of a conditionally convergent series of functions.* Mat. Sbornik **137** (1988) 114-127; Engl. transl.: Math. USSR Sb. **65**(1990) 119-131.

M.G. Krein (see I.C. Gohberg)

J.L. Krivine (see also D. Dacunha-Castelle)

[1973/74] *Théorèmes de factorisation dans les espaces réticulés.* École Polyt. Palaiseau., Sém. Maurey-Schwartz 1973/74, Exp. XXII-XXIII.

[1976] *Sous-espaces de dimension fini des espaces de Banach réticulés.* Ann. of Math. **104** (1976) 1-29.

[1977a] *Sur la constante de Grothendieck.* C. R. Acad. Sci. Paris **A284** (1977) 445-446.

[1977b] *Sur la complexification des opérateurs de L^∞ dans L^1.* C. R. Acad. Sci. Paris **A284** (1977) 377-379.

[1979] *Constantes de Grothendieck et fonctions de type positif sur les sphères.* Advances in Math. **31** (1979) 16-30.

W. Kurc (see H. Hudzik)

K.D. Kürsten

[1984] *Lokale Reflexivität und lokale Dualität von Ultraprodukten für halbgeordnete Banachräume.* Zeitschr. Anal. Anw. **3** (1984) 254-262.

S. Kwapień (see also T. Figiel)

[1968] *Some remarks on (p,q)-summing operators in ℓ_p-spaces.* Studia Math. **29** (1968) 327-337.

[1970a] *A remark on p-summing operators in ℓ_r-spaces.* Studia Math. **34** (1970) 109-11.

[1970b] *On a theorem of L.Schwartz and its applications to absolutely summing operators.* Studia Math. **38**(1970) 193-201.

[1970c] *A linear topological characterization of inner product spaces.* Studia Math. **38** (1970) 277-278.

[1972a] *Isomorphic characterizations of inner product spaces by orthogonal series with vector valued coefficients.* Studia Math. **44** (1972) 583-595.

[1972b] *On operators factorizable through L_p-spaces.* Bull. Soc. Math. France, Mém. **31-32** (1972) 215-225.

[1974] *On Banach spaces containing c_0. A supplement to the paper by J.Hoffmann-Jørgensen "Sums of independent Banach space valued random variables".* Studia Math. **52** (1974) 187-188.

S. Kwapień & A. Pełczyński

[1970] *The main triangle projection in matrix spaces and its applications.* Studia Math. **34** (1970) 43-68.

[1978] *Remarks on absolutely summing translation invariant operators from the disk algebra and its dual into a Hilbert space.* Mich. Math. Journ. **25** (1978) 173-181.

[1980] *Absolutely summing operators and translation invariant spaces of functions on compact abelian groups.* Math. Nachr. **94** (1980) 303-340.

S. Kwapień & S. Szarek

[1979] *Estimation of Lebesgue functions of biorthogonal systems with application to non existence of certain bases.* Studia Math. **66** (1979) 185-200.

Ky Fan

[1951] *Maximum properties and inequalities for eigenvalues of completely continuous operators.* Proc. Nat. Acad. Sci. USA **37** (1951) 760-766.

I. Labuda (see W. Filter)

H.J. Landau & L.A. Shepp

[1970] *On the supremum of a Gaussian process.* Sankhyà **A32** (1970) 369-378.

J.T. Lapresté (see also B. Beauzamy)

[1972/73] *Opérateurs se factorisant par un espace L^p, d'après S.Kwapień.* École Polyt. Palaiseau, Sém. Maurey-Schwartz 1972/73, Exp. XVI.

D.G. Larman & P. Mani

[1975] *Almost ellipsoidal sections and projections of convex bodies.* Math. Proc. Camb. Phil. Soc. **77** (1975) 529-546.

R. Latała & K. Oleszkiewicz

[1995] Studia Math. (1995)

M. Ledoux & M. Talagrand

[1991] *Probability in Banach Spaces.* Springer-Verlag, 1991.

P. Lévy

[1905] *Sur les séries semi-convergentes.* Nouv. Ann. de Math. **64** (1905) 506 - 511.

D.R. Lewis (see also Y. Gordon, H. König)

[1973] *Integral operators on $\mathcal{L}_p$-spaces.* Pacific Journ. Math. **46** (1973) 451 - 456.

[1975] *An isomorphic classification of the Schmidt class.* Compositio Math. **30** (1975) 293 - 297.

[1978] *Finite dimensional subspaces of L_p.* Studia Math. **63** (1978) 207 - 212.

[1979] *Ellipsoids defined by Banach ideal norms.* Mathematika **26** (1979) 18 - 29.

[1988] *An upper bound for the projection constant.* Proc. Amer. Math. Soc. **103** (1988) 1157 - 1160.

D.R. Lewis & C. Stegall

[1971] *Banach spaces whose duals are isomorphic to $\ell^1(\Gamma)$.* Journ. Funct. Anal. **12** (1971) 167 - 177.

D.R. Lewis & N. Tomczak-Jaegermann

[1980] *Hilbertian and complemented finite-dimensional subspaces of Banach lattices and unitary ideals.* Journ. Funct. Anal. **35** (1980) 165 - 190.

K .Lewis (see B. Cole)

P.K. Lin (see H. König)

W. Linde & A. Pietsch

[1974] *Mappings of Gaussian cylindrical measures in Banach spaces.* Teor. Veroj. i Primenen. **19** (1974) 472 - 487; Engl. transl.: Theory of Prob. and Appl. **19** (1974) 445 - 460.

J. Lindenstrauss (see also W.J. Davis, T. Figiel, R.C. James, W.B. Johnson)

[1963] *On the modulus of smoothness and divergent series in Banach spaces.* Mich. Math. Journ. **10** (1963) 241 - 252.

[1964] *Extension of compact operators.* Memoirs Amer. Math. Soc. **48**, 1964.

[1971] *On James' paper "Separable conjugate spaces".* Israel Journ. Math. **9** (1971) 279-284.

[1992] *Almost spherical sections; their existence and their applications.* Jber. Dt. Math.-Vgg., Jubiläumstagung 1990, (1992) 39 - 61.

J. Lindenstrauss & A. Pełczyński

[1968] *Absolutely summing operators in $\mathcal{L}_p$-spaces and their applications.* Studia Math. **29** (1968) 275 - 326.

J. Lindenstrauss & H.P. Rosenthal

[1969] *The $\mathcal{L}_p$ spaces.* Israel Journ. Math. **7** (1969) 325 - 349.

J. Lindenstrauss & C. Stegall

[1975] *Examples of separable spaces which do not contain ℓ_1 and whose duals are non-separable.* Studia Math. **54** (1975) 81 - 105.

J. Lindenstrauss & L. Tzafriri

[1977] *Classical Banach Spaces I.* Springer-Verlag, 1977.

[1979] *Classical Banach Spaces II.* Springer-Verlag, 1979.

J. Lindenstrauss & M. Zippin

[1969a] *Banach spaces with a unique unconditional basis.* Journ. Funct. Anal. **3** (1969) 115 - 125.

[1969b] *Banach spaces with sufficiently many Boolean algebras of projections.* Journ. Math. Anal. Appl. **25** (1969) 309 - 320.

M. Lindström & R.A. Ryan

[1992] *Applications of ultraproducts to infinite dimensional holomorphy.* Math. Scand. **71** (1992) 229 - 242.

J.E. Littlewood

[1930] *On bounded bilinear forms in an infinite number of variables.* Quart. Journ. Math. Oxford **1** (1930) 164 - 174.

H.P. Lotz

[1973] *Grothendieck ideals of operators on Banach spaces.* University of Illinois 1973.

G. Loupias (see C.H. Chu)

W.A.J. Luxemburg

[1969] *A general theory of monads.* Applications of Model Theory to Algebra, Analysis and Probability (1969) 18 - 56.

M.S. Macphail

[1947] *Absolute and unconditional convergence.* Bull. Amer. Math. Soc. **53** (1947) 121 - 123.

A. Makagon & H. Salehi

[1987] *Spectral dilation of operator-valued measures and its application to infinite dimensional harmonizable processes.* Studia Math. **85** (1987) 257 - 297.

B.M. Makarov

[1991] *p-absolutely summing operators and some of their applications.* Algebra and Analysis **3** (1991) 1 - 76; Engl. transl.: St. Petersburg Math. J. **3** (1992) 227 - 298.

R.P. Maleev & S.L. Troyanski

[1975] *On the moduli of convexity and smoothness in Orlicz spaces.* Studia Math. **54** (1975) 131 - 141.

P. Mani (see D.G. Larman)

P. Mankiewicz (see S. Heinrich)

A.M. Mantero & A.M. Tonge

[1979] *Banach algebras and von Neumann's inequality.* Proc. London Math. Soc. **38** (1979) 309 - 334.

[1980] *The Schur multiplication in tensor algebras.* Studia Math. **68** (1980) 1 - 24.

M.B. Marcus (see N.C. Jain)

M.B. Marcus & G. Pisier

[1981] *Random Fourier series with applications to harmonic analysis.* Annals of Mathematics, Studies **101**, 1981.

V. Mascioni (see also H. Hunziker)

[1988a] *Weak cotype and weak type in the local theory of Banach spaces.* Dissertation, Universität Zürich 1988.

[1988b] *On weak cotype and weak type in Banach spaces.* Note di Mat. **8** (1988) 67 - 110.

V. Mascioni & U. Matter

[1988] *Weakly (q,p)-summing operators and weak cotype properties of Banach spaces.* Proc. Roy. Ir. Acad. **88A** (1988) 169 - 177.

I. Masri

[1990] *Estimates for norms of multilinear Hankel operators and absolutely summing multipliers.* Kent State Ph.D. dissertation (1990).

C. Matsuoka

[1987] *Factoring functions of Maurey's factorization theorem.* Israel Journ. Math. **57** (1987) 318 - 326.

U. Matter (see also H. Jarchow, V. Mascioni)

[1985] *Absolutely continuous operators and super-reflexivity.* Dissertation, Universität Zürich (1985).

[1987] *Absolutely continuous operators and super-reflexivity.* Math. Nachr. **130** (1987) 193 - 216.

[1989] *Factoring through interpolation spaces and super reflexive spaces.* Rev. Roum. Math. Pures Appl. **34** (1989) 147 - 156.

R.D. Mauldin

[1981] *The Scottish Book.* Birkhäuser-Verlag, Boston-Basel-Stuttgart 1981.

B. Maurey (see also B. Beauzamy, B. Carl, N. Ghoussoub, W.T. Gowers, W.B. Johnson)

[1972/73] *Espaces de cotype p, $0 < p \leq 2$.* École Polyt. Palaiseau, Sém. Maurey - Schwartz 1972/73,Exp. VII.

[1973] *Sur certaines propriétés des opérateurs sommants.* C. R. Acad. Sci. Paris **A277** (1973) 1053 - 1055.

[1973/74a] *Une nouvelle caractérisation des applications (p,q)-sommantes.* École Polyt. Palaiseau, Sém. Maurey - Schwartz 1973/74, Exp. XII.

[1973/74b] *Type et cotype dans les espaces munis de structures locales inconditionelles.* École Polyt. Palaiseau, Sém. Maurey - Schwartz 1973/74, Exp. XXIV-XXV.

[1974a] *Théorèmes de factorisation pour les opérateurs à valeurs dans les espaces L^p.* Soc. Math. France, Astérisque **11**, Paris 1974.

[1974b] *Un théorème de prolongement.* C. R. Acad. Sci. Paris **A279** (1974) 329 - 332.

[1980/81] *Points fixes des contractions de certains faiblements compacts de L^1.* École Polyt. Palaiseau, Sém. d'Anal. Fonct. (1980/81) Exp. VIII.

B. Maurey & A. Nahoum

[1973] *Applications radonifiantes dans les espaces des séries convergentes.* C. R. Acad. Sci. Paris **A276** (1973) 751 - 754.

B. Maurey & A. Pełczyński

[1976] *A criterion for compositions of (p,q)-absolutely summing operators to be compact.* Studia Math. **54** (1976) 291 - 300.

B. Maurey & G .Pisier

[1973] *Caractérisation d'une classe d'espaces de Banach par les propriétés de séries aléatoires vectorielles.* C. R. Acad. Sci. Paris **A277** (1973) 687 - 680.

[1976] *Séries de variables aléatoires vectorielles indépendantes et propriétés géométriques des espaces de Banach.* Studia Math. **58** (1976) 45 - 90.

C.W. McArthur

[1956] *On relationships amongst certain spaces of sequences in an arbitrary Banach space.* Canad. Journ. Math. **8** (1956) 192 - 197.

C.A. McCarthy

[1967] c_p. Israel Journ. Math. **5** (1967) 249 - 271.

C.A. McCarthy & L. Tzafriri

[1968] *Projections in $\mathcal{L}_1$ and $\mathcal{L}_\infty$ spaces.* Pacific Journ. Math. **26** (1968) 529 - 546.

O.C. McGehee (see Y. Katznelson)

E. Merzbach (see S. Karni)

V.D. Milman (see also J. Bourgain, T. Figiel)

[1970] *The geometric theory of Banach spaces. I. The theory of basic and minimal systems.* Uspehi Mat. Nauk **25** (1970) 113 - 174.

[1971] *A new proof of the theorem of A. Dvoretzky on sections of convex bodies.* Funktsional Anal. i Prilozhen **5** (1971) 28 - 37; Engl. transl.: Funct. Anal. Appl. **5** (1971) 288 - 299.

[1992] *Dvoretsky's theorem – thirty years later.* Geom. and Funct. Anal. **2** (1992) 445 - 479.

V.D. Milman & G. Pisier

[1986] *Banach spaces with a weak cotype 2 property.* Israel Journ. Math. **54** (1986) 139 - 158.

V.D. Milman & G. Schechtman

[1986] *Asymptotic Theory of Finite Dimensional Normed Spaces.* Lecture Notes in Math. **1200** 1986.

B.S. Mitiagin & A. Pełczyński

[1966] *Nuclear operators and approximative dimensions.* Proceedings International Congress of Mathematicians, Moscow 1966.

[1976] *On the non-existence of linear isomorphisms between Banach spaces of analytic functions of one and several complex variables.* Studia Math. **56** (1976) 175 - 186.

N. Mittal (see A.I. Singh)

S.J. Montgomery-Smith

[1988] *The cotype of operators from $C(K)$.* Ph.D. Dissertation, Cambridge University 1988.

[1989] *The Gaussian cotype of operators from $C(K)$.* Israel Journ. Math. **68** (1989) 123 - 128.

L.C. Moore, Jr. (see C.W. Henson)

L. Nachbin

[1950] *A theorem of Hahn-Banach type for linear transformations.* Trans. Amer. Math. Soc. **68** (1950) 28 - 46.

[1965] *The Haar Integral.* Van Nostrand, 1965.

A. Nahoum (see also B. Maurey)

[1972/73] *Applications radonifiantes dans l'espace des séries convergentes. I. Le théorème de Menchov.* École Polyt. Palaiseau, Sém. Maurey - Schwartz 1972/73, Exp. XXIV.

H. Nakano

[1941] *Über das System aller stetigen Funktionen auf einem topologischen Raum.* Proc. Imp. Acad. Tokyo **17** (1941) 308 - 310.

C.M. Newman (see G. Bennett)

C.P. Niculescu

[1975] *Absolute continuity and weak compactness.* Bull. Amer. Math. Soc. **81** (1975) 1064 - 1066.

[1979] *Absolute continuity in Banach space theory.* Rev. Roum. Math. Pures Appl. **24** (1979) 413 - 422.

N.J. Nielsen (see also S. Heinrich)

[1973] *On Banach ideals determined by Banach lattices and their applications.* Dissertationes Math. **54** (1973) 1 - 62.

[1982] *The ideal property of tensor products of Banach lattices with applications to the local structure of spaces of absolutely summing operators.* Studia Math. **74** (1982) 247 - 272.

N.J. Nielsen & J. Szulga

[1984] *p-lattice summing operators.* Math. Nachr. **119** (1984) 219 - 230.

H. Niemi

[1984] *Grothendieck's inequality and minimal orthogonally scattered dilations.* Lecture Notes in Math. **1080** (1984) 175 - 187.

H. Niemi & A. Weron

[1981] *Dilation theorems for positive definite operator kernels having majorants.* Journ. Funct. Anal. **40** (1981) 54 - 65.

E.M. Nikishin

[1970] *Resonance theorems and superlinear operators.* Uspehi. Mat. Nauk **25** (1970) 129 - 191; Engl. transl.: Russian Math. Surv. **25** (1970) 125 - 187.

[1971] *Rearrangements of function series.* Mat. Sbornik **85** (1971) 272 - 286; Engl. transl.: Math. USSR Sb. **14** (1971) 267 - 280.

G. Nordlander

[1960] *The modulus of convexity in normed linear spaces.* Arkiv för Mat. **4** (1960) 15 - 17.

[1962] *On sign-independent and almost sign-independent convergence in normed linear spaces.* Arkiv för Mat. **4** (1962) 287 - 296.

E. Odell (see also D. Alspach)

[1980] *Applications of Ramsey theorems to Banach space.* Notes in Banach Spaces, University of Texas Press (Austin) (1980) 379 - 404.

K. Oleszkiewicz (see R. Latała)

A.M. Olevskiĭ

[1975] *Fourier series with respect to general orthonormal systems.* Springer-Verlag, 1975.

G.H. Olsen (see S. Heinrich)

W. Orlicz (see also R. Graślewicz)

[1929a] *Beiträge zur Theorie der Orthogonalentwicklungen.* Studia Math. **1** (1929) 1 - 40.

[1929b] *Beiträge zur Theorie der Orthogonalentwicklungen. II.* Studia Math. **1** (1929) 241 - 255.

[1933a] *Über unbedingte Konvergenz in Funktionenräumen (I).* Studia Math. **4** (1933) 33 - 37.

[1933b] *Über unbedingte Konvergenz in Funktionenräumen (II).* Studia Math. **4** (1933) 41 - 47.

P. Ørno

[1976] *A note on unconditionally convergent series in L_p.* Proc. Amer. Math. Soc. **59** (1976) 252 - 254.

H. Osaka

[1991] *Completely bounded multilinear functionals on C^*-algebras.* Math. Japon. **36** (1991) 363 - 369.

R. Ott (see H. Jarchow)

V.I. Ovčinnikov

[1976] *Interpolation theorems that arise from Grothendieck's inequality.* Funktsional Anal. i Prilozhen **10** (1976) 45-54; Engl. transl.: Funct. Anal. Applic. **10** (1976) 287-294.

R.I. Ovsepian & A. Pełczyński

[1975] *On the existence of a fundamental total and bounded biorthogonal system in every separable Banach space and related constructions of uniformly bounded orthonormal systems in L^2.* Studia Math. **54** (1975) 149-159.

A. Pajor

[1987] *Quotients volumique et espaces de Banach de type 2 faible.* Israel Journ. Math. **57** (1987) 101-106.

R.E.A.C. Paley

[1933] *On the lacunary coefficients of power series.* Ann. of Math. **34** (1933) 615-616.

R.E.A.C. Paley & A. Zygmund

[1930] *On some series of functions I.* Math. Proc. Camb. Phil. Soc. **26** (1930) 337-357.

R. Panzone (see A. Benedek)

P.L. Papini (see E. Casini)

A. Pełczyński (see also C. Bessaga, O. Blasco, E. Dubinsky, T. Figiel, M.I. Kadets, S. Kwapień, J. Lindenstrauss, B. Maurey, B.S. Mitiagin, R.I. Ovsepian)

[1960] *Projections in certain Banach spaces.* Studia Math. **19** (1960) 209-228.

[1961] *On the impossibility of embedding of the space L in certain Banach spaces.* Colloq. Math **8** (1961) 199-203.

[1962] *Banach spaces in which every unconditionally converging operator is weakly compact.* Bull. Acad. Polon. Sci. **10** (1962) 641-648.

[1966] *Uncomplemented function algebras with separable annihilator.* Duke Math. Journ. **33** (1966) 605-612.

[1967] *A characterization of Hilbert-Schmidt operators.* Studia Math. **28** (1967) 355-360.

[1968] *On Banach spaces containing $L_1(\mu)$.* Studia Math. **30** (1968) 231-246.

[1969] *p-integral operators commuting with group representations and examples of quasi-p-integral operators which are not p-integral.* Studia Math. **33** (1969) 63-70.

[1974] *Sur certaines propriétés isomorphiques nouvelles des espaces de Banach de fonctions holomorphes A et H^∞.* C. R. Acad. Sci. Paris **A279** (1974) 9-12.

[1976a] *All separable Banach spaces admit for every $\varepsilon > 0$ fundamental total biorthogonal systems bounded by $1+\varepsilon$.* Studia Math. **55** (1976) 295-304.

[1976b] *Banach Spaces of Analytic Functions and Absolutely Summing Operators.* Amer. Math. Soc., Providence R.I., CBMS **30** 1976.

[1980] *Geometry of finite dimensional Banach spaces and operator ideals.* Notes in Banach Spaces, University of Texas Press (Austin) 1980, 81-181.

[1985] Norms of classical operators in function spaces. Astérisque Soc. Math. France **131** (1985) 137-162.

[1992] *Non-isomorphism of the disc algebra with spaces of differentiable functions.* Proc. Center for Math. and Appl., Austral. Nat. Univ. **29** (1992) 183-195.

A. Pełczyński & H.P. Rosenthal

[1975] *Localization techniques in L^p-spaces.* Studia Math. **52** (1975) 263-289.

A. Pełczyński & K. Senator

[1986] *On isomorphisms of anisotropic Sobolev spaces with "classical" Banach spaces and a Sobolev type embedding theorem.* Studia Math. **84** (1986) 169-215.

A. Pełczyński & M. Wojciechowski

[1992a] *Paley projections on anisotropic Sobolev spaces on tori.* Proc. London Math. Soc. **65** (1992) 405-422.

[1992b] *Absolutely summing surjections from Sobolev spaces in the uniform norm.* North-Holland Mathematical Studies, **170** (1992) 423-431.

V. Peller

[1976a] *An analogue of J. von Neumann's inequalities for the space L_p.* Dokl. Akad. Nauk SSSR **231** (1976) 539-542; Engl. transl.: Soviet Math. Dokl. **17** (1976) 1594-1598.

[1976b] *Estimates of operator polynomials in an L_p space in terms of the multiplier norm.* Sem. Leningrad Otdel. Mat. Inst. Steklov (LOMI) **65** (1976) 132-148, 205-206

[1978a] *L'inégalité de von Neumann, la dilation isométrique et l'approximation par isométries dans L_p.* C. R. Acad. Sci. Paris **A287** (1978) 311-314.

[1978b] *Approximation by isometries and V.I. Macaev's conjecture for absolute contractions of the space L_p.* Funktsional Anal. i Prilozhen **12** (1978) 38-50; Engl. transl.: Funct. Anal. Applic. **12** (1978) 29-38.

[1982] *Estimates of functions of power bounded operators on Hilbert spaces.* J. Operator Theory **7** (1982) 341-372.

A. Persson

[1969] *On some properties of p-nuclear and p-integral operators.* Studia Math. **33** (1969) 213-222.

A. Persson & A. Pietsch

[1969] *p-nukleare und p-integrale Abbildungen in Banachräumen.* Studia Math. **33** (1969) 19-62.

B.J. Pettis (see also N. Dunford)

[1938] *On integration in vector spaces.* Trans. Amer. Math. Soc. **44** (1938) 277-304.

H. Pfitzner

[1992] *Weak compactness in certain Banach spaces.* Dissertation, Freie Universität Berlin (1992).

R.S. Phillips

[1940] *On linear transformations.* Trans. Amer. Math. Soc. **48** (1940) 516-541.

A. Pietsch (see also W. Linde, A. Persson)

[1963] *Zur Fredholmschen Theorie in lokalkonvexen Räumen.* Studia Math. **22** (1963) 161-179.

[1965] *Abbildungen von abstrakten Massen.* Wiss. Zeitschr. Univ. Jena **14** (1965) 281-286.

[1967] *Absolut p-summierende Abbildungen in normierten Räumen.* Studia Math. **28** (1967) 333-353.

[1968] *Hilbert-Schmidt-Abbildungen in Banachräumen.* Math. Nachr. **37** (1968) 237-245.

[1969] *p-majorisierbare vektorwertige Masse.* Wiss. Zeitschr. Univ. Jena **18** (1969) 243-247.

[1970/71] *Absolutely p-summing operators in L_r-spaces I, II.* École Polyt. Palaiseau, Sém. Goulaouic-Schwartz, 1970/71.

[1971] *Adjungierte normierte Operatorenideale.* Math. Nachr. **41** (1971) 189-211.

[1972a] *Nuclear Locally Convex Spaces* (3^d ed.) Springer-Verlag, 1972.

[1972b] *Eigenwertverteilungen von Operatoren in Banachräumen.* Theory of sets and topology. Akademie-Verlag, 391-402.

[1978] *Operator Ideals.* VEB Deutscher Verlag der Wissensch., 1978; North-Holland, 1980.

[1980] *Weyl numbers and operators in Banach spaces.* Math. Ann. **247** (1980) 149-168.

[1986] *Eigenvalues of r-summing operators.* Aspects of Mathematics and its Applications (North-Holland) (1986) 607-617.

[1987] *Eigenvalues and s-Numbers.* Akad.Verlagsges. Geest & Portig K.-G.; Cambridge University Press, 1987.

[1990a] *Eigenvalue criteria for the B-convexity of Banach spaces.* Journ. London Math. Soc. **41** (1990) 310-322.

[1990b] *Type and cotype numbers of operators on Banach spaces.* Studia Math. **94** (1990) 21-37.

G. Pisier (see also J. Bourgain, T. Figiel, U. Haagerup, J. Hoffmann-Jørgensen, M.B. Marcus, B. Maurey, V.D. Milman)

[1973] *Sur les espaces de Banach qui ne contiennent pas uniformément les ℓ_n^1.* C. R. Acad. Sci. Paris **A277** (1973) 991-994.

[1973/74a] *"Type" des espaces normés.* École Polyt. Palaiseau, Sém. Maurey-Schwartz 1973/74, Exp. III.

[1973/74b] *Sur les espaces qui ne contiennent pas de ℓ_n^1 uniformément.* École Polyt. Palaiseau, Sém. Maurey-Schwartz 1973/74, Exp. VII.

[1975] *Martingales with values in uniformly convex spaces.* Israel Journ. Math. **20** (1975) 326-350.

[1977/78a] *Les inégalités de Khintchine-Kahane (d'après C.Borell).* École Polyt. Palaiseau, Sém. Géom. Espaces de Banach 1977/78, Exp. VII.

[1977/78b] *Ensembles de Sidon et espaces de cotype 2.* École Polyt. Palaiseau, Sém. Géom. Espaces de Banach 1977/78, Exp. XIV.

[1978a] *Grothendieck's theorem for non-commutative C^*-algebras with an appendix on Grothendieck's constant.* Journ. Funct. Anal. **29** (1978) 397-415.

[1978b] *Une nouvelle classe d'espaces vérifiant le théorème de Grothendieck.* Ann. Inst. Fourier **28** (1978) 69-90.

[1978c] *Some results on Banach spaces without local unconditional structure.* Compositio Math. **37** (1978) 3-19.

[1979] *Some applications of the complex interpolation method to Banach lattices.* Journ. d'Analyse Math. **35** (1979) 264-281.

[1980] *Un théorème sur les opérateurs entre espaces de Banach qui se factorisent par un espace de Hilbert.* Ann. École Norm. Sup. **13** (1980) 23-43.

[1982a] *Holomorphic semi-groups and the geometry of Banach spaces.* Ann. of Math. **115** (1982) 375-392.

[1982b] *Quotients of Banach spaces of cotype q.* Proc. Amer. Math. Soc. **85** (1982) 32-36.

[1983] *Counterexamples to a conjecture of Grothendieck.* Ann. of Math. **151** (1983) 181-208.

[1985] *Factorization of Operators and Geometry of Banach Spaces.* Amer. Math. Soc., Providence R.I., CBMS **60** 1985.

[1986a] *Probabilistic methods in the geometry of Banach spaces.* Lecture Notes in Math. **1206** (1986) 167-241.

[1986b] *Factorization of operators through $L_{p\infty}$ or L_{p1} and non-commutative generalizations.* Math. Ann. **276** (1986) 105-136.

[1988] *Weak Hilbert spaces.* Proc. London Math. Soc. **56** (1988) 547-579.

[1989] *The Volume of Convex Bodies and Banach Space Geometry.* Cambridge University Press, 1989.

[1992a] *A simple proof of a theorem of Jean Bourgain.* Mich. Math. Journ. **39** (1992) 475-484.

[1992b] *Remarks on complemented subspaces of C^*-algebras.* Proc. Roy. Soc. Edinburgh **121A** (1992) 1-4.

G. Pisier & Q. Xu

[1987] *Random series in the real interpolation spaces between the spaces v_p.* Lecture Notes in Math. **1267** (1987) 185-209.

J. Puhl (see B. Carl)

F. Räbiger

[1991] *Absolutstetigkeit und Ordnungsabsolutstetigkeit.* Sitzungsber. Heidelberg. Akad. Wiss., Math. Nat.wiss. Kl. **1**, 1991.

G. Racher

[1980] *Über c_0-Modultensorprodukte.* Studia Math. **66** (1980) 201-211.

[1983] *A proposition of A. Grothendieck revisited.* Lecture Notes in Math. **991** (1983) 215-227.

H. Rademacher

[1918] *Über partielle und totale Differenzierbarkeit von Funktionen mehrerer Variabler und über die Transformation der Doppelintegrale.* Math.Ann. **79** (1918) 340-359.

[1922] *Einige Sätze über Reihen von allgemeinen Orthogonalfunktionen.* Math. Ann. **87** (1922) 112-138.

S.A. Rakov

[1976] *C-convexity and the "problem of three spaces".* Dokl. Akad. Nauk SSSR **228** (1976) 303-305; Engl. transl.: Soviet Math. Dokl. **17** (1976) 721-724.

F.P. Ramsey

[1929] *On a problem of formal logic.* Proc. London Math. Soc. **30** (1929) 264-286.

R.M. Redheffer & P. Volkmann

[1983] *Bilineare Funktionen mit Hilbertraum-Operatoren als Veränderlichen.* Math. Ann. **262** (1983) 133-143.

O. Reinov (see also E.D. Gluskin)

[1982] *Approximation properties of order p and the existence of non-p-nuclear operators with p-nuclear second adjoints.* Math. Nachr. **109** (1982) 125-134.

S. Reisner

[1979] *On Banach spaces having the property G.L.* Pacific Journ. Math. **83** (1979) 505-521.

J.R. Retherford (see also Y. Gordon, W.B. Johnson, R.J. Kaiser, H. König, C. Stegall)

[1972] *Operator characterizations of $\mathcal{L}_p$-spaces.* Israel Journ. Math. **13** (1972) 337-347.

R. Riedl (see also H. Jarchow)

[1994] *Composition operators and geometric properties of analytic functions.* Dissertation, Universität Zürich (1994).

B. Riemann

[1854] *Über die Darstellbarkeit einer Funktion durch eine trigonometrische Reihe.* Habilitationsschrift Universität Göttingen 1854 (Dover Publ. 1953).

R.A. Rietz

[1974] *A proof of the Grothendieck inequality.* Israel Journ. Math. **19** (1974) 271-276.

J. Ringrose

[1976] *Linear mappings between operator algebras.* Symp. Math. (Academic Press, London) **20** (1976) 297-315.

A.W. Roberts & D.E. Varberg

[1973] *Convex Functions.* Academic Press, 1973.

R. Rochberg (see R.R. Coifman)

C.A. Rogers (see A. Dvoretzky)

S. Rolewicz

[1972] *Metric Linear Spaces.* PWN, 1972.

M. Rosenberg

[1982] *Quasi-isometric dilations of operator-valued measures and Grothendieck's inequality.* Pacific Journ. Math. **103** (1982) 135-161.

H.P. Rosenthal (see also J. Bourgain, E. Dubinsky, W.B. Johnson, J. Lindenstrauss, A. Pełczyński)

[1966] *Projections onto translation invariant subspaces of $L_p(G)$.* Memoirs Amer. Math. Soc. **63**, 1966.

[1970] *On subspaces of L^p $(p>2)$ spanned by sequences of independent random variables.* Israel Journ. Math. **8** (1970) 273-303.

[1973] *On subspaces of L^p.* Ann. of Math. **97** (1973) 344-373.

[1974] *A characterization of Banach spaces containing ℓ^1.* Proc. Nat. Acad. Sci. USA **71** (1974) 2411-2413.

P. Rosenthal

[1987] *The remarkable theorem of Lévy and Steinitz.* Amer. Math. Monthly **94** (1987) 342-351.

Z.J. Ruan (see D.P. Blecher)

J.L. Rubio de Francia

[1985] *Fourier series and Hilbert transforms with values in UMD Banach spaces.* Studia Math. **81** (1985) 95-105.

[1987] *Linear operators in Banach lattices and weighted L^2 inequalities.* Math. Nachr. **133** (1987) 197-207.

J.L. Rubio de Francia & J.L. Torrea

[1987] *Some Banach space techniques in vector-valued Fourier analysis.* Colloq. Math. **54** (1987) 273-284.

W. Rudin

[1960] *Trigonometric series with gaps.* Journ. Math. Mech. **9** (1960) 203-227.

A.F. Ruston

[1962] *Auerbach's theorem and tensor products of Banach spaces.* Math. Proc. Camb. Phil. Soc. **58** (1962) 476-480.

D. Rutovitz

[1965] *Some parameters associated with finite dimensional Banach spaces.* Journ. London Math. Soc. **40** (1965) 241-255.

R.A. Ryan (see M. Lindström)

S. Sadosky (see M. Cotlar)

S. Saeki

[1992] *The Vitali-Hahn-Saks theorem and measureoids.* Proc. Amer. Math. Soc. **114** (1992) 775-782.

H. Salehi (see A. Makagon)

P. Saphar
[1970] *Produits tensoriels d'espaces de Banach et classes d'applications linéaires.* Studia Math. **38** (1970) 70-100.
[1972] *Applications p-décomposantes et p-absolument sommantes.* Israel Journ. Math. **11** (1972) 164-179.

Y. Sarantopoulos (see also C. Benítez)
[1986] *Estimates for polynomial norms on $L_p(\mu)$-spaces.* Math. Proc. Camb. Phil. Soc. **99** (1986) 263-271.
[1987] *Extremal multilinear forms on Banach spaces.* Proc. Amer. Math. Soc. **99** (1987) 340-346.

Y. Sarantopoulos & A.M. Tonge
[1995] *Extremal multilinear forms on real Banach spaces.*

J. Sawa
[1985] *The best constant in the Khintchine inequality for complex Steinhaus variables, the case $p = 1$.* Studia Math. **81** (1985) 107-126.

W. Schachermayer (see also N. Ghoussoub)
[1981a] *Integral operators on L^p spaces, I.* Indiana Univ. Math. Journ. **30** (1981) 123-140.
[1981b] *Integral operators on L^p spaces, II.* Indiana Univ. Math. Journ. **30** (1981) 261-266.

H.H. Schaefer
[1974] *Banach Lattices and Positive Operators.* Springer-Verlag, 1974.

R. Schatten
[1946] *The cross-space of linear transformations.* Ann. of Math.(2) **47** (1946) 73-84.
[1960] *Norm Ideals of Completely Continuous Operators.* Springer-Verlag, 1960.

R. Schatten & J. von Neumann
[1946] *The cross-space of linear transformations II.* Ann. of Math.(2) **47** (1946) 608-630.
[1948] *The cross-space of linear transformations III.* Ann. of Math.(2) **49** (1948) 557-582.

G. Schechtman (see also J. Bourgain, W.B. Johnson, V.D. Milman)
[1975] *Examples of $\mathcal{L}^p$-spaces.* Israel Journ. Math. **22** (1975) 138-147.

U. Schlotterbeck
[1969] *Über Klassen majorisierbarer Operatoren auf Banachverbänden.* Dissertation, Universität Tübingen 1969.

T. Schlumprecht
[1991] *An arbitrarily distortable Banach space.* Israel Journ. Math. **76** (1991) 81-95.

E. Schmidt
[1907a] *Zur Theorie der linearen und nichtlinearen Integralgleichungen I.* Math. Ann. **63** (1907) 433-476.
[1907b] *Zur Theorie der linearen und nichtlinearen Integralgleichungen II.* Math. Ann. **64** (1907) 161-174.
[1908] *Über die Auflösung linearer Gleichungen mit unendlich vielen Unbekannten.* Rend. Circ. Mat. Palermo **25** (1908) 53-77.

B.M. Schreiber (see J.E. Gilbert, C.C. Graham)

J. Schur
[1911] *Bemerkungen zur Theorie der beschränkten Bilinearformen mit unendlich vielen Veränderlichen.* Journ. Reine. Angew. Math. **140** (1911) 1-28.
[1920] *Über lineare Transformationen in der Theorie der unendlichen Reihen.* Journ. Reine Angew. Math. **151** (1920) 79-111.

C. Schütt
[1978] *Unconditionality in tensor products.* Israel Journ. Math. **31** (1978) 209-216.

J.T. Schwartz (see R.G. Bartle, N. Dunford)

L. Schwartz
[1969a] *Probabilités cylindriques et applications radonifiantes.* C. R. Acad. Sci. Paris **A268** (1969) 646-648.
[1969b] *Un théorème de dualité pour les applications radonifiantes.* C. R. Acad. Sci. Paris **A268** (1969) 1410-1413.

[1969c] *Applications des théorèmes de dualité sur les applications p-radonifiantes.* C. R. Acad. Sci. Paris **A268** (1969) 1612-1615.

[1970a] *Applications p-radonifiantes et théorèmes de dualité.* Studia Math. **38** (1970) 203-213.

[1970b] *Probabilités cylindriques et applications radonifiantes.* Journ. Fac. Sci. Univ. Tokyo **18** (1970) 139-286.

[1981] *Geometry and Probability in Banach Spaces.* Lecture Notes in Math. **852** 1981.

C.J. Seifert (see J. Diestel)

K. Senator (see A. Pełczyński)

J.H. Shapiro & P.D. Taylor

[1973] *Compact, nuclear and Hilbert-Schmidt composition operators on H^2.* Indiana Univ. Math. Journ. **23** (1973) 471-496.

L.A. Shepp (see H.J. Landau)

T.J. Shura (see P.G. Casazza)

N.G. Sidorenko (see S.V. Kisliakov)

B. Simon

[1979] *Trace Ideals and Their Applications.* London Math. Soc. Lect. Notes **35** 1979.

A.M. Sinclair (see also D.P. Blecher, E. Christensen, S. Kaijser)

[1984] *The Banach algebra generated by a derivation.* Operator Theory: Adv. Appl. **14** (1984) 241-250.

A.I. Singh & N. Mittal

[1991] *Complete finite representability in C^*-algebras.* Yokohama Math. Joun. **38** (1991) 83-94.

R.R. Smith

[1988] *Completely bounded multilinear maps and Grothendieck's inequality.* Bull. London Math. Soc. **20** (1988) 606-612.

M.G. Snobar (see also M.I. Kadets)

[1972] *p-absolutely summing constants.* Teor. Funktsij Funktsional Anal. i Prilozhen **16** (1972) 38-41.

T. Starbird (see L.E. Dor)

C. Stegall (see also D.R. Lewis, J. Lindenstrauss)

[1980] *A proof of the principle of local reflexivity.* Proc. Amer. Math. Soc. **78** (1980) 154-156.

[1981] *The Radon-Nikodým property in conjugate Banach spaces.* Trans. Amer. Math. Soc. **164** (1981) 507-519.

C. Stegall & J.R. Retherford

[1972] *Fully nuclear and completely nuclear operators with applications to $\mathcal{L}_1$- and $\mathcal{L}_\infty$-spaces.* Trans. Amer. Math. Soc. **163** (1972) 457-492.

E.M. Stein

[1970] *Singular Integrals and Differentiability Properties of Functions.* Princeton University Press 1970.

E.M. Stein & G. Weiss

[1971] *Introduction to Fourier Analysis on Euclidean Spaces.* Princeton University Press 1971.

E. Steinitz

[1913] *Bedingt konvergente Reihen und konvexe Systeme.* Journ. Reine Angew. Math. **143** (1913) 128-175.

J. Stern

[1978] *Ultraproducts and local properties of Banach spaces.* Trans. Amer. Math. Soc. **240** (1978) 231-252.

M.H. Stone

[1949] *Boundedness properties in function lattices.* Canad. Journ. Math. **1** (1949) 176-186.

L. Sucheston (see M.A. Akcoğlu, A. Brunel)

V.S. Sunder (see P.R. Halmos)

J.J. Sylvester

[1883] *On the equation to the secular inequalities in the planetary theory.* Philos. Mag. Ser. 5 **16** (1883) 267-269.

A. Szankowski

[1974] *On Dvoretzky's theorem on almost spherical sections.* Israel Journ. Math. **17** (1974) 325-338.

S.J. Szarek (see also S. Kwapień)

[1976] *On the best constants in the Khintchine inequality.* Studia Math. **58** (1976) 197-208.

[1979] *Bases and orthogonal systems in the spaces C and L^1.* Arkiv för Mat. **17** (1979) 255-271.

[1980] *Non-existence of Besselian basis in $C(S)$.* Journ. Funct. Anal. **37** (1980) 56-67.

B. Sz. Nagy (see also C. Foiaş)

[1947] *On bounded linear transformations in Hilbert space.* Acta Sci. Math. (Szeged) **11** (1947) 152-157.

J. Szulga (see also N.J. Nielsen)

[1983a] *On lattice analogues of absolutely summing operators.* Canad. Math. Bull. **24** (1983) 63-69.

[1983b] *On latticially summing operators.* Proc. Amer. Math. Soc. **87** (1983) 258-262.

[1984] *On p-lattice summing and p-absolutely summing operators.* Lecture Notes in Math. **1080** (1984) 292-298.

[1985] *On p-absolutely summing operators acting on Banach lattices.* Studia Math. **81** (1985) 53-63.

M. Talagrand (see also M. Ledoux)

[1988] *An isoperimetric theorem on the cube and the Kintchine-Kahane inequalities.* Proc. Amer. Math. Soc. **104** (1988) 905-909.

[1989] *The canonical injection from $C[0,1]$ to $L_{2,1}$ is not of cotype 2.* Contemp. Math. **85** (1989) 513-521.

[1992a] *Cotype of operators from $C(K)$.* Inventiones Math. **107** (1992) 1-40.

[1992b] *Cotype and $(q,1)$-summing norm in a Banach space.* Inventiones Math. **110** (1992) 545-556.

A.E. Taylor

[1947] *A geometric theorem and its application to biorthogonal systems.* Bull. Amer. Math. Soc. **53** (1947) 614-616.

P.D. Taylor (see J.H. Shapiro)

E. Thorp & R. Whitley

[1967] *The strong maximum modulus theorem for analytic functions into a Banach space.* Proc. Amer. Math. Soc. **181** (1967) 640-646.

H.S. Thurston

[1931] *On the characteristic equations of products of square matrices.* Amer. Math. Monthly **38** (1931) 322-324.

R.M. Timoney (see S. Dineen)

B. Tomaszewski

[1982] *Two remarks on the Khintchine-Kahane inequality.* Colloq. Math. **46** (1982) 283-288.

[1987] *A simple and elementary proof of the Khintchine inequality with the best constant.* Bull. Sci .Math. **111** (1987) 103-109.

N. Tomczak-Jaegermann (see also W.J. Davis, T. Figiel, D.J.H. Garling, R. Komorowski, H. König, D.R. Lewis)

[1970] *A remark on (s,t)-summing operators in L_p-spaces.* Studia Math. **35** (1970) 97-100.

[1974] *The moduli of smoothness and convexity and the Rademacher average of trace class S_p.* Studia Math. **50** (1974) 163-182.

[1979] *Computing 2-summing norms with few vectors.* Arkiv för Mat **17** (1979) 273-277.

[1989] *Banach-Mazur Distances and Finite-Dimensional Operator Ideals.* Longman Scientific & Technical, 1989.

A.M. Tonge (see also A.M. Mantero, Y.Sarantopoulos)

[1975/76] *Sur les algèbres de Banach et les opérateurs p-sommants.* École Polyt. Palaiseau, Sém. Maurey-Schwartz 1975/76, Exp. XIII.

[1976] *Banach algebras and absolutely summing operators.* Math. Proc. Camb. Phil. Soc. **80** (1976) 465-473.

[1978] *The von Neumann inequality for polynomials in several Hilbert-Schmidt operators.* Journ. London Math. Soc. **18** (1978) 519-526.

[1984] *Polarization and the two-dimensional Grothendieck inequality.* Math. Proc. Camb. Phil. Soc. **95** (1984) 313-318.

[1986] *Bilinear mappings and trace class ideals.* Math. Ann. **275** (1986) 281-290.

[1987] *Random Clarkson inequalities and L_p versions of Grothendieck's inequality.* Math. Nachr. **131** (1987) 335-343.

J.L. Torrea (see J.L. Rubio de Francia)

M.A. Triana (see B. Cuartero)

S.L. Troyanski (see also R.P. Maleev)

[1967] *On conditionally convergent series in certain F-spaces.* Teor. Funktsij Funktsional Anal. i Prilozhen **5** (1967) 102-107.

B.S. Tsirelson

[1974] *Not every Banach space contains ℓ_p or c_0.* Funktsional Anal. i Prilozhen **8** (1974) 57-60; Engl. transl.: Funct. Anal. Appl. **8** (1974) 138-141.

B. Turett (see A. Kamińska)

L. Tzafriri (see also T. Figiel, W.B. Johnson, J. Lindenstrauss, C.A. McCarthy)

[1974] *On Banach spaces with unconditional basis.* Israel Journ. Math. **17** (1974) 84-93.

[1979] *On the type and cotype of Banach spaces.* Israel Journ. Math. **32** (1979) 32-38.

J.J. Uhl, Jr. (see J. Diestel)

A. Ülger

[1988] *Arens regularity of the algebra $A\hat{\otimes}B$.* Trans. Amer. Math. Soc. **305** (1988) 623-639.

D.E. Varberg (see A.W. Roberts)

N.T. Varopoulos

[1972a] *Some remarks on Q-algebras.* Ann. Inst. Fourier **22** (1972) 1-11.

[1972b] *Sur les quotients des algèbres uniformes.* C. R. Acad. Sci. Paris **A274** (1972) 1344-1346.

[1974] *On an inequality of von Neumann and an application of the metric theory of tensor products to operator theory.* Journ. Funct. Anal. **16** (1974) 83-100.

[1975] *A theorem on operator algebras.* Math. Scand. **37** (1975) 173-182.

[1975/76] *Sous-espaces de $C(G)$ invariants par translation et de type $\mathcal{L}_1$.* École Polyt. Palaiseau, Sém. Maurey-Schwartz 1975-76, Exp. XII.

[1976] *On a commuting family of contractions on a Hilbert space.* Revue Roum. Math. Pures Appl. **21** (1976) 1283-1285.

P. Volkmann (see R.M. Redheffer)

J. von Neumann (see also R. Schatten)

[1937] *Some matrix inequalities and metrization of matrix spaces.* Tomsk Univ. Rev. **1** (1937) 286-300.

[1951] *Eine Spektraltheorie für allgemeine Operatoren eines unitären Raumes.* Math. Nachr. **4** (1951) 258-281.

A. Wald

[1933a] *Bedingt konvergente Reihen von Vektoren in $\mathbf{R}_\omega$.* Erg. Math. Koll. **5** (1933) 13-14.

[1933b] *Reihen in topologischen Gruppen.* Erg. Math. Koll. **5** (1933) 14-16.

A. Weil

[1940] *L'Intégration dans les Groupes Topologiques et ses Applications.* Hermann 1940.

G. Weiss (see R.R. Coifman, E.M. Stein)

J. Wermer (see also B. Cole)

[1969] *Quotient Algebras of Uniform Algebras.* Symposium on Function Algebras and Rational Approximation, University of Michigan 1969.

A. Weron (see H. Niemi)

H. Weyl

[1949] *Inequalities between the two kinds of eigenvalues of a linear transformation.* Proc. Nat. Acad. Sci. USA **35** (1949) 408-411.

R. Whitley (see E. Thorp)

M. Wojciechowski (see also A. Pełczyński)

[1991] *Translation invariant subspaces of Sobolev spaces on tori in L^1 and in the uniform norm.* Studia Math. **100** (1991) 149-167.

W.A. Wojczyński

[1975] *Geometry and martingales in Banach spaces.* Lecture Notes in Math. **472** (1975) 235-283.

[1978] *Geometry and martingales in Banach spaces, part II: independent increments.* Probability on Banach Spaces **4** (1978) 267-517.

P. Wojtaszczyk

[1977] *On Banach space properties of uniform algebras with separable annihilator.* Bull. Acad. Polon. Sci. **25** (1977) 23-26.

[1991] *Banach Spaces for Analysts.* Cambridge University Press, 1991.

H. Wolfson (see J. Bourgain)

T.K. Wong

[1971] *On a class of absolutely p-summing operators.* Studia Math. **39** (1971) 181-189.

K. Woźniakowski (see M.I. Kadets)

Q. Xu (see also G. Pisier)

[1987] *Cotype of spaces $(A_0, A_1)_{\theta 1}$.* Lecture Notes in Math. **1267** (1987) 210-212.

L.P. Yanovskii

[1979] *On summing and lattice summing operators and characterizations of AL-spaces.* Sibirskii Mat. Zh. **20** (1979) 401-408.

K. Ylinen

[1984] *Dilation of V-bounded stochastic processes indexed by a locally compact group.* Proc. Amer. Math. Soc. **90** (1984) 378-380.

[1988a] *The structure of bounded bilinear forms on products of C^*-algebras.* Proc. Amer. Math. Soc. **102** (1988) 599-602.

[1988b] *Non commutative Fourier transform of bounded bilinear forms and completely bounded multilinear operators.* Journ. Funct. Anal. **79** (1988) 144-165.

R.M.G. Young

[1976] *On the best possible constants in the Khintchine inequality.* Journ. London Math. Soc. **14** (1976) 496-504.

X.T. Yu (see T. Barton)

M. Zippin (see W.B. Johnson, J. Lindenstrauss)

A. Zygmund (see also R.E.A.C. Paley)

[1935] *Trigonometric Series.* PWN, 1935; Cambridge University Press, 1988.

Author Index

Subject Index